D1264755

LRC/LIBRARY

# SCIENCE,
# TECHNOLOGY,
## AND SOCIETY

---

## AN ENCYCLOPEDIA

EDITOR IN CHIEF
# Sal Restivo

*Rensselaer Polytechnic Institute*

## Editorial Board

Joan Leach, *University of Queensland*
Marcel LaFollette, *Independent Historian*
Susan Leigh Star, *Santa Clara University*

## Advisory Board

Karin Knorr-Cetina, *University of Konstanz, Germany*
Kenneth Manning, *Massachusetts Institute of Technology*
Emily Martin, *New York University*
Hilary Rose, *Independent Scholar*
Susan Cozzens, *Georgia Institute of Technology*

## Special Consultants

Linda Layne, *Rensselaer Polytechnic Institute*
Jennifer Croissant, *University of Arizona*
Wenda Bauchspies, *Pennsylvania State University*
Colin Beech, *Rensselaer Polytechnic Institute*
Evelynn Hammonds, *Harvard University*
Ellen Schieble, *Claremont Graduate University*
Rachel Dowty, *Rensselaer Polytechnic Institute*

# SCIENCE, TECHNOLOGY, AND SOCIETY

## AN ENCYCLOPEDIA

*Sal Restivo*

EDITOR IN CHIEF

Baker Taylor

2005 (6-27-2005)

147,750-L

LRC/LIBRARY

OXFORD

UNIVERSITY PRESS
2005

# OXFORD
## UNIVERSITY PRESS

Oxford University Press, Inc., publishes works that further
Oxford University's objective of excellence
in research, scholarship, and education.

Oxford  New York
Auckland   Cape Town   Dar es Salaam   Hong Kong   Karachi   Kuala Lumpur   Madrid   Melbourne
Mexico City   Nairobi   New Delhi   Shanghai   Taipei   Toronto

With offices in
Argentina   Austria   Brazil   Chile   Czech Republic   France   Greece
Guatemala   Hungary   Italy   Japan   Poland   Portugal   Singapore
South Korea   Switzerland   Thailand   Turkey   Ukraine   Vietnam

Copyright © 2005 by Oxford University Press, Inc.

Published by Oxford University Press, Inc.
198 Madison Avenue, New York, New York, 10016
http://www.oup.com/us

Oxford is a registered trademark of Oxford University Press

All rights reserved. No part of this publication may be reproduced, stored in a retrieval system, or transmitted,
in any form or by any means, electronic, mechanical, photocopying, recording, or otherwise, without prior
permission of Oxford University Press

**Library of Congress Cataloging-in-Publication Data**

Science, technology, and society an encyclopedia / Sal Restivo, editor in chief.
    p. cm.
    Includes bibliographical references and index.
    ISBN-13: 978-0-19-514193-1 (alk. paper)
    ISBN-10: 0-19-514193-8 (alk. paper)
    1. Technology—Social aspects. 2. Technological innovations—Social aspects.
    3. Science—Social aspects. I. Restivo, Sal P.
    HM846.S43 2005
    306.4'5—dc22

2004031121

Printing number: 9 8 7 6 5 4 3 2 1

Cover design by Joan Greenfield
Book design by Publication Services

Printed in the United States of America on acid-free paper

**Oxford University Press**
Sean Pidgeon, Acquiring Editor
Chris Collins, Acquiring Editor
Beth Ammerman, Development Editor
Joe Clements, Development Editor
Timothy J. DeWerff, Director, Editorial Development and Production
Casper Grathwohl, Publisher

*This volume is dedicated to the memories of five fellow presidents of the Society for Social Studies of Science: Nicholas Mullins, Derek J. de Solla Price, David Edge, Dorothy Nelkin, and Robert K. Merton.*

# CONTENTS

# Introduction

This work is designed for anyone interested in problems, issues, and questions related to the relationships between science, technology, and society (ST&S). It should also serve professionals in science and technology studies (S&TS) by summarizing basic tenets of the field and providing a concise and coherent overview of the achievements in their field. In this sense, it will also be a valuable resource for professionals in non-STS fields. This volume is focused on Science, Technology, and Society, not Science and Technology Studies. The distinction is meant to convey to potential readers and contributors that the book is not focused on bringing a high level of order and articulation to theory, method, and research in S&TS; rather, it is designed to illustrate the mutual shaping of science, technology, and society. The A–Z entries fall broadly into three main categories: Science and Society; Technology and Society; and Medicine and Society. High-school students should be able to come to this volume for accessible information on contemporary topics such as cloning or AIDS. At the same time, professional S&TS researchers should be able to find updates on the achievements and current state of fields such as the sociology of mathematics and the anthropology of information technology.

Authors were not limited by any Procrustean constraints in terms of perspectives or paradigms, and they bring a focused diversity to this volume. We have drawn on men and women from across the professional generations and from around the world to construct this work. They represent traditional disciplines such as history and political science, and interdisciplinary fields such as science and technology studies and cultural studies of science and technology. Our authors bridge the east-west divide from Japan and India to the United States and Canada, and the north-south divide from the Netherlands and the United Kingdom to Brazil and Africa. We have nonetheless chosen to follow a traditional separation of science and technology that is not supported by recent scholarship. The original organization of the book called for two sections, "Science and Society" and "Technology and Society." In the wake of editorial meetings, "Medicine and Society" was added to reflect the large number of studies in S&TS and ST&S devoted to issues and problems in medicine, biology, biomedicine, biotechnology, and sex and reproduction. Medicine is science and technology in the service of the public health and individual well-being. It is also a profession and thus is burdened by bureaucratic and economic interests, ideologies, and conflicts.

The organizing scheme that distinguishes science from technology from medicine works in part because we can still make sense of the distinction between Einstein's work on relativity theory and the formula $E = mc^2$ and the work of the scientists mobilized for the Manhattan Project that led to the building, testing, and use of the atomic bomb if we do not probe too deeply. We can still make some sort of sense out of the distinction between inventing the calculus and inventing the steam engine. Systematic studies of what we have traditionally understood and experienced as science and technology, however, do not support the ideas of "pure" or "basic" versus "applied" science, or even the distinction between science and technology. The concept of "technoscience" draws attention to the difficulty of sustaining the science/technology distinction in the face of what scientists and technologists actually do in the laboratory, the workshop, and the factory. The reasons for this will become clearer the more the reader engages the entries in this work. There was a rationale for titling this project *Technoscience and Society*, but the popular understanding of science, technology, and society made it advisable to stick to the more readily recognized terminology. In a period when the fundamental term of "society" itself has come under critical scrutiny, we thought it best to let the title carry tradition and the entries draw the reader into the complexities that characterize contemporary scholarship. Even then, in many cases the newer ways of thinking will have to be followed from the entries at least into the recommended readings.

The messages here uniformly echo themes and variations on one of the rallying slogans of the early years of ST&S: "Science is social relations" (now necessarily applied also to technology and medicine). What does this slogan point to? What is the significance of "science, technology, and society" and the more disciplined "science and technology studies" for our understanding of science, technology, and medicine? We are still embroiled in controversies over what should be the central dogma of social studies of science, variously stated as "science is social relations," "science is socially constructed," and "science is a social institution." Although some scholars have made serious efforts to heal the historical rift between philosophy of science and sociology of science, and between the physical and natural sciences on the one hand and social studies of science on the other, those rifts continue to generate conflicting views about what it means to say that science is social. I will focus on science here, as it is the "hard" case for the sociology of knowledge. There is a reason we have been through the "science wars" and not the "technology wars." The reason is that classical paradigms for formatting our views of science still have an extremely strong hold on the intellectual community and especially on philosophers, historians, and scientists. In spite of postmodernism and its excesses, in spite of relativism and its misrepresentations, in spite of the sociology of science and its uninformed detractors, positivism, scientism, and echoes of the Unity of Science movement are still abroad in our time. The idea that scientific knowledge is a through-and-through social phenomenon, that it is socially constructed, is still—in spite of the evidence—troubling even to some of the people who have contributed to demonstrating this idea. This project has not required that authors be wedded to this perspective. However, it is hard to imagine how we could have come up with the programmatic notions of science and technology studies—let alone the more general idea of science, technology, and society—without some commitment to the idea that science and technology are social relations.

My aim in this project was to encompass research in science, technology, and society as opposed to focusing on the field of science and technology studies. I have worked closely with my editorial board, advisers, and consultants, as well as with the Oxford University Press reference division, on establishing the topics to be covered and the people to invite as contributors. We have drawn heavily but not exclusively on leaders in science and technology studies as contributors and editors. The diversities and complexities we collectively represent are held together by the idea of reciprocal relationships between science and society, technology and society, and medicine and society. The idea that science and technology have impacts on society and culture surely is by now (especially in the wake of Hiroshima and Nagasaki, the Indochina War, and environmental disasters and degradations, and, on the positive side, penicillin and the Salk vaccine) a truism. This way of thinking about science and society, and science and values, is not new.

The nineteenth-century social philosophers and theorists who crystallized the social sciences wondered critically about the nature of science in their writings on society and culture as objects of scientific and philosophical inquiry. The period between the two World Wars also saw the rise of a science and society movement, captured in the title of J. D. Bernal's *The Social Function of Science* (1939). It is during this period that we find the sociology of knowledge emerging, provoked by the emerging political and economic engagements of the "democracies" with fascism. In the 1930s the sociology of science and the history of science came into their own, joining the philosophy of science as the three fundamental analytical approaches to the study of science.

What stands out about the period of the 1920s and 1930s, by contrast with the late twentieth and early twenty-first centuries, is the resistance to bringing social, historical, and cultural perspectives to bear on the content of science; on scientific knowledge itself. Even though early students of science and society such as Karl Marx, Émile Durkheim, Friedrich Nietzsche, and Oswald Spengler had suggested in different ways that scientific knowledge, mathematics, and logic could be studied sociologically, the prevailing attitude that emerged with the professionalization of the sociology of knowledge and science was that there could be no social theory of scientific knowledge or scientific facts. Scientific facts were considered to exist in a realm outside of the blood, sweat, and tears of our everyday sensual and material world, outside of history, outside of society and culture. It wasn't until the late 1960s that this idea began to be systematically challenged in a sustained manner by those researchers who founded and have sustained the field of science and technology studies. It is not incidental that many of these pioneers were originally trained and educated in science, engineering, or mathematics before turning their attention to social studies of science and technology. Those who did not come from a science, engineering, or mathematics background were well versed in philosophy and the philosophy of science. This development is exemplified throughout this book in the contributions of scholars such as David Bloor, Trevor Pinch, Steve Woolgar, Karin Knorr Cetina, and others. The science, technology, and society approach that we have adopted here means that there is a great deal of attention in this book given to the practical and social problems that we face in the world today, locally and globally. Thus, readers will find entries that deal with such issues as "upper limits" and "replication" (core problems for those scholars who worry about the very nature of science). Readers who are more concerned with the science-in-society nexus will find that we have addressed the social problems of science, technology, and medicine, including AIDS, the destruction of the rain forests, and biological terrorism.

## Science, Technology, and Society

Traditionally, our knowledge of how science works was based on second- and third-hand accounts offered up by historians, philosophers, and sociologists of science, reminiscing scientists, and journalists, and in anecdotal essays, biographies, and memoirs. On the surface, technology may have seemed to be the more transparent case because its inherent sensuality and materiality linked it more directly to our everyday lives. But invention and discovery in science, technology, and medicine were equally removed from critical and analytical inquiry. Critical inquiry was the first step away from anecdotal journalism, worshipful history of science, normative-functionalist sociology of science, positivist and logicist philosophies of science, and hagiographies of scientists, engineers, and physicians. Such inquiries are found throughout the modern period from Marx's critique of "bourgeois science" and Nietzsche's defense of a "joyous science" or "science of joy" to Wittgenstein's philosophical anthropology and Theodore Roszak's attack on the "wasteland." The real breakthrough for those interested in how science works came in the 1970s. What if it were possible actually to watch scientists at work, day to day, hour to hour? The development of S&TS was given a boost when scientific laboratories were discovered as sites for ethnographic research. Beginning in the middle of the 1970s, two major scientific and technological centers became transformed into sites for ethnographic study: the Stanford Linear Accelerator Center in (actually beneath) Menlo Park, California; and the Salk Institute in La Jolla, California. As early as 1972, anthropologist Sharon Traweek had taken a part-time job in SLAC's Public Information Office. By the time she had committed herself to exploring her surroundings anthropologically, philosopher Bruno Latour was already at work at the Salk Institute, and his work there in collaboration with Steve Woolgar (a social scientist with an engineering background) became the first published ethnography of science (*Laboratory Life: The Social Construction of Scientific Facts*, 1979). Traweek's work was not published in book form until 1988 (*Beamtimes and Lifetimes: The World of High Energy Physics*). Karin Knorr-Cetina's ethnography, *The Manufacture of Knowledge*, a study of a biology laboratory, appeared in 1981. The use of words like "construct" and "manufacture" when talking or writing about "facts" eventually migrated to the field of technology studies and gave birth to the SCOT (social construction of technology) program, associated with such researchers as Wiebe Bijker, Trevor Pinch, and Thomas Hughes.

By the early 1980s, the idea that science and technology were socially constructed was beginning to be taken for granted in the swiftly crystallizing field of social studies of science and technology (variously still referred to as "science studies," "technology studies," and "science and technology studies"). In short order, however, the social aspect of the construction process came under attack. Social construction seemed to suggest that facts were arbitrary inventions and were not given in nature. In the second edition of *Laboratory Life* (1986), Latour and Woolgar changed the subtitle from *The Social Construction of Scientific Facts* to *The Construction of Scientific Facts*. The postmodern "death of the social" theme was echoed in Latour and Woolgar's postscript for the new edition in the section headed "The Demise of the 'Social.'" Their explanation for this eulogy was that, now that we understood that the social was pervasive, and that all interactions were social, the term "social" was no longer needed. This view of the social has become increasingly prominent in Latour's writings, most notably and most recently in his books *Pandora's Hope* (1999) and *Politics of Nature* (2004). Along with some other colleagues in science studies, I have argued that the idea of the "death of the social" reflects a philosophical failure to understand the social at all.

If conflicts and confusions about the social dominate the internal politics of social studies of science and technology, the science wars dominate the external politics of the field. The two conflicts are linked in the sense that the science wars are grounded in the failure of certain scientists to grasp the nature of the social and to grant the scientific credibility of the social sciences. Readers should be aware of the active resistance some scientists have mounted against the very idea of a social or cultural theory of science. Briefly, their contention is that "social construction" implies an anything-goes relativism or that we can make of the world anything we want and call it true. No one at or near the center of the social studies of science and technology makes or has ever made this claim. Part of the problem is the place of the social sciences in our society and culture. In spite of what many of us would claim are significant discoveries and systematic research agendas in the social sciences, the credibility and legitimacy of the social sciences is overshadowed by the political powers that have accrued to the physical and natural sciences as a result of inventions and discoveries that have visibly and pervasively changed the world and changed our lives. This potential exists within the social sciences and has been actualized in many instances. Unfortunately, the successes of the social sciences are buried in an invisible revolution. Social norms, values, beliefs, and taboos have been impenetrable barriers to the critical and challenging discoveries of the social sciences. There are, to put it briefly, many unresolved fundamental conflicts over quite basic ideas in the social sciences (and humanities) generally and in S&TS and ST&S specifically. This work will not, and was not designed to, resolve these conflicts, but neither was it designed to facilitate or otherwise sustain them.

Studies of science, technology, and medicine have, like other areas of research and theory in the postmodern era, been plagued by problems of inclusion. Feminists, for example, have made major contributions to the field. The incorporation of their ideas and findings into the core paradigms of the field, however, has been systematically resisted. The same is true for the works of Third World scholars. The Other—nonwhite, female, nonwestern, southern rather than northern—is not only a challenge to students of science, technology, and medicine but also still something of a threat. The Other is viewed as a threat to science, and a threat to science is a threat to Western conceptions of truth and logic. The social theories of mathematics and logic notwithstanding, scientism, positivism, and an uncritical love of science sans politics and ideology still lurk in the shadows of science/technology and society studies. Feminist and Third World theorists of science, technology, and medicine hold the key to breaking the grip that the classical Westernist view of science still has on contemporary S&TS and ST&S research. Their views have not been welcomed into the mainstream of social studies of science, technology, and medicine because they tend to portray science as Western, objectivity as masculine, technology as nonprogressive, and the whole system of science, technology, and medicine as a tool of violent imperialist colonial ruling classes. This, then, is another arena of conflict explored within the pages of this book. Given the constructivist or constructionist wars, the science wars, the broader culture wars, and the wars of the Other, the format for this book could have been more self-consciously contentious, more provocatively polemical. While I have felt obligated to acknowledge and announce these conflicts in this introduction, the entries in this book demonstrate the positive potential of S&TS and ST&S for challenging and changing the ways in which we think about, make policy for, and practice science, technology, and medicine.

Scientists are quite prepared to acknowledge the social nature of science. Indeed, they seem to think that "revealing" this through social science research simply tells them something they already know. They are not so ready to admit or understand what we mean when we claim that scientific facts are social. The fact that science is social does not mean, they claim, that scientific knowledge is social. Let us look closely at just what it is that the science studies researchers are claiming. It is simply this—that scientific facts do not fall out of the sky, they are not "given" to us directly, we do not come to them by means of revelation. We do not simply look, see, name, and know. Human beings have to do work, and they have to work collectively, in order to construct facts. That work is embodied in the fact, just as the collective toil of the multitude of workers in Rodin's workshop is embodied in *The Thinker*. This is what it means to say that a fact is socially constructed. This does not mean, however, that the construction is arbitrary and can be carried out in any way we please. Clearly, constructing oars for rowboats is a product of human work, but oars cannot be and are not constructed in any way we please. Anthropologists refer to this as the principle of "limited number of possibilities." We are constrained to think, to do, and to make by the spatial and temporal contexts of our existence, by the degrees of freedom in our movements and our environments, and by the properties of our resources. In the end, it is not "nature" that we construct but the facts of nature formed out of the resources of our languages and our cultures. The examples of the oar and *The Thinker* suggest how we might think about those parts of our culture that are conventionally thought of as "abstract," "symbolic," "immaterial," or "transcendental." Abstractions and symbols are "things," they are things that we construct.

If we can understand that all of our thinking, doing, and making gives us all of the ingredients of culture, scientific and otherwise, we are still left with the problem of explaining the role of the individual in this process. Conventional wisdom in the West, influenced by philosophers from Plato to Descartes, credits individuals and especially geniuses with creativity and originality. Social and cultural influences and causes are minimized, ignored, or eliminated from consideration at all. Thoughts, original and conventional, are identified with individuals, and the special things that individuals are and do are traced to their genes and their brains. The "trick" here is to recognize that individual humans are social constructions themselves, embodying and reflecting the variety of social and cultural influences they have been exposed to during their lives. Our individuality is not denied, but it is viewed as a product of specific social and cultural experiences. The brain itself is a social thing, influenced structurally and at the level of its connectivities by social environments. The "individual" is a legal, religious, and political fiction just as the "I" is a grammatical illusion (as Nietzsche recognized). Less philosophical conceptions of the self or person developed in the writings of George Herbert Mead, C. Wright Mills, Erving Goffman, and Randall Collins provide empirically and theoretically credible defenses of this idea.

My colleagues and I are often asked by scientists to explain what difference our work makes to what they do. In some cases, this question has an aura of anti-intellectualism about it—why bother studying and doing research if it is not going to make an immediate and visible difference in our lives? There is no reasonable answer to the question framed in this way.

Our defense of what we do must be that it needs no defense beyond the need to know, the intrinsic value of satisfying our curiosity. There is, however, a more positive answer to the question. The "science of science" will give us a more complete understanding of science and facilitate our ability to carry out research with a better understanding of what we are doing. It should be obvious to skeptical laboratory scientists that the more perspectives we can bring to bear on a phenomenon, the more we will know about it. This could even serve as a definition of objectivity: the more perspectives we bring to our study, the more objective our understanding becomes (Nietzsche already recognized this in his proposal to bring as many "eyes" to bear on a phenomenon as possible). Perhaps we need go no further than to contemplate Dr. Jonas Salk's remarks in the wake of the study of his laboratory by Latour and Woolgar. "Such studies" (in this case, an anthropological study of Salk's own Salk Institute for Biological Studies), Salk wrote in his introduction to the first laboratory ethnography, "can help scientists understand themselves through the mirror provided, and help a wider public understand the scientific pursuit from a new and different and rather refreshing point of view." Can we expect such research to change the way we do physical and natural science, and social science? The answer, if we take science itself and its history seriously, must surely be *yes*.

Can we already know what kinds of changes to expect? I do not think so. Perhaps, however, it is already clear that greater self-consciousness about science as a social process, greater scrutiny of science, and a more critical approach to understanding science can bring and indeed has brought ethical and value issues embedded in science to the surface. It has also influenced what we *call* science, as our investigations have taken us into the knowledge systems of other cultures and even into the knowledge systems of subcultures in our own society. We can perhaps get a sense of the sort of influence science and technology studies can have on how we *do* science and technology by considering the impacts of university internal review boards, ethical oversight committees, science shops, the impact of AIDS activists on AIDS research, and (for a few years at least) the former Office of Technology Assessment of the U.S. Congress. This may not be the kind of "change" that physical and natural scientists are thinking about when they query us. It may be that what is at stake here is the very nature of human inquiry. Science is just a systematized variant of our general capacity for inquiry, and science and technology studies is just another science. The same factors that have given rise to S&TS and ST&S and their influence on scientific practice are all beginning to change how we do science, who does science, in what contexts, and toward what ends. But then, changing agents, contexts, and ends is in fact changing how we do science. In the meantime, it seems that we can reasonably expect to improve our understanding of and self-consciousness about science, technology, and medicine as social processes and social institutions.

If we broaden our perspective to consider the science, technology, and society movement, we can see how some of the fundamental ideas being studied and refined in science and technology studies owe a debt to the ST&S pioneers who stressed policy and educational issues and problems. That debt has roots in the programs that developed in the 1960s and 1970s in the United Kingdom at the University of Sussex, and in the United States at Pennsylvania State University under the leadership, respectively, of Christopher Freeman and Rustum Roy. Their ideas and programs continue to resonate in many of our undergraduate college programs and departments in Science, Technology, and Society and in the annual meetings of the International Association for ST&S (IASTS, founded in 1988 as the National Association for Science, Technology, and Society). If the Society for Social Studies of Science (4S) represents the "high church" in S&TS/ST&S, IASTS represents the "low church." In this context, the "high church" focuses on research and theory (the interests of a professional discipline), whereas the "low church" focuses on social justice, participatory democracy, technological literacy, and the impact of science and technology on society. The low church has been an important conduit for moving high-church ideas, concepts, and theories into the schools and colleges and into the arenas of public discourse on science, technology, and medicine. Perhaps this book can be one of the steps toward what Robin Williams and David Edge once described as "broad church."

*Sal Restivo*

## Science and Society

Science and society are best seen as co-constructed: each has been built up with the other, in an endlessly recursive process. The interchanges between science and society are so thick and deep that finding a line to separate one from the other may be impossible. Societal forces are implicated in scientific practice at every turn, just as the course of human history has been steered consequentially by scientific knowledge. Sorting out the reciprocal causes and effects of science and society is useful as an analytical tool, but it should not obscure the reality of a relationship that is blurred, overlapping, and mutually constitutive.

Science provides knowledge that is then more or less taken up by people in the rest of society, for their own diverse purposes. However, scientific knowledge does not enter society as the only epistemic game in town. An explanation of the origin of species grounded in evolutionary theories of Charles Darwin continues to compete with creationist accounts of human origins grounded in a literal reading of the Christian Bible. Science provides knowledge that some may consider superior or authoritative, but only if rival understandings—religion, ideology, art, practical wisdom, common sense—are made to be different and delegitimated as less worthy or reliable. The transit of knowledge from a small community of scientific practitioners out to a huge society of potential consumers rests on complicated judgments about the credibility of scientists and the utility of scientific knowledge, as measured by ordinary people against competing nonscientific understandings and beliefs. Scientific truth does not shine by its own lights: if people in society accept $E = mc^2$ or the double-helical structure of DNA as the best account of natural reality, they have already made a prior decision that the methods and institutions of science stand above other available means to produce knowledge.

Science enters society via technologies that continue to transform the human condition. Medical research discovers cures for dread diseases, physical and astronomical calculations enable travel to the moon, biotechnology enhances the productivity of rice and wheat needed to feed hungry children. Much hinges on the belief that science has made for a better world. The cultural authority of scientists in society may be traced back to people's assumption that technological progress is dependent upon scientific research and knowledge. But is it so? The boundary separating science from technology is itself blurred and overlapping, with arrows of causation running in both directions. Historians tell us that progress in physics depended upon the steam engine more significantly than the invention of the steam engine depended upon abstract theories of physics. Engineers, technicians, scientists, craft workers, entrepreneurs, inventors, investors, managers, and many others contribute to the development of salutary machines and devices, often in a seamless way, and with no particular sequence (science may, or may not, be the fount). Science often gets the credit for technologies when they work, but less often it gets the blame for failures. When the *Challenger* space shuttle exploded, responsibility for the accident was placed on managers and engineers embedded in an organizational culture at NASA that allowed signs of impending disaster to be ignored—but not on imperfect science.

Science also enters society as a means to ratify legislative policies and legal decisions. Science is widely assumed to be a refuge of objectivity in a sea of self-interest and parochial opinions—in part because its methods for producing knowledge (experiment and quantification, for example) are assumed to be governed by procedural rules that reduce or eliminate personal discretion. Scientists turn up in courts and hearing rooms, where they provide expert testimony thought to provide a reliable cognitive basis for decisions involving environmental impact or medical risk. However, scientific knowledge is rarely decisive in settling things once and for all. Often, experts disagree on the state of scientific understanding or on the pertinence of some theory or fact for a particular case at hand—giving rise to the "hired gun," the expert who is paid to bring in science to support one side of an adversarial deliberation, in the process raising doubts about the supposed objectivity of scientific method and belief.

Society both constrains and enables the pursuit of science. Scientific research depends on copious material and organizational resources that cannot be generated by scientists on their own. Whether patronage is provided by the Medici dynasty or the National Science Foundation, the financial wherewithal for laboratories, equipment, salaries, expeditions, and publications must come from somewhere in society, and it usually does so with strings attached. Scientists working under the Nazi regime were expected to produce not only evidence of Aryan superiority but the technological means to ensure military victory. Scientists working under a capitalist regime in the pharmaceutical industry are expected to limit their research to lines of inquiry likely to have beneficial consequences for the corporate bottom line. Even the growth or decline of pure (or basic) science depends upon the willingness of people in society to allow scientists the autonomy and freedom to pursue knowledge wherever they wish. Such license is not automatic: animal rights activists, for example, have bombed laboratories in hopes of eliminating research that uses dogs or monkeys as experimental subjects.

Near-future relations between science and society will depend largely on whether scientific knowledge is defined as a *public* or a *private* good. Who owns science, and who shall benefit from it? As a public good, the practice of science is both accessible and accountable to the needs and interests of society at large, and scientific knowledge is available to all. As a private good, scientific knowledge becomes a commodity, produced for the benefit of those who pay for it and owned by them. Eventual resolution to current debates over human cloning and intellectual property will say much about whether the future of science is steered by democratic negotiation or by the market.

## BIBLIOGRAPHY

Gieryn, Thomas F. *Cultural Boundaries of Science: Credibility on the Line.* Chicago: University of Chicago Press, 1999. An exploration of the competition between science and its rivals for cultural authority.

Jasanoff, Sheila. *The Fifth Branch: Science Advisers as Policymakers.* Cambridge, MA: Harvard University Press, 1990. How scientific experts get involved with democratic decision making.

Kloppenburg, Jack Ralph, Jr. *First the Seed: The Political Economy of Plant Biotechnology, 1492–2000.* New York: Cambridge University Press, 1988. An analysis of how corporate interests have steered the research agenda, with help from a compliant government.

Ravetz, Jerome R. *Scientific Knowledge and Its Social Problems.* New Brunswick, NJ: Transaction Publishers, 1996. Synoptic account of how society insinuates itself in science, with a focus on the difference between pure and applied research.

Vaughan, Diane. *The Challenger Launch Decision.* Chicago: University of Chicago Press, 1996. On the relationships between science and technology, observed in an episode of failure.

*Thomas F. Gieryn*

# Technology and Society

What is the rationale for studying technology and society? It is a commonplace that we live in an advanced technological world. Not a day goes by without the announcement of a new innovation or invention, without concerned reports about the adverse effects of some new technical system, or without a breathless expression of the likely virtues of some new technology.

The *Oxford English Dictionary* offers the definition of technology as "a discourse or treatise on an art or arts; the scientific study of the practical or industrial arts." The origins of the word stem from the Greek word *techne*, so we might expect the term to have a long history. But in fact the first recorded use of the term is in the early seventeenth century. Significantly, this coincides with the emergence of eras of specialization and industrialization in modern Western societies. So the appropriation of the Greek term at that time signals an emerging sense of specialized expertise or knowledge in relation to particular areas of skill—the very idea that some aspects of life and activity can be thought of as "technical." This in turn has provided an abiding and profound legacy for our current common conceptions of the relation between "technology" and "society." In particular, the latter term is taken to denote those residual, nonspecialist areas of life that are thought to somehow exist separate from and outside the technical. In an important sense, recent research on technology and society offers an important challenge to this conceptual separation between technology and society. It suggests that we need to look beyond the task of merely illuminating connections and relations between technological and societal aspects of modern life; it instead proposes that the very separation between them, understood as the contingent outcome of a particular historical period, needs now to be rethought.

Since the first emergence of the term, technology has come to encompass a vast range of areas and kinds of application. Tools and instruments of many kinds—although often predating the industrial revolution, for example, those in early primitive societies—are now spoken of as technology. We can think of areas as diverse as industrial engineering and production management; energy conservation and management; farming and agricultural technology; chemical process industries; medical technologies from birth control to diagnosis and treatment to the control of the spread of infectious diseases; military technologies and weapons; climate control and weather prediction systems; building construction and architectural technologies; transportation technologies; and so on. In recent years, particular attention has been given to two areas: biotechnologies, especially those involved genetic engineering and genetically modified organisms; and information and communications technology. The latter category itself encompasses a vast range of activities, including broadcasting, information systems and processing, photography, video and films, telecommunications, computer systems, Internet and World Wide Web–based technologies, and mobile telephony.

Unsurprisingly, given this vast range of technological applications, the ranges of questions about them are legion. In almost every instance, the development and implementation of any technology raises questions about the possible effects on existing social arrangements. Thus, for example, just within the sphere of business and management, the possible introduction of a new technology invites concern about how this will generate organizational change. Will the new system help reduce costs, will it provide new business opportunities, and will it facilitate and support strategic change? Faced with (usually much hyped) descriptions of the advantages of the new technology, managers need to decide whether they actually need the new technology. Should the business purchase it? Will it lead to competitive advantage? How should the new technology be installed and deployed? What will be the effects on the workforce? How should technological change be managed?

## Technology and Technocracy

From another perspective, the very idea of technology has been associated with particular views about how societies might be organized. In particular, the term *technocratic*—often used loosely in contemporary usage as a kind of denigration of officialdom, which is insensitive to the human dimensions of organizations—in fact stems from the deliberate attempt to apply principles of technological progress to modes of government in the United States of the 1930s. Building on the enthusiasm for Frederick W. Taylor's ideas of scientific management, supporters of the principle of technocracy proposed a form of government that would be run by technicians guided solely by the imperatives of their technology. The argument was that because businessmen were proving themselves inept organizers and managers and because they were unable to bring about reforms of industry in the public interest, the task of reorganizing should be given to engineers. Organizations such as the Committee on Technocracy, in 1932, were formed to promote the principles of organizations and business better founded and more efficiently run on the principles of technical expertise rather than by long-standing custom and practice. This enthusiasm for, and optimism about, the application of the principles of technology to the problems of societal organization echoed some of the contemporary moves in Soviet Russia to implement economic and societal organization based on rational and "scientific" principles of planning and control.

But if the idea of technology was taken by such movements as exemplary of the principles of positive societal organization, the polar opposite view was taken by critics, such as by the Frankfurt School of "critical theory." This German philosophical and sociological movement, dating from the 1920s, promoted the idea that the current ills of society were due in large part to the unacknowledged ideological assumptions and interests that pervade science and technology. So the application of principles of scientific and technological rationality to problems of societal organization could be tantamount to embracing nontheoretical and repressive interests. Critical theorists especially objected to the idea of value freedom in the Weberian conception of science, insisting instead that aspirations for value neutrality were nothing more than ways of stifling debate about the true underlying interests of those promoting ideals of rationality and efficiency. Endemic value judgments were concealed by the cloak of what purported to be self-evident reasonableness. In this view, a central, objectionable, yet unchallenged assumption has been the desirability of the technological domination of nature.

## Technology and Science

As can be seen from the discussion thus far, depictions of technology frequently embrace an implicit view of the connection between science and technology. In particular, it is assumed that science provides the rules and laws of nature that are then the basis for their instantiation in the shape of technologies. The pervasiveness of this view is signaled by the fact that definitions of technology commonly portray it as "the application of science to the practical aims of human life or to the change and manipulation of the human environment," offered in the *New Encyclopaedia Britannica, Micropaedia,* vol. 11. The same source notes the more recent elision of the terms: by the late twentieth century, the term *technology* "had become a global term connoting not only the tangible products of science but also the associated attitudes, processes, artifacts, and consequences." However, in recent years, the view that science and technology are related in this way has been challenged. It is now thought preferable to understand science and technology as intimately connected and, in particular, to regard scientists and technologists as two largely overlapping social groups or institutions. This reflects more accurately the interpenetration of the two domains, the constant blurring and redefining of what counts as "pure" and "applied," the transition of personnel between the two areas and the bidirectional flow of ideas and applications. This change also has consequences for the extent to which our research perspectives on technology can fruitfully adopt and extend analytic issues in the social basis of science. The term *techno-science* has emerged to denote this fusion of both substantive and analytic concerns.

## Definitions and Usages

The conceptualization of "technology" as largely ancillary to or separate from "society" has given rise to a variety of definitions and understandings of technology. Technology is used to refer to the inner workings of some mechanism or machine. At a larger scale, it refers to the same workings of a system of devices and mechanisms. Less commonly, it is used to encompass a whole series of social and other arrangements.

At the core of these ambiguities in the definition of technology is the extent to which various understandings of "society" are allowed in; in other words, the extent to which the social is thought to be involved with technology. Thus at one extreme, technology seems to refer strictly to those aspects of operation deemed not to involve the human element. For example, one is familiar with the reference in common parlance, to "technical problems": "We apologize for the late arrival of the train due to technical problems." The phrase can be heard as defining and circumscribing possible engagement with the phenomenon by human agents. It is being claimed as one of those problems that is amenable to the skills and knowledge of only a specific group of skilled individuals. The situation is only understandable and (hence) solvable by this restricted group of experts; it is not something that the nontechnical ordinary passenger (in this case) would understand or need know about. "Cars" are sometimes spoken of in these terms, with an emphasis on the mechanical components, as if such components were not themselves shot through with considerations of the human and social elements involved in their production and use.

In another usage, "technology" admits the involvement of humans (and hence the psychological and the sociological), as when, for example, one speaks of the technology (machinery) of government, or of society as a machine. On the whole, this metaphorical kind of usage is usually intended to draw attention to the (remorseless) workings of a system of devices and machinery (usually) to the detriment of the mere humans involved. The metaphor has also recently been applied to business phenomena. For example, researchers have spoken of the technology of management techniques, or the technology of strategy. With a more sophisticated usage of the metaphor, drawing on Foucauldian themes to understand the emergence and growth of such disciplines as psychology and psychiatry, Nikolas Rose uses the term *technologies* to refer to "the technical assembly of means of judgment, the techniques of reformation and cure, and the apparatuses within which intervention is to take place."

It is important to consider the utility of applying metaphors about technology to ostensibly nontechnical phenomena. On the one hand, the application of the metaphor draws attention to the systematic arrangements involved in social phenomena. Thus, to speak of "the school system" or "clinical examinations" as technologies, is to suggest that these systems have been designed with specific intended purposes, that they operate according to preconceived means, and that they are anticipated to have particular effects. One interesting aspect of this metaphorical usage is that it implies that a full understanding of the rationale, the operation, and the effects of this "technology" is restricted to a specific group or groups. In other words, the description of this social system as a technology highlights the ways in which knowledge and power are socially distributed. One part of this is the suggestion that only some people know and are (possibly) capable of changing the situation and that they possess the skills and expertise to determine whether and to what extent the system is working effectively or failing.

Nonetheless it could be argued that this kind of usage remains metaphorical in the particular sense that the school system is not commonly thought of as technological in the same way as, say, the assemblage of electronic components on the motherboard of a PC. Notwithstanding the advantages of exposing the technological features of a relatively (perceivedly) "soft" social system (the school system), it arguably remains more difficult and challenging to expose the societal dimensions of a "hard" technical system (the computer motherboard). To get this balance right, we need to review the various available perspectives on the relation between technology and society.

**BIBLIOGRAPHY**

Agar, J. *The Government Machine: A Revolutionary History of the Computer.* Cambridge, MA: MIT Press, 2003.
Barry, A. *Political Machines: Governing a Technological Society.* London: Athlone Press, 2001.
Feenberg, A. *Questioning Technology.* London: Routledge, 1999.
Rose, Nikolas. *Governing the Soul.* London: Free Association Books, 1999.

*Steve Woolgar*

# Medicine and Society

Medicine and society are inextricably intertwined. Medicine—the complex of meanings and practices established to understand, classify, study, manage, and eradicate disease—is an inherently social phenomenon. It comprises a constellation of interconnected institutions, ranging from public health authorities and hospitals to medical schools and professional organizations. The production of biomedical knowledge is constituted with and through the technological innovations, social structures, and cultural priorities of a given community. Framed by developments in technoscience and by the sociocultural significance imparted to disease, enacted through the governance of disease by bureaucracies and health professionals, medicine is a unique site at which to understand human societies. Social concerns are evident in medical theory and practice, and medicine has profoundly influenced the social world in turn.

The spread of contagious disease has been linked to changes in social organization. At the beginning of the historic era, for example, the transformation from diffuse hunter-gatherer groups to centralized, agriculturally based communities created the conditions necessary for waterborne diseases to develop. In the nineteenth-century United States the interlinked social processes of immigration, industrialization, and urbanization concentrated human populations in dense, poorly sanitized metropolitan living and working arrangements that often became seedbeds for the transmission of infectious diseases. The U.S. cholera epidemics of 1832, 1849, and 1866 were thus born not only of the bacterium *Vibrio cholerae* (which would be identified by Robert Koch in 1884) but also of these economic and demographic shifts in which it thrived. During the epidemic waves of the 1830s and 1840s, American cities were ravaged by cholera; the available therapies to combat it proved highly ineffective. City officials mandated the quarantine of actual and perceived cholera sufferers in the interest of public safety—a medically justified measure that had extramedical implications for those groups stigmatized as carriers of the disease. Specifically, moral meanings were imputed to this segregation; groups such as immigrants and the impoverished, who were more likely to be exposed to cholera for environmental reasons such as overcrowding, polluted water, and inadequate sanitation, were unfairly stigmatized as predisposed, innate carriers of the disease. In this way, fears of disease were overlaid with fears about the changing nature of American life in the nineteenth century.

Medical researchers, health reformers, city planners, and politicians together marshaled their efforts to help combat the disease. Combining insights from the application of scientific techniques to medicine as well as microscopy and population statistics, a nascent public health system was eventually able to slow the transmission of cholera dramatically by the epidemic wave of 1866. The conditions for the three waves of cholera epidemic arose from a confluence of socially produced transformation; communities responded to these changes with the creation of institutions to foster the public's health but also with the construction of categories of social stratification around disease status. These "cholera years" provide one illustration of the ways in which medicine and society together form a feedback loop.

The rise of scientific medicine—in particular, the introduction of the experimental method and laboratory technology—afforded communities some independence from the capricious ebb and flow of epidemics such as cholera throughout history, as well as the loss of human capital and social order that accompanied them. From the 1870s to the turn of the century, the microorganisms responsible for most infectious diseases were identified. Subsequently, the majority were contained with antibiotics and vaccines in the late nineteenth and twentieth centuries. These treatments—often called "miracle cures" because of their dramatic results and far-reaching social impact—transformed the rhythms of human life, extending average life expectancy, decreasing infant mortality rates, and stimulating population growth. (Of course, the extent of these improvements varied by race, socioeconomic status, gender, and region.)

## Germ Theory, Antisepsis, and Anesthesia

The trajectory of medical inquiry has been marked by several milestones. Among the most important of these was the discovery that microorganisms were agents of contagious disease. "Germ theory" (the theory and science of disease etiology via the entry of microorganisms into the body) and bacteriology proceeded from a series of important observations in the nineteenth century by leading scientists including the French chemist and microbiologist Louis Pasteur and the German physician Robert Koch. In the second half of the nineteenth century, based on studies that demonstrated that airborne microorganisms were responsible for the contamination (fermentation and putrefaction) of wine and beer, Pasteur hypothesized a causal link between bacteria and infectious disease. Building on Pasteur's work, Koch later identified the bacteria that caused pulmonary tuberculosis and cholera. He also refined Pasteur's "germ theory" by establishing a set of standards—known today as Koch's postulates—by which a microorganism could be proved to cause disease. In the moral universe of cultural meanings, "germ theory" meant that disease was not the outward manifestation of innate, internal defects (often attributed to African Americans) or a sinful disposition (often attributed to non-Christians) but rather the result of external invaders. Thus, the "germ theory" of disease, in some measure, undermined forms of medical racialism that stigmatized some social groups as more inherently prone to disease than others. In its aftermath, diseases were seen in somewhat less explicitly moral terms, linked to the failings of a particular ethnic or class group, and more as opportunistic and universal scourges. (Moralizing and accusatory language about infectious disease would again be encountered over a century later in the case of the highly stigmatized disease HIV/AIDS, which was once referred to as both the "gay plague" and the "Haitian plague.")

Germ theory also had an impact on treatment techniques. Owing to ineffective sanitary measures, it was not uncommon for patients to become more ill during hospitalization than they were before they were admitted. Pasteur's observations led him to encourage the sterilization of instruments and linens in French hospitals by exposure to high temperatures through steaming and boiling. British surgeon Joseph Lister refined sterilization by introducing systematic antiseptic techniques to surgery, including the use of carbolic acid, which significantly decreased postoperative infection. Lister's antisepsis system was among the first direct applications of bacteriology (although the Austrian physician Ignaz Semmelweis had reduced hospital infections by introducing antiseptic practices some twenty years earlier).

Another advance in surgical method was the development of anesthesia, which diminished the experience of pain related to surgical procedures for patients. Antisepsis and anesthesia increased patients' confidence in medicine's ability to cure illness and eliminate pain. They also advanced the cause of medical science by enabling more complex and time-intensive surgical procedures.

## The Changing Status and Expectations of the Physician

The introduction of scientific methods into medical practice transformed the profession as well as its object. Until the late nineteenth century, doctors were not required to have studied medicine and were relied on mainly to provide comfort and guidance to their patients. As the practice of medicine shifted from cure to prevention, doctors were now expected to provide results based on scientific evidence. As a result of this access to forms of knowledge beyond the understanding of the general public, more authority and power was granted to the medical profession, and the nature of the doctor/patient relationship changed. Once the source of a disease was ascertained, patients expected that doctors should be able to cure them. Additionally, those doctors with scientific training were now distinguished from a range of alternative healers, from homeopaths to midwives to quacks, resulting in an elevation in the eyes of the public of the status of the profession vis-à-vis other healing practices, which persists today.

## Vaccines and Society

With the causal agents of many of the leading infectious diseases identified, researchers began to seek treatments capable of destroying illness-causing microorganisms. The development of immunizations for bacterial and viral disease had begun long before the mechanics of infectious disease was widely known. In the first half of the nineteenth century, Pasteur had success, following the prior work of British physician Edward Jenner, vaccinating sheep against anthrax and chickens against cholera. Bacteriological principles prompted researchers to link this early work on immunization theories of germ transmission. Based on the theory that, although significant exposure to a microorganism could cause disease, limited exposure to a weakened form of a germ afforded some protection, or immunity, vaccines against bacteria and viruses would subsequently be developed.

Few vaccines had as dramatic an impact on modern society as those developed by Jonas Salk and Albert Sabin in the 1950s to prevent the spread of poliomyelitis. In the early twentieth century, polio was a leading cause of death in the United States. Although the microorganism that causes the paralyzing and crippling disease was identified in 1908, when the epidemic hit its high point some four decades later there were still no effective diagnostic techniques, vaccines, or therapies for the disease. Unsurprisingly, when the U.S. polio epidemic reached its height in 1954, killing and disabling thousands, there was a national panic. In the face of this crisis, American communities fell back on long-held assumptions about the connections between uncleanliness—literally and figuratively—and ill health. Thus, as with earlier epidemics of tuberculosis and cholera, the poor, immigrants, and racial minorities were cast as vectors of infectious disease; disease status was linked to economic status, presumed moral status, and citizenship status. Such commonly held beliefs in professional and lay circles about the association between social status and disease status impeded the epidemiological investigation into the transmission of polio. Medical researchers were not immune to commonly held assumptions that the poverty and foreignness were determinants of disease. It would take the impeccably patrician President Franklin Delano Roosevelt's revelation of his bout with polio and increased rates of the disease in middle-class communities, especially suburban children, as well as new epidemiological insights, to alter this dynamic. The change in perspective about disease susceptibility opened minds to the possibility of other causes of polio. In 1955, the Salk vaccine was introduced, and within a year, polio mortality rates had declined by 50 percent. Albert Sabin's more effective oral vaccine was released in 1957. Today, polio has been all but eradicated in the United States.

## The Antibiotic Revolution

The discovery of penicillin by Alexander Fleming in 1929 heralded the promise of another medical "magic bullet": antibiotics. Before these drugs became known, infection of a small injury could lead to serious illness and even death. Observing that a plate culture of *Staphylococcus* bacteria had been contaminated by a mold, *Penicillium,* and that bacteria adjoining the mold dissipated, Fleming was led to the conclusion that a fungal substance had the ability to eliminate bacteria that cause disease. In a publication of the results of his subsequent investigations into the mold, Fleming noted that the active substance, which he named penicillin, might have medical usage if it could be distilled and produced in large quantities. During World War II, medical scientists had limited success with small amounts of penicillin used to treat battlefield injuries, a feat that transformed the war effort by eliminating the threat of death once posed by infection. Soon after, British scientists working with U.S. manufacturers successfully took up Fleming's challenge, refining methods for producing the antibiotic on a large scale. Penicillin was soon hailed as a "miracle cure," capable of curing a wide range of ills from blood poisoning and strep throat to sexually transmitted diseases. Penicillin was without a doubt one of the most important medical discoveries of the last century. With several chemical varieties, it remained the most widely used antibiotic as the twenty-first century began.

## Challenges for Medicine in the Twenty-first Century

The nineteenth and twentieth centuries ushered in many impressive techniques capable of precisely identifying disease agents, developing vaccines to prevent their transmission, and producing antibiotics to eradicate infectious illness; in the process, they transformed the tenor of human life and the social world. Yet the emergence of scientific medicine has also brought unexpected challenges in the form of new diseases and chronic illnesses. Within a few years of the successful mass production of penicillin, some quickly mutating bacteria developed resistance to it. The situation worsened in ensuing decades, and presently there are numerous pathogens—many commonly found in hospitals—that demonstrate multiple resistance to penicillin and other antibiotics. For example, streptomycin, the antibiotic treatment long used to treat tuberculosis, has proven ineffective in some instances, in the face of drug-resistant strains of the bacteria that cause TB. Tuberculosis is on the rise in poor and disenfranchised communities in the United States and abroad because of inadequate access to health care access and the persistence of living conditions that increase the opportunity for disease transmission as well as the evolution of drug-resistant strains of the disease.

The longer life spans made possible by the many successes of modern medicine have produced unexpected consequences in the form of chronic and degenerative conditions such as cancer, heart disease, and Alzheimer's disease, conditions that did not exist in such high frequency when humans lived for fewer years but that now are among the leading causes of death. We have also seen the emergence of autoimmune diseases such as lupus. Unlike the infectious diseases of prior decades, there are currently no cures for these diseases.

**BIBLIOGRAPHY**
Farmer, Paul. *AIDS and Accusation: Haiti and the Geography of Blame.* Berkeley: University of California Press, 1993.
McBride, David. *From TB to AIDS: Epidemics among Urban Blacks since 1900.* Albany: State University of New York Press, 1991.
Rogers, Naomi. *Dirt and Disease: Polio before FDR.* New Brunswick, NJ: Rutgers University Press, 1992.
Rosenberg, Charles E. *The Cholera Years: The United States in 1832, 1849, and 1866.* Chicago: University of Chicago Press, 1990.
Starr, Paul. *The Social Transformation of American Medicine.* New York: Basic Books, 1982.

*Alondra Nelson*

# SCIENCE, TECHNOLOGY, AND SOCIETY

## AN ENCYCLOPEDIA

# A

## ABORTION/ANTI-ABORTION CONFLICT.

The U.S. Supreme Court's 1973 ruling in *Roe v. Wade* legalized induced abortion in the first trimester of pregnancy "free of interference by the State" (paras. 18, 126); in the second under regulation that "reasonably relates to the preservation and protection of maternal health" (paras. 19, 125); and in the third "when it is necessary to preserve the life or health of the mother" (paras. 20, 127). In *Roe* the court interpreted women's right to terminate pregnancies and physicians' right to perform abortions as within constitutional rights to privacy; abortion in the United States is framed as a "negative" right to act without government interference, rather than an "affirmative" right guaranteed by the state.

Legal abortion preserves women's lives from unsafe pregnancies and nonmedical abortions. For feminists, the "right to choose" abortion also symbolized women's liberation from patriarchal control. Although "pro-choice" rhetoric is consistent with American values of self-determination, "choice" may be a misguided symbol for reproductive rights; many women experience abortion not as desirable but as an unfortunate necessity following failed contraception, forced intercourse, or "positive" diagnosis of fetal abnormality. Further, women's ability to "choose" an abortion in the United States remains subject to some federal and state-level legislative restrictions. The 1976 Hyde Amendment restricts Medicaid-funded elective abortions, and in 2000, 87 percent of U.S. counties lacked abortion providers. Reproductive rights groups have run into conflict with disability rights activists over the possible eugenic use of elective abortion following prenatal screening, such as for Down syndrome, to limit the range of acceptable human life.

*Roe* ultimately gives physicians, not pregnant women, the ability to determine whether and when abortion is warranted. In the nineteenth century, women of all social classes could legally procure abortion, often using herbal abortifacients. As "regular" physicians distinguished themselves from midwives and homeopaths, many lobbied state legislatures to criminalize induced abortion. Shortly after its formation in 1847, the American Medical Association (AMA) declared human life to begin at conception and not, as women apparently believed, at "quickening," midway through gestation, when a woman first feels fetal movement in the womb. In taking an anti-abortion stance, physicians not only professionalized but moralized their practice through association with saving lives. By end of century, abortion was criminalized throughout the United States and recognized to be a medical issue. It is an historic irony that abortion was medicalized to restrict its practice, only to be legalized a century later precisely based on its status as medical procedure, a private matter between patient and doctor.

Since *Roe*, "pro-life" activists redefined the question of "life" in the abortion controversy as fetal, rather than maternal, right to life. Legalization galvanized opposition from pro-life activists, for whom abortion eroded traditional gender roles and women's moral standing as childbearers. Right-to-life activists return to early AMA formulations of distinct life beginning at conception. Appropriating prenatal medical imaging technologies, they produce propagandistic displays. Photographic and ultrasound images of

1

free-floating fetuses are used to portray fetal life as not only viable but autonomous, suggestive of personhood and rights. Pro-life campaigns project images of intact, well-developed fetuses despite the fact that 90 percent of abortions in the United States occur during the first trimester. Radical pro-lifers view abortion as a holocaust and a symbol of America's moral degeneracy. Activists commit property crimes including arson at abortion facilities, and during the late 1980s and 1990s several abortion providers were murdered by extremists who championed a Christian nation at millennium's end where God's law would prevail over human law.

Following the second Bush Administration's 2001 signing into effect of the Mexico City Policy (called the "global gag rule" by reproductive rights advocates), developing countries have felt the effects of American abortion controversies. This policy, cutting off U.S. international aid money to organizations directly or indirectly engaged in abortion-related services including referral, hampers the promotion of contraceptives and HIV awareness. Illegal abortions continue to be a significant cause of morbidity and mortality among women of reproductive age throughout the nonindustrialized world.

The legal status and availability of abortion is often determined by government interest in population control. The People's Republic of China's one-child policy, which is not evenly enforced across the country, has been implemented in part through mandatory abortions; Ceausescu's regime in Romania, on the other hand, attempted to raise population growth by banning abortion and contraception. Under state socialism, reproduction has figured as a form of production subject to government regulation. The Soviet Union provided workers with medical abortion as routine birth control; in impoverished 1990s Russia, abortions, having been routinized, outnumbered live births by two to one.

Japan legalized abortion in 1948, when families and government shared interests in reducing family size. Women in Japan and southeastern Europe turned to abortion during World War II to cope with wartime poverty. Initially used by mothers to limit family size but now practiced by women of all ages, abortion in countries like Japan and Greece is not regarded as posing a symbolic threat to motherhood. Where abortion is medicalized as routine birth control, a backup to nonmedical contraceptive methods, it is medical contraception rather than abortion that symbolizes women's reproductive rights. Elsewhere, legalization restricts women's access to abortion. Turkey's 1983 Population Planning Law allows elective abortion through ten weeks of pregnancy, but a married woman's husband must consent. Germany requires counseling before medically approved abortion.

*See also* **Birth Control: Male Contraception; Birth Control: The Pill.**

**BIBLIOGRAPHY**
Ginsburg, Faye D. *Contested Lives: The Abortion Debate in an American Community.* Berkeley: University of California Press, 1989. An ethnography of pro-choice and pro-life activists in Fargo, North Carolina.
Luker, Kristin. *Abortion and the Politics of Motherhood.* Berkeley: University of California Press, 1984.
*Roe v. Wade,* 410 U.S. 113 (1973). Louisiana State University Law Center Medical and Public Health Law Site, http://biotech.law.lsu.edu/cases/reproduction/roe_v_wade.htm. The full text of the ruling, with paragraph numbers added as cited in this article.
Solinger, Rickie, ed. *Abortion Wars: A Half Century of Struggle, 1950-2000.* Berkeley: University of California Press, 1998. Includes a timeline of major legislative and social developments concerning abortion in the United States and United Kingdom.

*Heather Paxson*

# AGRICULTURAL SCIENCES.

The agricultural sciences include a wide range of disciplines. At the center are what are often known as *production sciences,* in which the object of research is to increase agricultural productivity. These fields include the animal and plant sciences, soil science, and agricultural engineering. Related fields focus on pest management (entomology, plant pathology), postharvest issues (food science, nutrition), and natural resource management (fisheries, forestry). In addition, agricultural economics, home economics, and rural sociology are often found in organizations that engage in agricultural research.

Although the agricultural sciences have their origins in the knowledge accumulated by millions of farmers over thousands of years, the agricultural sciences only recently became the subject of careful examination by scholars. Before the 1970s, most studies of the agricultural sciences were unabashedly celebratory in character, as they were often commissioned by public entities engaged in agricultural research. Much of the critical work in philosophy, history, economics, and sociology emerged independently of science and technology studies (STS) because early STS scholarship was based on physics as the model for research. As such, STS scholars were relatively uninterested in agricultural science.

## Philosophy

Philosophical analysis of the agricultural sciences emerged out of a renewed interest within philosophy over what is commonly referred to as *applied ethics*. Given the advent of the new agricultural biotechnologies, especially as used in genetically modified crops, philosophers have given considerable attention to questions of risk—to humans and to other living organisms. One particularly important finding of ethical analysis is the difference between scientific notions of risk (statistical probability of harm) and social notions of risk (something arising out of human volition). Thus, scientists may argue that a genetically modified crop is harmless because the level of risk to human health is extremely low, whereas consumers may argue that however low it might be, they should be the ones to decide whether or not to take that risk. On a practical level, the latter position would suggest mandatory labeling to ensure that consumers have informed consent.

Philosophers have also noted some dilemmas associated with the distribution of risks from agricultural practices. For example, in most industrialized nations, environmental laws prohibit the entry of products that carry pests and diseases not found in those nations. However, in an effort to comply with such laws, farmers in developing nations sometimes spray exports with pesticides that rapidly degrade such that little or no residue is left when the product is consumed. However, those same products may bring about acute pesticide poisoning among farm workers in developing nations. Such questions of distributive justice are commonplace in agricultural settings.

Ethicists have also posed questions about the impact of agricultural practices on the environment, particularly with respect to the conservation of biological diversity in both domesticated crops and animals, as well as those plants and animals displaced by agricultural production. In particular, philosophers have attempted to define *sustainability*, a term commonly used today within the agricultural sciences, and to determine what ethical approaches could be applied to sustainability.

Finally, ethicists have grappled with the many questions surrounding the treatment of farm animals. Although there is widespread agreement that maltreatment of farm animals is unethical, what constitutes adequate treatment is the subject of considerable debate. Practices such as the debeaking of chickens, the use of small cages for farm animals, and the use of special diets to ensure rapid weight gain have often been justified based on scientific studies of farm animal stress. However, philosophers point out that, even if tests show no undue stress, farm animals behave quite differently when left to their own devices. Some ethicists argue that managed production should, as much as possible, conform to what the animals would do if left in their natural state. Others argue that, given the environmental damage and suffering inflicted by animal production, we would do best by abandoning animal agriculture and switching to a largely vegetarian diet.

## History

Historians have noted that the agricultural sciences emerged out of a combination of peasant trial and error as well as highly specialized knowledge usually developed by a priestly class. For example, peasants quickly learned how to select animals for their health and docility as well as how to select plants for the hypertrophy of the edible parts. At the same time, the priestly class in ancient civilizations often predicted seasonal change and set planting dates by careful observation of the stars. For example, ancient Egyptian agriculture depended on maintaining accurate prediction by priests of the annual flooding of the Nile Valley. In contrast, the Roman poet Virgil, undoubtedly summarizing the common wisdom of his day, provided his contemporaries (ca. 30 B.C.) with detailed commentary on how to farm well, as had the Greek poet Hesiod some six centuries earlier.

Even though certain scientific developments in agriculture can be traced to antiquity, there is general agreement among historians that modern agricultural science began with the creation of botanical gardens in the sixteenth century. Although they were initially intended to re-create the Garden of Eden, the religious motivation for establishing them soon gave way to more prosaic aims. In particular, botanical gardens permitted the classification and comparison of plants from around the world. Medicinal properties of plants were explored first because it was widely believed that all plants had such qualities. Later, the gardens were used to identify economically useful plants and to transfer them to locations where they could be farmed on large plantations. Perhaps the most famous of these gardens is that founded at Kew, not far from London, in 1759. Like other botanical gardens, Kew served to provide education to the general public, to classify useful plants, to serve as a collection point for plants from throughout the British Empire, and to increase the wealth of the empire as well. Plants such as coffee, cocoa, rubber, and tea, in addition to the most effective methods of cultivating them, were spread round the world by the gardens.

In the nineteenth century, botanical gardens were gradually eclipsed by agricultural experiment stations. Unlike the gardens, experiment stations applied experimental methods to cultivated plants. Initially founded in Germany, experiment stations became commonplace in the United States after the passage of the Hatch Act of 1887, which provided funds to set up at least one in each state. The experiment stations combined field, greenhouse, and laboratory methods in their quest for increased productivity. Until the end of World War II, agricultural experiment stations were often the sole recipient of public largesse for nonmilitary scientific and technological research. They were also pioneers in the development of applied statistics, using them to determine the effectiveness of various agricultural practices. Using statistical methods, experiment station scientists helped to correct a long held misunderstanding about yield. Farmers and the general public had always thought that yield was highest when the size of the edible parts of a plant was largest. Experiment station scientists demonstrated that yield was better defined as the total units of weight of edible parts harvested per unit area of land. Measured in this way, it became apparent that large size did not correlate well with yield.

With the help of extension services, which were public agencies founded to help farmers adopt new practices, the experiment stations played an important role in transforming agriculture in industrialized nations. In particular, the stations provided improved varieties of plants and animals that yielded more and were more resistant to pests and diseases. Until the mid-twentieth century, stations in industrial nations focused on food crops (in part to keep industrial wages low by providing cheap food), while those in what were then colonies focused on the production of crops for export to the industrial nations (such as rubber, coffee, cocoa, and cotton). With the advent of independence, most experiment stations in what became known as the *developing world* began to engage in research on increasing food production for local consumption. They were helped by the creation of International Agricultural Research Centers, formed during the Cold War in part to offer a "Green Revolution" as an alternative to "Red" (Communist) revolutions. The Green Revolution rapidly increased yields of rice, wheat, and other crops in developing nations. However, critics argue that this was done at the expense of considerable environmental degradation and the displacement of rural populations.

## Sociology

Sociological studies of the agricultural sciences began with an interest in the diffusion and adoption of agricultural innovations. This is hardly surprising given that many rural sociologists were employed by agricultural experiment stations. Diffusion theorists accepted the production-oriented position of their colleagues in the natural sciences; hence, they focused on the adoption of innovations by farmers and viewed those innovations as highly desirable. The model of communication commonly used by diffusion theorists was adopted from engineering. *Messages* were *transmitted* from *sender* to *receiver*. Later, the concept of *feedback* was added, also borrowed from engineering. It was intended to determine whether the message received was interpreted correctly by the receiver. Adoption of these innovations was found to be based on the sociopsychological characteristics of the receivers: the farmers who were better educated, less risk-averse, more cosmopolitan, and more willing to invest in new technologies were found to be early adopters of innovations. In contrast, those who adopted late were labeled by the pejorative term *laggards,* suggesting a certain traditionalism and even irrationality. By the 1970s, critics of the diffusion model began to argue that the characteristics of the innovations were ignored by those involved in the study. In their enthusiasm for the products of agricultural research, diffusion theorists had failed to note that the innovations studied tended to be costly and large. Moreover, they required considerable skill to operate and maintain. Not surprisingly, those farmers lacking the necessary education, capital, and land made the rational decision not to adopt.

Critics also challenged the application of the diffusion model to Green Revolution varieties. They observed that the packages of innovations developed around the improved seed often involved capital-intensive investments for irrigation, fertilizers, and pesticides. They noted that not all farmers had the resources necessary to adopt. Moreover, they also documented the turmoil that took place in many rural areas. This included the displacement of farm workers and small farmers to urban slums, a decline in the status of women, pollution of groundwater by pesticide runoff, and a drop in water tables as a result of irrigation overuse.

Others challenged the engineering model central to diffusion theory, drawing on the works of philosophers Hans-Georg Gadamer and Jürgen Habermas, to propose a perspective drawn from the hermeneutic-dialectic tradition. In particular, they argued that communications between scientists and farmers had to be in the form of a dialogue. Put differently, farmers had to be able to challenge the central assumptions of industrial agriculture and to play a greater role in the design and formulation of research questions.

During the 1980s, the sociological research agenda broadened. Sociologists began to study problem formulation in agricultural research. Drawing on the work of historians, they suggested that science and commerce were inextricably linked in agricultural research. Such linkages extended from the choice of research problems, to the relations between public and private research, and to the values held by agricultural scientists. Following a trend in STS research generally, they also began to examine more carefully the content of the agricultural sciences.

One particularly productive line of questioning involved the study of entire agricultural commodity chains—from farm to fork. Unlike the diffusion theorists, proponents of commodity studies examined the entire series of events from the formulation of the research problem, to the empirical research needed to perfect it, to its use in production, and to the consequences of its widespread adoption. For example, a study of the development of the tomato harvester by public university researchers began by examining the engineering necessary to develop a mechanical harvester. It continued by noting how breeders had to develop a tomato sufficiently hard to withstand rough treatment by the machine. The study further noted that the machine designed was profitable for use only on tomato farms far above the average size. The study also observed the consequences of the widespread use of the harvester: widespread labor displacement without compensation and increased farm size. The study concluded by questioning the wisdom of using public funds in a manner that largely benefited the wealthy at the expense of the poor.

Another sociological research trajectory has been the study of the new agricultural biotechnologies, and especially the creation of genetically modified plants. Scholars have noted that, in principle, these new technologies permit the insertion of any gene into any organism, thereby bypassing sexual reproduction processes. Moreover, they truncate the time needed to develop new varieties and reduce the space needed for research by partially substituting Petri dishes for experimental fields. In addition, unlike older biological technologies, the new biotechnologies have been subject to utility patents—in part due to a series of court rulings that expanded the scope of patent law. These new biotechnologies, combined with patent law, have transformed both agriculture and agricultural research. On the one hand, they have shifted much of the agricultural research agenda from the public to the private sector, as large chemical and pharmaceutical companies have invested heavily in search of (sometimes illusory) greater profits. On the other hand, they have transformed the seed and agrochemical industries. Before the advent of the new technologies, the seed industry consisted mostly of small companies. Many of the varieties they released were developed in public experiment stations. Now, the seed industry more closely resembles other oligopolistic industries, with a few leading firms dominating the industry and licensing their genetically modified products to smaller companies. In contrast, there has been a decline in insecticide sales as insect-resistant crops have been introduced. Similarly, the rapid success of herbicide-tolerant crops, most of which are tolerant to the herbicide glyphosate, has reduced considerably the market for other herbicides. In addition, there is some concern about declining biodiversity as fewer firms breed new crops and concern about selection pressure on insects

that can permit them to overcome the narrow base of insect resistance in genetically modified crops. However, to date, the evidence of either is scant.

Finally, sociologists have recently begun to study the role that science plays in the formulation, maintenance, and enforcement of food and agricultural standards. Standards for food safety, environmental management, worker health and safety, geographical origin of products, intellectual property, organic products, color, and nutrition, among others, are ubiquitous. Nearly all of them are in some fashion science based. Science can be implicated in the tests performed to meet a standard, in the definition of the product or process, and/or attempts to convince a potential purchaser of the value of a particular product. Such research has shown that several aspects of standards are of critical importance: who participates in the negotiations leading to the acceptance of a particular standard is relevant because the standard may contain hidden biases against particular groups. Standards may themselves determine who can participate in a given market, by imposing costs required for entry. Finally, standards have distributive consequences; some will win and some will lose when a particular standard is adopted.

## Economics

As in sociology, most early economic studies of agricultural science took the products of research as undiluted goods and emphasized "adjustment" by farmers and others in the agrofood system. However, within the last 25 years, economists have focused more on the social rates of return on investments in public agricultural research. The economists' studies have been used effectively to lobby legislators for additional funds for public research. In contrast, critics of the approach (often from other disciplines more concerned with equity issues) have noted the tendency to count only the benefits (which are easily quantified) and to exclude difficult-to-measure environmental and health costs.

A number of strands of economic research have emerged from studies of rates of return.

Some economists have examined comparative studies of the rates of return across various major commodities. Others have examined what is known as spillover. Specifically, public research is often not protectable by patents or copyrights. Thus, research conducted in a given nation can be easily transferred to other nations with similar agroclimatic features. Not surprisingly, economists have argued that developing nations should not invest in research with high spillover, but should rely on the International Agricultural Research Centers for that research. In contrast, they argue that developing nations should invest in research not easily transferred across national boundaries. However, the emergence of stronger intellectual property rights in agriculture is probably reducing spillover effects worldwide.

Still other economists have attempted to discern the appropriate mix of public and private funds in agricultural research. Some note that given strong intellectual property rights, the private sector should take on a greater share of research, thus leaving the public sector with agriculturally related social science and environmental research. In fact, the United Kingdom has privatized much of its agricultural research with some limited success. Critics argue that letting the suppliers of farm inputs such as chemicals and seeds set the research agenda will force abandonment of technological changes that might benefit farmers or the general public but that yield no profits to those companies. They also note that public sector research allows an independent voice to be heard in debates about the desirability of particular research products.

A key contribution of economics has been the development of the "induced innovation" theory to explain broad national research directions in agricultural research. According to the theory, innovations are induced by the relative scarcity of land, labor, and capital. For example, in the United States there is abundant land yet labor is relatively scarce. Therefore, research has tended to increase yields per unit of labor. In contrast, in Japan, where land is scarce and labor is relatively cheap, research has tended to focus on increasing yields per unit of land. In addition, proponents of the theory argue that public research is responsive to the demands of farmers as voiced in legislative politics. In that sense, innovations are induced by farmer demand for public support for certain types of research. Critics have noted that this is likely to be true only in democratic regimes where farmers have a voice in politics.

Unfortunately, studies of the agricultural sciences remain fragmented. Disciplinary barriers remain high. In addition, most scholars of the agricultural sciences remain within agricultural colleges.

## BIBLIOGRAPHY

Anderson, Edgar. *Plants, Man and Life.* Berkeley: University of California Press, 1967.

Brockway, Lucile H. *Science and Colonial Expansion: The Role of the British Royal Botanic Gardens.* New York: Academic Press, 1979.

Busch, Lawrence, William B. Lacy, Jeffrey Burkhardt, and Laura R. Lacy. *Plants, Power, and Profit: Social, Economic, and Ethical Consequences of the New Biotechnologies.* Cambridge, MA: Blackwell, 1991.

Friedland, William H., Amy E. Barton, and Robert J. Thomas. *Manufacturing Green Gold: Labor, Capital, and Technology in the Lettuce Industry.* Cambridge U.K.: Cambridge University Press, 1981.

Fuglie, Keith, Nicole Ballenger, Kelly Day, Cassandra Klotz, Michael Ollinger, John Reilly, Utpal Vasavada, and Jet Yee. *Agricultural Research and Development: Public and Private Investments under Alternative Markets and Institutions.* Washington, DC: U.S. Department of Agriculture. 1996. A review of the rates of return literature.

Hayami, Yujiro. *Development Economics: From the Poverty to the Wealth of Nations.* Oxford: Clarendon Press, 1997. An application and explanation of induced innovation theory.

Jamieson, Dale. *Morality's Progress: Essays on Humans, Other Animals, and the Rest of Nature.* New York: Oxford University Press, 2002.

Kloppenburg, Jack R., Jr. *First the Seed: The Political Economy of Plant Biotechnology, 1492–2000.* New York: Cambridge University Press, 1988.

Perkins, John H. *Geopolitics and the Green Revolution.* New York: Oxford University Press, 1997.

Rogers, Everett M. *Diffusion of Innovations.* New York: Free Press, 1995.

Thompson, Paul. *The Spirit of the Soil: Agriculture and Environmental Ethics.* London: Routledge, 1995.

*Lawrence Busch*

# AGRICULTURE AND TECHNOLOGY.

As bread continues to be "the staff of life," agriculture abounds with emotive meanings, linked with political-economic agendas. These agendas promote conflicting accounts of the problem to be solved. Agriculture exists ostensibly to feed people, yet its organization and methods are driven by other aims: to earn a living at various stages of the agro-food chain, to sell inputs to farmers, to control the commodity trade, and to "add value" that can be realized as profit.

Any agricultural technology embodies a choice of such agendas. Its design process involves models of nature, efficiency, and development. Dominant innovations generally frame the problem as a deficiency (such as in animals or seeds) that must be "remedied" through a laboratory-based "improvement." This problem definition favors some values and purposes over others.

Such agendas generally remain beyond democratic accountability. Even within so-called "liberal democratic" societies, regulatory decisions revolve around "risk-benefit" analysis as the official grounds for a government to restrict (or not) an innovation. Often this framework has been contested through public protest. For examples of agricultural technology as culture, this brief article surveys a hormone supplement, seeds, and development issues.

## Recombinant BST: Industrializing Cows?

Conflicting agendas are exemplified by the controversy over bovine growth hormone (BGH), a product designed to increase milk yields in dairy cows. The manufacturers use a more technical term, *bovine somatotropin* (BST), to avoid the emotive term *hormone*. According to Monsanto, BST is a naturally occurring protein supplement that enhances the cow's effi-

ciency. According to opponents, Monsanto is pushing a drug that forces cows to work harder on the factory farm and that pollutes natural processes. Thus, each side appeals to a different concept of what is "natural."

Early criticisms warned that BST would aggravate familiar problems. When dairy cows were selectively bred for higher milk yields, some showed increases in infectious disease. Opponents argued that BST injections could cause similar effects. In turn, this would lead to increased use of antibiotics, which could pose direct or indirect risks to human health. BST would encourage farmers to intensify factory-farming methods, already driven by competitive pressures to produce cheaper milk.

Such debates led to opposite outcomes across the Atlantic. Despite mass opposition in the United States, BST was approved as safe by the U.S. Food and Drug Administration (FDA). Facing greater opposition, the European Union authorities banned BST, initially on grounds that greater milk production would contradict EU policy of reducing agricultural surpluses. In the decade since the U.S. approval, the FDA has belittled evidence of harm as merely an "animal welfare" problem, whereas the European Commission has funded research on risks to justify its ban. More generally, the U.S. political system has promoted agricultural technologies that claim to increase productive efficiency and market competitiveness, while the EU has more readily accommodated demands to protect farmers from such pressures. Such divergences underlie many trans-Atlantic conflicts over agricultural products.

## Redesigned Seeds

Seeds have been a site of struggle over the entire agro-food chain. Given the inherent reproducibility and variability of seeds, their natural characteristics have provided an opportunity for farmers to improve varieties through selective breeding, while developing their skills in cultivation methods. Capitalist strategies have sought to uncouple seeds from farmers' control, to commoditize natural resources, and thus to make farmers dependent upon a separate supply industry.

Already by the mid-nineteenth century, noted Karl Marx (p. 527), "Agriculture no longer finds the natural conditions of its own production within itself, naturally arisen, spontaneous, and ready to hand, but [rather] these exist as an independent industry separate from it."

As a major step in commoditizing nature, breeders developed hybrid seeds that cannot breed true, so that farmers have to buy them anew each season. Citing the supposed benefits of "hybrid vigor," U.S. agricultural research prioritized such varieties rather than improving open-pollinated ones. For seeds that could still be bred and improved by farmers, proprietary control was extended by various laws restricting farmers' rights to sell (or even to reproduce) their own grain for seed. Especially in the United States, farmers have also been locked into dependency through other purchased inputs, grain contracts, debt, and other means. According to the sociological Jack Kloppenburg, farmers have become reduced to "propertied underlabourers," though this dependency has provoked opposition.

*The Pesticide Umbrella.*   After World War II, agrochemical pesticide usage was greatly expanded, especially in the countries of the global North. Formerly, plant breeding had given priority to pest resistance traits and genetic variation. Increasingly, however, these were marginalized in favor of yield criteria as breeders depended upon a "pesticide umbrella" to protect crops from pests.

Agriculture became more dependent upon a chemical fix that conceptualized all other life as a threat. Pests then developed resistance to agrochemicals, and farmers increased sprays in response. Environmental damage became a major public controversy in the United States, especially with the 1962 publication of Rachel Carson's book *Silent Spring*. The agrochemical industry sought to discredit her warnings, though the dual agronomic and environmental crisis stimulated a search for alternative methods.

Alternatives were developed through ideas of working with natural forces and local resources. *Integrated pest management*

(IPM) manages the cultivation system to control pests through such means as crop rotation, fertilizer application, soil preparation, and time of sowing. IPM was later incorporated into *integrated crop management* (ICM), which selects components of the farm system to avoid pests. Such components include soil management (to enhance crop health), resistant cultivars, natural predators, and limits on pesticide or mineral residues.

Such alternatives were potentially profitable for farmers but remained marginal, at least until the 1990s. ICM methods needed regional cooperation for several reasons: for example, to build learning networks among farmers and to ensure that agrochemicals do not eliminate local predators. Government programs failed to provide such measures, so farmers were locked into an agrochemical dependency.

*High-Response Varieties.*   Agrochemical dependency was extended to the global South through the "Green Revolution." So-called "high-yielding varieties" (HYVs) were really "high-response varieties." Designed for more intensive cultivation methods, their higher yields depended upon agrochemicals, irrigation, and other purchased inputs. Yet their name suggested that they were "high-yield" because of a genetic trait.

Systems using HYVs increased grain yields of wheat and rice, as in India. This increase could count as greater efficiency only by measuring a single commodity, while ignoring previous benefits that were now lost. Higher grain yield meant less straw, used locally as animal feed. Many farmers had done intercropping; for example, rows of sorghum and wheat alternating with legumes, whose combination helped to renew soil fertility while providing other nutrients. Those benefits were lost in the switch to HYVs. More generally, land use shifted away from cultivating oilseeds and legumes, which had been a cheap protein source, widely known as the poor person's meat. Eventually India had a shortage of oilseeds and legumes, thus requiring imports.

The inherent violence of the Green Revolution has been documented by the Indian activist Vandana Shiva. HYVs favored those farmers who could obtain loans for the purchased inputs. Financial dependency and market competition drove many into debt, even out of business, leading some to commit suicide. Landless peasants became wage laborers for the successful farmers or migrated to cities. Some moved to live or work near the Union Carbide plant at Bhopal, which supplied agrochemicals for the HYVs, scene of the poison-gas release disaster of 1984.

Those outcomes logically followed from the agribusiness agenda that guided HYV research, largely funded by the Rockefeller Foundation. According to its chief, "agriculture is a business and, to be successful, it must be managed in a business-like fashion" (Ross). As such language indicates, the Green Revolution redefined agricultural efficiency in terms of calculable commodities, while devaluing any resources or benefits that did not fit such a model.

*Agbiotech: Fetishized Metaphors.* A commoditization agenda was extended through agricultural biotechnology, in turn originating from the science of molecular biology. This science reconceptualized "life" in physical-chemical terms. Through a computer metaphor, DNA became coded "information," which could be freely transferred across the species barrier. "As technology controlled by capital, it is a specific mode of the appropriation of living nature—literally capitalizing life," argued the prophetic writer Edward Yoxen (1981).

Indeed, biotechnology recasts nature in terms of computer codes, combat, and commodities. Through social metaphors, nature is redesigned in the image of biotechnology, while human qualities are fetishized as properties of "smart seeds"—genetically modified (GM) crops. According to Monsanto, genetic engineering is "a natural science" that simply rearranges a universal "genetic code." GM crops have "in-built genetic information" that provides natural defenses against pests or other adverse conditions. Biotechnologists also speak of a clean, surgical precision, such as redesigning seeds to attack pests or to withstand herbicides (which are often sold by the same company selling

the seeds). Some also speak of "value-added genetics": searching for genetic changes that can enhance the market value of agricultural inputs or products.

This technological trajectory has intersected with a wider debate over sustainable agriculture. At issue is how to diagnose and remedy the systemic hazards of intensive monoculture, such as agrochemical pollution, pest epidemics, and pest resistance. Agbiotech research attributes these problems to deficient seeds, which must be corrected by editing their genetic information, thus making agriculture more efficient and clean.

GM crops are redesigned mainly to benefit the agro-food industry, argues the activist Henk Hobbelink. Through expectations for greater efficiency, farmers are thrown into greater competition with each other while becoming more dependent upon input suppliers and grain traders. Ironically, such dependency is portrayed as liberation from natural threats to crops.

## Disputing Development Models

Amid a wider debate over sustainable development (for example, how to reduce harm from agrochemical usage), different diagnoses of the problem underlie stances toward GM crops. These promise to enhance agronomic "control" and "efficiency," conceptualized within a purely external control of crop pests, whereby inserted genes function in ways analogous to the chemical pesticides that they may replace or supplement. Such a solution marginalizes efforts at internal control, which would restructure agronomic-ecological relationships so as to minimize the need for pest control agents.

According to its supporters, biotechnology offers essential tools for environmentally friendly products. Within their eco-efficiency model of sustainability, GM crops provide a tool for integrated crop management (ICM): more productive agriculture, more soil conservation, less insecticide use, less energy, better habitat protection. Companies promote such products as essential tools for the "sustainable intensification of high-input agriculture" and thus greater productivity, especially as a means to increase food production in the South. Society faces the risk of forgoing these benefits, they argue.

By contrast, critics attribute our agro-environmental problems to intensive mono-cultural methods, which make crops more susceptible to pests and disease. In their view, GM crops aggravate the problems of high-input agriculture, stimulate a "genetic treadmill," impose unknown hazards, and preclude more beneficial alternatives. As a different solution, sustainability would mean more extensive methods, dependent upon human skills.

Also at issue are consequences for agro-biodiversity. According to biotechnology proponents, the greater variety of genetic combinations will increase biodiversity, which thus is redefined as laboratory simu-lations. According to critics, GM crops threaten the diverse crop varieties that have been developed by farmers in the field. Sup-posed benefits of pesticide reductions have little relevance in places where most farmers have not used agrochemicals anyway or have experienced their use as a problem of soil degradation and economic dependence, as in the Green Revolution.

Humanitarian claims about biotechnology depend on assumptions about world hunger and malnutrition. According to some propo-nents, biotechnology will be essential for "feeding the world." In the simplest version of this argument, greater world hunger may result from inadequate agricultural inputs and from population growing faster than food production. GM crops can solve the problem by increasing production, resisting pests, or tolerating inhospitable climates (dry or saline soils). In the more subtle ver-sion of the argument, land degradation re-sults from overintensive cultivation, which can be avoided by using GM crops to in-crease productivity in environmentally be-nign ways.

According to critics, however, the threat of world hunger results mainly from unjust forms of land use and food distribution. In recent years, more and more land has been converted from staple food crops to bulk commodities used as animal feed or to spe-cialty crops as export commodities. GM

versions of the latter will accelerate the loss of staple crops. "High-yield" varieties have already displaced traditional crop va-rieties while making farmers more dependent upon chemical-intensive methods and other purchased inputs. Herbicide-tolerant crops may further displace the multi-cropping methods that previously protected farmers from pests and provided diverse nutrients such as vitamin A. By holding patents on GM crops, moreover, companies can charge royalties to farmers who replant seeds and thus discourage independent breeding.

In Mexico maize yields have been in-creased by Green Revolution-type "modern varieties," with more purchased inputs, to produce maize mainly for animal feed. This development has marginalized peasants who grow maize in rotations with other crops, which helped control pests. Neverthe-less, 2.5 million households still cultivate maize on small-scale, low-input, rain-fed farms. Their livelihoods are further threat-ened by cheap U.S. imports, some of them from higher-yield GM maize. This threat in turn increases the pressure on Mexican farm-ers to adopt GM maize varieties and more in-tensive methods in order to compete in the market.

By analogy to development issues in the South, GM crops intersected with a debate over agricultural futures in Europe too. Trans-Atlantic biotech companies designed GM crops to replace agrochemicals or to sub-stitute less harmful ones than before. Yet such products were widely criticized as means to extend industrialization and con-trol over the seed market. Widespread com-mercialization went ahead in the United States but was blocked or delayed in Europe by commercial boycotts in the late 1990s. These obstacles undermined the "Life Sci-ences" strategy, which had aimed to inte-grate research across the three sectors of crops, pesticides, and pharmaceuticals.

The trans-Atlantic trade conflict arose partly from different cultural meaning of agriculture. U.S. farms are seen as analogous to factories, sharply demarcated from wilderness and nature conservation areas. Although European agriculture also uses intensive methods, this model has undergone challenge. Farmland is widely regarded as an integral part of the environment—as an aesthetic landscape, a local heritage, a livelihood for traditional-style farmers, and a wildlife habitat. In such ways, the European risk debate and regulatory controls have emphasized environmental issues that are hardly considered risks by most people in the United States.

*Biopiracy by Whom?* GM techniques have been used also as a symbolic means to privatize seeds, going a step beyond hybrid seeds. According to advocates of greater patent rights, GM crops are inventions, involving a significant contribution by scientists. According to opponents, however, such products are discoveries (or simulations) of common resources that have been already selected and cultivated by farmers over many generations.

To qualify for patent rights, a product or process must have contributed an "inventive step." The United States and European Union have interpreted that criterion to accept broad patent claims on GM crops. Patents have encompassed substances derived from plants traditionally cultivated in developing countries, such as pesticidal agents from the neem tree (*Azadirachta indica*). Even some non-GM seeds have been subjected to royalty payments charged to farmers in Southern countries, thus threatening their livelihoods. The U.S. government has sought to extend its patent criteria to other countries by using the TRIPS (Trade-Related Aspects of Intellectual Property Rights) rules under the WTO.

Amid that conflict, the term *biopiracy* has arisen and acquired two opposite meanings. For advocates of greater patent rights, biopiracy means violating those rights, such as by using patented materials without a license agreement or without paying royalties. For opponents of such rights, the real biopiracy is the patents themselves, because they make plant material private that should remain freely reproducible as a common resource, or because Southern farmers should be reimbursed for their plant-breeding skills.

*Alternatives: South and North.* Problems of farmer dependency have generated fierce protests against GM crops. Many Indian farmers had previously been abandoning mixed farming systems in favor of cotton monoculture based on hybrid seeds, thus intensifying their dependence upon purchased seeds, which sometimes led to crop failures. They regarded Monsanto's GM insecticidal cotton as a further step toward privatizing seeds, and in 1999 many Indian farmers publicly "cremated" field trials. As a greater threat to livelihoods, the Andhra Pradesh government adopted a program, "Vision 2020," which aims to increase agricultural efficiency but, in effect, displaces most current farmers through intensification and greater dependence on purchased inputs such as GM seeds. This "modernization" program has been scrutinized in a participatory exercise on possible futures for agriculture and livelihoods (Pimbert and Wakeford, 2002).

Alternatives to both agrochemicals and GM crops are being developed in the South. Through various nongovernmental organizations (NGOs), farmers learn to substitute their own skills and local materials for purchased inputs. Farmers replant their own seeds to preserve diverse cultivars in the agricultural field and to develop independent methods of crop protection. For example, when Brazil's *Movimiento de Trabalhadores sem Terra* (Landless Workers' Movement) expropriated land and started farming, at first they imitated the chemical-intensive methods of large landowners. Ensuing problems led them to develop organic cultivation methods, recovering and building on earlier traditions.

In Europe, too, public protest against GM crops has given further stimulus to alternative agricultural innovations that were already being developed. Food retail chains require, and help, farmers to adopt cultivation methods that avoid pest problems and so reduce the need for agrochemicals. They promote integrated crop management (ICM), which enhances knowledge of how best to use various methods and inputs. For example, retail chains fund research on soil-management methods that strengthen plant resistance to pests and disease. Organic breeding institutes develop pest-tolerant seeds, which may be more durable in the face of novel pests. More opportunities arise to develop and apply agroecology techniques.

Thus the agro-supply industry undergoes pressure to change not only the characteristics of products but also the concept of innovation. The outcome will depend upon the socio-natural models and political-economic forces that drive change.

## BIBLIOGRAPHY

Branford, Sue, and Jan Rocha. *Cutting the Wire: the Story of the Landless Movement in Brazil.* London: Latin American Bureau, 2002.

Carr, Susan. ed. "Innovation Strategies in the European Agricultural Life Sciences" (special issue). *Science and Public Policy* 29.4 (2002): 241–312.

Cowan, R., and P. Gunby. "Sprayed to Death: Path Dependence, Lock-in and Pest Control Strategies." *The Economic Journal* 106 (1996): 521–542.

Greens/European Free Alliance. *Agroecology: Toward a New Agriculture for Europe*, Brussels: Greens/EFA in the European Parliament, 2001.

Hobbelink, Henk. *Biotechnology and the Future of World Agriculture: The Fourth Resource.* London: Zed Books, 1991.

Kloppenburg, Jack. *First the Seed: The Political Economy of Plant Biotechnology, 1492–2000.* Cambridge, U.K.: Cambridge University Press, 1988.

Krimsky, Sheldon, and Roger Wrubel. *Agricultural Biotechnology and the Environment: Science, Policy and Social Issues.* Chicago: University of Illinois Press, 1996.

Levidow, Les. "Simulating Mother Nature, Industrializing Agriculture." In *FutureNatural: Nature, Science, Culture*, edited by George Robertson, Melinda Mash, Lisa Tickner, Jon Bird, Barry Curtis, and Tim Putnam, 55–71. London: Routledge, 1996.

Marx, Karl. *The Grundrisse* (1850–1859). Translated by Martin Nicolaus. London: Penguin, 1973.

Parrott, Nicolas, and Terry Marsden. *The Real Green Revolution: Organic and Agroecological Farming in the South.* London: Greenpeace Environmental Trust, 2002, http://www.greenpeace.org.uk/MultimediaFiles/Live/FullReport/4526.pdf.

Pimbert, Michel P., and Tom Wakeford. *Prajateerpu: A Citizens Jury/Scenario Workshop on Food and Farming Futures for Andhra Pradesh, India.* London: International Institute for Environment and Development, 2002, http://www.iied.org/sarl/pubs/otherpubs.html.

Ross, E. B. *The Malthus Factor: Poverty, Politics and Population in Capitalist Development.* London: Zed Books, 1998.

Shiva, Vandana. *The Violence of the Green Revolution: Third World Agriculture, Ecology and Politics.* London: Zed Books/Third World Network, 1991.

Shiva, Vananda. *Monocultures of the Mind: Perspectives on Biotechnology and Biodiversity.* London: Zed Books/Third World Network, 1993.

UK Food Group. *Hungry for Power: The Impact of Transnational Corporations on Food Security.* London: UK Food Group, 1999, http://www.ukfg.org.uk/docs/Hungry%20For%20Power.pdf.

Yoxen, Edward. "Life as a Productive Force: Capitalizing the Science and Technology of Molecular Biology." In *Science, Technology and the Labour Process: Marxist Studies*, Vol. 1, edited by Les Levidow and R. M. Young. London: CSE Books/Atlantic Highlands: Humanities Press, 1981; reissued by Free Association Books, 1983: 66–122.

*Les Levidow*

# AIDS AND SOCIETY.

Now in its third decade and reported in most countries and populations of the world, the HIV/AIDS epidemic is widely recognized as a social as well as a biological crisis. First officially reported in 1981 in the United States, acquired immunodeficiency syndrome (AIDS) continues to spread globally with devastating effects for millions of people each year. Caused by a virus (human immunodeficiency virus, HIV) that attacks the human immune system and renders its victims vulnerable to a wide range of life-threatening conditions and diseases, AIDS first struck urban communities of white homosexual men in the United States and other industrialized nations. This initial appearance led to the perception of AIDS as a "gay plague" brought on by some feature of gay men's lifestyles, deviant sexual practices, or moral deficiencies. Although the epidemic now looks radically different and is, worldwide, increasingly reported among heterosexuals and people of color, the early history of AIDS continues to shape public and even professional perceptions of the epidemic. The stubborn persistence of this stereotype despite scientific evidence illustrates just one of the ways in which HIV/AIDS is a social phenomenon as well as an infectious disease. More generally, because HIV is transmitted by way of specific human activities and behavioral practices that often involve sex, illegal drugs, or treatment for a preexisting disease (such as hemophilia), the stigma attached to these practices not uncommonly migrates to the persons or groups engaging in them. Thus, a diagnosis of HIV infection is sometimes called a "double diagnosis" because it reveals both a serious and incurable medical condition and a socially stigmatized identity or pattern of behavior.

The HIV/AIDS epidemic is uniquely social; indeed, its social dimensions are as defining as the genetic makeup of the virus. To take this view is not insignificant, for as Jonathan Mann and Daniel Tarantola argue, "how a problem is defined determines what will be done about it" (p. xxxiii). We can map the epidemic's social importance in several distinct domains.

### The Individual

Individuals diagnosed with HIV/AIDS encounter a host of social problems that are often as difficult to manage as the disease itself. Whatever a society's experience with the epidemic, an individual diagnosed with HIV or AIDS may still evoke fear, panic, denial, hatred, discrimination, loss of employment and benefits, misinformation, gossip, stigma, psychological and physical abuse, and expulsion from the community. In the first decade of the epidemic, prevention and intervention efforts focused largely on the individual. Defining HIV/AIDS as a problem of individual behavior, this traditional public health

approach aimed at inducing and sustaining individual behavioral change. The basic template for prevention and intervention programs worldwide has been the "three-stage model," which seeks to 1) provide individuals with information and education (about HIV, modes of transmission, preventive measures such as the use of condoms, and so on), 2) provide individuals with health and social services (for instance, affordable antiretroviral drug therapy), and 3) ensure nondiscrimination toward people with HIV and AIDS (for instance, by prohibiting job loss). The vast majority of AIDS education, intervention, and behavior modification efforts—whether grassroots or professional, whether in Hong Kong or Hoboken—employ some version or other of individually based behavioral change and service provision for individuals. Unquestionably, these efforts have helped educate the public, reduce hysteria and misinformation, slow the epidemic, and save lives.

### Communities, Populations, and Societies

But just as HIV/AIDS is more than a problem of individual behavior, its social dimensions go beyond its impact on individuals. Wherever the epidemic appears, it exacerbates already existing tensions, weaknesses, and unresolved problems; traveling along a society's "fault lines," HIV/AIDS tends to reinforce prevailing social divisions—between dominant and marginalized groups, between "us" and "them," between the "normal" and the "deviant." Following the template of past epidemics, HIV/AIDS may provoke calls for quarantine, mandatory testing, some form of public identification of those found to carry the virus, termination of protections and benefits for the infected, and restrictions on civil liberties. At the same time, societies have responded to the epidemic with compassion, creativity, and incisive political criticism. Art, literature, popular culture, community organizing, church initiatives, and social activism all reflect the extraordinary social impact of the epidemic.

## The State

When HIV/AIDS arrives, government officials likewise swing into action, even if the action is a forceful denial that HIV is present or that such a thing as AIDS even exists. Called upon by fearful citizens to "do something," they may initially propose draconian legislation and far-reaching policy changes, raising the stakes on quarantine and a host of other traditional public health measures. In Illinois, for instance, a state law that was passed in the early years of AIDS required every couple who applied for a marriage license to be tested for HIV; a year later, when only a handful of cases had been identified, but the state had lost millions of dollars in revenue to neighboring states, the expensive screening procedure was discontinued. On a larger scale, journalist Randy Shilts's 1987 book *And the Band Played On* paints a devastating portrait of denial, neglect, and panic in the U.S. government in the years when AIDS was seen largely as a "gay disease." Indeed, both the U.S. and the U.K. governments passed measures in the 1980s that prohibited the state from supporting any education and prevention programs or materials that could be construed as promoting homosexual activity. While many nations have responded in more enlightened ways and have developed excellent anti-AIDS programs, the global pandemic now requires innovative and coordinated multinational approaches to intervention and control. This in turn may require redefining the epidemic.

## Social Definition of the Epidemic

Even though the AIDS epidemic has identifiable social effects on individuals and creates predictable responses by communities, societies, and the state, the history of the epidemic reveals a fourth critical domain in which society matters. As Mann and Tarantola put it, "the lesson of the modern pandemic is that certain features of society are fundamental determinants of the epidemic's natural history" (p. 4). Whereas the HIV/AIDS epidemic spreads from one individual to another by way of specific behavioral and social practices, a growing body of evidence makes clear that the broad social and structural characteristics of a society facilitate or discourage the occurrence of those practices and set the stage for individual vulnerability and resistance. The course of the epidemic, in other words, can be significantly influenced by the broad social and structural character of a society, such as by its overall stability; political, material, and spiritual resources; fundamental stance toward human rights and dignity; and its commitment to the physical and economic well-being of its citizens. Recent approaches to the epidemic thus seek to link individual behaviors with their societal context, attend more closely to the broad social determinants of health and disease, and develop action and intervention measures that integrate HIV/AIDS work with social, cultural, and economic studies and interventions.

### BIBLIOGRAPHY

Mann, Jonathan, and Daniel J. M. Tarantola, eds. *AIDS and the World II: Global Dimensions, Social Roots, and Responses.* New York: Oxford University Press, 1996.

Shilts, Randy. *And the Band Played On: People, Politics, and the AIDS Epidemic.* New York: St. Martin's Press, 1987.

*Paula A. Treichler*

# ALTERNATIVE MEDICINE.
*See* COMPLEMENTARY AND ALTERNATIVE MEDICINE

# ARTIFICIAL INTELLIGENCE.

Artificial intelligence is the attempt to make machines that think like humans. In this article the abbreviation AI will be used to refer to this enterprise, and the phrase *artificial intelligence technologies* (AIT) will be reserved for new technologies that initially grew out of AI but that mimic only some aspects of human abilities, such as a subset of speech recognition or pattern recognition, while avoiding the deep problems. AIT and AI are often confused with one another.

AI proper has an important bearing on sociology in general, and social studies of science in particular, because of the light it can shed on the notion of the social. Most sociologists believe that most of a person's capacities are gained through the person's embedding in social groups. If machines could succeed in mimicking human reasoning, then either humans would have learned to "socialize" machines or there would be something wrong with the idea of "the social." At the moment, humans have no idea how to socialize machines; there are no machines that can be raised from birth and learn language within a family, nor any that are imprinted with a ready-made set of social abilities and a capacity to continue to build them through social interaction. From time to time such abilities have been claimed for machines as they have evolved. For example, neural nets appear to be capable of learning by themselves, but only in a crude, behavioristic way, as one might train a pigeon or the like, so the deep problem of socialization has not been approached. This means that any real successes in AI would threaten the sociologist's idea of the social.

This is not the only kind of relationship between artifical intelligence and the social sciences. Sociologists are interested in the way new technologies change society, and the changes brought about by AIT are one such area of inquiry. One might think of machines of all sorts as already an integral part of society, but this is to use the notion of "the social" in a way that bears less directly on sociology as an enterprise. Also, social studies of science have a legitimate concern with the development of AITs of various kinds, and especially their relationship to military projects.

Returning to AI proper, the attempt to automate scientific discovery is a hard case for those who would wish to maintain the idea of the social. At the heart of the sociology of scientific knowledge (SSK) is the idea that even scientific knowledge is deeply invested with the social, whereas dominant models of science take it to be a paradigm of universality divorced from social influence. Thus, imagine a human community that has developed in isolation from other communities: There would be no grounds to expect such a community to develop, say, the English language, still less the nuances of any particular dialect spoken at a given moment in history. One accepts that such capacities would not develop in the absence of social contact between the community and the social group that embodied the dialect. On the other hand, the dominant view of science would lead us to be less surprised should such an isolated community rediscover many of our scientific and mathematical laws. It is this view of science that is challenged by SSK, which treats any body of scientific knowledge as very much like a dialect in a natural language. If the current generation of asocial machines could rediscover scientific laws on their own, this would support the dominant view and challenge the sociological view of science. The sociology of scientific knowledge is, then, a hard case for the larger argument about AI and the social. If the sociologists can hold their ground in the case of science, then the ground can be held much more easily in the case of social activity as a whole.

Attempts to develop AI can be seen, then, as an expensive experiment to test the deep ideas of sociology in general and of SSK in particular. Like all experiments, however, this one suffers from indeterminacies in its outcome associated with the experimenter's regress and the like. One important source of confusion is the confusion between AIT and AI, which is amplified by humans' ability to "repair" the deficiencies in others' communication and attribute far more competence to partners in discourse that they deserve. The need for repair is crucial to ordinary communication, because speech is normally indistinct, broken, overlaid with other sounds, and invested with allusion to shared but unspoken contexts. It is only by "reading" within context and repairing the "mistakes" that we are able to make sense of others' speech and action. On these tendencies depends the success of confidence tricksters and fraudsters of various kinds, who can rely on the "mark: to do most of the work necessary to see what they do as a competent performance.

Thus, supposed tests of artifical intelligence as they are normally designed, such as versions of the Turing test, which are used in much-publicized computer challenges, trade on humans' ability and willingness to repair the broken discourse of computers, whereas the computers deal with only complete or near-complete language input. A fairer test would involve more symmetry between what the computer and the human have to accomplish in the way of context-dependent understanding. The asymmetry that continues to exist is revealed in attempts to build speech transcribers and such devices, which continue to founder on the inability of machines to understand contexts that they could grasp only through immersion in society. To provide a simple example, the following is a true statement: "My spell checker will correct *weerd processor* but not *weird processor.*" This is a difficulty for my spell checker, but not as big a difficulty as the fact that in the context of this essay, both spellings should remain uncorrected (which the human editor, understanding the context, immediately sees). Careful attention to the deficiencies of automated speech-processing devices, from spell checkers to speech transcribers, shows that the world is still a safe place for the idea of the social as conceived by sociology as a whole.

Turning back to the hard case of science, the issue is more complex and turns on a careful separation of what is being done by the humans and what by the machines (Collins and Kusch talk of *polymorphic* and *mimeomorphic* actions). For example, a pocket calculator appears to be a competent arithmetic machine, though the sociologist of science would like to say that even arithmetic is a social activity. We can resolve the problem by pointing out that arithmetic consists of a mechanical part—something that can be likened to stacking pebbles—and a part that fits this kind of activity into human life including, say, approximation. The number of decimal points retained in a calculation depends on whether we are converting, say, a person's height from one kind of unit to another or calculating the course of a spacecraft, and these continually changing subtleties are, once more, like the subtleties of dialect. The problem is visible even inside arithmetic in the buildup of errors within calculations. Thus, simple calculators will usually make a mistake at about $7.0000/11.0000 \times 11.0000$, and are even more likely to get it wrong if the computation is done in two stages. (Though some calculators will get this simple calculation right, the deeper problem of automation of arithmetic recurs in more complex tasks.)

Computers can be programmed to discover systematic relationships between variables given a database, and it has been said that this means that scientific discovery is an individual process. Once more this is easy to understand so long as one thinks of discovery as the mechanical part of science (as the spellings of words in a dictionary are the mechanical part of language). But before any computer can do this, it has to separate data from noise, and in science this is a socially approved judgment. In the same way, the output of such discovering computers must be filtered. Not just any relationship is scientifically interesting, or science would be no more than "data dredging." Once more, a close examination of supposed independent discovery machines shows that they depend on humans to filter their output in the same way as a modern spell checker generates a number of potential corrections but leaves it to the human to make the appropriate choice. In sum, even the hard case of science turns out to support the sociologists' view; the supposed discovering machines can help, but not replace, the humans in the scientific process.

The continuing development of various generations of intelligent machines continues to refine our understanding of the social, but it comes no closer to resolving its deep problems.

**BIBLIOGRAPHY**

Collins, Harry M., and Martin Kusch. *The Shape of Actions: What Humans and Machines Can Do.* Cambridge, MA: MIT Press, 1998.

Holton, Gerald. *The Scientific Imagination: Case Studies.* Cambridge, U.K.: Cambridge University Press, 1978.

Kuhn, Thomas S. "The Function of Measurement in Modern Physical Science." *ISIS 52* (1961): 162-176.

Langley, Pat, Herbert A. Simon, Gary L. Bradshaw, and Jan M. Zytkow. *Scientific Discovery: Computational Explorations of the Creative Process.* Cambridge, MA: MIT Press, 1987.

Maurer, David W. *The Big Con: The Story of the Confidence Man and the Confidence Game.* New York: Bobbs Merrill, 1940.

H. M. Collins

## ASTRONOMY.

*See* PHYSICS AND ASTRONOMY

## ATOMIC AGE.

No other age in human civilization has as definitive a beginning as the Atomic (or Nuclear) Age. Whether it began with the first self-sustaining chain reaction in the atomic pile under the squash court at the University of Chicago (December 2, 1942), or with the detonation of the first atomic device in the Trinity test at Alamogordo, New Mexico (at 11:29:45 Greenwich Mean Time, July 16, 1945), we can chart the Atomic Age from a specific time during World War II.

Such origins in a time of global conflict have shaped the trajectory of nuclear technology through the present. Peaceful uses of nuclear energy are intertwined with military ones, and both entail global considerations of its deployment and long-term effects.

The Atomic Age finds its origins in relation to Germany in the 1930s. When Otto Hahn and Fritz Straussmann split the uranium atom in 1938, and Lise Meitner discovered the experimental possibility of a fission chain reaction, the relation between nuclear research and Nazi ambitions worried scientists elsewhere. Physicists who, like Albert Einstein and Leo Szilard, had fled Nazi Germany because of fear of persecution were particularly anxious. On the brink of war in 1939, Einstein wrote a letter to President Roosevelt outlining the potential problem. The ominous specter of a Nazi atom bomb dictated a series of tactical decisions to delay its development in Germany and strategic decisions to accelerate nuclear research in the United States and Great Britain. With a wartime cost of $2 billion, the Manhattan Project involved thousands of personnel filling hundreds of roles.

The Manhattan Project was a mammoth task. Yet, within three years of his appointment in 1942, Brig. Gen. Leslie Groves, supported by his chief science advisor, J. Robert Oppenheimer, melded disparate groups from coast to coast, from northern Canada to Oak Ridge, Tennessee, into a unit that produced the first atomic bombs.

Much had changed, however, between August 1939 and July 1945. The Allies had known for more than a year that the Nazi bomb program was unsuccessful. By the time of the Trinity test, Germany had surrendered, America mourned the death of Roosevelt, and Harry S. Truman had become President. No one thought that Japan, the remaining enemy, had a nuclear research program, much less a bomb.

Whether a different solution could have ended the war or not, the troubling question remains whether racial prejudice made it easier to use the atomic bomb on Japan. What is certain is that short-term political interests have driven the development, deployment, and use of nuclear weapons from the outset. Nowhere is this more evident than the Cold War nuclear arms race between the United States and the Soviet Union.

During the years following Hiroshima and Nagasaki, the United States began a series of atmospheric tests, especially following the Soviet detonation of an atomic bomb in 1949. One year after the first U.S. thermonuclear device upped the ante in 1952, the Soviets tested their own "layer cake" bomb (less powerful than the American design), followed by their first full-fledged two-stage hydrogen bomb in 1955. Once missile technology and guidance systems improved the odds of intercontinental delivery, within a few years multiple warheads, nuclear submarines, and hardened missile silos extended the race. In strategic terms, both Americans and Soviets adhered to the policy of MAD (Mutually Assured Destruction), while also proceeding with the development of tactical weapons for the battlefield.

Although the global effects of nuclear testing were limited by the adoption of the Nuclear Test Ban Treaty in 1963, the environmental effects of nuclear war were largely ignored until the 1980s, when early computer simulations tested scenarios of a nuclear war. Carl Sagan was among the group that identified "nuclear winter," the catastrophic result of the atmospheric debris, even from a limited nuclear exchange, that would plunge the planet into subzero temperatures for more than a year.

This information fueled public pressure for limitations on the numbers of nuclear weapons and strengthened long-standing demands for their elimination. Less successful were attempts to limit the number of states with nuclear capability. The dangers of nuclear proliferation were illustrated by recent nuclear programs in India and Pakistan as well as those rumored in Israel and Arab states in the Middle East.  While the Cuban missile crisis might have brought the world to the brink of nuclear destruction in 1962, the breakup of the Soviet Union and the loss of control over the Soviet nuclear arsenal brought with it the potential for such technology to fall into new hands.

Quite apart from the geopolitical implications of nuclear weaponry is the long-term environmental effect of radioactive contamination. Whether a result of catastrophic accidents (Three Mile Island in 1979 or Chernobyl in 1986), bioconcentration of radioactivity released in the production or testing of nuclear weapons, or problems associated with the storage and disposal of nuclear reactor waste, nuclear technology threatens future generations.

## BIBLIOGRAPHY

Bailyn, Bernard. *The Making of the Indian Atomic Bomb: Science, Secrecy, and the Postcolonial State.* London: Zed Books, 1998.

Ambio Magazine Staff. *Aftermath: The Human and Ecological Consequences of Nuclear War.* New York: Pantheon Books, 1983.

Rhodes, Richard. *The Making of the Atomic Bomb.* New York: Simon and Schuster, 1987.

Rhodes, Richard. *Dark Sun: The Making of the Hydrogen Bomb.* New York: Simon and Schuster, 1996.

Wyden, Peter. *Day One: Before Hiroshima and After.* New York: Simon and Schuster, 1984.

*Peter Denton*

# AUTOMOBILE, THE.

The invention of the first self-propelled vehicle is usually attributed to Nicholas Cugnot, who in 1760 built a steam-powered three-wheeler that was intended to haul artillery pieces for the French army. The unwieldy vehicle soon crashed and overturned, and the army lost interest. Powered road vehicles then enjoyed a brief vogue in England, where a few steam-powered coaches transported passengers during the middle of the nineteenth century.

Steam-powered automobiles were produced in respectable numbers during the late nineteenth and early twentieth centuries, as were electric cars. Each type had characteristic deficiencies. Steamers were fast, but they required a long heating-up period, and they consumed water rapidly. Electrics were clean and quiet but suffered from limited range and a low top speed. Consequently, the development of the automobile is inextricably linked to the four-stroke internal-combustion engine, which was invented in Germany by Nikolaus August Otto in 1876. These engines were at first used as stationary power sources, but in 1885 Otto's assistant, Gottfried Daimler, fitted one to a wooden-framed two-wheeler, thereby creating the first motorcycle. Credit for building the first automobile is usually given to another German, Karl Benz, whose three-wheeler took to the road in 1886. Daimler followed with a four-wheeled automobile a year later, but during the last decade of the nineteenth century, France supplanted Germany as the center of automotive progress as a result the efforts of such firms as Panhard et Lavassor, Darracq, and Peugeot. It is significant that the latter firm began with the manufacture of bicycles (which are still being produced under that brand name), because the early automobile industry made considerable use of parts and techniques initially developed for bicycle use, such as tubular frames, chain drive, tension-spoke wheels, and pneumatic tires. Beyond the technologies that it provided, the bicycle stimulated the development of the automobile by inculcating the idea of personal mobility, while at the same time giving an impetus for the improvement of roads and highways.

Bicycle firms were prominent in the early days of automobile manufacture, but it was Henry Ford, a former electrical power plant engineer, who became the dominant figure in the emerging industry. Ford's company was not the first to mass-produce automobiles, but it took volume production to levels previously unimagined. Rigid standardization, interchangeable parts, and special-purpose machine tools complemented Ford's introduction of the assembly line in 1913, resulting in vastly increased production volume and relentless cost reduction. Lower costs allowed dramatically lower prices and huge sales increases, and by the mid-1920s half of the cars on America's roads were Ford Model Ts. At this time the ratio of cars to people in the United States was about 1 to 5.

The United States was the premier car culture, and even during the severe economic depression of the 1930s the percentage of the population owning automobiles dropped only slightly. The diffusion of the automobile lagged in other parts of the world, and it was not until the 1950s and 1960s that Europe and then Japan replicated the American experience of mass automobile ownership. By the end of the twentieth century automobile ownership was nearly universal in the industrialized world.

Personal preferences have stimulated widespread automobile ownership, but public policies have also encouraged it, especially in the United States. Gasoline has been only modestly taxed, and for many years tax revenues were exclusively devoted to highway construction and maintenance. In 1956 the federal government embarked on what was to become the largest civil engineering project in human history: the Interstate Highway System. Encompassing more that 42,000 miles and still under construction, this network of high-speed, limited-access roadways greatly facilitated travel. It also encouraged the acceleration of suburbanization and the consequent decline of central cities.

The federal, state, and local governments of the United States did little to regulate the automobile until the 1960s, when a number of car-related issues began to enter the political arena.

Automobile-generated air pollution had become a serious problem in several parts of the United States, most prominently in Southern California. The reduction of smog-causing emissions required not only technological fixes, from crankcase ventilation valves to catalytic converters and computerized engine-management systems, but also the enactment and enforcement of governmental regulations that mandated their use. Regulation at the federal level began with the Motor Vehicle Air Pollution Control Act of 1965. By the end of the century these efforts resulted in cars that were far cleaner, but some of the benefits were offset by a relentlessly expanding car population.

Safety was the next issue that elicited the involvement of the federal government, as consumers became increasingly aware of the deficiencies of contemporary car design. The creation of the National Highway Traffic Safety Administration (NHTSA) in 1967 initiated substantial governmental involvement in making cars more crashworthy and protecting their occupants.

With cheap, lightly taxed gasoline, fuel economy had not been much of an issue in the United States until the temporary losses of Middle Eastern oil in 1973-1974 and 1979 constricted supplies and produced long lines at gas stations. Raising fuel taxes to reduce consumption was deemed politically unacceptable, and the federal government has instead mandated that manufacturers must meet a Corporate Average Fuel Economy (CAFE) standard of 27.5 miles per gallon for their automobiles and 20.7 mpg for their light trucks.

Cars are much safer today, but motor vehicle accidents still account for around 42,000 deaths annually in the United States. Similarly, automobile exhausts are much cleaner, but they are still a significant source of air pollution, and fuel economy standards have only slightly diminished the voracious consumption of petroleum products. There are approximately 550 million cars and light trucks on the world's roads today, a number that will increase substantially as the people of China, India, and other developing countries acquire more and more private automobiles. Addressing the problems engendered by the automobile while enjoying its many benefits will be one of the key technological challenges of the twenty-first century.

## BIBLIOGRAPHY

Berger, Michael. *The Devil Wagon in God's Country: The Automobile and Social Change in Rural America, 1893–1929.* Hamden, CT: Archon Books, 1979.

Dunn, James A., Jr. *Driving Forces: The Automobile, Its Enemies, and the Politics of Mobility.* Washington, DC: Brookings Institution Press, 1998.

Flink, James J. *The Automobile Age.* Cambridge, MA: MIT Press, 1988.

Foster, Mark S. *Nation on Wheels: The Automobile Culture in America since 1945.* Belmont, CA : Thomson, Wadsworth, 2003.

Lay, M. L. *Ways of the World: A History of the World's Roads and the Vehicles That Used Them.* New Brunswick, NJ: Rutgers University Press, 1992.

McShane, Clay. *Down the Asphalt Path: The Automobile and the American City.* New York: Columbia University Press, 1994.

Rae, John. *The Road and Car in American Life.* Cambridge, MA: MIT Press, 1971.

Volti, Rudi. *Cars and Culture: The Life Story of a Technology.* Westport, CT and London: Greenwood, 2004.

Wik, Reynold M. *Henry Ford and Grass-Roots America.* Ann Arbor: University of Michigan Press, 1972.

*Rudi Volti*

# B

## BIOENGINEERING.

Although there is much interest in genetic therapy and genetic "engineering," to date very little has been successfully offered. This article will concentrate on two of the most widely used types of bioengineering technologies (although there are considerable areas of overlap between them): first, the technologies used to overcome infertility; then, the technologies of screening and selection to create a particular kind of embryo— technologies that some think are eugenic in their design or impact.

### Overcoming Infertility via Bioengineering

Reproductive technologies are often viewed as instruments of choice that help individuals or couples have children with whom they share a genetic tie. However, choices regarding matters of reproductive technologies are often not real choices, and access to these services is limited to a privileged class. Still, the number of reproductive technologies available to candidate individuals and couples has multiplied since 1983, when the United States saw its first live birth via *in vitro* fertilization (IVF), and research and development move at a brisk pace. These new reproductive technologies vary in their levels of invasiveness, in their success rates, in their risks, and in their financial costs. Yet all have important social implications for women, for mothers, for fathers, for children, for families, and for society.

The socially reinforced desire among Americans for children with whom parents share a genetic tie has helped fuel the rapid expansion and economic success of the reproductive technology industry. In addition, the framing of infertility as a disability or a

disease has legitimated research, spending, and investment in these technologies. In 1995 new reproductive technologies were an approximately $350 million per year industry. If spending on infertility care, more broadly, is included, that figure jumps to about $2 billion per year. Despite these dollar estimates, the valuation of the genetic or biological connection is difficult to overstate in American culture; the belief in the strength of relation by blood (or seed) is a strong one. But by placing such emphasis on biological and genetic connections, social relationships and social processes are devalued. It is a modern phenomenon that procreation can be separate from sex and that conception can occur outside of the human body. Yet, while our technology moves at a breakneck pace, our ideas about the mothers, fathers, and families that result from these modern-day pregnancies lag far behind.

The far-reaching capabilities of new reproductive technologies make it possible for a male and a female donor to contribute the sperm and the egg (gametes), for the egg to be fertilized by the sperm in a laboratory and the resulting embryo implanted in another woman (sometimes called a *surrogate*), and for two other individuals (the intended parents) to raise the child. And though only 40 percent of infertility problems are attributed solely to the woman, and an equal percentage are attributed solely to the man, infertility treatment disproportionately plays out in the medical treatment of women's bodies. For these reasons, reproductive technologies affect men and women in very different ways.

*Assisted Reproductive Technologies (ARTs).* ARTs involve the manipulation of the egg produced by a woman. Three technologies are categorized as ARTs: *in vitro* fertilization (IVF), gamete intrafallopian transfer (GIFT), and zygote intrafallopian transfer (ZIFT).

*In vitro* fertilization most often begins with the artificial (chemical) stimulation and priming of the ovaries of the *genetic mother* (the woman who is to provide the eggs). Eggs are extracted from her ovaries and are fertilized with known or donor sperm in a petri dish. Some or all of the resulting embryo(s) are then transferred through the cervix of the *gestational mother* or *birth mother* (the woman who is to carry the pregnancy) to her uterus, with the hopes of yielding a viable pregnancy that ultimately results in a successful birth. Multiple embryos are usually implanted into the uterus in an effort to increase the likelihood of pregnancy. Because multiple embryos are usually implanted, twins and even triplets are common with this procedure, introducing serious health risks for the fetuses.

IVF is a highly invasive procedure, particularly in light of its relatively low and controversial success rate of approximately 15 percent, not to mention its extremely high cost (usually about $10,000 per round of treatment), which is rarely covered by insurance. Clinics have been criticized for their reporting procedures. Some included only fully completed cycles of treatment in their figures and left out cycles that had to be stopped midway, while others counted twins as two successes instead of just one. In response, Congress passed the Fertility Clinic Success Rate and Certification Act (FCSRCA) in 1992 to mandate and standardize the reporting practices of operating fertility clinics in the United States; all statistics are now reported to and compiled by the Centers for Disease Control (CDC), which then makes the data available to the public. Critics cite the length of the treatment, the invasiveness of the procedure, the cost of the service, and the low success rate of the method.

In gamete intrafallopian transfer a laparoscope (a tube for guiding operating instruments and viewing the bodily environment, inserted through an incision into a patient's abdomen) is used to transfer robust unfertilized eggs and sperm (gametes) to the intended gestational mother's fallopian tubes, where they can combine to form the desired embryo.

Zygote intrafallopian transfer resembles GIFT except that the eggs are fertilized outside of the body, in a laboratory setting, and the resulting zygotes (fertilized eggs, or single-celled embryos) are transferred to the gestational mother's fallopian tubes through the laparoscope, rather than to her uterus via her cervix as in IVF.

*Contract Pregnancy.* Contract pregnancy, or commercial surrogacy, begins with the assumption that one woman, the so-called surrogate, carries the fetus(es) to term in her womb and a different woman will act as the "intended mother," or mother who raises the child. Unlike adoption, in which a woman becomes pregnant and perhaps delivers the child and *then* decides to enter into an adoption agreement, surrogacy arrangements are entered into *prior* to conception and with the presupposition that the surrogate will not be the intended mother following the birth.

The egg may come from the intended mother, the surrogate, or a donor. In many cases, the sperm is donated by the "intended father," or father who will raise the child. However, like eggs, sperm may also be provided by a donor. When the surrogate is also the genetic mother, she is inseminated with sperm from the intended father or donor. When the surrogate mother is not the donor of the egg, the surrogate is made pregnant using ARTs (with embryos resulting from egg and sperm donations that may or may not be from one or both of the intended parents), and she then carries the fetus(es) to term. After the birth, the surrogate relinquishes her parental rights in the child(ren) to the intended parents. There may or may not be any future relationship between the surrogate and the child, depending on the legal arrangement, usually entered into between the surrogate and the intended father.

Contract pregnancy enables people to have a genetic tie to their children even in cases where the intended mother is unable to carry a pregnancy to term. However, surrogate agreements are usually drawn up to favor the rights of the intended parents, or the people who pay for the surrogacy "service." As just described, the surrogacy arrangement can be quite complicated, depending on who contributes the genetic material for the pregnancy. There have been some high-profile cases (most notably the *Baby M* case in 1987) involving women who served as surrogates and then fought for rights to the children they carried in their wombs. In the *Baby M* case, Mary Beth Whitehead acted as a surrogate for a couple, the Sterns. Whitehead realized that she had a strong bond to the baby she carried; she could not give the baby up to the Sterns. The case garnered national and international media attention, and ultimately Whitehead was determined a noncustodial legal mother and given visitation rights. This case and other cases come to the fore because the parties could not come to a settlement or agreement on their own, outside of the courts.

But even when contracts are adhered to, ethical concerns remain. In surrogacy contracts, it is likely that the "intended parents" are white and have money, whereas the surrogate mother is a woman without significant financial resources and may also be a woman of color. Not surprisingly, a so-called surrogate earns a pittance for her services—about $10,000, or just over $250 per week of pregnancy. Proponents of contract pregnancy claim that acting as a surrogate is a sort of gift to a couple, done primarily for altruistic reasons, without a monetary incentive. But considering that there is often a significant disparity between the financial resources of the intendeds and those of the surrogates, it is quite clear that serving as a surrogate is unlikely a purely altruistic act. And given that the actual work involved in pregnancy, labor, and delivery is considered unskilled, and that the racial background of the surrogate does not affect the physical traits of the baby, this is precisely the kind of work that low-income women and women of color are considered "qualified" to do.

Critics argue that commercial surrogacy commodifies the female body, reducing it to a vessel or container. That is, this arrangement devalues the effort and the experience of pregnancy as well as the social relationship that develops between a pregnant woman and her fetus. With the stroke of a pen and the help of some very clear laws protecting "intended parents," such as those in the state of California, so-called surrogates are reduced to low-skilled workers whose work itself is not valued, despite the fact that they produce a product that is among the most sacred and treasured in American culture.

## Screening and Selection

Screening can be done by testing an already implanted fetus or by testing a zygote produced *in vitro* before implantation. Screening procedures are used to eliminate the risk of some medical condition deemed a defect, to ensure the presence of some quality deemed desirable, or to select the sex of the child.

*Prenatal Screening.* The technologies of prenatal screening and diagnosis are far more widespread than are the technologies known as ARTs. But these too, from ultrasound diagnosis to the more sophisticated genetic tests, raise questions about eugenics—of deciding which lives are, in the Nazis' words, "lives unworthy of life." Prenatal screening has been increasingly routinized in maternity care in the United States, with ultrasound examinations and maternal serum screening now standards. These routine tests, however, raise very difficult questions and can create very difficult situations. Routine screening, including ultrasound visualizations and maternal blood tests, have high rates of false positives. Women thus may find themselves facing more invasive prenatal diagnosis following routine testing, without adequate counseling or preparation.

Amniocentesis is the most common of the definitive diagnostic techniques used. It involves the insertion of a large needle through the mother's abdomen into the amniotic sac to collect a sample of the amniotic fluid, which surrounds the fetus. This fluid is then analyzed for certain genetic abnormalities, most commonly Down syndrome. Amniocentesis is used after standard screening tests such as ultrasound or maternal serum testing, in women with a family history of a genetic disease, and with women over the age of 35—a somewhat arbitrary age designation. The rationale behind the routine use of amniocentesis for this last group of women is that the incidence of Down syndrome increases as maternal age goes up. and at age 35 the risk for Down syndrome surpasses the risk of spontaneous abortion from amniocentesis itself. In addition to the risk of miscarriage, women may experience cramping, discomfort, spotting, and leakage of amniotic fluid following the procedure.

Criticisms of this technology include that results of the test often take weeks. During this period of "tentative pregnancy," women are in limbo, unsure whether they are having a baby or an abortion. All told, 95 percent of women who have amniocentesis get good results—no genetic abnormalities. For the other 5 percent the question of what to do with the information looms, particularly given that results may come back as late as the twenty-fourth week of pregnancy. Almost none of the conditions diagnosable are treatable *in utero;* for women not considering abortion, there is little reason to have prenatal diagnosis. However, this prospect is not something generally discussed with women when an ultrasound examination or maternal serum screening test is offered.

*Embryo Screening.* Preimplantation diagnosis, or embryo screening, involves diagnosing the genetic material for desirable and undesirable traits prior to implantation and then, using ARTs, implanting only those zygotes that possess desirable traits. An enormous variety of genetic diseases can be identified.

The engineering of human genetic material is a controversial subject because engineering for so-called "better" genetic outcomes means that certain traits and characteristics are valued over others. Those valued traits, from sex to ability/disability issues, are reflective of cultural values and therefore subjectively singled out as preferred characteristics.

*Sex Selection.* A far more controversial use is for sex selection. That is, the sex of each zygote is determined and individuals then select which zygotes to implant, based on preferred sex. Sex selection, once considered to be the clear line in the sand between the ethical and unethical use of prenatal genetic information, is now available at fertility clinics throughout the United States. Couples undergo screening and counseling to determine whether they are "appropriate" candidates for sex selection. This screening process is intended to determine whether sex selection is being used for the "right" reasons. Having two boys, only being able afford three children in total, and really wanting a daughter passes muster at some clinics. Many ethicists worry that preimplantation diagnosis is a slippery slope. One may also question whether sex selection is fundamentally different from screening for any other condition, a point made by the disability rights movement activists. In both cases, the quality of life of the prospective child and of its parents rests primarily on social conditions.

## Reproductive Cloning

The interest in reproductive cloning, in which the nucleus of an adult cell is used to replace the nucleus of an egg cell, growing a child who will be an identical twin of the adult donor, is perhaps the most controversial of the new reproductive technologies. Although claims have been made, it is not, at this writing, ascertained that any human being has been cloned. Given what we know about identical twins, and the enormous variability they show even when grown in the same pregnancy and raised in the same circumstances, we know that cloned individuals will not be "identical" reproductions, and the fantasies of armies of clones, or cloned Hitlers or Mozarts, remain just that: fantasies.

## On Pregnancy, Choice, and Society

Pregnancy is about making a baby, but it is also about building a family, forming relationships, making the woman a mother. Even if the pregnancy is not successful in producing a baby, it may very well be successful in its other products. Death and grief can also build a family. By nurturing the woman, her relationship with her partner and her family and friends, and her feelings about herself and her lost child, midwives do a good job constructing success, satisfaction, and even family out of death. Modern culture, however, often fails to see that positive things can be born from grief, especially when such situations can be avoided. And it is in this context that so-called "choice" becomes compulsory and hardly choice at all.

From the beginning, advances in genetic testing have made things more complicated, not less. Tay-Sachs disease was one of the first genetic conditions for which prenatal testing and selective abortion were made available. Other options, including donor insemination, are less likely to be offered. It has also been noted that when a prenatal test is made available, funding and interest may shift away from research on cures; this has been the case with Tay-Sachs.

Infertility itself is a social construction of its times. It used to be that a couple had to be actively trying to become pregnant for two years before being considered for infertility treatment. Today, a year is the benchmark, and six months for women older than 35, according to Resolve: the National Infertility Association. This same organization declares infertility a medical disease. and, so it goes, we must treat it medically, aggressively.

While many view new reproductive technologies as a miracle, a gift, the technologies are not available to everyone who wants them or "needs" them. The technology is expensive, and someone has to pay for it. Considering that about 40 million Americans are without health insurance, and most people have to pay for these technologies out of their own pockets anyhow, it is clear that new reproductive technologies are accessible to only a select population. Simply put, in the United States people with money have access to them, and people without the disposable income do not.

Economic class is closely linked to race, further underscoring the divide between the "haves" and the "have-nots." In a society divided by racial categorization, procreation is experienced in race-specific ways. Commodifying procreation and reproduction affects whites and nonwhites differ-

ently. The price of socially desirable white eggs has doubled in the same time that the fee for surrogacy, where the race of the surrogate does not affect the child, has remained the same and even fallen in real dollars. And although the incidence of infertility in women of color is almost double that of white women, the costs of infertility treatments render treatment unaffordable for most people of color. White babies are highly desirable; there is no shortage of adoptive parents who want a white baby. The same cannot be said for children of color in need of a home.

These and other issues remind us that, although there are ethical dilemmas associated with all reproductive technologies, the real issues involve access to these services and choice in the context of the knowledge these services provide. It is too easy to forget that given their price tags, these services ultimately benefit so few people. According to Resolve, fewer than 5 percent of the people undergoing fertility treatment even need the advanced reproductive technologies that are currently available—and yet hundreds of millions of dollars per year are spent on developing such technologies.

When access is limited, choice is illusory. For instance, is there really a "choice" of selective abortion if one has the option of amniocentesis but not access to the social services necessary to raise and nurture a child who is born with a disability?

It becomes increasingly clear that the issue is power, not technology per se. Individual choice, in the hands of a relatively select few, underscores rather than undermines social inequality. While the promises of the new reproductive technologies are alluring, the realities raise complex social questions regarding gender, race, class, mothers, fathers, and families.

**BIBLIOGRAPHY**

Atwood, Margaret. *The Handmaid's Tale: A Novel.* New York: Anchor Books, 1998.

Becker, Gaylene. *The Elusive Embryo: How Women and Men Approach New Reproductive Technologies.* Berkeley: University of California Press, 2000.

Bowring, Finn. Science, *Seeds and Cyborgs: Biotechnology and the Appropriation of Life.* London: Verso, 2003.

Ginsburg, Faye D., and Rayna Rapp, eds. *Conceiving the New World Order: The Global Politics of Reproduction.* Berkeley: University of California Press, 1995.

Huxley, Aldous. *Brave New World* (1932). New York: Perennial, 1998.

Kass, Leon, and James Q. Wilson. *Ethics of Human Cloning.* Washington, DC: American Enterprise Institute Press, 1998.

McGee, Glenn. *The Perfect Baby: Parenthood in the New World of Cloning and Genetics.* New York: Rowman and Littlefield, 2000.

Ragone, Helena. *Surrogate Motherhood: Conception in the Heart.* Boulder, CO: Westview Press, 1994.

Rich, Adrienne. *Of Woman Born: Motherhood as Experience and Institution.* New York: W. W. Norton, 1986.

Roberts, Dorothy. *Killing the Black Body: Race, Reproduction, and the Meaning of Liberty.* New York: Pantheon, 1997.

Rothman, Barbara Katz. *The Tentative Pregnancy: How Amniocentesis Changes the Experience of Motherhood.* New York: W. W. Norton, 1993.

Rothman, Barbara Katz. *Recreating Motherhood.* New Brunswick, NJ: Rutgers University Press, 2000.

Rothman, Barbara Katz. *The Book of Life: A Personal and Ethical Guide to Race, Normality, and the Implications of the Human Genome Project.* Boston: Beacon Press, 2001.

*Bari Meltzer, and Barbara Katz Rothman*

# BIOLOGICAL TERRORISM.

*Biological terrorism* and *bioterrorism* are terms that came into widespread use in the United States during the 1990s. Both refer to the use of *biological agents* by *nonstate actors* to frighten, injure, or kill people (including attacks on crops or livestock). Biological terrorism is distinguished by *who* terrorists are, *what* sort of agents they use, and *why* they use those agents.

## Biological Agents

Biological agents include pathogens (living microorganisms that cause disease) and toxins (nonliving poisons extracted from living microorganisms, plants, or animals). Pathogens that have been developed into biological weapons (BW) include the *bacteria* that cause anthrax, tularemia, and plague; *viruses* such as smallpox and Marburg; and *rickettsiae* such as *Coxiella burnetii*, which causes Q fever. Toxins such as botulinum and ricin have also been used as biological weapons. Theoretically, any pathogen could be used as a biological weapon, but doing so poses considerable challenges.

Biological agents can be used as weapons in two ways. First, an individual might infect himself or herself with a pathogen such as smallpox that can be transmitted from person to person, in the hope that this would spark a wider outbreak of the disease. This method would likely work only on a small scale, would be difficult to distinguish from a naturally occurring epidemic, and would almost certainly result in the death of the person who had originally been infected.

For a biological agent to be effective on a large scale, it must be "weaponized." A weaponized agent must have several characteristics. It must be highly likely to cause disease in humans, or *virulent*. Since most naturally occurring pathogens are not especially virulent, they have to be selected, bred, or genetically modified to increase their virulence. A biological agent must also be relatively stable so that it can survive being stored, transported, and delivered to its target. Since pathogens tend to be fragile and unstable, they must be manipulated in a laboratory to improve their stability. Finally, a terrorist must have a relatively large quantity of an agent in order to produce a weapon that can affect large numbers of people. Naturally occurring pathogens are difficult to collect in quantity, so they must be cultivated and mass-produced in a laboratory.

In nature, pathogens infect humans through breaks in the skin or by being ingested or inhaled. Exposing large numbers of people to pathogens through ingestion, which would require the contamination of food or drinking water, is difficult. Modern sanitation methods such as chlorination and filtration, as well as safety precautions such as cooking and pasteurization, are effective in killing pathogens. Contaminating a large water source such as a reservoir is unfeasible because of dilution effects. Inhalation is thus the most likely route for a biological weapon. However, creating an *aerosol* (a cloud of infectious droplets or particles that remains suspended in the air long enough to be inhaled) requires extensive manipulation of the pathogen. Effective delivery of an aerosol is also difficult, because most pathogens break down quickly in sunlight, and any wind will rapidly disperse an aerosol cloud, rendering it harmless.

Manufacturing biological weapons is expensive and requires considerable scientific expertise, and until recently only large nations had the financial resources, technological sophistication, and laboratory infrastructure necessary to do so. The United States, Russia, and other nation-states have maintained BW programs since World War II and have successfully weaponized a number of agents.

## Nonstate Actors

In recent years it has become feasible for nonstate actors (individuals or groups not directly affiliated with a nation-state) to manufacture and use BW. Developments in biology and biotechnology have given humans unprecedented ability to manipulate and control microrganisms. This has led to the "dual use" dilemma: Much legitimate biomedical research aimed at curing infectious diseases can also be used to construct BW. Scientific expertise and technology are also more widely available. The equipment necessary to weaponize pathogens exists in countless academic institutions and private pharmaceutical and biotechnology companies, and it can often be purchased on the open market. The number of nations conducting BW research has also multiplied, and many of these might provide expertise or actual weapons to nonstate actors.

Unlike nation-states, which have used or threatened to use BW against enemy combatants during war, nonstate actors might target civilians at any time, and for many reasons. Politically motivated terrorist groups seek to publicize a cause by creating widespread fear or panic; millenarian sects seek to hasten an imminent apocalypse; hate groups such as nationalist extremists or racial or religious supremacists seek to destroy a perceived enemy or minority group. Terrorists might also be affiliated with or sponsored by individual nations seeking to attack an enemy covertly.

## Historical Examples

Although a number of individuals and groups have attempted to obtain or manufacture BW, only three confirmed cases of biological terrorism have ever occurred. In 1984, members of a religious sect in Oregon called the Rajneeshees attempted to sicken citizens in order to prevent them from voting in local elections. As an experiment, they contaminated salad bars in ten restaurants with *Salmonella enterica*, an organism that causes food poisoning. Over the next two weeks, at least 751 people became sick; none died.

During the early 1990s, a millenialist Japanese religious sect called Aum Shinrikyo attempted to develop BW. They conducted research into anthrax, botulinum toxin, and Q fever and reportedly sent medical missionaries to Zaire in order to obtain samples of the Ebola virus. On ten separate occasions between 1990 and 1995, they attempted to disseminate botulinum toxin from trucks in Tokyo and a U.S. Navy base at Yokohama and to release anthrax from a device atop a building in downtown Tokyo. None of these attempts resulted in any casualties. In 1995 the group disseminated sarin gas, a chemical weapon, in the Tokyo subway; this attack sickened more than 1000 and killed 12.

In October and November 2001, five letters containing *Bacillus anthracis* spores were mailed to news organizations and government offices in Florida, New York, and Washington, D.C. This resulted in 22 cases of anthrax, five of which were fatal, and the probable exposure of up to 400 others to the bacillus. The majority of cases and exposures occurred among postal workers and other mail handlers. In the following weeks at least 4000 anthrax hoaxes were reported, and approximately ten million Americans purchased antibiotics. Although the individual or group responsible for this attack has not been identified, the anthrax used in this attack was later determined to be the Ames strain, developed by the American BW program during the 1980s.

*See also* **Military, The, and Technology**

**BIBLIOGRAPHY**

Miller, Judith, Stephen Engelberg, and William J. Broad. *Germs: Biological Weapons and America's Secret War.* New York: Simon and Schuster, 2001.

Stern, Jessica. *The Ultimate Terrorists.* Cambridge, MA: Harvard University Press, 1999.

Tucker, Jonathan B., ed. *Toxic Terror: Assessing Terrorist Use of Chemical and Biological Weapons.* Cambridge, MA: MIT Press, 2000.

*Nicholas B. King*

# BIOMEDICAL TECHNOLOGIES.

The widely publicized Human Genome Project has been a massive undertaking requiring enormous computing power. Its goal has been to produce an accurate map of the chemical structures that make a person a person—the blueprint of human life, involving several billion pieces of information. Along the way, the project has also become a test case for weighing the benefits of private versus public science represented on each side by highly visible scientists with charismatic personalities, with the largely government-funded public project at the U.S. National Institutes of Health led by Francis Collins competing head-to-head with a for-profit commercial operation led by J. Craig Venter of Celera Genomics to crack the code first.

Each group has claimed victory on several occasions as new milestones have been reached; most observers seem to accept the proposition that the project would not have advanced as quickly as it did were it not for the intensity of the competition. The rivalry may have peaked during a joint presentation at the 2001 American Association for the Advancement of Science meeting at which Venter's Celera group, publishing in AAAS's prestigious *Science,* and Collins' NIH group, publishing in an issue of *Nature* that was literally hot off the presses when it arrived at the meeting, simultaneously revealed similar "draft" human genome maps. (Actually, neither map was entirely complete, and the work of both groups continues.) While the rivalry may indeed have spurred both groups to work harder and faster, it also spurred a heated debate about the conflict between the preservation of commercial patent rights, based on keeping details proprietary, and the advancement of public science, based on a policy of open information sharing.

Knowing most of the genetic code, or even all of it, does not, however, mean understanding it and does not translate directly into effective therapies for genetics-related problems. Genes with specific known effects must still be identified, defined, and distinguished from the amorphous chunks of code. Many traits are believed to be the result of the interaction of multiple genes and often reflective of environmental influences as well. Even identical twins do not always have the same personalities, problems, or diseases. Nevertheless, the project is an important and highly visible step in the direction of linking genetic heritage with a variety of conditions. This step has also raised people's awareness of the profound social and ethical issues associated with the complete mastery of human genetics that the project appears to promise in the not-so-distant future. The issues include the following:

- Genetic testing and privacy: What will employers and insurance companies do with the information about individuals' susceptibility to particular diseases?
- The question of "designer babies": Is it right for parents to choose their children's gender, height, weight, coloration, athletic ability, or intelligence?
- The essential nature of human individuality and identity: Should the code ever be duplicated to produce a new human, and if it is, will this clone be the same person, or a new one?
- The relations between genetics and ethnic identity, genetics and personality, and genetics and human behavior: How much of our decision making is based in biology and how much is actually a matter of choice?

Against this backdrop, public controversies have raged about the heritability of homosexuality (the "gay gene" idea), of obesity (the so-called "fat gene"), and of individual predispositions to mentally disturbed, aggressive, or criminal behavior. The idea that so much of human behavior might be "in the genes" represents an assault on the Western legal system (and some Western theology) by undermining the presumption that humans make behavioral choices that are free and that may be rationally determined. Some worry that the kind of genetic determinism that this line of research seems to reinforce will blind us to the social and environmental determinants of behavior, such as learned social values, economic influences, and family dynamics. At the same time the eventual promise of the project may be to enable us to transcend the tyranny of biology, making possible the achievement of human control over human evolution and destiny to an unprecedented degree. But the minute we have the capability to correct defective genes, we will be faced with the dilemma of having to choose which human characteristics actually fall in the "defect" category. As the old saying goes, we must be careful what we ask for—we might get it.

Despite the ethical challenges that a complete knowledge of human genetics may eventually engender, there are some cases where social consensus on the right course of action is more likely than in other cases. If diseases such as cystic fibrosis or diabetes can be treated with gene therapy, if dysfunctional organs can be replaced with substitutes from modified animals, or if nerve cells or other critical human tissues can be made to regenerate themselves, the benefits would appear overwhelming. In fact, most people in both North America and Europe (the areas where opinion data are generally available) are much more supportive of medical biotechnology than of agricultural biotechnology, most likely because the benefits to the quality of human life are so readily apparent for medical interventions. Yet in each of these examples—gene therapy, xenotransplantation, and the use of stem cells—substantial ethical controversy has arisen.

## Gene Therapies

Once knowledge of human genetics progresses to the point that specific genetic defects can be identified and located on a human chromosome, can they also be compensated for by somehow substituting compatible genetic material that lacks the defect? Can animal organs be bioengineered to be compatible enough for transplant into human beings? Many researchers are pursuing the development of these and other therapies based on manipulating DNA, the genetic blueprint of life, although these have not provided much in the way of easy medical answers. What they have provided is new public attention to issues of medical ethics.

When teenager Jesse Gelsinger died in a gene therapy experiment at the University of Pennsylvania in 1999, the director of the university's Institute for Human Gene Therapy was reported to have been indefinitely banned from using human subjects in subsequent experiments. News reports also indicated doubt about whether Gelsinger and his family were given adequate information about the potential risks of the highly experimental therapy. To make matters worse, the director was alleged to have had a financial interest in a gene therapy company that had supported the Institute's research (and presumably was in a position to benefit from the knowledge gained by the work that resulted in the boy's death).

The gene therapy research community has responded with increased attention to the ethics governing such situations, but profound questions remain that are still disturbing to some observers. Who will get to decide which characteristics are "defects" requiring compensation, and which are normal variations in appearance or ability? For some observers, gene therapy raises some of the same disturbing questions as genetic engineering of human beings. If today's athletes self-medicate with performance-enhancing drugs, will not tomorrow's be tempted to seek ways to improve their own genomes? If today's infertile couples seek sperm donors of unusual intelligence or a particular physical appearance, will not tomorrow's seek to obtain these qualities by "fixing" their children's perceived deficiencies through gene therapy? It is easy to say that responsible members of the medical community will not let such things take place, but in a society in which plastic surgery to improve appearance—not simply to correct self-evident defects—is commonplace, it is not easy to believe there will be no demand.

Present gene therapy research focuses on restoring or replacing the function of cells affected by a genetic defect. These modifications, known as *somatic modifications,* cannot be passed on to the patient's children. Future gene therapy might eventually involve heritable changes as well, although this *germ line modification* would be substantially more difficult. Genetic alteration of this kind could conceivably eliminate hereditary illnesses in a family's future generations, but any unanticipated side effects would also be passed on. The prospect of making permanent modifications to the human gene pool, even in the interest of eliminating disease, is disquieting. Evolution, an imperfect process, has brought us this far. Do we really have the wisdom to know when to redirect its course?

## Xenotransplantation

The transplant of an organ or tissue from a nonhuman species into a human being has its own challenges and opportunities. Heart disease is a leading killer in the developed world, and liver and kidney disease take other lives. Healthy human organs available for transplant are in short supply. Attempts to transplant baboon hearts into people have failed, and artificial hearts are not yet adequate. As with gene therapy, experiments with these techniques unleashed a flurry of criticism, in part because of the probability of failure. Yet if xenotranplantation is ever to become the hoped-for alternative to death by progressive organ failure, further experimentation on human patients, as well as potential donor species, is inevitable. Meanwhile, bioengineering may make it possible to produce animals with specific organs that stand a better chance of functioning in human bodies.

The prospect of xenotransplantation carries not only significant risks but significant cultural barriers to acceptance. Anecdotes about heart transplant recipients who develop strong feelings about their donors are common. Even if it proves practical, the idea of keeping a human patient alive through a transplanted pig's liver is unlikely to engender easy acceptance. In addition, the development of animals with transplantable organs might raise animal welfare concerns, especially during the period when the capacity to make functional and safe transplants to human bodies remains an experimenter's vision rather than a reality.

## Stem Cell and Embryonic Tissue Research

While research on gene therapies and xenotransplantation continues, basic research involving human embryonic stem cells has become even more controversial, and unusually swift government action in the United States has effectively halted the development of new embryonic stem cell lines by restricting federal funding, by order of President Bush, in 2001. Embryonic stem cell research is closely associated with both abortion and cloning, two of the most controversial medical interventions of the last century.

Stem cells are cells that have the capacity to differentiate into one or more types of mature tissue, such as muscle, brain, blood, and skin. Stem cells are therefore the source of tissues that are not naturally self-regenerating when differentiated, such as nerves. For this reason, the use of stem cell material is believed to have potential for treating spinal cord injuries and has also been suggested as a possible treatment for other diseases ranging from juvenile diabetes to Parkinson's and Alzheimer's. In addition to differentiating, stem cells can continue to reproduce themselves as stem cells. Once isolated, a *line* of stem cells can apparently be replicated almost indefinitely.

The most potent forms of stem cells—those that can *each* develop into all of the various specific types of cells found in complete adults—are those that constitute embryos and early fetuses. The products of abortion, along-side tissues from naturally occurring miscarriages, are one possible source from which these cells can be extracted. Embryos that were created for implantation into otherwise infertile women but that have not been used have also been identified as a key source of this most powerful form of stem cells; it is not really necessary to use tissue from implanted and aborted fetuses, though an embryo is still destroyed in the process. Some have suggested that cloning techniques might be used to increase the availability of necessary material. Again this is not essential, but it creates another close association of this technology with an even more controversial one. These associations may have helped create an image of deliberate "test tube" production of individual humans for purposes of research or, worse, as sources of medical "spare parts" for living individuals who have had themselves cloned. Some people do object, on religious and other grounds, to the intentional destruction of human embryos under any circumstances, from wherever they are derived. Those who are conservative on the issue of abortion or repelled by the prospect of human cloning are also distressed by the prospect of using any embryonic or fetal material for medical research, and in the United States these are often vocal and politically powerful voices.

Research continues into stem cells found in adults, a less controversial line of inquiry that has already produced clinical trials and may yield substantial medical contributions, even though adult stem cells have been thought to be much more limited in their flexibility.

Research involving embryonic stem cells taking place elsewhere in the world is expected to receive a significant boost because of the restrictions that currently apply within U.S. borders, which prohibit federally funded researchers from creating new embryonic stem cell lines and thus confines them to working with a limited selection of tissues that already exist. Some fear the loss of U.S. scientific leadership in this area as a result.

## Issues for the Long Term

Despite the controversies surrounding embryonic stem cell research, and despite the fact that the "big science" approach of the Human Genome Project has not directly resulted in many practical new therapies as yet, research on medical treatments growing out of the current explosion of knowledge of human genetics is still expected to yield big results. In cultures that value human well being, medical applications of science and technology are likely to continue to receive broad public support despite ethical reservations. Unfortunately, as both the Celera-NIH rivalry and the case of Jesse Gelsinger illustrate all too well, medical research is often driven as much by the profit motive and the ego of the researcher as by humanitarian considerations.

Long after the key ethical questions of what uses and modifications of human genetic material are appropriate have been answered, the question of differential access to what are likely to be enormously expensive medical alternatives will probably remain. Who will receive gene therapy, organ xenotransplants, or the benefits of stem cell research—both within the industrialized countries that are developing these alternatives and, more generally, around the world? Most likely, those at the top of the economic pyramid will be the primary beneficiaries of these discoveries, although the development of these technologies as public versus private commodities, the adequacy of regulations regarding physician and researcher conflicts of interest, and the impact of international agreements governing access to intellectual property may all influence their distribution.

See also **Human Subjects in Medical Experiments; Medical Technologies; Medical Values and Ethics**

## BIBLIOGRAPHY

Celera. http://www.celera.com. The commercial genome project company founded by J. Craig Venter that publicly competed with the public Human Genome Project and has expanded its mission to the pursuit of therapeutic applications.

Chapman, Audrey R., Mark S. Frankel, and Michele S. Garfinkel. *Stem Cell Research and Application: Monitoring Frontiers of Bio-*

*medical Research*. Washington, DC: American Association for the Advancement of Science; Institute for Civil Society, 1999. http://www.aaas.org/spp/sfrl/projects/stem/report.pdf. Careful review of ethical concerns; clarifies religious objections and how views on stem cell research may relate to those on cloning and abortion.

Coalition of Americans for Research Ethics. *Do No Harm*. http://www.stemcellresearch.org. Dedicated to promoting adult stem cell research, as opposed to embryonic, and publicizing its results.

Hubbard, Ruth, and Elijah Wald. *Exploding the Gene Myth: How Genetic Information is Produced and Manipulated*. Boston: Beacon Press, 1993. Readable essay presentation introducing concerns, especially the assumption of genetic determinism.

National Reference Center for Bioethics Literature. "Human Gene Therapy." *Scope Note* 24. http://www.georgetown.edu/research/nrcbl/scopenotes/sn24.htm. Last updated August 2004. Excellent history and extensive annotated bibliography on subject, including various position statements.

The National Human Genome Research Institute. http://www.genome.gov. Informative source for information on both the scientific project and the associated research on ethical, legal and social implications.

Nelkin, Dorothy, and M. Susan Lindee. *The DNA Mystique: The Gene As a Cultural Icon*. New York: W. H. Freeman, 1996. Provocative discussion by a well-known sociologist of the cultural significance of human genetic heritage.

*Susanna Hornig Priest*

# BIOTECHNOLOGY.

The "brave new world" of genetically engineered plants and animals promises everything from plant-based edible vaccines, to sheep and goats that produce pharmaceutical drugs in their milk, to pigs whose organs can be transplanted into people, to yellow rice designed to save millions in the developing world from serious nutritional deficiency. Which of these imagined futures (and hundreds of others) will prove practical on the basis of economics, environmental and health impacts, and popular acceptance remains to be seen. Meanwhile, a startup company in Texas, Genetic Savings and Clone (GSC), has promised (for a substantial fee) to preserve a specimen of your dead pet's DNA until the day your beloved and much-missed Rover can be cloned to create a brand-new and genetically identical dog—an enterprise that neatly exemplifies the broadly entrepreneurial character of this new area of science. How much is emerging promise, and how much science fiction or simple hype? (After exhausting a substantial research grant, gene scientists at Texas A&M University, funded by GSC, were unable to clone a dog, although they reported success in the case of a cat. In June 2004 GSC produced two cloned kittens and promised "dog cloning available in 2005.")

*Biotechnology* means different things to different people. Broadly construed, the term refers to the modification or exploitation of naturally occurring biological processes (whether in plants, in animals, or in microorganisms) to meet perceived human needs. By this widest definition, most agriculture and medicine—not to mention wine and cheese making—could be described as biotechnology. Currently the term most often refers specifically to active manipulation of the genetic code of life, the deoxyribose nucleic acid (DNA) molecule—sometimes resulting in *recombinant DNA* (rDNA) products in which the genes of one species are combined with those of another to produce a desired trait (genetic engineering).

## Food Fights

Few developments in modern science and technology have engendered as much controversy as the introduction of genetically modified foods. Public concerns first became widely visible in the United States in 1994 with the introduction of a new supply of the dairy cow growth hormone, bovine somatotropin (BST), produced by genetically modified bacteria. The resultant *rBST* (recombinant BST) became the first bioengineered item to be broadly adopted in U.S. food production; it is administered to dairy cows to increase their milk production. Even though neither the dairy cows nor their milk were actually being directly modified, the protest over rBST's use in dairy herds became the first public opinion and policy test of the new gene technologies, and it was partly the dairy farming community in small-farm states such as Vermont and Wisconsin that led the charge against it. Concern quickly spread to consumers, however, and because labeling dairy products produced with rBST could mean expensive segregation of the milk supply system and could possibly exacerbate these fears further, the technology also touched off a controversy about labeling biotech food products that is still unresolved.

The original dairy industry fear seemed to be that this product, which quickly became widely utilized in the United States though still banned in neighboring Canada, would accelerate the trend toward fewer and larger farms and help drive smaller dairy operations out of business. Animal welfare advocates also feared increased stress on cows being chemically forced to produce more milk, and human health concerns were raised about the heavier use of antibiotics that could result and the perception that milk quality might be lowered. As these concerns spread from farmers to various advocacy groups to consumers, the biotechnology industry assumed that public fears were attributable to misinformation and miseducation. In most cases, however, the correlation between knowledge and opinion in this area has proven to be weak; values, beliefs and attitudes seem to be more important.

For consumers to accept biotechnology, its benefits have to be clear. For this reason, medical biotechnology is often more broadly accepted than food biotechnology in parts of the world where there is no food shortage, such as North America and Europe. Several major bioengineered crops have been introduced as environmental positives, not just moneymakers for industry or agriculture, because they require fewer applications of chemicals. Roundup Ready® soybeans, for example, were developed to be treated with a relatively benign herbicide at an early stage of growth without harming the crop. (These soybeans appeared during a period when their developer, Monsanto Corporation, was about to see its patent on the Roundup herbicide itself expire, leaving it searching for a new source of revenue to replace lost profits.) But others fear the emergence of herbicide-resistant "superweeds" with some of the same properties as the engineered soybean plants, and they worry that the doses of herbicide that these plants can now withstand may not be benign from an environmental point of view.

Similarly, so-called Bt cotton and Bt corn have been engineered to express a mild pesticide of their own as a result of incorporation of genes from a naturally occurring soil bacterium, *Bacillus thuringiensis*. These bacteria had been a staple of "organic" gardeners who wanted to get rid of insect pests without relying on industrially produced chemicals that they saw as environmentally problematic. The engineered-in Bt element might be able to replace some use of these chemicals in conventional agriculture, resulting in an engineered crop it was felt that the environmental community could embrace. Some fear, however, that broader use of Bt materials in commercial agriculture will lead to loss of effectiveness as insect resistance develops through natural selection. That outcome would eventually mean an end to the element's applicability in smaller-scale organic agriculture, not just in larger-scale conventional agriculture.

Controversy after controversy has followed the introduction of genetically modified agricultural species. In 2000 StarLink® corn, a Bt corn that had been approved only for animal feed, turned up in the human food supply in supermarket taco shells. Although this corn itself was not seen as a significant health risk, the incident further galvanized U.S. public opinion on the issue, because it suggested that restrictions and guidelines on these crops may be difficult to enforce. Tests on Bt corn at Cornell University in 1999 suggested that monarch butterflies might be harmed by the pollen of these engineered crops falling on the milkweed leaves on which they feed as caterpillars. A controversial study published in the widely respected science journal *Nature* in 2001 suggested that modified corn genes had un-expectedly cross-pollinated with naturally occurring stock across much of Mexico, threatening the genetic integrity of tradi-tional crops. Debate raged over whether modified fish capable of outcompeting natu-rally occurring species might escape into open waters, with unpredictable environ-mental and biodiversity consequences. Ru-mors have surfaced of partial failures of Bt cotton crops, causing suicides among poor farmers in India.

Currently, "second-generation" biotech-nology is producing crops engineered to con-tain chemicals used in pharmaceutical production and in other industrial applica-tions, using food plants as a base. Although such crops are supposed to be grown only under special conditions and scrupulously kept separate, news reports have stated that another Texas company, ProdiGene, was fined millions of dollars in 2002 for carelessly mixing remnants of an experiment test crop of this kind with other crop material.

Most scientists involved in developing agricultural biotechnology say that they see this technology as having substantial poten-tial for good. It may be possible to create more nutritious foods, more drought-resistant foods, foods that can actually help fight human disease, and foods that can withstand the unique environmental conditions charac-teristic of poorer areas of the world such as parts of Africa and Asia. Others, however, argue that the environmental risks are too great, that the enormous investment in re-search required to bring some of these dreams to reality would be better spent else-where, that consumers are too suspicious of these foods, and that for the most part the GMO (genetically modified organism) crops that we are actually growing are designed for large-scale farming and do not have any of the envisioned benefits for poor people living in marginal agricultural environments.

According to the latter view, biotechnology, rather than being used to shape food prod-ucts tailored to conquer specific problems—whether health-related or environmental—in the developing world, is mostly being used to enhance the efficiency of large-scale monoculture (production of a single crop on very large acreages) within the United States and a handful of other countries. Simultane-ously, the use of this technology further con-centrates the economic power of a handful of huge multinational agribusiness concerns. Because GMO seed is patented, new license fees must be paid by farmers to these com-panies every year, both within and outside the United States; farmers can no longer save a portion of their crop for the following year's seed, a limitation that is particularly critical for subsistence farmers in the developing world (and that has been used to bring legal action against North American farmers).

Further, no mechanisms are in place to en-sure that biotechnology products designed to enhance human health reach their in-tended targets. Perishable edible vaccines, like conventional vaccines, will be difficult to administer in areas of the world underserved by transportation and health care systems and lacking refrigeration. Nutrition-enhanced "yellow rice" may not actually be planted, grown, and, most importantly, eaten (rather than sold) by those who most need it, even if the seed is made readily available. In short, such products may be promising, but their eventual impact is still far from certain.

## Biotech as Industrial Science

In some ways biotechnology is the ultimate "big science" system. Its proponents are university-based scientists, as well as those in industrial laboratories. Biotechnology build-ings and departments are springing up at universities around the world, especially in the United States; substantial investments have been made in highly trained technicians and expensive equipment, and an entire cadre of scientists now have careers—not to mention, in many cases, stock portfolios—that depend on biotechnology's future suc-cess. Scientists face increasing competition for a shrinking federal research dollar; state dollars are likely to be earmarked for projects with industrial promise, such as biotech. The result is that a new coalition of biotechnology advocates is being formed within the aca-demic community, a coalition with a clear vested interest in identifying problems for which biotechnology can be the solution. Their enthusiasm is also fueled by sincere belief in the value of science and the usefulness of technology; these beliefs are both professional and cultural.

In one unusual but illustrative case, that of the University of California at Berkeley's Department of Plant and Microbial Biology, a multimillion-dollar grant from the Swiss biotechnology company Novartis is part of a complex and largely unprecedented arrangement through which Novartis gets "first rights" to a share in any patentable discoveries the department makes, in exchange for generous financial support. This is a very different arrangement from the more limited research grants and contracts conventionally awarded to university-based researchers, which do not obligate entire academic departments. While an attempt was made to hammer out an agreement that preserves academic freedom, its success in this regard is still being evaluated. At a minimum, the arrangement will almost certainly shift the department's research agenda in the direction of biotech-related products that are commercially viable.

Beyond this impact on the academy, and in addition to the substantial list of scientific uncertainties associated with what proponents believe is biotechnology's potential promise, this technology poses a variety of challenges that are more clearly social. These include ethical, regulatory, and public opinion considerations.

## Ethical Arguments

Academic ethicists divide ethical arguments into *utilitarian* ones, concerned with intended and unintended consequences (the thinking behind both cost-benefit analysis and "ends justify means" argumentation), and *rule-based* or *rights-based* ones, based on principles of action or behavior that are seen as inviolate (such as the physician's admonishment to "first do no harm," the U.S. Bill of Rights provisions, or the Old Testament injunction against killing).

Many of the criticisms, as well as the potential benefits, of biotechnology in agriculture can be classified as utilitarian in nature.

Both the concerns over negative effects that might result from gene pool alteration, changes to the ecological balance, and health and economic impacts, and the visions of possible positive benefits in terms of human health, nutrition, food availability and price, and environmental protection, are essentially utilitarian. These arguments are concerned with the good and bad consequences of gene technology, whether they are related to the environment, economs, health, nutrition, or the general safety and security of the food supply.

However, when asked about the morality of biotechnology or genetic engineering, many people respond in other ways entirely. Their first reaction often has to do with the inherent "wrongness" (or "rightness," in some cases) of these technologies. For religious people, objections are often expressed as a matter of "playing God." For others, including many environmentalists, reservations may involve the idea of "messing with Nature" or "tampering with life." People express reluctance to interfere with the "natural order" or "balance of nature," even if no specific negative consequences can be predicted. Still others may have a vague sense that crossing species boundaries is "wrong" but no clearly formed idea as to why. In other words, the thinking of some people in this area seems to be rule-based, not utilitarian, whether framed in religious, environmental, or more general terms.

Rule-based thinking may be more likely to lead to criticism of genetic manipulation rather than support for it because the issue is framed as "breaking a rule" or "violating nature" through the use of biotechnology, rather than a matter of weighing risks against benefits. But it is also possible for positive reactions to be rule- or rights-based; for example, some people may believe on religious grounds that we were given the capability for this technology to be used for our benefit, and that therefore we have a right and even an obligation to exploit it. Those who believe scientific progress is always inherently beneficial may also perceive an ethical obligation to pursue biotechnology's development, even if the obligation is not phrased in religious terms.

Clearly, most of these arguments—especially those that are rule-based, whether positive or negative, religious or scientific—are largely matters of value and belief. Although one perspective on risk management holds that purely utilitarian criteria could and should be used to decide whether to develop and adopt technologies and that cost-benefit determinations should always drive social policy with respect to technology, this idea ignores both scientific uncertainties and rule-based ethical arguments, and it raises new questions about the ethics and values underlying what are perceived as "benefits," who will receive them, and who will be subjected to associated "risks." In other words, the utilitarian perspective on biotech's risks leaves many values-based questions unanswered. Nevertheless, biotech's promoters argue the risks are small and the potential benefits enormous, even while critics continue to argue that the environmental and health effects are unknown and the envisioned benefits not sure enough to outweigh them.

## Regulatory Environment

Regulations governing food biotechnology are enormously complex. As an example, as the early world leader in the development of biotechnology and bioengineered products, the United States was faced with an emerging regulatory issue for which few precedents were available. To this day, biotech regulation in the United States falls under the administration of three separate federal agencies: the U.S. Department of Agriculture (USDA), the U.S. Environmental Protection Agency (EPA), and the U.S. Food and Drug Administration (FDA). Confusion over the respective roles of these three powerful agencies of the U.S. government persists, both within and outside the United States.

The USDA is the primary agency charged with actively monitoring food safety; it is also the agency charged with nurturing and promoting U.S. agriculture. (Some critics charge that these two missions are in conflict.) The EPA oversees environmental quality and enforces regulations affecting it, for example, by exercising oversight on crops

engineered to produce their own pesticides (such as the Bt crops). The FDA is directly concerned with regulating what reaches consumers and how they are told about benefits and risks. A fourth agency, the U.S. Federal Trade Commission (FTC), regulates the content of advertising.

The FDA has not required labeling of bioengineered food products. Its "product not process" rule means that so long as the foods are not demonstrably different in terms of chemical or nutritional properties from conventional foods, they should be treated no differently. Labeling of biotechnology-based or GMO food products remains a contentious issue, domestically and internationally.

Is the regulatory glass half full or half empty? Critics claim that this alphabet soup of regulators was not developed to respond to the "brave new world" of genetic modification, is poorly coordinated, and lets too many things slip through the cracks. Biotech's backers claim that the existence of these multiple overlapping agency responsibilities ensures that nothing is left to chance and produces a regulatory system on which U.S. citizens and others who consume U.S. food products can rely, a system that produces the safest food in the world. Which argument average citizens accept likely depends more on which institutions are trusted to act in their interests than on specific scientific or legal details.

Not only does each country in the world have a different set of agency responsibilities on issues like these; when these foods cross national borders, a number of multinational agencies also have jurisdiction. On the international level, to name only a few of the most visible players, the World Trade Organization sets import and export rules for its members, and the Food and Agricultural Organization and the World Health Organization share responsibility for setting food safety standards.

## National Opinion and International Politics

Recent figures from a Michigan State University study show that about 69 percent of the U.S. public believe that biotechnology will improve the quality of life, and only 11 percent believe that it will make the qualify of life worse. Other data, however, generated a few years ago at Texas A&M University using similar questions, showed that about 30 percent of the U.S. public believes that genetic engineering (using this specific phrase) will make the quality of life worse, a figure nearly comparable to that obtained for nuclear energy production. It is fair to say the U.S. public is quite positive about biotechnology, consistent with our cultural perspective on science and technology generally. However, it is also fair to say that a substantive minority is pessimistic about particular applications, such as genetic engineering, with controversial implications. The rest of the world is often less optimistic than the U.S. majority, resulting in heated international debates. While risk analysts and regulatory agencies struggle to resolve the scientific questions, the disagreements persist, because they are often based on ethics and values, differences in national agricultural policy, and patterns of trust in institutions and so cannot be resolved by science.

In Europe in particular, the importation of food GMOs under World Trade Organization (WTO) rules has emerged as a major and very persistent international controversy that illustrates how trust, values, and beliefs affect responses to GM foods across national and cultural borders. The growing and marketing of GMO products in Europe may have become controversial because these came to represent broader-based U.S. and multinational corporate economic and cultural interference. But there were other reasons. More Europeans than Americans may consciously value food as an important element of culture, expect proactive government intervention to protect agriculture, are cynical about official reassurances in the post-mad cow era, and take the objections of nongovernmental organizations (NGOs) such as Greenpeace and other environmental groups seriously.

Fundamental policy differences between the precautionary approach of many European governments (in which potentially harmful technologies must be established to be safe before being adopted) and what is called "sound science" policy making in the United States (in which hard evidence of harmfulness is required before a technology

will be restricted) also exist. As a result, GMO exportation from the United States to Europe has become a major stumbling block to more cooperative and friendlier trade relations between the United States and the European Union. However, it is important to remember that a substantial proportion of the U.S. public also has concerns; these are not confined to Europe, nor are they confined to a particular political or religious subgroup in North America. Although there is some relationship between knowledge of genetics and support for biotechnology, both in the United States and across Europe, it is a weak one.

To date, biotechnology has not become especially controversial in China or much of the rest of Asia, where population pressures and the desire to adopt progressive Western science dictate that any promised benefit in terms of productivity is likely to be embraced, but it has become controversial in India. India has an emerging scientific sector and a biotechnology industry of its own; of course, it also has an enormous population of people to feed, many of whom are engaged in small-scale or subsistence agriculture. On the one hand, like the earlier "Green Revolution" involving the widespread adoption of fertilization, hybrid seeds, and pesticide use, biotechnology adoption may help produce more food, and it may also give a boost to Indian science. On the other, some Indian farmers are already hurt by agricultural overproduction, smaller farmers may fear being disadvantaged by biotechnology-based approaches, and U.S. industry may get more economic benefit from biotech adoption than Indian science will. Even the Green Revolution was controversial. Patent issues associated with biotech development have also politicized attempts to commercialize processes making use of traditional Indian foods and medicines. Finally, the role of religious perspectives that may influence popular reactions in India remains to be explored.

Most recently, several African countries, perhaps most visibly the Democratic Republic of the Congo (the former Zaire), have objected to being sent GMO crops as food aid. Such reactions, widely seen in the United States as springing from scientific ignorance fueled by a misunderstood European precedent, are actually reflective of uniquely African conditions and concerns. Devastated economies, hungry populations, the burden of HIV-related health problems, and very high awareness of the historical contribution of colonialism to current conditions in Africa all mean that African leaders are understandably resistant to accepting any risks, even those that U.S. experts may see as trivial, to their food supply or their economic independence. These utilitarian concerns may or may not prove justified, but asking questions about them is not necessarily irrational or "anti-technology." They reflect not only recent European influence but the historical influence of colonialism and a variety of significant internal stresses.

Currently, biotech-related policy efforts in Africa, as in much of the developing world, center around finding mechanisms to allow nations to make use of these technologies in agriculture without paying what some see as unaffordable patent fees. There is also widespread concern with protecting commercial rights to indigenous crop species that may yield valuable products, whether traditionally identified as useful or discovered more recently through "bioprospecting" in resource-rich areas such as rain forests. The biotech industry, to whom the benefits of trade with Africa are limited by the health of African economies as well as the vagaries of international politics, now seem anxious to reap a public relations benefit by demonstrating their willingness to cooperate, including making investments in crops they can adapt to African conditions to solve African problems.

See also **Agriculture and Technology; Medical Values and Ethics**

### BIBLIOGRAPHY

Gaskell, George, and Martin Bauer, eds. *Biotechnology 1996–2000: The Years of Controversy.* London: Science Museum Press, 2001. Technical but an excellent source of comparative opinion and media data from Europe and North America.

Pew Initiative on Food and Biotechnology. http://pewagbiotech.org/. Pew Charitable Trust project site with news, debate, analysis, and opinion data regarding food biotechnology.

Priest, Susanna Hornig. *A Grain of Truth: The Media, the Public, and Biotechnology.* Lantham, MD: Rowman and Littlefield, 2001. Outlines the early evolution of the conflict, with emphasis on media's role.

The Rockefeller Foundation. http://www.rockfound.org. Respected foundation offers balanced information on prospects for using biotechnology on behalf of the world's poor.

Thompson, Paul B. *Food Biotechnology in Ethical Perspective.* Aspen, CO: Aspen Publishers, 1997. Applied ethicist with a longstanding interest in agricultural issues discusses food biotechnology.

Council for Biotechnology Information. http://www.whybiotech.com. Arguments in favor of agricultural biotechnology from organization of biotechnology companies and trade associations.

*Susanna Hornig Priest*

# BIRTH CONTROL.

This entry includes two subentries

Male Contraception
The Pill

## Male Contraception

For most of modern history, contraception research and development has focused mainly on women, who literally bear the most immediate consequences of technological failure. This is not the only reason for the gynocentric history of the technology, however. As Nancy Alexander expressed it in 1994, "Turning off the continuous production of 200 million sperm per day presents tougher problems than stopping the development of one egg a month."

The twin obstacles to development of effective male contraception are that effective methods have not proved to be reversible and that reversible methods have not proved to be effective. The preindustrial remedies of *coitus interruptus* (withdrawal before orgasm), spreading ointments on the penis, or encasing it in a sheath of specially prepared animal intestine or waxed fabric, were well known to be unreliable means of preventing conception and were regarded as uncomfortable by both partners.

Despite its defects, the condom emerged as the dominant male contraceptive technology in Western culture by 1700. First attested in 1564 in Western medicine in the work of Gabriel Fallopius (1523?–1562), the condom was described by Madame de Sévigné (1626–1696) as "an armor against enjoyment, and a spider web against danger." The sheath was slipped on over the erect penis and then secured with tape or ribbon tied around the man's back. The device may have been named after a Dr. Condom, who was the chief medical luminary of the notoriously libertine court of Charles II of England in the second half of the seventeenth century, although there is no evidence that he was associated with its invention. Condoms have had many popular names. Whereas the British called them "French letters," Giacomo Casanova (1725–1798) knew them as *redingotes d'Angleterre* (English riding coats), and in modern French they are still called *capotes anglaises* (English cloaks). In twentieth-century America they have usually been called "rubbers."

The crepe rubber condom, invented by Charles Goodyear and Thomas Hancock in 1844, and its successor, the modern latex condom, introduced in the 1930s, have a somewhat better record of preventing conception than their sheep-intestine predecessors and are more comfortable to wear, but they share with earlier models their tendency to leak, burst, or come off the penis during intercourse. Most men also report that they interfere with sensation.

The condom's virtues are three: first, it is a fairly dependable method of preventing transmission of disease in penile-vaginal intercourse; second, as a means of contraception it is readily reversible; third, it is inexpensive and easy to distribute. Unfortunately, the device remains unpopular both for contraception and for prevention of disease transmission. As the worldwide AIDS epidemic continues, the World Health Organization and many other agencies, national and international, have embarked on massive public education and condom distribution campaigns, especially in the Third World.

The condom remains the only widely used reversible form of male contraception. Vasectomy—severing and ligature of the two vasa deferentia, the tubes that deliver spermatozoa from the testicles to the prostate—is highly effective but, despite research efforts on surgical techniques, has proved irreversible in nearly every case. It is, however, often recommended as an option for a man who already has several children and does not intend to have more. In some cases, husbands choose this option when their female partners have in the past borne most of the burden of contraception. Both men and women report in surveys that they would prefer that men share more of the responsibility for contraception than they currently do, and vasectomy allows a man in a stable relationship to take on this responsibility.

Inventors, aware that a vast market exists for reliable and reversible male contraception, have patented hundreds of products of all three known types: barrier, mechanical, and oral contraceptives. Barrier devices are typically variations on the condom or sheath theme, with innovations only in shape or material. Mechanical means include such imaginative schemes as electrically heating the scrotum to discourage sperm production, but the majority are methods of occluding either the urethra, the urethral canal, or the vasa deferentia.

One of the most elaborate of the American designs for vas occlusion was David R. Pressman's "Corporal Fluid Control Using Bistable Magnetic Duct Valve," for which U.S. Patent 3,731,670 was issued in 1973. A thermoplastic housing on the outside of the vas held in place a metal valve on the inside, of which all but the binary duct was to have been of gold. The binary duct was a small ferrous metal ball which, in the closed position, entirely occluded the vas. Since it could be opened and closed by external action with a magnet, it was possible to open the valve every six months or so reduce fluid pressure. This device, which received a rave review in *Esquire* magazine (which went so far as to suggest that it be implanted in all young men at puberty), proved to have the same defect as others of its type: Any device placing enough pressure on the walls of the vas to form an effective occlusion was also tight enough to cause tissue necrosis. This problem with occlusion devices remains unsolved.

Finally, some experiments with oral contraceptives for men have been successful in reducing fertility, particularly those involving gossypol, which is derived from cottonseed oil. The effects, however, are not reliably reversible, and oral contraceptive methods for men remain experimental.

## BIBLIOGRAPHY

Bradbury, Jane. "Male Contraceptive Pill Would Be Acceptable to Men and Women." *Lancet* 355.9205 (2000): 727.

Brodie, Janet Farrell. *Contraception and Abortion in Nineteenth-Century America*. Ithaca, NY: Cornell University Press, 1994.

Himes, Norman E. *Medical History of Contraception*. New York: Gamut Press, 1963.

McLaren, Angus. *A History of Contraception from Antiquity to the Present Day*. Oxford: Basil Blackwell, 1990.

Riddle, John M. *Contraception and Abortion from the Ancient World to the Renaissance*. Cambridge, MA: Harvard University Press, 1992.

Schwartz, Jill L., and Henry L. Gabelnick. "Current Contraceptive Research." *Perspectives on Sexual and Reproductive Health,* 34.6 (2002): 310–316.

Skerrett, P. J. "Sperm Busters." *Technology Review* 97.1 (1994): 19.

Tone, Andrea. *Devices and Desires: A History of Contraceptives in America*. New York: Hill and Wang, 2001.

*Rachel Maines*

## The Pill

The birth control pill has an intriguing history that illustrates important aspects of the cultural and social dynamics of technological development. To start with, the history of the contraceptive pill shows how technologies are not simply the results of the laboratory work of a few male scientists who discover the secrets of nature. Women and social movements can be important actors in technological innovation. The feminist-inspired birth control movement, with Margaret Sanger as one of its most prominent members, played a crucial role in getting scientists to work on the first oral contraceptive for women. Margaret Sanger, a women's rights activist and pioneer for birth control in the United States, lobbied extensively for years for laboratory research on contraceptives. In 1951, she accomplished her goal when she convinced Gregory Pincus, a biologist at the Worcester Foundation for Experimental Biology working in the field of hormones, to reorient his research on hormones toward contraception. He eventually became one of the "fathers" of the birth control pill. Sanger also provided Pincus with the required funds. Sanger raised $150,000, largely contributed by her friend Katherine Dexter McCormick, to get Pincus started on research toward the development of oral contraceptives for women. McCormick, a feminist birth control advocate and one of the first women graduates of the Massachusetts Institute of Technology, eventually financed the entire research effort that brought the pill into being.

Furthermore, the history of the birth control pill illustrates how the shape technologies take cannot be understood in terms of intrinsic properties of the technology. The drug profile of the first oral contraceptive for women resulted from carefully designed strategies to overcome moral and religious resistance toward the new technology. Pincus and John Rock, a gynecologist at Harvard who conducted the first clinical studies on hormonal contraceptives, selected the regime of medication in such a way that women would still have their menstrual periods. Because progestin, the major compound of the early oral contraceptives, has the capacity to prevent menstrual bleeding as long as it is taken, Pincus and Rock could have designed a contraceptive that produced a menstrual cycle of any desired length or even no menstruation at all. However, they deliberately chose a regime of administration in which the pill should be taken for twenty days of the menstrual cycle to produce a "normal" cycle length in which bleeding occurred once every four weeks. This strategy of mimicking nature was used to convince the Catholic opponents and other critics of the pill that this new contraceptive was not going against nature and therefore, Rock hoped, not in conflict with Catholic Church teaching, which opposed all "unnatural" means of preventing conception. Possible protests against the birth control pill also shaped the way in which the first oral contraceptives were put on the market. In the United States, The Netherlands, and Spain, the new technology was initially approved by drug regulatory authorities and marketed as a drug for the treatment of menstrual irregularities, carrying a "warning" that women would not become pregnant while taking this medication. This practice illustrates how the "activity" of a drug-that is, which effect counts as the intended effect and which as the side effect-is not dictated by nature but is, in this case, the result of a selection process aimed to diminish religious and moral protests towards birth control technologies.

Finally, the birth control pill exemplifies how a technology can have manifold and contradictory impacts on society. Technologies can be considered as the great liberators as well as oppressors of our time. The history of the birth control pill clearly illustrates these two contrasting faces of technology. On the one hand, the history of the birth control pill is a story about the empowerment of women: Oral contraceptives for women have definitely contributed to the increasing liberation of women. The pill enabled women to reduce the health risks and social burdens of unwanted pregnancy. The pill also provided women with a mechanism to determine whether or when they wanted to try to get pregnant. On the other hand, the history of the contraceptive pill is also a story about delegating risks and responsibilities for contraception to women rather than men. Because of the innovation in female birth prevention methods—including the hormonal contraceptive pill, the IUD, and hormonal methods such as Norplant—women's methods have come to predominate practices of family planning. In Foucauldian terms, contraceptives are "disciplinary technologies": the predominance of modern contraceptive drugs for women has disciplined men and women to delegate responsibilities for contraception largely to women. Consequently, the risks of taking hormonal contraceptives have been put principally on the shoulders of women. Ever since the pill was introduced, physicians, scientists, and women's health advocates have reported serious side effects of hormonal contraceptives. The history of the birth control pill thus shows the ambiguity and tensions of the transformative capacity of technology in society.

### BIBLIOGRAPHY

McLaughlin, Loretta. *The Pill, John Rock, and the Church: The Biography of a Revolution.* Boston: Little, Brown, 1982.

Oudshoorn, Nelly. *Beyond the Natural Body. An Archeology of Sex Hormones.* New York and London: Routledge, 1994.

Oudshoorn, Nelly. *The Male Pill. A Biography of a Technology in the Making.* Durham, NC: Duke University Press, 2003.

Vaughan, Paul. *The Pill on Trial.* Harmondsworth, U.K.: Penguin Books, 1972.

*Nelly Oudshoorn*

# BRAIN AND MIND SCIENCES.

Cognitive neuroscience dominates the enterprise of mind-brain studies. Its two sources of authority are cognitive science, which approaches the mind as a computational or information-processing device, and brain research. This article reviews some analyses of cognitive neuroscience by participants in this area of research and then turns to sociological studies. Succeeding sections cover individual topics, such as consciousness, emotion, and social cognition, and the related area of mental illness is briefly considered.

## Perspectives on Mind-Brain Studies

Cognitive neuroscience's objective is to map the mind, which is considered to be an internal, private thing, onto the brain. However, the only areas of the brain that are even somewhat understood at present are those that receive direct sensory information or that generate movement. Although the neural basis of other mental functions, such as memory, emotion, awareness, language, planning, and so forth, are being intensively investigated, results remain elusive.

*Participant Analyses.* Participants (psychologists, neuroscience researchers, and philosophers of science) operate within the internalist worldview, which holds that minds are contained in individual brains: their commentaries focus on procedural problems and remedies within that context. Criticism, when voiced, is conservative and it often amounts to the observation that, since the brain is very complex, it is too early to expect good mind-brain accounts. For example, philosophers of science point out that neuroscientists do not reduce mental entities to neural entities because they import psychological accounts into their descriptions of neural activity in advance. Although this method of connecting psychology and neuroscience would seem to invite arbitrary, ad hoc theorizing about mind-brain relations, it is upheld as promising.

Occasionally the tone is more critical. William Uttal argues there is no evidence to support *any* extant descriptions of psychological entities. Localizationist accounts assume that psychological functions arise from activity in specific brain regions. Uttal shows that apparently localized brain activations seen in imaging studies may be artifacts of the machines and techniques used to capture them and not independent evidence of localized mental operations.

Despite such concerns, the number of imaging studies is growing rapidly, thus prompting the publication of journals dedicated to brain imaging. Human brain-imaging techniques (fMRI, functional magnetic resonance imaging, and PET, positron emission tomography) have come to overshadow traditional lesion studies and neurophysiological recording in animal and human subjects. Consistent with Uttal's critique, there is some disquiet concerning technical and interpretational difficulties in the field and concerning the fact that researchers do not generally share their raw data. It is acknowledged that the relations between neuronal activity, energy metabolism, regional blood flow, and the imaging signal are still undefined and that the statistical analyses used to generate colored "maps of the mind" from these signals vary widely in the assumptions they incorporate. Researchers are beginning to admit that apparent localizations of mental functions to discrete anatomic brain regions may be artifacts of the statistical packages used to analyze the data. These weaknesses are construed as technical problems to be managed: in the imaging community, underlying localizationist and internalist assumptions prevail.

Internalist assumptions are being challenged by a few philosophers and neuroscientists, however. Andy Clark and Marco Iacoboni argue that cognition is to be found not "inside the head" but in the organism's embodied interaction with the world. A detailed interactive account of vision has been put forward by J. Kevin O'Regan and Alva Noë. Even with promising beginnings, a comprehensive account of how brain and cognition engage the social world remains to be developed.

*Social Studies of Cognitive Neuroscience.* Susan Leigh Star documented the practices of neuroscience researchers, who, beginning in the second half of the nineteenth century, tended to promote a localizationist account of the mind and brain. For at least a century, neuroscientists rejected the opposing diffusionist (distributed) account, despite significant empirical anomalies. As Star showed, rather than prompting the rejection of localization, contradictory findings were used to create a more plastic yet durable theory. Localizationist views of the brain today have a sophisticated counterpart in cognitive theory, namely, the idea that there are isolable subprocesses, or modules, of mind. Although debates about modularity abound, and localization's definitions have expanded to include large-scale networks of neural activity, evolving versions of both modularity and localization are intertwined and lie at the heart of contemporary cognitive neuroscience.

Star emphasized that localization is not an idea, but practical work. Neuroscientists struggled then as they do today to transform the problem of mentality with its many contingencies into something simple enough to produce workable goals, images, and tasks. What they choose to work on must meet the requirements of their careers and the complex local settings in which they work. Practicality often dictates a focus on an isolated aspect of mentality or on the results available from a particular machine. However, as these contingencies get deleted from descriptions of scientific work, ideas about the mind are abstracted away from their practical origins and reified into theories.

Uttal's recent work illustrates Star's point about practicality. He reviews the evidence for and against isolable subsystems (modules) of mind as articulated by the eminent neuropsychologist Timothy Shallice, showing that Shallice lays out the logical weaknesses of the arguments for modularity but then makes a purely pragmatic decision to adhere to the notion anyway.

## Topics in Mind-Brain Science

Limitations of approaches to the study of the mind other than the sociocultural approaches are particularly apparent in the cognitive neuroscience debates about consciousness that are dominated by the concept of a pure, isolated subject.

*Consciousness.* "What it's like" to have first-person experience is considered on the one hand, the isolated brain is considered on the other, and arguments about how to link them ensue. There are several proposed solutions to the conundrum. One is the belief that with continued research, the neural correlates of pure subjective experience will be discovered. More specifically, it has been suggested that consciousness derives from perceptual interactions with the physical environment. Still another view is that the very idea of "what it's like" is an empty abstraction considered apart from the dynamic flux of brain states and that consciousness is simply any brain state with enough transient "clout" to determine future states and responses.

Sociocultural approaches to the problem of consciousness have been influenced by interpretations of Wittgenstein's private language argument. Inner states first occur in social contexts, where they are primitively expressed through physical behavior; thus, the inner state and its public expression are intrinsically related. Subsequently, the child learns to use language as an elaboration of primitive expressions, allowing it to gear its behavior into the collective ways of its group. This works well in everyday life, but for theorists, linguistic reference terms generate the mistaken notion that inner states are objects that can be treated logically like other nonpersonal objects in the world. Wittgenstein's position was that subjective feeling-states are logically approachable only through observing how sensation-language operates, not by mistakenly construing them as objects. Conventional language—by stripping away the embodiment and relatedness of unfolding social interaction—brings us to believe mistakenly in an inner entity called *consciousness.*

Psychoanalysts with an intersubjective orientation are revising traditional notions of consciousness. According to them, children learn to articulate their inner states in a relational milieu; some interpersonal environments either may not provide a matrix for certain articulations or may actively discourage them. Unarticulated states do not become part of ongoing, adult conscious experience. This view contrasts with traditional Freudian accounts in which strictly internal mechanisms relegate certain mental contents to the unconscious. Other social approaches to subjective experience also emphasize its interpersonal nature.

*Emotion.* Throughout the history of psychology, emotion has proved impossible to define in a way that sets it apart from other states of the organism. Nevertheless, efforts to define it, account for it somatically and neurally, and retain a taxonomy of basic emotions persist. Emotion is currently a popular topic in cognitive neuroscience, both in the research literature and in books for lay audiences. The neurologist Antonio Damasio has supplied an influential account referred to as the *somatic marker hypothesis.* It connects emotional bodily states and their neural representations with decision making, holding that "rational thought" is influenced by past emotional experience.

Until recently, a number of studies in humans seemed to prove that a brain structure called the *amygdala* is critical for emotional experience. However, patients with damage to the amygdala report the same patterns and intensity of emotional experience as others, leaving the brain basis of subjectively felt emotion in doubt. What is certain, though, is that the amygdala—or a brain system of which it is part—is essential for interpreting social signals emanating from faces. These findings, taken together, would seem to encourage a social approach to emotions over an isolated-mind approach, but the idea that emotions are essentially private and must exist somewhere in the brain is still dominant.

LeDoux (1996) has taken a different approach by explicitly studying behavior and not phenomenal experience, using an animal model of fear involving anticipated foot shocks. That these studies give different neural results than studies using other fear-inducing stimuli suggests that fear as a general concept and prototypical basic emotion may not map well onto the brain. LeDoux addressed the poor fit between the psychological concept under study and the laboratory findings by proposing revisions to the psychological concept, a remedy frequently used by neuroscientists as they wrestle with anomalies created by the standard cognitive neuroscience approach.

The underlying reason that the general term *fear* and the even more general term *emotion* cannot be captured by such experiments is that their usage and expression derive from complex social practices. Cognitive neuroscientists generally ignore the social origins of psychological concepts and attempt to reify them directly in brain activity. Furthermore, behavior and the underlying mentality that it is assumed to express are often attributed directly to an asocial brain, that is, without considering the extensive social experience that has shaped brain, behavior, and mentality in individual subjects.

Neuroscientific accounts of emotion resemble their lay counterparts in taking emotion to be an internal phenomenal state, existing apart from its social aspects. The persistence of belief in emotion despite theoretical arguments against it derives from an awareness of somatic, especially visceral, changes evoked by situations. To the extent that such changes are suddenly mobilized, as opposed to being tonic, they are likely to be labeled as emotional. Organisms continually adjust to the environment; therefore, no meaningful distinction exists between emotional and nonemotional states apart from linguistic convention.

The amygdala receives highly processed information in all modalities from sensory brain areas and is directly connected to endocrine, autonomic, and motor centers. Thus it is a neural link in the chain between incoming information and bodily adjustments. That the amygdala is crucial in social signaling situations, whereas its connection with emotion per se is problematic, may eventually encourage neuroscientists to adopt a view of the human brain/body as a node in social networks and to discard isolated-mind thinking.

*Social Cognitive Neuroscience.* Social cognitive neuroscience is emerging as a subfield of cognitive neuroscience. Its central principle is that human beings use specific neural-cognitive mechanisms to respond to the social signals of others. It seeks to identify the components of social cognition and localize them to specific brain circuits. By studying the social abilities and brains of nonhuman primates, it hopes to discover how human social cognition evolved. It generally assumes the brain has some inborn social potential; it asks how young brains interact with the environment during development to produce adult social capacities. Developmental psychology, clinical neuroscience, comparative anatomy, and primatology contribute.

Eye gaze direction and certain facial muscle configurations are signals used by many primates. An evolutionarily old neural system, present in the normal human brain at birth, may prepare the human infant to respond to such signals. This putative neural system is scaffolding for the accumulation of subsequent social experiences, which are built upon early, simple responses to the sights and sounds of faces. Elaborated in response to what the environment provides, it ultimately produces complex, finely tuned responses to such entities as belief ("cold" theory of mind), irony, and faux pas and such signals as flirtation or threat ("hot" theory of mind).

Comparative anatomists have discovered correlations in various primate species between social group size and the size of neocortex and some portions of the amygdala, which receives neocortical inputs. They argue that social stimuli are cognitively demanding; therefore, it makes sense that neocortex size and social complexity have coevolved. The neocortex may have undergone selection pressure to store memories of identities and past actions of others so as to match them appropriately to the current setting, thus allowing for social manipulation through simple but extensive social learning. For example, tactical deception in primates does not necessarily imply insight into others' knowledge and thus need not be related to a theory of mind. To explain the origins and neural basis of human social cognition, evolutionary accounts invoke prolonged development and large brains with great capacity for learning in complex social environments.

Relative preservation or deficits in social abilities of certain children have attracted the attention of developmentalists. Williams syndrome, autism, and the autistic variant Asperger's syndrome are subjects of intense research effort. Although the spectrum of autistic disorders is heterogeneous and requires careful specification, autism clearly deserves its paradigmatic status in the field of social cognition. Autistic persons appear to have an inborn defect in elementary social processing, as shown by a number of studies of face and eye gaze processing, whereas Williams syndrome is in many respects the converse, with relative sparing of social abilities in contrast to profound deficits in most other areas.

Controversies in the field include the roles of frontal cortex versus limbic regions in the theory of mind; what counts as theory of mind in primates and other species; to what extent naturalistic observations versus controlled experiments are admissible in the study of a primate theory of mind; whether the neural substrates of face perception are in fact face-specific; and more generally whether social cognition is a "module" with specially adapted brain mechanisms or instead uses general-purpose mechanisms. An alternative to the modularity concept is the idea that numerous strands of specialized cognition develop in complex interaction: these may be bundled together when the environment elicits them in concert.

A potential pitfall of social brain studies is that some ideas about sociality, just like ideas about emotion and other aspects of mentality, may be reifications. As in other areas of cognitive neuroscience, there may be a tendency to abstract away the social history of the individual when studying his or her competencies, even though developmental psychologists have correctly emphasized the importance of the social milieu in shaping brain activity.

*Mental Illness.* Psychiatry has been the object of many sociological studies. Some have analyzed its institutions from the perspective of social control. Others have emphasized that psychiatric diagnoses and treatments are context-dependent, pragmatic, and ad hoc—in short, the result of practical, situated work. Although the era of psychosurgery is past, there is once again an overwhelming focus on brain treatments in the form of psychoactive medications, despite the absence of context-independent criteria for psychiatric diagnosis.

The pharmaceutical industry's stake in ascribing disordered behavior to disordered brain chemistry is obvious. How these ascriptions have become entrenched in both the professional and popular realms, despite flimsy and contradictory evidence, is addressed in works by Valenstein (1998) and Healy (2002). They recount the emergence of receptor theories of mental illness in the 1950s, the confluence of academic and pharmaceutical interests, and how technologies such as brain scans and receptor binding studies were construed to confirm the biochemical basis of mental illness. The economic power of the pharmaceutical companies allowed them to take over academic psychiatry, with the result that the profit motivations of the industry constitute the single most important factor shaping training, research, and publication in psychiatry at present. Diagnoses are invented in order to market drugs that appear to reduce certain symptoms; they are imported into official diagnostic manuals because psychiatric departments, organizations, and committees are funded by pharmaceutical interests. Furthermore, cognitive neuroscientists appear to be adopting categories of psychiatric diagnosis uncritically in areas such as emotion research.

## Future Directions

Despite cognitive neuroscience's powerful presence in contemporary culture, there are few social science studies of the field. Just as Star showed that localizationist theory is a practical achievement, a similar sociological approach could be used to show that internalism is achieved through the work practices of cognitive neuroscience. Such an approach could show how anomalies that might undermine internalism are dealt with by creative elaborations of internalist theory and how alternative externalist, or social, views of mentality are rendered invisible. Institutional arrangements and work practices that underpin cognitive neuroscience as an internalist, nonsocial enterprise could be fruitfully examined.

A movement toward externalism on the one hand—the idea that the mind is not "in the head" but in the individual's embodied interactions in the world—and social aspects of brain function on the other, also deserve study by sociologists of science. New coalitions and interests that unite these approaches may be emerging. As they develop, the ways in which mind-brain relations are conceptualized and studied are likely to be transformed.

## BIBLIOGRAPHY

Clark, Andy. *Being There: Putting Brain, Body, and World Together Again.* Cambridge, MA: MIT Press, 1997.

Damasio, Antonio. *Descartes' Error: Emotion, Reason, and the Human Brain.* New York: G. P. Putnam's Sons, 1994 .

Dennett, Daniel. "Are We Explaining Consciousness Yet?" *Cognition* 79.1-2 (2001): 221–237.

Healy, David. *The Creation of Psychopharmacology.* Cambridge, MA: Harvard University Press, 2002.

Iacoboni, Marco. "Attention and Sensorimotor Integration: Mapping the Embodied Mind." In *Brain Mapping: The Systems,* edited by A. Toga and J. Mazziotta. San Diego: Academic Press, 2000: 463–490.

LeDoux, Joseph. *The Emotional Brain: The Mysterious Underpinnings of Emotional Life.* New York: Simon and Schuster, 1996.

O'Regan, J. Kevin, and Alva Noë. A sensorimotor account of vision and visual consciousness, *Behavioral and Brain Sciences,* 24 (2001): 939–973.

Star, Susan Leigh. *Regions of the Mind: Brain Research and the Quest for Scientific Certainty.* Stanford, CA: Stanford University Press, 1989.

Stolorow, Robert, Donna Orange, and George Atwood. "World Horizons: A Post-Cartesian Alternative to the Freudian Unconscious." *Contemporary Psychoanalysis* 37.1 (2001): 43–61.

Uttal, William. *The New Phrenology: The Limits of Localizing Cognitive Processes in the Brain.* Cambridge, MA: MIT Press, 2001.

Valenstein, Elliot S. *Blaming the Brain: The Truth About Drugs and Mental Health.* New York: The Free Press, 1998.

*Leslie Brothers*

# C

## CHEMICAL AND BIOLOGICAL SCIENCES.

The influence of chemistry and biology can be found throughout the world around us. All activities involving naturally occurring processes, from food production to animal husbandry to protection of the environment, rely on knowledge of biology. In medicine, biology is essential in the analysis of disease, preparation of vaccines, and understanding of genetics and birth defects. Chemistry is similarly omnipresent and is necessary for everything from the manufacturing of silicon chips used in computers to the synthesis of drugs for conditions ranging from high blood pressure to erectile dysfunction. In fact, every synthetic product that drives our culture of consumption, such as gasoline, plastics, cosmetics, paints, dyes, and medications, comes from chemical processes and is informed by knowledge of chemistry.

There is an increasing amount of public awareness about chemistry and biology, Information from these sciences is often reported by the news media (for example, forensic data in celebrity trials) and often figures in detective or medical novels, TV shows, and movies. This increased awareness has also created an increased level of scrutiny. Whereas worries about chemistry are about the same today as they were in the 1960s, involving concerns about pesticides, environmental toxins, and chemical warfare, the same cannot be said for biology. Advances in biology have raised new moral and pragmatic questions about cloning, transgenic plants and animals, and antibodies in foods, causing a great deal of concern in the public sphere. As legal and ethical regulations fall behind the pace of scientific progress, the research itself has become immersed in a level of politics that would have seemed impossible even twenty years earlier. We now have presidential candidates weighing in on whether they think stem cell research is a good idea, as government and the general population try to decide the course of scientific advances.

In order to understand the true impact of chemistry and biology on our lives, we must understand how the research in these fields gets done and thus how the discourse in these fields has been shaped. To do so we must look at the ways in which these fields are structured, funded, and operated.

### The Institutional Structures of the Chemical and Biological Sciences

Chemistry has been divided into the fields of physical, organic, inorganic, and analytical chemistry for a hundred years and still uses these distinctions, with the addition of the field of biochemistry in the last fifty years. Biochemistry has meanwhile grown into a distinct discipline in its own right. Biology has evolved from botany, zoology, physiology, microbiology, animal behavior, ecology, and evolutionary biology to include the very active and important fields of genetics, enzymology, virology, cell biology, bioinformatics, neurobiology, developmental biology, and immunology. These new fields in biology rely heavily on chemistry and biochemistry as well as biology proper for the tools needed for research. Whereas every chemistry department will have the fields just listed, these different fields in biology may well have their own departments at different sites in the university.

Biochemistry has become a central science to both chemistry and biology. It is very hard to tell from a description of a faculty member's research whether they the person is a chemist, biochemist, or biologist. Likewise, the separation of fields within disciplines has become less defined over time. For example, it was announced in a recent issue of *Chemical and Engineering News* ("the *Time* magazine for chemists") that Angela M. Belcher, 37, an associate professor of materials science and engineering at MIT, had won the MacArthur Foundation "genius award." She was acclaimed as a chemical scientist who "uses genetically engineered viruses to serve as templates for the synthesis of submicroscopic conductors and semiconductors. She has genetically modified viruses to interact with solutions of inorganic semiconductors, yielding self-assembling metal films and wires" (Raber, p. 11). Thus she is a virologist, engineer, chemist, physicist, and computer scientist in her work, and a specific academic label cannot hope to encompass the scope of her research.

A botanist today may well be using genetics and the analysis of DNA to classify ferns, whereas a microbiologist may need to use the principles of enzymology. In fact, the interdisciplinary fields have become the most exciting and productive areas of research today. It is interesting to note that these areas are also the most welcoming to diversity. It may be a result of the flexibility necessary to operate at the boundaries of a discipline also brings an ability to value and accept the differences in people.

Aside from divisions between different fields, chemistry and biology communities are also divided into academic (which places the highest value on doing pure research) and industrial (which needs and values applied research). Not only are different kinds of research valued in academia and industry but there is also divergence in the ways research is conducted and whether collaborative work is encouraged. Academic research has a more archaic structure, where individual professors have their own labs and have to obtain funding for their research and for the equipment and supplies it requires by writing competitive grant proposals. They recruit graduate students and hire postdoc-

toral fellows to do this research. These professors may collaborate with professors in their discipline or in another discipline, but they each maintain their own fiefdoms. An entire department may also submit grant proposals, usually to the National Science Foundation (NSF) or the National Institutes of Health (NIH), to obtain very expensive instruments, which are then used by many members of the department. Equipment obtained by individual professors is normally only used only by their own groups. This method of organizing research actually discourages academic researchers from attacking really big or complex problems, which are addressed quite efficiently by government and private research foundations. When working on interesting problems that are interdisciplinary, at the interface of disciplines, one often needs knowledge in several disciplines. Collaboration among faculty in several disciplines in exploring interesting questions greatly facilitates such research. However, collaboration in chemistry is often frowned upon, and questions such as "Whose research is this really?" abound in both panels that fund research and department committees that decide on tenure for young faculty members. Assistant professors in chemistry are often advised to wait to develop collaborations until after they have obtained tenure and developed their own research presence. Academic biologists are not nearly as resistant to collaboration as chemists are, but this varies by the subdiscipline.

In contrast, Government labs and the research labs of companies and foundations such as the Howard Hughes Medical Institute (HHMI) value collaboration and have found that throwing together mixed groups of scientists from different disciplines increases the likelihood of truly innovative results. This appears to be true for both applied and pure research. Government labs are powerful, well-funded and important sources of research, which sometimes bridge the gap between pure and applied research. The government labs at the NIH have nineteen separate research institutes working on cancer, aging, heart disease, and other fields.

Other very important government labs include the Centers for Disease Control and Infectious Diseases (CDC), the National Aeronautics and Space Administration (NASA), Army and Naval Research Laboratories, the National Institute of Standards and Technology (NIST), and Environment Research Laboratories.

The processes of deciding on what is an important research question also differ between academia, government labs, and industry. In academia the decisions on what research is truly important and should be studied are made by professors at the top twenty or so research universities and the funding agencies. In industry the marketplace drives the topics funded for research. The selection of topics for research in government labs is influenced by those chosen in academia and industry and is also influenced by political considerations. The graduate students who receive doctorates from the top twenty research universities provide most of the university faculty for all of the research universities as well as liberal arts colleges in the country as well as many of the government research scientists. Thus a very small population decides the course that most research will take, a sample that turns out to be quite skewed toward one gender and ethnicity. The chemistry professors at the top fifty research universities are 88 percent white males, while those in biology departments are about 80 percent white males. One somewhat controversial explanation for the reason science has become so male-dominated is given by David Noble, who believes that the scientific culture of today was developed from the all-male monastic asceticism of the late medieval period, and the mores and boundaries of the monks are still reflected in the present day.

Today funding for research for the academic community is almost entirely governmental, with foundations such as HHMI or companies providing the rest. The majority of the governmental funding comes from the NSF, and the NIH, with the Department of Energy (DOE), Department of Agriculture, Department of the Navy, Department of the Army, and smaller government agencies providing the rest. The same research project may receive funding from several funding sources. The government funding organizations convene panels made up of academic professors who review the grant proposals and decide which faculty and ideas get funded. The grant proposals must contain information about the significance of the research and the preliminary results. Many also require information about intended outreach to the community or plans for recruiting women and minorities into research positions.

The composition of the panels of scientists who review grant proposals and advise the NSF or NIH as to which grant proposals should be funded has changed. In the past twenty years, the directors and their staffs at NSF and NIH have created panels that have included professors from second-tier research universities as well as from research-oriented liberal arts colleges, along with those from the top 50 research universities. These panels have shown increased participation by both women and minorities. Surprisingly, the professors from the top twenty research universities very rarely participate on these panels today, although funding is essential to their research.

## Racial and Gender Diversity in the Chemical and Biological Sciences

Despite increasing diversity in the grant review panels, there has not been much progress in creating a more diverse academic environment. Of the 1637 tenured/tenure track faculty in chemistry at the top fifty research universities, only 22 were Hispanic (1.3 percent), 18 were African-American (1.1 percent) and 3 were Native American (0.2 percent). Another 27 departments had no professors in these minority groups. There were no African-American assistant professors, and of the 12 full professors, most were nearing retirement. Of these professors, 197 were women (12 percent), with 8 percent at the full-professor level. Mary Frank Fox, the NSF Advance Professor of Sociology at Georgia Institute of Technology, where there are two women in a department of 32), says (Marasco, p. 59):

[W]omen's doctoral degrees are not translating into expected academic rank over time. Despite the number of women with doctoral degrees earned in the 1970s and 1980s, and the passage of years allowing these women to mature in professional experience, the proportion of women who attain academic rank as full professor has not kept pace with the growth of women holding doctorates. Even assuming up to 15 years from receipt of doctorate to rank of full professor, women's degrees are not translating into expected rank over time.

The change in the numbers and percentages of doctorates awarded to women is shown in Table 1.

Despite considerable funding by NSF and NIH to increase diversity in all the sciences, there has been very little increase in doctoral degrees awarded to minorities in Natural Sciences (Physical, Earth, Atmospheric, Biological and Agricultural) since 1977 (see Table 2).

While the question may be asked whether minorities want to become scientists, it is clear that they increasingly arrive at college with this career path in mind (see Table 3).

**Table 1.  Doctoral Degrees Awarded to Women**

|  | 1970 | | 1999 | |
|---|---|---|---|---|
|  | Number | Percent | Number | Percent |
| Physical Sciences | 227 | 5.8 | 831 | 23 |
| Biological and Agricultural Sciences | 538 | 12.9 | 2680 | 40.8 |

Source: National Science Foundation.

**Table 2.  Doctoral Degrees Awarded in Natural Sciences to Minority U.S. Citizens**

|  | 1977 | | 1999 | |
|---|---|---|---|---|
|  | Number | Percent | Number | Percent |
| African American | 85 | 1.3 | 220 | 2 |
| Hispanic | 76 | 1.2 | 256 | 2 |
| Native American | 4 | 0.06 | 38 | 0.35 |

Source: National Science Foundation.

**Table 3.  Minority Percentages of Freshmen Intending to Major in Science and Engineering**

|  | 1977 | 2000 |
|---|---|---|
| African American | 6.0% | 11.5% |
| Hispanic | 0.5% | 7.1% |

Source: National Science Foundation.

Industry and government labs, on the other hand, have been much more welcoming to women and minorities in the sciences, and the percentages of women and minorities are much higher than in academia. Companies have been much more successful at recruiting and retaining minorities and at establishing a culture that values diversity.

## The Making of a Chemical or Biological Scientist

Whatever a student's gender or ethnicity, the process of working through graduate school itself is a long and daunting affair. Students are expected to spend ten to twelve hours a day in the lab every day of the week. Today these graduate students spend five to seven years completing their doctorates and are then expected to spend at least two more years working as senior research fellows in another lab if they intend to become faculty. In both graduate school and postdoctoral work, the research project is developed and owned by the research director, who writes the grant proposals and obtains the money to pay the graduate and postdoctoral fellows.

The average time spent as a graduate student has increased from four years in the 1970s to over six years today.

## Ethics and Implications

Unlike scientists in industry or government labs, academic scientists do not take courses on teamwork, communication, management, diversity, or ethics. Very little in the way of training or even discussion about ethics is available to graduate students in chemistry, but it is present in some biology curriculums. Ethics courses tend to focus on integrity in reporting results and the importance of not altering data.

Today the public is increasingly concerned with what could be problematic and dangerous research, which can occur without structural mechanisms for regulation. Whereas legislation in Europe has forbidden the use of transgenic plants and animals for food, it has had little effective opposition here in the United States. One example of the dangers of transgenic plants has been provided by research on the genetically engineered bent grass developed by Monsanto and Scotts as a turf for golf courses. This grass is resistant to the chemical herbicide Roundup. A recent study by Lidia S. Watrud, research ecologist at EPA's National Health and Environmental Effect Research Laboratory (a government lab), found that pollen from the Roundup Ready® creeping bent grass, traveling up to 13 miles, could transfer the herbicide resistance to wild creeping bent grass and to related species as well. This study validates fears that genetic alteration of some crops may not be self-contained and could spread widely to other domestic and wild plants. Previous studies had found that crop genes spread out to only a 1400-foot radius. When the gene transfer research and cloning research began, many scientists as well as the public were very concerned about the implications and dangers. In talking to colleagues in biology, they said that regulating gene research and cloning would be next to impossible because it is so easy to do and inexpensive.

We have reached the point in history at which we have already made enough scientific advances to end life on earth, either slowly with global warming or quite rapidly with nuclear weapons or with transgenically produced biotoxins. The explosion of scientific knowledge over the past fifty years has left the development of social controls for the expanding technological advances in the dust. The evolution of social development has not kept pace with the dangers involved, and there is an urgent need for all scientists to receive training that will allow them to be responsible and ethical in their work. Currently these scientists may be the only ones who have the knowledge necessary to evaluate the risks in their work.

*See also* **Bioengineering; Biotechnology**

## BIBLIOGRAPHY

Commision on Professionals in Science and Technology online database, *http://www.cpst.org*

Hileman, Bette. "Gene Flow from Transgenic Grass." *Chemical and Engineering News*, Vol. 82, No. 39 (2004): pp 5. http://pubs.acs.org/cen/news/8239/8239notw1.html.

Howard Hughes Medical Institute. *Janelia Farm Research Campus.* http://www.hhmi.org/janelia.

Marasco, Corinne A. "No Change in Numbers of Women Faculty." *Chemical and Engineering News* 82.39 (2004): 32–33.

Marasco, Corinne A. "Numbers of Women Nudge Up Slightly." *Chemical and Engineering News* 81.43 (2003), 58–59.

National Science Board. *Science and Engineering Indicators 2002*, Vol 1. Arlington, VA: National Science Foundation, 2002. *http://www.nsf.gov/sbe/srs/seind02/start.htm.*

Noble, David F. *A World Without Women: The Christian Clerical Culture of Western Science.* New York: Alfred A Knopf, 1993.

Raber, Linda. "Funding Genius, Four of 23 MacArthur Fellows Named for 2004 Are in Chemistry-Related Fields." *Chemical and Engineering News* 82.40 (2004): 11.

Watrud, Lidia S., E. Henry Lee, Anne Fairbrother, Connie Burdick, Jay R. Reichman, Mike Bollman, Marjorie Storm, George King, and Peter K. Van de Water. "Evidence for landscape-level, pollen-mediated gene flow from genetically modified creeping bentgrass with CP4 EPSPS as a marker." *Proceedings of the National Academy of Sciences* 101.40 (October 5, 2004): 14533–14538. http://www.pnas.org/cgi/content/full/ 101/40/14533.

Yarnell, Amanda. "HHMI's Janelia Farm Sets Scientific Focus." *Chemical and Engineering News* 82.41 (2004): 46.

Sheila E. Browne and David T. Browne

# CLONES AND CLONING.

Biologists and sociologists have different interests in clones and cloning. Many biologists study cloning in terms of agriculture (better animal- and plant-breeding practices), medicine and research (as for regenerating injured tissues, preserving endangered species, or even bringing back extinct species), and how cloning can increase knowledge about humans. Sociologists ask what definitions of *clone* stem from interactions between scientist-experts and non-experts. They examine such topics as agricultural impacts (such as economic consequences, effects on systemic societal configurations, or influences), genetic identity, the nuclear family, individuality, creativity, procreative liberty, eugenics, and scientific progress.

Biologists recognize several definitions of *clone*. For example, many plants easily form clones from a severed part of the parent plant. Some invertebrates, such as worms, can also regenerate parts if their bodies are severed in certain places. If each survives independently, they are genetic clones. Cloning of vertebrates, however, is generally limited to multiple births (identical twins). Technology produced Dolly, the famous "cloned" sheep, although this process was originally referred to only as "nuclear transplantation" and not as "cloning."

In nuclear transplantation the nucleus of a haploid germ cell (sperm or egg) is removed and replaced with the nucleus of a diploid somatic cell. Human conception typically occurs when the sperm (haploid germ cell) fertilizes the egg (haploid germ cell), and a diploid cell forms. In nuclear transplantation, however, the genetic material of the haploid egg is removed and replaced with the nucleus of a diploid cell. The nucleus of the diploid cell (usually from an early embryo) thus introduces its full genetic complement into the egg, which then has the potential to develop into a "clone" of the organism. Since embryonic cells lose qualities that make them suitable for cloning as they get older, the number of animals that can originate from a single embryo is limited. Cloning has other practical limitations. For example, some scientists suspect that Dolly's premature aging and progressive lung disease resulted from nuclear transplantation shortcomings. These difficulties pose obstacles to human cloning that are seen more readily in scientific than in popular circles.

Two myths in particular are repeated in films, television, and science fiction. The first is that a cloned animal is an exact replica. The fact is that a cloned animal may not act or even look like the individual from which it is cloned. Cloning is not duplication. For example, the clone of a domestic calico cat at Texas A&M University in 2002–2003 neither looked nor acted like the cat from which it was cloned. The other recurrent myth suggests that research such as the Human Genome Project enables scientists to duplicate a human genome technologically. "Coding the human genome" only means identifying genes that code for the production of proteins. Genes that code for proteins constitute less than half a human genome. In fact, a large part of the genome was until recently referred to as "junk DNA." The term fell out of use because, whatever its functions, which are still a mystery, that DNA is not "junk." So, projects such as the Human Genome Project only aim at mapping less than half the human genome—all that our understanding currently allows.

Sociologists distinguish between propagating the human race and propagating human values and knowledge. Social institutions—that is, rules and regulations that define social interactions in given contexts—help perpetuate knowledge and values. These include connotations of the term *clone*. Do our genes, from "nature," confer our identity? Or is it upbringing and environment, "nurture," that makes us who we are? Is it both? For the sociologist, complex social interactions link an individual to a community and a community to a society. These connections can only be reproduced through social institutions. For example, stories of technologically reproduced humans have been around a lot longer than molecular genetics. Such stories, such as Mary Shelley's *Frankenstein,* still play a role in current understandings of cloning technology.

So for the sociologist the question is not "Can a human be cloned?" but "Can a social environment be cloned?" Organisms develop in specific contexts, during specific times at specific places. Taken out of those contexts, identity changes. For example, the social context of agriculture, family, economics, individuality, creativity, and procreative liberty changes through time and among disciplines and thus can change the meanings of these concepts. When a plant physiologist says the word *clone,* she can mean a plant cutting, whereas a member of the Raelian sect who believes that Jesus was resurrected by aliens with cloning technology may attach a very different meaning to the word. A "clone" represents a category of life, recognized as a copy of a part or whole organism, but social contexts give this category applications and uses that drive cloning debates.

*See also* **Bioengineering; Biomedical Technology; Biotechnology**

**BIBLIOGRAPHY**

Franklin, Sarah. "Animal Models: An Anthropologist Considers Dolly." Lancaster, U.K.: Department of Sociology, Lancaster University, 2003. http://www.comp.lancs.ac.uk/sociology/papers/franklin-animal-models.pdf.

National Bioethics Advisory Commission. "The Science and Application of Cloning." In *Clones and Clones: Facts and Fantasies about Human Cloning,* edited by Martha C. Nussbaum and Cass R. Sunstein. New York: W. W. Norton, 1998.

*Rachel Dowty*

# COMMUNICATION IN THE SCIENTIFIC COMMUNITY.

The deeply social endeavors of science and technology are held together by communication that is both cooperative and competitive. To carry out their work, scientists and technologists need the knowledge, thought, and techniques developed by others, but to contribute to knowledge and advance technology they must identify, develop, and argue for some novelty of discovery, thought, or invention, distinctive from prior advances. Furthermore, they must make their novelty visible, persuasive, and useful to others—enlisting those others in the propagation of their thought. This work is carried out through communication.

## Communication in Knowledge Creation and Dissemination

Formulating knowledge is itself a process of articulating a communicative statement, by which our knowledge is crystallized and shared. Studies of laboratory interaction have repeatedly observed how, through talk and gesture, scientific teams looking at a representation of data notice, identify, label, and attribute meaning to an interesting phenomenon. Natural phenomena, as they are observed or manipulated, are inscribed into data by the recording scientists or by various instruments that serve as inscription devices. Sociologists Bruno Latour and Steve Woolgar describe the work of a biological laboratory as a process by which living specimens are labeled and turned into samples, which are then, for example, centrifuged and assayed, resulting in graphic representations of the contents. These graphic representations are then numerically valued and charted into graphs and tables, which are in turn analyzed and embedded into scientific papers. Although much work is done through the physical manipulation and observation of objects in the laboratory, the results of this material experience are transcribed into communicative symbols that continue to be transformed in the world of calculation and communication.

But the story does not end with the production of papers. Papers are presented to audiences of scientists at conferences and seminars, where they are the topic of discussion. Articles are submitted to journals, where they are read and evaluated by referees and editors, who communicate their evaluations back to the authors, who, if so requested, may then revise. Articles, once published, go to libraries and private subscribers. Readers then search journals for findings that will help them advance their own knowledge and research. Further research articles then cite previous work to build on them, and review articles sum up earlier findings to point to new research directions. An article may become so associated with a particular idea that reference to the article comes to represent the idea. People whose work is closely related to each other's cite many of the same people as well as each other's work. Analysis of shared patterns of citation, called *cocitation analysis*, can identify emerging areas of scientific work.

Further from the research front, articles may be cited in handbooks and textbooks that help familiarize students with areas of research. If the knowledge claim presented in the original article becomes widely accepted and relied on, it may become a shared piece of common knowledge that it is no longer identified with any article or author. As it has become fully incorporated into shared scientific knowledge, its origins have become obliterated. Textbooks themselves then become embedded within the complex communicative systems of the classroom with syllabi, discussions, lectures, other readings, assignments, and exams. Further dissemination of scientific knowledge may occur through journalism and other forms of popular science writing, sometimes to provide practical knowledge, sometimes to satisfy public curiosity, and sometimes to help increase public support for scientific programs.

Funding and support for science foster additional systems of communication. The most visible edge of this is the proposal and reviewing process, but there are many other documents by which proposals are called for, support solicited, capabilities of research teams represented, and grants and contracts administered and monitored. This communication may be as much legal, budgetary, and bureaucratic as it is overtly scientific, but all of it is the communication that makes science possible. There are differences among the sciences in the groups of people with whom they communicate. People in the environmental sciences, for example, often communicate with government agencies and community groups; people in geological sciences often have regular connections with extractive industries and their regulators as well as groups concerned with monitoring and preparing for volcanic and earthquake disasters. Engineers and other developers of technology may have even more complex patterns of communicative patterns. Their work often circulates within corporations and must be attentive to legal, regulatory, risk, financial, production, sales, and consumer issues, which are often mediated by other specialists such as patent lawyers, consumer relations and marketing departments, managers, and technical writers. As the development and deployment of technology regularly involve mobilizing many forms of interest and power, power relations are typically enacted within communications surrounding technology.

## Spoken, Written, and Electronic Communication

Much of communication within science and technology is spoken. Ideas are shared, interpreted, and argued, and experiments designed, over informal chat. Technology design teams thrash through ideas and alternative models; development teams communicate to turn design ideas into workable practice. Over the lab bench, people coordinate through talk to make the experiment work, identify and interpret phenomena visible in data traces, and consider the implications and uses of their projects. Conference and seminar presentations are also often more talked than written, supported by outlines of key points, data charts, graphs, and other visual displays.

Informal communication can also extend into written text. From the earliest periods of natural philosophy, investigators shared their results through letters. Scientist and inventors as early as Leonardo da Vinci (1452–1519) recorded their ideas, designs, observations, and results in notebooks. As collaboration became more common, particularly in the later nineteenth century, notebooks and other laboratory records served as well to coordinate work and attention.

Laboratory notebooks also have an archival function: to substantiate findings and claims that appear in the public literature. As industrial laboratories grew, the coordinating function of laboratory notebooks became increasingly important, along with other internal reports and records. With the nineteenth-century elaboration of patent laws, notebooks also came to serve as legal evidence for priority of invention.

In the twentieth century, as technology for duplicating papers became convenient, informal circulation of drafts, or *preprints,* of scientific articles to colleagues became common. The Internet now supports the distribution and archiving of preprints along with sharing of large data sets. Through communications technologies, scientists in different locations can even work simultaneously, observing data being collected in real time at a third location. The groups of people who regularly communicate with each other informally have been called "invisible colleges," a term first applied to the seventeenth-century membership of the Royal Society.

However, the more enduring and formal communications of published work in journals and books are reproduced many times, shipped to distant places, and archived in research libraries. Although laboratory talk accompanies experiments, data collection, and interpretation, and although informal documents help coordinate complex, multiperson, multisite projects, formal written documents are significant in the social presentation of knowledge, development of extended argument, creation of shared knowledge and archives, and structuring science as communal activity.

## The Resources of Language

Because knowledge emerges within communication and takes on the form of communication, it employs the many resources of language, although with a particular selection. One of the most obvious elements is the emergence of scientific terms and words. The linguist Michael Halliday has examined the nominalization process, whereby actions are turned into nouns, which then become more abstract and combine with other abstractions in multiword noun phrases (an example is the phrase "nominalization process" itself). This process occurs both within the course of a single text and over the history of research areas. Another linguistic feature often associated with scientific writing, the avoidance of the first person by the use of the third person and passive voice, upon closer investigation turns out not to be so simple. Although objects of investigation and theory-based abstractions are often the subjects of sentences, the first person appears regularly to indicate roles appropriate to the work of the scientist, such as choosing and carrying out procedures, adopting assumptions, and drawing conclusions. Disciplines also vary in the use of the first person. Douglas Biber, Dwight Atkinson, and Kok Cheong Lee have each examined the linguistic resources used in scientific writing more comprehensively.

From the earliest period of modern science the difficulties of using language to carry out the work of science have been recognized. The Renaissance philosopher Francis Bacon (1561–1626) noted that language was plagued by four kinds of idols. The "idols of the tribe" are the limitations of human mind, sense, and perception. The "idols of the cave" are the idiosyncrasies of each person's separate experience, character, education, reading, and experience, which lead us each to perceive things differently. The "idols of the marketplace" are the word meanings and associations that grow out of common experience rather than philosophic investigation. Finally, the "idols of the theatre" are the residue of the philosophic dogmas that have influenced people's minds and perceptions. Bacon never suggests that we can totally expunge these idols from our minds and discourse, but only that we attempt to address their ill effects.

Some of Bacon's followers in the later seventeenth century more optimistically thought. as did Thomas Sprat (1635–1713), that they could fully cleanse language of the ornamental colors of rhetoric or, as did Bishop John Wilkins (1614–1672), that they could develop a more philosophic language with a univocal correspondence between things and signs. Such a total cleansing of the language proved impossible, however, for the tools of language are what make communication possible. Metaphor, for example, uses similarities to communicate new ideas and experiences to audiences, building on what they are already familiar with. Even more, it allows investigators to draw on what they know as they attempt to discover and formulate knowledge about the previously unknown. Metaphor is an invaluable cognitive tool and is inescapable in language, for new meaning grows by building on former meanings. Scientific writing, similarly, uses many of the other traditional figures of speech and thought (though they may be subsumed into scientific analytic method), such as antithesis, serial or graded ordering, and repetition.

The practice of hedging and modulating claims is also central to making arguments and assessing certainty about claims, as well as protecting the integrity of arguments against overgeneralization or unwarranted certainty. In addition, scientific writing, because it requires constantly challenging the views of others in order to advance new or competitive claims, has been found to use the linguistic mechanisms of politeness.

## Scientific Communication as Argument

Even more fundamentally, science has not been able to escape rhetoric, argument, and advocacy, because each scientific paper, when first written, is not immediately an unquestioned truth. In a scientific paper an author attempts to convince colleagues of some claim or set of related claims. At issue may be the identification of a phenomenon, the accuracy and veracity of data, the appropriateness of method, the general usefulness of a set of methods, the interpretation of data, or a new theoretical position. Scientific claims and methods must present themselves in competition with other claims, so an article must present good reasons why the its claims should be taken as more accurate and more important than other claims in other articles. Although standards of scientific

argument have been refined over the centuries, they are always directed at making claims persuasive. Standards of argument are often identified as methodological rather than rhetorical, because they have to do with finding and producing the most relevant and precise data or making the connection between data and ideas, but these nonetheless are issues of persuasiveness. The current standards have emerged because they have proven to be the most persuasive, trumping weaker arguments resting on less powerful data and reasoning.

The life of a scientific claim also lies in its usefulness to others so that it gets cited and regularly repeated. The great majority of articles get few or no citations, so, whether they are true or false, they lie unused on the back shelves of a library. A smaller number of works, however, greatly influence methods, problem formulations, and theoretical explanations, as well as our knowledge of phenomena.

## History of Scientific and Technical Communication

The history of science and technology goes hand in hand with the history of communicative technologies and practices along with the emergent social organization built on communication. The introduction of the printing press into fifteenth-century Europe facilitated the precise reproduction and distribution of observations and knowledge, allowing for comparison and collection at many sites, advancing the communal character of science. The printing press also fostered the recording and distribution of craft knowledge, so that technology spread more widely and accumulated more rapidly. Much of the practice and craft of science and technology, nonetheless, required (and continues to require to today) direct personal transmission, particularly in laboratory skills of manipulating physical objects and events and in developing the habits of thought to produce, elaborate, and substantiate innovations. Nonetheless, the shared knowledge and practical guidelines available in books helped to align and expand attention to common practices and problems.

The colonial expansion of Europe during the fifteenth to nineteenth centuries, facilitated by advances in maritime, military, and navigational technology, fostered new practices and forms of information gathering and communication. Governance and economic exploitation of distant lands required extensive transmission of information about geography, climate, mineral wealth and extraction, and flora and fauna.

In the latter half of the seventeenth century the development of scientific societies, along with journals, represented another major step forward in scientific communication. The Royal Society of London, founded in 1660, and the Académie Royale des Sciences, founded in 1666, were early associated with journals to spread more widely the presentations, demonstrations, and arguments that occurred at their meetings. *The Philosophical Transactions of the Royal Society* and *Le Journal des Sçavans* were both founded in 1665.

The founding of journals raised many social, intellectual, and communicative issues whose solutions set precedents that would influence the operations of scientific journals that followed. The editor of the journal had a strong shaping hand in soliciting, selecting, and setting standards for articles; but within a century the modern system of expert referees emerged to make publication decisions. Whereas *Le Journal des Sçavans* presented articles anonymously as the product of a collective enterprise, individual authorship was prominent in the *Philosophical Transactions* from the beginning. Publication allowed scientists to claim priority and credit for their discoveries. This individual credit model came to dominate publication practice.

Also, the various genres of scientific article needed to emerge gradually from prior genres. The earliest issues of the *Philosophical Transactions* were written in the voice of the editor reporting on correspondence he had received. Soon, however, the letters themselves were published directly. Over time the articles lost the trapping of letters and became accounts of what was observed, in increasing detail. Challenges led to quantification of detail, detailed presentation of circumstances of observations, and substantive accounts of experimental and observational methods. By the mid-eighteenth century, articles in English journals turned into narratives of inquiry, and by the end of the century articles were organized around general claims supported by experimental or observational evidence. Modern practices of citation and reviewing the literature became standard practice only in the nineteenth century. French articles from the beginning were more object-oriented, theoretical, and argumentative than English articles. German journals, on the other hand, emerged later and seemed to lag behind stylistic and theoretical tendencies in English and French journals. In the twentieth century English-language journals came to dominate science, and now most major journals in most fields are published in English, disadvantaging scientists and scientific communities in non-English-speaking countries, particularly less developed countries.

The increasing number of journals and professional societies communicating knowledge escalated rapidly in the nineteenth century, supported by social and technological changes, including less expensive printing and paper, regularized mail services, increasing economic importance of technological practice, increasing literacy, and expanding schooling. In the latter nineteenth century, universities became associated with scientific research, and technological professions became associated with university training. As research and training became more specialized and academic, scientific publications became more esoteric, theoretical, and filled with specialized vocabulary. Currently the Web of Science, a major indexing service, reports on over 5700 major scientific journals.

This enclosure of science required substantial rhetorical work to define and maintain the social boundaries of science that allowed science to be internally regulated, to gain public authority over its areas of interest, and to garner public and commercial support for its endeavors. To convey the esoteric work of science to popular audiences, new journals emerged. As science and technology also became more important for other social, economic, and governmental institutions, many other new genres and pathways of communication developed, from the environmental impact statement and testimony before Congressional committees to pharmacological advertisements touting the latest drug breakthroughs and online instruction manuals embedded within computer programs. To assist technologists in communicating with their peers, their corporate co-workers, and the public, technical writing instruction emerged as a university field in the twentieth century, and the technical writer became a major employment title in many organizations. As newspapers increased reporting on science and technology in the latter twentieth century, scientific journalism also became a recognized specialty.

The rise of the Internet and other electronic communications technologies have facilitated the rapid and extensive transmission and availability of communication of science at the research front, including preprints (early versions of articles presented for discussion), electronic journals, and virtual forums and online conferences. The Internet has also facilitated new means of carrying out the internal business of journals, societies, and research projects as well as the popular dissemination of scientific knowledge and educational communications. These new opportunities are serving to reorganize both the social arrangements and the form and content of communications. Many ideas and arrangements are being tested out, but their implications are unclear. Whatever will emerge as the next set of regularized communicative practices will no doubt use the Internet in conjunction with talk, informal text, print, electronic file, and other emergent communication technologies.

## BIBLIOGRAPHY

Ammon, Ulrich (ed.). *The Dominance of English as a Language of Science: Effects on Other Languages and Language Communities.* Berlin: Mouton de Gruyter, 2001.

Atkinson, Dwight. *Scientific Discourse in Sociohistorical Context: The Philosophical Transactions of the Royal Society of London, 1675–1975.* Mahwah, NJ: Lawrence Erlbaum, 1999.

Baake, Ken. *Metaphor and Knowledge: The Challenges of Writing Science.* Albany: State University of New York Press, 2003.

Bazerman, Charles. *Shaping Written Knowledge: The Genre and Activity of the Experimental Article in Science.* Madison: University of Wisconsin Press, 1988.

Bazerman, Charles. *The Languages of Edison's Light.* Cambridge, MA: MIT Press, 1999.

Biber, Douglas. *Variation Across Speech and Writing.* Cambridge U.K.: Cambridge University Press, 1988.

Eisenstein, Elizabeth. *The Printing Revolution in Early Modern Europe.* Cambridge, U.K.: Cambridge University Press, 1983.

Fahnestock, Jeanne. *Rhetorical Figures in Science.* New York: Oxford University Press, 1999.

Fleck, Ludwik. *Genesis and Development of a Scientific Fact.* Trans. Fred Bradley and Thaddeus Trenn. Chicago: University of Chicago Press, 1979.

Gieryn, Thomas F. *Cultural Boundaries of Science: Credibility on the Line.* Chicago: University of Chicago Press, 1999.

Gilbert, G. Nigel, and Michael Mulkay. *Opening Pandora's Box.* Cambridge, U.K.: Cambridge University Press, 1984.

Gross, Alan G. *The Rhetoric of Science.* Cambridge, MA: Harvard University Press, 1990.

Gross, Alan G., Joseph E. Harmon, and Michael Reidy. *Communicating Science: The Scientific Article from the 17th Century to the Present.* Oxford: Oxford University Press, 2002.

Halliday, Michael A. K., and James Martin. *Writing Science.* Pittsburgh, PA: University of Pittsburgh Press, 1994.

Johns, Adrian. *The Nature of the Book: Print and Knowledge in the Making.* Chicago: University of Chicago Press, 1998.

Knorr-Cetina, Karin. *The Manufacture of Knowledge.* Oxford: Pergamon Press, 1981.

LaFollette, Marcel C. *Making Science Our Own: Public Images of Science, 1910–1955.* Chicago: University of Chicago Press, 1990.

Latour, Bruno. *Science in action.* Cambridge, MA: Harvard University Press, 1987.

Latour, Bruno, and Steve Woolgar. *Laboratory Life: The Construction of Scientific Facts.* Beverly Hills, CA: Sage, 1979.

Lee, Kok Cheong. *Syntax of Scientific English.* Singapore: Singapore University Press, 1978.

Lynch, Michael. *Art and Artifact in Laboratory Science: A Study of Shop Work and Shop Talk in a Research Laboratory.* London: Routledge and Kegan Paul, 1985.

Meadows, Arthur Jack. *Communicating Research.* San Diego, CA: Academic Press, 1998.

Montgomery, Scott L. *The Scientific Voice.* New York: Guilford Publications, 1996.

Myers, Greg. *Writing Biology: Texts in the Social Construction of Scientific Knowledge.* Madison: University of Wisconsin Press, 1990.

Nelkin, Dorothy. *Selling Science: How the Press Covers Science and Technology.* New York: W. H. Freeman, 1987.

Pera, Marcello. *The Discourses of Science.* Trans. Clarissa Botsford. Chicago: University of Chicago Press, 1994.

Slaughter, Mary M. *Universal Languages and Scientific Taxonomy in the Seventeenth Century.* Cambridge, U.K.: Cambridge University Press, 1982.

Van Nostrand, A. D. *Fundable Knowledge: The Marketing of Defense Technology.* Mahwah, NJ: Lawrence Erlbaum, 1997.

Winsor, Dorothy A. *Writing Power: Communication in an Engineering Center.* Albany: State University of New York Press, 2003.

*Charles Bazerman*

# COMMUNITY AND TECHNOLOGY.

One of the apocryphal tales in the science and technology studies literature is the story of a communal water pump in the village of Ibieca, Spain, told by Richard Sclove in *Democracy and Technology.* According to Sclove, the village lived a peaceful and harmonious existence so long as the women of the community assembled at the communal water supply to participate in the daily social exchanges that sustained the community's solidarity. However, in succumbing to the myth of technological progress, the male elders of the village replaced the communal water pump with a new municipal water supply that brought water to individual homes. While this project supposedly made the women's work easier, it unintentionally disrupted the social fabric of the community.

There is reason to doubt the veracity of this tale—the social networks that constitute our perceptions of community are often more extensive, and fragmented, than in Sclove's story. The tale also seems to speak more to the myth of the isolated, American suburban housewife rather than from the historical realities of a rural village in Spain. Nevertheless, the story raises several issues of direct significance to this essay, including how civic concerns are embedded in technological choice and design; that technology is integral to a sense of community; that gender and other cultural identities mediate this relationship; and that the mundane technologies of everyday life contribute as much to a sense of community as the latest information and communication technologies. This last point is especially important, given the hope people place in the Internet for reviving communities, both virtual and real. If the extensive literature in science, technology and society says anything, it is that all of these issues must be held in careful balance if we are to assess a complex relationship such as that between technology and community.

Given the very different ways in which people use the word *community*—the "community of West Philadelphia," the "gay and lesbian community," or even the "American community" (Putnam, 2000)—it is important to begin with a precise definition of the term, along with an account of the relationship between community and civil society. Formally, a community can be defined as any identifiable group that has a shared set of values and interests, where this commonality is sustained through some form of social interaction. (Though this definition is sufficient as a sociological definition, some also add that a community must have members who work actively to uphold the values of the community.) However, other characteristic differences in a community can affect how technology mediates its interactions. Thus, communities may refer, at one level, to physically contiguous communities, such as the Italian-American neighborhood in North Boston or the African-American community in West Philadelphia. It may also refer to less contiguous groups, such as the gay and lesbian community or the international community of nuclear physicists. Also, as invoked in the story of Ibieca, the term *community* may transcend the interests of specific groups and can be equated more with a sense of communitarian interests and civic obligation. By this definition, "community," at this higher level, refers to a civil society that is able to mediate differences and sustain social harmony amidst a more complex pattern of shared interests and differences. While this essay is organized more along the lines of major bodies of science and technology studies scholarship that examines the relationship between technology and community, it will address questions of community at all three levels; it also closes with a specific look at the relationship between information technologies, and the promise of virtual communities.

Sclove's *Democracy and Technology* is a good place to start this overall discussion. Sclove's writings are based on a critical tradition in political philosophy, with ties to Jacques Ellul, Hannah Arendt, Lewis Mumford, and others. His work draws most directly on Langdon Winner's article, "Do Artifacts Have Politics?" published in 1980 in *Daedalus*. Winner's article was an early response to the constructivist turn in the history and sociology of technology, where social groups with identifiable interests were seen as driving the process of technological design and stabilization. While hailing this as a necessary advance beyond the *technological determinist* position, Winner criticized what he saw to be an equally misguided *social* determinist approach in which, when taken literally, "technical *things* do not matter at all" (Winner, p. 21). Using examples such as the difference between nuclear reactors and solar energy, Winner set out to document that there are inherently democratic and authoritarian technologies, following earlier arguments made by Mumford. The authoritarian measures required to prevent nuclear proliferation threatened the character of the civil societies that chose to adopt nuclear energy. Drawing on established traditions within political philosophy, the critical edge to Winner's writings lies with the

attention he gives to political process. Whereas social constructivism, as understood by Winner, is purported to describe how new social and technological arrangements emerge out of existing social interests, Winner emphasized that the very act of technological choice and design are political acts that could themselves transform social interests and commitments.

Sclove's *Democracy and Technology* amounts to a careful development of Winner's thesis, with many examples of particular value to a discussion about community and civil society. The story of Ibieca simply opens the volume. Subsequent examination of artifacts such as classrooms and housing, and institutions and laws, such as the Americans with Disabilities Act, all document that the choice and design of technology contribute to very different forms of social order. Sclove also allows no clear separation between technologies of the workplace and those pertaining to community. He suggests that different arrangements for industrial, and now postindustrial, production have direct implications for the character of civil society. Those forced into the drudgery of assembly line work, or highly constrained service jobs such as telemarketing, are less likely to have the requisite characteristics of a democratic citizenry. There are scholars who would contest this interpretation, suggesting that oppressive workplace technologies can do as much to foster solidarity and political engagement; they would reject the tacit model of the "over-socialized individual," a model that views individuals as highly constrained by social (and technological) forces, that is implicit to Sclove's analysis. Nevertheless, this larger literature in social history and "labor process" studies, not to mention Marxist theory itself, shares with Sclove the ultimate goal of promoting the civil discourse necessary for the social regulation of technology.

Indeed, at the heart of both Winner and Sclove's argument is the idea that technological design and choice ought themselves to be a site for political action. Using transitive reasoning, Sclove argues that, "[i]f citizens ought to be empowered to participate in determining their society's basic structure, and

technologies are an important species of social structure, it follows that technological
design and practice should be democra-
tized" (p. 26). It is important to note that neither Sclove nor Winner cares that this
position is incompatible with the existing
structures of authority and interest within
capitalist democracies. In approaching the
matter from their particular branch of political philosophy (technically known as "sub-
stantivism," a branch of moral philosophy),
their goal has been to develop an ethical
foundation for challenging oppressive institutions that lie beyond the din of present
civil discourse.

Included in Winner's early essay is also a
famous argument about New York city plan-
ner and engineer Robert Moses and the
bridges on the Long Island Parkway. Drawing on comments collected by Moses' biographer, Robert Caro, Winner describes how
Moses arranged for low-arch bridges to be
built across the Long Island Parkway in such
a way so as to preclude the use of public
transit buses. The claim is that this was a
racist policy designed to create segregated
urban spaces, especially by excluding poor,
inner-city blacks from using a new public
beach built by Moses on Long Island for the
moral uplift of the city's burgeoning middle
class. Understood historically, Moses' actions
can be tied to prevailing views of social order
and social evolution that should sound biased
and amoral to readers today. They also represent a technocratic outlook, namely the faith
that technology and central planning could be
mobilized to achieve a harmonious society.

Winner's argument has recently been criticized by Bernward Joerges through a close
and reflexive reading of Winner's text, in
which the text is revealed to be itself a kind
of political artifact. However, the fact that
there are alternate routes to Jones Beach, as
pointed out by Joerges, does not alter the social geography constructed by the relative
ease and difficulty of access created by different modes of transportation. More importantly, Moses' bridges contributed neither
more nor less than any other element of culture in producing perceptions of social sanc-
tion and prohibition that sustained the

semiotic constructions that constituted Jones
Beach as a "white" and "middle-class" beach
in 1930s, Depression-laden United States.
Both the technological and nontechnological
aspects of culture all contributed, say, to the
kind of personal evaluations that a black
mother, fearful of social scorn or the consequences of that scorn as applied to her three-
year-old son, might have made in choosing
instead to go to the more accessible and socially permissible amusements of Coney Island. Although Winner may be accused of
placing untoward emphasis on technology
as an element of social order, he cannot be
faulted for drawing attention to the role that
technological design and material culture
play in its construction. The very fact that
Moses' actions seem out of place to contemporary observers also points indirectly to the
efficacy of moral reasoning and political philosophy. In any case, the built environment
has played a fundamental role in partition-
ing urban spaces into communities defined
by their various differences.

## Community and Urban Infrastructure

This point of view is backed by a rather robust set of classic studies in the history of
transportation, including many studies of
the "streetcar suburb." But to resist, once
again, the temptation to impart too much
causality to specific technologies, it is important to consider that suburbanization
occurred amidst the backdrop of massive
demographic movements caused by industrialization and the mechanization of agriculture. Especially in the United States and
other countries that permitted a high rate of
immigration, strong ethnic and racial communities formed within urban spaces in response to new technologies of mass
production. The identity of these communities resulted from complex, and sometimes
violent social and cultural negotiations involving ethnicity, class, race, religion, and
gender. These negotiations have continued
with the global extension of industrial capitalism and the rise of a postindustrial
economy.

Generations of sociologists, beginning
with those at the University of Chicago, have

studied the phenomenon of urban community formation, superimposing, in many cases, an ecological metaphor on social geography. What remains important in this literature is the significance given to center-periphery relations and how they describe the relationship between the suburbs and the city, including dysfunctional relationships that contributed to the economic decline of the inner-city core. Transportation technologies and the politics behind them have been deeply implicated in this decline. Whether through placement decisions for the urban portions of U.S. interstate highways or through the Federal Housing Administration's system of guaranteed loans, explicit policy has contributed to the extension of a social ideology of segregation and of "single-family homes"—itself a package of domestic technologies and ideology. Cities in the United States, as elsewhere, have been shaped through intricate relationships between market forces, government intervention, technology, and questions of social identity and difference. Sadly, both market forces and federal policies have left the poorest urban communities straddled with an aging urban infrastructure that they can ill afford.

Recent studies of urban infrastructure have moved away from the "big" technologies of urban transportation, and have turned instead to the more mundane technologies of everyday life—what one scholar has referred to as "pedestrian" technologies in an intentional double entendre. Those drawn to Michel Foucault's postmodernist studies of the technologies of social order—Jeremy Bentham's "panopticon" prison design, in which the cells are arranged in a circle around a central tower from which a hidden guard can see into all of them, for instance—have delighted at the effects that simple objects such as speed bumps, street lights, and the physical barriers of a gated community can have on urban spaces and their sense of community. As these scholars would immediately acknowledge, these subtle technologies of social order have a long history, as with the use of electric lighting during the 1893 Columbian Exposition to recast Chicago into a city of light devoid of the crime and malfeasance attributed to the city's darker (and poorer) immigrant communities. The Foucauldian obsession with power and oppression, however, can be misleading. There are those who have applied the same insights in a more positive direction. By envisioning bike paths, pedestrian right-of-ways, and wheelchair-accessible structures, including the graduated curbs now found at almost every street corner, scholars in science and technology studies, as well as a larger community of urban planners, have demonstrated how relatively minor changes in urban infrastructure can have profound effects on the civic character of a city. There have also been various efforts to apply such perspectives on the scale of the urban neighborhood in shaping its built environment to foster social exchanges and the economic vitality of the neighborhood. Many of these efforts unfortunately have a distinct flavor of gentrification, which is to say improvements geared towards the already affluent, and ignore and sometimes even displace poorer ethnic communities that also occupy the same neighborhood. Nevertheless, there have been quite a few urban renewal initiatives that have successfully drawn on the principles of a carefully built environment to restore economic vitality to the inner city.

In this article so far, technology has played several different roles in constituting communities and civil society. It has been portrayed as contributing to the overall character of civil society, most notably at the level of a city, which can influence how communities and neighborhoods emerge around social differences. Technology has also been portrayed as something that mediates the interactions within a geographically contiguous community—the effect, for instance, that a carefully planned, built environment has in enhancing social exchanges within a specific neighborhood. Finally, technology can facilitate communication and cultural identity among geographically distended communities whose shared interests are sustained through the use of communications media.

## Information Technologies and Community

With all three of these different aspects of the relationship between technology and community in mind, it is possible to consider to what extent recent information technologies have added to or detracted from communities and civil society. There is no doubt that the Internet and its Web-based technologies have strengthened existing communities and contributed to the proliferation of new subcultural identities. The Internet has offered unique characteristics of interactivity, anonymity, searchability, and user-initiated inquiry, on which different groups, whether *Star Trek* fan clubs, artists' groups, religious organizations, skinhead groups, or new on-line communities such as The WELL, have selectively drawn to build a particular identity. Recent scholarship on the Internet points out, correctly, that one of the distinguishing characteristics of the Internet is its potential for providing users with the means to generate "relevant content"—content relevant to a specific community—making it a many-to-many medium of communication quite different from the one-to-many mode of communication of traditional broadcast media.

Still, claims of "revolution" should be weighed against historical evidence. Race records—record labels that catered to the African American community—and the immense popularity of ethnic radio stations during the early history of broadcast radio suggests that earlier technologies played a vital role in strengthening existing communities and their cultural identities. They also document the coproduction of content between audience and producer that belies common assumptions about broadcast media. Especially when weighed against the practical challenges many groups face in generating relevant content for the Internet, the use of the Web by contemporary subcultures should be seen as an extension of, rather than a radical departure from, prior community uses of media. In general, the Internet has proven to be of greatest interest and value to geographically distended communities such as those with rare hobbies, political and religious interests, or illnesses such as Hodgkin's disease, where the Internet has helped draw together a large enough audience to justify generating substantial content on a matter of common interest.

The proliferation of radical, conservative, ultranationalist, and religious fundamentalist sites has meanwhile generated substantial concern about the Internet as an instrument of social disintegration. The underlying concern is that the stronger the allegiance individuals hold to a specific subculture, the weaker their commitment to society as a whole. Such concerns have a long history, and tend to be based on an impoverished notion of civil society. The basic premise for any democratic vision of social order is that there will always be groups with politicized differences. Order is nevertheless sustained so long as there is a commonly accepted and enforced political process for mediating those differences. The civil strife that unfolded in Bosnia and Rwanda, fomented in part by ethnic radio broadcasts, demonstrates the real dangers of tools of cultural production that accentuate differences by amplifying real and imagined communities. Nevertheless, the conservative commitment to social order in most industrialized nations ensures that current uses of the Internet will have little effect on the overall character of civil discourse. This criticism applies equally well to the opposite hope that the Internet is the harbinger of direct democracy. The U.S. system of representative democracy, as driven by interest group politics, is far more likely to be strengthened through campaign finance reform. On the other hand, political activism and engagement using the Internet, including the recent, Internet-based fundraising strategies that have a distinctly populist orientation, have brought notable changes to character of civil discourse.

There have also been serious and admirable efforts to use new digital technologies to enhance existing geographically based communities. Successful sites such as Charlotte's Web in Charlotte, North Carolina, and the LibertyNet in Philadelphia, have specifically built upon the existing cultural, religious, and institutional resources of established urban communities. Generally known as Asset-Based Community Development, these projects have given real-world communities a presence in cyberspace, often to complement urban renewal efforts. Those developing such sites have uniformly found that the only means of generating relevant content is to draw on existing community organizations, because they have the defined interests and organizational capacity to produce and sustain current content of value to the community. Even then, more traditional modes of communication, such as a local alternative weekly, remain by far the dominant means of communicating local events and cultural resources, even as established social networks remain the dominant mode of communication within neighborhoods and urban subcommunities.

The greatest impact of information technology on urban communities, however, has not been in strengthening communities but in laying anew the structural foundations upon which community are created. The industrial carnage visible across the Northeast and Midwest and its contribution to the decline of the inner city attest to the global capital flows made possible through new information technologies. The resulting shift to service-sector employment in Western countries has contributed to the phenomenon of the "digital divide," which current research suggests is much deeper than simply differential access to computers. The latest research in fact has begun to ask more fundamental questions about how information technologies, in all of their different guises, contribute to new patterns of social stratification—socioeconomic advantage and disadvantage—especially in the United States. Doing so places the social study of information technology on the same footing as studies of industrialization during the previous century. The predominant character of urban communities has to be understood on such grounds.

Nevertheless, IT-based community networking initiatives have emerged as a real resource for those displaced by the very spread of information technologies. The most effective "community networks," in this sense, have worked with well-identified populations—working-class ethnic and low-income African-American neighborhoods, senior centers, battered women's shelters, and the like—to help them navigate through the effects of a structural shift to "the new economy." New computing and network access facilities, established as a "community technology center," have often been set up through the efforts of an existing community organization, such as a YWCA or a public housing facility, often aided by the initiative and voluntary assistance of information technology professionals. These multiuse facilities aim, simultaneously, to promote digital literacy, foster technological interest, and offer skills development and retraining as specifically suited to regional economic opportunities. An especially valuable feature of these community networks has been current job listings and enhanced access to social services made possible by the use of information technology. Still, these programs constitute but a limited response to a structural problem, and they reflect equally the paucity of public services in the United States and its lack of social commitment to deal with the disruptive aspects of technological change.

## BIBLIOGRAPHY

CTCNet. Community Technology Centers' Network. http://www.ctcnet.org. A social and online network for community technology centers.

Engwicht, David. *Reclaiming Our Cities and Towns: Better Living with Less Traffic.* Philadelphia: New Society Publishers, 1993.

Etzioni, Amitai. *The Spirit of Community: Rights, Responsibilities, and the Communitarian Agenda.* New York: Crown Publishers, 1993.

Jackson, Kenneth. *Crabgrass Frontier: The Suburbanization of the United States.* New York: Oxford University Press, 1987.

Norris, Pippa. *Digital Divide: Civic Engagement, Information Poverty, and the Internet Worldwide.* Cambridge, U.K.: Cambridge University Press, 2001.

Plugged In. http://www.pluggedin.org. A model community technology center set up in East Palo Alto, California.

Putnam, Robert. *Bowling Alone: The Collapse and Revival of American Community.* New York: Simon & Schuster, 2000.

Rheingold, Howard. *The Virtual Community: Homesteading on the Electronic Frontier*, rev. ed. Cambridge, MA: MIT Press, 2000.

Sclove, Richard. *Democracy and Technology*. New York: Guilford Publications, 1995.

The WELL. http://www.well.org. A "literate watering hole for thinkers" founded in 1985 as the Whole Earth 'Lectronic Link.

Winner, Langdon. *The Whale and the Reactor: A Search for Limits in an Age of High Technology*. Chicago: University of Chicago Press, 1986.

*Atsushi Akera*

# COMPLEMENTARY AND ALTERNATIVE MEDICINE.

Conventional therapeutic systems—that is, systems of diagnosis, etiology (explanation), and treatment—are certified as valid by scientific research communities, declared legal by state regulators, covered by major state or private insurers, and utilized by the medical profession and its auxiliary health care providers. The term *alternative medicine* refers to therapeutic systems that are used instead of conventional medicine, and *complementary* refers to therapeutic systems used alongside conventional medicine. Increasingly, the term CAM (complementary and alternative medicine) has been used to refer to this field.

In addition, the term *experimental medicine* refers to therapeutic systems that are not yet accepted as conventional but are undergoing evaluation with conventional status likely in the foreseeable future, such as new drugs. The three categories of conventional, complementary/alternative, and experimental medicine are widely used but nevertheless not completely distinct from one another. For example, some doctors may prescribe therapies that expert research communities no longer consider valid or that insurers do not cover. Likewise, some CAM therapies are undergoing intensive evaluation and increasing utilization, and therefore they might be considered to be experimental.

The term *traditional medicine* also warrants some explication. In biomedical circles the term often refers to conventional medicine, which is "traditional" in the sense that the medical profession considers it established or accepted. However, the World Health Organization defines traditional medicine as indigenous to different cultures, and it distinguishes *traditional medicine* from CAM, which it defines as neither indigenous nor accepted as conventional medicine. In this essay *CAM* will be used as the broader term that includes what the WHO is calling *traditional* or *indigenous medicine*.

In general, conventional medicine is based on the biomedical model of disease, that is, a model that draws on the materialist and mechanistic assumptions found in the modern natural sciences. This model is often amplified by inclusion of psychological and social levels of explanation, so it might be more properly termed "sociopsychobiomedical." However, conventional medicine is often practiced in the absence of a fully articulated biomedical model— that is, etiologies may be uncertain, diagnoses may be imprecise or craftlike, and treatment may be by trial and error. Nevertheless, even where biomedicine faces the unknown, the preferred methods of explanation are based on the principles of materialism and mechanism.

In contrast, CAM systems may adopt a biomedical model, a nonbiomedical model, or a mix of the two. An example of the former is the model of cancer as a bacterial infection that can be treated with bacterial vaccines or even antibiotics. Although unconventional, the therapies are based on recognizable mechanisms from a biomedical perspective. Medical researchers in the major universities and research centers may consider bacteria-and-cancer CAM to be unfounded, experimental, or even bad science, but they can recognize the bacterial theories and therapies as part of the family of existing biomedical concepts and methods. Other CAM systems are based on nonbiomedical concepts or mixes of biomedical and nonbiomedical concepts, such as humors (quasi-biological entities that were found in the ancient biologies of Old World societies), vital energies, spiritual beings, or spiritual development. Thus, as a conceptual system the field of "CAM" runs from scientific systems that may approximate experimental medicine to the world of religion. One common denominator among the many flavors of CAM is that there tends to be a greater emphasis on treating the person as a whole (holism) than in conventional medicine.

Many other classifications or ways of conceptualizing CAM have been developed. For example, the National Center for Complementary and Alternative Medicine of the U.S. National Institutes of Health has classified CAM into five areas: alternative medical systems (such as Chinese medicine or naturopathy), mind-body therapies, biologically based therapies (such as herbs and food supplements), manipulative therapies, and energy therapies. The broader field of CAM conceptualized here would also include both rejected-drug therapies (such as laetrile) and magico-religious therapies.

## Utilization Patterns and CAM Social Movements

A substantial survey literature has now documented the extensive use of CAM and its substantial growth in recent decades. CAM utilization appears to have grown in many countries, including the advanced industrialized countries, where surveys typically show that about half the population has had recourse to at least one CAM therapy during the previous year. Users in industrialized countries tend to be more well educated, more often women, and more often suffering from chronic disease or pain than the population average. A substantial social science literature explores the driving factors behind CAM utilization, such as the aging of the population, changes in diet and lifestyle, and the successes of biomedicine in preventing and curing many infectious diseases. The latter has resulted in a transition in mortality toward chronic diseases that are generally considered noninfectious, such as most forms of cancer, cardiovascular disease, diabetes, and arthritis. The treatments offered by conventional medicine for those and other chronic diseases tend to have higher levels of toxicity and lower levels of efficacy than, for example, the antibiotic treatment of the classic acute infectious diseases. As a

result, CAM emerges not only where biomedicine is not available or where religious and cultural preferences are oriented toward nonbiomedical systems but also at the frontiers or limits of the safety and efficacy of conventional medicine, such as for the treatment of many types of cancer.

The growth of patient utilization of CAM challenges the narratives of progress sometimes encountered among the leaders of conventional medicine. Anti-CAM editorials in medical journals sometimes argue that there is no such thing as CAM and that there is (or should be) only good and bad medicine based on good and bad science. From this perspective there should be no deviation from scientific medicine (that is, therapies that are theoretically consistent with biomedicine and evidentially supported by clinical trials), and any therapeutic systems that are not part of scientific medicine should be eradicated. This view is usually associated with a policy position of "quackbusting" or suppression of CAM practitioners.

A contrasting view is found among some of the patient-based CAM social movement organizations and among some CAM healthcare professionals and researchers. Perhaps the most well-developed expression of this view is found for CAM cancer therapies in the United States from the 1950s through the 1980s, but similar views are found for smaller networks of patient advocates, clinicians, and researchers for other diseases. Social movement leaders reversed the rhetoric aimed at CAM cancer therapies to argue that the medical establishment and the pharmaceutical industry were blocking the access of patients to more safe, efficacious, and less expensive CAM treatments because the supporters of conventional medicine were more concerned with their own profits than with the advancement of scientific knowledge and the public good. CAM advocates pointed to the credibility of their own biomedical theories, to the low toxicity of most CAM therapies, and to their possible efficacy as documented in best cases, clinically based data sets, subclinical experiments, and the occasional clinical trial. They also pointed to a variety of cases in which the evaluation of

CAM therapies by conventional medical researchers was plagued by bias.

## Integrative Medicine and Epistemic Politics

During the 1990s and into the first decade of the twenty-first century the rather pitched rhetoric from both sides tended to subside, although occasional flare-ups continued to occur. In their place, medical research leaders increasingly argued that CAM should be evaluated rather than debunked, and the CAM professions increasingly developed their own journals for peer-reviewed publication of CAM-oriented research. As a result, the epistemic politics shifted somewhat away from the dismissal and suppression of alternatives to selective acceptance based on the legitimating concept of "evidence." In turn, the CAM field underwent restructuring as it was selectively incorporated into conventional medicine. The more alternative therapies tended to be marginalized as the more complementary uses of CAM became integrated. Likewise, there was increasing opportunity and pressure for CAM practitioners to undergo licensing and professionalization. The more professionalized CAM systems—such as, in the United States, chiropractic, naturopathy, acupuncture and Chinese medicine, and massage—were increasingly incorporated into insurance plans and some conventional medical practices in the form of complementary or auxiliary therapeutic systems.

Changes in insurance policies are one of several faces of the integration process. Widespread knowledge of the growth in demand for CAM spurred both private and public insurers to increase coverage since the early 1990s. However, a decade later in the United States most CAM expenditures continued to be out-of-pocket, and insurance coverage was generally limited, often only to chiropractic but sometimes extended to naturopathy, acupuncture, and massage. Most U.S. states required chiropractic coverage, and some states, such as Washington, had enacted much more extensive mandates. Private insurance company coverage in competitive markets is determined less by evidence in the sense of "clinical efficacy" than by evidence in the sense of profitability calculations. Those calculations rest on evaluations of cost savings relative to conventional therapies, and they are made with increasing precision. Although insurance coverage increases CAM utilization, when it is offered on an experimental basis to patients, only a small percentage of covered patients tend to access CAM (less than 10 percent in the studies reviewed here). Insurers would need enormous savings on such a small percentage of patients, or a large impact on demand, for the CAM coverage to have a significant impact on profitability. Meanwhile, expansion of insurance coverage may be having a more significant impact on CAM providers, who must standardize diagnosis and treatment in order to receive reimbursement.

Another dimension of the integration of CAM occurs in the medical profession's selective embrace of CAM in the clinical setting. The development of physician-controlled "integrative medicine" in the clinical setting has benefited patients by increasing access to some forms of CAM. Likewise, coordination among diverse therapeutic systems could also eliminate potentially dangerous interactions, such as drug-herb incompatibilities. However, as with insurers, clinical integration has tended to filter out the more alternative uses of the CAM. For example, patients have access to the CAM systems mostly as adjuvants to their conventional cancer treatments and usually from CAM systems that have undergone professionalization.

## Research and Evaluation

The research arena involves a similar selective ordering of the CAM field. Biomedical researchers generally maintain a hierarchy of research methods, with randomized clinical trials (RCTs) at the top. RCTs randomize patients into two or more arms, one of which is usually a control group that receives a placebo (or, in some cases, the conventional standard of care) and the other of which receives the experimental treatment (here the CAM treatment). A step down the rung of evidence is retrospective methodology (such as retrospective cohort studies); this methodology usually compares existing records of patients with CAM treatments to matched control populations or published success rates for comparable patients who received conventional therapies. Although most medical researchers consider retrospective methods to be less reliable than RCTs, the power of observational studies with respect to RCTs is gaining increasing recognition. The next step down the ladder of evidence is usually the best case series. Usually it is dismissed as "anecdotal" evidence, but a best case series can be considered impressive if it demonstrates dramatic improvements for diseases that have poor prognoses under conventional medicine, such as series that have generated long-term survivors for pancreatic cancer. Finally, subclinical methods also provide contributing evidence; they include animal studies, *in vitro* studies of the effects of therapies on tissue systems or cell cultures; and biochemical analyses of substances used in CAM, such as herbs.

In general, conventional medicine is disease-oriented, and RCTs are set up to test a standardized therapy or a small group of therapies for a particular disease on a sample of patients who represent a fairly broad segment of the population. However, CAM systems are frequently individual-oriented, and they tend to target multiple areas in need of improvement in the patient's mind, body, and emotional state. As with the insurers' need for standardization in the clinical setting, in the research setting CAM systems often do not fit well into the framework of evaluation methodologies for standardized conventional therapies. Although one can design RCTs at a very broad level to randomize patients to conventional treatments versus a CAM clinic that provides an individualized complex therapeutic program, the interpretation of the results is difficult because so many variables are left uncontrolled. For example, if a CAM cancer clinic substantially outperforms the standard therapy for lung cancer patients, is this because of the doctor-patient relationship, the ambience of the clinic, the dietary program, the mind-body program, the pharmacological program, the attitude changes in patients, or other aspects of the CAM clinic? Because the holistic approach requires so many variables to be "black-boxed," the results of a clinical trial of this sort are so ambiguous that it is difficult to know what to recommend as a change that would benefit patients.

In conventional medical research, druglike portability of therapies is favored over the comparative evaluation of specific clinics or clinicians. Consequently, clinical research on CAM therapies tends to simplify total programs to test one or a small number of therapeutic interventions. This research design tends to filter out both nonbiomedical concepts (such as humoral typing, subtle energies, and spirituality) and person-centered treatment in favor of substances or techniques that can be replicated and made portable across a wide variety of settings for a wide range of patients with the same disease or symptoms. Furthermore, RCTs also tend to select complementary therapy designs (such as chemotherapy plus high-dose vitamin supplements that reduce the side effects of chemotherapy) rather than fully alternative therapies (such as complete dietary programs that are used instead of chemotherapy). Although neither the filtering-out of nonbiomedical concepts and holism nor the bias toward complementary care is a necessary feature of clinical research design, in practice clinical research design tends to shape the CAM field in ways that reinforce the same shaping effects of insurance programs and integrative clinical practice.

## Regulatory Politics

The position of RCTs at the apex of the methodological hierarchy is due in part to the fact that RCTs are required in many countries in order for a therapeutic substance to achieve the legal regulatory status of drug. However, the high cost of both RCTs and the drug approval process in general means that pharmaceutical companies will invest money only in substances that can be patented and that can achieve a profit volume great enough to compensate for the hundreds of millions of dollars invested in the drug approval process. In the United States natural substances such as herbs and food-derived supplements can be used as foods in health care settings, but only structure-and-function claims can be attached to them (such as "the substance promotes colon health"). If a disease claim is attached to the substance (such as "it successfully treats colon cancer"), the substance in effect becomes a drug and must pass through the regulatory process of RCTs for such claims to be made legally. However, private firms are unlikely and unable to invest the hundreds of millions of dollars required to bring an unpatented natural substance to market as a drug.

Although the regulatory systems of some countries, particularly Germany and some of the Asian countries, provide alternative models that make it easier for unpatented substances—such as herbs that have long been used—to achieve a status beyond that of structure-and-function claims, in general the substances used in CAM tend to have a secondary legal status with respect to the patented drugs of the international pharmaceutical industry. The lower or more ambiguous legal status tends to reinforce the shaping of the CAM field toward the use of treatments that are complementary to drugs and surgery, not alternatives to them. The lack of capitalization due to the lack of patentability means that the more expensive forms of research, such as RCTs, tend to remain part of the body of "undone science" for CAM, thus contributing to its inability to gain the status of insurability, physician acceptance, scientific acceptance, and regulatory approval that characterize conventional medical therapies. Indeed, CAM advocates sometimes quip that RCTs are aptly named the "gold standard" because it takes a lot of gold to set the standard.

Although the public sector could in theory compensate for the market failure created by the lack of patentability for the unpatented natural substances that are commonly used in CAM (such as herbs, foods, and food supplements), the cost of running clinical trials on even a small percentage of CAM therapies would be prohibitive. In the United States and some other countries, there is some government funding for the evaluation of CAM therapies, but the funding is minuscule in comparison both with total health-care research spending and with the universe of studies needed based on patient utilization patterns. As a result, CAM practitioners and patients make decisions based on levels of evidence that they often consider inadequate. To some extent the lack of evidence is compensated for by the relative safety of most CAM therapies. Because CAM therapies often either have a long history of safe usage or are based on dietary, mind-body, or lifestyle modifications, concerns with side effects and toxicity tend to be lower in general than for conventional therapies such as drugs, surgery, or radiation treatment. As a result, even when CAM practitioners and patients adopt a rigorous model of utilization decisions based on evidence, they tend to be more willing to try treatments that have not undergone extensive evaluation through RCTs or large retrospective studies: There are two reasons for their willingness to use CAM therapies even in the absence of full documentation: The risks are assumed to be lower, and the conventional therapies are, in some cases, unacceptable because of high side effects or low efficacy. In sifting through the evidence for specific CAM therapies, CAM-oriented health-care providers and patient advocacy leaders display greater willingness to look at the entire body of evidence (including best cases and subclinical studies) and make a holistic assessment of potential safety and efficacy. This approach to evaluation might be termed "methodological pluralism," in comparison with both the standard model of the hierarchy of evidence in medical research and the better recognized phenomenon of "medical pluralism," that is, the coexistence of conventional and CAM practitioners. Thus, the integration of CAM is playing a role in the pluralization of research methodologies that is another aspect of the epistemic modernization of medicine in the contemporary period.

In summary, the relationship between conventional medicine and CAM is dynamic and undergoing rapid change. Beginning in the 1980s but increasingly during the 1990s, the leaders of conventional medicine shifted their approach to CAM from debunking and suppression toward evaluation and integration, although the older strategy has far from disappeared. The more complementary therapies and more biomedically oriented concepts are being selected rather than the more alternative and nonbiomedical side of the CAM field. However, conventional medicine is also undergoing changes. It is paying greater attention to toxicity and side effects of drugs and other conventional therapies (particularly for chronic disease), the role of nutrition and mind-body relationships in establishing the terrain for disease and health, person-oriented treatment rather than disease-oriented treatment, and the evidential power of clinical methods other than RCTs. Those are four examples of areas in which CAM is having some influence on conventional medicine.

*See also* **Biomedical Technologies**

**BIBLIOGRAPHY**

Baer, Hans. *Biomedicine and Alternative Healing Systems in America*. Madison: University of Wisconsin Press, 2001. A thorough overview of the CAM field in the United States by a well-known medical anthropologist.

Benson, K., and A. J. Hartz. "A comparison of Observational Studies and Randomized, Controlled Trials." *New England Journal of Medicine* 342 (2000):1878–1886.

Eisenberg, David, T. L. Delbanco, S. L. Ettner, S. Appel, S. Wilkey, M. Van Rompay, and R. C. Kessler. "Trends in Alternative Medicine Use in the United States, 1990–1997: Results of a Follow-up National Survey." *Journal of the American Medical Association* 180 (1998): 1569–1575. Eisenberg's work is probably the most highly cited among the many surveys that document the growth and extent of CAM utilization in the United States.

Hess, David. *Can Bacteria Cause Cancer?* New York: New York University Press, 1997. An approach to one strand of CAM that utilizes a four-field STS analysis of history, social science, philosophy, and policy.

Hess, David. *Evaluating Alternative Cancer Therapies*. New Brunswick, NJ: Rutgers University Press, 1999. A survey of how the leaders of the CAM cancer therapy movement in the United States viewed issues of evidence and evaluation during the mid-1990s.

Hess, David. Home Page. http://home.earthlink.net/~davidhesshomepage. More on the author's work on CAM.

Kelnor, Merrijoy, Beverly Wellman, Bernice Pescosolido, and Mike Saks, eds. *Complementary and Alternative Medicine: Challenge and Change*. Amsterdam: Harwood, 2000. A good collection of essays that introduces some of the major social science researchers in this field.

Moss, Ralph. *The Cancer Industry*. 2nd ed. New York: Equinox Press, 1996. The classic exposÈ of scientific bias in CAM cancer therapy research, by the person who has become the leader of patient advocacy in this field. See the extensive Website at http://www.ralphmoss.com.

National Center for Complementary and Alternative Medicine, U.S. National Institutes of Health. *What Is CAM?* http://nccam.nih.gov/health/whatiscam/. 2003. Accessed January 1, 2003. Website of the official CAM research organization of the U.S. government.

Pelletier, Kenneth, and John Astin. "Integration and Reimbursement of Complementary and Alternative Medicine by Managed Care and Insurance Providers: 2000 Update and Cohort Analysis." *Alternative Therapies in Health and Illness* 8.1 (2002.): 38-39, 42, 44. Ongoing research program on CAM and the U.S. insurance industry.

Rosner, Anthony. "Fables and Foibles: Inherent Problems with RCTs." *Journal of Physical and Manipulative Therapy* 26.7 (2003): 460–467. A careful analysis of epistemic politics in chiropractic by one of the leading researchers in the field.

Tillman, Robert. "Paying for Alternative Medicine: The Role of Health Insurers." *Annals of the American Academy of Political and Social Science* 583 (2002): 64–75. A special issue of a leading social science journal devoted to social studies of CAM.

World Health Organization. *Essential Drugs and Medicines Policy. Definitions*. 2003. http://www.who.int/medicines/organization/trm/orgtrmdef.shtml.

*David J. Hess*

# COMPUTERS.

Computers are a major focus of contemporary fascination with technology. Computing tools of one kind or another have a long history, especially if one includes early and primitive devices used for measurement and calculation around the world. Such devices include the tally stick, which has been used since prehistoric times to record quantities (such as the amount of cargo being loaded onto a ship) in the form of carved notches; the abacus; and double-entry bookkeeping. It is important to include such devices in one's analytical ambit even though these superficially bear little resemblance to modern electronic computer systems. Instead of being dazzled by the specificity of the modern, we need to understand computers in terms of the ways in which technological systems in general are promoted, are designed, and occupy a central place in the social consciousness.

A major and remarkable change in the identity of computers has been their transformation from systems of calculation to systems of communication and connection. Computers have metamorphosed from calculators to connectors. Originally built with a view to mechanizing computations far beyond the power of humans, computers have been turned into entities that enable unforeseen capacities for communicating and accessing information. As with many stories in the history of technology, this particular characteristic of computers is the contingent outcome of unanticipated developments rather than a straightforward extrapolation from known features of existing systems. Thus the World Wide Web (WWW), by far the most commonly encountered way to access information and do business over the Internet today, was originally conceived as a way of sharing research results between hard-pressed particle physicists.

## Computer System Failures

Large-scale computer-based systems are notoriously ambitious, costly, and prone to spectacular failure. In recent years the United Kingdom has witnessed major failures of computer systems associated with passport provision, air traffic control, the London Ambulance Service, the government Child Support Agency, and the government Department of Work and Pensions. In 1994 OASIG, a special interest group of the UK government Department of Trade and Industry concerned with the organizational aspects of information technology (IT), estimated that 80 to 90 percent of investments in new technology were failing to meet their objectives in full, with about half of these projects being abandoned completely; and that only 20 to 30 percent of new systems were delivered on time and to budget. OASIG found that it was thought rare for failures to be due solely to technical factors and that human and organizational factors were believed to be the major sources of problems. In particular, many changes were seen as being technology-led, with insufficient consideration of business goals and human and organizational factors; and project management methods and structured case tools were criticized as being too technically oriented and neglecting critical human and organizational issues.

Arguably the problem here is that our understanding of technology stills lacks a throughgoing conception of the mutually implicative nature of technology and society. This means, in particular, that it is always inadequate to speak of the "human and organizational factors" as if they were just another, residual party to the phenomenon, to be also "taken into account" alongside "technical and technological factors," because this form of apparently beneficent analytic pluralism still construes the realm of the human, social, and organizational as essentially distinct from the realm of the scientific and technical. (Worse still is the interpretation that the phenomenon of technical failure comes down to a problem of "culture," since this seems to claim that the origins of the difficulty are not just separate from, but somehow also mystically beyond the understanding of, those involved.)

## Projections and Realities

Computers represent one of two key areas (the other is biotechnology) that have attracted special attention in recent years. In particular, the massive recent growth of Internet technologies and their applications has generated huge amounts of speculation about the extent and significance of the impact these technologies have upon "society." A good indicator of the centrality of computers to these wider concerns about the impacts of technology is the phenomenon of dismally incorrect predictions about their likely future growth and usage. Less common are predictions that underestimate future trends. Famous examples of these latter include the prediction made by Thomas Watson, chairman of IBM, in 1943 that the world would need perhaps five computers and, more recently, the remark attributed to Microsoft chairman Bill Gates in 1981 that 640 kilobytes (the maximum amount of memory available under Microsoft's operating system for the IBM Personal Computer at that time) "ought to be enough for anybody." Perhaps much more common are overoptimistic projections such as the recent dot-com boom (and subsequent crash) resulting from speculation about the significant transformative effects of Internet-based technologies.

As with other technologies, it is important to our understanding of computer-based technological systems that we deploy an analytic perspective that is clear about its presumptions on key matters, such as the "actual" capacity of the system in question.

One particular contentious summary claim about the transformative effects of new electronic technologies is the idea that we are experiencing a transition to a "virtual society." At the core of this vision is the notion that electronic technologies can enable communication via computers (and other electronic devices) that can replace face-to-face interaction. One aspect of this vision was that, with the onset of virtuality, people would spend as much time in an imaginary virtual world as in their real world, if not more. Among other profound consequences, this would herald the much-vaunted "death of distance." Social and psychological interaction, economic transactions, and political relations could proceed unimpeded by the need for physical proximity. More generally, we could anticipate fundamental shifts in how people behave, organize themselves, and interact as a result of the new technologies. We could expect significant changes in the nature and experience of interpersonal relations, communications, social control, participation, inclusion and exclusion, social cohesion, and trust and identity-all of which would have effects on the policy agenda, the nature of commercial and business success, the quality of life, and the future of society.

## The Five Rules of Virtuality

Against the foregoing heady visions of societal transformation through computer systems, most research stresses the unexpected, counterintuitive outcomes of Internet technologies. These outcomes can be expressed in terms of five "Rules of Virtuality," organized around initial anticipations of a move from the real to the virtual.

*Rule 1. The uptake and use of new computer technologies depend crucially on local social context.* Although widespread growth in the uptake and use of computer systems is frequently commented upon, this overall pattern conceals some interesting subvariations. For example, one study discovered that in the midst of the boom, large numbers of teenagers were ceasing to use the Internet. It seems that this cohort had quickly exhausted what they found attractive about the Internet and had returned to other (offline) pursuits. This is not to say, of course, that such users will not "come back" to the Internet, perhaps with the full integration of Internet and mobile telephony, but the trend gives food for thought. The more general point is that uptake and use do not stem straightforwardly from the "actual character" of the technology itself. Instead, the nature and capacity of computer technologies are not given but rather are contingent on their apprehension and use in a specific local context.

*Rule 2. The fears and risks associated with computer technologies are unevenly socially distributed.* Reception and use cannot then stem straightforwardly from the "actual" technical character of computer technologies. Instead, views about new computer systems and the anticipations, concerns, and enthusiasm (or otherwise) are unevenly socially distributed. For example, in research into the impact of surveillance-capable technologies on social relations at work, it was unexpectedly found that respondents accorded a markedly low priority to the question of privacy at work, whereas it is part of the agenda of certain social agencies to emphasize the capacity of computer systems to have a detrimental impact on privacy. The research reveals a variety of counterintuitive usages of computer systems that are not easily classifiable as either conformity or resistance to surveillance-capable technologies.

*Rule 3. Virtual computer technologies supplement rather than substitute for "real" (noncomputer) technologies.* Research shows that new computer systems tend to supplement rather than substitute for existing practices and forms of organization. The "virtual" thus sits alongside the "real," which, in popular imagings of a world transformed by computers, it is usually supposed to supplant. The iconic example is the much vaunted "paperless office." Against expectation, the computerization (and other automation) of office practices did not make obsolete the use of paper. Instead, the new forms of computer-mediated communication sit alongside the continued use of memos, notes, phone calls, faxes, and face-to-face interaction. This gives rise to interesting new forms of interrelationship between the virtual and the real and to the modification of both modes of communication. Thus, for example, a study of the computer mediation of changing patterns of emotional support shows that virtual social life provides an added dimension to a person's social life, not a substitute for it. Computer-mediated, virtual sources of emotional support were used together with other resources and became enmeshed into people's social lives, in some cases transcending the boundaries between "real" and "virtual" life.

*Rule 4. The more virtual, the more real!* This rule is an extension of the previous one. Not only do new, computer-mediated, "virtual" activities sit alongside existing "real" activities, but the introduction and use of new computer technologies can stimulate more of the corresponding "real" activity. Thus, an unexpected finding of one study of telecommuting work is that telecommuters end up physically traveling more than they had done previously. Why did this happen? Computer-mediated communication enabled the telecommuters to make many more contacts with prospective clients rapidly and efficiently, yet the accepted mode of then relating to these new clients was still to meet and deal with them face to face, requiring increased physical travel on the part of the telecommuters. In a similar way there has been speculation that the very large increases in (especially intercontinental) business air travel over recent years owe much to the increased use of e-mail between businesses; that increases in the numbers of museums online has led to an increase in the number of people physically visiting those museums; or that a person's use of computerized means of communication, such as e-mail lists, leads to further or increased use of more traditional communicative media such as telephone conversations and letter writing.

The precise nature and extent of these kinds of "Rule 4" crossover effects are not yet well known. For example, it is unclear whether these are merely transitory effects that apply only during a period of gradual adoption of new computer systems, or whether they constitute more permanent ways of acclimatizing to and inculturating these new systems. The answers clearly have profound implications for weighty questions about the "impact" of computers on society. The kinds of processes suggested by Rule 4 raise key questions about the implications of computer systems for the supposed realignment of work and leisure activities and for new forms of interrelation between social organization, social relations, and citizenship.

*Rule 5. The more global, the more local.* Computer systems and technologies are famously implicated in the much-discussed phenomenon of globalization. Computer technologies are widely regarded as facilitating the rapid movement and spread of symbolic and financial capital. They facilitate the rapid traffic in communication, the instantiation of activities and institutions at diverse widespread locales, and the insinuation of standardized identities and imagery (especially brands) in multiple locations. Globalization is quintessentially about the death of distance. In contrast to these claims, Rule 5 draws attention to research showing how instantiations of global communication and identity depend critically on attention to the local setting. The very effort to escape the local, to promote one's transcendent global or virtual identity, actually depends on specifically local ways of managing and using the computer system. For example, a study found that the implementation of a new system for accessing remote databases in a major retail bank involved considerable local fixes and "workarounds" before its (global) communication potential could be realized. Participants spoke of needing to "make the system at home." As another example, users of mobile phones are familiar with declaring their locality (such as "I'm on the train!") almost as a prerequisite to the beginning of a remote conversation.

## Expecting the Unexpected

The insights into the operation and effects of computer systems suggested by the above five rules have important implications for the problems in the use and deployment of computer systems, mentioned earlier in this article. The most important lesson is to expect the unexpected, which means to anticipate outcomes that run against expectation. In broad order, it is a safe bet that computer systems will not be used to the extent we imagined, nor by the people we anticipated, nor in the ways we expected. Of course, the source of our expectations is important in this. How, and by whom and what, are our expectations about computer systems shaped in the first place? Suffice it say that we need to be aware of the ways in which "hyped" expectations can derive in part from the collusion of different social institutions in promoting the prospect of radical change, such as governments, for whom technical fix solutions and social change and improvement are part of the political rhetoric; commercial firms, who, with advertising agencies, promote their products in the best possible light; and the

media, for whom stories about dramatic effects are a core currency of their operation.

By eschewing summary claims about sweeping effects in favor of attending to the details of their actual contingent social construction and use, we may hope to engender a more measured response to the claims and problems associated with computer systems.

*See also* **Internet, The; Technology in Culture**

**BIBLIOGRAPHY**
Castells, Manuel. *The Internet Galaxy: Reflections on the Internet, Business, and Society.* Oxford: Oxford University Press, 2001.
Hine, Christine M. *Virtual Ethnography.* London: Sage, 2000.
Miller, Daniel, and Don Slater. *The Internet: An Ethnographic Approach.* Oxford: Berg, 2000.
Sellen, Abigail, and Richard Harper. *The Myth of the Paperless Office.* Cambridge, MA: MIT Press, 2001.
Wellman, Barry, and Caroline Haythornthwaite, eds. *The Internet in Everday Life.* Oxford: Blackwell, 2002.
Woolgar, Steve, ed. *Virtual Society? Technology, Cyberbole, Reality.* Oxford: Oxford University Press, 2002. This volume sets forth the Five Rules of Virtuality.
Woolgar, Steve. "Reflexive Internet? The British Experience of New Electronic Technologies." In *The Network Society: A Cross-Cultural Perspective,* edited by Manuel Castells, 125–142. Cheltenham, U.K.: Edward Elgar, 2004.

*Steve Woolgar*

## CONTRACEPTION.

*See* BIRTH CONTROL

## CULTURE AND SCIENCE.

A revolution in how we understand sciences in their cultures has been underway since the 1960s. It has always been clear that political, social, and cultural values and interests can serve as "prison houses" of knowledge, obstructing attempts to understand how nature and social relations actually work. Now we can also see how cultures act as "toolboxes," providing distinctive resources for science projects in different cultural contexts.

Almost a century ago, the British anthropologist Bronislaw Malinowski observed that the scientific impulse flourishes in every culture. Yet for many it has remained difficult to see that the knowledge systems of non-Western cultures, too, make valuable contributions to the human knowledge enterprise. Moreover, it has proved equally difficult, especially for Westerners, to accept the fact that modern Western sciences are as deeply embedded in their cultures as are the knowledge systems of non-Western cultures and that features that look culture-free to Westerners are in fact distinctively Western. Westerners resist the idea that the modern West is also a specific, local culture, or, rather, a distinctive set of such cultures. Thus, there is a gap between, on the one hand, our increasing awareness of the ways that sciences, "ours" and "theirs," are always culturally embedded and, on the other hand, the traditional legacy in the West of representing as the only "real sciences" ones that are culture-free. It turns out that the absence of cultural fingerprints is neither possible nor desirable; it is not a defensible criterion for the empirical adequacy of a knowledge system.

This gap has unfortunate consequences. For one thing, significant questions remain relatively unaddressed. Moreover, in the Net Society, as sociologists name an emerging aspect of global social relations, Western and non-Western projects for the production and management of information are linked in increasingly complex ways. Accounts of knowledge production that ignore non-Western sciences, and that ignore their links to Western science projects, lose touch with the reality of how sciences function in global society today. Worse, attempts by powerful groups to control the production of knowledge in undemocratic ways flourish all the more in a context of ignorance about the relations between scientific and technological change in the West and the non-West (the "North" and the "South," in recent parlance). We need a better map on which to chart scientific change around the globe past and present.

## Enabling and Obstructing Sciences

Let us begin by recollecting how cultural interests and values have advanced the growth of scientific knowledge. There have been many histories of scientific projects of such complex cultures as those of China, India, and the Arabic world. More recently, anti-Eurocentric reports have appeared that detail the wealth of sophisticated scientific knowledge developed in so-called simple societies. Many cultures developed accurate calendars to assist agricultural projects; astronomical knowledge as an aid to navigation over oceans, deserts, and ice floes; engineering knowledge to terrace and irrigate Andean mountainside agriculture, to build seaworthy ships, and to construct aqueducts, palaces, churches, bridges, and dams. Effective medical, surgical, and pharmacological knowledge has had to be produced in every culture because staying healthy is always a challenge. Astonishing mathematical achievements appeared in many non-European cultures. Modern European sciences have always borrowed information, methods, technologies, and conceptual aids (such as Arabic numerals) from other cultures, as those cultures, in turn, have borrowed from Europe.

On the one hand, such information has been used to support a universal claim. Malinowski used such information in this way. Indeed, it is hard to imagine how a group not interested in discovering nature's regularities could survive. On the other hand, what aspects of nature's order have seemed important to study, how they have been examined (the methods used), the theories and assumptions through which observations have been reported, as well as what has motivated particular scientific engagements and how the results of research have been disseminated-all of these concerns have resulted in strikingly different useful understandings of nature's order and of sciences' social relations.

One can begin to make sense of such patterns by noting five well-documented ways in which cultures function as "toolboxes" for knowledge systems, thereby enabling the growth of empirical knowledge in ways useful to a culture's ability to survive and even flourish.

First, cultures settle in or travel through distinctive locations in nature's order. They live on deserts and in rain forests, in the Arctic and the tropics, and on mountaintops and shorelines,  and they survive even while being vulnerable to volcanoes, earthquakes, avalanches, tornadoes, and hurricanes. They travel from Micronesia to New Zealand, across the Sahara, from Genoa to the Caribbean, and from Cape Kennedy to the Moon. So to survive and flourish each must develop systematic knowledge of the aspects of nature's order with which it interacts. Thus, different cultures will produce knowledge about various parts of heterogeneous nature and social relations.

Second, even in the same physical location, cultures can have different interests in the world around them. For example, living on the borders of the Atlantic Ocean, one culture will develop knowledge necessary for successful fishing; another, for using the ocean as a coastal trading route. A third will be interested in using the ocean as a dump for toxic wastes, a fourth in mining oil and gas under the ocean floor, and a fifth in desalinating ocean water for drinking. About the very same location in nature each will develop bodies of systematic knowledge. Notice, also, that in acquiring distinctive patterns of knowledge, we could say that it simultaneously acquires even larger bodies of systematic ignorance. For example, those who intend to use the ocean as a toxic dump will tend to be ignorant of fishing knowledge because the two interests exclude each other. Yet even when this is not the case, focus on one set of questions tends to exclude interest in others that could have been asked.

Third, cultures bring different discursive traditions to the way they see nature, and such traditions help to focus scientific impulses in some directions rather than others. Consider, for example,  the medieval Europeans' representations of nature as being the creation of the Christian God, or as a living organism; then, in the early modern period, as a mechanism; and, recently, representations of the Earth as a spaceship or lifeboat wherein scarce resources must be carefully tended and administered. In each case, different scientific questions have arisen about how humans could survive and flourish in a nature constituted within such distinctive constraints and possibilities. In each case, other potentially useful questions were obscured by the particular cultural lens through which nature was viewed.

Fourth, cultures tend to organize the production of knowledge in the ways they organize other kinds of work. Gaining pharmacological knowledge can occur in prestigious medical universities with their laboratories and clinics, in transnational corporate pharmaceutical laboratories, or by the work of indigenous healers in rainforest cultures. Demographers, obstetricians, and midwives acquire different knowledge about pregnancies, births, and maternal health. As modern Western sciences themselves correctly insist, how one interacts with nature always both enables and limits what one can know about it.

Finally, a culture's position in local and global political and economic relations both enables and limits what it can know about nature's order. Poor people, like other politically marginalized groups, develop distinctive bodies of knowledge about the aspects of nature's order with which they must deal, ones that the dominant groups do not find scientifically interesting. As biologist Anne Fausto-Sterling argues, feminist approaches to biology, especially, have put neglected questions, new answers, and productive new biological paradigms into mainstream biological and medical sciences. Political and economic relations shape where groups live in nature's order (near toxic dumps or in economically marginal environments?), what their interests will be in the world around them, the narratives about nature's order that they will tell, and the ways they organize the knowledge that they need. Many cultures around the world do not have access to modern science's research results. They must learn how to survive without health insurance, in desiccated environments, and with few material resources.

## Looking at Science with Both Eyes Open

Both historians of European sciences and of the sciences of other cultures now agree that it is a mistake to understand culture only as a "prison house of knowledge"; thus, we should additionally understand now that any culture will both enable and obstruct the growth of knowledge. This insight opens up interesting new issues, especially if we look at sciences with our eyes open both to Western and non-Western inquiry. Here are just four of such challenges that await us.

1. How do pieces of one culture's science—a method, a central concept, a representation of some part of nature, a set of observations—travel into another culture's science? What is lost in the journey? For example, what traveled and what was left behind in the recent journey of acupuncture from Asian medicine into modern Western biomedicine?

2. If we cannot reasonably conceptualize any science as culture-free, on what grounds can we justify competing knowledge claims? Representations of nature frequently seem strange to observers from another culture. In what ways do the values of scientific practices override the importance of sciences' representations of nature?

3. When they emerged, modern Western sciences directly embedded democratic values and interests in their procedures. For example, the social status of the observer was to carry no weight when evaluating scientific claims. Scientific methods were to be public processes, not secrets protected from public scrutiny. The results of research were to be available to all, not the possession of privileged groups. Yet today the most powerful sciences seem contained by only powerful interest groups. In what ways can modern Western sciences be directed to engage more productively with the kinds of prodemocratic, social justice values and interests that marked them so distinctively at their origins?

4. As the multicultural science movements demonstrate, the greater participation of groups with different values and interests can advance the growth of knowledge. How should objectivity, rationality, "good method," and the universality of science be reconceptualized to take account of this phenomenon?

## BIBLIOGRAPHY

D'Ambrosio, Ubiratan, and Helaine Selin. *Mathematics across Cultures.* Dordrecht, The Netherlands: Kluwer Academic, 2000.

Fausto-Sterling, Anne. *Myths of Gender: Biological Theories about Women and Men* New York: Basic Books, 1985.

Goonatilake, Susantha. *Aborted Discovery: Science and Creativity in the Third World.* London: Zed Books, 1984.

Harding, Sandra. *Is Science Multicultural? Postcolonialisms, Feminisms, and Epistemologies.* Bloomington, IN: Indiana University Press, 1998.

Hess, David J. *Science and Technology in a Multicultural World: The Cultural Politics of Facts and Artifacts.* New York: Columbia University Press, 1995.

Nader, Laura, ed. *Naked Science: Anthropological Inquiry into Boundaries, Power, and Knowledge.* New York: Routledge, 1999.

Needham, Joseph. *The Grand Titration: Science and Society in East and West.* London: George Allen & Unwin Ltd., 1969. Essays comparing Chinese and European sciences.

Selin, Helaine, ed. *Encyclopedia of the History of Science, Technology, and Medicine in Non-Western Cultures.* Dordrecht, The Netherlands: Kluwer Academic, 1997.

*Sandra Harding*

# D

## DISEASES, EMERGING.

The term *emerging diseases* was adopted during the 1990s by American scientists and public health experts to describe communicable diseases whose incidence in humans has increased during the past twenty years. They may be caused by viruses, bacteria, or other micro-organisms. Some of these, such as the Ebola virus, have recently "emerged" as a result of interspecies transfer, crossing over from animals into the human population. Others, such as the tubercle bacillus, *Mycobacterium tuberculosis,* have "re-emerged" as a result of mutations that have made them resistant to antimicrobial drugs. Most diseases have emerged or re-emerged as a result of technological, environmental, or social changes.

### History of the Term

Concerns over the appearance of new diseases are centuries old, and the phrase "emerging diseases" can be identified in the medical literature at least as far back as the 1960s, when some scientists used it to refer to the chronic ailments that were becoming increasingly prevalent in the industrialized West. However, not until the 1990s did it appear as a coherent concept and the intellectual kernel of a broad public health campaign. The person generally credited with originating it is Rockefeller University virologist Stephen S. Morse, who chaired a 1989 conference, "Emerging Viruses: The Evolution of Viruses and Viral Disease," and published a series of articles on the topic during the early 1990s.

Writing in the shadow of HIV/AIDS, an epidemic that had taken the public health community and the biomedical sciences by surprise and had resisted effective intervention for more than a decade, Morse was concerned with predicting and tracking the appearance of new diseases. He distinguished between "truly new" viruses, resulting from major evolutionary changes via mutation or recombination, and existing viruses that were transferred unchanged or with slight variations into the human population. These latter "emerged" through a two-step process, first crossing over from an animal host into humans and then infecting and spreading within the human population. Morse argued that, although it was almost impossible to predict the evolution of new viruses, anticipating and tracking their emergence was feasible given a proper understanding of the basic mechanisms of interspecies transfer and viral spread. He called these mechanisms the "rules of viral traffic" and called for a program of "viral traffic studies," incorporating virology, molecular genetics, field and evolutionary biology, ecology, and the social sciences.

In 1992 the Institute of Medicine (IOM) of the National Academy of Sciences published a report, *Emerging Infections: Microbial Threats to Health in the United States.* The IOM report was written by a committee of infectious-disease researchers and public health experts, chaired by Nobel laureate and molecular biologist Joshua Lederberg and epidemiologist Robert Shope. Written after decades of budget cuts to American public health institutions and to infectious-disease research in particular, it sought to convince policymakers that infectious diseases presented an immanent threat to American health and security. It identified more than fifty emerging viruses, bacteria, and other microorganisms and discussed a broad range of factors contributing to their emergence and spread.

It also presented specific recommendations for methods of detecting, tracking, and combating emerging infections.

During the 1990s the IOM's vision of emerging diseases became the centerpiece of a broad campaign to attract resources for public health and infectious-disease research. Advocates held numerous conferences, issued reports and publications reiterating the IOM's arguments, developed independent institutes and funding streams, and established a journal dedicated to the topic. Later in the decade, they began stressing the danger of bioterrorism as well, arguing that natural and human-created outbreaks presented similar threats. They were largely successful in attracting funding from American and international public health institutions as well as attention from the mass media.

Emerging diseases were an object of keen interest for the mass media and entertainment industries. Nonfiction books, including best-sellers such as Richard Preston's *The Hot Zone* and Laurie Garrett's *The Coming Plague: Newly Emerging Diseases in a World Out of Balance,* addressed the topic, as did numerous stories on television news programs and in newspapers, magazines, and journals. Movies such as *Outbreak* and Robin Cook's *Virus* presented fictionalized accounts of outbreaks of new infections, and widespread coverage of "killer germs," such as the O157:H7 strain of the common intestinal bacillus *Escherichia coli* and new pathogens such as the Ebola and West Nile viruses, contributed to a sense that Americans were increasingly vulnerable to infectious disease.

## Causes of Emergence

The factors contributing to disease emergence can be divided into six broad categories: changes in human demographics and behavior; modern medical, food-processing, and agricultural technologies; economic "development" and environmental change; increased international transportation and commerce; inadequate public health measures; and microbial adaptation and change. Each of these factors, or several in combination, can alter the "vectors" via which

pathogens are transmitted from animals to humans and within human populations.

Human demographic changes such as population growth, migration, and urbanization contribute to emergence by creating conditions in which pathogens can thrive and be easily transmitted. Large cities, in which people often live in overcrowded conditions and have inadequate access to sanitation and clean water, are especially vulnerable to epidemics of infectious disease. For example, the viruses that cause dengue fever and yellow fever are transmitted to humans by the *Aedes aegypti* mosquito, which lays its eggs in standing water. Recent outbreaks of dengue fever in Latin America and Southeast Asia are attributable in part to the lack of effective mosquito control in densely populated urban centers. As more people migrate to cities, such outbreaks are expected to increase.

Individual human behaviors contribute to disease emergence by altering vectors and by affecting individuals' immune systems. Sharing of needles by intravenous drug users is one of the vectors for HIV, as is sexual activity, which is also the vector for several other diseases including syphilis and hepatitis B and C. Larger political, economic, and social changes, such as increased poverty or war, produce undernourished and marginalized populations, who are more vulnerable to infections.

Modern medical technologies contribute to emergence as well. Widespread and incomplete use of antimicrobial drugs leads to the development of resistance among microorganisms. Modern hospitals, which concentrate the sickest members of the population in a small space and use large quantities of antimicrobials, provide an ideal setting for the transmission of infectious diseases. Invasive surgeries increase potential exposure to infections, and use of drugs that suppress the immune system, following transplant surgery and as part of chemotherapy against cancer, make individuals more vulnerable to infections. *Nosocomial* (hospital-acquired) infections kill at least 20,000 Americans each year. Industrial management techniques in private hospitals, which seek to maximize the ratio between patients and doctors and nurses, also amplifies the likelihood of infection by decreasing time for sanitation and increasing the number of contacts between health care workers and patients.

Recent outbreaks of food- and water-borne illnesses are in part attributable to modern agricultural and food-processing technologies. Large-scale industrial farms, which use huge quantities of antibiotics and often recycle parts of butchered animals as feed, have been implicated in outbreaks of *E. coli*, *Salmonella*, and bovine spongiform encephalopathy (also known as "mad cow disease"). Increased consumption of imported fresh fruits and vegetables has multiplied the opportunities for transmission of food-borne diseases.

Environmental changes associated with economic "development" have a major impact on disease vectors. Outbreaks of Rift Valley fever in Egypt and Mauritania have been associated with the building of dams, which provided standing-water breeding grounds for mosquitoes that carry the virus. Increases in Lyme disease, which is transmitted from mice to humans via deer-borne ticks, have been linked to the effects of deforestation and reforestation on the deer population in the United States.

International travel and commerce have long been associated with disease emergence. The growing quantity of people and materials and the increasing speed of travel have amplified the diseases to be transmitted between different human populations. Local outbreaks of disease now have more opportunities to ignite international or even global epidemics. The biology of microorganisms also contributes to disease emergence through microbial adaptation. Microbes can evolve extremely quickly and can adapt to environmental changes and develop resistance to antimicrobial drugs and pesticides in a relatively short time.

Finally, inadequate public health measures contribute to disease emergence. Inadequate sanitation and water processing have been implicated in outbreaks of cholera and other diseases. Lack of immunizations have lead to increasing rates of preventable childhood illnesses such as measles. Finally, decreasing rates of infectious disease brought on by the mid-century epidemiologic transition in industrialized nations has led to a greater focus on chronic diseases and a sense of complacency about communicable diseases.

## Responding to Emerging Diseases

The IOM report outlined four broad areas of intervention against emerging diseases: surveillance, training and research, vaccine and drug development, and behavioral change. This framework has been followed by numerous other reports and publications and has been adopted by the U.S. Centers for Disease Control and Prevention (CDC), which identified it as the source of "a new consensus."

Epidemiologic surveillance refers to the systematic monitoring and collection of information regarding health and illness. The IOM report recommended the expansion of current efforts and the development of Internet-based, computerized surveillance networks. These would enhance scientists' and public health officials' ability to rapidly identify and track signs of disease emergence: increasing numbers of cases of a disease, changes in the geographic distribution or mortality rate of a disease, or the appearance of a novel *syndrome* (a cluster of symptoms) with no known cause.

The IOM report also encouraged financial support for training programs and research in epidemiology and infectious-disease research, disciplines that had suffered from declining interest and poor funding in the preceding decades. Epidemiologists and other public health experts would be needed to predict, track, and respond to emerging diseases. In addition, scientific experts in virology, immunology, genetics, and other biomedical sciences would be needed to conduct research into the basic mechanisms by which microorganisms evolve, propagate, and transfer between species.

Prevention of emerging diseases depends upon basic public health measures such as vector control, sanitation, and immunization; the latter in turn depends upon the development of effective vaccines, and treatment depends on the development of effective therapeutic drugs. In recent decades, diminishing attention has been paid to the development of pharmaceuticals for infectious diseases, and many basic public health programs have seen their budgets slashed or have been eliminated altogether. Emerging-diseases advocates recommend that these programs be revised. They also urge pharmaceutical companies to devote more research toward development of vaccines and antimicrobial drugs, recommend increased public funding for drug development, and call for changes that would streamline the approval and regulation of new vaccines and drugs.

**BIBLIOGRAPHY**

Drexler, Madeline. *Secret Agents: The Menace of Emerging Infections.* Washington, DC: Joseph Henry Press, 2002. A more recent popular account of emerging diseases.

Centers for Disease Control and Prevention. *Preventing Emerging Infectious Diseases: A Strategy for the 21st Century.* Atlanta: U.S. Department of Health and Human Services, 1998.

*Emerging Infectious Diseases.* http://www.cdc.gov/ncidod/EID/index.htm. Online journal of the United States Centers for Disease Control and Prevention.

Farmer, Paul. *Infections and Inequalities: The Modern Plagues.* Berkeley: University of California Press, 1999. An anthropological account of infectious diseases, including a critical piece on emerging diseases.

Garrett, Laurie. *The Coming Plague: Newly Emerging Diseases in a World Out of Balance.* New York: Farrar, Strauss and Giroux, 1994. A best-selling account of global infectious diseases.

Gladwell, Malcolm. "The Plague Year." *The New Republic,* July 17 & 24 1995, 38–46. A critical review of Preston's and Garrett's books, and media coverage of emerging diseases more generally, by a leading science journalist.

Henig, Robin Marantz. *A Dancing Matrix: How Science Confronts Emerging Viruses.* New York: Vintage Books, 1993. An early popular account of research on emerging diseases.

*Journal of the American Medical Association* 275.3 (January 17, 1996). Special issue of this leading American medical journal, devoted in part to emerging diseases. Thirty-five other journals in twenty-one different countries have focused such special issues on this topic.

Lederberg, Joshua, Robert E. Shope, and Stanley C. Oaks, Jr. *Emerging Infections: Microbial Threats to Health in the United States.* Washington, DC: National Academy Press, 1992. The influential IOM report on emerging diseases.

McGee, Daniel E. "Emerging Meanings: Science, the Media, and Infectious Diseases." *Journal of the American Medical Association* 276.13 (1996): 1095–1097. An excellent critical review of representations of emerging diseases in scientific and popular texts.

McNeill, William H. *Plagues and Peoples.* New York: Doubleday, 1976. A classic historical account of the factors leading to the appearance and spread of infectious diseases.

Morse, Stephen S. "Emerging Viruses: Defining the Rules for Viral Traffic." *Perspectives in Biology and Medicine* 34.3 (1991): 387–409. One of Morse's earliest articles.

Morse, Stephen S. *Emerging Viruses.* New York: Oxford University Press, 1993. A collection of essays from the 1989 conference on emerging viruses.

Preston, Richard. *The Hot Zone.* New York: Random House, 1994. A popular account of the outbreak of Ebola hemorrhagic fever among a group of monkeys in a primate quarantine unit in Reston, Virginia.

Tomes, Nancy. "The Making of a Germ Panic, Then and Now." *American Journal of Public Health* 90.2 (2000): 191–198. A critical piece by a historian comparing concern over emerging diseases in the 1990s with a similar "germ panic" earlier in the century.

*Nicholas B. King*

# DRUGS AND SOCIETY.

At the turn of the twenty-first century, criminal drug users in the United States were robbing pharmacies, breaking in through layers of pharmaceutical infrastructure, to get to the deeply protected and tightly controlled heroin-like high released from the center of the OxyContin pill. Even the innermost shield, the drug's time-release mechanism, was being broken in pursuit of the synthetic opiate it contained. Though we often think of criminal drug use and pharmaceuticals as operating in separate spheres, the laws and crimes surrounding the synthetic opiates remind us that these spheres are never really separable. Not surprisingly, opiates were, a century earlier, the drugs that were central to the creation of the  divide between illegal and legal drugs. The constellation of events that led to the illegalization of nonmedicinal opiates and the regulation of medicine in the United States provide a glimpse into the entangled forces that continue to shape the social interactions of drugs the world over.

Opiates were widely used in the United States in the 1800s; their use was spurred by the development of the medical syringe, pain treatment during the Civil War, the growth of the patent medicine industry, and opium smoking in immigrant labor communities. An addiction tended to be perceived as a medical problem rather than as a social menace, and drug sales were not restricted by law. The international trade of opiates was extremely lucrative, subject to heavy taxation and to international struggles for control.

The first law that made drug use a crime in the United States was enacted in 1875, when San Francisco established an ordinance prohibiting opium smoking. The ordinance was directed at the opium-smoking practices centered in the Chinese immigrant community, a community that had grown with the industrialization of the U.S. West. Following the gold rush and the construction of the transcontinental railroad, Chinese laborers became identified as a threat to the job market, and their strong urban presence challenged American cultural norms in the growing western cities. Opium dens became a focal point of anti-Chinese sentiments. City leaders argued that the dens were centers of moral and cultural corruption, often citing examples of young white women being led by Chinese opium dealers to promiscuity and prostitution.

The San Francisco ordinance, mired in economic, racial, and cultural tensions, marked the rise of an antiopiate movement. The movement was, in turn, sustained by the moralizing currents of the temperance movement. In 1909, the U.S. government passed the 1909 Smoking Exclusion Act, which banned the purchase of opium for smoking while enforcing the legitimacy of the medicinal market. The act did more to protect the medical market than it did to combat illicit opiate use; many users turned to morphine and other opiates, and underground drug cultures flourished in response.

At the same time, the medical profession in the United States was becoming increasingly standardized and regulated, with the American Medical Association playing a central role in standardizing and legitimating medical schools and medical practice. One of the tougher challenges to the authority and control of American doctors had been the patent medicine industry. Doctors had been in competition with medicines that contained opiates, cocaine, or alcohol and that were marketed as remedies for a wide range of diseases and ailments. The manufacturers, without revealing the ingredients of the medicines, acquired a large market through mail-order catalogs and national advertisements, offering medical advice and encouraging consumers to self-medicate rather than pay the expense of physician visits. In 1905, having accumulated political and financial resources, the American Medical Association officially took on the patent medicine industry, establishing the Council on Pharmacy and Chemistry, which evaluated drugs and established standards and advertising policies designed to sever any ties between patent medicines and the medical profession.

The growing power of the American Medical Association coincided with an increasing public trust in scientific and professional expertise and with the federal government's increasing power to reform institutions and regulate industries. Critiques of the patent medicine industry merged with journalistic critiques of large industry, especially the meat-packing industry, and resulted in the Pure Food and Drug Act of 1906. The act did not prohibit the sales of narcotics in medicines but required labels identifying the ingredients of all medicines. The act was later amended to restrict fraudulent claims of effectiveness on labels and in advertisements. Regulating the claims on labels remains the collective focus of pharmaceutical manufacturers, government agencies, and the medical industry, which together form the powerful international pharmaceutical infrastructure.

Efforts to regulate pharmaceuticals converged with efforts to regulate international trade and to control the importation of opiates. Struggles for control over the lucrative international trade of opiates had been going on for a century. But growing international moral and ethical concerns over opiates—fueled largely by Christian missionaries in British- and American-controlled territories—gave the United States the impetus to call for the formation of the International Opium Commission, consisting of thirteen nations at its first conference in Shanghai in 1909 and twelve nations at its second conference at The Hague in 1911. The conferences laid the groundwork for international controls of drug trafficking, established a definition of opiates in need of international control, and declared the use of opiates for anything other than medicinal purposes to be an international problem.

## BIBLIOGRAPHY

Abraham, John. *Science, Politics, and the Pharmaceutical Industry: Controversy and Bias in Drug Regulation.* New York: University College London/St. Martin's Press, 1995.

Courtwright, David T. *Dark Paradise: A History of Opiate Addiction in America.* Cambridge, MA: Harvard University Press, 2001.

Gootenberg, Paul. *Cocaine: Global Histories.* New York: Taylor & Francis Ltd, 1999.

South, Nigel, ed. *Drugs: Cultures, Controls, and Everyday Life.* Thousand Oaks, California: Sage Publications, 1999.

Starr, Paul. *The Social Transformation of American Medicine: The Rise of a Sovereign Profession and the Making of a Vast Industry.* New York: Basic Books, 1982.

*E. P. Shea*

# E

## ECONOMY, THE, AND SCIENCE.

As various combinations of tacit and codified knowledge become the basis for the formation of new firms and the enhancement of existing industry, scientists play a new role in society as entrepreneurs. Intellectual property is becoming as important as financial capital as the basis of future economic growth, as indicated by the inadequacy of traditional models for valuing firms primarily in terms of their tangible assets. The creation of "knowledge management" as an academic discipline and the role of knowledge manager in industry also exemplify this transformation.

As the production of scientific knowledge is transformed into an economic enterprise, the economy is also transformed because it increasingly operates on an epistemological base. An increasing number of firms, generated from the university and/or attracted to the university, are being built within the penumbra of the campus. Governments at the multinational, national, regional, and local levels encourage the growth of science and technology research as a basis for firm formation in regions that formerly lacked these capabilities. Industry is acting like a university in sharing knowledge not only with universities but with fellow firms as they collaborate with each other through strategic alliances in a quasi-academic mode.

### Science and Industry

The creation, dissemination, and utilization of knowledge are increasingly involved in industrial production and governance. Industry traditionally advanced through incremental innovation based on existing products and competencies. As industries such as railroads matured, they rationalized the innovation process by introducing calculative methods derived from science even as they resisted discontinuous innovation that might disrupt routines. Thus, the relationship of the large corporations that had been created by the late nineteenth century to science was contradictory.

Science was initially utilized in industry as a handmaiden to invention, supporting its systematization in Edison's "invention factory" and similar venues. Discontinuous innovations are likely to come out of advanced research, such as that typically carried on at universities but also in some companies. Some firms established well-funded laboratories and produced advances in basic science, and their discoverers—such as Irving Langmuir of General Electric—were awarded the Nobel Prize. Some of these advances were translated into industrial use, such as Bell Laboratories' transistor inventions in solid-state physics, which were used in telephone switching systems.

Nevertheless, many companies find it difficult to take advantage of the innovation potentials that they have created from research. Hindrances include organizational rigidities, accumulation of technical expertise, and tacit knowledge based upon traditional products. Short-term financial comparisons typically show greater profitability from older products than from new ones that might also draw customers away from existing lines. IBM was not among the first to commercialize RISC (reduced instruction set computing) although its laboratories were responsible for early advances. Vacuum tube makers rarely made the transition to transistors.

### Firm Formation

The growing role of the university in the economy is the new factor in innovation, at least on the scale with which it is emerging. On the one hand, only a small fraction of university innovations, based upon dollars of R&D spending, are actually utilized by industry, to date. In addition to enhancing the receptivity of existing firms, an alternative strategy is to form new firms to realize innovations made on campus. Entrepreneurship training programs and *incubator* facilities, directly tied to the research and teaching activities of the university, extend them in the direction of industrial innovation.

Moreover, the incubator itself is often more than not a creature of the university. Quite often an incubator project is a cooperative venture between a university, a local government authority, and a consortium of companies interested in enhancing the local innovation environment. This constellation differs dramatically, in its functions and roles, from the innovation model that existed prior to the emergence of knowledge-based economic development.

## Innovation in Innovation

What is peripheral and what is central to innovation have been transformed in recent years. Knowledge-producing institutions have developed the organizational capacity not only to recombine old ideas and to synthesize and conceive new ones, but also to translate them into use. Innovation has assumed a broader meaning than the development of new products in firms. Instead of focusing only on individual technologies, there is a broader concern with the support structure for innovation, especially encouraging an enhanced relationship between university and industry.

There is a shift from a sole focus on technological innovation to organizational innovation and from what happens in a single organization or institution to what happens across the institutional spheres, not only in producing products but also in developing new organizational formats to promote innovation. Design incubators have been established in Brazil to help existing firms create new products and business units. Entrepreneurial training and incubation have been

linked in Sweden, creating a staged process of firm formation from academia.

## The Enhanced Role of the University and Government

A triple helix of university-industry-government networks creates a seamless web between science and the economy. The triple helix is realized through hybrid organizational mechanisms that promote innovation. Institutional spheres previously had a single purpose and direction. Although they maintain this primary purpose, they also take the role of another, often in association with one or two other institutional spheres. Thus, a university produces scientific results and translates them into new technologies and firms. A government is expected to provide rules and regulations, but it may also provide venture capital.

As institutions take on some of each other's roles, each maintains its primary role and distinct identity. Nevertheless, even traditional tasks are fulfilled in new ways. Thus, government sets requirements for technological innovation by mandating safety and environmental standards. MIT's relations with industry in the early twentieth century resulted in a series of organizational innovations such as the one-fifth rule regulating consultation and the utilization of the legal concept of the contract to regularize and expand hitherto informal ties.

In the United States, the university-government relations constructed from these models initially developed for relations with industry; in other societies, the movement has been in the opposite direction. For example, the growth of university-government relations was intertwined with the formation of a national identity in Germany in the early nineteenth century, with the Humboldtian University integrating teaching and research. Apart from agricultural research, university-government relations in the United States emerged from the World War II military research projects, undertaken at the behest of academic scientists who saw the potential to develop advanced weaponry through the application of science to military problems, on the one hand (radar), and as an ultimate outcome of theoretical advance on the other (the atomic bomb).

## Multifunctional Institutional Spheres

University-government relations transcended the wartime emergency as academics realized that theoretical advance could arise from problem-oriented research as well as vice versa. Heretofore, the institutional spheres were either separate, or, when contiguous, one was directly in control of another. During World War I the military organized R&D and employed scientists under its direction; in World War II the scientists convinced government to supply resources to support projects that they created and directed. After World War II the military also developed agencies to support and direct science to its ends, but it realized that supporting basic research was in its interest as well.

The military realized that practical uses can come with undirected basic research even as the scientists realized through their wartime experience that they were deriving basic research questions when working on practical (military, for example) problems such as radar. Vannevar Bush's (1945) *Endless Frontier* provided a framework for the understanding of the role of science in society induced by the World War II experience and extended into the postwar period by addressing civilian as well as military issues. A new organizational and theoretical framework was required to link government and university.

## The Origins of the Endless Frontier

An individualistic ideology of scientific accomplishment, created in the late nineteenth century, obscured the collaborative and practical nature of science from its founding in the seventeenth century. Before the existence of autonomous disciplines, research with both theoretical and practical implications was the original format of science. Moreover, the original substrate of the medieval university was based on the unity of knowledge and the inherent interdependence of all disciplines.

An ideology was constructed to justify autonomy for science. In the United States, it was feared that the founding of universities by holders of large industrial fortunes could influence the direction of research, including the hiring and firing of professors as well as what topics were acceptable to be studied. The ideology of basic research was strengthened by sociologist Robert K. Merton's positing of the normative structure of science during World War II. This initiative arose from the need to defend science from external attack by proponents of the Nazi racialist ideas of science, as well as from Lysenko's suppression of genetics in the Soviet Union.

The third element in establishing science as an isolated enterprise was the 1945 U.S. government report *Science: The Endless Frontier.* The success of science in supplying practical results during World War II in one sense supplied its own legitimation. But with the end of the war, a rationale was needed that would, on the one hand, accommodate scientists' wishes to control the disbursement of funds with the need to promise practical results in the civilian arena, equivalent to what had been achieved for the military in wartime in order to secure those funds. The report contained an implicit concept of science as a self-regulating mechanism, operating according to a linear progression: put in the money at one end and the results will flow out at the other in 50 years' time.

## The Endless Transition

There is a movement away from an assumption that there is a single starting point of research and an end point of the economy: a linear model that works automatically. To facilitate technology transfer, the United States has established a series of assistance programs and has revised its intellectual property regime in order to reap the benefits of munificent research funding. Other countries, such as Sweden, with high rates of R&D spending and relatively low rates of economic return, are currently undertaking parallel steps. It is also realized that it is necessary to start from the standpoint of problems in society, the reverse linear model, and seeing how knowledge can be used to address them.

An interactive model of innovation is created by joining a linear model proceeding from research with a reverse linear format proceeding from problems in industry and society, with each reinforcing the other. The objective is not only to transfer knowledge or even knowledge embodied in intellectual property and patents, but also—tied together with an entrepreneur—to move it out into new firms. A series of intermediary mechanisms, such as incubator facilities and science parks, actually make linearity work as an assisted linear model of innovation.

This often involves scientists' taking entrepreneurial roles, beyond the laboratory or even research management tasks in a large firm. For example, researchers observed in a California firm, developing solar photovoltaic technology in the late 1970s, did not fit the existing sociological models of academic or industrial scientist. On the one hand, they were not interested in publication; their goal was patents. On the other, they did not appear to suffer the tensions and "role strain" identified among scientists in large corporate laboratories, making the transition from academic self-direction to taking direction.

The solar firm's scientists were not ordinary employees; they were also financial backers of the startup and participated in the decisions affecting the project. Financially independent from stock options received in the semiconductor industry, they were using some of those monies to help fund the solar firm. This led to conceptualization of the photovoltaic researchers as "entrepreneurial scientists" and then to research on their academic counterparts, also involved in startups. The new study took into account the effects of changes in the U.S. patent law on university–industry relations in the early 1980s. Overnight, or at least over the summer, universities that had not previously been involved in technology transfer established offices for this purpose.

Even without the new involvement in patenting on the part of many universities, an entrepreneurial framework was already embedded in the infrastructure of the research university. An entrepreneurial ethos was initially induced as an unintended consequence of the necessity for professors to raise their own research funds in the mid- to late nineteenth century. It was then reinforced during the twentieth century by the experience of managing research groups that had firmlike characteristics. These quasi firms were but a step away from becoming actual firms, given a perceptual change, seeing research findings as potentially commercializable as well as publishable, the availability of a source of startup capital and business opportunities.

## The Capitalization of Knowledge

Technology transfer from academia currently makes a significant contribution to the U.S. economy. Academia–industry connections have, of course, long been present. They were expressed in the mid-nineteenth century in the development of academic chemistry and in the founding of technological universities such as MIT. They can be traced back to seventeenth-century pharmaceutical science and the early development of a pharmaceutical industry in Germany. Nevertheless, until quite recently this modality existed only in an embryonic stage. For example, although Marx perceptively cited Perkins's research as the basis for the dyestuffs industry, he was at a loss to offer additional instances in the mid-nineteenth century.

During the first academic revolution, the theoretical and specialized outlook of the graduate schools was conveyed throughout the academic institutional order. In the course of the second academic revolution, the valorization of research is integrated with scientific discovery, returning science to its original seventeenth-century format prior to the appearance of an ideology of basic research in the mid-nineteenth century. However, scientific and industrial development have been reintegrated at a new level.

## The University's Comparative Advantage

The comparative advantage of the university is its flow of students in comparison to the research institute or corporate laboratory, where the staff is relatively stable except under conditions of rapid expansion. New students bring new ideas to the research process not only by themselves, but also through interaction with older generations of students and professors. The regular entry and departure of persons is important, but also especially important is the reordering of social relations in the laboratory that regularly takes place owing to the rhythm of academic life.

Nor is the consulting firm likely to displace the university as the prime node of the knowledge society. Such firms typically work by recycling a template to satisfy successive clients. Economic pressures rarely allow time for reflection on the accumulated data. Although there are notable exceptions, these typically occur in collaboration with academic partners. Indeed, such firms may be extensions or spin-offs of academic research groups. They operate in a symbiotic relationship, drawing upon students as a labor pool and bring real-world experience into academia for theoretical analysis.

## The Third Mission

The establishment of universities has been always a strategy for latecomers or lagging regions to build up industrial clusters. One element of the latecomer strategy is designing a new entrepreneurial university; another is restructuring existing universities accordingly. As the university takes up a new role in promoting innovation, its educational and research missions are transformed. As the university expands its role in the economy, from a provider of human resources to a generator of economic activity, its relationship to industry and government is enhanced. When academia is a strong institutional sphere, it has the ability to set its direction without the need for strong external boundaries.

The academic development process is from teaching college to research university (the first academic revolution) and then from research university to entrepreneurial university (the second academic revolution). The first phase in the emergence of entrepreneurial science is the development of academic research groups as quasi firms based upon a system of competitive research funding. The second phase is academic participation in the transfer of technology to enterprises, through purpose-built intermediary mechanisms. Finally, academics become directly involved in entrepreneurial activities and firm formation. Research groups and firms exist on a continuum as the boundaries between the university and other institutional spheres overlap.

Business expertise, formerly localized within the university in extra-academic functions, is extended to traditional academic fields. A growing number of U.S. universities are willing to use a small portion of their endowment funds to capitalize new firms, typically in association with other investors. This can be seen as the latest stage in a long-term movement of endowment managers to a more risk-intensive investment strategy, having previously shifted from a concentration on preferred stocks in the prewar period to common stocks in the postwar. An entrepreneurial academic ethos that combines an interest in fundamental discovery with application is reemerging as new and old academic missions persist in tension.

Universities are being founded by economic development projects to further their goals. The science park, which has been a static unit, is now evolving into a dynamic center incorporating firm formation. Instead of the university being the base of the science park, the science park is becoming the base of a new technological university in this emerging model of regional economic development. Thus, the IT University at the Kista Science Park in Stockholm completes the loop by providing intellectual capital. In this instance, knowledge-based economic activity precedes the development of academic work, which then builds upon and remains closely tied to its source. Flow-through of students and research projects is expected to encourage entrepreneurial activity and firm formation on a regular basis.

Rather than being subordinated to either industry or government, the university is emerging as an influential actor and equal partner in an innovation promotion and industrial policy regime. By taking the lead in developing new relationships with government and industry, the university upgraded itself from a secondary to a primary institution in society. The contemporary university brings together people of different backgrounds, thus creating networks that can be activated later. In a society with fixed functions and roles, this potential of the university is less important than in a society where functions and roles must periodically be recreated. This is a fundamental difference between pre- and postmodern eras.

*See also* **Economy, The, and Technology**

**BIBLIOGRAPHY**

Altbach, Philip, Robert Berdahl, and Patricia Gumport, eds. *American Higher Education in the Twenty-first Century.* Baltimore: Johns Hopkins University Press,1999: 63.

"AUTM Releases 2003 Licensing Survey," *Innovation Matters* 1.1 (2003): 8.

Braczyk, Hans Joachim, Philip Cooke, and Martin Heidenreich, eds. *Regional Innovation Systems.* London: University College London Press, 1999.

Etzkowitz, Henry. "Entrepreneurial Scientists and Entrepreneurial Universities in American Academic Science." *Minerva.* 21 (1983):198–233.

Etzkowitz, Henry. "Entrepreneurial Science in the Academy: A Case of the Transformation of Norms." *Social Problems* 36 (1989): 1.

Etzkowitz, Henry. *MIT and the Rise of Entrepreneurial Science.* London: Routledge, 2002.

Etzkowitz, Henry, and Ashley Stevens. "Inching Toward Industrial Policy: The University's Role in Government Initiatives to Assist Small Innovative Companies in the U.S." *Science Studies* 8.2 (1995): 13–31. A discussion of the methodological bases of how the calculations in *Technology Access Report,* 13 No. 2 (2000) translate into economic growth.

Etzkowitz, Henry, Magnus Gulbrandsen, and Janet Levitt. *Public Venture Capital.* New York: Aspen/Kluwer, 2001.

Faulkner, Wendy, and Jacqueline Senker. *Knowledge Frontiers: Public Sector Research and Industrial Innovation in Biotechnology, Engineering Ceramics and Parallel Computing.* Oxford: Oxford University Press, 1995.

Feingold, Mordechai. "Science as a Calling: The Early Modern Dilemma." *Science in Context* 15.1 (2002): 79–120.

Goncharoff, Katherine. "Tiny Technologies, Big Risk." *The Daily Deal,* August 15, 2002, 9.

Graham, Loren, ed. *Science and the Soviet Social Order.* Cambridge, MA: Harvard University Press, 1990.

Hollinger, David."Science as a Weapon in the United States during and after World War II." *ISIS* 86 (1995): 440–454.

IBM Director of Academic Relations, Presentation at Academic-Industry Conference, SUNY, Purchase, New York, 1993.

Jencks, Christopher, and David Reisman. *The Academic Revolution.* New York: Doubleday, 1966.

Kevles, Daniel. *The Physicists: The History of a Scientific Community in Modern America.* New York: Knopf, 1978.

Kornberg, Arthur. *The Golden Helix: Inside Biotech Ventures.* Herndon, VA: University Science Books, 2002.

Kornhauser, William. *Scientists in Industry.* Greenwich CN: Greenwood Publishing Group, 1982.

Lamoreaux, Naomi, Daniel Raff, and Peter Temin. *Learning by Doing: In Markets, Firms and Countries.* Chicago: University of Chicago Press, 1999.

Machlup, Fritz. *The Production and Distribution of Knowledge in the United States.* Princeton, NJ: Princeton University Press, 1962.

Mansfield, Edwin. *Innovation, Technology and the Economy.* London: Edward Elgar Publishing Ltd, 1995.

Merton, Robert. K. *Sociology of Science.* Chicago: University of Chicago Press, 1979.

Milland, Andre. *Edison and the Business of Innovation.* Baltimore: Johns Hopkins University Press, 1990.

Mingers, John. "Can Social Systems Be Autopoetic?" *The Sociological Review* 50.2 (2002): 300–310.

Nelson, Richard. "The Problem of Market Bias in Modern Capitalist Economies." *Industrial and Corporate Change* 11 (2002): 207–244.

Nonaka, Ikujiro, Hirotaka Takeuchi, and Hiro Takeuchi. *The Knowledge Creating Company: How Japanese Companies Create the Dynamics of Innovation.* Oxford: Oxford University Press, 1995.

Oleson, Alexandra, and John Voss, eds. *The Organization of Knowledge in Modern America: 1860–1920.* Baltimore: Johns Hopkins Press, 1977.

Rivette, Kevin, and David Kline. *Rembrandts in the Attic: Unlocking the Hidden Value of Patents.* Boston: Harvard Business School Press, 1999.

Slaughter, Sheila, and Larry Leslie. *Academic Capitalism: Politics, Policies and the Entrepreneurial University*. Baltimore: Johns Hopkins University Press, 1997.

Storr, Richard J. *Harper's University*. Chicago: University of Chicago Press, 1968.

Storr, Richard J. *The Beginnings of Graduate Education in America*. Chicago: University of Chicago Press, 1953.

"U.S. Universities and Research Institutions Ranked by Adjusted Income per 100 Faculty." *Technology Access Report* 13 (2000): 11.

Usselman, Steven. *Regulating Railroad Innovation: Business, Technology and Politics in America: 1840–1920*. Cambridge, U.K.: Cambridge University Press, 2002.

Wang, Jessica. "Merton's Shadow: Perspectives on Science and Democracy." *Historical Studies in the Physical Sciences* 30 (1999): 280–306.

Henry Etzkowitz

# ECONOMY, THE, AND TECHNOLOGY.

From the emergence of the first tools, developed for hunting and for working the land, to the devices that first enabled transoceanic exploration, to the microchips and personal computers of today, technological change has both influenced and been influenced by economic change. Although technology is sometimes posited as an "engine" of economic transformation, recent scholarship in science and technology studies encourages us to think about this relationship in a more dynamic way. Specifically, this scholarship cautions us against inadvertently conceptualizing technology as an autonomous historical force outside of society or envisioning social change as flowing from technological change in a deterministic way. Not only could things have happened otherwise, but one must always consider the social factors at play in shaping how and why given technologies are developed and gain popularity at a given time, while others are not.

This essay will focus on the relationship between technological change and economic change through a consideration of three time periods: the late eighteenth to the late nineteenth century; the early twentieth century to the 1950s; and the 1950s to the present. Technology has been an important dimension of human life since sticks, stones, or bones were adapted to become the first tools; and much has been written about the emergence of, for example, early water technology, windmills, armor, and gunpowder. Rather than attempt a comprehensive cataloguing of technology and the economy over the span of all time, this framework allows us to focus on the transformations that are most salient in understanding issues of today.

## The Late Eighteenth to Late Nineteenth Century: The Industrial Revolution

The most significant modern economic change associated with technological innovation was the industrial revolution, occurring first in Britain and somewhat later in Europe, North America, and elsewhere. Prior to the industrial revolution, work and leisure followed the rhythms of daylight and the shift of the seasons. Work and home life blended together, and both food and consumer goods were produced and consumed in more or less the same place. The invention of the steam engine, patented by James Watt in 1769, and the expansion of the factory system and the capitalist mode of production through the nineteenth century, led to qualitatively different kinds of work and workplaces from what had come before. Through the rise of capitalism, states became largely allied with what we today call the "business system," and laws evolved to support ownership and labor discipline. Because workers had to adapt themselves to the new spaces and temporal regimes of the industrial system, we can even say that this transformation produced different kinds of bodies.

In contrast to working with individual machines in small shops, factories constituted geographically centralized sites of production that housed large numbers of workers. In them, both the workday and the production process came to be increasingly standardized. The spread of industrialism and the factory system also led to an engendering of different kinds of labor, such that by the mid-nineteenth century wage work (production) came to be understood as something that took place outside the space of the home and was associated with men, while the home was viewed as women's "sphere" and the space for reproduction of the workforce—that is, the unpaid care-work of raising families and keeping homes running. Despite the power of this gendered binary in the Western cultural imaginary, it is important to note the extent to which it is a racialized, classed ideal. Single women, working-class women, and almost all (non-slave) women of color were actively engaged in the wage workforce since the beginning of the industrial revolution (although at lower wages than white males). In addition, certain kinds of factory labor were conducted primarily or exclusively by women, such as textile manufacturing, expanded through the invention of the cotton gin in 1793 and the automatic loom in 1894.

Labor requirements of the factory system coupled with improvements in farm technologies led to the replacement of farming by manufacturing as the basis of the economy. In turn, this led to a rise in urbanization, such that by the end of the nineteenth century most Europeans lived in cities, as did most Americans by the 1920s. Trains, first developed in the early nineteenth century and powered by the steam engine, increased the flow between population centers and opened up new territory to economic development, especially in North America. Together with the telegraph and, by the turn of the century, the telephone, trains accelerated the rate at which people, information, and resources could move from place to place, making the world feel smaller and more connected.

These technologies enabled the creation of commodity chains linking resources with ever more distant consumers, and they served to reinforce the importance of cities as loci for trading and processing natural resources gathered from outlying agricultural and wilderness areas. By the end of the nineteenth century larger cities in the industrialized world were being lit up by electric streetlights and served by intraurban electric street rail lines (developed in 1888). Such innovations heightened the productive capac-

ity of the workforce by allowing the workday to stretch into the evening and allowed workers to live farther from their places of employment. They also changed urban culture itself by expanding the range of after-dark activities such as theaters, variety shows, and dance halls, and public transportation greatly increased the number of the people who could partake of such amusements.

Shaped by the Enlightenment and the Scientific Revolution, the industrial revolution was informed by a faith in the capacity of science and technology to function as agents of economic prosperity and positive social change. However, there have always been communities and persons hurt by technological change, and there have been those who have challenged technological development. In England between 1811 and 1816 the Luddites resisted technological change by destroying the machines they saw as threatening their values of social reciprocity, mutual obligation, and craft standards. Though for some the factory system served as an instrument of production, it also reduced craftspeople to deskilled wage workers. A spontaneous movement of resistance (that ultimately failed), the Luddites tried to resist the primitive capitalism that threatened their way of life.

Like slavery and, to some extent, colonialism, industrialization led to the concentration of wealth for those who owned (or helped manage) the means of production but to poverty and economic oppression for those who had to sell their labor power, as observed by Marx and Engels. From the mid- to late nineteenth century the realities of factory work and urban poverty that industrialism produced began to generate sharp criticism, such as through Engels's account of Manchester's "satanic mills" and Jacob Riis's exposé of tenement life in New York City. Discontent within sectored labor forces led to the expansion of trade unions, and demands on the part of these organizations (coupled with middle- and upper-class fear of uprisings, as well as large-scale industrial disasters such as the Triangle Shirtwaist factory fire in 1911) led to certain urban and social reforms.

## The Early Twentieth Century to the 1950s: Expansion and Consolidation

By the early twentieth century, ideas of modernity, standardization, and efficiency had come to be viewed as social goods within the industrialized West. Globally, time itself had become standardized with the institution of time-zones in 1884 and the agreement on a common global calendar and in the mid-1910s. The new value placed on standardization manifested itself in economic life through technologies designed to rationalize both time and the working body. Examples include time clocks, designed to track worker punctuality (or its absence); workplaces designed to maximize visual surveillance of the workforce; and time and motion studies (the practice of timing how long it took workers to complete a given task in order to rationalize and increase production, pioneered by Frederick Taylor in the 1880s). By the 1920s a critique of the industrial workplace as technological dystopia had emerged, as evidenced by such works Fritz Lang's 1926 German film *Metropolis*, in which workers are depicted as subterranean slaves nearly incapable of separating themselves from the machines that drive them.

In many ways the early twentieth century also marked the beginning of the contemporary information economy. The invention and diffusion of the typewriter in the 1860s and 1870s meant that greater volumes of information could be generated more easily at less cost. This lead to the expansion of the financial industries (especially banking and insurance), as well as the expansion of the amount of information that governments could collect and store regarding the citizenry (such as through the census). The development of interior steel-frame architecture in the late nineteenth century allowed office buildings to become larger and more diversified, reinforcing the importance of cities as center-points for information processing and economic decision making.

In addition to the information economy; the invention of the internal combustion engine and, by the 1910s, the assembly-line mode of production led to significant changes in economic patterns, land use patterns, and the environment itself. The assembly line was first developed by Henry Ford to produce low-cost automobiles within reach of the working class, but this technology came to be applied to the manufacture of a wide range of consumer goods. The assembly line streamlined the production process by bringing together all the elements, both human and mechanical, necessary to turn separately manufactured component parts into a finished product. While workers remained at their workstations, the product itself moved from one workstation to the next. Reorganizing the production process in this way resulted in lower costs, higher productivity, and, of course, a highly standardized product. Though developed in the United States, Fordism was enthusiastically adopted in Russia, Europe, and elsewhere and was formative in shaping twentieth-century industrial landscapes worldwide.

By the early twentieth century the United States had become a net exporter of technology, with companies such as General Electric and Westinghouse serving not only as producers of consumer durables but as loci of technological research and development.

Yet as technological innovation expanded, access to technical knowledge was narrowing. The machines people engaged with at work and at home were growing more complex, harder to understand, and more difficult to repair. Social sanctions and structures of racism and sexism constrained technical knowledge even more: the number of doctoral degrees awarded to women in science and engineering fields in the United States decreased between 1920 and 1960, and technological innovation was widely viewed as the preserve of white men.

World Wars I and II provided graphic demonstrations of the effect of technological innovation on modern warfare. Machine guns, employed for the first time in a large-scale conflict in World War I, proved that even death could be mechanized, and nuclear bombs and the gas chambers of Nazi concentration camps, first deployed in World

War II, showed all too graphically the scope and scale of technology's destructive potential. For some, the devastation these devices left in their wake led to a loss of innocence about technology's supposed promise as an instrument of emancipation. The contesting narratives about the capacity of nuclear fission, arguably the most destructive technology known to humanity in 1950, illustrate how views on the appropriate role of high-risk technologies in economic development differed in the postwar world. In nuclear fission some saw the awesome potential of this energy source as an engine to fuel economic development. For others, however, nuclear capacity was marred forever by the destructive power of nuclear bombs, which held the power, for the first time, of global destruction. To many, the possible economic benefits of nuclear energy did not outweigh the risks, and as early as 1949 the Educational Foundation for Nuclear Science was founded and began publishing the *Bulletin of the Atomic Scientists,* a not-for-profit publication devoted to educating the public about the dangers of nuclear technology.

## 1950s to Today: International Development, Ecological Economics and the Rise of the Information Economy

Until the mid-twentieth century most countries outside Europe, Russia, North America, and the former British Empire (as well as many rural places within the aforementioned list) had not experienced industrialization. After the end of World War II this began to change; and places such as China, India, South America, and parts of Africa began to undergo (sometimes rapid) economic development. The postwar era saw the emergence of a variety of multinational organizations founded to encourage international development, such as the World Health Organization and the U.S. Agency for International Development (USAID); and to stabilize global markets, such as the World Bank and International Monetary Fund. Some of the tenets underlying postwar economic development schemes included the idea that the path to prosperity lay in modernization, and, as articulated by economist

Walt Whitman Rostow, that a progression toward advanced capitalism was part of the "natural" development of all societies. Modernization was to be achieved was by replacing traditional forms of architecture, agriculture, transportation, and energy production with larger-scale alternatives that were more highly technical in nature and did not typically acknowledge local history or existing cultural forms.

One of the most graphic ways in which this ideology reshaped the economies of the developing world has been through changes to farming and agricultural technologies. Under the banner of the "Green Revolution," a variety of postwar development schemes sought to encourage farmers to grow just one kind of crop (instead of rotating among several different crops); move to large-scale farms requiring large capital investments; and adopt Western farm technologies such as the tractor, genetically modified seeds, and chemical-based pesticides. Though producing greater amounts of food at lower cost, such techniques created financial reliance on Western companies such as John Deere and Monsanto for the machine parts, seeds, and fertilizer required of large-scale farming. Thus, in some respects postwar international economic development has had the effect of making certain countries in the developing world more, not less, economically dependent on the West. In this way, socio-technical systems that position the global South as a market for Western products and position advanced capitalism as the desired goal for all countries, recall the cultural relations of imperialism.

The economic landscape of the postwar world was also fundamentally shaped by the Cold War and the arms race. Within this system the United States and the Soviet Union justified high levels of state funding for military research and development as well as space programs, both of which are heavily reliant on high-end technological development. For the major players in the Cold War the development and production of weapons and other kinds of military technologies became a central pillar of the economy by the 1960s. In industrialized nations, especially the United States, the postwar era was marked by an expansion of the manufacturing sector and a rise in consumerism. Both manufacturing jobs and the service jobs maintaining the large infrastructural systems that manage transportation, energy, water, and waste were associated with strong unions and high wages in the postwar era. Consumerism was bolstered through television and radio, which, in addition to providing news, serve as a means of delivering product advertising and, in the case of television, programs featuring appropriately consumptive domestic settings. In the home, domestic work itself was becoming increasingly technologically sophisticated by the second half of the twentieth century. At the same time, as a result of their contributions to the wartime economy as emergency workers and in the context of a particularly strong economy, women in the United States, Canada, and elsewhere began working in many kinds of jobs from which they had barred before the war been, including operating heavy machinery and working in highly technical workplaces.

As a result of a spike in the birth rate after World War II (the "baby boom"), the introduction of mass production in the home construction business, and public policy that favored cars and roads over streetcars and other forms of public transportation, postwar North America saw an explosion of suburban tract development zoned for solely residential use and an increasing reliance on automobiles. In addition to intensifying economic activity on the geographic fringe of traditional centers of population, in many places, a car has become a virtual requirement in order not only to get to and from work but also to shop, get to school, or visit entertainment centers—basically, to do anything outside the home. Especially since the oil shock of the early 1970s, commentators from many quarters have expressed concern about the consumptive suburban lifestyle and the centrality of the automobile and fossil fuels in sustaining it. Critiques have ranged from global warming due to the greenhouse effect of increased carbon dioxide emissions, to the press on natural resources resulting from low-density development on a mass scale, to the loss of economic bases in many urban neighborhoods due to migration to the suburbs and beyond. In response, the last quarter of the twentieth century saw the emergence of various efforts to promote technological forms that are less damaging to the environment, such as through sustainable development, "green architecture," "smart growth," and "ecological economics."

Finally, one of the most significant legacies of the military-industrial complex has been the expansion of the information economy, as seen graphically in the rise of Silicon Valley, Microsoft, and the dot-com bubble of the late 1990s. The rise of the information sector has been enabled through telecommunications technologies of the microprocessor and the Internet, the latter of which was originally developed for military use and with government funding. With the capacity for information to flow ever more rapidly across greater distances, the collection, management, and analysis of digitally recorded information has come to occupy a significant role in the economies of industrialized nations in the last third of the twentieth century and into the present day. Especially for those countries that industrialized in the eighteenth and nineteenth centuries, the rise of the information sector has coincided with a decline in manufacturing, as companies began taking advantage of newly relaxed trade regulations in the 1990s by relocating production to sites in the developing world where labor is cheaper. At the same time, over the last thirty years of the twentieth century Asian countries have borrowed Western technologies and transformed their own economies as well as the economies of the West by manufacturing and exporting manufactured goods, from cars to DVD players. As with the industrial revolution, the economic impacts of the information economy have been highly uneven, both between industrialized and nonindustrialized nations and within industrialized nations, creating digital and economic divides between information-rich and information poor.

To conclude, understandings about the relationship between technology and the economy are multiple and contested. They are structured, in part, by social location, one's relationship to the labor market, and amount (or lack) of social advantage. Over the course of history, technological change has sometimes brought with it unforeseen negative consequences for humankind and the environment. In an increasingly globalized world, ownership and wealth are becoming increasingly concentrated, and capital is becoming increasingly free of geographic constraints (or accountability to any given community). Although technological innovation often creates new economic and social opportunities, it is always important to consider what price is being paid, and by whom, in these contexts.

See also **Economy, The, and Science**

**BIBLIOGRAPHY**
Berman, Marshall. *All That Is Solid Melts into Air: The Experience of Modernity.* New York: Penguin Books, 1988.
Bijker, Wiebe, Thomas Hughes, and Trevor Pinch. *The Social Construction of Technological Systems: New Directions in the Sociology and History of Technology.* Cambridge, MA: MIT Press, 1987.
Castells, Manuel. *The Rise of the Network Society.* Malden, MA: Blackwell Publishers, 1996.
Cowan, Ruth Schwartz. *More Work for Mother: The Ironies of Household Technology from the Open Hearth to the Microwave.* New York: Basic Books, 1983.
Cronan, William. *Nature's Metropolis: Chicago and the Great West.* New York: W. W. Norton, 1991.
Fouché, Rayvon. *Black Inventors in the Age of Segregation: Granville T. Woods, Lewis H. Latimer, and Shelby J. Davidson.* Baltimore: Johns Hopkins University Press, 2003.
Graham, Stephen, and Simon Marvin. *Splintering Urbanism: Networked Infrastructures, Technological Mobilities and the Urban Condition.* New York: Routledge, 2001.
Jacobs, Jane. *Dark Age Ahead.* Toronto: Random House, 2004.
Schumacher, Ernst Friedrich. *Small Is Beautiful: A Study of Economics As If People Mattered.* New York: Harper and Row, 1973.
Shiva,Vandana. *Staying Alive: Women, Ecology, and Development.* London: Zed Press, 1988.
Staudenmaier, John M. *Technology's Storytellers: Reweaving the Human Fabric.* Cambridge, MA: MIT Press, 1985.
Winner, Langdon. *The Whale and the Reactor: A Search for Limits in an Age of High Technology.* Chicago: University of Chicago Press, 1986.

*Kate Boyer*

# EDUCATION AND SCIENCE.

Discourses about science and education have changed markedly since the 1950s, becoming more diverse and sophisticated, focusing on the global situation rather than on only Europe and North America, and turning to the education of girls and women. Historical, feminist, antiracist, and anticolonial views of science have found their way into the writings and classrooms of a few countries as well as into United Nations documents and policies. Science education continues to be promoted and funded as central to economic growth, national self-definition, and globalization. But at the same time, liberally minded skeptics note that scientific literacy is low and that this undermines democracy because citizens should be able to participate in scientific decision making. Radical critics suggest that science education itself is inequitable and that science must be taught differently in order to engage the interest of more than just an elite. Still others challenge the notion of global capitalist scientific education, arguing that capitalist science not only threatens indigenous knowledges of the world, particularly the activities of First Nations women, but also the earth itself.

## U.S. Government Views

One dominant voice on matters of science education is the U.S. government in its various branches, whose calls for reform of how science is taught date back to the Cold War, between the United States and the former USSR, in particular, to the USSR's launch of the satellite *Sputnik* in 1957. This event was interpreted as a threat that the United States was falling behind scientifically and that U.S. science education had not kept pace with science education in other countries, which could translate into a lack of competitiveness economically. The solution was to introduce National Science Foundation (NSF) curricula designed by scientists and intended to allow the student to experience "discovery" as scientists do.

In the boom period from 1965 to 1975, which overlapped with the American civil and women's rights movements, science education was declared elitist, and Lyndon Johnson's government poured billions of dollars into education at every level. Social class issues were addressed by a radical increase in the number of students entering postsecondary institutions and studying science.

## Views of Mainstream Philosophers, Historians, and Educationalists of Science

By 1975, U.S. achievement scores in science were found to be lagging behind those of other countries, and both the extremes of "top-down" (theory-driven) and "'bottom-up" (experiment-driven) curricula were faulted. The "rote memory of facts" or "lists of discoveries" in teaching elementary and secondary school science was rejected along with the lecture method in postsecondary science teaching (at least on paper). This "banking" pedagogy, or the idea that the student was an empty vessel into which wisdom could be deposited, was later contrasted by Sheila Tobias with the more contextual and critical approach often used in humanities and social sciences teaching. But "inquiry" or "discovery" learning in the extreme was also found wanting, since students do not have the institutions of science at their disposal in the classroom.

The failures of the early NSF curricula and similar efforts in Britain were blamed by science historians and philosophers on a variety of ills. Perhaps the top-down limited curricula used from about 1750 to 1950 were not appropriate, but neither was the science curriculum designed by scientists in the 1960s. Those who established university programs for the study of the history and philosophy of science described the U.S. science curriculum in those years as boring in its discipline-based theoretical abstractness and generally devoid of context—produced by practitioners who themselves failed to appreciate the cultural context of science. These historians and philosophers then produced a spate of books popularizing stories about the great men of science as well as some documenting that there were also great women in science. Also faulted in poor science achievement was the excessive television watching of American children, claimed to translate into anti-intellectualism and consumer fetishism instead of the pursuit of knowledge. There were also complaints about the underfunding of science programs and equipment, lack of knowledge and/or interest of teachers in science, and lack of training and support for the teaching of science.

Yet more calls for scientific literacy, notably in the American Association for the Advancement of Science (AAAS) document *Science for All Americans*, were followed by yet more curricular reform launched in the 1990s. Project 2061 (the year Halley's comet is due to return) was designed to clear the science curriculum of its crowded lists of discoveries, to interweave the history and philosophy of science into high school teaching, and to make it more interesting to students.

The regular reforms advocated for the teaching of science have led to a pedagogical approach called *social constructivism* that attempts to thread its way between the extremes of top-down and bottom-up teaching in the science classroom. Science is assumed to be a social construction of human beings, and learners are introduced to the "discourse community" of scientists in order to master its content. This approach is now filtering into undergraduate university science teaching, after having influenced American elementary and secondary education.

## International Testing Research on Science Achievement

Eventually, perennial U.S. concern about the ineffectiveness of science education translated into the creation of massive efforts at achievement testing, which verified that most American teenagers stopped taking science as soon as they could and ranked below many other countries in science achievement. The Third International Mathematics and Science Study (or TIMSS) was the largest comparative international study of education ever undertaken. These data, collected in 1994/1995 in 41 countries and again in 1999 (and 2003) by the U.S.-based International Education Association (IEA), have been the focus of a good deal of controversy. Although there is sizable variation by grade level, the results generally show that science and mathematics achievement is highest in Chinese Taipei (Republic of China), Singapore, Korea, and Japan, and is lowest in such countries as the Philippines, Morocco, and South Africa. Despite high achievement, large proportions of students in Japan and Korea report that they do not enjoy their science and mathematics education, finding it stressful. Thus, even for those countries showing above-average performance, press reports following releases of the data have emphasized the negative, and have sparked a redoubling of efforts to standardize and improve science curricula. For example, in some of the high-achievement countries, deficiencies in environmental education are being addressed.

In the United States, where the data show a strong performance by Grade 4 students but a much poorer performance by the end of secondary school, concern led to a large videotape study of science classrooms, the results of which showed that American teachers are more isolated than those in Japan. Both the economic and social indicators published for the countries involved in 1999 and the book *The Impact of TIMSS on the Teaching & Learning of Mathematics and Science* suggest that there is so much variation in cross-cultural educational structures and curricula that these types of comparisons cannot be interpreted without taking context into account. In the case of South Africa, for example, one in eight schools have no sanitation facilities, a quarter have no water within walking distance, and less than half had an adequate supply of textbooks. Further, most TIMSS items do not match the South African curriculum, and English was not the home language of the majority of students taking the test. These structural disparities between rich and poor countries could have been pre-

dicted to be even more pronounced if more countries had met the strict inclusion criteria of the study.

A perusal of TIMSS reports shows only a beginning awareness of the complex societal- and international-level differences in power that underlie the differential performance of various social classes, sexes, nations, and races. Nor do TIMSS researchers question the relevance of what Dawn Gill and Les Levidow label capitalist science literacy for regions outside of North America and Europe. The same can be said for the TIMSS interpretation of American results. Critics have pointed out that any reform would be a challenge in view of the massification and decentralization of U.S. education. With over 50 million students in over 80,000 schools in 50 states, and over 3,000 colleges and universities, the American educational system is quite diverse compared with other systems. This diversity could be seen as a good thing, since U.S. society is more diverse than most other societies, and a multicultural science reflecting a diversity of worldviews could be argued to be inherently suitable for a democracy. Instead, the diversity of curricula has been seen by TIMSS analysts and others as a "cacophony" or "splintered vision" in which there is a lack of concordance between the standardized curriculum materials developed with NSF funding, what is actually taught in the classroom, and what students know.

## Antiracist, Anticlassist, and Feminist Science Education

In addition to the critiques of the American science curriculum made by mainstream historians and philosophers (that it was too abstract and discipline-based), a gender component of poor science achievement was identified: so-called math phobia among American girls. Whereas in the December 1980 issue of *Scientific American* researchers at Johns Hopkins University were still suggesting a naive biological explanation for this finding, feminist investigation has shown that mathematics and science education was *rejected* by a large majority of American girls as irrelevant to their lives. Feminist proposals for reform of science education came in three waves, according to Angela Calabrese Barton: The first, in the 1960s and 1970s, focused on the structural and institutional barriers to recruiting women into science and keeping them interested. Second-wave feminism, prominent in the 1980s, emphasized that women were different from men and proposed a "gender-inclusive science" education that promoted cooperative rather than competitive student-centered learning and multiple ways of knowing and doing science. Often repeated were the comments of the well-known scientist Barbara McClintock that doing science involved developing a "feeling for the organism" rather than the "cold rationality" of traditional scientific methods. In addition, the biology of women and of apes was shown to be distorted by male, middle-class heterosexual values.

Black feminists and others warned about the grouping of all women into one category in second-wave feminist science education, which paved the way for a third wave of feminist critique of science education. Science itself was identified as the problem. All ways of knowing science are *situated*, that is, proposed from a particular race, gender, and class *standpoint*, say contemporary feminist critics of science. There is no "neutral" science or "view from nowhere." What does this mean? For third-wave feminists, it means that teachers and learners need to question their own location in teaching and learning science as well as to actively pursue constructing new sciences and new ways of doing science. For example, feminist scientists such as Lynn Margulis, to mention only one, have reinterpreted evolutionary theory as *symbiotic* rather than competitive.

Proposals for antiracist and anticlassist science education also came in waves. Among the earliest were studies revealing the class-based and race-based biases in educational testing in Britain. Thus, the claim by Arthur Jensen that black–white IQ differences were hereditary was discredited, as was the rationale for tests that tended to stream working class youth into vocational training whereas middle-class youth went on to uni-

versity. Later, British and American reports explicitly documented classroom racism, and NSF reports quantified differences in ethnoracial educational attainment in U.S. science. Whereas some advocated that multicultural sensitivity or content could be "added on" to science curricula unproblematically, antiracists have emphasized, as have third-wave feminists, that specific curricula must be constructed anew. For example, Gill and Levidow argue convincingly that *all* ability testing is based on racist assumptions about ability in that they privilege Eurocentric ways of being and knowing. When the results of such tests are compared, the racist or sexist assumption is made that the individuals with lower scores have a deficit of some kind, either in ability or self-image. These authors provide examples of what an antiracist science education would look like by discussing the nature of race, analyzing grading in science education, and revealing the role of corporate and environmental racism in the thousands of deaths in Bhopal, India, that occurred as a result of a gas leak in 1984.

## Incorporation of Science Education Critique into Mainstream Sources

The realization that there are gender differences in science achievement has also become a central focus for those who make international comparisons of science achievement. For example, TIMSS published gender comparisons across 33 countries in 2000. The 150 pages of tables, accompanied by an abbreviated text, do not find gender differences, overall, at the fourth and eighth grades between the achievement of boys and girls in these countries, but huge differences emerge by the end of high school. The report urges caution in interpreting these results, speculating that boys and girls may not really be exposed to the same curriculum, formally or informally, and that girls may have done better with "free response" questions rather than multiple-choice questions. However, there is little evidence that advanced feminist critique has been studied by those assembling the results of these studies.

Some problems with how these international educational achievement scores are obtained have, however, been emphasized in United Nations documents. For example, the United Nations Educational, Scientific, and Cultural Organization (UNESCO) has produced a book series about various aspects of science education since 1984, with the most recent focusing on women and Africa and reviewing the problems with international and gender comparisons. UNESCO's *World Education Report* (1995) and its *Medium Term Strategy* (1996–2001) focused on science education of women and girls, particularly in Africa, proposing to take the lead in eliminating differences in the education and training opportunities available to men and women (a view that also does not recognize advanced feminist critique). Labeling education as gender-biased, UNESCO's Project 2000+ was established to help realize the *World Declaration of Education for All*, an obvious extension of the 1990 AAAS document *Science for All Americans.*

The most recent UNESCO volume advocates the "scientific advancement" of Africa and describes African girls as "human resources" for industrial development. However, it shows more sophistication than the original AAAS document or the TIMSS publications in assessing gender differences. It has at least partially incorporated second-wave feminist science views, for example, in suggesting that the original IEA 1970/1971 studies comparing girls' and boys' achievement in 19 countries had used multiple-choice questions, which advantaged the boys. It describes substantial differences in how boys and girls solved some of the problems; questioned the role of teachers in discouraging girls, particularly in Africa; and linked these to knowledge and power relations. Feminist literature is cited that describes a "culture of violence" in which girls at school are harassed by boys and by teachers, and rote memorization and corporal punishment, in some countries, are the norm. Solutions posed include providing excursions outside the classroom, providing female role models, some single-sex education, education of teachers about gender issues,

and the reconstruction of alienating science courses as more socially relevant. These documents also deal with each country or region separately, recognizing that social conditions vary so markedly from region to region in the world. But they stop short of making generalizations about global gender or race relations.

## The History of Education as Exclusionary Practice vs. Education as a Route toward New Sciences

Meanwhile, contemporary historians, philosophers, and sociologists have begun to sketch the history of science education as comprising an important set of exclusionary practices. Feminist historians have shown that the whiteness and maleness of natural science can be traced back to the exclusionary practices of European monks who guarded the educational routes into science during the Middle Ages. For example, European women were largely excluded from the development of sciences, such as physics, because they were not allowed to acquire higher education in the monastery schools. Accounts of science generally describe science as born after the Renaissance and Reformations of Europe, a secular vision of "salvation" using man's reason to better society and tame unpredictable Nature. This secular attitude, called *positivism,* unfortunately championed the ideas of only one region of the world as well as only one subset of the societies in which it arose. Touting their success in taming nature, the United States and Europe have been attempting, in what indigenous scholars call *neocolonialism,* to export this "monocultural" way of knowing to the rest of the world. But signs of global warming, environmental degradation, overpopulation, and global poverty have sparked a reexamination of the original project, particularly by environmental educators and advocates for indigenous knowledges. Is Western science all it is cracked up to be? Aren't indigenous farming practices in Africa and India more friendly to the earth? And isn't education about solving societal problems more than just a "transfer" of the ideas of a privileged global elite into the heads of those who are less privileged?

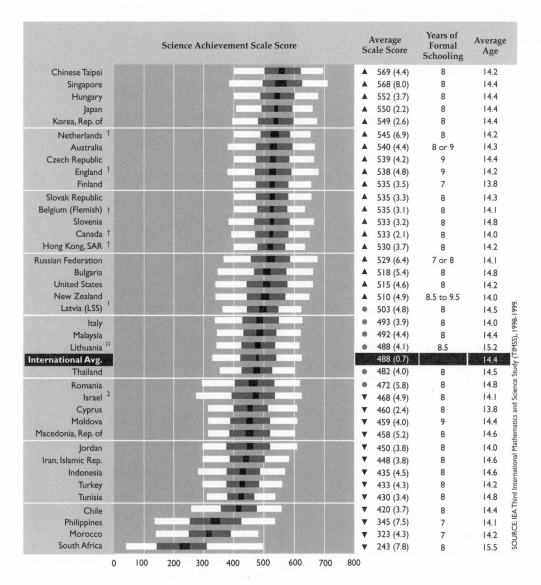

| | Science Achievement Scale Score | | | | | | | | | Average Scale Score | Years of Formal Schooling | Average Age |
|---|---|---|---|---|---|---|---|---|---|---|---|---|
| Chinese Taipei | | | | | | | | | | ▲ 569 (4.4) | 8 | 14.2 |
| Singapore | | | | | | | | | | ▲ 568 (8.0) | 8 | 14.4 |
| Hungary | | | | | | | | | | ▲ 552 (3.7) | 8 | 14.4 |
| Japan | | | | | | | | | | ▲ 550 (2.2) | 8 | 14.4 |
| Korea, Rep. of | | | | | | | | | | ▲ 549 (2.6) | 8 | 14.4 |
| Netherlands † | | | | | | | | | | ▲ 545 (6.9) | 8 | 14.2 |
| Australia | | | | | | | | | | ▲ 540 (4.4) | 8 or 9 | 14.3 |
| Czech Republic | | | | | | | | | | ▲ 539 (4.2) | 9 | 14.4 |
| England † | | | | | | | | | | ▲ 538 (4.8) | 9 | 14.2 |
| Finland | | | | | | | | | | ▲ 535 (3.5) | 7 | 13.8 |
| Slovak Republic | | | | | | | | | | ▲ 535 (3.3) | 8 | 14.3 |
| Belgium (Flemish) † | | | | | | | | | | ▲ 535 (3.1) | 8 | 14.1 |
| Slovenia | | | | | | | | | | ▲ 533 (3.2) | 8 | 14.8 |
| Canada † | | | | | | | | | | ▲ 533 (2.1) | 8 | 14.0 |
| Hong Kong, SAR † | | | | | | | | | | ▲ 530 (3.7) | 8 | 14.2 |
| Russian Federation | | | | | | | | | | ▲ 529 (6.4) | 7 or 8 | 14.1 |
| Bulgaria | | | | | | | | | | ▲ 518 (5.4) | 8 | 14.8 |
| United States | | | | | | | | | | ▲ 515 (4.6) | 8 | 14.2 |
| New Zealand | | | | | | | | | | ▲ 510 (4.9) | 8.5 to 9.5 | 14.0 |
| Latvia (LSS) [1] | | | | | | | | | | ● 503 (4.8) | 8 | 14.5 |
| Italy | | | | | | | | | | ● 493 (3.9) | 8 | 14.0 |
| Malaysia | | | | | | | | | | ● 492 (4.4) | 8 | 14.4 |
| Lithuania [1‡] | | | | | | | | | | ● 488 (4.1) | 8.5 | 15.2 |
| **International Avg.** | | | | | | | | | | 488 (0.7) | | 14.4 |
| Thailand | | | | | | | | | | ● 482 (4.0) | 8 | 14.5 |
| Romania | | | | | | | | | | ● 472 (5.8) | 8 | 14.8 |
| Israel [2] | | | | | | | | | | ▼ 468 (4.9) | 8 | 14.1 |
| Cyprus | | | | | | | | | | ▼ 460 (2.4) | 8 | 13.8 |
| Moldova | | | | | | | | | | ▼ 459 (4.0) | 9 | 14.4 |
| Macedonia, Rep. of | | | | | | | | | | ▼ 458 (5.2) | 8 | 14.6 |
| Jordan | | | | | | | | | | ▼ 450 (3.8) | 8 | 14.0 |
| Iran, Islamic Rep. | | | | | | | | | | ▼ 448 (3.8) | 8 | 14.6 |
| Indonesia | | | | | | | | | | ▼ 435 (4.5) | 8 | 14.6 |
| Turkey | | | | | | | | | | ▼ 433 (4.3) | 8 | 14.2 |
| Tunisia | | | | | | | | | | ▼ 430 (3.4) | 8 | 14.8 |
| Chile | | | | | | | | | | ▼ 420 (3.7) | 8 | 14.4 |
| Philippines | | | | | | | | | | ▼ 345 (7.5) | 7 | 14.1 |
| Morocco | | | | | | | | | | ▼ 323 (4.3) | 7 | 14.2 |
| South Africa | | | | | | | | | | ▼ 243 (7.8) | 8 | 15.5 |

0   100   200   300   400   500   600   700   800

SOURCE: IEA Third International Mathematics and Science Study (TIMSS), 1998-1999.

Percentiles of Performance

5th   25th   75th   95th

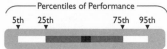

Average and 95% Confidence Interval (±2SE)

▲  Country average significantly higher than international average

●  No statistically significant difference between country average and international average

▼  Country average significantly lower than international average

Significance tests adjusted for multiple comparisons

† Met guidelines for sample participation rates only after replacement schools were included (see Exhibit A.8).

1 National Desired Population does not cover all of International Desired Population (see Exhibit A.5). Because coverage falls below 65%, Latvia is annotated LSS for Latvia–Speaking Schools only.

2 National Defined Population covers less than 90 percent of National Desired Population (see Exhibit A.5).

‡ Lithuania tested the same cohort of student as other countries, but later in 1999, at the beginning of the next school year.

( ) Standard errors appear in parentheses. Because results are rounded to the nearest whole number, some totals may appear inconsistent.

Androcentric white science—say an increasing number of scholars working in the area of science, technology, society, and the environment—has been exalted for too long as the "indigenous knowledge of Europe." Perhaps it was useful prior to the twentieth century for stabilizing humanity's food supply, relationships with other species, and in labor saving so that time could be set aside for leisure. However, in view of the fact that there are now more human beings per area of earth than there are of most other species and in view of the massive air, water, and soil diversion and pollution caused by our dominant science, many ethicists and historians have called for a limit to its growth. It is time for a new beginning, they say, and education can be the way toward that new beginning. This call comes from environmental educators, such as David Orr, who advocates a world that takes its environment seriously, as well as from scholars in indigenous knowledges, such as Gregory Cajete, Vandana Shiva, and George Dei, who have contributed to university-level institutions for the teaching of an indigenous science that respects the earth.

*See also* **Education in the Sciences; Education, Philosophy and Science of**

## BIBLIOGRAPHY

Barton, Angela Calabrese. *Feminist Science Education.* N.Y.: Teacher's College Press, 1998. This book is part of the publisher's Athene series, and it provides a history of feminist influence on science and science education, followed by a personal account of feminist science teaching.

Cajete, Gregory. *Look to the Mountain. An Ecology of Indigenous Education.* Skyland, NC: Kivaki Press, 1994. An introduction to the idea of indigenous science education. A pragmatic extension is included in the author's extension of this work, which is entitled Igniting the Sparkle, an Indigenous Science Education Model.

Dei, George J. Sefa, Budd L. Hall, and Dorothy Goldin Rosenberg, eds. *Indigenous Knowledges in Global Contexts: Multiple Readings of Our World.* Toronto: University of Toronto Press, 2000. A collection of articles by 16 authors describing the nature of indigenous knowledges, indigenous knowledges and the academy, and transforming practices, such as Aboriginal healing and spirituality.

Gill, Dawn, and Les Levidow, eds. *Anti-Racist Science Teaching.* London: Free Association Books, 1987. A collection identifying racism in science, demonstrating antiracist curriculum change and providing provocative examples of antiracist teaching.

Jenkins, Edgar W., ed. *Innovations in Science and Technology Education,* vol. 7, Paris: UNESCO Publishing, 2000. Focusing on Africa and on women, this volume contains many useful country-specific chapters, as well as an introduction that attempts to deal briefly with feminist critiques.

Lynch, Sharon J. *Equity and Science Education Reform.* Mahwah, N.J.: Lawrence Erlbaum Associates, 2000. A useful book for exploring issues of inequity in U.S. science education, presenting detailed descriptions of groups in American society, numerous comparative tables, and discussion of how various dimensions of discrimination intersect and interlock.

Margulis, Lynn. *Symbiotic Planet: A New View of Evolution.* Amherst MA: Basic Books, 1998. A classic in feminist science.

Matthews, Michael R. *Science Teaching. The Role of History and Philosophy of Science.* New York: Routledge, 1994. One of many histories and critiques of American science education that advocates providing context in science teaching. Ignores early feminist and antiracist critique of science education and seems to miss the point of social constructivist pedagogy.

Mullis, Ina V. S., Michael Martin, Edward Fierros, Amie Goldberg, and Stephen Stemler. *Gender Differences in Achievement: IEA's Third International Mathematics and Science Study.* Chestnut Hill MA: Boston College, 2000.

Orr, David W. *Earth in Mind: On Education, Environmentalism, and the Human Prospect.* Washington DC: Island Press, 1994.

Robitaille, David, Albert Beaton, and Tjeerd Plomp. *The Impact of TIMSS on the Teaching and Learning of Mathematics and Science.* Vancouver, BC: Pacific Educational Press, 2000. An attempt by the TIMSS group to place its work within a its country-specific interpretation of how the results of international science education have been received by each country.

Rutherford, F. James, and Andrew Ahlgren. *Science for All Americans.* New York: Oxford University Press, 1990.

Schmidt, William H., Curtis C. McKnight, and Senta A Raizen. *A Splintered Vision: An Investigation of U.S. Science and Mathematics Education.* Dordrecht, The Netherlands: Kluwer Academic Publishers, 1997. A conservative analysis of U.S. science curricula.

Shiva, Vandana *Monocultures of the Mind: Perspectives on Biodiversity and Biotechnology.* London: Zed Books, 1995. One of many books produced by the Indian physicist, philosopher of science, and feminist, arguing that Western science is monocultural and that it is bringing about a dangerous reduction in biological diversity on earth.

Taylor, Peter C., Penny J. Gilmer, and Kenneth Tobin, eds. *Transforming Undergraduate Science Teaching. Social Constructivist Perspectives.* New York: Peter Lang, 2002. Contains four concluding essays that explore spirituality and dreams in science education.

Tobias, Sheila. *They're Not Dumb, They're Different: Stalking the Second Tier. An Occasional Paper on Neglected Problems in Science Education.* Tucson, AZ: Rand Research Corporation, 1990.

Wyer, Mary, Mary Barbercheck, Donna Giesman, Hatice Orun Ozturk, and Marta Wayne, eds. *Women, Science and Technology.* New York, 2001.

*Linda Muzzin*

# EDUCATION IN THE SCIENCES.

Science and technology are often considered central for expanding and sustaining the economies of developed countries as well as for underdeveloped countries seeking to join the ranks of the more industrialized. Science and technology's role for national economic well-being is widely recognized, as will be explained in the following. The need for increasing the pool of future scientists and those in technologically intensive occupations is becoming a more widely recognized part of the increased importance of science and technology to society.

## The Problem of the Pool of Future Scientists

The European Council meeting of March 2000 in Lisbon saw a goal set for the European Union of becoming "the most competitive and dynamic knowledge-based economy in the world, capable of sustainable economic growth with more and better jobs and greater social cohesion." That goal was to be reached in 2010 (Commission of the European Communities, 2002). The European Council meeting in Barcelona in 2002 reviewed progress toward attaining that goal and set a further goal necessary to attain the first—an amount approaching 3 percent of the gross domestic product of the European Union was to be devoted to investment in research and technological development by 2010. The report announcing this goal highlighted the fact that the Lisbon goal was at risk owing to a "large and growing gap" in research and development investment between the European Union and the United States. The report also noted that "[European] Community policies already recognize the importance of having sufficient research scientists and engineers with appropriate qualifications. Research and Development are particularly labor intensive, and available data show that the lack of human resources is a major constraint on the EU's capacity to deliver on the 3% objective" (Commission of the European Communities, p. 8).

The European Commission's High Level Group on Human Resources for Science and Technology (hereafter HLG) presented a preliminary report entitled "Increasing Resources for Science and Technology in Europe" in Brussels, April 2004. It stated that an additional workforce of about 500,000 entering into science, engineering, and technology (SET) careers was needed to reach a target "on the order of 1.2 million additional research personnel" to provide the resources thought needed to reach the 3 percent goal by 2010. It indicated that in 2001 the EU's 15 countries had only about 5.2 full-time equivalents of researchers per 1,000 of workforce and that this was growing annually at a rate of about 2.6 percent. This "researchers per 1,000 in the workforce" number increased rapidly during the 1990s in some EU countries (such as Greece and Portugal, where it increased by more than 100 percent). It was further noted that this figure, unfortunately, compared to 9.14 for Japan and 8.08 for the

United States (High Level Group on Human Resources, 2004). The preliminary report touched off immediate expressions of concern and alarm.

Though focusing mainly on the EU, clearly this is evidence of a growing worldwide demand for people trained in the sciences, engineering, mathematics, and technology. Shortages involved in meeting this demand have created an international "brain chase," in which countries seek to solve their shortages by importing their talent from several Southeast Asian countries. The demand is great, increasing, and involves cross-national competition.

This raises several questions. What size will the pool of trained workers in science, engineering, and technology (SET) be in future years? Will it be adequate for each nation's goals of economic development to be reached and sustained? What can be done to ensure that each country attains adequate scientific workforces? What can be done to ensure that an adequate "pool" exists among all countries so that cross-national competition can help each country attain its needs for a scientific workforce?

This is not a single country's problem. It is a problem of the overall pool from which countries can draw desired workers. From the EU's point of view, the HLG noted, "From a supply perspective, it can be argued that on the present trajectory of those entering SET careers, EU ambitions will not be met. There is need for a . . . change in recruitment into SET at all levels. Dramatically increasing the number of women entering SET careers would go a long way toward helping to solve the problem" (High Level Group on Human Resources, 2004).

The report above cites the United States as comparatively well off in developing an adequate scientific workforce. This is far from true in absolute terms. Science and engineering employment is expected to increase about three times faster than the rates for all U.S. occupations from 2000 to 2010. This translates into some 2 million new jobs—an increase of almost 50 percent. A trend began in 1980, and, over the 20-year period since, science- and engineering-related occupations

outside of academia and excluding technicians and programmers have increased over 150 percent—almost a 5 percent average annual growth.

The issue confronting the United States, the European Communinty, and other countries is whether enough people can be found to fill these jobs without importing them from other countries. Will the national pool (as well as the overall pool if we include importing workers among countries) be sufficient for each country's economic needs? Cross-national competition for SET workers will become steeper as more countries enter the competition and as demand increases for talent-exporting countries.

## Education and the Pool of Future Scientists

Beyond these questions about the scientific workforce and pool adequacy, there are also questions about how to attain such workforce increases. The HLG preliminary report mentioned inducting more women into the scientific workforce. This is one of the many issues that face policy experts in science, technology, and society issues. In addition to gender issues, there are also questions of how such workforce increases are related to educational adequacy. This includes primary, secondary, and even tertiary (university or vocational training) concerns. Herein the discussion turns to educational issues and mainly to those of secondary education.

For instance, the U.S. science and engineering workforce includes many individuals educated in fields not considered part of science or engineering. Many picture a worker in science occupations as an individual with a doctorate. However, even for the United States, where a large proportion of citizens receive some form of tertiary education, only about 6 percent of its science workforce held doctorates in 2000. Similar findings seem certain where even fewer typically pursue tertiary education. This intensifies the educational questions confronting the countries. If one only had to increase doctorates, more attention could focus on tertiary education. However, again using the United States as an example, slightly less than 50 percent of the science workforce have only baccalaureate degrees and 25 percent do not have any college degree.

This stresses lower-secondary (middle school) and secondary school experiences, both in educating the future workforce and in ensuring the development of an adequate workforce (without restrictive and expensive demands for on-the-job training). The lack of tertiary education in the workforce increases the importance of course taking and attainments in science and mathematics in secondary school.

## A Snapshot of Secondary Science Curricula

Of what does secondary science education typically consist? A brief "snapshot" of what is common seems helpful and is provided by curriculum data collected in the Third International Mathematics and Science Study (TIMSS), which was a cross-national comparison of mathematics and science education for almost 50 countries conducted under the auspices of an international consortium with funding from the participating countries. It tested students in mathematics and science at five grade levels in 1995, tested science literacy at the end of secondary school, investigated comparative curricula, and collected extensive data from teachers and students.

Much of the developed countries' lower-secondary curriculum was about chemistry, physics, and the biochemical aspects of life. Comparable U.S. curriculum mostly was about descriptive biology and geology. The United States covered little, if any, physical science, and that little amount focused on descriptive aspects of matter.

Most TIMSS countries' lower-secondary curriculum in mathematics mostly dealt with algebra and geometry, especially for the highest-achieving countries. The U.S. curricula were again typically less demanding— around 80 percent of U.S. students continued the study of elementary arithmetic and very little algebra and geometry. Problems in the United States were exacerbated by its curriculum's lack of coherence, the repetitiveness, and the fragmentation.

## The Prospective Science Workforce: Size and Characteristics

Here the pool of the future science workforce for a country is taken as those students who in the last year of secondary school indicated that they planned to continue their education at some level and that they planned mostly studies related to science, mathematics, or engineering. This means only that they had such plans, not that they would carry them out. The percentages of students at the end of secondary school who planned study in one of nine science-related areas for a representative sample of 19 TIMSS countries are given in Table 1. The distribution of percentages among countries is shown in Table 2. Chemistry, earth science, and physics plans varied little across countries, averaging about one percent each. Biology's variation was larger but only averaged about four percent. Few students indicated interest in pursuing these areas further. Health sciences and engineering had the largest percentages (about 9 percent).

## Evolving Interest

Might it be that this small interest in pursuing science evolved during schooling? Might it continue to evolve past secondary schooling? There is little data to answer these questions. One source—but only for one country, the United States—is the Longitudinal Study of American Youth (LSAY). This study traced the educational history of a nationally representative cohort of U.S. Grade 7 students in 1988 through their 1993 end of secondary schooling.

The LSAY included asking what job students expected to have when they were 40. This was asked each grade from Grade 7 through Grade 12 (the end of secondary school). This allows estimating how the percentages of intended science study changed over the grades—"science," here including all science- and engineering-related interests.

Table 3 presents estimated percentages of students' desire to enter science professions and how they change over the grades. The diagonal entries (upper left to lower right) are percentages of U.S. students at each grade who wanted to become employed in

**Table 1. Percentages of Students Planning Further Study in Nine Different Science Areas for 19 Selected Countries**

| Percentage of Students | Biological Sciences | Chemistry | Computer and Information Science | Earth Science | Engineering | Health Occupations | Health Sciences | Mathematics | Physics | Average |
|---|---|---|---|---|---|---|---|---|---|---|
| United States | 3.4 | 1.6 | 5.2 | 1.2 | 10.8 | 5.5 | 15.7 | 1.8 | 0.5 | 5.1 |
| Australia | 7.6 | 1.4 | 5.8 | 0.9 | 9.3 | 2.5 | 11.8 | 0.8 | 0.6 | 4.5 |
| Canada | 4.3 | 1.3 | 4.8 | 1.9 | 10.0 | 3.6 | 11.1 | 1.8 | 0.7 | 4.4 |
| Switzerland | 4.4 | 1.3 | 5.4 | 1.3 | 8.4 | 7.5 | 5.4 | 1.0 | 1.4 | 4.0 |
| Czech Republic | 5.9 | 0.7 | 14.3 | 2.5 | 7.5 | 2.5 | 3.9 | 2.2 | 0.9 | 4.5 |
| Cyprus | 1.4 | 1.3 | 7.6 | 0.2 | 4.0 | 3.9 | 9.5 | 2.9 | 0.7 | 3.5 |
| Germany | 6.4 | 1.6 | 2.7 | 1.3 | 15.0 | 4.7 | 8.0 | 1.0 | 1.0 | 4.6 |
| Denmark | 3.6 | 2.0 | 6.4 | 0.7 | 6.5 | 1.1 | 9.7 | 1.9 | 0.8 | 3.6 |
| Hungary | 5.7 | 0.8 | 8.9 | 2.0 | 8.0 | 3.9 | 3.8 | 3.1 | 0.6 | 4.1 |
| Iceland | 2.5 | 1.6 | 4.6 | 1.0 | 8.8 | 3.7 | 14.6 | 0.9 | 1.7 | 4.4 |
| Israel | 4.6 | 2.4 | 11.6 | 1.3 | 6.9 | 1.7 | 10.5 | 1.0 | 0.9 | 4.5 |
| Italy | 4.3 | 1.5 | 3.8 | 1.4 | 12.4 | 4.2 | 12.4 | 0.7 | 0.9 | 4.6 |
| Lithuania | 4.0 | 0.6 | 4.1 | 1.0 | 3.5 | 0.9 | 3.0 | 1.4 | 3.7 | 2.5 |
| Netherlands | 2.3 | 1.7 | 2.9 | 0.7 | 14.8 | 13.1 | 4.4 | 0.9 | 0.7 | 4.6 |
| Norway | 3.5 | 1.0 | 3.9 | 0.5 | 11.8 | 4.5 | 13.5 | 1.0 | 1.2 | 4.5 |
| New Zealand | 3.5 | 1.3 | 4.9 | 1.5 | 12.9 | 2.1 | 8.8 | 1.2 | 0.8 | 4.1 |
| Russia | 3.5 | 1.1 | 10.5 | 0.7 | 6.8 | 6.1 | 4.9 | 2.6 | 2.1 | 4.2 |
| Slovenia | 5.7 | 1.8 | 6.0 | 1.5 | 9.1 | 1.9 | 7.8 | 2.8 | 0.6 | 4.1 |
| Sweden | 3.7 | 1.6 | 9.5 | 0.8 | 13.0 | 3.7 | 7.6 | 1.3 | 1.0 | 4.7 |
| Average | 4.2 | 1.4 | 6.5 | 1.2 | 9.4 | 4.1 | 8.8 | 1.6 | 1.1 | 4.2 |

**Table 2. Distributions of Percentages of Students Planning Further Study in 19 Selected Countries by Science Area**

| Area | 5th Percentile | 1st Quartile | Median | 3rd Quartile | 95th Percentile |
|------|---------------|--------------|--------|--------------|-----------------|
| Biological Sciences | 2.2 | 3.5 | 4.0 | 5.2 | 6.5 |
| Chemistry | 0.7 | 1.2 | 1.4 | 1.6 | 2.0 |
| Computer & Information Science | 2.9 | 4.3 | 5.4 | 8.2 | 11.8 |
| Earth Science | 0.5 | 0.8 | 1.2 | 1.4 | 2.0 |
| Engineering | 3.9 | 7.2 | 9.1 | 12.1 | 14.8 |
| Health Occupations | 1.1 | 2.3 | 3.7 | 4.6 | 8.1 |
| Health Sciences | 3.7 | 5.1 | 8.8 | 11.5 | 14.7 |
| Mathematics | 0.8 | 1.0 | 1.3 | 2.0 | 2.9 |
| Physics | 0.6 | 0.7 | 0.9 | 1.1 | 2.2 |

**Table 3. Transitional Percentages Students Retaining Their Interests in Science Careers through U.S. Secondary School (1988 US Grade 7 Cohort through 1993 End of Secondary; Longitudinal Study of American Youth)**

| | | | Grade | | | |
|------|------|------|------|------|------|------|
| Grade | 7th | 8th | 9th | 10th | 11th | 12th |
| 7th | 13.2 | 51.8 | 46.3 | 39.1 | 35.2 | 26.2 |
| 8th | 6.4 | 12.4 | 61.9 | 53.0 | 44.9 | 36.0 |
| 9th | 8.9 | 7.1 | 13.9 | 66.7 | 53.0 | 39.3 |
| 10th | 11.6 | 9.9 | 7.0 | 15.3 | 62.9 | 48.2 |
| 11th | 12.0 | 10.9 | 9.0 | 6.5 | 15.1 | 58.6 |
| 12th | 10.3 | 9.1 | 8.1 | 6.0 | 4.2 | 12.4 |

science. For example, 13.2 percent of seventh graders indicated that intention, 12.4 percent of twelfth graders, and so on.

A fairly stable proportion of approximately 12 to 15 percent of students indicated such interests over the grades. This Grade 12 number can be compared with TIMSS data, but the TIMSS students were asked about intended further study whereas LSAY asked what occupation students intended to follow when age 40. Summing appropriate TIMSS areas gives an estimate for science of 22.7 percent, somewhat higher than LSAY and probably reflecting the difference between planning science study and planning science careers.

Were the nearly 13 percent at Grade 7 who planned careers in science the same students as the 13 percent at Grade 12? No. Only about one-fourth (26.2 percent) of those students had that same desire by Grade 12. About three-fourths of interested Grade 7 students in Grade 12 indicated they were not

likely to enter the science workforce. About 4 percent of Grade 7 students sustained their interest in a science career, but some seventh graders who expressed no interest in a science career later became interested before the end of secondary school. At Grade 8, around 6 percent indicated this change; this increased to more than 10 percent by Grade 12.

The off-diagonal values in Table 3 are percentages of students choosing science careers given other choices in other grades. Below and to the left of the diagonal are estimated student percentages in a grade that chose science careers at a later grade but not at an earlier grade; that is, the grade level given by the row will indicate the grade during which a choice of science career was made, given they did not express such a choice at the grade shown by the column. For example, 6.4 percent of those who did not choose a science career at Grade 7 did so by Grade 8, 8.9 percent did so by Grade 9, and so on.

Entries above and to the right of the diagonal are the estimated student percentages expressing a science career choice at an earlier grade and doing so again later; that is, given that the student expressed a science choice at the grade indicated by the row, the percentage indicates the proportion who continued to do so at the grade given by the column. For example, only 51.8 percent of those students who indicated the choice of science career at Grade 7 continued to do so in Grade 8; only 46.3 percent continued to do so in Grade 9, and so on.

What do these data say about U.S. students' initial and evolving interest in science careers? Only about 13.2 percent of U.S. seventh graders saw themselves in a science job at age 40. Of this 13.2 percent, only about one in four (26.2 percent) sustained the choice in twelfth grade. In contrast, only about 10.3 percent of twelfth graders who expressed no interest in a science career in seventh grade did so by twelfth grade. Overall the data indicate an evolution away from interest in science careers.

These data were only available for the United States. However, it is unlikely that the comparable data will show markedly different patterns in other countries although the absolute levels may be higher. Such data make it clear that one of the things that must be addressed to ensure an adequate pool of future scientists is how to change this evolution away from science careers to an evolution toward science careers. How does a country increase the pool of those who initially choose science careers? How do they keep them interested in such choices and even entice others to join them in such a choice?

Two possible approaches seem reasonable in light of the results presented in Table 3. The proportion of students expressing such an interest is relatively stable over the six years, around 12 to 15 percent. One might seek to increase the twelfth-grade commitment level by either increasing retention of this base interest level or by increasing the percentage of other students not initially interested who become interested at later grades. One could, of course, attempt to do both.

Can these data show us something about these two strategies? Table 4 attempts this. The first column of Table 4 represents the proportion retaining interest in a science career from

Grade 7 through 12 based on the data in Table 3; this might be called the *retention percentage.* The second column represents those who gain a new interest in a science career in a subsequent year and can be called the *gain percentage.* The third column represents the odds in favor of a gain strategy rather than a retain strategy. For example, at twelfth grade, the percentage of students who will express an interest in a science career is about 12.4 percent. Of these, 3.5 percent continued interest from seventh grade and 8.9 percent gained interest later. The odds in favor of a newly gained interest at twelfth grade rather than a retained interest are 2.59. That is, the odds favor a strategy of gain by about 2.5 to 1 (or 5 to 2).

Table 4 suggests that at seventh grade the best "lever" to increase the end-of-secondary-school pool of students interested in a science career is to increase interest over the next few grades. The data show that the odds favor a gain strategy in Grades 8 and 9 but a retain strategy at Grades 10 and 11. No one strategy is best over all of schooling. Although based on U.S. data, the pattern seems likely similar for other countries.

## The Challenge of Developing a Scientific Workforce

The potential pool of people interested in the future study of science and presumably likely to seek a career in a science-related field is relatively small, as seen in these data. The percentage of students who indicate at the end of secondary school that they are planning to study more in the four basic sciences is only around 2 to 3 percent. How well these students are prepared academically in the sciences also varies across the countries.

How these students develop, retain, or lose interest in science careers is particularly important given the small pool and, in the United States in particular, given the relatively weak curriculum surrounding their study. The LSAY data showed that the best strategy to increase the pool likely varies among the grades. For some countries this is more problematic than for others. For instance, in the United States, the grades critical for science recruitment are the very grades in which the U.S. curriculum is

**Table 4. Proportions of Retained Science Career Choices Compared with the Proportions of Gaining New Science Career Choices and the Odds in Favor of Retention Rather Than Gaining New Choices for Seventh Grade through End of Secondary U.S. Students (1988 U.S. Grade 7 Cohort through 1993 End of Secondary; Longitudinal Study of American Youth)**

| Grade | Percentage of Students Retaining Interest from Grade 7 to Grade 12 | Percentage of Students Gaining Interest in Subsequent Years | Odds Favoring Gaining Newly Interested Students over Retaining Previously Interested Students |
|---|---|---|---|
| 7th | 3.458 | 8.940 | 2.59 |
| 8th | 4.464 | 7.972 | 1.79 |
| 9th | 5.463 | 6.974 | 1.28 |
| 10th | 7.355 | 5.084 | 0.69 |
| 11th | 8.849 | 3.566 | 0.40 |

weak by international standards. Just at a point in time when increasing the pool of students interested in a science career is best done by enticing them, U.S. schooling seems to bore students with weak, unchallenging, poorly structured, and highly repetitive curricula.

All countries must face these issues if they see a science-intensive workforce as necessary for economic growth. Key questions must be faced in setting workforce and educational policy. How do students become interested in careers in science? What keeps them interested?

*See also* **Education and Science; Education, Philosophy and Science of**

## BIBLIOGRAPHY

Commission of the European Communities. *More Research for Europe: Toward 3% of GDP.* Communication from the Commission. Brussels, 2002.

Department of Education, National Center for Education Statistics. *Pursuing Excellence: A Study of U.S. Eighth-Grade Mathematics and Science Teaching, Learning, Curriculum, and Achievement in International Context.* NCES 97–198. Washington, DC: Government Printing Office, 1996.

Department of Education, National Center for Education Statistics. *Pursuing Excellence: A Study of U.S. 4th-Grade Mathematics and Science Achievement in International Context.* NCES 97–255. Washington, DC: Government Printing Office, 1997.

Department of Education, National Center for Education Statistics. *Pursuing Excellence: A Study of U.S. Twelfth-Grade Mathematics and Science Achievement in International Context.* NCES 98-049. Washington, DC: Government Printing Office, 1998.

Glanz, James. "Trolling for Brains in International Waters," *New York Times,* April 1, 2001, D-3.

High Level Group on Human Resources for Science and Technology. *Increasing Resources for Science and Technology in Europe.* Brussels, Belgium: European Commission, 2004. Preliminary report presented to the European Commission Conference, "Europe Needs More Scientists," April 2004.

Mullis, I., M. O. Martin, A. E. Beaton, E. J. Gonzalez, D. L. Kelly, and T. A. Smith. *Mathematics and Science Achievement in the Final Year of Secondary School.* Boston: Boston College, 1998.

National Science Board. *Science and Engineering Indicators 2002.* Arlington, VA: National Science Foundation, 2002. http://www.nsf.gov/sbe/srs/seind02/.

Pincock, Stephen. "EU Seeking Scientists." *The Scientist.* Daily News, April 1, 2004. http://www.biomedcentral.com/news/20040401/03.

Rowsey, Robert E. "The Effects of Teachers and Schooling on the Vocational Choice of University Research Scientists." *School Science and Mathematics* 1 (1997): 20–26.

Schmidt, W. H. "Too Little Too Late: American High Schools in an International Context." In *Brookings Papers on Education Policy,* edited by Diane Ravitch, 253–307). Washington, DC.: Brookings Institution Press, 2003.

Schmidt, W. H., R. Houang, and L. Cogan. "A Coherent Curriculum: The Case of Mathematics." *American Educator* (Summer 2002):, 10–26, 47–48.

Schmidt, W. H., C. C. McKnight, R. T. Houang, H. C. Wang, D. E Wiley, L. S. Cogan, and and R. G. Wolfe. *Why Schools Matter: A Cross-National Comparison of Curriculum and Learning.* San Francisco: Jossey-Bass, 2001.

Schmidt, W. H., C. C. McKnight, and S. A. Raizen. *A Splintered Vision: An Investigation of U.S. Science and Mathematics Education.* Dordrecht, The Netherlands: Kluwer Academic Publishers, 1997.

Schmidt, W. H., C. C. McKnight, G. A. Valverde, R. T. Houang, and D. E. Wiley. *Many Visions, Many Aims: A Cross-National Investigation of Curricular Intentions in School Mathematics.* Dordrecht, The Netherlands: Kluwer Academic Publishers, 1997.

Schmidt, W. H., S. A. Raizen, E. D. Britton, L. J. Bianchi and R. G. Wolfe. *Many Visions, Many Aims: A Cross-National Investigation of Curricular Intentions in School Science.* Dordrecht, The Netherlands: Kluwer Academic Publishers, 1997.

Schmidt, W. H., H. A. Wang, C. C. McKnight "Curriculum Coherence: An Examination of US Mathematics and Science Content Standards from an International Perspective." *Journal of Curriculum Studies* (in press).

U.S. Department of Education. *The Digest of Education Statistics 2000.* NCES 2001- 034. Washington, DC: U.S. Department of Education, Office of Educational Research and Improvement. http://www.nces.ed.gov/pubs2001/digest.

*William H. Schmidt and Richard T. Houang*

# EDUCATION, PHILOSOPHY AND SCIENCE OF.

A philosophy of education is a set of ideas about the nature and purposes of education. At its heart is a set of aims, including a view of what is valuable to know and learn. It is also based on ideas about what knowledge is, which aspects are most important, and how it is best taught and learned.

All major societies and civilizations have had systems of teaching and learning and educational philosophies since the days of Mesopotamia, where the earliest writing and mathematics have thus far been discovered. The aims in those days were to produce a small number of scribes who could record and calculate taxes and trade and write down the glorious achievements of the rulers.

Over 2,000 years ago, Plato developed a philosophy of education for the free citizens in Athens. This was intended to develop the abstract powers of the mind through the study of language, including grammar, rhetoric, and logic, and mathematics, including astronomy and music. Only pure subjects were chosen for study, as practical knowledge and manual crafts were considered degrading and suitable only for the slave classes. Here we have the origins of both the three R's (reading, 'riting, and 'rithmetic) and the idea that different classes of people need different types of education.

Private schools and academies for the sons of the rich (and a few daughters, too) have been available for a number of centuries, as well as some church schools. However, it has been only in the past hundred years or so that general elementary education became available to all in developed Western countries.

Central to any educational philosophy is a set of educational aims. These encompass more than the end product to which teaching is directed, for they also express the underlying values that education is intended to embody and communicate. But educational aims cannot considered in isolation from their social context. They represent the values, interests, and ideologies of individuals and groups. These interests and ideologies are often in conflict. Social groups not only differ in their aims and ideas about attaining them, but also in the different groups of students to which the aims apply.

## Social Groups and Their Educational Aims and Philosophies

In 1961, Raymond Williams published a historical analysis of educational philosophies and identified three social interest groups who emerged in the late nineteenth century. He argues that these groups, with widely contrasting ideologies and educational philosophies, exerted a powerful influence on the foundations of modern education systems and have a continued impact. This analysis has been updated, and the different interest groups identified in modern education have been expanded to five, based on sociological, historical, psychological, and philosophical data and theories.

1. The *Industrial Trainers* are made up of managers, skilled and white-collar workers, and religious fundamentalists at all levels with a "bourgeois" ideology, who prioritize the utilitarian and disciplinary aspects of education. The aims of the Industrial Trainers are utilitarian, concerned with the training of a suitable workforce in basic skills, but there is also a powerful social training dimension to their aims, concerned with "teaching the required social character—habits of regularity, 'self-discipline', obedience, and trained effort" (Williams, p. 161–162). This can be traced back to the Protestant work ethic and associated puritanical values. Industrial Trainers are the proponents of "back-to-basics," drill, rote learning, and strictly enforced authoritarian discipline. Their aims are to develop mastery of basic numeracy and literacy skills, and fact-based knowledge, as well as training in obedience and subservience inculcated through repeated drilling and strict discipline.

2. The *Technological Pragmatists* represent a pragmatic and utilitarian tradition in education, valuing practical skills, technological progress, and the certification of learning, all as a means of furthering economic progress in society, without the backward-looking authoritarianism of the Industrial Trainers. This group represents the interests of industry, business, and the public sector and is concerned with the acquisition and development of a broad range of knowledge and personal skills, those that are useful in modern employment. This group sees social progress as following from the advance of industrialization and technology, and expects more of education than the mastery of basic skills. Their aims for education include mastery of aspects of computers, science and technology, and communication and problem-solving skills. Their emphasis is on utilitarian knowledge learned through practical projects and applications.

3. The *Old Humanists* represent the conservative professional and cultured classes. They value the traditional humanistic studies, and their intended product—the cultured, well-educated person—for its own sake. Their educational aim is "liberal education," that is, the transmission and conservation of the cultural heritage, which is made up of pure (rather than applied) knowledge in traditional forms, including the human, social, and physical sciences. This group endorses the "canon" of great English and American traditional literature and the values of the Ivy League universities in the United States. They value knowledge for its own sake, emphasizing its purity and aesthetics as opposed to utilitarian knowledge, which is viewed as inferior. Thus they hark back to Plato's ideals.

4. The *Progressive Educators* are liberal professionals who value education for the sake of the child, to allow the individual to flower, develop, and reach his or her full potential. They are the modern representatives of the progressive tradition, whose proponents have included Rousseau; Montessori; and, in part, Dewey. Progressive Educators are associated with child-centered teaching, emphasizing active learning and creativity. This group endorses discovery learning, problem solving, and protecting the child from failure. This group is representative of the reform movement in modern American education, such as the National Council of Teachers of Mathematics's Mathematical Standards movement, emphasizing mathematical problem solving, the individual construction of understanding, and positive attitudes to learning mathematics.

5. The *Public Educators* represent a more radical reforming and activist group, concerned with democracy, equity, and social justice. Their aim is to empower all sectors of society, including women, peoples of color, and other minority groups to participate in the democratic institutions of society and to share more fully in the prosperity and opportunities of modern industrial society. The Public Educators see all education as political, as a means to empower students. Their aim is to empower individuals to take con-

trol of their lives, to fully participate in economic life and in the democratic decision making of modern society and, ultimately, to facilitate social change to a more just society. Public Educators tend to be political activists or committed teachers and professionals working with minority students and adult returners to college. They typically view knowledge as socially constructed.

## A Model of Educational Ideologies

The term *ideology* is sometimes understood to have negative connotations, including fanatical or irrational theories of society. Karl Marx first used the term for "false consciousness," in which a thinker "imagines false or apparent motives" (Meighan, p.174). But he later used it in the sociological sense intended in which an ideology is an overall, value-laden philosophy or world-view, comprising a broad interlocking system of ideas and beliefs. In this section, ideologies are understood to be competing belief systems, combining both epistemological and moral value positions. No pejorative meaning is intended in the use of the term. Meighan describes educational ideologies as comprising sets of beliefs operating in various contexts and at various levels with several layers of meaning. The model outlined here is based on these notions, and reflects the degree of complexity involved at two levels. At the primary level are the deeper elements of the ideology, comprising the underlying epistemological and ethical system of the ideology.

At the secondary level are derived elements pertaining to education.

The example detailed in Table 1 concerns ideologies of mathematics education. However, this model can be easily adapted to other social and physical science disciplines in education.

The primary level of the ideology of mathematics education includes the overall epistemological and ethical position, a philosophy of mathematics (a specialized part of the epistemology) and a set of values. If applied to another discipline, only the specialized part of the epistemology, the philosophy of the discipline itself, needs to be changed. Further elements are a theory of the child, which is the special part of a theory of the person pertaining to education, and a theory of society. The last component is a set of aims of education.

At the secondary level are derived elements pertaining to mathematics education. First, there is the theory of school mathematics. Mathematical knowledge is central to the whole process of mathematics education, so a theory of school mathematical knowledge is needed, in addition to a philosophy of mathematics. Second are the aims of mathematics education. Third, the means of attaining these aims are also required, and these include a theory of teaching of mathematics, including the role of the teacher. The teaching of mathematics has been transformed during its history, by developments in the resources for teaching and learning, including texts, electronic calculators, and microcom-

**Table I. The Mathematics Education Ideologies of Five Interest Groups**

| | INTEREST GROUP | | | | |
|---|---|---|---|---|---|
| TENETS | INDUSTRIAL TRAINER | TECHNO-LOGICAL PRAGMATIST | OLD HUMANIST | PROGRESSIVE EDUCATOR | PUBLIC EDUCATOR |
| Political ideology | Neoconservative fundamentalists | Meritocratic and conservative | Traditional conservative | Liberal | Political activist and radical |
| Set of values | Authoritarian values of choice, effort, self-help, work, moral weakness | Utilitarian, pragmatism, expediency, wealth creation, technological development | "Blind" Justice, objectivity, rule-centred, hierarchical, paternalistic, "Classical" view | Person-centered, caring, empathy, human values, nurturing, maternalistic, "Romantic" | Equity, social justice and awareness, engagement and critical citizenship view |

*(Continued)*

**Table 1. The Mathematics Education Ideologies of Five Interest Groups** *(Continued)*

| TENETS | INDUSTRIAL TRAINER | TECHNO-LOGICAL PRAGMATIST | OLD HUMANIST | PROGRESSIVE EDUCATOR | PUBLIC EDUCATOR |
|---|---|---|---|---|---|
| | INTEREST GROUP | | | | |
| Theory of society | Rigid hierarchy, market-place | Meritocratic hierarchy | Elitist, class stratified | Soft hierarchy, state support for needy | Inequitable hierarchy, needing reform |
| Theory of the child | Elementary school tradition: child "fallen angel" and "empty vessel" | Child "empty vessel" and "blunt tool," future worker or manager | Paternalistic discipline builds, character, culture tames | Child-centered progressive: view child "growing flower" and "innocent and savage" | Social Conditions view: "clay molded by environment" "sleeping giant" |
| View of school math | Set of truths and rules | Unquestioned body of useful knowledge | Body of structured pure knowledge | Process view: personalized mathematics | Socially constructed |
| Math aims | "Back-to-basics": numeracy and social training in obedience | Useful mathematics to appropriate level and certification (industry-centered) | Transmit body of mathematical knowledge (mathematics-centered) | Creativity, self-realization through mathematics (child-centered) | Critical awareness and democratic citizenship via mathematics |
| Theory of learning | Hard work, effort, practice, rote | Skill acquisition, practical experience | Understanding and application | Activity, play, problem solving, exploration | Questioning, decision making, negotiation |
| Theory of teaching math | Authoritarian transmission, drill, no 'frills' | Skill instructor motivate through work-relevance | Explain, motivate, pass on structure | Facilitate growth experiences, problem solving positive attitudes | Discussion, conflict, questioning of content and pedagogy |
| Theory of resources | Chalk and talk only, anti-calculator | Hands-on and microcomputers | Visual aids to motivate and communicate | Rich environment to explore | Socially relevant, authentic data |
| Theory of assessment in math | External testing of simple basics, avoid cheating | External tests and certification, skill profiling | External examinations based on hierarchy | Teacher led assessment, failure avoidance | Various modes. use of social issues and content |
| Theory of math ability | Fixed and inherited, realized by effort | Inherited ability | Inherited cast of mind | Varies, but needs cherishing | Cultural product: not fixed |
| Theory of social diversity | Differentiated schooling by class, crypto-racist, mono-culturalist | Vary curriculum by future occupations | Vary curriculum by ability only (math neutral) | Humanize neutral maths for all: use local culture | Accommodation of social and cultural diversity a necessity |

puters, and a theory of resources for mathematics education is included. Teaching is instrumental to learning, the intended outcome of mathematics education, so a theory of learning mathematics, including the role of the learner, is needed. This relates to both epistemological assumptions and moral and ethical views concerning the individual's responsibilities, and hence derives in part from theories of the child and society.

A theory of the assessment of mathematics learning is included because of its vital social role. Further important elements include a theory of intelligence and ability (and its fluidity or fixity) because concepts of ability are of special importance in mathematics, and last, a theory of social diversity in mathematics education. This encompasses ideas about the impact of race, ethnicity, socioeconomic status, gender, and so forth, on mathematical attainment.

The model of educational ideology is applied to the interest groups outlined in Table 1. Although very compressed, Table 1 indicates interesting connections within and across ideological groupings. For example, the Industrial Trainer neoconservative fundamentalists are shown as having authoritarian values prioritizing choice, effort, self-help, and work, and they attribute moral weakness to those unable, for whatever reason, to manifest these virtues. Their theory of the child is based on the punitive elementary school tradition, which sees children as "fallen angels" and "empty vessels," i.e., susceptible to sin and wrongdoing unless strictly disciplined and guided, and who will only fill up with the right values and knowledge as a consequence of hard work, effort, practice, and rote learning. These ideas are reflected in the mathematical aims focusing on "back-to-basics" numeracy and social training obedience. The theory of teaching mathematics is one of authoritarian transmission, using drill, and no "frills," and is also anticalculator, because it is perceived to allow students to skip the mental effort or work of calculations. By the same token, it opposes the Progressive Educator emphasis on play, discovery, and child-centeredness, which is directly antithetical to its emphasis or work and discipline. The theory of assessment in mathematics emphasises the external testing of simple basic skills, and precautions to avoid children's natural tendency to cheating. Mathematical ability is understood to be fixed by inheritance, but can only be realized by student efforts to memorize facts and master skills. However, superior talents will be capable of mastering more than the basics.

## Ideologies in the British National Curriculum

In the late 1980s, the ideology of the Industrial Trainers played a key role in the development of the British National Curriculum under the Conservative government led by Margaret Thatcher, who was identifiably a member of this group. This development was contested by different groups and resulted from a series of alliances and compromises, in which this ideology ultimately dominated.

The legally binding National Curriculum in mathematics embodies the aims of three of the groups. It is a course of study of increasing abstraction and complexity, providing a route for future mathematicians, meeting Old Humanist aims. It is a technologically oriented but assessment-driven curriculum, meeting Technological Pragmatist aims. It is one component of a marketplace approach to schooling, and an assessment-driven hierarchical curriculum with traces of the Progressive Educators expunged, meeting (some) Industrial Trainers aims. The overall range of mathematical content specified exceeds the basic skills deemed necessary by the Industrial Trainers. However, the underlying assessment framework ensures that below average–attaining students will study little more than the basics, in keeping with these aims. Overall, the outcome was largely one of victory for the Industrial Trainers (and their allies), despite the progressive climate of professional opinion prevailing among mathematics education professionals. Nowhere in the end result or the complex processes of negotiation and compromise leading up to it were the aims or ideology of the Public Educators seen in evidence.

The development of the British National Curriculum in Science followed a similar course, although the ideologies differ in detail. Practical work is traditional in science, and without the connotations of progressive child-centeredness in mathematics. Thus, the Industrial Trainers position in science does not view practical activity by students as anathema, as it does in mathematics, where it signifies child-centeredness and permissiveness. Nor does it see "social training in obedience" as important in science as it is in mathematics and the three R's in general. The Old Humanist position in science does not oppose practical applications of science and technology. The pure-applied dichotomy does not have the same force as in mathematics, where it has strong value connotations, and was traditionally associated with class distinctions ("pure" associated with the leisured and professional classes, "applied" with the working and mercantile classes). Likewise, appreciating the social context of science is not seen as having the same ideological significance as it does in mathematics, where it may be seen to represent a radical, social constructivist, or leftist position. For although the social and value neutrality of mathematics is assumed by all positions other than that of the Public Educator, this is not the case with science. The National Curriculum Council stated that science is a human construction and refers to its origins in other cultures, as well as to the moral and ethical issues its applications raise.

In contrat to mathematics, the committee charged by the government with the development of the National Science Curriculum had a major Public Educator strand in their philosophy, as well as Progressive and Old Humanist strands. In their final report, they proposed 20% of the science curriculum was to be devoted to social, historical, philosophical, and ethical dimensions of science; 40% to process aspects of science; and only 40% to traditional knowledge and understanding. These can be seen as roughly corresponding to the Public Educator aims, a compromise centering on the Progressive Educator aims, and a coalition of the other three sets of aims combined, respectively.

The government's response was to require that 70% of the curriculum be devoted to Knowledge and Understanding, and 30% be devoted to Exploration of Science. The latter is practical, largely skills-based scientific work, a traditional part of the science curriculum, with the more Progressive Educator-oriented components on investigation and communication proposed in the final report stripped away. This represents a shift toward the aims of the three more conservative groups (Old Humanist, Technological Pragmatist, and Industrial Trainer). The component corresponding to the Public Educator dimension was completely cut away.

Throughout the development of the British National Curriculum in all of the social and physical science areas, the Public Educator aims and ideology have been expunged and the only traces of the Progressive Educator aims and ideology remaining are those that are a compromise with the Technological Pragmatists. Practical work is only included where it can be justified as being centered on employment and industry, rather than on the child. The resulting framework is a centrally imposed hierarchical model of knowledge that is assessment driven, emphasizing basic skills, with mandatory, externally imposed tests at ages 7, 11, 14, and 16 in all state schools. This represents a victory for the three conservative group ideologies, with overall dominance for the Industrial Trainers. Even with the change of government to Labour in the late 1990s, this pattern remains, with increased emphasis on utilitarian knowledge and skills (contra the Old Humanist ideology) and testing. The only slight shift is the emphasis on improving achievement of minority, low socioeconomic status, and low attaining students, suggesting the increased impact of the Technological Pragmatists, rather than the Public Educator ideology.

## The U.S. "Math Wars" and "Science Wars"

Opposing ideologies and the groups that support them have been clearly evident in other recent controversies concerning knowledge and education. The "science wars" have concerned defenders of the traditional positivistic scientific worldview and social commentators on science. This can be read as Old Humanist versus Public Educator ideologies, as the former adherents are deeply committed to the objectivity of scientific knowledge or at least of the scientific method, whereas the latter regard science as a social construction, inseparable from the real world of political struggles between competing groups. Such controversies have also been widespread in the world of education.

An ideological struggle has been going on in the teaching of English between the supporters of phonics versus a whole-language approach. The most vociferous proponents of phonics are typically Industrial Trainers with their "back-to-basics" philosophy, supported by Old Humanists. They regard knowledge of English as based on atomic facts about spelling and grammar, and only when this has been mastered can fluency develop. The groups supporting the whole-language approach are typically Progressive Educators whose philosophy emphasises children's experience as a whole and thus favors the whole-language approach. There is also strong support from Public Educators, who endorse social theories of learning that foreground apprenticeship and view the learner as immersed in social practice, which fits naturally with the whole-language approach.

Currently, the hottest controversy of this type is the so-called math wars. The U.S. Secretary of Education Richard W. Riley (1998) acknowledged this when he called for a "cease-fire" and a focus instead on improved standards of attainment in mathematics. He recognized the differences in ideologies between protagonists but referred to them as *opinions*. In response, about 200 mathematicians and scientists led by David Klein and Hung-Hsi Wu, in an open letter, urged him to withdraw federal government endorsement of 10 mathematics programs for schoolchildren it had designated "exemplary" and "promising." These programs, including MathLand and Connected Math, were castigated as being "horrifyingly short on basics," for not teaching standard arithmetic algorithms, the division of fractions, and for doing away with textbooks as too hierarchical. These attacks are typical of a coalition of Industrial Trainers and Old Humanists, because of the emphasis on basic skills and mathematics-centeredness, and the critical stance towards child-centered learning with problem solving and the active construction of meaning.

The ten critiqued mathematics programs are based the National Council of Teachers of Mathematics's proposals for a reformed math-

ematics curriculum and teaching principles (see bibliographical entry herein and earlier publications). This proposes a shift away from basic skills, direct instruction, and memorization toward problem solving, active learning, and the understanding of central mathematical concepts based on constructivist learning theory. It also emphasises equity and claims that mathematical understanding and success is attainable by all, not just a gifted elite. The document reflects the widespread views of mathematics education professionals and teachers and evidently embodies Progressive and Public Educator philosophies of education. These proposals are also the subject of strong attacks in the math wars.

One of the chief battlegrounds in this and similar ideological struggles concerns educational testing and credentialing. Educational credentials are regarded as the objective yardstick of what students achieve in schools and college. These are vital in the world of employment and in societal views about the success of educational systems. International comparisons in mathematics and science have had powerful effects in industrialized nations, and have provided ammunition for critics of education policies, some warranted and some less so. However, the nature of testing regimes owes much to the underpinning ideological presuppositions about the nature of knowledge, learning, and the aims of education. The different groups described here have widely differing views in these domains. As these competing groups know all too well, testing has another important function: testing regimes have a powerful controlling effect on the content and pedagogy of education. Accountability in the terms laid down by the all-important testing regimes necessitates "teaching to the test," unless these important indicators of education effectiveness are ignored.

*See also* **Education and Science; Education in the Sciences**

### BIBLIOGRAPHY

Ernest, P. *The Philosophy of Mathematics Education*, London: Falmer Press, 1991.

Meighan, R. A *Sociology of Educating*, Eastbourne, U.K.: Holt, Rinehart, and Winston, 1986.

National Council of Teachers of Mathematics. *Principles and Standards for School Mathematics*, Reston, VA: National Council of Teachers of Mathematics, 2000.

Riley, R. W. "The State of Mathematics Education: Building a Strong Foundation for the 21st Century." *Conference of American Mathematical Society and Mathematical Association of America*, 1998.

Williams, R. *The Long Revolution*. London: Penguin Books, 1961.

*Paul Ernest*

## ELECTRICITY.

*See* NETWORKS AND LARGE-SCALE TECHNOLOGICAL SYSTEMS

## ENERGY SYSTEMS.

Oil wells, pipelines, refineries, gas stations, coal mines, generating plants, transmission wires, and related technologies together make up the planet's largest and most expensive technological system. Based primarily on oil, coal, and natural gas, energy services structure everyday life both directly and by enabling computing, air-conditioning, telecommunications, and global transport. For nearly half a century, environmentalists and technological innovators have been urging reduced reliance on fossil fuels, but technological momentum remains dominant.

### Conventional Energy Systems

The contemporary energy system has two basic components, electricity and fluid fuels. In the year 2000, the world had a total installed electrical capacity of 3,365 gigawatts (billions of watts), producing nearly 14 trillion kilowatt hours (kWh). The United States consumes about a quarter of the total at a cost of more than $220 billion annually. This is supplied by some 100,000 electric power plants and 10 million miles of power lines, controlled increasingly by investor-owned electric utilities, with some governmental ownership and some onsite generation by industry. Coal-fired power plants account for about 38 percent of electricity generation in the United States, followed by natural gas, oil, nuclear, and hydro power. Most other countries also rely heavily on fossil-powered generating plants, although Japan and France use far more nuclear.

In the mid-1960s, dozens of nuclear plants were ordered annually, as utility executives jumped on a bandwagon for "light water" reactors, using ordinary water as the coolant and moderator. Popular protests, lawsuits, and eventually pressure from legislators forced reactor manufacturers and their customers to retrofit existing plants, with round after round of expensive safety devices. This slowed construction and licensing, and costs escalated to as much as $4 billion per plant, far above the cost of fossil-powered electricity. Some utilities made bad management decisions in trying to implement nuclear power, scaling up the size of reactors very quickly and without having a clear idea of their costs. With interest rates as high as 18 percent during the 1970s, there came a point at which no utility executive was willing to take the risks of ordering a plant that might bankrupt the company. One especially costly case was that of the Shoreham plant on Long Island, built but never operated, partly because local, state, and federal authorities could never agree on a feasible evacuation plan from that crowded region in the event of a major accident. These economic forces, combined with the Three Mile Island accident in 1979, *The China Syndrome* movie, and the 1986 accident at Chernobyl, put the finishing touches on the demise of nuclear power.

If the organization most responsible for pushing nuclear power, the congressional Joint Committee on Atomic Energy, had proceeded more slowly, both proponents and opponents might have had time to realize that it is technically feasible to build tiny nuclear reactors that cannot suffer catastrophic accidents. Controversies still might have arisen regarding radioactive wastes, like those that continue to swirl regarding the Yucca Mountain repository in Nevada. Without the threat of catastrophe, however, nuclear power might have been more acceptable psychologically and politically, and, hence, more feasible financially. Small reactors may one day be reconsidered, but for now the only nations not phasing out nuclear power are those where citizen groups are effectively kept out of courtrooms and legislative proceedings.

Fluid fuels are the second main component of the energy system. The world oil industry produces over 75 million barrels of oil per day (27 billion barrels annually), transported by fleets of huge tankers and by perhaps a half-million miles of pipeline. World natural gas consumption was 84.2 trillion cubic feet (tcf) in 2000, with Russia and the United States together using nearly half. Demand is increasing faster than supply, with projections for 2020 ranging to 174 tcf worldwide. The fastest growing use is for electric power generation, because the fuel burns cleanly and plants are relatively cheap to build. Natural gas is not as far along toward depletion as is petroleum, but the midpoint of world reserves probably will be passed some time in the next generation, and certainly by the middle of this century.

Conventional energy sources are vulnerable to terrorism. Oil pipelines, tankers, and refineries are typical targets in war. Nuclear reactors are a target for the theft of plutonium, as well as for direct attacks. Even with redundancies built into electrical supply networks, transformers and switching equipment are easy to attack, and the supply of back-up parts is miniscule. In the aftermath of 9/11, some observers began wondering about the desirability of moving to more decentralized, less vulnerable energy sources.

## The Conventional System's Momentum

The fossil fuel–based energy system is coextensive with contemporary life. More than a trillion dollars globally is invested in power plants, transmission lines, oil wells, tankers, and pipelines, as well as furnaces, automobiles, water heaters, and air conditioners. Some of these items have functional lifetimes of half a century, and changeover would be costly. Social aspects of the system also resist change. Bankers are not eager to fund houses utilizing unconventional technologies, for example. Engineers mostly work for utility companies, automobile manufacturers, and other conventional firms, not for smaller companies doing research and development on fuel cells, making plastics out of corn, or developing photovoltaics. Even university professors consult disproportionately for the backward-looking companies; for example, no department of nuclear engineering emphasized tiny reactors. Technologically oriented professional associations issue standards for training, accreditation, and licensing based on last decade's knowledge. And fossil-based industries donate heavily to electoral campaigns, and have excellent access to elected officials. Unfortunately, the main governmental efforts to change the fossil energy system have proven misguided, especially encouraging manufacturers and utility companies to move into nuclear power by providing massive research and development funding and limiting liability for accidents. Governments have provided tax credits for solar, but much of that proved wasted. Some state public utility commissions have required utilities to purchase electricity generated in innovative ways by small-scale producers, but the effect is small. Encouraging energy efficiency has worked out fairly well, but that obviously does not change the basic energy system. Despite these exceptions, the general pattern is that the social components of the energy system magnify rather than counteract technological momentum.

It has long been apparent that industrialized countries depend for oil on politically volatile regions, yet the United States imports more than ever, partly because its oil production has been gradually declining since the peak in 1970. Despite the oil embargo of 1973, recurrent Middle East crises, and OPEC struggles, political debates focus less around geopolitical dependency and oil depletion than around environmental and safety concerns, in part because policy makers have no short-term solutions to this dependency. NIMBYism in Japanese energy siting, Green Party efforts to influence the German bureaucracy, and rebuffs by the court system of France resemble politics more generally in those nations. In the western United States, energy politics involves land and water controversies. Leading-edge areas such as California and Denmark emphasize nuclear opposition and focus on renewables. Local smog from automobiles, acid rain from power plants, and the prospect of global warming from the burning of carbon-rich fossil fuels have stimulated both international treaties and local governmental regulations. Citizens groups generally have joined governments and businesses in focusing on limiting emissions, modest conservation efforts, and incremental technical refinements to improve efficiency and reduce pollution "at the end of the pipe." Although Greenpeace and other groups have sought more fundamental alterations in the energy system, they have met with little success. The United States has withdrawn entirely from the Kyoto Protocol for reducing greenhouse gas emissions, and it remains to be seen how effectively even those countries that have ratified it will implement it.

Momentum in energy systems also derives from institutionalized ideas in government agencies, professional associations, and even universities. Energy decision making works within a set of problem definitions, standard operating procedures, ways of approaching problems, and institutional cultures that are more accepting than challenging of the status quo. This is especially curious given the fact that the possibility of resource wars rarely is far from the front pages, coupled with oil supply predictions since the 1950s that can be summarized as, "More than half of the world's reserves will be used by x date"—although the date keeps changing. Nevertheless, in the United States, a substantial majority believes that Americans need not move toward buses and trains, that gas taxes should not be increased to discourage consumption, and that motor vehicles need not become radically more fuel efficient. Ultimately, getting energy-related institutions to embrace major changes will require changing deeply ingrained ideas.

Such changes as do occur tend to be ill considered and elite dominated. Business-oriented ideologues have been pressing to privatize state-owned energy companies, to separate electricity generation from distribution and marketing, and to deregulate prices. This "neoliberal" approach is championed on grounds of efficiency, but advocates for the poor perceive it as a ploy to drive up profits at the expense of nonaffluent consumers. In the late 1990s, many legislators passed laws they only dimly understood, while interest groups, media, and the public were pretty much asleep during the process, and relevant experts did little to help. The short-term results included a doubling of San Diego electricity prices in a matter of months, and over the longer term has come a loss of reserve "spinning capacity"—backup electricity-generating capacity, formerly required by regulators. With the split of the industry into component pieces, none has incentive to invest in expensive equipment that sits idle except during emergencies. Some parliamentary systems have managed to formulate and implement an energy strategy, with France and Japan held up as models of deliberate choice in favor of nuclear and Denmark and Germany among the leaders in alternative energy. Most nations have a more coherent energy policy and lower per capita energy usage than the United States, as indicated by the high European gas tax and the predominance there of high-mileage diesel vehicles. Nevertheless, a uniform momentum appears everywhere in the sense that fossil fuels remain dominant.

## Alternative Energy Systems

Alternative energy advocates long have believed that renewable energy from sun, wind, and biomass could supply the planet's needs. Electricity generated by wind turbines is now the largest renewable source, with installed capacity of 31,000 megawatts in 2002, equivalent to about 31 large coal or nuclear power plants. Huge wind turbines allow electric utilities to add a "green" panache to their corporate images while leveraging their capacities for generating, distributing, marketing, and billing for electricity. Wind power is expected to triple within a decade, partly because it is the least-expensive renewable source, producing electricity profitably at less than $0.05 per kWh at good sites. Wind energy advocates, mostly outside the conventional energy system, have successfully pushed the system to respond, making wind almost mainstream.

Solar thermal energy for space heating, once considered the most promising alternative source, received a substantial push from governmental tax credits in the 1970s. But it so far has proven too fussy and too expensive for widespread use, although passive solar (e.g., special south-facing windows) remains a good method of reducing the heating/cooling required for buildings. Few architects and housing developers have adapted passive solar design and construction practices, however, partly because they have not been trained and have no incentive to do so. Discussions arise periodically about high-technology, space-based solar systems, and some utilities have experimented with giant solar arrays using mirrors to focus sunlight on tanks of molten salts in deserts. Photovoltaics—semiconductor devices that generate electricity when hit by light—were stimulated by space and military purchases, and sales increased tenfold in the 1990s as research and development and economies of scale brought prices down. Solar photovoltaics already have overtaken solar thermal, with 80 percent of world capacity in Germany, Japan, and the United States. Photovoltaic electricity thus far remains too expensive to use except for very high value applications, such as for small devices like calculators and in remote settings off the conventional electric grid. Both the European Union and the United States in 2002 announced research programs into using hydrogen as a fuel. Not a source of energy but a means of storing it in convenient form, hydrogen is produced by running an electric current through water ($H_2O$) and by then separating out the hydrogen, which can be stored, distributed through pipelines, or used in fuel cells to generate electricity. Hydrogen also can be made from natural gas, and technologists are currently working on other ways of producing it. However, critics argue that hydrogen actually would yield less net energy than direct utilization of whatever fossil fuel sources are used to generate the hydrogen; they also argue that there would be no net reduction in pollution.

Researchers are exploring other alternative energy sources, such as bio-oil, a liquid derived from low-grade wood waste. It potentially could be used for space heating and electricity generation as a substitute for fuel oil, and could serve as a feedstock for a "greener" chemicals industry. Bio-oil raises the same environmental questions as any other use of forests, but it could be a renewable, carbon-neutral energy supply not dependent on geopolitical conditions in oil-producing regions.

## Modifying Energy Systems

If the price of renewables drops while gas and oil become more costly, market choices may begin to replace fossil fuels with renewables. However, modifying systems with technological momentum is more complex than simply unplugging one technology and plugging in another as prices change. Moreover, price itself is determined not solely by market forces: there may be a "world" price for a barrel of crude oil, but the prices consumers pay depend also on patterns of taxation and government subsidies.

Social preferences also shape energy outcomes, as in American preferences for larger vehicles. For another example, very efficient compact fluorescent light bulbs are cheaper overall than conventional incandescent bulbs but have not been widely embraced. People are reluctant to spend $10 on a single bulb, even if the long-term savings in electricity more than offsets the initial investment. Some do not like the color of the light, have fixtures or shades of the wrong shape, or may simply resist switching to the unfamiliar. Others prefer convenience to conservation or simply are ignorant about alternatives. For similar reasons, most homes still lack programmable thermostats, three decades after the 1973 OPEC oil embargo led to widespread awareness of conservation. Hybrid cars from Honda and Toyota have not flown off U.S. car lots, despite their equivalent performance to other small cars and despite excellent ratings from automotive experts, suggesting that cars running off more exotic technologies may be even more difficult to sell. These issues obviously are challenges for technological research and development, but also point to the social components of technological momentum.

## Momentum and Barriers to Change

"*Hydrocarbon Man*, now virtually the sole surviving human subspecies, will certainly be extinct by the end of this century, meaning that the very future of mankind is therefore at stake. This perhaps explains why oil depletion is such a sensitive subject, which many people, especially those in government, prefer not to know about."—The energy analyst Colin Campbell

Despite the conventional energy system's momentum, changes are inevitable as the world passes the peak of petroleum extraction in the coming generation, with the natural gas peak soon following. Neither extraction rates nor prices necessarily will react quickly, but, as relative scarcity looms, oil and gas will begin to command higher prices. This will lead producers to invest in more sophisticated extraction technologies to get a higher fraction of the resources out of the ground. Higher prices will also stimulate research and development on energy efficiency and on presently marginal sources, such as tar sands; prices will possibly reduce industrial activity via recession; and they will probably encourage behavioral changes (e.g., adjusting thermostats, reducing driving). Meanwhile, climate change from greenhouse gases probably will become increasingly evident.

So the real question is not whether the energy system will change, but whether the changes will be deliberate, farsighted, and equitable. To date, the fossil system's momentum remains profound, and disruptive social, economic, and environmental consequences seem likely, with burdens falling disproportionately on poorer countries and on poorer people within the affluent regions. Without the extensive physical, financial, and social investments that create momentum, however, poor nations do have an opportunity for creating alternative energy systems: because it would be very costly to duplicate conventional energy systems in these countries, advocates of alternative energy in some respects have an easier task there than where they must try to displace a huge existing system.

In sum, despite decades of warnings coupled with insightful efforts by alternative energy advocates, a majority of energy decision makers have been sleepwalking during the past generation, carried along by fossil-energy momentum in a pleasant sort of technological drift, denying impending realities. Most government officials, business executives, and consumers appear to be hoping for the best rather than planning prudently. This adherence to the past contrasts sharply with the innovation juggernauts in consumer electronics, telecommunications, and biotechnology, where innovation arguably is moving far too fast to allow sensible learning from experience. Although opposites in certain respects, the failure to reduce reliance on fossil energy and the failure to modulate other innovative trajectories can be seen as flip sides of the same problem: humanity has not yet created institutions, learning processes, and incentive systems capable of governing technological change equitably, farsightedly, and at an appropriate rate of change.

## BIBLIOGRAPHY

Energy Information Administration. *Electric Power Annual 2000; Oil Market Basics, Trade, Tankers and Pipelines,* and *Renewable Energy Annual.* http://www.eia.doe.gov.

Hubbert Peak of Oil Production. http://www.hubbertpeak.com.

Hughes, Thomas Parke. *Networks of Power: Electrification in Western Society, 1880–1930* Baltimore, MD: Johns Hopkins University Press, 1983.

International Energy Agency. http://www.iea.org.

Laird, Frank N. *Solar Energy, Technology Policy, and Institutional Values,* Cambridge, U.K.: Cambridge University Press, 2001.

Morone, Joseph G., and Edward J. Woodhouse. *The Demise of Nuclear Energy?: Lessons for Democratic Control of Technology.* New Haven, CT: Yale University Press, 1989.

Taffetas, Kenneth S. *Hubbert's Peak: The Impending World Oil Shortage.* Princeton, NJ: Princeton University Press, 2001.

*World Electric Power Plants Data Base,* September 2002. Platts UDI publishing. http://www.platts .com/udidata.

*Frank N. Laird and Edward J. Woodhouse*

## ENGINEERING CULTURES.

Each summer in Paris, an enormous military parade commemorates Bastille Day, July 14, 1789, when commoners stormed the royal fortress and wrested power from the king, formally initiating what later became known as the *French Revolution.* The parade is led each year by second-year students from the École Polytechnique, the top engineering school in France. At the key moment on the key day, when the entire nation is focused on itself and its accomplishments under the leadership of a republican government, France makes its elite engineers visible to an extent found nowhere else in the world. Those engineering graduates who make it into the state administration in fact constitute the highest ranked occupation in the country. Importantly, both those engineers and the majority of engineers employed in lower-status locations in provincial governments or in the private sector demonstrate through their work and careers that advanced mathematical knowledge is valued above all else in French engineering training and practice.

In sharp contrast, engineers in the United Kingdom, especially in England and Wales, have struggled with relatively low status throughout their more than 200-year history. A nineteenth-century emphasis on training through an apprenticeship system established a focus on the value of practical engineering knowledge that continues to this day, even amid the twentieth-century emergence of school-based training and the increased attention paid to theoretical knowledge. Also, with the exception of work within the military, engineers have tended to seek employment and build their careers in the private sector, outside of government. Struggling for higher status in a hierarchical class system that placed greatest value on classical training in the liberal arts, engineers have relied on a type of practice-based membership group, the professional society, in an effort to advance engineering as a legitimate profession alongside law, medicine, and the clergy.

Germany offers yet another pattern in the knowledge and social positioning of engineers. As suggested by the pervasive cultural icon, the BMW motor vehicle, German engineers have developed over time a primary focus on the production of "quality technics," where *quality* entails the realization of engineering precision, measured in close tolerances, and *technics* refers to both the outputs and the mechanisms of technological production. Quality technics has served as a marker of advancement in German society since the late nineteenth century, becoming especially important during the 1930s under National Socialism. Developing in parallel, an engineering focus on precision has involved actively developing what some historians have called *scientific technology,* or forms of scientific knowledge specifically designed to help solve technical problems. Through the linkage between technics and advancement, engineers have come to know that by applying engineering precision in quality processes of technological development, they also contribute directly to advancing both the German nation state and humanity in general. In other words, although German engineers have tended to work in the private sector, similar to British engineers, they have also had the opportunity to be identified directly with national development, similar to French engineers.

Although the number of engineers in the world is comparable to the number of scientists, engineers have received far less attention in studies of science and technology in society. Distinct patterns among French, British, and German engineers call attention to what has been a key barrier to the understanding of engineers and engineering in society: great diversity in what counts as an engineer and engineering knowledge. Wide variations exist in different countries in the central concerns and social locations of engineers amid other knowledge workers, including scientists, technicians, government officials, business managers, and so on. Different types of people are called "engineers."

Sometimes the term *engineer* refers to the holder of a degree and sometimes to a job title that one can occupy. Sometimes engineers experience high status, sometimes low status, and often ambiguous status. Important differences in engineering careers lie not only between countries but within countries as well.

One pathway to understanding contingent developments and persistent patterns in engineering knowledge and personhood is to examine the cultural meanings that challenge engineers in their work. However, doing so effectively requires considering what is happening when engineers experience and respond to configurations of cultural challenges. A focus on the professional identities of engineers provides a way of following links between engineering knowledge and engineering personhood and, hence, understanding how engineers have been active agents of their own positioning in different countries.

## Mapping Engineering "Up" from Society

One reason that engineering knowledge and engineers have received far less attention than science and scientists is dominance of the view, both inside and outside of academia, that engineering is located "downstream" of basic science. In this view, engineering consists of forms of knowledge and collections of activities associated in some way with the application of scientific knowledge to practical problems. To the extent that engineering knowledge lives or gains form in this derivative sense, one should seek to understand its essential elements by first looking "upstream," sorting out the defining features of the relevant scientific knowledge. Only after such prior work is completed can one begin to understand and find order in the application of this knowledge to solve problems. In this way, the downstream location of engineering thus also appears to be an accurate indicator of its apparent subordinate level of importance. Along with the applied sciences, engineering gains both content and significance fundamentally through its links to basic science.

However, strong reasons exist for pulling engineering epistemology, or theorizing about the content and positioning of engineering knowledge, out from under the shadow of scientific knowledge. In particular, consider the fact that the key mathematical activity of engineering analysis always works with the goal of somehow making society "better." As historian Ken Alder (p. 60) observes in describing the emergence of engineering analysis in eighteenth century France, its operational method is "to describe quantitatively the relationships among measurable quantities, and then to use these descriptions to seek a region of optimal gain." Today, school curricula for engineering training tend to present engineering analysis as a disciplined activity of mathematical problem solving. For students in the United States, such problem solving involves drawing a boundary around a given problem, abstracting its features out into mathematical worlds (such as statics, dynamics, fluid mechanics, thermodynamics), solving it in the mathematical terms of those worlds, and then applying the solution back to the original problem, all to facilitate a gain for whoever is faced with the problem in the first place. In other words, where scientists have long been charged with bringing new objects into the purview of society, as discovery, with the ultimate goal of benefiting society, the key expectation confronting engineers has been to improve society directly by developing and improving the performance of human constructs.

Born in the European Enlightenment, this mandate to improve society also brings design methodology and practice into the epistemological mix of engineering. The activity of design embeds engineering deeply in everyday life, generating outputs that are supposed to count as solutions to everyday problems and making it difficult to distinguish *a priori* engineering knowing from engineering doing.

In parallel with other developments in science and technology studies, researchers are increasingly inquiring into engineering epistemology by attempting to map engineering "up" from society rather than "down" from science. When scrutinized from the bottom up, engineering knowledge begins to appear so diverse because it is mapped so closely onto the diverse societies that engineers serve and in which they function. In other words, the activity of serving, of making society better, depends upon what counts as better at particular times and in particular places. Perhaps engineers, rather than functioning as mere disseminators of basic knowledge, are actively engaged in selecting, adapting, and developing the forms of scientific knowledge they need in order to successfully intervene in and, hopefully, improve everyday life. In this way of thinking, diversity in the knowledge and social positioning of engineers appears as an accomplishment, a product of successful struggle in diverse circumstances, rather than an unfortunate limitation. The challenge is to show how the epistemological value of engineering knowledge is linked to the wider social value of engineering work and the professional identities of engineers. Here is where culture comes in.

## Responding to Culture

How do cultural meanings act on, or influence, engineers? The issue comes down to the question, what do engineers share? Consider the cases introduced above of France, United Kingdom, and Germany. It would be both grossly misleading and analytically unhelpful to claim that these countries have distinct and independent cultures that shape their engineers in uniform ways.

In the first place, the cultural contrasts traditionally thought to distinguish countries from one another are by no means sharp. With dramatic increases in both the virtual and physical movements of engineers around the world using emergent transportation and communications technologies, not only are such cultural boundaries between countries increasingly porous, but perhaps it also no longer makes sense to posit them at all. Such is especially the case for Europe, where much evidence exists of engineers actively working in the context of perspectives from other countries. Secondly, the life trajectories and experiences of engi-

neers vary greatly within a given country, such that differences among engineers within a given country might actually be greater than differences among engineers from different countries. In short, no one-to-one correspondence can be said to exist between the culture of a country and the knowledge and personhood of engineers. It makes little sense to argue that French, British, and German engineers share distinct national cultures.

But the idea that cultural meanings that have become national in scale influence engineers still has merit. One way to elaborate this idea is to think about engineers not as passively sharing cultures, as underlying sets of beliefs or assumptions, but as actively "responding" to cultures as "codes of meaning" that challenge them as people. For example, rather than saying that all engineers in a given country share a value in mathematics or believe in the importance of practical knowledge, one can say that all engineers trained or working in a given country may have to respond to a code of meaning that places value on mathematical or practical knowledge, respectively.

How engineers actually respond to a code of meaning affects its future scale, depending on whether their actions reproduce, transform, and/or replace the code. In this sense, national engineering cultures are not membership groups but codes of meaning that have scaled up and, hence, become dominant at the level of the nation state. So when engineering schools in France emphasize mathematical theory over everything else in their texts, teaching, and exams, and when "successful" engineering students demonstrably accept this as what counts as engineering knowledge, then we can say that mathematical theory has become part of a dominant code of meaning challenging people who study and practice engineering in France.

The key question for engineering epistemology in this way of thinking lies in how patterns in engineering knowledge and personhood might emerge amid historical contingencies. In their lives and work, engineers regularly respond to all sorts of meanings from inside and outside of engineering that have different contents and live at different scales, depending upon unique life experiences with family, religion, education, travel, friendship, and so on. Accordingly, one must treat differences among engineers as the expected norm, the default setting. Yet patterns do emerge, especially at national scales. Since the identity of the engineer emerged along with the Enlightenment concept of advancement in society, perhaps national patterns in engineering knowledge and personhood may have developed as engineers responded to distinct popular images of national progress.

## Responding to Popular Images of Progress

In the French context, a fundamental focus on mathematics and the positioning of elite engineers in the executive branch of government suggests that engineers have long responded to a teleological image of national progress, that is, as advancement toward a potential future state of perfection. As eighteenth-century French *philosophes* argued in advocating greater rationality for social life— because nature, having been created by God, is organized by a principle of regularity— human society could become more godlike by emulating the principled order of nature, abolishing obsolete privileges and making itself more orderly. Both before and after the French Revolution, the state administration has been the major agent for increasing social order. A key vehicle has been the development of a national infrastructure of transportation, communications, and energy technologies. Private industry has been of far less interest. The authoritative position of the state administration remained stable even as the French battled over defining the titular leadership, producing a dizzying mix of three monarchies, two empires, and five republics in a two-hundred-year period.

Engineering analysis based in abstract mathematics gained legitimacy as the crucial national tool for theorizing and enacting the march toward perfection through higher states of social order. The seventeenth-century philosopher René Descartes had established the idea that nature could be seen as a huge mechanism, analyzable in mathematical terms. Working to fulfill the popular image of progress, French administrators constructed parallel hierarchies of education and employment, with engineering schools and state employment at the top. Higher ranked than universities, the elite engineering schools, or *grandes écoles*, have been limited to top performers on an extremely difficult math-based examination, *les concours*; have consistently placed greatest emphasis on training in advanced mathematics; and have provided graduates with direct pathways to top positions in the state. In other words, French engineers have responded to a national image of progress in a patterned way by privileging mathematical capabilities and activities. Accordingly, they have successfully built and managed a large state apparatus that has made them the envy of engineers worldwide.

In contrast, the commitment of British engineers to practical knowledge and their struggles with lower-status positions suggests a patterned response to an image of social progress as material improvement over the past. In the British context, successful improvement within society has been defined competitively as the increasing material welfare and comfort of individuals, or self-improvement, measured in part by the distance individuals achieve from manual labor and by the quality of their education. Class status became an indicator of one's level of social advancement. Importantly, the English industrial revolution was a phenomenon of lower-status crafts, transforming artisans and craftsmen into wage laborers while enabling some new industrialists to climb into gentlemanly status alongside traditional agricultural elites and gentry. But in contrast with the United States, English industry focused not on mass production for mass use and benefit but on batch processing oriented toward society's upper echelons. In contrast with France, the state adopted a laissez-faire stance toward national progress, limiting itself mainly to authorizing charters to private companies.

Responding to the dominant code of self-improvement, British engineers explicitly rejected the French emphasis on mathematical theory and assessed their work in terms of its practical benefit. Insisting on training through the apprenticeship system, engineers both emerged from the ranks of craftsmen and emphasized hands-on experience and craft knowledge. Engineers sought gentlemanly status by emulating the traditional professions of law, medicine, and clergy and organized themselves into professional societies, however, with uneven success. In a country in which material success was indicated by distance from manual labor, the idea of an elite engineer was almost a contradiction in terms. Even the children of relatively prominent nineteenth-century engineers tended to avoid careers in engineering. The polytechnic institutes that eventually emerged to provide school training were located below universities, and professional societies were never able to displace employers in the control of engineering work. In short, responding to a call for progress through material improvement, engineers produced a pattern that established a passionate attachment to practical knowledge and a permanent struggle for higher status.

In the German case, in placing high value on the production of quality technics, engineers have responded to an evolving image of progress as emancipation of the human spirit, a freeing of something that is naturally internal to the human essence. During the German Enlightenment, progress in society came to be known as the unfolding of reason, articulated by philosophers in universities and enacted by bureaucrats in rationalized governments, especially in Prussia. The concept of engineering was not indigenous to Germany but was borrowed from Great Britain and, especially, France. Engineers emerged among the lower-status guilds of artisans, which had long been known for their conservatism. Responding to the idea of progress through reason, activist engineers sought higher status through education, establishing Higher Technical Institutes for engineering education. Their efforts met with increasing success after unification of the German states in 1870, as industry became a new site for marking human and,

hence, German progress. The unfolding German spirit could now be found in the physical and material existence of quality technologies and products. Late-nineteenth- and early-twentieth-century Germany is a story of the rapid rise of high-quality German industry, especially the steel and chemical industries.

Beginning in the twentieth century, especially during and after the Weimar Republic, engineering leaders responded to the idea of progress through industry by promoting the production of quality technics, working up from practice to theory only to the extent necessary to produce a quality outcome. They organized a second tier of engineering education, in which "gaining a feel" for materials became a defining activity and work producing quality products in private industry simultaneously brought national significance because it demonstrated German advancement. The engineering emphasis on quality as precision became prominent after World War II as engineers gained stable status as an important category of German society. In sum, German engineers were able to gain increasing credibility for themselves and their forms of knowledge by responding strategically to the national shift from reason to technics as the main site for emancipation of the German spirit.

## Research Questions

One key category of questions to pursue involves following how European patterns in engineering knowledge and personhood have traveled through colonial relations to challenge people in other parts of the world. How, for example, do engineers respond to differing configurations of what counts as national progress, advancement in society, and/or improvement? Engineers in the United States appropriated patterns from British and French sources while responding to an emergent national image of progress as increased standard of living for the masses, a concept and challenge that may indeed be central to the contemporary experience of what is today called "globalization." Engineers in former British and French colonies have often struggled with mixes of challenges that are of indigenous, as well as colonial, origin. The distinct patterns that have emerged presumably depend upon unique national trajectories and participation in ongoing international interactions.

A second, crucially important, set of questions concerns how engineers have responded to codes of meaning that extend beyond the boundaries of nation states. A key example is the capitalist organization of the private firm. Might examining the patterns through which engineers in different countries tend to position themselves as mediators between management and labor be a practical strategy for examining and assessing the so-called structural effects of capitalism on engineers? Might national similarities and differences lie precisely in configurations of challenges from codes of meaning living inside capitalist organizations and codes of meaning originating in everyday life?

Third, what sorts of factors distinguish more successful from less successful engineers at different times and in different places? Historians of engineering tend to focus on the small number of engineers that have achieved prominence. How might we understand the populations of engineers who have had routine careers? What about engineers who failed, or who have left before becoming engineers? Might following engineers as active agents of identity formation provide a means for providing a more complete mapping of similarities and differences among engineers?

Lastly, what counts as reform in engineering? By asserting that engineers actively respond to cultural codes of meaning, we are implicitly describing engineers as engaged in "cultural projects," for their activities affect those codes of meaning. By thinking about codes of meaning as posing challenges to people rather than grounding their assumptions, cultural change becomes an attainable goal, requiring one to develop and scale up alternative meanings. Serious study of how engineers have in fact built identities for themselves by responding to cultural codes of meaning may make it easier to identify and scale up better alternatives in the future.

## BIBLIOGRAPHY

Alder, Ken. *Engineering the Revolution: Arms and Enlightenment in France, 1763–1815.* Princeton, NJ: Princeton University Press, 1997. Shows how artillery engineers laid the foundation for the French Revolution and the subsequent authority of state engineers.

Bucciarelli, Louis L. *Designing Engineers.* Cambridge, MA: MIT Press, 1996. Advances the concept of "object worlds" to show how engineering design involves an interaction of different perspectives.

Canel, Annie, Ruth Oldenziel, and Karin Zachmann, eds. *Crossing Boundaries, Building Bridges: Comparing the History of Women Engineers 1870s–1990s.* Amsterdam: Harwood Academic Publishers, 2000. Compares women's struggles for acceptance in engineering schools in different national contexts.

Fox, Robert, and Anna Guagnini, eds. *Education, Technology and Industrial Performance in Europe 1850–1939.* Cambridge, U.K: Cambridge University Press. 1993. Compares forms of engineering education in Europe.

Hutton, Stanley, and Peter Lawrence. *German Engineers: Anatomy of a Profession.* Oxford: Clarendon Press, 1981. A study by British sociologists based on a survey of 1000 German engineers.

Kranakis, Eda. *Constructing a Bridge: An Exploration of Engineering Culture, Design, and Research in Nineteenth-Century France and America.* Cambridge, MA: MIT Press, 1997. Analyzes the emergence of engineering cultures in nineteenth-century France and United States.

Meiksins, Peter, and Chris Smith. *Engineering Labour: Technical Workers in Comparative Perspective.* New York: Verso, 1996. Cross-national comparison of the positioning of engineers and engineering work.

Picon, Antoine. *French Architects and Engineers in the Age of Enlightenment.* Translated by Martin Thom. Cambridge, U.K.: Cambridge University Press 1992. Analyzes the separation between architects and engineers in eighteenth-century France.

Reynolds, Terry S., ed. *The Engineer in America: A Historical Anthology from Technology and Culture.* Chicago: University of Chicago Press, 1991. A collection of articles by historians of technology on engineers in the United States.

Vincenti, Walter G. *What Engineers Know and How They Know It: Analytical Studies from Aeronautical History*. Baltimore: Johns Hopkins University Press, 1990. An anatomy of engineering design knowledge drawn systematically from five detailed case studies.

Weiss, John Hubbel. *The Making of Technological Man: the Social Origins of French Engineering Education*. Cambridge, MA: MIT Press, 1982. Explores the early history of the École Centrale des Arts et Manufactures, founded in 1829 as the first private engineering school in France.

Whalley, Peter. *The Social Production of Technical Work: The Case of British Engineers*. Albany: State University of New York Press. 1986. Analyzes the work lives and occupational positioning of British engineers, drawing on data from two factories.

Zussman, Robert. *Mechanics of the Middle Class: Work and Politics among American Engineers*. Berkeley: University of California Press. 1985. Analyzes the work lives and occupational positioning of American engineers, drawing on data from two factories.

*Gary Lee Downey and Juan C. Lucena*

# EPISTEMIC CULTURES

Everyone knows what science is about: It is about knowledge, the "objective," and perhaps "true," representation of the world as it really is. The problem is that no one is quite sure how scientists and other experts arrive at this knowledge. The notion of an *epistemic culture* is designed to capture these interiorized processes of knowledge creation. It refers to those sets of practices, arrangements, and mechanisms bound together by necessity, affinity, and historical coincidence that, in a given area of professional expertise, make up how we know what we know. Epistemic cultures are cultures of pursuing and warranting knowledge. This is what the choice of the term *epistemic,* rather than simply *knowledge,* suggests. The notion of the epistemic here builds upon earlier studies' findings and methodological orientation in the new sociology of science, whose defining concern has been to open up the black box

that has constituted scientific inquiry and make sense of the various activities observed.

Several other things are important in the understanding of epistemic cultures. First, the concept of epistemic cultures is motivated by the epistemic diversity and fragmentation of the "golem" (to use Harry Collins's and Trevor Pinch's term) of science. It implies a disunity in science and knowledge, which runs counter to the notion that there is only one scientific method, only one scientific rationality. Science, technology, and expertise are obvious candidates for cultural divisions; they are pursued by specialists separated off from other specialists by long training periods, stringent division of labor, distinctive technological tools, particular financing sources, and so on. The notion of an epistemic culture takes up where this assessment leaves off. It brings into focus the specific object relations regimes, the construction of the epistemic subject and the referent, the meaning of "empirical," the ethno-methods of consensus formation, the forms of engagement or disengagement with the social world manifest in an area. Nor are individual scientific cultures homogeneous, either. A second point is that epistemic environments are merged realms of existence and forms of life. They bring natural objects and their contexts together with social contexts and processes. They also draw on different background knowledges, which become linked in knowledge work. How the mergers are accomplished, what reconfigurations they bring about, and what fault lines still run between the inner and outside realms of existence of all entities involved, are essential parts of the description of an epistemic culture. Third, epistemic cultures become increasingly important with the rise of knowledge societies. There is a widespread consensus today that knowledge has become a productive force and a determining factor in the "reflexive modernization" (to use the phrase of Ulrich Beck) of contemporary Western societies. Knowledge and expert systems account for the transition from industrial to postindustrial societies, which are permeated with contexts of technical accomplishment and professional expertise. If these arguments are right, what we call "society" is to a significant degree now constituted by knowledge settings. Epistemic cultures are the cultures of such settings. If we wish to understand postindustrial societies, we will have to understand the cultures that are specific to these societies.

## Investigating Epistemic Cultures

What are typical areas for investigation in relation to epistemic cultures? A first question surely is "Who are the entities in an epistemic culture?" For example, who or what are the epistemic subjects, those we traditionally think of as the agents in scientific practice and the authors of scientific findings? We can assume that *agency* is constituted differently in different cultures, as are *personhood, subjectivity, collectivity,* and other cognates of the term. For example, high-energy physics experiments, which are today among the largest and longest-lasting worldwide (approximately 200 physics institutes, collaborating for 20 years and more), develop strong post-traditional communities as a practical accomplishment and normative choice. These communal socio-technical forms erase the individual scientist as an epistemic subject. What emerges instead is a distributed form of agency that involves the human collective, the detector itself, which may be said to be the real producer of experimental effects, and the "experiment," which comprises the human collaboration and the detector and the processes that link them. The notion of an epistemic culture also brings into focus questions of the cultural definition of "non-agents," the "objects" of knowledge—though putting it this way is misleading, because objects and subjects can both have active and passive elements. In particular, objects of knowledge—for example, the microbes Pasteur investigated, the chromosomes in McClintock's research on the cytogenetics of maize, the space shuttle studied by Vaughan, a detector in high-energy physics, or even a mathematical object—tend to be "doers" in scientific research; they have powers, produce effects, may have their own internal environments, mold perception, and shape the course of an experiment. Furthermore, what is at stake in epistemic cultures is not simply the definition of subjects and objects but their reconfiguration in relation to the natural and social orders as they exist outside expert systems and in relation to each other—and this is one way in which we can give concrete sense to the merging of lifeworlds articulated in epistemic cultures. When objects are reconfigured—for example, when subatomic particles become signs that particle collisions leave in detector materials and that are prone to false appearances, fading, "ghost" production and the like—a specific intramural reality is created that brings physics close to semiotics, complete with interpretation problems, long chains of meaning reconstruction, a certain loss of the empirical, and the proliferation of virtual life (through simulation, for instance).

Another aspect of epistemic cultures concerns object relations regimes. How, for example, does one, in a given "knowledge area," gain access to the "referent," the targeted object, and to stability in outcomes? How does a science understand and enact empirical inquiry, and what strategies does it use? High-energy physics can again serve as an example. It is intensely focused on negative knowledge. Negative knowledge is not non-knowledge but rather knowledge of the limits of knowing, which is gained from the disturbances, distortions, errors, and uncertainties of research. In theology (primarily Eastern Christian but also Jewish and Hindu) there is an approach called the apophatic or *via negativa*, which advocates the study of God in terms of what He is *not* rather than what He is, since no positive assertions can be made about His essence. High-energy physics experiments show similar preferences. One narrows down the regions of positive knowledge by developing what one might call liminal knowledge. One delimits the properties of the objects of interest by measuring the properties of the objects that interfere with them and distort them.

Thus, an improved data run in the experiments observed may lead to a longer list of error terms and more refined error measurements rather than to an elimination of errors. Another illustration of the liminal approach is limit analyses, which appear to be the most frequently produced results of the various experiments. A limit analysis identifies the boundaries of a domain within which one has searched the terrain but "not found" the physical process of interest—hence, a domain within which the process is unlikely to occur. The liminal approach is a culturally specific elaboration of an object relations regime. The author did not find this approach, for example, in the field of molecular biology; nor did she find there the emphasis on reflexive self-observation and self-understanding in which the liminal approach is embedded.

## Macro-Epistemics

The notion of an epistemic culture "cuts the idea of culture down to size," relocating it in the micropractices of laboratories and other habitats of knowledge practices. It follows Clifford Geertz's recommendation either to make culture a delimited notion or else to abandon it (Geertz, p. 13). Not all places of knowledge, however, are bounded spaces. There is a case to be made for including in the empirical agenda more distributed locations, some of which are linked up over vast distances. For example, consider the international networks of programming wizards who, between them, assemble and upgrade software code, which they may make freely available. These networks are made possible by electronic connections, and they have global reach. Still, these are delimited groups of specialists working in one facility, that of the electronic space, and drawn together by code, which becomes their centering object. A more decentered setting involving macroepistemic actors (institutions that take on specific knowledge-related tasks) has emerged in economic areas. One example consists of the circuits of information and regulation that focus on the *global financial architecture*, which the International Monetary Fund (IMF) defines as "the framework of markets, practices and organizations that governments, businesses and individuals use when they carry out economic and financial activities." In this area, the IMF and other overseeing agencies, working groups, and "conferences" of banks and governments, as well as statistical offices and rating agencies that rate a country's economic "worth" (thereby validating knowledge that circulates about it), observe, analyze, and process information that flows between governments and economic institutions. A recurring theme in this context is transparency of information, which is held to be necessary to prevent economic crisis, to guarantee the stability of the system, and to grow global markets cooperatively. Thus, on one level, the international financial system is an epistemic system: It rests on the architecture of observation rules and strategies, on the work of national and transnational units that generate and validate the observations, and on the information flows that circulate between these units. The epistemic system safeguards the economic system. Information rules and strategies seek to discover the "truth" of units' states and other relevant developments in order to detect possible problem areas, anticipate ripple effects, and spot signs of turns in economic cycles. The units in the circuit do not produce a single outcome, and no one may fully publish their results. Nonetheless, we are confronted with a knowledge system; it is the design of the system, for example the epistemic rights and procedures of the respective units, that is at stake in discussions of the global financial architecture. Merging lifeworlds in this case is a formidable task that includes negotiating compatibilities between different nation-state administrations and their political cultures.

## Effects of Epistemic Cultures

Characteristics of epistemic settings may be reproduced in the context of these settings, where they may reinforce independent tendencies and trends. One area where this is manifest are forms of binding between human beings and nonhuman objects. Expertise depends on object relations. Object worlds make up the embedding environments in which expert work is carried out, thus constituting something like an emotional home for expert selves. As a consequence, object environments define individual identity and situate and stabilize selves. Experts also develop intimate relationships with objects of expertise. They learn to handle and observe them and they also imagine their interior states as they attempt to understand them. Experts are linked to objects through libidinal sequences of wantings projected on objects, which continually pose new questions and transform themselves into new objects, as pointed out by Hans-Jörg Rheinberger. The libidinal, reciprocal, and in other ways binding components of experts' object involvements make it plausible to construe these involvements as forms of *postsocial relations* rather than simply as work or instrumental action. Object relations of the sort exemplified are also present in general social life, where they reproduce aspects of epistemic cultures. Part of the epic character of the transformations many authors observe may have something to do with "objectualization," an increased orientation toward objects as sources of the self, of relational intimacy, of shared subjectivity and social integration. One driving force behind this may be the declining functionality of human relationships. In this scenario, objects may simply become the risk-winners of human relationship risks and failures, and the entities towards which attachments are redirected. But another driving force may the objects themselves, which, in everyday life, are no longer fixed entities in the way simple instruments such as a hammer are. As objects in everyday life become (high) technological devices, some of the properties these objects have in epistemic settings carry over into daily life: their unfolding nature, their intelligence, their relational demands, and the relational possibilities they offer (see, for example, the study by Sherry Turkle). The transfer onto society of a new definition of objects sustains and enhances object relationships as forms of binding the self with the other.

*See also* **Knowldege Construction**

## BIBLIOGRAPHY

Beck, Ulrich, Anthony Giddens, and Scott Lash. *Reflexive Modernization.* Stanford, CA: Stanford University Press, 1994.

Collins, Harry M., and Trevor J. Pinch. *The Golem: What Everyone Should Know About Science.* 2nd. ed. Cambridge, U.K.: Cambridge University Press, 1998.

Fox Keller, Evelyn. *A Feeling of the Organism: The Life and Work of Barbara McClintock.* San Francisco: W. H. Freeman, 1983.

Galison, Peter, and David Stump, eds. *The Disunity of Science.* Stanford, CA: Stanford University Press, 1996.

Geertz, Clifford. *Available Light: Anthropological Reflections on Philosophical Topics.* Princeton, NJ: Princeton University Press, 2000.

International Monetary Fund. *Progress in Strengthening the Architecture of the International Financial System: A Factsheet.* Washington, DC: International Monetary Fund, July 2000. http://www.imf.org/external/np/exr/facts/arcguide.htm.

Knorr Cetina, Karin. *Epistemic Cultures. How the Sciences Make Knowledge.* Cambridge, MA: Harvard University Press, 1999.

Knorr Cetina, Karin. "Culture in Global Knowledge Societies: Knowledge Cultures and Epistemic Cultures." In *The Blackwell Companion to the Sociology of Culture,* edited by Mark Jacobs and Nancy Weiss Hanrahan. Oxford: Blackwell Publishers, 2004.

Latour, Bruno. *The Pasteurization of France.* Cambridge, MA: Harvard University Press, 1988.

Restivo, Sal. "The Social Roots of Pure Mathematics." In *Theories of Science in Society,* edited by Thomas Gieryn and Susan Cozzens, 120–143. Bloomington: Indiana University Press, 1990.

Rheinberger, Hans-Jörg. *Toward a History of Epistemic Things.* Stanford, CA: Stanford University Press, 1997.

Turkle, Sherry. *Life on the Screen.* New York: Simon and Schuster, 1995.

Vaughan, Diane. *The Challenger Launch Decision: Risky Technology, Culture, and Deviance at NASA.* Chicago: University of Chicago Press, 1997.

*Karin Knorr Cetina*

# F

## FAMILY, THE, AND SCIENCE.

The family is a core social institution. While marriage and children have historically defined a family, today single parents living with children, heterosexual couples remaining without children, and same-sex couples with or without children have required redefinition of the American family.

Science itself is woven into the changing family. The average life expectancy rose from 47 years in 1900 to 77 years in 2000, so people marry later in life and have more years to spend together. Women are older when they have children. They have fewer births, and the children who are born are more likely to live. In 1900 the infant mortality rate (the number of deaths to infants less than one year of age per 1000 births) was well over 100; by 2000 the rate had dropped to 10. In 1900 the most common household size was seven. Between 1940 and 2000, the most common household size was two. The "sandwich generation" composed of those in their fifties will spend as many years caring for aging parents as they did caring for their own children. The following sections will explore just a few of the discoveries in the world of science and their relationship to changes in the family.

### Sex and Sexuality

While sex and sexuality were American taboo in the earlier part of the 1900s, growing knowledge about human biology, sexual reproduction, and birth control have contributed to the decoupling of sex from procreation and marriage. Whereas 51 percent of men and 87 percent of the women born before 1890 were virgins at the time of marriage, less than 10 percent of Americans today are still virgins by the time they reach 20 years of age. Marriage is no longer the frontier for exploring sexuality.

Between 1938 and 1963 Alfred Kinsey and his colleagues collected sexual histories from more than 18,000 research participants. The famous "Kinsey Reports," *Sexual Behavior in the Human Male* and *Sexual Behavior in the Human Female,* along with widespread public lectures, set the stage for the *sexual revolution.* Though his methods have come under criticism, Kinsey's work challenged the Freudian psychoanalytic model of sexuality and replaced it with a biological and social science-based paradigm.

Kinsey's work and a boom in health movement publications such as *Our Bodies, Ourselves* (Boston Women's Health Book Collective 1973) had people thinking and talking about their bodies and sexualities in ways that would have shocked earlier generations. Men and women were encouraged to explore their own physical bodies on their own and in partnerships. Couples put their own sexual relationships under the microscope for inspection, comparison, and experimentation.

Although homosexuality appears in anecdotes throughout history, Michel Foucault, a French social historian and theorist, suggests that the construction of an identity or a category of people who are identified as "homosexual" is a recent phenomenon constituted through Westphal's 1870 article on "contrary sexual sensations" (Foucault, p. 43). Kinsey's work also contributed to a growing knowledge of the commonality of homosexual encounters, as well as a range of other sexual practices including oral sex, anal sex, and masturbation.

While social science and science critique continues to make the case for the social shaping and social constructedness of sexuality (see, for example, the works by Anne Fausto-Sterling and by Judith Stacey and Timothy Biblarz), biological and genetic research has tried to assert linkages between sexual orientation and biology. Framing sexual orientation as biological has been part of the political framework for the gay rights movement. A 1989 study found that societies that believe gays "were born that way" are less homophobic. Yet gay rights activists and others are aware of a sordid history in which biological arguments have been the source of genocide. Nazi Germany used biological rationales to support the extermination of Jews, gays, and other groups of designated undesirables. Yet science headlines that have suggested a biological basis for sexuality, whether founded or not, have been a part of changing attitudes about gays and gay families. As science impinges on public opinion regarding gay families, it also affects a myriad of other family issues and laws including domestic partner rights for same-sex couples and rights to fertility treatment, adoption, and child custody.

## Reproductive Control

Reproductive science and technologies have been central to our ability to structure and plan for the inclusion or exclusion of children in the family. Family size, timing of children, and the options for gay couples and singles to have children have greatly expanded in the last century.

The history of the practice and science of birth control in and out of family arrangements is woven through the archives of many societies. The *Petrie Papyrus* is an Egyptian medical guide dated to 1850 B.C.E. and contains references to birth control methods. Some methods handed down through the ages were effective. Some were not. Others caused complications.

Andrea Tone documents early correspondence between newlywed couples and their friends indicating a joint endeavor by married men and women to control conception. In 1858 we see early descriptions of rubber caps and in 1869 of rubber condoms, years after Charles Goodyear's 1839 discovery of the vulcanization of rubber and its many uses. Although many women knew otherwise, physicians until the 1920s continued to insist that women, like other mammals, ovulated during their periods, so physicians instructed them on a "safe period" during the middle of their cycle.

Poor families and families of color have a dark history with reproductive control. Although eugenics began to be discredited as bad science in the 1940s, it is estimated that more than 70,000 sterilizations were performed under U.S. state sterilization statutes that were implemented using deceit and force. These statutes served as models for Nazi Germany's sterilization law and practice. More recent social policy that funded the insertion of Norplant birth control devices under the skin for women on public support, but did not fund their removal, indicates an ongoing social dilemma regarding the application of science to control select American families.

On the other end of the spectrum, science provides more options for conceiving children, particularly for those American families with the resources to pay. Artificial insemination dates back to antiquity. *In vitro* fertilization for humans was introduced in 1978 in England when Robert Edwards, an embryologist, and Patrick Steptoe, a gynecologist, performed the first successful such procedure. By 2003 more than 20,000 babies had been born worldwide using this practice. Fertility science has been a godsend for many couples with difficulties conceiving. Some show up on the doorstep of fertility specialists as early as three months after first trying to conceive. The drugs and procedures, however, come with their own physical, psychological and emotional costs. Women submit themselves to months of often costly and sometimes debilitating treatments to fulfill a dream of family that privileges biological connection to children.

## Childbirth and Family

Birth used to be a family event that took place in the home. In 1900 midwives still "caught" 50 percent of babies born in the United States. Yet this period also marked the rise of medicine and the practice of hospital-based obstetrics. With newly found knowledge of germs and sanitary techniques, including simple hand washing, England trained its midwives on the new advances. In the United States, however, midwives were construed as dirty and ignorant, and a new cadre of white, male, and inexperienced medical practitioners maneuvered birth into the institution of the hospital.

The new obstetricians introduced more "civilized" techniques such as laying down laboring women. This made birthing more difficult for mothers but easier to "manage" for physicians. This position and drug interventions created opportunities for greater intervention through instruments such as forceps and procedures such as episiotomies. Such practices sped up the birth process but often caused other complications.

The Natural Childbirth Movement began in the late 1940s and was promoted in books such as the British obstetrician Grantly Dick-Read's *Childbirth without Fear* (1944). Other prominent names that later entered the movement included Ferdinand Lamaze of France and Robert Bradley of Colorado. Each helped to renormalize pregnancy, return greater control to women, and bring attention back to pregnancy and birth as family events. Bradley's *Husband-Coached Childbirth* (1965) moved fathers off the waiting room floor and put them in a central role for birthing. This move was symbolic of the larger role that men would later come to play in the raising of children.

At the close of the twentieth century, family planning moved to the genetic level. Surveillance of pregnancy and birth became expected with an ever-expanding list of tests and monitoring procedures. Technologies such as ultrasound provide windows into the developing life that have speeded up the bonding process for family members. Techniques such as "embryo selection" allow an anticipating family to select and implant only those embryos without a particular genetic disease, providing greater assurance of the creation of healthier families.

Yet today women and their partners no longer simply hope to deliver a "healthy" baby; they are experiencing an increased pressure *not* to deliver a "disabled" baby. Although some women feel better gathering as much information as possible about their developing babies, other women find that the process makes an already stressful situation worse. At least part of this heightened anxiety is due to the fact that most women and their partners who discover fetal abnormalities must choose between delivering and aborting. *In-utero* treatment of a human fetus, as well as germ line (egg and sperm) genetic alteration, is now in scientific infancy. According to the National Human Genome Research Institute, human germ line gene transfer was not being actively investigated as of late 2004 and remained controversial on multiple fronts. Decisions around the development and application of these technologies will add further complexity to the legal, ethical, and social reproductive issues facing our changing families.

## The Raising of Children

In a child we see the future. In the family we see the future of the child. For better or worse, it is to science that we entrust the future of both. During the twentieth century we have seen a shift from a focus on family and immediate needs, to a focus on the needs and interests of the child. Psychology's efforts to be recognized as a science played an important role in transforming the field of child raising into a science. The early American child was perceived in many ways as a miniature, savage adult. Yet, beginning at the end of the nineteenth century, early psychologists, mother's movement leaders and even the leaders of domestic science began to frame childhood as serious business that must not be left to the untrained.

In the early part of the twentieth century, research institutes and government conferences became centers for studying and doing something about the child. The early tone was one of restraint and regularity. Women's magazines and child-raising books and manuals became central mechanisms for spreading the new behaviorist theories on the subject.

That tide turned when Benjamin Spock popularized a brewing child-centered approach with the publication of *The Common Sense Book of Baby and Child Care* (1946). Against the backdrop of behaviorism, Spock's ideas were revolutionary. If sales are an indicator of reception, it appears that parents around the world were pleased with the new message from the science of child development. The book went through seven editions, was translated into 39 languages, and sold more than 50 million copies. Earlier parents had been told that holding a crying child only spoils him or her. Spock encouraged parents to go with their instincts, cuddle the crying child, be flexible and more focused on the individual needs of each child, and have fun parenting.

The struggle between child-centered, permissive parenting and more restrictive parenting did not end with Spock. Parents today often find themselves caught between the experts. They might read "attachment"-oriented books such as *The Baby Book* by Martha and William Sears. This approach draws on non-Western child-rearing practices such as the "family bed" that emphasize physical and emotional closeness between parent and child. At the same time, parents may reference Richard Ferber's *Solve Your Child's Sleep Problems* (1986), which emphasizes child independence and self-comfort. This formula provides for progressively increased lengths of letting the child cry in order to train her to fall asleep on her own.

The fields of pharmacology and nutrition science have had a tremendous impact on improving childhood health and all but eradicating diseases such as polio and measles. Babies are more likely to be carried to term, are born healthier, and are much less likely to die. Yet, within the overall wave of medicalization, we see boundaries extended to include an increasing array of behaviors and symptoms for medical treatment and drug treatments.

As just one example, the psychiatric profession developed what is now known as Attention Deficit Hyperactivity Disorder (ADHD) as a diagnostic category in the 1950s. Ritalin (methylphenidate), the drug most prescribed for ADHD, is estimated by the World Health Organization to be given to "10 to 12 percent of boys between the ages of 6 and 14 in the United States," according to Peter Breggin. As a practicing psychiatrist, parent, and author of *Talking Back to Ritalin* (1998), Breggin has been critical of the extensive use of the drug as well as of families and physicians who, in too many cases, have not examined for change the very context of the child's life.

Science, therefore, is never in isolation in these endeavors. As in other areas in which science and medicine have extended their reach, the culture is ripe for their "magic wand." Sometimes the magic reduces disease and human suffering. Sometimes the magic creates controversy and a need for greater reflection on the context that created the need in the beginning.

## Home Sweet Home: The Rise of Domestic Science

More than 90 percent of the population in the early 1800s lived in rural areas. The preindustrial home was the center of production for daily life. Family members helped with the production of vegetables, bread, clothing, soap, candles, and even medicines. The industrial revolution and urbanization shifted many types of production outside the home and created new forms of home production. Some feminists such as Oliver Schreiner and Charlotte Perkins Gilman suggested that the transfer of production out of the home freed women to join men in the public sphere. Yet others such as Ellen Swallow Richards struggled to reinvent the home as women's domain and the family as her central project.

The foundations for the field of "domestic science" had been bubbling throughout the nineteenth century, with a stream of advice manuals and books such as Lydia Maria Child's *American Frugal Housewife* (1828) and Catherine Beecher and Harriet Beecher Stowe's *The American Woman's Home* (1869). Yet it was Richards who moved the field to new levels. In 1871 she became the first female student and then faculty member to negotiate a place for herself at the Massachusetts Institute of Technology. Having trained at Vassar in laboratory techniques then in their infancy, Swallow made a place for herself in academy and for science in the home by applying these techniques and others to domestic matters: sanitation, product testing, food science, ventilation, and systematic cleaning arrived on the heels of urbanization, which had brought with it all the health issues of human crowding.

Domestic Science legitimated higher education for women and professionalized the role of the homemaker. Swallow's focus on the home was more of a concession than a driving vision. She trained secondary education teachers (many of whom were also women) in domestic science to make the presence of women in the sciences palatable in the halls of men. Swallow's own work domesticated earlier work in epidemiology and germ theory. Cleaning and sanitation became a moral responsibility of every good homemaker.

Cleaning and homemaking took on a routinization endemic to modern time. The scientific management of production promoted in industry by Frederick W. Taylor and others had similar effects in the home. Instead of freeing time for homemakers, it just increased the standards of production. "Wash day" became every day instead of once a month as when laundry was done by hand. Barbara Ehrenreich and Deirdre English argue that despite all the emphasis on cleaning, it is not clear that the more intense standards improved family health. Clearly the key scientific contributions to improving the health of family members were water sanitation, waste management, and immunizations, all of which were logical extensions of germ theory. Nonetheless, with new standards of homemaking based on science, the American home was forever changed.

Domestic science home visits became a method for assimilating urban poor families into middle-class American desires, if not luxuries of life. A new sense of order for family was conveyed; family schedules, cleanliness standards, food choice, and food preparation were all conveyed under the auspice of "right living." The movement also paved the way for "time saving" appliances that have ironically increased the amount of time that American families spend on household labor.

Test kitchens such as the "Good Housekeeping Experiment Station," established in 1900, put the science of food chemistry into corporate practice. Yet the influence of these kitchens was as much about marketing and sales, with the "Good Housekeeping Seal of Approval," as it was about science. Photos of female home economists dressed in lab coats in kitchens gave scientific credibility to products such as Corning Ware and helped convince homemakers to use them in their own kitchens.

## The Family in Motion

The institution of medicine and the underlying science rose to glory in the late nineteenth and twentieth centuries with discoveries in germ theory, sanitation, immunization, reproductive technologies, and psychology. Many of the changes made in the name of science eased human sufferings. Yet some of those changes increased hardships instead as women, people of color, and other marginalized groups became victims of the premature discarding of traditional methods and the introduction of harmful new practices, experiments, and related social policies. The field of genetic science is in the current spotlight, with both utopian and dystopian promise. "How will science change the family?" is perhaps less the question than "What will tomorrow's family do with the science?"

*See also* **Abortion/Anti-Abortion Conflict; Biomedical Technologies; Birth Control; Sex and the Body in Science; Sex and the Body in Technology**

### BIBLIOGRAPHY

Boston Women's Health Book Collective. *Our Bodies, Ourselves.* Boston: New England Free Press, 1973.

Bradley, Robert. *Husband-Coached Childbirth.* New York: Harper and Row, 1965.

Breggin, Peter. *Talking Back to Ritalin.* Monroe, ME: Common Courage Press, 1998.

Breggin, Peter. 2000. Testimony before the Subcommittee on Oversight and Investigations Committee on Education and the Workforce. U.S. House of Representatives, Washington, DC. http://www.breggin.com/congress.html.

Bullough, V.L. "Reviewer Alfred C. Kinsey" (book review). *Journal of Sex Research* 36 (1999): 309–311.

Clarke, Robert. *Ellen Swallow: The Woman Who Founded Ecology.* Chicago: Follett Publishing Company, 1973.

Conrad, Peter. "The Discovery of Hyperkinesis: Notes on the Medicalization of Deviant Behavior." *Social Problems* 23 (1975): 12–21.

Ehrenreich, Barbara, and Deirdre English. *For Her Own Good: 150 Years of the Experts' Advice to Women*. New York: Anchor Books, 1978.

Fausto-Sterling, Anne. *Sexing the Body: Gender Politics and the Construction of Sexuality*. New York: Basic Books, 2000.

Ferber, Richard. *Solve Your Child's Sleep Problems*. New York: Simon and Schuster, 1986.

Foote, R. H. *The History of Artificial Insemination: Selected Notes and Notables*. American Society of Animal Science. 2002. http://www.asas.org/jas/symposia/esupp2/Footehist.pdf.

Foucault, Michel. *The History of Sexuality*. Vol. 1: *An Introduction*. Translated by Robert Hurley. New York: Pantheon Books, 1978.

Georgia Reproductive Specialists. *In Vitro Fertilization* (IVF-ET). 2003. http://www.ivf.com/ivffaq.html.

Gladue, Brian A., Richard Green, and Ronald E. Hellman. "Neuroendocrine Response to Estrogen and Sexual Orientation." *Science* 225 (1984): 1496–1499.

Grantly, Dick Read. *Childbirth without Fear: The Principles and Practice of Natural Childbirth*. New York: Harper and Brothers, 1944.

Hanna, Kathi E. *Germline Gene Transfer*. Bethesda, MD: National Human Genome Research Institute, National Institutes of Health. September 2004. http://www.genome.gov/10004764.

Hobbs, Frank, and Nicole Stoops. *Demographic Trends in the 20th Century*. Washington, DC: U.S. Census Bureau, 2002. Report No. CENSR-4.http://www.census.gov/prod/2002pubs/censr-4.pdf.

Jensen, Mari N. "Heterosexual Women Have Noisy Ears." *Science* 153 (1998): 151.

Kinsey, Alfred C., Wardell B. Pomeroy, Clyde C. Martin, and Paul H. Gebhard. *Sexual Behavior in the Human Female*. Philadelphia: W. B. Saunders, 1953.

Kinsey, Alfred C., Wardell B. Pomeroy, and Clyde E. Martin. *Sexual Behavior in the Human Male*. Philadelphia: W. B. Saunders, 1948.

Leavitt, Sarah. *From Catharine Beecher to Martha Stewart: A Cultural History of Domestic Advice*. Chapel Hill: University of North Carolina Press, 2002.

LeVay, Simon. "A Difference in Hypothalamic Structure between Heterosexual and Homosexual Men." *Science* 253 (1991): 1034–1037.

Lund, Dale A. "Caregiving." In *Encyclopedia of Adult Development*, edited by R. Kastenbaum, 57–63. Phoenix, AZ: Oryx Press, 1993.

Ogburn, W. F., and M. F. Nimkoff. *Technology and the Changing Family*. Boston: Houghton Mifflin Company, 1955.

Puterbaugh, Geoff. *Twins and Homosexuality: A Casebook*. New York: Garland, 1990.

Rapp, Rayna. "Real-Time Fetus." In *Cyborgs and Citadels: Anthropological Interventions in Emerging Sciences and Technologies*, edited by G. L. Downey and J. Dumit. Santa Fe, NM: School of American Research Press, 1997.

Risman, Barbara, and Pepper Schwartz. "After the Sexual Revolution: Gender Politics in Teen Dating." *Contexts* 1 (2002): 16–24.

Roberts, Dorothy. *Killing the Black Body*. New York: Pantheon, 1997.

Rothman, Barbara Katz. *In Labor: Women and Power in the Birthplace*. New York: W. W. Norton and Company, 1982.

Rubenstein, William B. *Lesbians, Gay Men, and the Law*. New York: The New Press, 1993.

Sears, William, and Martha Sears. *The Baby Book: Everything You Need to Know About Your Baby from Birth to Age Two*. Boston: Little, Brown, 2003.

Shilts, Randy. *And the Band Played On*. New York: Penguin Books, 1987.

Stacey, Judith, and Timothy Biblarz. "(How) Does the Sexual Orientation of Parents Matter?" *American Sociological Review* 66 (2001):159–183.

The Dr. Spock Company. *Dr. Benjamin Spock, 1903–1998*. 2002. http://www.drspock.com/about/drbenjaminspock/0,1781,,00.html.

Tone, Andrea. *Devices and Desires: A History of Contraceptives in America*. New York: Hill and Wang, 2001.

U.S. Census Bureau. *Selected Historical Decennial Census Urban and Rural Definitions and Data*. Washington, DC: U.S. Census Bureau, 2002. http://www.census.gov/population/www/censusdata/ur-def.html.

Wajcman, Judy. *Feminism Confronts Technology*. University Park: Pennsylvania State University Press, 1991.

*Mary Virnoche*

# G

## GENDER AND GLOBALIZATION.

Popular commentators, like many social scientists, believe that the invention and diffusion of digital technologies are fundamentally transforming our society. Much emphasis is placed on major new clusters of scientific and technological innovations, particularly the widespread use of information and communication technologies (ICTs), and on the convergence of ways of life around the globe. The increased automation of production and the intensified use of the computer are said to be revolutionizing the economy and the character of employment. Globalization gurus stress the historically unprecedented intensity, extensity, and velocity of global flows, interactions, and networks embracing all social domains. In the global, networked society, the dominant form of work becomes information- and knowledge-based work. At the same time, leisure, education, family relationships, and personal identities are seen as molded by the pressures exerted and opportunities arising from the new technical forces.

Theorists of the global society have, for the most part, failed to consider whether this technological revolution might have a differential impact on women and men. The common theme is that everything in the digital future will be different, but it is not clear whether gender relations will also be different—in fact, the question is seldom raised. Although the optimistic commentators on the digital revolution promise freedom, empowerment, and wealth, rarely do they show interest in the relationship between technology and gender. Their over-sight is particularly disappointing given the contribution that feminist scholarship has made to our understanding of the gendered nature of both work and technology. The key issue to be considered here is the extent to which the traditional hierarchies of the gender order are being disrupted by the digital global economy.

### The Network Society

The notion that we are living in a postindustrial era has been absorbed into everyday language. Several different schools of thought exist, but the recurring theme is a claim that theoretical knowledge has taken on a qualitatively new role. One of the best-known commentators of such change is Manuel Castells, who argues that the revolution in information technology is creating a global economy, the product of an interaction between the rise in informational networks and the process of capitalist restructuring. In the "informational mode of development," labor and capital—the central variables of the industrial society—are replaced by information and knowledge. In the resulting "network society," the compression of space and time made possible by the new communication technology alters the speed and scope of decisions. Organizations can decentralize and disperse, with high-level decision making remaining in major cities while lower-level operations, linked to the center by communication networks, can take place virtually anywhere. For Castells, the information age, organized around "the space of flows and timeless time," marks a new epoch in the human experience.

Although optimistic and pessimistic visions of the information age exist, all such visions focus on the assumed outcomes for employment. The optimists see the expansion of the information-intensive service sector as producing a society based on lifelong learning and a knowledge economy. This implies that a central characteristic of work will be the use of expertise, knowledge, judgment, and discretion in the course of producing a product or service, requiring employees with high levels of skills and knowledge. The pessimistic approach, by contrast, sees growing technology-induced unemployment and increased vulnerability to global capital. It argues that automation standardizes work tasks and diminishes the need to exercise analytical skills and theoretical knowledge.

What is common to most of these understandings of the new social order is their tendency to adopt an evolutionary and technologically determinist stance. The idea that technology, specifically information and communication technology, is the most important cause of social change permeates Castells' analysis of the network society. Similarly, there is a tendency to conceptualize these technologies in terms of technical properties and to construct their relation to the social world as one of implications and impacts. The result is a rather simplistic view of the role of technology in society. In this, Castells is typical of most scholars of globalization, who fail to engage with the burgeoning literature in science and technology studies (STS) that has developed over the last two decades.

This is not to deny the transformative potential of technology. Rather, it is to emphasize that technological change is itself shaped by the social circumstances within which it takes place. Such a perspective prevents a purely technological interpretation and recognizes the embeddedness of technology in society and the variable outcomes of these technologies for different social groups. Technology and society are inextricably bound together, meaning that power, contestation, inequality, and hierarchy inscribe new technologies. Although ICTs can indeed be constitutive of new social dynamics, they can also be derivative of or merely reproduce older conditions. The sexual division of labor, however, has always been central to technological development and the organization of work. What then are the implications of digitalization for gender equality?

## The Service Economy

One of the most important social changes of the twentieth century was the feminization of the labor force. The sharp separation between work and home has been eroded by this development, at least for some social strata. Most striking is the sharp increase in the labor-market participation of married women and women with young children. The causes of this feminization are complex, but clearly they link to the substantial growth in service-sector activity and employment. In most advanced capitalist economies, the manufacturing sector has declined, while new jobs have been created in services. In one sense, this advantages women, since they have long been associated with service work, especially jobs in clerical work, retail, catering, and the health and education professions.

However, accompanying the feminization of the labor force has been a dramatic growth in economic inequality between different groups of women. This phenomenon is especially striking in the United States, where feminist demands for gender equality have been more potent than elsewhere and where inequality in wealth and income has increased sharply in recent years. Although this trend to feminization has not brought about a major breakdown of gender segregation, there has been a significant movement of women into traditionally male professions such as law, medicine, and managerial occupations. These elite women have unprecedented access to well-paid, high-status occupations, while at the bottom of the occupational hierarchy, women have expanded their share of already femininized lower-skilled or lower-paid occupations. There has been an enormous growth among part-time workers, temporary workers, and at-home "independent contractors," the majority of whom are women.

The increase in temporary or casual work that characterizes this era of economic globalization has been made possible by the proliferation of ICTs that support it. For example, the financial service and telecommunications industries rely heavily on information technology (IT) for service and sales delivery. Business transactions ranging from personal retail banking to transnational financial market deals are increasingly mediated by IT. Automation not only results in the replacement of routine manual labor; IT systems also generate detailed information about the work process. Moreover, company-wide intranets can coordinate work tasks, disseminate information, and facilitate the exchange of opinions among employees in different ranks and functional departments. All these features of IT have dramatically changed the way work is organized.

Overall, then, the shift to a postindustrial society cannot be viewed simply as progress for women. On the one hand, women are more fully integrated into the paid labor force and are unlikely to be relegated to the domestic sphere of yesteryear. On the other hand, many of the new jobs created are temporary and part-time. Although these jobs offer women more flexibility, the use of IT by employers to fine-tune their labor requirements can cost women dearly in terms of pay, working conditions, and training opportunities. The skill requirements for much service-sector employment tend to be social and contextual, making them less amenable to formal measurement. The issue of how skills or competencies are perceived, labeled, accredited, and rewarded is critical for women's ability to participate in and benefit from the "knowledge-based economy."

The failure to regard women's social and communicative skills as knowledge-based and reward them accordingly has strong echoes with the way in which women have been traditionally defined as technically unskilled, and thus excluded from well-paid work. The association between technology, masculinity, and the very notion of what constitutes skilled work was and is still fundamental to the way in which the gender division of labor is reproduced. Historically, this pattern was a consequence of the male domination of skilled trades that developed

during the Industrial Revolution. Male craft workers actively resisted the entry of women to areas of technological work to protect their own conditions. Men's traditional monopoly of machinery, and the subsequent formation of engineering as a male profession, have been identified as key to maintaining the taken-for-granted association of technical work with men. The fact that the culturally dominant form of masculinity in contemporary Western society is still strongly associated with technical prowess and power is a legacy of this history.

For all the rhetoric about women prospering in the emerging digital economy, the signs are that men's domination of science and technology has continued. Women are making few inroads into technology-related courses in the information technology, electronics, and communications sector, and they face considerable barriers when they attempt to pursue a professional or managerial career in this sector. Indeed, the percentage of undergraduate degrees in computer science awarded to women in the United States nearly halved between 1984 and 1999 (37 percent of all degrees in 1984 compared to 20 percent in 1999). The result is that women are chronically underrepresented in precisely the jobs that are key to the creation and design of technical systems in the new economy. Increasingly, these technical systems constitute the world we inhabit.

### Changing Location

All visions of the global network society place great emphasis on the way ICTs allow for an increasing disassociation between spatial proximity and the performance of paid work. The idea is that with the advent of technological innovations, production no longer requires personnel to be concentrated at the place of work. In this scenario, using the home as workspace liberates people from the discipline and alienation of industrial production. Computer-based homework or telework offers the freedom of self-regulated work and a reintegration of work and personal life. Moreover, an expansion of teleworking will allegedly lead to much more sharing of paid work and housework, as men and women spend more time at home together. Mothers are particularly seen as the beneficiaries of this development, as working from home allows much greater flexibility to combine employment with childcare.

Futurologists commonly assume a dramatic increase in teleworking, but the number of those employed regularly to work online at home is rather small. Nevertheless, teleworking has important implications for the way women's work is understood. We need to distinguish between skilled or professional workers who work from home and the more traditional "homeworkers" who tend to be semiskilled or unskilled low-paid workers. The former certainly do have more choices about how they schedule their work to fit in with the rest of their everyday lives. However, these teleworkers, who tend to work in such occupations as computing and consultancy, are typically men. Women who telework are mainly secretarial and administrative workers. Thus, a rather conventional pattern of occupational sex segregation is being reproduced in this new form of work.

Indeed, women and men are propelled into teleworking for very different reasons. While women's main motivation is childcare and domestic responsibilities, men express strong preferences for the flexibility, enhanced productivity, convenience, and autonomy of such working patterns. The common media image of a woman working while the baby happily crawls across a computer is misleading. There is an important difference between being at home and being available for childcare. Women continue to carry the bulk of responsibility for domestic work and childcare, and, for them, telework does not eliminate their double burden. Even among the minority of professional women who work from home, few are able to separate the demands of motherhood and domesticity from paid work. For men, who can more easily set up child-free dedicated "offices" at home, telework often leads to very long and unsocial hours of work. These long hours tend to militate against a more egalitarian and child-centered way of life.

More significant than telework, although it has received much less attention, is the capacity of ICTs to facilitate and encourage people to bring work home from the office. Not only has there been a general increase in the number of hours worked at the workplace for managers and professionals, but the expectation of availability has also been greatly extended with the advent of mobile phones, email, and fax machines. In this sense, the boundaries between the public world of work and the private home have become blurred. However, this almost always happens in such a way as to facilitate the transfer of work into the home rather than the transfer of home concerns into the workplace. This has made the balance between work and home at senior levels even more difficult. ICTs may have raised productivity, but they have certainly not reduced working time.

## Globalization: Divisions and Unity

Any discussion of the geographical relocation of work sites must address the use of female labor in the developing world. Although the international division of labor is not a new phenomenon, innovations in ICTs allow a spatial flexibility for a growing range of tasks. The use of Third World female labor by multinational manufacturing industries offering poorly paid assembly jobs is well documented. Garment assembly and seamstress work is subcontracted to small offshore companies in the South, while the process of design and cutting is carried out in the North.

With increasingly automated means of coordinating marketing, production, and customer demand on a daily basis, however, garment companies have begun to reverse their reliance on Third World labor. Western and Japanese companies alike are increasingly intent upon "close-to-market" strategies that involve subcontracting work to smaller companies in the West. In this move back to their host countries, companies mainly employ women from immigrant and ethnic minority groups, ensuring a captive, regional labor force that is compelled to accept low wages and exploitative working conditions. Despite the high levels of capital investment and advanced ICTs used by these firms, there is little transfer of technical skills and expertise to the women who work in these manufacturing jobs.

White-collar, professional, and clerical jobs are also moving to developing countries, underpinned by the diffusion of ICTs. The emergence of the Indian IT industry is a notable case. Skilled software design and development projects are sited in the West, whereas the lower-level programmers employed are located in offshore companies, maintaining consistently high production rates for very low wages. Countries such as India have a ready pool of female labor available for software work—women who are well educated, English speaking, and technically proficient. It is estimated that in India women constitute over 20 percent of the total IT workforce, which is higher than women's participation in the Indian economy as a whole. Despite the repetitive nature of the work and the lack of job security, Indian women's income, their authority in household matters, and their social mobility have improved as a result.

Alongside the export of software jobs to countries such as India, Mexico, and China, the proliferation of mobile phones, the Internet, and cyber cafes is providing the means for women's groups to organize networks and campaigns to improve their conditions. Indeed, the combination of IT with telecommunications, particularly satellite communications, provides new opportunities and outlets for women. Although affluent women in highly industrialized countries are better placed than other groups of women to take advantage of these technologies, the Internet and the mobile phone may ultimately have even greater significance for women in low-income households and communities in the global South. Pay-as-you-go mobile phones have enabled hundreds of millions in Africa, Asia, and the former Soviet Union to bypass the financial and bureaucratic obstacles of landline phones and to get connected. Around the world, although women still account for a lower proportion of Internet users than men, their share is rapidly rising.

Indeed, early concerns that the globalization of communications would lead to homogenization and reduce sociability now seem mis-

placed. On the contrary, all the signs are that new electronic media can help to build local communities and project them globally. The expansion of cyberspace makes it possible for even small and poorly resourced non-government organizations (NGOs) to connect with each other and engage in global social efforts. These political activities are an enormous advance for women who were formerly isolated from larger public spheres and cross-national social initiatives. Just as the car increased women's mobility and capacity to participate in public space, so the new ICTs have expanded women's horizons and capacity to connect with networks and campaigns to improve their conditions. To this extent, women are reinterpreting the technologies as a powerful tool for political organizing and the means for creation of new feminist communities.

Throughout this article, I have stressed the need to keep a skeptical eye on purely technological interpretations of the effects of globalization. Instead, we need to recognize the embeddedness of technology within social relations and the variable, sometimes contradictory, outcomes of digital technologies for different groups of women. ICTs can be constitutive of new gender power dynamics, but they are also derivative of preexisting conditions of gender inequality at work. While the rise of the information economy has led to a feminization of the labor force, these new forms of work to some extent replicate old patterns of sex segregation. The skills that are exercised in predominantly female jobs are still undervalued, and women are making slow inroads into the upper echelons of ICT occupations. Although the flexibility and spatial mobility afforded by ICTs have expanded opportunities for women, the male cultures associated with technical expertise still serve as a brake on progress toward global gender equality.

*See also* **Gender, Race, and Class in Science; Gender, Race, and Class in Technology**

**BIBLIOGRAPHY**
Castells, Manuel. *The Rise of the Network Society.* Oxford: Blackwell, 1996.
Castells, Manuel. *The Internet Galaxy.* Oxford: Oxford University Press, 2001.

Cockburn, Cynthia. *Machinery of Dominance: Women, Men and Technical Know-How*. London: Pluto, 1985.

Felstead, Alan, and Nick Jewson. In *Work, At Home: Towards an Understanding of Homeworking*. London: Routledge, 2000.

Govind, Kelkar, and Dev Nathan. "Gender Relations and Technological Change in Asia." *Current Sociology* 50.3 (2002): 427–441.

Held, David, Anthony McGrew, David Goldblatt, and Jonathan Perraton. *Global Transformations: Politics, Economics and Culture*. Cambridge, UK: Polity Press, 1999.

Mitter, Swasti, and Sheila Rowbotham, eds. *Women Encounter Technology: Changing Patterns of Employment in the Third World*. London: Routledge, 1995.

National Council for Research on Women. *Balancing the Equation: Where Are Women and Girls in Science, Engineering, and Technology?* New York: 2001.

Oldenziel, Ruth. *Making Technology Masculine: Men, Women, and Modern Machines in America, 1870–1945*. Amsterdam: Amsterdam University Press, 1999.

Rubery, Jill, Mark Smith, and Colette Fagan. *Women's Employment in Europe*. London: Routledge, 1988.

Sassen, Saskia. *Losing Control? Sovereignty in an Age of Globalization*. New York: Columbia University Press, 1996.

Wajcman, Judy. *Feminism Confronts Technology*. University Park, PA: The Pennsylvania State University Press, 1991.

Zuboff, Shoshana. *In the Age of the Smart Machine: The Future of Work and Power*. New York: Basic Books, 1988.

*Judy Wajcman*


# GENDER, RACE, CLASS IN SCIENCE.

New social movements of the 1960s and 1970s played an important role in creating public recognition of how deeply sciences are embedded in the social, political, and economic projects of their cultures. During the same period, historians, sociologists, and ethnographers were arriving at a similar conclusion. These researchers had set out to explore how central aspects of Western sciences have represented assumptions and preoccupations of their particular time and place, and continue to do so today. Sciences are part of their cultures. Moreover, both kinds of movements revealed how politics and culture sometimes stimulate and direct the growth of scientific knowledge in valuable ways. This came as a surprise to those who assumed that social values and interests can only block the growth of knowledge.

This account focuses on how social projects of gender, race, and class discrimination and oppression have shaped modern Western natural sciences and how the sciences in turn have provided resources for such oppression. It also identifies some ways in which the recipients of such discrimination have learned to fight back. First, the terms *gender, race*, and *class* must be defined, and we will note the influential roles that the sciences have played in such definitions.

## *Gender, Race,* **and** *Class* **Defined**

In Plato's *Republic*, he claimed that an ideal society would be a class society because individuals are born with different qualities of "souls." In an ideal society, those with souls of gold, as he described them, should be the rulers; citizens with souls of silver should be assigned to a warrior class; and the largest class, with souls of lead, were destined to serve as craft laborers and artisans. Plato articulated a form of biological determinism, although, of course, he could not draw on modern biological theories to support it.

Belief in some kind of biological determinism, often taken to be of divine origin, has persisted in many cultures. Today it motivates searches for biological causes of existing differences between racial groups and between the genders. Yet, in the last three decades, biologists and social scientists have finally succeeded in convincing many people that the causes of such differences are at least in large part social, or, rather, that the effects of biology and culture cannot in principle be distinguished in the ways that biological forms of determinism require. (Public recognition of the causes of class inequality underwent such a change 50 to 100 years earlier.) Moreover, the practices of national and international agencies, such as the U.S. Bureau of the Census and United Nations organizations, have begun to disseminate around the world the new social accounts of inequality. Thus, one can see that the sciences have helped to stabilize social inequalities, but that they also have provided resources for delegitimating them. Cultural change and scientific change have always highly influenced each other.

Gender, race, and class discrimination and oppression have different histories and direct different kinds of social practices (including scientific and technological practices). Yet they share significant features also. Let us start off by considering some less obvious aspects of gender differences, as these are understood today. First, whereas gender is a feature of individuals, in that at birth each of us is assigned to a gender, it also characterizes social structures and symbolic systems. Each individual is at birth inserted into a particular gendered position in our culture's complex social structure. Such positions carry distinctive privileges, obligations, and burdens: A culture's social structure itself is gendered. In various cultures, gender structures differ in the quantity and quality of gender they contain or require. Modern Western societies turn out to be much more highly gendered than many non-Western societies are: the division of labor and other human activities by gender tends to be relatively high in modern Western cultures. For example, the job of physician has been until recently reserved mostly for men in societies where medical research is of high status. In societies with few such research opportunities, however, such as Russia, doctoring has been a low-status, women's occupation. Similarly, many Eastern European and Third World countries can boast a higher percentage of women university professors of physics than the United States and most Western European countries can. The former societies have a lower gender division of labor in this respect than do the latter.

Furthermore, gender appears also as a symbolic system that is used by cultures to organize the meanings of humans' experiences of the blooming, buzzing confusion of the world around us. Thus, the capacity for objectivity, rationality, and the highest forms of ethical behavior have usually been thought of as uniquely masculine capabilities, as Evelyn Fox Keller and others have pointed out. Devalued behaviors, such as exhibiting vanity or weeping, and devalued activities, such as child care, house cleaning, and everyday responsibility for food preparation, are represented as feminine.

Race and class make similar appearances as attributes of individuals, social structures, and symbolic systems, as Donna Haraway and others report. This last may be less obvious, yet Western discourses are full of references to the "noble character" of great scientists and to the way truly scientific work "advances civilization" and thus contrasts with the merely technical achievements of hand laborers and with so-called primitive thought. Race and class, like gender, function in symbolic as well as individual and structural ways.

Next, focusing on the structural and symbolic features of gender, race, and class highlights how all three phenomena are fundamentally social relations between groups rather than sets of isolatable features. Achieving what are regarded as riches depends upon the existence of people who lack them, so in order to understand what makes the lives of the poor so miserable, one must examine also the lives of the rich, where wealth and luxury can accumulate only because the poor are not permitted a fair share of such resources. Similarly, women's and men's lives are explicable only in terms of the structural relations of women and men and symbolic relations of womanliness and manliness to each other in any particular cultural context. Ruling-class men, who have the jobs of administrators and managers in contemporary Western cultures and are usually mostly white men, have the privilege of not having to do the labor of servicing their own or anyone else's bodies—such as cooking, cleaning, maintaining clothes, or tending to the bodies of their relatives who are sick, elderly, or children. This privilege is created by the labors of others, who are mostly women and/or poor people and people of color. Insistence on hierarchically organized gender, race, and class differences is a way of justifying the unequal distribution of scarce resources, including desirable values or meanings (including, for example, "scientific"). The degree and character of status and power differences within gender or race or class relations varies from culture to culture, but the fact of their hierarchical organization does not, according to historians and anthropologists.

Moreover, gender, race and class are "intersectional"; that is, what a woman is and symbolizes, the responsibilities and privileges she bears, vary according to her race and class. Thus, the scientific and technological experiences, needs, and desires of women or men in the educated classes in the modern West are not the same as those of women or men in other classes and cultures around the globe. Race, class, and gender are historically dynamic, continually changing over time as each is shaped by and in turn shapes each other in new ways. Other cultural projects, such as those focused on sexuality, nationalism, religion, and empire, also interact with these three powerful structures of inequality. Scientific and technological changes always redistribute resources differently by race, class, and gender wherever such hierarchical systems exist.

These resources enable us to grasp some of the main scientific issues for the social justice movements that have struggled on behalf of groups disenfranchised by their class, gender, and/or race. The three are next discussed, in that order.

## Class

Class relations unequally distribute social status and economic/political power. Only dominant classes have had the social status and power to define what can count as an important research problem and what counts as evidence-based empirical knowledge. For example, dominant groups have thought it important to address only some of the health needs of the poor. Enabling the poor to resist contagious diseases has received relatively high attention because such diseases can easily spread to those with higher class status, such as the masters, bosses, and rulers. Also, keeping workers healthy enough to perform dangerous and arduous work, such as most

mining, manufacturing, and soldiering, has received at least some attention. Yet the gap in longevity between the have-nots and the haves indicates how much more attention the health needs of the poor require to meet social justice goals.

Moreover, because the production and distribution of the results of scientific research today are controlled largely by governments and profit-hungry corporations, there is little incentive in cultures with great class inequality to fund research that would primarily benefit poor people. It is difficult to make high profits from sales of inexpensive drugs or technologies. Recently, Third World governments and nongovernmental organizations have had some success in forcing pharmaceutical firms to make AIDS remedies available more cheaply and to refrain from punishing generic producers of AIDS drug "cocktails" in poor countries. So the control of scientific information and technologies through patents and copyrights owned by governments and profit-driven enterprises restricts access to the results of scientific knowledge to already economically and political advantaged groups. Finally, class systems keep us all ignorant in additional ways. For example, research directed toward finding safe ways to dispose of toxic wastes, or toward producing products that generate fewer such wastes, probably would have proceeded faster were it not permissible to dump such wastes in impoverished areas of the United States or to pay impoverished countries to dispose of them. Similarly, the availability of cheap labor removes motivation to develop effective technological labor.

Who gets access to careers in science and engineering? Traditionally these have been a route to upward mobility in the modern West as well as in other countries around the globe where people want Western lifestyles (though, of course, not everyone does!). Becoming a scientist or engineer offers opportunities to enter an international community that is perceived to transcend the limitations of local cultures. Yet entrance to lengthy and expensive scientific training requires an encouraging cultural climate, access to preprofessional math and science training, and freedom from family responsibilities and even from the need to support oneself during training. Such requirements block scientific and technological careers from most women and men around the globe.

Who bears the costs of scientific research and of its applications and technologies? As we have already seen, the least advantaged groups bear most of such costs. Additionally, critics point out that funds assigned to maintaining a space station or to physicists' or biologists' efforts to "discover the origin of life" could have been assigned to supporting universal health insurance, underwriting the cost of prescription drugs for the indigent elderly or for other poor people, or for securing safe supplies of water and food around the globe. In this sense, it is poor people who have disproportionately paid for such elite scientific projects. The scientific and technological interests of elites always can be pursued only at the cost of neglecting other possible social projects. Moreover, the poor, and especially poor women and racial minorities, have often disproportionately become science's experimental subjects in slave quarters, hospitals for indigents, and in non-Western cultures around the world. Furthermore, the installation of "scientific" agriculture, forestry, animal breeding, and fishing consistently has eliminated access of the world's poorest peoples to the land, seeds, forests, animals, marine resources, and the labor that they have needed to stay alive. Whereas the "green revolution" no doubt has had some favorable consequences for some groups of the poor, it mainly profited the transnational corporations that sold the fertilizers and pesticides required by "scientific agriculture" and those who could gain rights to the larger plots of land necessary to make such farming a financial success, as argued by Rosi Braidotti and coauthors, Susantha Goonatilake, Patrick Petitjean and coauthors, and Ziauddin Sardar. Also, scientific management of industrial production—the quest for greater "efficiency" defined in terms of higher production at lower costs—has consistently moved the control of labor from the worker to management and workers' financial rewards to the investing classes around the globe.

Since the 1960s, class analyses have also tended to be articulated through protests against such other social hierarchies as those of gender and race. Gender and race are always also class issues, and vice versa. Only brief additions to the kinds of such analyses already indicated can be given in the following.

## Gender and Race

Gender and race analyses have questioned five aspects of the sciences. The class analyses cited in the preceding section also expand and deepen the gender and race foci. First, where have women (and men), blacks, whites, and other racial groups been located in scientific institutions in different countries, different sciences, and different historic periods, and what have been the obstacles to the entrance, retention, and advancement of the "minorities" in each case? How have the minorities overcome such obstacles? Second, where are there sexist and androcentric (male-centered), racist, and Eurocentric distortions and silences in the research results of biology, the social sciences, and engineering? The work on sexist and racist reproductive theories and on sexist and racist biological determinisms are just the most visible of such studies. Third, how have misuses and abuses of applications of sciences and sciences' technologies disadvantaged women and racial minorities in such fields as reproduction and health, domestic technology, urban design, workplace, and military sciences and technologies? Fourth, how has science education—in different eras, different sciences, and different cultures—enacted androcentric and Eurocentric pedagogies and curricula, and promoted such goals? Fifth, how have the sciences produced and promoted androcentric and Eurocentric conceptual frameworks, meanings, and philosophies of sciences? What have been the particular consequences for the content of knowledge in each scientific field of the exclusion or restriction of minorities in the field's social structure? How do the presence of androcentric and Eurocentric sciences and distortions in the results of scientific work reflect back on the standards for objectivity, rationality, and "good science" in each case?

Because faulty individualist assumptions are so well established in the natural sciences, it is sometimes hard to grasp that "bad science," and the bad intent of individual scientists, are not the issue in such criticisms. Rather, their target is *standard* scientific assumptions and practices. Feminist and antiracist complaints cannot be dismissed on the grounds that sexist and racist sciences are not "good science" (though they are not) or that they are not "real science" (which they are). The issue is not that ignorant or bad-intentioned scientists have produced biased results of research. Rather, the issue is that scientific institutions, with their standards and practices, have been complicit with attempts of the most powerful groups in modern Western cultures to maintain and improve their power over other groups. The new social movements of the 1960s and 1970s helped to change the social climate within which scientific research occurred, making it possible to detect cultural aspects of science that before then had been invisible and to imagine alternative frameworks for understanding nature's order.

One might be tempted to think that the questions of feminist, antiracist, and other social justice groups introduce gender, race, or class values or politics into sciences and their philosophies that had been value-neutral before the critics spoke. However, as argued above, the class, gender, and race dimensions of the sciences were already there. What these critics do introduce is a method of detecting class, gender, and race oppression in the sciences that has otherwise escaped the notice of the scientists and most of the rest of us. Determining where and how such aspects of the sciences function is an empirical matter, requiring the same kinds of high rigorous standards that the sciences have promoted for other kinds of projects. Indeed, because the sciences' standards were too low to detect the class, gender, and race distortions, it would be reasonable to say that these critics introduce higher standards of empirical adequacy—of objectivity, rationality, and good method. Thus, scientific approaches are used in these accounts to improve both the accuracy and the comprehensiveness of sciences as well as democratic social practices.

## The Pursuit of More Competent and Just Western Sciences

The last few decades of social and intellectual change have enabled us to understand how class, gender, and race relations in any culture influence every aspect of its processes of knowledge production and dissemination. Cultures that insist on such hierarchical organization of social relations lose valuable opportunities to gain empirically and theoretically more evidentially supported understandings of themselves and the world around them, not to mention the opportunity to advance social justice. The social movements that resist such gender, race, and class discrimination—supported by the sciences themselves and by new directions in the history, ethnography, and sociology of sciences—have been able to produce and inspire more accurate and comprehensive accounts of nature and social relations, which can also advance social justice. Thus they retrieve and renew important founding values of modern Western sciences.

*See also* **Gender and Globalization; Gender, Race, and Class in Technology**

**BIBLIOGRAPHY**

Braidotti, Rosi, et al. *Women, the Environment, and Sustainable Development*. Atlantic Highlands, NJ: Zed Books, 1994. How modern Western sciences have acted complicitously with the disempowerment of women, degradation of environments, and dedevelopment and maldevelopment of societies in the Third World.

Fausto-Sterling, Anne. *Myths of Gender: Biological Theories about Women and Men*. Second Edition. New York: Basic Books, 1994.

Goonatilake, Susantha. *Aborted Discovery: Science and Creativity in the Third World*. London, Zed Books, 1984. A survey of Third World cultures' scientific achievements and the decline of these sciences after encounters with European cultures.

Gould, Stephen Jay. *The Mismeasure of Man*. New York: W. W. Norton, 1981.

Haraway, Donna. *Primate Visions: Gender, Race, and Nature in the World of Modern Science*. New York: Routledge, 1989.

Harding, Sandra. *Whose Science? Whose Knowledge? Thinking From Women's Lives.* Ithaca, NY: Cornell University Press, 1991.

Harding, Sandra, ed. *The "Racial" Economy of Science: Toward a Democratic Future.* Bloomington, IN: Indiana University Press, 1993.

Harding, Sandra. *Is Science Multicultural? Postcolonialisms, Feminisms, and Epistemologies.* Bloomington, IN: Indiana University Press, 1993.

Keller, Evelyn Fox. *Reflections on Gender and Science.* New Haven, CN: Yale University Press, 1984.

Lewontin, R. C., Steven Rose, and Leon J. Kamin. *Not in Our Genes.* New York: Pantheon Books, 1984.

Petitjean, Patrick, et al., eds. *Science and Empire: Historical Studies about Scientific Development and European Expansion.* Dordrecht, The Netherlands: Kluwer Academic, 1992. Proceedings of a UNESCO-sponsored conference.

Rossiter, Margaret. *Women Scientists in America: Struggles and Strategies to 1940.* Baltimore: Johns Hopkins University Press, 1982.

Sardar, Ziauddin, ed. *The Revenge of Athena: Science, Exploitation, and the Third World.* London: Mansell, 1988. Proceedings of an early conference of Third World science intellectuals.

Schiebinger, Londa. *Nature's Body: Gender in the Making of Modern Science.* Boston: Beacon Press, 1993.

Selin, Helaine, ed. *Encyclopedia of the History of Science, Technology, and Medicine in Non-Western Cultures.* Dordrecht, The Netherlands: Kluwer Academic, 1997. The most comprehensive collection of reports and analyses available.

*Sandra Harding*

# GENDER, RACE, AND CLASS IN TECHNOLOGY.

Technology offers an ambivalent promise. On the one hand we can hardly imagine the passage of daily life without it. Tasks as mundane as closing a door with hinges and as unfathomable as eradicating a cancer with chemotherapy are delegated to technological agents. In other words, humans use objects to do tasks for them, and though objects often can be helpful, it is also crucial to remember that people remain only in partial control of how things will act. In this sense, objects are social actors that factor into politics, economics, and social life. Thus, certain kinds of hinges ensure that three-year-olds will not enter dangerous boiler rooms; certain combinations of chemicals seem to defer (if only for a short time) the spread of various malignancies in mice and humans. On the other hand, as we know too well, hinges also squash fingers, and chemotherapy drugs induce side effects that can include secondary cancers. Within cultures of interrelating human and nonhuman sets of agencies, differences—some of which fissure into relations not only of difference but of inequality—materialize. The material objects of everyday life—from concrete curbs that delineate roads but obstruct wheelchairs, to cigarettes that provide bursts of pleasure but instigate deadly disease—offer a way to explore how our social categories, such as race, class, and gender, are meaningful. In this sense, technology in relation to gender, race, and class is about the creation of differences, and these differences are not benign.

Rather, critical scholarship has demonstrated the ways in which technology can make it seem as though differences that fall under categories such as race, class, and gender were natural and not created by social values, categories, and understandings. Thus, if women are consistently seen as naturally good nurturers, or Latinas seen always as passionate, or African Americans as good athletes, the problem is not that each of these attributes is necessarily culturally and economically less valued (although in many cases they are), but that technology can supplement these social valuations and limit the opportunities that people have.

One example of this is the way in which, through the twentieth century, mechanical ability was thought to be associated with masculinity. Historians have shown how this marked a distinct change from the nineteenth century, in which women took an active part in designing technologies. In part this had to do with the growth of engineering as a field and the strategies used to exclude women. These strategies include such things as teaching girls that their natural attributes were better suited to nurturing, cooking, cleaning, and homemaking; limiting educational opportunities for girls and women; presenting images of engineers as muscular masculine heroes; and diminishing women's contributions to science and technology. Often this has meant that women's' contributions have been misattributed to men (some examples include the erasure of Rosalind Franklin's work leading toward the discovery of the helical structure DNA, or Emily Roebling's work in the construction of the Brooklyn Bridge). Women have used different strategies to try to get around the roadblocks. These have included overqualification, marrying members in the same academic field in order to create a research team, and publishing under male pseudonyms. Since the 1960s, when engineering schools opened their doors to women, more women have been pursuing these careers. However, enrollments in engineering majors are still low (under 10 percent of classes in some subfields), and salaries for women are still quite significantly below those for men.

An extension of this point can be seen by examining the use of agricultural technology in California. The short-handled hoe, in combination with racial stereotypes, reinforced the powerful assumption, which prevailed in California after World War II until at least the 1970s, that Mexican-American workers were particularly good at stoop labor. Since low-paying agricultural labor was virtually the only job opportunity for these immigrants, race discrimination and the material technologies of the labor itself recursively highlighted the tautology of the assumptions that they were good at certain kinds of work *because* of their race. Thus, in the legal suit that was launched in the 1970s in an effort to have the short hoe banned, growers argued both that Mexican Americans were susceptible to injury because of their genetic make-up (which a Palo Alto doctor, Oakley Hewitt, testified was similar to that of "Eskimos"), and also that they were uniquely suited to use the hoe because of their dexterity and height. These contradictory versions of what traits were attributed to nature indicate some of the politics of nature and how they relate to race, class, and gender.

While in this case stoop labour was particularly racialized, class is also a central category in how technology is understood as useful or as socially good. The history of technology provides many examples in which the overall good of the society was perceived to be worth the cost of lost lives and limbs. The growth of the assembly line offers a clear and nonapologetic window on this view.

Henry Ford, acknowledged as one of the founders of assembly line work, wrote in his autobiography, "Of [7,882 different jobs at the factory] . . . 3,595 jobs could be performed by the slightest weakest sort of men [or] satisfactorily filled by older women or children. [Of these,] 670 could be filled by legless men, 2,637 by one-legged men, 2 by armless men, 715 by one-armed men and ten by blind men" (Ford, p. 108). This quote is important in illustrating the class dimensions of technology for several reasons. First, it demonstrates the way in which working class people's bodies were understood as extensions of the machinery, useful only insofar as that they could physically make up for the shortcomings of the assembly line with their arms, fingers, or leg. Second, this example shows an early version of the way in which consumption in the United States came to act as a ruse for access.

Ford widely voiced the notion that every family should own an automobile. Rather than marketing specifically for wealthy people, as other manufacturers did, he mass-produced the car and brought prices down in each consecutive year during the first two decades of the century and set up loan programs in an effort to sell cars as widely as possible. In this way he demonstrated a point that Adam Smith had also made in 1776 in his treatise, *The Wealth of Nations*. Smith maintained that, while the division of labor necessary for industrial production, which the assembly line takes to such an extreme degree, requires work that is far from enlightening, it is more than made up for by the ability that workers have to consume the goods that they participate in making. Production and consumption were understood to operate as two sides of the same coin toward the wealth of nations.

Gender, race, and class can never be extricated as categories. Each of us carries intersecting identities. To look back at Ford's factory then, women bore an economic brunt of the category of gender, because women were paid significantly less than men for the same work. Not only these economic factors, but social considerations, such as the exclusion of most women (with a few exceptions for wealthy women) from automobile clubs; racing; and auto-related jobs such as sales and mechanics, have led to diminished mobility and economic opportunity for women throughout the century.

Many women have been able to elude some of the effects of these barriers through heterosexual relations, particularly through marriage. The assumption that women would marry in many cases excused women's low pay, since they were understood to be temporary or secondary workers. In many cases this was false, but these economic and social conditions also led to women's dependence on men. In the airline industry until the 1980s, women who married were typically required to leave their jobs.

Twentieth century "women's work" has overwhelmingly included either domestic labor, caretaking work such as nursing and teaching, or clerical work. In the early century, the typewriter became a means by which women could enter the previous all-male space of offices, and by 1920, clerical labor was virtually completely feminized. As women joined the clerical labor force, salaries began to decline (at the same time as factory wages increased), and whereas clerical work had previously held the possibility for rising in the corporate hierarchy, women experienced a glass ceiling that, again, was not only economic but cultural. This becomes clear in looking at the injuries that resulted from new digital technologies such as scanners and computer keyboards in the 1980s. Despite widespread and serious injuries, no national design standards or broad press coverage was offered until white-collar journalists and computer programmers complained of the injury. Contemporary litiga-

tion indicated that the legal system was unable to address the problems of design of these objects; rather, management understood the issue as an expected and acceptable part of repetitive activities such as typing and scanning. This demonstrates the overlap between gender and class where one can offer terms to excuse the other.

Thus, technology, and its social filters such as policy, law, and design, provides us with an opportunity to understand better how technology both amplifies and alters social relations: how risk is distributed, how tasks are distributed and valued. In short, whose bodies and lives count?

In this sense the move of African Americans from nonpersons to persons can demonstrate the extent to which technology can illustrate social paradoxes. Even after the Fourteenth Amendment of 1866, African Americans were mistaken by others as not fully human. Thus, African American consumers would bear the brunt of disrespect and overt racism. This could take the form of having to look at fingers and toes of lynching victims on display at local shops, being sold second-class goods, and having to wait until shops were empty before being served. In this sense, the growth of chain supermarkets and their lack of personal service came as a relief to many African Americans, and the right to consumption became a primary focus of the civil rights movement of the 1960s. These movements included not only picketing racist establishments for jobs and setting up school and other programs but also lobbying for equal access to lunch counters and other consumer venues.

At the same time, during the 1950s and 1960s African Americans were "discovered" by marketers as a new, and potentially lucrative, niche market group. Niche magazines such as *Ebony* and *Tan* began to carry advertisements for cigarettes and liquor, and tobacco companies began targeting African Americans even as the popularity of their products in the white market declined because white Americans were starting to understand the health risks of smoking. Since African Americans tended to be less well educated and tended to have less access to health resources, they were vulnerable to the later health consequences of smoking. When they brought these complaints—framed in terms of racism—to the legal system, courts ruled that the question of targeting was beside the point. African Americans were offered the same cigarette for the same price. Thus, social distinctions offer a contradictory promise: On the one hand, everybody wants to be a part of the promises of technology, and on the other, the costs are not distributed equally and everyone does not have an equal opportunity to know what these costs will be.

It is also crucial to note that race, class, and gender are not the only ways to configure human relations to technologies, and in some ways they can be too limiting or even misleading. People identify in all kinds of ways, from patients to survivors to experts. People are also identified by other people and institutions as being in particular kinds of relationships with technology. One way to see how this happens is in relation to the way in which the growth of the automobile industry not only required a population of drivers but forced virtually everyone into some relationship with the automobile. For many urban dwellers, and for all of us at one time or other, cars force us to inhabit the role of the pedestrian. Pedestrians have very little social or material power. Despite the facts that pedestrians amount to approximately 20 percent of all car-related deaths in the United States and that small design changes could drastically lower this proportion, car designers do not need to take pedestrians into account when designing the front ends of cars. In fact, current sport utility vehicle designs are some of the most dangerous for pedestrians and other drivers in the history of automobility. In order to bring a legal complaint and change an issue, a group has to be recognized as a legal entity. Thus, people who want to claim that they have been injured by, say, silicone breast implants have been able to establish themselves as a legal class of persons and then on that basis, argue for compensation. In most cases, these groups hold in common a mode of consumption or more recently, exposure to toxic substances such as asbestos or Agent Orange. Pedestrians have

been unable to have their complaints against auto manufacturers recognized and, therefore, have not been able to make the legal argument that car designs should be altered to increase pedestrian safety. At the same time, as pedestrians have very little legal and social power to effect design changes, they have been virtually unable to hold drivers responsible for criminal negligence in courts.

Much of this current situation is a result of very early assumptions in automobility that the car was a tool that could be fully controlled by its driver. Cars were understood as ordinary (not dangerous) objects, so crashes were perceived as "accidents." Perfect driving was assumed to be humanly possible. These assumptions, and the way they have played out in policy and law, have led to an extremely vulnerable class of people—even those simply walking across the parking lot between their cars and the Wal-Mart. While this category has overlapping relations with race, class, and gender, it is also a distinct class, born out of a particular relation to technology. The pedestrian remains a class of people both defined by, and defined out of, automobility.

Scholars in the first wave of a critical engagement with technology, in particular with the design of the material world, focused on the ways in which difference was physically encoded and enforced through the built environment. Langdon Winner provided one of the first articulations of this way of thinking about technology in his aptly titled article, "Do Artifacts Have Politics?" For Winner, technical artifacts and systems do have politics in two crucial ways. First, they code prior social inequalities—in particular, pernicious ones such as racism. To illustrate this point he discusses the way in which road overpasses in New York were built such that public buses could not pass underneath them. Although, once built, an overpass looks like a more or less normal part of the environment, Winner discusses the way in which the lowered height meant that African Americans and poor people, who more often used public transit, could not access ostensibly public beaches. Second, Winner discussed how technical artifacts can require certain types of social organization; he calls these "inherently political technologies." Nuclear power, he suggests, simply cannot be a democratic technology because of the requirements for its use and containment, which include a "centralized, rigidly hierarchical chain of command closed to all influences that might make its workings unpredictable." Nuclear power implies and necessitates the existence of a complex bureaucracy because of the secrecy required and danger inherent to such as system; thus, it is at odds with democracy.

From another angle, one might examine the purportedly labor-saving devices introduced as commodities to the post–World War II American market. Technologies such as washing machines and irons actually increased the labor for women when the time saved by these devices were replaced by higher expectations and increased workload for women. Further analyses of the way that formal social institutions such as law understand technology further substantiates these claims. Cars have been a main object for rendering invisible and logical the subordination of women. The cultural understandings of femininity and the automobile rendered violence against women in automobiles virtually illegible in legal suits brought by women raped in automobiles. The legal system has consistently viewed the entering of a car (even, in one case, a taxicab) as consent to sexual relations.

Second-wave theorists built on these understandings of hierarchical categories of human difference, primarily by extending these analyses of how the supposedly natural categories of inequality, such as race, class, and gender, are built into the very projects of science and personhood itself. In this way, the unequal effects of technology are understood as symptoms of deeper roadblocks to human flourishing. Women were constructed and defined as women through their relationships with objects such as typewriters, household cleaning products, washing machines, and automobiles—objects that, as noted, did not ultimately contribute to women's benefit. It was these activities that rendered persons legible, as a correlate of sexual organs. In other words,

recognizable gender is a precondition for personhood, and the behaviors and practices that are accepted within those gender roles are recursively recognized and then performed in ways that bring gender into being. Rather than being a preset category into which people are born, gender is always being made and remade through the people and performances that inhabit the category.

To bring these reflections and insights back to the subheading of this entry, then, is to realize the full force of the notion that social differences are coded into technologies. It is only because of these differences that technology can be made legible as a social good. The politics and economics of human flourishing simply cannot be understood outside of technology and the ways in which it has been used to define and amplify certain political goals over others. Social divisions are necessary to the success of "things," from the factory workers who enabled the industrial revolution, to the millions of housewives who—as consumers and as homemakers for corporate employee husbands—to allow the explosive growth of corporations through the 1950s and 1960s, to the African Americans who provided at one historical end the slave labor to develop the American South and at the other the consumer power to maintain a variety of markets in the 1970s and 1980s.

*See also* **Gender and Globalization; Gender, Race, and Class in Science**

**BIBLIOGRAPHY**

Butler, Judith. *Undoing Gender.* New York and London: Routledge, 2004.

Cowan, Ruth Schwartz. *More Work for Mother: The Ironies of Household Technology from the Open Hearth to the Microwave.* New York: Basic Books, 1983.

Cockburn, Cynthia. "The Material of Male Power." In *Social Shaping of Technology,* edited by Donald MacKenzie and Judy Wajcman, 125–146. Philadelphia: Open University Press, 1999.

Cohen, Lizabeth. *A Consumers' Republic: The Politics of Mass Consumption in Postwar America.* New York: Alfred A. Knopf, 2003

Ford, Henry. *My Life and Work.* New York: Doubleday, Page, 1923.

Haraway, Donna J. *Simians, Cyborgs, and Women: The Reinvention of Nature.* New York: Routledge, 1991.

Oldenziel, Ruth. *Making Technology Masculine: Women, Men, and the Machine in America, 1880–1945.* Amsterdam and Ann Arbor: University of Amsterdam Press, 1999.

Pursell, Caroll W., Jr. "Toys, Technology, and Sex Roles in America, 1920–1940." In *Dynamos and Virgins Revisited: Women and Technological Change in History,* edited by Martha Moore Trescott, 252–267. Metuchen, NJ, and London: Scarecrow Press, 1979.

Rossiter, Margaret W. *Women Scientists in America: Struggles and Strategies to 1940.* Baltimore: The Johns Hopkins University Press, 1982.

Rossiter, Margaret W. *Women Scientists in America: Before Affirmative Action, 1940–1972.* Baltimore: The Johns Hopkins University Press, 1995.

Sanger, Carol. "Girls and the Getaway: Cars, Culture, and the Predicament of Gendered Space." *University of Pennsylvania Law Review* 144.2 (December 1995): 705–756.

Terry, Jennifer, and Melodie Calvert, eds. *Processed Lives: Gender and Technology in Everyday Life.* London and New York: Routledge, 1997.

Winner, Langdon. "Do Artifacts Have Politics?" *Daedalus* 109.1 (Winter 1980): 121–136.

*Sarah S. Jain*

# GENETICS AND GENETIC ENGINEERING.

In 1906 the English biologist William Bateson gave the name *genetics* to a newly emerging scientific field whose practitioners sought to understand the fundamentals of biological heredity. This new field built upon the findings of an obscure Austrian monk named Gregor Mendel regarding the properties of heredity in peas published forty years earlier. In April 2003 scientists announced the completion of a "map" of the human genome, the 3.1 billion base pairs that make up the complete genome, or genetic material, in the nucleus of each cell of a human being. This map is widely believed to provide a foundation for a new medical practice oriented around genetics. The histories and social practices that produced and continue to elaborate upon these moments in time offer insights into the complex ways that social values, cultural ideals, and human action are woven into every level of the production, circulation, and uptake of scientific knowledge and its application in medical technologies.

Genetics offers an exceptional object for inquiries into the interrelationships among science, technology, and society, because its histories and contemporary practices repeatedly illustrate how people simultaneously produce their understandings of social and natural order. As such, it offers a particularly rich example of how both social and natural truths are folded into what we understand as scientific knowledge. Here a brief review of genetics, eugenics, and the Human Genome Project will precede a discussion of major themes in social studies of contemporary genetics.

## The Founding of a Scientific Discipline

The founding of genetics as a field of scientific inquiry came as a part of Victorian era fascination with natural history in general and with heredity in particular. In this environment, the naturalist Charles Darwin and the philosopher Herbert Spencer developed their respective theories of biological and social evolution. Darwin's theory of "descent with modification" provided a theoretical foundation for the field of genetics—the study of biological heredity. The themes of evolution, progress, and development elaborated by these English intellectuals were circulating among naturalists but also in numerous other domains, such as literature and art, during their lifetimes. The English literary critic Gillian Beer argues that Darwin imported Victorian social ideals into his 1859 descriptions of nature in *On the Origin of Species.* The English anthropologist Marilyn Strathern illustrates how one sees the very essence of Victorian kinship in Darwin's attention to biological relationships among various kinds. Ideas about branches of families, lineages, genealogies, and degrees of relatedness had powerful significance in Victorian England; they determined laws of inheritance and succession as well as rules of daily social activity. Ironically, Victorian kinship was not about nature at all. Rather, the Victorians saw as one of their great achievements—one of the very things that made them civilized in relation to those they dominated in their colonies—their ability to rise above nature, to repress what they viewed as distasteful urges of nature, and to organize life not around the baseness of nature but around ideals of society. Karl Marx observed that "it is remarkable how Darwin recognizes among beasts and plants his English society with its division of labor, competition, opening of new markets, 'inventions' and the Malthusian 'struggle for existence.' " (cited in Yanagisako and Delaney, p. 5). In Darwin's work we find a fine example of how the social is embedded in nature in the very founding of this new approach to understanding the natural world. What was so radical about Darwin's work at the time was his insistence on locating humans in nature. Darwin challenged Victorian ideals in this regard, in part, by adopting Victorian understandings of kinship to describe what he was in nature. Yet we now tend to understand these concepts as transparent scientific descriptions of nature.

## Eugenics

Sir Francis Galton, a cousin of Darwin, brought together the ideas of social and biological evolution, coining the term *eugenics* in 1883 to indicate active interest in improving the human population through a combination of encouraging some particular human matings and discouraging others. Eugenics, undergirded by ideological concerns about social and natural order that brought together ideas about social and biological progress, exploded as a worldwide social movement from 1900 to 1940. Perhaps the most fully elaborated and horrifying form of eugenics was that implemented in the World War II period by the Nazi regime as part of its program of "racial hygiene," through which particular persons—"Aryans"—were encouraged to reproduce while others—such as Jews, homosexuals, and the disabled—were defined as public health risks, a definition that facili-

tated their sterilization, incarceration, and extermination. Although the Nazi project is frequently viewed as an example of bad science, bad medicine, or modernity gone awry, it is important to recognize that eugenic ideas and policies were widespread and were considered cutting-edge scientific practice throughout the world during the interwar period. For example, many U.S. states had eugenic laws, such as statutes permitting forced sterilization, and eugenic ideals powerfully shaped American immigration policy during this period, with well-known disastrous consequences for those attempting to escape Europe. In his history of eugenics, Daniel J. Kevles quotes an American eugenicist as complaining that the "Germans are beating us at our own game" (Kevles, p. 116) because Germany had been so successful in implementing eugenic policies. In its heyday as a social movement eugenics was viewed as a scientific approach to the improvement of human health and thus society. Among its proponents were respected and accomplished physicians and scientists.

## Post–World War II Genetics

In the post–World War II period the word *eugenics* took on a pejorative connotation. In light of revelations about how eugenics had been implemented by the Nazis, geneticists sought to separate the science of genetics from public policy. In spite of its widespread repudiation, the conceptual framework of eugenics has survived intact in many medical and popular practices and theories up to the present, where it remains attached to widespread ideologies about social and natural order as well as to an ideal of improving human health and society. Postwar genetics continued to build upon theories and techniques developed earlier in the century, particularly the use of "model systems"—the flies, mice, and worms that have assisted in developing knowledge about both humans and other organisms.

For contemporary genetics, the most significant midcentury finding came with the 1953 publication of James Watson and Francis Crick's Nobel Prize–winning description of the structure of deoxyribonucleic acid

(DNA). Their description relied on photos taken by the English X-ray crystallographer Rosalind Franklin, who died before Watson revealed that it was upon seeing one of Franklin's X-rays of DNA without her knowledge that he successfully began formulating the double helical structure of DNA. The American philosopher Evelyn Fox Keller suggests that Watson and Crick's finding engendered an entirely new form of analysis, replacing the practices of classical genetics with the techniques of molecular genetics. These techniques have supported the rapid development of new genetic knowledge and technique, in part through the organized effort known as the Human Genome Project. These new scientific practices also bear a dynamic relationship to new forms of social organization, including changes in institutional structures, biomedicine, education, government funding, corporate biotechnology, pharmaceutical investment, regimes of ownership, and highly mediated popular representations of heroic scientific practice.

## The Human Genome Project

In the late 1980s scientists in Japan, the United States, and Europe, with the support of their governments, undertook a major scientific initiative known as the Human Genome Project (HGP). The U.S. government earmarked $3 billion for this attempt to delineate the billions of base pairs making up the genetic material in human beings, promising, along with the scientists who promoted the project, dramatic new interventions into human health. The nearly constant popular and professional media barrage of genetic findings since that time points not only to the rapid proliferation of genetic knowledge but also to a public discourse in which genetic causality is increasingly used to explain human diversity and affliction. Many of these claims are based on commitments to reductive genetic determinism that attach enormous import to the human genome. For instance, the Nobel Prize–winning molecular biologist Walter Gilbert describes the human genome as the "Holy Grail" of biology and suggests that learning the base pair sequence

of the human genome will uncover the very essence of what it is to be human. He makes this point in lectures by pulling out a compact disc and telling the audience, "This is you." His point is that eventually genetic research will allow every individual to have her or his entire genome encoded on compact disc. What is perhaps more remarkable than such a technological feat is the idea that such a representation could indeed actually "be" a person. These kinds of claims resonate powerfully in the United States, where ideologies regarding innate ability and biological determinism have long histories. At the same time that scientists and others herald molecular biology as the royal road to human health, a diverse array of individuals—from bench scientists to clinicians, activists, lay people, historians, anthropologists, and others—work to challenge any too-easy biological reductionism about the power and meaning of genetics and genetic difference.

Scientists working in this area of biology have made impressive discoveries, including finding the genes for many heritable conditions and numerous potential links to some of the more common complex conditions, all of which are increasingly viewed as having a genetic substrate. In June 2000 human genome scientists announced the completion of a rough draft of the human genome, and in 2003 they announced completion of the entire map. The story of the Human Genome Project and the completion of this map is one of politics, ideology, altruistic and arrogant personalities, scientific competition, collaborations crossing both public/private and lay/expert divides, money, and hope. In other words, it is a story of science as usual in our time. With the completion of the map, the project's emphasis has shifted from structural to functional genomics—work that focuses not on mapping DNA sequences but on understanding the complex functioning of genetic material.

With the explosion of new scientific knowledge and technique associated with the Human Genome Project, the gene itself has become a cultural icon, integrated into popular and material culture in the form of such phenomena as DNA-Man comic books, DNA perfume sold in helical-shaped glass bottles, and double helix silverware patterns. This fascination with the gene bears a dynamic relationship to media coverage of genetic practices.

Pervasive media coverage of emerging genetic practices and findings often exaggerates the implications of the findings themselves. Peter Conrad has described this kind of reporting as "genetic optimism," a phenomenon whereby genetic findings are enthusiastically covered in the popular press, but results that refute those findings are rarely brought to the public's attention. Some suggest that this genetic optimism is a function of the media, science reporting, and the problem of translating scientific knowledge for a popular audience. This model of science popularization understands science reporting as a mere translation of new knowledge for a popular audience as that knowledge flows out of science laboratories. This view, however, does not adequately account for the way social values, cultural ideals, and human action work at every level of scientific inquiry and reporting or for the fact that scientists increasingly use and control what the media have access to through press releases, interviews, news conferences, public relations experts, and staged media events.

Although delineation of specific genes linked to specific conditions with obvious inheritance patterns enables the development of genetic tests that can be used for both pre- and postnatal testing for genetic conditions, these findings have, to date, not translated into successful treatment for such conditions. At the same time, most of the widespread common conditions frequently linked to genetics—diabetes, cancer, heart disease—are related to the interactions among multiple genes and between genes and the environment, making the finding of and testing for those genes still not practical. We can describe this state of affairs, which was predicted early in the genome project, as a "therapeutic gap" (Kevles and Hood). Enormous energy and resources are now being poured into closing this gap.

## Social Studies of the New Genetics

Writing in 1991 about the impact of genetic technologies in light of new knowledge associated with the Human Genome Project,

Canadian epidemiologist Abby Lippman introduced the concept of geneticization. Lippman uses this concept to emphasize that genetics is the dominant explanatory model in today's stories of health and disease. This concept highlights the way new genetic knowledge intensifies connections between perceived heritability of such complex social traits as intelligence or criminality and their assumed explanation at the level of individually carried DNA. These connections, in turn, underlie powerful beliefs in both genetic determinism and the value of new biotechnologies of genetic improvement. Lippman thus suggests the persistence and potential expansion of eugenic thinking in North America today across a broad spectrum of social groups who consider the genome the site at which the human future can or must be negotiated. The tension-fraught knowledge and practice associated with genetics do not simply develop or seep into collective consciousness by themselves; they must be taught, learned, and experienced before they can operate as a worldview and as social practice. This expanding genetic worldview among many constituencies is now being elaborated in numerous contexts around the globe.

Perhaps the most immediate context in which people engage new genetic knowledge is in prenatal genetic testing, a practice that has become routinized in many locales, including both highly developed and developing nations. One of the primary procedures for prenatal diagnosis involves extracting fetal cells from the amniotic fluid of pregnant women through amniocentesis, culturing them, and examining them for the presence or absence of particular genetic anomalies. The emergence and routinization of prenatal genetic testing is grounded in a convergence of ideology, social practice, law, technique, and knowledge. In the United States, for example, the technique for amniocentesis and chromosome analysis had been known earlier, but it was not until after the Supreme Court expanded legal access to abortion in its 1973 *Roe* v. *Wade* decision that clinicians worked to develop its use in routine practice.

This expansion of reproductive rights coincided with the social practice of women entering the workforce in new ways and delaying childbearing. By 1978, scientific publications were reporting a correlation between what is described as "advanced maternal age" and chromosome anomalies. As women were referred for or sought out prenatal genetic testing, they provided the raw materials upon which to perfect and expand both knowledge and technique in this area of science and clinical practice. The desire for these services was premised not only on widely shared though diversely defined ideals of producing "healthy" babies but also on an ideology of human perfectibility. In her work on the social impact of amniocentesis in the United States, anthropologist Rayna Rapp illustrates the complex ways that women and men encounter, create, and contest the power and meaning of genetic knowledge and practice as they imagine and produce a new generation.

People also produce genetic knowledge and practice through their own experiences with genetic conditions and through the social networks individuals and families bearing specific genetic conditions often establish. These networks include the research scientists, clinicians, lay support groups, and more general populations that people affected by genetic conditions regularly encounter. In this context a genetic worldview is constituted in interaction, through the robust collaborations of clinicians, individuals and families bearing genetic conditions, and researchers that develop as people with very different subject positions all seek to develop and engage genetic knowledge. Some of those actors may use their new and multiple locations to contest a too-easy determinism, or to develop interventions—molecular and otherwise—they consider choice-enhancing. This author has used the phrase "flexible eugenics" to describe how powerful American ideologies regarding knowledge, individualism, and choice are embedded in a free-market approach to genetic technologies. It is precisely this choice orientation to genetics that the American sociologist Troy Duster has suggested opens a "back door to eugenics."

The U.S. case contrasts with the situation in other nations. In the Netherlands, where access to genetic services is highly regulated around a concept of medical necessity, these services are employed to ensure citizens' abilities to achieve the Dutch social ideal of fitting in. In China and India the use of amniocentesis for sex selection also suggests an appropriation of genetic technologies rooted in specific cultural commitments, particularly desires for male babies. In the Indian case, Indian feminists have mounted vigorous campaigns to control the use of amniocentesis, including successfully campaigning to outlaw its use for sex selection.

A shared desire to produce genetic knowledge is part of what creates the networks of association described in the previous examples. Researchers, clinicians, policy makers and government agencies are putting significant energy and resources into moving genetics beyond its past focus on rare genetic conditions to the more widespread commonconditions such as cancers and heart disease that affect large numbers of people throughout the world but especially those in wealthy nations. Although geneticists repeatedly stress the great genetic similarity of all humans, and most of the minute differences between people have no known biological significance, some of those minute differences do make people who have them sick. The effort necessary to understand conditions involving multiple genes and gene-environment interaction requires large-scale epidemiological studies and the willingness of large numbers of ordinary people to participate in genetic research. Such research projects allow scrutiny of the similarities and differences among people that may lead to understanding the risks and susceptibilities for the common conditions so many seek to address. Such efforts, of course, may lead to new forms of categorization. Duster reminds us to be wary of how such efforts might play into old taxonomies such as those reflected by racial thinking and practice. As part of the effort to engage people in contemporary genetics, a variety of groups with diverse interests are now working to educate ordinary people about genetics through an extraordinary array of genetic education efforts. These projects include

- An effort sponsored by a British nongovernmental organization to educate impoverished indigenous and *dalit* ("untouchable") rural Indian farmers in order to prepare them to conduct a "citizen's jury" that will evaluate the use of genetically modified seed as part of a development strategy in the state of Uttar Pradesh
- A program instructing American judges in genetics so that they can evaluate the use of such knowledge in their courtrooms
- A federal govermnent-sponsored undertaking to engage virtually all Vermont citizens in a conversation about genetics
- A project spearheaded by senior faculty and deans in elite medical schools that seeks to train future physicians in genetics
- A number of projects associated with academic health centers that aim to educate people about genetics to encourage participation in genetic research aimed at closing the therapeutic gap

All of these projects seem to offer an understanding of the world, the body, the self, and the relatedness of all life forms in which significant truths are considered to reside at the level of the molecule. In each of these sites one can expect the power, value, and meaning of genetic knowledge and practice to be engaged and contested as people with diverse backgrounds and experiences encounter new knowledge and forms of categorization. Indeed, if one applies the insights offered by the previous examples to the emergence of these efforts, one might understand these educational projects as sites in which the precise meaning of genetics now and in the future will be worked out.

## Genomic Futures

Scientists, policy makers, and elected officials have persistently promised the translation of new genetic knowledge into dramatic medical interventions. The state, academic institutions, venture capitalists, and established biotech and pharmaceutical companies are now putting vast resources into this effort. At the same time, many people are personally and professionally invested in the development of new medical treatments. As individuals and groups engage each other to develop genetic medicine, diverse knowledge and experiences will further, challenge, andor appropriate the power and meaning of genetics. These circumstances make genetics an ongoing site for exploring the ways social values, cultural ideals, and human action are at work in the production of contemporary and emerging understandings of social and natural order.

*See also* **Genetics and Society**

### BIBLIOGRAPHY

Beer, Gillian. *Darwin's Plots: Evolutionary Narrative in Darwin, George Eliot, and Nineteenth-Century Fiction.* 2nd ed. Cambridge, U.K.: Cambridge University Press, 2000.

Conrad, Peter. "Genetic Optimism: Framing Genes and Mental Illness in the News." *Culture, Medicine, and Psychiatry* 25.2 (2001): 225–247.

Darwin, Charles. *The Origin of Species* (1876). Washington Square, NY: New York University Press, 1988.

Duster, Troy. "Buried Alive: The Concept of Race in Science." *Chronicle of Higher Education* 48.3 (September 14, 2001): B11–B12.

Duster, Troy. *Backdoor to Eugenics.* New York: Routledge, 2003.

Heath, Deborah. "Locating Genetic Knowledge: Picturing Marfan Syndrome and Its Traveling Constituencies." *Science, Technology, and Human Values* 23.1 (1998): 71–97.

Hubbard, Ruth, and Elijah Wald. *Exploding the Gene Myth: How Genetic Information is Produced and Manipulated by Scientists, Physicians, Employers, Insurance Companies, Educators, and Law Enforcers*, rev. ed. Boston: Beacon Press, 1997.

Keller, Evelyn Fox. "Nature, Nurture, and the Human Genome Project." In *The Code of Codes*, edited by Daniel Kevles and Leroy Hood, 281–299. rev. ed. Cambridge, MA: Harvard University Press, 1993.

Keller, Evelyn Fox. *The Century of the Gene.* Cambridge, MA: Harvard University Press, 2000.

Kevles, Daniel. *In the Name of Eugenics.* Cambridge, MA: Harvard University Press, 1985.

Kevles, Daniel, and Leroy Hood, eds. *The Code of Codes: Scientific and Social Issues in the Human Genome Project*, rev. ed. Cambridge, MA: Harvard University Press, 1993.

Lippman, Abby. "Prenatal Genetic Testing and Screening: Constructing Needs and Reinforcing Inequities (The Human Genome Initiative and the Impact of Genetic Testing and Screening Technologies)." *American Journal of Law & Medicine* 17.1–2 (Spring–Summer, 1991): 15–50.

Mendel, Gregor. *Experiments in Plant Hybridization* (1865). Translated by the Royal Horticultural Society of London. Cambridge, MA: Harvard University Press, 1965.

Nelkin, Dorothy. *Selling Science: How the Press Covers Science and Technology*. rev. ed. New York: W. H. Freeman, 1995.

Nelkin, Dorothy, and M. Susan Lindee. *The DNA Mystique: The Gene as a Cultural Icon*. New York: W. H. Freeman, 1996.

Paul, Diane. *Controlling Human Heredity: 1865 to the Present*. Atlantic Highlands, NJ: Humanities Press, 1995.

Petersen, Alan. "Biofantasies: Genetics and Medicine in the Print News Media." *Social Science and Medicine* 52 (2001): 1255–1268.

Rapp, Rayna. *Testing Women, Testing the Fetus: The Social Impact of Amniocentesis in America*. New York: Routledge, 1999.

Rapp, Rayna, Deborah Heath, and Karen-Sue Taussig. "Genealogical Dis-Ease: Where Hereditary Abnormality, Biomedical Explanation, and Family Responsibility Meet." In *Relative Values: Reconfiguring Kinship Studies*, edited by Sarah Franklin and Susan McKinnon, 384–409. Durham, NC: Duke University Press, 2001.

Sayre, Anne. *Rosalind Franklin and DNA*. New York: W. W. Norton, 1975.

Strathern, Marilyn. *After Nature: English Kinship in the Late Twentieth Century*. Cambridge, U.K.: Cambridge University Press, 1992.

Taussig, Karen-Sue. "Normal and Ordinary: Human Genetics and the Production of Dutch Identities." PhD diss., The Johns Hopkins University, 1997.

Taussig, Karen-Sue, Rayna Rapp, and Deborah Heath. "Flexible Eugenics: Technologies of the Self in the Age of Genetics." In *Genetic Nature/Culture: Anthropology in the Age of Genetics, Genetics in the Age of Anthropology*, edited by Alan Goodman, Deborah Heath, and M. Susan Lindee, 58–78. Berkeley: University of California Press, 2003.

Yanagisako, Sylvia, and Carol Delaney, eds. *Naturalizing Power: Essays in Feminist Cultural Analysis*. New York: Routledge, 1995.

*Karen-Sue Taussig*

# GENETICS AND SOCIETY.

On July 3, 2000, *Time* magazine asserted that "history books will mark this week as the cer-emonial start of the genomic era." The remark is typical of how many journalists, scientists, business leaders, and politicians describe the significance of genetic research and genetic technology. That is, genetics is said to bring about changes that are so profound and so far-reaching as to amount to the beginning of a new era, no less world-altering than the beginning of the industrial era in the late eighteenth century. Such claims are not necessarily exaggerations. The financial and institutional resources committed to genetic research, the results of genetic research, and the promise of genetic research are together dramatically transforming medicine and health care, the pharmaceutical industry and the life sciences, education, criminal investigations, and agriculture and food production, as well as the way that we understand and explain human behavior and social relationships.

Still, to pinpoint the start of an era—even if it is the ceremonial start of an era—to a specific week is a bit of a dramatic exaggeration. The impetus for *Time*'s grand pronouncement (and many similarly enthusiastic news headlines) was a press conference, held in Washington, D.C., featuring President Bill Clinton, the leaders of the international Human Genome Project (HGP), and representatives of the private company Celera Genomics. Together, they were announcing the completion of the first draft of the sequence of the human genome. Marking the rather arbitrary milestone of a draft (or a project that was *almost* done), the announcement itself was not nearly as newsworthy as the alliance that had been formed by the political, scientific, and business leaders who made it.

The alliance marked the resolution of an intense (and publicity-soaked) competition between the HGP (an international consortium of public research institutions) and Celera (a private, for-profit company). The two competitors agreed to share the credit for the sequence of the human genome—the sequence that each had been rushing to finish before the other. The intensity of the "race" and the well-orchestrated media event (complete with the President of the United States) announcing its resolution should remind us that it is not genes and genomes themselves that are so world-altering and history-making. Genetic material is not simply a research object; it is a site of intense struggle for economic and political control.

## Genetics and Medicine

Genetic material and genomic information are especially valuable to the medical and pharmaceutical industries, for whom the identification of genes associated with particular diseases and health conditions can be translated into medical treatments and drugs, which can in turn be owned and patented. The prospect of new medical treatments and drugs is economically valuable; the promise of curing disease is politically valuable. Medical genetics, however, also raises large and complicated ethical questions in several areas. First, research that involves collecting people's genetic information comes with challenges to those people's privacy, their rights as research participants, and their claims on the research results. Second, health care and medicine that are genetically tailored to individual patients raise more concerns about privacy and, ironically, about individuals' rights to choose and control their own approaches to health care. Finally, even more complicated ethical and moral questions arise when researchers seek valuable genetic information from distinct populations of people who are not assimilating in Western society (that is, either culturally or geographically isolated peoples). Questions about the ownership, control, and usage of genetic information of an entire population (especially when that population does not benefit from the research) expand the complexity of the ethical and moral dilemmas associated with medical and pharmaceutical research.

## Genetics and Agriculture

Tensions between the promise of progress and ethical uncertainties are not unique to the medical arena. In agriculture and food development, genetic engineering technology has enabled the creation of new strains of crops designed to resist pests, the manipulation of animal and plant DNA to meet consumer preferences, and the introduction of specialized growth hormones to increase the efficiency of food production. Although genetic engineering and genetically modified foods are promoted by government-funded research and international biotechnology corporations that purport to be striving to expand and improve the world food's supply, these products and technologies meet with resistance from many sectors of society. Citizens and consumer groups across the world resist genetically modified foods, expressing concern about the safety of food supplies and the politics of multinational corporations controlling the production and distribution of food. Adding to the concern over food safety and international food-market politics (which quite literally reshape plant and animal genomes), environmentalists express concern about the loss of genetic variation, the irreversibility of many genetic experiments, and the unforeseen effects that designer crops will bring to the ecological balance of plants, insects, animals, and diseases. Organizations of independent farmers express concern over the loss of local autonomy and diversity in farming practices as a result of both the economic power of biotechnology corporations and the problems of "genetic pollution," including the pollen spread of genetically modified crops and ecological changes.

## Genetics and Human Behavior

Public conflict about genes and genetics also extends to debates about the degree to which genetics can explain human behavior and social phenomena. Debates are often framed in terms of questions of "nature versus nurture," with genes symbolizing the power of "nature." In the framework of nature versus nurture, genetic determinism suggests that we cannot change who we are as individuals; social or cultural determinism suggests that people can influence the forces that shape who we are. The claims of genetic determinism that capture the most public attention tend to be claims about socially contentious issues, especially issues that are sustained by the tensions of race, class, sexuality, and gender. For example, research linking memory and other brain functions with particular genes has been used to argue that intelligence is genetically determined. The genetic determinism of intelligence is then extended to justify differences in academic performance and professional successes that tend to coincide with differences in race and economic class. Emphasis on genetic explanations of intelligence diminish the accountability of public institutions in the face of race- and class-based inequalities.

Critics note that media attention on questions of nature versus nurture tend to draw attention away from changes in social policy. Whether it is about nature-nurture debates, races to sequence the genome, claims about the start of a new era, or the promise of medical research, gene talk captures attention. The dramatic publicity surrounding genetics should not be seen as trivial or exaggerated responses to "real" changes taking place. The promise of genetics and the persuasive power of gene talk are themselves major driving forces of social change.

See also **Genetics and Genetic Engineering**

E. P. Shea

# GOVERNMENT AND SCIENCE.

The allocation of public monies in the "public interest" is a thoroughly *political* process. American democracy ensures that the executive *proposes* and the legislative *disposes.* Compromise is the name of the game. So it is with the relationship between science and government, a "fragile contract" (in the words of David Guston and Kenneth Keniston) that the architect of the National Science Foundation (NSF), Vannevar Bush, called a "science and society compact." The so-called Bush Report, *Science—The Endless Frontier,* advanced a partial blueprint for a post–World War II U.S. science policy focused on the support of research and education in nonprofit institutions (mainly universities).

Together with another, more empirical report issued by the President's Scientific Review Board, *Science and Public Policy,* the Bush Report defined a set of expectations for the administration and nurturance of science and technology (S&T) that have been at the core of American life and, indeed, dominant in the world of scientific research and technological innovation, for the past half-century.

As the National Science Foundation grew, "policy for science" began to clarify the federal role around issues of funding levels and sources, research priorities, and the development and utilization of human resources for science and engineering. Complementary "science for policy" issues—the uses of scientific knowledge and capabilities for governance in the service of the larger society—received considerably less attention.

Today, the science-society compact is affirmed in the institutional apparatus of the U.S. government and the laws, mandates, and programs it authorizes and funds through an elaborate dance between the executive and legislative branches (and increasingly, the judiciary). As the National Science Board, the governing body of NSF, puts it, "The process of making choices constitutes the heart of policymaking—determining priorities and investment levels, nurturing long-standing programs, and responding appropriately to emerging research opportunities. . . . Inevitably, S&T are among the factors tied to choices made in the strategic allocation of scarce resources" (National Science Board, 2000b).

To many both inside and outside of science, however, the policy-making process remains a "black box," overshadowed by a protracted annual budget process and beset by contentiousness about "competing goods." How to sustain and manage the federal research system is an ongoing challenge to the relationship between science and government, especially the policy apparatus that has evolved over more than half a century.

In meeting the challenge, national goals are negotiated, funded, and pursued by the myriad participants who advise, compete within, and staff the policy process. Space exploration, a cleaner environment, safer and more fuel-efficient cars, and mapping the human genome to understand disease all emanate from a base of scientific discovery. Each entails the application of organized knowledge, ostensibly to improve the social condition (both at home and around the globe).

## The Policy Apparatus

Who decides what should be funded, developed, and applied to advance the human condition through S&T? Many countries have ministries of science—a single office for the whole country that determines what science should be funded, by whom, and at which institutions. In the United States many agents, from the President's Science Advisor in the Office of Science and Technology Policy, to expert witnesses called before congressional committees, to scholars sitting on committees and issuing reports, play prominent roles in a highly decentralized advisory structure. Yet other "stakeholders" are notable participants, such as ordinary citizens invited to comment on proposed rule changes in the *Federal Register*, think tanks that offer their interpretation of policies gone awry or issues deserving more urgent attention, and professional associations that lobby for budget increases in the departments and agencies that make scientific decisions daily in the name of their respective missions.

About two dozen executive agencies and departments include S&T research and development (R&D) in their portfolios. Before the September 11, 2001, terrorist attacks and the creation of a Department of Homeland Security, six of them—the Departments of Agriculture, Defense, Energy, and Health and Human Services (home of the National Institutes of Health), the National Aeronautics and Space Administration, and the National Science Foundation—accounted for over 90 percent of the federal R&D budget. Their allocations are meted out by 13 appropriations subcommittees of the U.S. Congress after scores of authorizing committees and subcommittees have reviewed strategic plans and recent program outcomes. Some departments, such as the Department of Energy (DOE), are subject to multiple committee jurisdictions because their mission ranges from unrestricted basic research to classified national security (weapons, intelligence) facilities. Even agencies with a science mission and budget per-

haps more modest in scope, however, such as the Environmental Protection Agency and the Department of the Interior (which includes the National Park Service), are just as vital to the national interest.

The discretionary federal budget resides within a perplexingly complex environment. The national investment in science and technology R&D has shifted over the last three decades, so the federal proportion represents just one-third of the total. It can be argued that private industry now drives the agenda. A perennial question about relations between science and government centers on what the appropriate federal role is. Does the government fill gaps formed when the private sector fails to support worthy projects that have long time horizons and weak prospects of commercialization? Or does federal R&D consciously distribute its risks over many performers to seed a diversity of projects with the expectation that some small fraction will yield profound discoveries—methods, treatments, unforeseen connections—that advance both knowledge and quality of life for scores of citizens here and worldwide?

Such questions are raised by those who advocate for achieving particular goals. Of particular concern, given government's prominent role in the funding of basic research, is its posture toward institutions of higher education where most such research is conducted. Universities occupy a unique historical place in the science-government relationship. They are not only the chief performer of basic research, but also the institution responsible for renewing the science and engineering workforce through the conferral of baccalaureate and advanced degrees.

At the doctoral level the federal government invests heavily—through fellowships (awarded directly to students), traineeships (awarded to departments and programs), and research assistantships (awarded through research grants to faculty)—to ensure a U.S. technical workforce that remains unrivaled in diverse talent, innovative capability, and versatility to respond to emerging market opportunity. Three of five doctorates conferred annually by U.S. universities are awarded in a science or engineering discipline. Clearly, these specialists have been deemed a national resource. If they enter academia, they are likely to compete for federal research grants and pursue a career path rewarded as much for entrepreneurial skills as for research, teaching, and service to various communities. If they enter the private sector, they are destined to fuel the engine of innovation and ascend to ranks of corporate leadership. Either way, they become international, as well as U.S., resources, disproportionately responsible (given their small numbers) for a technology-driven global economy.

More than half the doctoral degrees awarded in the physical sciences and engineering today are earned by foreign citizens attracted to premier programs in American research universities. Throughout the twentieth century, immigration policy welcomed this flow of talent. With heightened national security concerns, the United States may constrict the flow at the same time that countries significantly develop their high-tech infrastructure, such as those of the Pacific Rim, thereby intensifying the competition for world talent. As more science faculty near retirement, and more new Ph.D.s either opt to return to their nations of origin or find themselves barred from working in positions that would utilize their technical skills, the United States is unlikely to hold the edge as the nation of choice for training or employing future scientists and engineers. The impact of graduate-degree knowledge carriers as an innovative resource that contributes to and shapes the twenty-first century cannot be underestimated. This is a policy issue precisely because it calls into question the scope and leadership of the federal government in preserving the science-society compact.

## Spirits in a Material World . . . and the Political Arena

In the policy world, the adage goes, "where one sits determines where one stands." Examples of conflicting perspectives on S&T issues abound. Congressional "oversight" feels like "micromanagement" to a program manager in a federal R&D agency; the relaxation of regulatory protocols signals ideological exertion;

and appointments to presidential advisory committees are predicated on passing stated or unstated "litmus tests," the values—as inferred from one's published record and public pronouncements on issues such as abortion and cloning—of those selected to advise and staff the policy process.

Policymaking occurs in a world beyond the public display of lawmaking and budgeting. Each branch of government has a repertoire of tools for advancing its partisan will: The executive branch has executive orders and pocket vetoes (both of which defy congressional approval). The legislative branch wields congressional "earmarks" (which appropriate funds for local political purposes but simultaneously spurn executive policy, department/agency priorities, and authorization committee wisdom). The judicial branch hands down judicial decisions (which occasionally rise to the Supreme Court) that can negate policies (as in the University of Michigan affirmative action admissions cases) considered desirable by the two other branches of government. These are the realities of a dynamic political system that subjects all bidders to the strains of disparate values and scarce resources.

Science may represent what a popular song of the 1980s called "spirits in a material world." While surely not immune from these deliberations, science remained above the fray far longer and more effectively than other social institutions. Like any other interest group, its lofty and altruistic aspirations notwithstanding, S&T must compete for the sustenance that the system doles out.

Not surprisingly, earmarks for science-related bricks-and-mortar as well as research projects on campus have grown to a $2 billion a year enterprise, though they substitute raw power for agency priorities. Likewise, the patenting of intellectual property (almost every university now has a "technology transfer" office) signals the privatization of research over the open sharing of findings among academic researchers, while it restricts public access to and affordability of the invention in question. These are but two illustrations that "merit" (what one knows and can do, instead of who one is or whom one knows) coexists, and is indeed sometimes subordinated to, the overriding political (and business) culture that prevails in various sectors, organizations, and branches of government.

Scientists and many stakeholders lament developments such as these, citing (1) a compromise of values that are the hallmark of science—openness, selflessness, and knowledge as a public good; (2) increased conflicts of interest, inescapable in a system reliant on peer judgments, where reviewers are often competitors for scarce resources (whether federal project grants or corporate consulting dollars); and (3) requirements that institutions of higher education investigate and report allegations of misconduct involving researchers, the ownership of data, and the claims that federal monies enabled commercialization and private (corporate) gain. These are knotty issues of ethics and disclosure. The frequency of cases indicates how high the stakes; how heavy the burden on sponsors, performers, and stakeholders alike; and how quickly trends can become chronic conditions.

No wonder that foremost in the institutional relations of science and government, nestled within a "checks and balances" system, accountability for actions and outcomes has mounted. Legislation enshrined in the Government Performance and Results Act (GPRA) of 1993 was a response to the "Reinventing Government" campaign of the 1992 presidential election. In a decade, it has transformed how agencies present and justify their annual budget requests, using performance metrics. Whether appropriations rise or fall with these report cards again depends on where one sits and who is doing the grading.

Inevitably, the interplay of interests in the policy process engenders controversy and conflict. The resolution of conflict, through the passage of laws and the implementation of policy, is thus endemic to how science and government negotiate the next big issue. Curbs on research—given threats to health, privacy, or security—are particular lightning rods. The continuing furor over dangers posed by stem cell research and nanoscale science and engineering are two recent ex-

amples. They also dramatize how the advance of knowledge pushes the boundaries of what is known, believed, and valued.

## Beyond Myths

The generation of policy through new knowledge should prompt decision makers to act, or act differently, as science migrates through the policy apparatus. Vetting and refining new findings into fundable initiatives reflects the use of what we know today that we did not know yesterday. If the new science supports a broad constituency need while encompassing a national priority, it is likely to gain more adherents and momentum, regardless of cost, than if there is opposition due to labeling or charges of politically motivated scientific claims. Sometimes the cost of building a "big science" instrument is deemed excessive for a single nation to bear. So it was with the particle accelerator known as the Superconducting Supercollider (SSC): The U.S. Congress changed its mind about the priority of this investment, to be located in Texas, and withdrew its support, sending physicists, jobs, presumed Nobel Prizes, and a sizable boost to the state economy elsewhere. This decision established a precedent for multinational support of such facilities.

The politics of science play out in other forms as well. Describing a finding or prediction as "voodoo science" or "fuzzy math"—phrases that suggest not merely a benign absence of logic but a politically motivated claim without real support from the scientific evidence—may dog it or doom it despite the approval that the researcher's work may have within the expert community. Take, for instance, emerging data on global warming. If the results clash with a sitting Administration's position as stated in international treaties, then they are likely to be greeted with political skepticism, deemed unwelcome and unwarranted, and ignored as the basis for a science policy change. In short, the expert community, however extensive its access and credibility, may not carry the day. Typically, members of that community are likely to disagree, so a political actor can invoke the lack of consensus as grounds for not acting to change policy.

In this sense, experts, like other stakeholders, are resources. Scientists sometimes forget that their set of values, steeped in the scientific method of experiment and the shaping of theory under the guidance of data, is only one among many value sets that may not be widely shared by citizens outside of science. Religion, to be sure, provides certain values that require faith, which cannot be tested as a scientific hypothesis can. There remain many imponderables in crafting policy—different priorities and alternative explanations for observations and experiences. In the political arena, anecdotes count. Putting a human face on an issue can often be more compelling than statistical "causes and effects" in carrying the vote and the day.

How to act, and when, is the crux of policy making. Those who participate in the process—elected and appointed officials, individual and collective experts, policy staff, the public attentive to particular issues—confront the challenge of formulating and recommending policy for science through a willingness to act under uncertainty. Science is a tentative business because the empirical record keeps growing. Thus, the movement from analysis to action is not definitive. There are no "right" and "wrong" answers. Public policy is a world of options in which we must constantly ask who will benefit and who will be placed at risk.

In addition, policy for science is served by a small cadre of practitioners. Relatively few professionals combine science degrees with advanced policy degrees, so policy specialists with relevant experience (via a fellowship on Capitol Hill, for instance) hardly dominate the federal apparatus. Yet they draft the legislation and help to mobilize support for it. Those who staff the process elsewhere, in R&D agencies, nongovernmental organizations, and universities, are a special category of stakeholder. They may conduct policy-related research (on a discipline-anchored problem), execute policy analysis (on a topical issue), serve on expert committees (at, say, the National Academy of Sciences), testify at hearings on proposed legislation or agency/program performance, advise a court on the scientific soundness of claims in a product liability case, or all of the above.

As geologist and former congressional aid Daniel Sarewitz describes, scientists have perpetrated various myths that assert their authoritativeness while sustaining governmental support for research expenditures that promise endless benefits with minimal accountability to the nonscientific public. Sarewitz's critique is more a commentary on the role of science in U.S. culture—revered but not well understood—than a damning of institutional relations. He argues that science is a metaphor for optimism, adventure, economic growth, and social well-being. He also wonders whether the notion that what is good for science is good for society really holds, why the quality of life of our own population is so uneven (much less inhabitants of the underdeveloped world). The universal promises of science remain unrealized. The existence of knowledge does not ensure or hasten its distribution to those who could benefit from it.

Concerns such as those of Sarewitz are not necessarily an indictment of science policy, but they do raise reasonable questions about priorities, investments, and the politicization of knowledge for advancing certain interests and retarding others. Science is a publicly funded institution in U.S. society. Therefore, public expectations cannot be dismissed. Likewise, citizens' fears about catastrophes derived from scientific knowledge—dirty bombs, sabotaged planes, contaminated mail—should not be disparaged. Science as a problem-generating, as well as a problem-solving, activity must continue to be articulated.

## Whither the Compact?

In the simplest terms, most of science policy is negotiated between two expert communities, one scientific, the other political. As political theorist David Guston (1999) observes, these communities "collaborate at the boundary of politics and science." Forty years earlier, physicist Alvin Weinberg dubbed this policy space "trans-science." This is where policy issues live—shepherded through a labyrinth of organizations, each with a distinctive rationality, yet "helping government think" (in the words of Carol Weiss). The process requires mediation, translation, and stewardship—in efforts to inform the policy-making process, effect outcomes, and give transparency to the public interest.

In this arena, science and government engage in political debate about technical issues and what is "scientifically right." Open debate is essential in a participatory democracy, where a spectrum of views can be heard and challenged. Determining what is scientifically right can translate into "policy wrongs." All participants draw on an ever-changing pool of evidence and are steeped in partisan interpretations of what it all means and, moreover, what should be done about it. "Resolution" comes in the form of legal challenges, new precedents, and the passage of new laws. Out of this maelstrom, what unites participants in the exploitation of the process are the concepts of "majority rule" and "minority rights."

Still, it is unimaginable today that social progress can be made without scientific progress. That fact is a testament to the success of the compact between science and government. Accountability is the public demand for evidence that social progress is the chief aim of investments in scientific research. How to measure returns on that investment—intangible, workaday ways as well as in abstract and lofty rhetoric—is an imperative that both institutions in collaboration must fulfill. Indeed, the overarching challenge to science and government as intersecting institutional spheres is not funding (a fiction entertained by most researchers) but rather a bigger vision of the organization and pursuit of U.S. and global scientific needs.

Whither the science and society compact? It must be adapted if it is to be preserved. As political scientist Roger Pielke, Jr., states, "Historically, we've tried to uphold the notion that science is value-free and politics deals with values, and we try to keep them separate. But reality is much more complicated. You can't build a wall between them; the real action is how we connect and put them together" (quoted by Agres).

*See also* **Government and Technology**

## BIBLIOGRAPHY

Agres, Ted. "Science, Policy, and Partisan Politics." *The Scientist Daily News*, Aug. 13, 2003.

http://www.biomedcentral.com/news/20030813/04.

Bush, Vannevar. *Science—The Endless Frontier: A Report to the President on a Program for Postwar Scientific Research* (1945). Washington, DC: National Science Foundation, 1990.

Chubin, Daryl E., and Ellen W. Chu. *Science Off the Pedestal: Social Perspectives on Science and Technology.* Belmont, CA: Wadworth, 1989.

Chubin, Daryl E., and Jane Maienschein. "Staffing Science Policy-making." *Science* 290 (2000): 1501.

Chubin, Daryl E., and Willie Pearson, Jr., eds. *Scientists and Engineers for the New Millennium: Renewing the Human Resource.* Washington, DC: Commission on Professionals in Science and Technology, 2001.

Friedman, Sharon M., Sharon Dunwoody, and Carol L. Rogers, eds. *Communicating Uncertainty: Media Coverage of New and Controversial Science.* Mahwah, NJ: Lawrence Erlbaum Associates, 1999.

Golden, William T., ed., *Science and Technology Advice to the President, Congress, and Judiciary.* Piscataway, NJ: Transaction Publishers, 1988.

Guston, David H. *Between Politics and Science: Assuring the Productivity and Integrity of Research.* Cambridge, U.K.: Cambridge University Press, 1999.

Guston, David H. and Kenneth Keniston, eds. *The Fragile Contract: University Science and the Federal Government.* Cambridge, MA: MIT Press, 1994.

Kalil, Thomas. "A Broader Vision for Government Research. *Issues in Science and Technology* (Spring 2003): 29–33.

National Science Board. *Science and Engineering Indicators—2000.* Arlington, VA: National Science Foundation, 2000a.

National Science Board. "Science and Technology Policy: Past and Prologue. A Companion to *Science and Engineering Indicators—2000."* Arlington, VA: National Science Foundation, 2000b.

Sarewitz, Daniel. *Frontiers of Illusion: Science, Technology, and the Politics of Progress.* Philadelphia: Temple University Press, 1996.

Steelman, John R. *Science and Public Policy* (1947). New York, Arno Press, 1980.

U.S. Congress, Office of Technology Assessment. *The Regulatory Environment for Science.* Springfield, VA: National Technical Information Service, February 1986.

U.S. Congress, Office of Technology Assessment. *Federally Funded Research: Decisions for a Decade*. Washington, DC: U.S. Government Printing Office, May 1991.

Waxman, Henry. *Politics and Science in the Bush Administration*. U.S. House of Representatives, Committee on Government Reform – Minority Staff, Special Investigations Division, August 2003. www.henrywaxman.house.gov/ and www.house.gov/reform/min/politicsandscience/pdfs/pdf_politics_and_science_rep.pdf.

Weinberg, Alvin M. *Reflections on Big Science*. Cambridge, MA: MIT Press, 1966.

Weiss, Carol H., ed. *Organizations for Policy Analysis: Helping Government Think*. Newbury Park, CA: Sage Publications, 1992.

*Daryl E. Chubin*

# GOVERNMENT AND TECHNOLOGY.

Governments have been involved with technology throughout human history, at least since the days of the Pharaohs in ancient Egypt. Military technology was essential to the power of medieval governments, and modern governments employ technology to stimulate economic growth, improve their citizens' standards of living, and pursue missions in many other areas, including environment, public health, and defense. The involvement of governments with technology can be described under three broad headings: promotion or development of technology; regulation of technology; and use of technology and technical information in government policy making.

## Government's Role in Technology Development

Governments support the development of technology both directly and indirectly. Direct funding of research and development (R&D) provides government with a means of shaping technologies according to public policies. In fiscal year 2003 the U.S. federal government spent an estimated $117 billion, or about 5.4 percent of the overall federal budget, on R&D.

Some government R&D provides the foundation for development of products intended to be manufactured and sold by industry, such as hydrogen cars or improved airport screening devices. Other government investments help build the nation's long-range scientific and technological strength; examples are the National Nanotechnology Initiative in the United States and the Sixth Framework Programme of the European Union. Direct funding of R&D is a major function of government in virtually all industrialized countries and has grown sharply in the past half-century, in concert with the increasingly technological nature of society and the growing role of technology in the world economy.

In the United States, direct federal government support for development of commercial technologies is politically controversial. The history of federal efforts to promote industrial technology goes back at least to the 1920s, but virtually all such initiatives have become enmeshed in disputes between political leaders who advocate an activist role for government in the nation's economy and those who see government's role as more limited and subordinate to the market. During the past decade, one focus of controversy has been the Advanced Technology Program of the Department of Commerce, a relatively small program that provides subsidies for firms to do R&D on technologies whose chances of succeeding are too small to attract private investors but whose potential payoffs are great. Republican presidents and congressional leaders have sought to kill the program at every opportunity, while Democrats have stood fast in its defense.

Programs of this nature are more readily accepted in many other nations, whose governments are less concerned with issues of what is or is not appropriate for government. In France, for example, ANVAR (Agence Nationale de Valorisation de la Recherche, called the French Agency for Innovation in English), promotes and finances innovation in French industry, particularly in small and medium-sized enterprises and facilitates the emergence of new products and processes.

Many state governments in the United States, eager to attract or retain high-paying jobs in technology-based industries, also subsidize commercial technological development. Michigan is using funds acquired from the settlement of a huge multistate lawsuit against the tobacco industry to leverage investment in biotechnology businesses in order to build a "life sciences corridor" in the southern part of the state. Pennsylvania's Ben Franklin Partnership has invested more than $300 million of state funds in technology development since its founding in 1982.

A large share of government investment in R&D in most countries is primarily intended to serve government's own needs, for example, strengthening military capabilities, building better roads and bridges, or providing a scientific basis for regulating food and drug safety or the environment. This kind of R&D can also help to strengthen technology-based industry. In some cases, the resulting technologies may be adapted for commercial uses or may be "dual-use," directly serving commercial as well as government users. In the United States, the Department of Defense (DOD) is an important source of such funding. Slightly more than half of all federal funding for R&D in the United States typically comes from the defense portion of the federal budget in any given year. Among the civilian technologies that originated or were supported in early stages by DOD R&D funds are the Internet, microelectronics and integrated circuits, optical fibers, and many aeronautical and space technologies. Other U.S. agencies that support technology development for government use are the National Aeronautics and Space Administration (NASA), the Department of Energy, and the Department of Homeland Security.

Governments also provide indirect support for technology development. Some of this support takes the form of subsidies to the private sector through tax codes. Most industrial nations encourage industrial firms to invest in R&D by allowing 100 percent of R&D expenses to be deducted from a firm's income tax. Australia, where industrial investment in R&D has long been regarded as insufficient for the nation's long-term economic growth, allows a 125 percent deduction. Some nations, including the United States, Canada, Japan and France, also allow firms that increase their R&D spending to take a certain percentage of the increase (20 percent in the case of the United States) as a tax credit. With this mechanism, govern-

ments can increase industrial investment in R&D while allowing individual firms to choose technologies and R&D projects according to their own needs and judgment.

Other indirect mechanisms that governments use to support technology development include regulation and creation of a suitable environment for technological innovation—for example, by providing protection for intellectual property. To force the development of technology, government may establish a future regulatory standard that it knows cannot be met with existing technology. Such a standard puts pressure on firms in the affected industry to conduct R&D and develop technology that can meet the standard. The best-known American example of technology forcing is the Clean Air Act Amendments of 1970, which required automakers to reduce the exhaust emissions of new vehicles by 90 percent in four to five years. This legislation, and the mandate three years later for eliminating leaded gasoline, led to the implementation of the catalytic converter—although this was not accomplished within the time frame specified in the law.

If they hope to profit from their innovations, firms and organizations that conduct R&D, as well as individual inventors, need to establish property rights in them. Governments provide the means for protecting such rights through patents and copyrights in order to create an environment that will encourage technological innovation. The U.S. Constitution provides the basis for patent protection, and the legal system allows innovators to defend their patents when others threaten to infringe on them. In recent years, developments in information technology as well as in the life sciences have posed new challenges to traditional concepts of intellectual property, raising issues such as whether life forms or DNA sequences can be patented, what kinds of rights apply to computer code, and whether it is possible or desirable to protect the rights of intellectual property owners in a world where those properties (such as music, movies, and books) can be readily duplicated by almost anyone with a computer and shared worldwide on the Internet.

Another important way in which governments are involved with technology development is through the establishment of standards. From keeping exact time to ensuring that one person's or company's inch, centimeter, or kilogram is exactly the same as another's, or that a bolt with a certain thread specification fits into a nut with that thread, or that a nation's television receivers are compatible with its broadcasters, standards are fundamental to technology, commerce, and trade. In the United States, the National Institute of Standards and Technology (NIST) is responsible for these functions. NIST laboratories develop new physical standards and measurement methods. It also provides basic reference standards and works with industry to calibrate measuring instruments to those standards and with international organizations to ensure compatibility with other nations' standards.

## Regulation of Technology

As they develop and as they affect society, technologies raise issues that create demands for government regulation. These issues may involve human health and safety or the environment; allocation of technological resources, such as the electromagnetic spectrum); protection of consumers from fraudulent or defective products; social goals, such as bringing the benefits of technology to diverse populations or geographic regions; or political goals, such as preventing the acquisition of technologies by foreign enemies, economic rivals, or terrorists. At times, the issues may also involve ethics, as in the case of research involving human embryos and reproductive cloning.

Virtually all countries have government agencies that regulate technology. In the United States, several federal agencies are responsible for regulation to protect human health and safety, including the Food and Drug Administration, the Occupational Safety and Health Administration, the Environmental Protection Agency, and the Nuclear Regulatory Commission. In Canada, the Health Products and Food Branch of Health Canada performs functions similar to the U.S. Food and Drug Administration. In Germany, the analogous agency is the Federal Institute for Drugs and Medical Devices.

Among the products that the U.S. Food and Drug Administration (FDA) regulates are vaccines and other biologics, such as blood and blood products; cosmetics; medical devices, such as pacemakers and contact lenses; foods, with regard to processing and contamination; and prescription, as well as over-the-counter, drugs. Before a prescription drug can be marketed in the United States, it must go through an elaborate approval process. Based on animal studies and human clinical trials, the FDA must determine (1) whether the drug is safe and effective in its proposed use(s), and whether its benefits outweigh its risks; (2) whether the drug's proposed labeling is appropriate, and what it should contain; and (3) whether the methods used in manufacturing and quality control are satisfactory. The process often takes years and costs millions of dollars. The considerations are similar in most other countries, although the judgments of risks and benefits may differ among them.

Because the regulatory process often creates long delays between the discovery or invention of a new drug and its availability to patients, and because of the high costs involved, drug regulation is often the subject of controversy. Controversies also arise in regulation of the environmental effects of technologies or in regard to safety issues. The U.S. Environmental Protection Agency and its counterparts in other nations, such as Environment Canada and the Swedish Environmental Protection Agency (Naturvårdsverket), are charged with the enforcement of laws intended to prevent or remediate environmental pollution. Preventing pollution of rivers, lakes, and other bodies of water imposes costs on firms whose production processes create potential pollutants. The same is true of sources of air pollution, such as that produced by coal-burning power plants, oil refineries, and automobiles. Those who are asked to pay these costs quite naturally resist, generally arguing that the presumed benefits do not justify the costs. The disputes often focus on uncertainties in the data on which the regulations are based.

Probably the most significant controversy of this nature in the United States in the early twenty-first century relates to the problem of global climate change. A strong consensus has developed among scientists that the earth's surface temperature is rising; that it is rising at an accelerating rate; that the rise is being caused by human activities, especially the increase of atmospheric carbon dioxide due to the burning of fossil fuels; and that the increase is likely to cause serious problems in the next 50 to 100 years, including greater climate instability and a significant rise in sea level. Most of the nations in the world have agreed to take action to slow or halt the rise. This action will be very costly, but failure to act could be even more costly. The United States, which has the world's largest economy and is the largest emitter of carbon dioxide and other greenhouse gases, would bear the largest share of the costs. The federal government and much of the U.S. business community have resisted taking action, arguing that there are too many uncertainties—regarding how much the climate is changing, what the potential impacts are, and what may be causing the changes. The dilemma is a central one in the regulation of technology: How certain must one be in order to take a regulatory action, especially when that action is likely to be costly?

The controversy over genetically modified organisms (GMOs), especially in foods, is similar. Guided by the "precautionary principle," which holds that "when an activity threatens harm to human health or the environment, precautionary measures should be taken even if some cause-and-effect relationships are not fully established scientifically," most European nations assert that GMOs should not be allowed in foods. United States regulators, arguing that there is no persuasive scientific evidence that GMOs are harmful, and citing many benefits, have generally rejected the European position.

## Use of Technical Information in Government Decision Making

As the global change and GMO issues illustrate, technology is an increasingly important factor in the programs that governments undertake, in the problems that confront them, and in the policies that they develop in response to or in anticipation of these problems. These policies require specialized knowledge that governments must either maintain internally or obtain from outside sources. The government agencies that use technology or fund R&D have engineers and scientists on their staffs. Some, like NASA, have a substantial number; others have relatively few. Virtually all of these agencies, however, have technical advisory committees that supplement their in-house capabilities.

At the top levels, governments of most industrialized countries have bodies that provide science and technology advice to their leaders. Since World War II, the president of the United States has had a science and technology advisor, situated throughout most of the period in the Executive Office of the President. In the administration of President George W. Bush, the advisor became a special assistant to the president, serving as director of the White House Office of Science and Technology Policy and also as co-chair of an external advisory group known as the President's Council of Advisors on Science and Technology. The influence of the advisor and the advisory councils and committees have waxed and waned depending on the nature of the issues facing the president, the power structure within the White House, and the president's personality and degree of interest in scientific and technological matters.

Departments and ministries often have their own science and technology advisory committees as well. In the United States, the Department of Energy, the Environmental Protection Agency, the Department of Defense, and many other agencies have such committees. United States government agencies also obtain technical advice from the National Research Council (NRC), which is the operating arm of the National Academies of Science and Engineering and of the Institute of Medicine. With a staff of over 1,000, the NRC conducts hundreds of studies each year to provide government agencies with authoritative advice on topics ranging from supercomputing to smallpox. Agencies also obtain information and advice through studies conducted by universities, nonprofit institutions (such as the RAND Corporation), and private firms under contracts and grants.

Although executive branch agencies and their counterparts have many resources for science and technology information and advice, the situation is different for legislative bodies. Members of parliament in most nations have very limited personal and committee staffs. With relatively few positions available, staff members must be generalists: those with technical expertise are rare, except on specialized committees. Supplementing these staffs are such bodies as the Library of the House of Commons in the United Kingdom. The Library produces research papers, a small fraction of which are devoted to science and technology. The United Kingdom, however, also has a small Parliamentary Office of Science and Technology (POST), which provides more specialized advice on technical topics. Since its establishment in 1989, POST has produced reports for Parliament on subjects ranging from broadband Internet access to stem cell research.

Things are different in the U.S. Congress. It is not uncommon for senators to have personal staffs ranging from 50 to 100. Representatives' staffs are smaller, but they are still considerably larger than their parliamentary counterparts in other nations. Senators and representatives who have committee assignments relating to technology or science, or who have major research facilities in their constituencies, can therefore afford to hire some staff with technical qualifications. In addition, science and engineering associations have since 1973 maintained fellowship programs that make PhD-level scientists and engineers, selected through national competitions, available to members of Congress for one-year assignments. Other nations are beginning to initiate such programs as well. Switzerland has established one on a small scale, and the United Kingdom, Australia, Japan, and South Africa are all considering doing so.

The Office of Technology Assessment (OTA), created by the U.S. Congress in 1972, was the first governmental body devoted exclusively to conducting studies and providing advice to a legislature on the impacts of technology on society. The agency conducted hundreds of studies for Congress on subjects ranging from advanced networking technology to AIDS to the lumber industry. It was unique among agencies serving the U.S. Congress not just in the range of its work, but in the depth of its technical expertise and in the long-term view it brought to its studies.

In 1995, in what was officially called an economy move, the U.S. Congress eliminated funding for OTA. Other nations (such as the United Kingdom), which had followed the U.S. example in establishing parliamentary offices of technology assessment, did not follow the U.S. lead in abolishing these offices. Today, at least ten European nations plus the European Parliament are served by technology assessment organizations. These bodies conduct policy studies of issues raised by scientific and technological developments, many of which seek to anticipate the future impacts of technologies under development or the impacts of current societal trends. Some conduct "participatory assessments," engaging members of the public in their analyses.

## Outlook

The vast amount of resources that governments, industrial firms, universities, and other organizations worldwide are investing in research and development and the growing demand for technological products and technological solutions to the problems of poverty, illness, environmental degradation, and global insecurity virtually ensure that the rate of technological change will only accelerate in the twenty-first century. The changes will continue pose new challenges for government. Some of these challenges may be foreseen; most will not. When researchers for the U.S. Department of Defense developed a means of linking mainframe computers at a few select laboratories in the late 1960s, no one involved foresaw the immense changes their innovation, forerunner of the Internet, would engender in publishing, commerce, entertainment, and many other fields. And no one imagined that this innovation would provide a means for bored teenagers to wreak havoc on computers halfway around the world or that industry and government would need to spend millions of dollars protecting against such threats. The course that emerging technologies—nanotechnology, biotechnology, robotics, and more—will take in coming years is uncertain. What is certain, however, is that the future of government and the future of technology are inseparably linked.

See also **Government and Science**

**BIBLIOGRAPHY**
Branscomb, Lewis M., and James H. Keller, eds. *Investing in Innovation: Creating a Research and Innovation Policy That Works.* Cambridge, MA: MIT Press, 1998. An attempt to defuse the ideological arguments about government support of commercial technology.
Flamm, Kenneth. *Creating the Computer: Government, Industry, and High Technology.* Washington, DC: The Brookings Institution, 1988. Traces the roots of the various technologies that converged to make modern computers and information technology possible; documents the role of several governments, including the United States, the United Kingdom, Japan, and Germany in the development of these technologies.
Greenberg, Daniel S. *Science, Money, and Politics.* Chicago: University of Chicago Press, 2001. Although nominally about science, its interactions with government, and its ethical transgressions, this book is also relevant to government's technology policies.
Hart, David M. *Forged Consensus: Science, Technology, and Economic Policy in the United States, 1921–1953.* Princeton, NJ: Princeton University Press, 1998. Makes a strong case that the events and policy decisions of the period about which he writes continue to have a strong influence on contemporary science and technology policy.
Heimann, C. F. Larry. *Acceptable Risks: Politics, Policy, and Risky Technologies.* Ann Arbor, MI: University of Michigan Press, 1997. The politics and policy of dealing with high-risk technologies and the disasters they sometimes cause, using examples from space flight and pharmaceuticals.
Kleinman, Daniel, ed. *Science, Technology, and Democracy.* Albany, NY: State University of New York Press, 2000. A collection of essays on citizen involvement in government policy making for science and technology.
Mani, Sunil. *Government, Innovation, and Technology Policy: An International Comparative Analysis.* Cheltenham, U.K.: Edward Elgar, 2002. Analysis of the role of government in commercial technology development in eight countries.
Morgan, M. Granger, and Jon Peha, eds. *Science and Technology Advice for Congress.* Washington, DC: RFF Press, 2003. Papers from a conference on Congress's need for technical information and whether the Office of Technology Assessment should be resurrected to fill it.
National Research Council. *The Digital Dilemma: Intellectual Property in the Information Age.* Washington, DC: National Academies Press, 2000. An attempt to clarify the issues surrounding digitized intellectual property by a blue-ribbon committee convened by the U.S. National Academy of Sciences; includes policy and research recommendations.
Sarewitz, Daniel. *Frontiers of Illusion: Science, Technology, and the Politics of Progress.* Philadelphia: Temple University Press, 1996. A sharp critique of the myths that guide science and technology policy in the United States, by a former congressional staff member.
Stokes, Donald E. *Pasteur's Quadrant: Basic Science and Technological Innovation.* Washington, DC: Brookings Institution Press, 1997. A seminal work on the interaction of science with technology.
U.S. National Science Foundation, Division of Science Resources Statistics. *Science and Engineering Indicators–2004.* Arlington, VA: National Science Foundation, 2004. Published every other year, this is the most comprehensive authoritative source of statistics on science and technology in the United States; the full text and tables are available online at http://www.nsf.gov/sbe/srs/seind/ start.htm.
Vig, Norman J., and Herbert Paschen. *Parliaments and Technology: The Development of Technology Assessment in Europe.* Albany, NY: State University of New York Press, 1999. A unique attempt to compare technology assessment mechanisms in a number of different countries.

*Albert H. Teich*

# GRASSROOTS SCIENCE.

Grassroots science is science done by people outside the mainstream of professional science. Grassroots science includes research by amateurs and lay people as well as some dissident work by professional scientists.

*Science*—both professional and grassroots—involves the creation and use of systematic knowledge, using standard procedures, within a community of practitioners, involving both theory and experiment. Scientific knowledge is public knowledge, within the practitioner group and often beyond, thereby excluding some proprietary knowledge generated within corporations and some secret knowledge generated within intelligence agencies.

*Professional science,* in addition to these general features, is built around practitioners who have extensive formal training and work full-time as scientists. Professional science also often involves use of expensive equipment. Editors, referees, and prestigious scientists defend the boundaries of professional science, imposing definitions of what is and what is not science. Orthodox science is not as open to new or outside ideas as many people think.

*Grassroots science,* in contrast, is done by amateurs or by professionals separately from their main paid work, and it usually involves much less expensive equipment. Some people become grassroots scientists because they love to learn about nature but have no opportunity or no desire to undertake a professional career in science. Others want to challenge orthodox theories. Yet others believe that professional science is biased toward corporate and government priorities and that grassroots science provides a way to truths that are otherwise ignored or obscured by vested interests.

The boundaries between grassroots and professional science are blurry and changeable, and so are the boundaries between science and nonscience.

## Amateur Science

There is a thriving community of amateur astronomers, ranging from beginners who look at stars and planets with small telescopes to experienced observers who systematically seek to observe new objects in space. There seems to be a de facto division of labor, with professional astronomers mainly using massive instruments to look into deep space, such as faraway galaxies, and amateurs using smaller instruments to look at planets, moons, comets, variable stars, and other more accessible phenomena. Because there are so few professional astronomers and so many possible astronomical objects to observe, amateurs can make important contributions, especially now with the availability of cheap powerful computers, the Internet, and lower-cost high-tech telescopes and other instruments. Unlike most sciences, astronomy does not involve experimentation. Amateurs are also prominent in botany and zoology, other fields where observation, either directly or using low-cost instruments, retains a central importance.

The primary differences between amateur and professional astronomy are that the latter involves formal training, full-time paid work, and expensive equipment. The two groups largely agree on the goals and methods of science. The sophistication of new tools used by amateurs is making many professionals appreciate their contributions. Amateurs sometimes can support professional scientific endeavors by providing labor or resources, such as computer users who contribute spare computing power to the Search for Extraterrestrial Intelligence (SETI).

*Ufology* is the study of unidentified flying objects (UFOs). Mainstream scientists mostly reject the view that UFOs are manifestations of alien intelligence, and the whole field of ufology is often dismissed as pseudoscience. Although a few professionals are involved with ufology, the field is dominated by amateurs, with a thriving community of magazines, meetings, and communication networks. Ufology is thus a facet of grassroots science that is stigmatized by mainstream science. Mainstream scientists take various actions to ensure that their own work is kept separate from ufology, such as preventing UFO research from being published in mainstream journals. This process of distinguishing and separating mainstream science from what is labeled pseudoscience is called *boundary work.*

Independent inventors could be called grassroots technologists. James Lovelock, who made many important scientific contributions and developed the Gaia hypothesis, was a home-based inventor. Independent inventors typically work alone; many of them make a living from other activities. Computer hackers—in the original sense of building or modifying computers and software—are another species of grassroots technologist.

Much of science conducted before World War II was small in scale. Through the 1800s, amateurs played a major role in science. So it might be said that until the twentieth century, most science was grassroots science in that it was less dependent on governments, large corporations, or universities.

Amateur astronomy and independent invention can be highly sophisticated. These forms of grassroots science and technology are not easy or obvious for most people. Just as some members of amateur theater groups can be of professional caliber but not receive any payment, some grassroots science can be of professional standard but be undertaken out of pure interest.

## Dissident Science

Dissent is central to science: the formulation of new ideas and the discovery of new evidence is the driving force behind scientific advance. At the same time, certain theories, methods, and ways of approaching the world—often called paradigms—are treated as sacrosanct within the professional scientific community. Those who persist in challenging paradigms may be treated not as legitimate scientists but as renegades or outcasts. Some of these dissidents could be said to be doing grassroots science because they operate outside the normal system of training, employment, and major equipment.

For example, there are many individuals who have developed challenges and alternatives to relativity, quantum mechanics, and the theory of evolution, three theories central to modern science. Some of these are amateurs who have jobs outside of science. Others are professional scientists who have degrees, publications, and honors but who undertake their dissident work as an adjunct to their mainstream careers. These dissident individuals, though they may espouse incompatible theories, are brought together through meetings, networks, and organizations such as the Natural Philosophy Alliance.

Linus Pauling is an example of a professional scientist who became a dissident: He won a Nobel Prize in chemistry but later developed unconventional ideas about vitamin C and cancer, in an area in which he had no formal training.

Dissident scientists usually agree with mainstream scientists about the aims of science—namely, the advancement of knowledge—and about the methods of science—namely, critical examination of theory and evidence—but disagree about what theories are correct.

Mainstream scientists sometimes ignore dissidents, sometimes attack them, and sometimes seek to incorporate their ideas into the mainstream. Cold fusion—nuclear fusion at room temperature—started out as a dramatic challenge to orthodoxy by established scientists. When initial results could not be widely reproduced, cold fusion research was attacked and then forgotten. In the aftermath of the original attention, many scientists continue to explore cold fusion, some with funding from corporations, but their findings are ignored by the mainstream. Cold fusion has elements of grassroots science, though professionals play a significant role in it.

Acupuncture is a method of healing long used in China as part of a non-Western understanding of the body. Traditional Chinese acupuncturists were often full-time healers, but the practice itself is inexpensive, so it can be said to be a form of grassroots medicine. Western medical authorities at first rejected acupuncture as unscientific but, following demonstrations of its effectiveness, eventually accepted or tolerated it as a practice under the canons of Western biomedicine, rejecting its associations with non-Western concepts of the body. Acupuncture thus is an example of grassroots science that has been incorporated into Western professional practice, in part severing its links with the grassroots. A similar process has occurred with some other parts of complementary medicine such as meditation.

## Science and Social Movements

A social movement involves many people acting in concert to create a preferred image of society; familiar examples are the environmental, peace, and feminist movements. Social movements typically have a core of activists (professionals, volunteers, or both), a wider group of occasional participants, and a still wider group of passive supporters. Social movements can contain great internal diversity but agree about general goals. Social movements are natural homes to grassroots science, especially when a movement challenges an establishment that has the backing of professional science.

*The Alternative Health Movement.*   The alternative health movement emphasizes prevention and, for dealing with disease, concentrates on nutrition and natural products and methods. The alternative health movement includes many trained professional practitioners but also encourages popular learning and self-help, whereas conventional medicine, in contrast, emphasizes cure through intervention by medical professionals.

Many individuals make observations about their own health and the health of family and friends, noting that a certain food, herb, or method of behavior has beneficial effects. When these observations are shared with others, some become widely adopted, becoming "folk medicine."

Some proponents of alternative health keep a close eye on conventional medical research findings, noting those that are relevant to nutritional healing and other movement interests. Participants in the movement sometimes recommend to mainstream researchers that they test particular substances or methods. Thus there can be a mutually supportive relationship between the movement and portions of mainstream medical research.

At the same time, some mainstream medical practitioners and researchers are hostile to alternative health. This is apparent in pronouncements that taking vitamin supplements is a waste of money or in police raids on alternative cancer therapists, the raids being encouraged by mainstream opponents.

Many proponents of alternative health say that mainstream medical science is distorted by corporate, government, and professional pressures. In this context, grassroots medical science presents itself as being truer to the ethos of science as a search for truth unsullied by vested interests.

*The Feminist Movement in Medicine.*   The feminist movement has also developed a grassroots challenge to orthodox medicine, which the movement sees as patriarchal, both dominated by men and oriented to male concerns. Reproduction has been a key area of contention, with mainstream (or "malestream") medicine alleged to have medicalized reproduction through contraceptives and Caesarean sections, ignoring the practical knowledge that women have of the operation of their own bodies.

With the second wave of the feminist movement beginning in the 1960s, women's personal knowledge was collected and circulated by feminist collectives, most famously the Boston Women's Health Book Collective. Movement activists thus served to codify the knowledge of individual women, also drawing on mainstream research, some of it by researchers sympathetic to the movement.

One practice promoted by the women's health movement has been breastfeeding, which went into decline in many industrialized countries through pressure by medical authorities for rigid weight gain targets linked to a bottle-fed norm. Breastfeeding advocates have developed great understanding of what is needed to make breastfeeding successful (for example, how to deal with cracked nipples) and promoted this alternative model to both new mothers and medical professionals. Grassroots science in this instance involves development of knowledge about breastfeeding in a community of practitioners who are in an ongoing challenge to mainstream practice.

*AIDS Activism.*   From the time AIDS first became recognized as a disease in 1981, it was apparent that gay men were prime targets. By that time, the gay movement was well organized in many western countries. Gay men's health activists learned as much as they could about AIDS, studying virology, immunology, epidemiology, and other specialized fields relevant to the disease. As well as drawing on conventional biomedical science, AIDS activists also knew a great amount about the behavior of gay men, knowledge they could use to recommend interventions against the spread of AIDS. AIDS activists thus combined their knowledge of sophisticated professional science with grassroots behavioral knowledge.

Some activists intervened in orthodox medical policy making, influencing the design of trials of potential AIDS drugs and pushing for speedier release of drugs, thus becoming de facto adjuncts of mainstream biomedicine—a case of grassroots science being partially incorporated in professional science. Other AIDS activists have taken a more alternative route, investigating substances such as dinitrochlorobenzene (DNCB)—of little interest to pharmaceutical companies because it is not patentable—and distributing them via so-called guerrilla clinics.

*Community Epidemiology.* In the 1970s in Japan, *minamata* disease was causing devastation in some communities. Teams of community volunteers, aided by sympathetic scientists, used simple techniques, including interviewing members of local communities, to track down the source of the disease, which was poisoning by mercury pollution from industry. They did this more effectively than well-funded teams of professional scientists using sophisticated methods for analyzing samples and running computer models.

Since then, there has been a considerable expansion of community research, especially in the United States, with much of the work being done in environmental and health areas, where mainstream research is often influenced by corporate and government agendas. In terms of doing good science, community research follows the model of conventional scientific research, but there are several differences. Community researchers are usually unpaid, often without formal re-

search training. They pick topics relevant to local community concerns, often challenging corporate or government agendas. They largely communicate with their local communities and other community researchers, not necessarily seeking publication in conventional scientific journals.

In professional epidemiology—the study of the incidence and transmission of disease—researchers may dismiss anecdotal evidence as unworthy of attention (sometimes in a selective fashion that may or may not be related to its relevance to corporate sponsors). In popular epidemiology, in contrast, the same anecdotal evidence can provide the inspiration for a more detailed investigation.

*The Appropriate Technology Movement.* The appropriate technology movement pursues development and use of technology that is appropriate to people's needs, especially poor people in poor countries. It includes organic gardening, biogas generators, inexpensive water filtration techniques, passive solar design, and a host of other technologies and techniques that can be locally used and controlled. Many of the developers and promoters of such technology are grassroots technologists.

*The Free Software Movement.* Free software (which may or may not be distributed free of charge) is open source (the code is openly available for inspection) and is licensed so that it cannot be turned into proprietary software. In constructing and improving free software, suggestions are taken from anyone who is interested, with a core group making decisions about implementing changes. The most famous free software is the operating system Linux, but there are many other programs.

The free software movement is a type of gift economy, in which participants do not seek financial gain but instead the respect of peers and the satisfaction of contributing to a worthwhile product. Some contributors are computer professionals who help with free software separately from their jobs; others are committed students or amateurs. The movement has begun to expand to other domains; for example, there has been an open-source cola with the ingredients listed on each can, and the recipe posted on the Internet.

The free software movement has several similarities to grassroots science. Both harness tremendous voluntary efforts by participants through the satisfaction of contributing to a collective enterprise, and both contribute to alternative knowledge or products that challenge professional systems.

*The Scientific Establishment as a Social Movement.* Professional science is itself a social movement. There are proselytizers and lobbyists for professional science, necessary to win support for massive expenditures to train and pay scientists and support their research. In the early days of modern science, struggling in the face of hostility from churches and an indifferent public, its social movement characteristics were more obvious. Today, professional science is a social institution, but it still must maintain its struggle for resources and credibility.

## Indigenous Knowledge

In order to survive, people in nonindustrial societies must learn a great deal about nature, for example about weather and the seasons, edible and medicinal plants, the behavior of animals, and human physical and mental capacities. This sort of indigenous knowledge has characteristics of grassroots science: it is systematic knowledge, publicly available in a community of practitioners, and involves theory and practice. Professional scientists have often ignored indigenous knowledge or dismissed it as unscientific because practitioners do not follow the scientific method and because the knowledge itself, for example interpreting disease as caused by spirits that need to pacified through rituals, is wrong.

On closer inspection, it is not so easy to dismiss indigenous knowledge as unscientific. First, scientific knowledge is open to revision: Just because knowledge is wrong, according to the judgment of today's scientists, does not mean that it is unscientific. Second, what is called "the scientific method" is, on closer inspection, more variable and situation-specific than commonly imagined. Some anthropologists say that the scientific method does not readily distinguish indigenous belief systems from modern science.

In indigenous societies, there are no paid knowledge-makers with access to major resources. If indigenous knowledge is a form of science, it is definitely grassroots science.

There is a similar type of knowledge found in everyday life in industrial and post-industrial societies, which can be called everyday knowledge. Within groups active on certain topics, there is a lot of folk knowledge that is developed, shared, modified, and used for practical activities. Examples include farmers' knowledge of local weather, carpenters' knowledge of materials and construction techniques, bodybuilders' knowledge about drugs, nutrition, and training techniques, and musicians' knowledge about instruments and ways of playing. There are a great many groups in which insider knowledge is cultivated. Some of this knowledge is articulated in handbooks and formal training; sometimes professional scientists take an interest and undertake observations and experiments that recast some of this everyday knowledge in terms of the frameworks of scientific disciplines. But much everyday knowledge remains embedded in the communities that cultivate it. In this sense, very many people are part of grassroots scientific communities.

## The Future of Grassroots Science

The label "science" is normally restricted to scientific knowledge that is produced by credentialed professionals using sophisticated, expensive equipment, with other forms of knowledge being relegated to the status of being "unscientific." The idea of grassroots science is a challenge to this conventional image of science.

Grassroots science, in its clearest forms, is similar to professional science in creating systematic public knowledge through standard methods in a community of practitioners. The key difference is that grassroots science is far less dependent on formal training, professional employment, and expensive equipment.

The trend in the past century has been towards professional science, with the professionals usually ignoring, rejecting, or incorporating grassroots knowledge. Yet there remains interest and capacity, especially among amateurs and social movements, for embedding knowledge making into everyday lives. The future of grassroots science lies in the future of this resistance to the professionalization of every realm of activity.

**See also Indigenous Knowledge**

## BIBLIOGRAPHY

Boston Women's Health Book Collective. *Our Bodies Ourselves for the New Century.* New York: Touchstone, 1998. A new edition of the classic text.

Epstein, Steven. *Impure Science: AIDS, Activism, and the Politics of Knowledge.* Berkeley: University of California Press, 1996.

Ferris, Timothy. *Seeing in the Dark: How Amateur Astronomers Are Probing Deep Space and Guarding Earth from Interplanetary Peril.* New York: Simon and Schuster, 2002.

Fischer, Frank. *Citizens, Experts, and the Environment: The Politics of Local Knowledge.* Durham, NC: Duke University Press, 2000. The case for citizen participation in science policy.

Johnston, Robert, ed. *The Politics of Healing: Histories of Alternative Medicine in Twentieth-Century North America.* New York: Routledge, 2003.

Moody, Glyn. *Rebel Code: Linux and the Open Source Revolution.* New York: Perseus Books Group, 2002.

Sheldrake, Rupert. *Seven Experiments That Could Change the World: A Do-It-Yourself Guide to Revolutionary Science.* London: Fourth Estate, 1994. An invitation to grassroots research that challenges orthodoxy.

Simon, Bart. *Undead Science: Science Studies and the Afterlife of Cold Fusion.* New Brunswick, NJ: Rutgers University Press, 2002.

Verran, Helen. *Science and an African Logic.* Chicago: University of Chicago Press, 2002. An examination of indigenous knowledge.

*Brian Martin*

# H

## HEALTH CARE, ACCESS TO.

Unlike all other developed and some developing countries, the United States does not guarantee universal access to a range of health care services for its citizens. At present, there is no universally accepted definition of *access to health care*, but in 1993, the National Academy of Sciences's Institute of Medicine proposed a definition that includes both the use of health services and health outcomes. It defined access as "the timely use of personal health services to achieve the best possible health outcomes" (Institute of Medicine 1993, p. 4). Others have defined access as the ability to receive regular, high-quality medical care in the most appropriate health care setting.

It is generally accepted that access to and utilization of health care services are particularly important for people with existing health problems (those with needs for ameliorative and curative medical care) and apparent risks for new or more complicated health problems (those with needs for preventive care). It is also generally accepted that access to health care in the United States is influenced by the organization and financing of the health care system, the characteristics of individuals, and the fit between individuals in particular circumstances and the existing health care infrastructure.

### Primary Measures of Access

Having health insurance and a usual source of care are the two most commonly used indicators of access to health care. According to the behavioral model of health services utilization originally developed by Ronald Andersen and refined over several decades, having health insurance and a usual source of care enable entry into the health care system when acute or ongoing care is needed—the former by providing financial access and the latter by promoting familiarity, convenience, confidence, and satisfaction with the system.

The development of new medical services (including pharmaceuticals, surgical procedures, and imaging technologies) has contributed to increasing life expectancy and better management of both acute and chronic medical conditions. These technological advances have also contributed to rapid increases in health care costs. In 2002, spending on health care in the United States accounted for 14.9 percent of gross domestic product (GDP); it is projected to increase to 18.4 percent by 2013. As the scope and cost of health care services has increased over several decades, the primary mode of financing health care has shifted from direct payment by consumers to third-party payment, with the majority of health care financed through private (mostly employer-based) or government-sponsored health insurance plans.

In the absence of health insurance coverage, access to care is restricted to what can be paid out of pocket and to reliance on emergency rooms and free clinics that provide care to the indigent. In many research and policy arenas, having "access" to health care services has become synonymous with being covered by a health insurance plan. Reducing financial barriers to care has been the primary focus of federal and state policy initiatives aimed at expanding access to health care in the United States. Such initiatives include offering free health care to veterans through the Veterans Administration hospital system. Additionally, in 1965, the federal government established Medicare, the program that provides health insurance to the majority of Americans over the age of 65 years and some persons with specific, disabling health conditions (for example, end-stage renal disease) and Medicaid, which provides health insurance to some low-income Americans who are younger than 65 years old. More recent initiatives include the expansion of the State Child Health Insurance Program, which aims to expand access to Medicaid for low-income children.

Despite these and other efforts to expand access to health insurance and health care for specific segments of the U.S. population, the majority of Americans have no guaranteed health insurance coverage. Most must purchase it individually, which is particularly expensive, or obtain it through an employer-sponsored plan (either through payroll deduction for participation in a group plan or as an employment benefit). According to the U.S. Census Bureau, in 2003, approximately 45 million (or 15.6 percent of) Americans did not have health insurance coverage, including 8.4 million (or 11.4 percent of) American children. Many more persons were *underinsured*, which means that they had health insurance but lacked coverage for specific health problems.

There are large sociodemographic differences in health insurance coverage, indicating that access to care in the United States is not equitable. For example, in 2003, 19.6 percent of African Americans and 32.7 percent of Hispanics lacked health insurance, whereas only 11.1 percent of non-Hispanic whites lacked coverage. Individuals without health insurance are found at every income level, although being uninsured is more common for those who have low incomes: in 2003, 24.2 percent of those living in households with a total income of less than $25,000 were uninsured, whereas 8.2 percent of those in households with an income greater than $75,000 were uninsured. Low-wage workers have particularly high rates of uninsurance because their employers often do not offer insurance as a benefit and because they earn too much to qualify for Medicaid and too little to purchase health care individually or

through their employers. The vast majority of Americans who are uninsured are in working families; in 2002, 69 percent of the uninsured were in households with a full-time worker and 12 percent were in households with a part-time worker.

Persons with health insurance are more likely to have a usual source of care, and this affords an additional advantage with respect to access to appropriate care in an appropriate location. Moreover, having health insurance and a usual source of care are generally among the strongest predictors of health services utilization, and they have been shown consistently to enhance the timely use of medically necessary health services, increase the use of preventive health care, increase the continuity of care for chronic conditions, and reduce costly emergency room utilization. By increasing access to and utilization of medically necessary care, health insurance and having a usual source of care reduce *unmet need for health care*, which can be defined as medically necessary care that is not obtained.

According to data from the National Center for Health Statistics (2004), 86.4 percent of Americans reported that they had a usual source of care in 2004. Although the vast majority of respondents to the survey indicated that they had a usual source of care, there were significant differences reported by age, race, and sex. Children (under the age of 18 years) and the elderly (over the age of 65 years) were most likely to report having a usual source of care. Whites were more likely to have a usual source of medical care than either African Americans or Hispanics, and women were more likely than men to report a usual source of care. The length of time that individuals are uninsured also affects the likelihood of having a usual source of care. According to the Kaiser Commission on Medicaid and the Uninsured, in 2002, 8 percent of those who were insured for the full year reported no usual source of care, compared with 17 percent of those who were uninsured for less than 6 months, 19 percent of those who were uninsured for 6 to 11 months, and 30 percent of those who were uninsured for 12 or more months.

Among working-age adults with a regular source of care, where they receive care varies by health insurance status. According to the 1993 Access to Care Survey, 91 percent of adults with private health insurance reported using a private doctor's office as their regular source of care, compared with 62 percent of adults with public health insurance and 76 percent of adults with no health insurance. Clinics were the regular source of care for 6 percent of adults with private insurance, 30 percent of adults with public insurance, and 16 percent of persons with no health insurance. Although only 2 percent of working-age adults reported using the hospital emergency room as their usual source of care, when compared with adults having private insurance, adults with public insurance and the uninsured were respectively 5 and 4 times more likely to report using the emergency room as their regular source of care.

Studies have consistently documented that uninsured individuals have less access to a regular source of care in the health care system and are at risk of worse short-term and long-term health outcomes than are insured individuals. According to the Kaiser Commission on Medicaid and the Uninsured, in 2002, 22 percent of those who were insured for the full year did not have a physician visit in the past year, compared with 27 percent of those who were uninsured for less than 6 months, 35 percent of those who were uninsured for 6 to 11 months, and 52 percent of those who were uninsured for 12 or more months. The IOM reports that uninsured children use medical and dental services less frequently than children with health insurance, and are less likely to have a regular source of care, and thus miss opportunities for preventive health services and the early identification and treatment of medical problems. This is significant because disparities in access to health care early in life likely contribute to poorer health over one's lifetime.

## Other Indicators of Access to Health Care

Unmet need for health care is a particularly important indicator of access because it com-

bines two dimensions: medical need and a lack of health care utilization. By definition, people who need care but do not use it have some barrier to health care that impedes their access substantially. Thus, estimates of unmet need provide a means of assessing problems with access to health care.

According to the National Center for Health Statistics, the number of Americans with unmet health care needs increased from 1997 to 2004. Adults aged 18 to 64 years were more likely to report unmet needs than either children or the elderly. African Americans and Hispanics were more likely than whites to go without care, and, at any age, women were more likely than men to report unmet needs. Not surprisingly, unmet need for health care is strongly correlated with lack of financial access. The Kaiser Commission on Medicaid and the Uninsured reports that uninsured Americans are more likely than the insured to postpone seeking care because of cost (47 percent versus 15 percent); need care but not get it (35 percent versus 9 percent); not fill a prescription because of cost (37 percent versus 13 percent); have problems paying medical bills (36 percent versus 16 percent), and be contacted by a collection agency about medical bills (23 percent versus 8 percent).

Geographic differences in access to health care resources currently exist in the United States. Health care providers (including physicians, hospitals, and other health providers) tend to be clustered in and around urban areas, and rural areas are often left without an adequate supply. Geographic access to health care can be measured by the number and type of physicians or hospitals within a defined geographic area (for example, the number of obstetricians practicing within a county), or within a specified mileage radius of a location (for example, the number of primary care physicians within a 10-mile radius of a ZIP code). The ratio of practicing physicians to residents is also used to measure geographic access. Hospital or facility access within a geographic area is often measured by identifying the number of inpatient beds available to serve the residents of that geographic area.

Geographic access is complicated in some circumstances by the type of health care service that is needed. For example, in 2000, 87 percent of counties in the United States did not have an abortion provider, despite the fact that abortion is a legal medical procedure. Lack of access to local abortion services is nearly total in rural counties, of which 97 percent do not have a single provider; the proportion of unserved counties has been increasing since 1978. Studies also document that persons living with HIV/AIDS who live outside of major urban areas often travel long distances to receive care and that some eventually move for care.

Many researchers describe access in terms of how long it takes to receive necessary care. For example, the number of days between when a person calls for an appointment with their health care provider and when they receive care is commonly measured. The availability of health care services may also be measured by the number of times that individuals receive services. For example, the average number of visits to a primary care physician in a year is often used to assess both the availability of primary care providers and the appropriateness of the level of primary care services being received by consumers.

## Additional Influences on Access to Health Care

The ability to access health care is influenced by the characteristics, experiences, and preferences of individuals, over and above the financial access to care afforded them by health insurance or other economic resources. Individuals' personal resources, including their knowledge of where to seek care and their trust in the health services available, have been identified as important variables affecting access to health care. For example, researchers have studied the level of satisfaction that individuals express about their encounters with health care providers, and how satisfaction influences access to care for subsequent medical problems. Language barriers create significant problems with access to and experiences with health care, and thus concern about linguistic access for those who do not speak English as a primary

language and those with low levels of literacy is substantial.

Discrimination may affect access to health care for some racial/ethnic minorities in the United States. Compared with whites, African Americans are sometimes offered and receive different treatments for the same medical conditions. Members of racial/ethnic minority groups often report difficulties in their interactions with health care providers, which contributes to their perception that their needs are not being met. Similarly, lesbian, gay, bisexual, and transgender persons often report negative experiences with health care providers, which may affect their subsequent decisions about seeking care, disclosing information to providers, and accessing further care at a later point. Studies suggest that negative experiences with allopathic health care providers push some individuals toward alternative health care providers and that the use of alternative health care is substantial and increasing. General discussions of access to traditional health care systems may be inadequate for fully capturing the complexity and dynamics underlying the process of accessing care.

## Access to Health Care for Communities

The consequences of unequal access to care for the health care institutions in the communities in which uninsured individuals live, and these communities themselves, have begun to be the focus of research. The IOM reports that those who lack health insurance and live in a community with high rates of uninsurance have worse health outcomes than their peers who live in communities with low overall rates of uninsurance. While the reasons for this synergy are not clear, it has been proposed that health care providers in areas of high uninsurance, whose care may be uncompensated or incompletely compensated, are forced to cut back on services for all patients.

The IOM also found that communities with a higher than average number of uninsured residents have less access to certain health services. Local hospital emergency departments, which are required by law to provide care for all persons regardless of their ability to pay, may become a regular source of primary acute and chronic care for those with limited access to regular sources of health care. The increase in patient volume in emergency departments for minor illness or injury care, and for the management of chronic care, may adversely affect access to and the delivery of health care services for all patients by diverting resources from other settings.

Communities with high numbers of uninsured residents face financial uncertainty and lost productivity. In any given year, the amount of uncompensated care delivered to individuals in the United States by health care providers is approximately $35 billion; local taxpayers fund subsidies provided to hospitals to cover uncompensated care. The community-level financial impact of inadequate access to the health care system includes more than just the direct costs of medical care paid by taxpayers. Lost productivity and diminished quality of life because of poorer health outcomes among the uninsured have been estimated to cost between $65 billion and $130 billion annually.

## Ensuring Access to Health Care

Ensuring more equitable access to health care does not necessarily require an increase in national spending on health care services. According to the World Health Organization, a branch of the United Nations, more money is spent for medical care in the United States than any other country in the world. In 2001, the United States spent $4,887 per capita on health care services. In contrast, countries such as Switzerland, Japan, Germany, and France, who all provide a basic level of health services to their citizens and have better health outcomes than the United States, spent significantly less for health care. In spite of higher levels of spending on medical care, in 2002, the United States ranked 29th in the world on healthy life expectancy, a measure that takes into account length and quality of life.

A recent assessment of the quality of health care in America suggests the extent to which access to care in the United States is compromised. This study, known as the Community Quality Index Study, estimated the extent to which recommended care was provided to a representative sample of the population (Kerr et al. 2004) and found that adults received only about half of recommended preventive, acute, and chronic medical care services. This suggests that the problems with access to health care in America cannot simply be reduced to a lack of insurance and that solving the complex access to health care problem in the United States will require concerted efforts by individuals and health care providers to improve access to appropriate care.

Several strategies have been proposed over the last 50 years to offer a basic level of access to health care services to all citizens of the United States, although none have been implemented successfully. In assessing the issue of access to health care, the IOM proposed the following principles in 2004, around which they argued any further efforts to equalize access to health care should be organized (Institute of Medicine 2004, p. 8–9):

1. Health care coverage should be universal
2. Health care coverage should be continuous
3. Health care coverage should be affordable to individuals and families
4. The health insurance strategy should be affordable and sustainable for society
5. Health insurance should enhance health and well-being by promoting access to high-quality care that is effective, efficient, safe, timely, patient-centered, and equitable

Problems with access to health care in the United States are longstanding, more complex than the lack of insurance alone, and have consequences for society, communities, and institutions, as well as individuals. Following the principles outlined by the IOM would bring financial access to health care in the United States in line with that of all other more developed, Western nations and many developing countries, and would shift the focus of policy debates from questions about how to expand access for the uninsured to defining which health care services are to be guaranteed for all citizens.

**BIBLIOGRAPHY**

Andersen, Ronald M. "Revisiting the Behavioral Model and Access to Medical Care: Does it Matter?" *Journal of Health and Social Behavior* 36 (1995): 1–10.

Bloom, B., G. Simpson, R. A. Cohen, and P. E. Parsons. "Access to Health Care. Part 2: Working-Age Adults." *Vital and Health Statistics,* Series 10, No. 197. Washington, DC: Department of Health and Human Services, 1997.

DeNavas-Walt, Carmen, Bernadette D. Proctor, and Robert J. Mills. *Income, Poverty, and Health Insurance Coverage in the United States: 2003.* U.S. Census Bureau, Current Population Reports, P60-226. Washington, DC: U.S. Census Bureau, 2004.

Institute of Medicine, Committee on Monitoring Access to Personal Health Care Services. *Access to Health Care in America.* Washington, DC: National Academies Press, 1993.Institute of Medicine, Committee on the Consequences of Uninsurance, Board of Health Care Services. *Care Without Coverage: Too Little, Too Late.* Washington, DC: National Academies Press, 2002.

Institute of Medicine, Committee on the Consequences of Uninsurance, Board of Health Care Services. *Health Insurance is a Family Matter.* Washington, DC: National Academies Press, 2002.

Institute of Medicine, Committee on the Consequences of Uninsurance, Board of Health Care Services. *A Shared Destiny: Community Effects of Uninsurance.* Washington, DC: National Academies Press, 2003a.

Institute of Medicine, Committee on the Consequences of Uninsurance, Board of Health Care Services. *Hidden Costs, Value Lost: Uninsurance in America.* Washington, DC: National Academies Press, 2003b.

Institute of Medicine, Committee on the Consequences of Uninsurance, Board of Health Care Services. *Insuring America's Health: Principles and Recommendations.* Washington, DC: National Academies Press, 2004.

Kaiser Commission on Medicaid and the Uninsured. *The Uninsured and Their Access to Health Care.* Washington, DC: The Henry J. Kaiser Family Foundation, 2003. http://www.kff.org/uninsured/1420-05.cfm.

Kaiser Commission on Medicaid and the Uninsured. *Lack of Coverage: A Long-Term Problem for Most Uninsured.* Washington, DC: The Henry J. Kaiser Family Foundation, 2004.

Kerr, E. A., E. A. McGlynn, J. Adams, J. Keesey, and S. M. Asch. "Profiling the Quality of Care in Communities: Results from the Community Quality Index Study." *Health Affairs* 23 (2004): 247–256.

National Center for Health Statistics. *Early Release of Selected Estimates Based on Data from the January-March 2004 National Health Interview Survey.* Hyattsville, MD: National Center for Health Statistics, 2004.

World Health Organization. *World Health Report 2004: Changing History.* Geneva, Switzerland: World Health Organization, 2004.

*Nancy A. Myers and Andrew S. London*

# HISTORY OF SCIENCE.

The word *scientist* was coined in the early nineteenth century by William Whewell (1794–1866), one of the original historians of science. The occasion for giving practitioners of the sciences this new name for themselves was a growing belief that the progress of humanity depended upon their efforts. Whewell chose deliberately; the Latin word *scientia* means "knowledge," and a scientist is "one who makes knowledge." Our idea of science—past and present—is prefigured by the word's literal meaning, a meaning underscored in the past two centuries by histories in which science is routinely depicted in triumphalist narratives as the domain of rational heroes who dispel dark ignorance by illuminating the hidden forces of nature.

By the time academic professionals undertook the history of science in the twentieth century, scientists themselves had crafted historical accounts of their disciplines. By the late nineteenth century, histories of astronomy, geology, chemistry, physics, and various emerging disciplines were appearing with growing frequency, many of them written by leading scientists and intellectuals such as Adam Smith, Charles Lyell, Wilhelm Ostwald, Marcelin Berthelot, and Ernst Mach. As scientists (hitherto referred to in English as "natural philosophers") specialized, they looked to the past for not only an understanding of the origins of their specialties but a confirmation that their specialties were valuable and, indeed, essential to society. Such histories of science tended, therefore, less toward dispassionate inquiries into the complex origins and development of a discipline and more toward advertisements for its social importance. Abetting this approach was the fact that the professionalization of history in general, and of the history of science in particular, was still far in the future. In the latter part of the nineteenth century, a leading historian, Leopold Ranke, could still claim with no hint of irony that the historian's goal was to relate the past *wie es eigentlich gewesen* (how it essentially was). Still below the horizon of consciousness was today's growing insight that a historian must practice rigorous self-scrutiny to eliminate even a portion of the innumerable subjective influences—patriotism, religious conviction, ethnocentrism, political ideology—that tend to make the most high-minded scholar render the past in a form uncannily consistent with his personal values and cherished worldview.

## Science: Revised and Edited

Our stereotypical and anachronistic image of science was promulgated in accord with the nineteenth-century scientist-historians' emphasis on the social benefits of their disciplines. Accordingly, histories of science have stressed what might be termed the "internal" aspects of science. For science to appear socially beneficial it had to appear reliable, and to appear reliable it was made to seem independent of the subjective values that prejudice most social activities.

Consider, for instance, the scientific undertakings of Nicolaus Copernicus (1473–1543 C.E.), still the characteristic starting point in histories of modern science. Only retrospectively is it possible to look back at Copernicus and imagine that his goal was to upset the geocentric model and replace it with the heliocentric. It is erroneous to ignore his religious passion and make him retrospectively into a rational iconoclast as he is portrayed in modern accounts, part of the elite company of heroes of science. His motive for asserting his abstract, heliocentric thesis was not—as we might, in our haste to retrospectively

claim the past, wish to believe—to reveal objective truth by defeating unscientific dogma but rather to *maintain* dogma. Central to the cosmological and astronomical ideas inherited from Aristotle was the belief that the celestial realm was characterized by perfection—that objects moved in circular, uniform motion. The phenomena of the cosmos did not, however, always conform to Aristotle's explanation. Thirteen hundred years years before Copernicus the astronomer Ptolemy, working in the north African city of Alexandria, noted many imperfect celestial phenomena, which he did his best to explain in terms of Aristotle's model. For example, each year the outer planets stop their eastward motion against the stars, go back westward for a while, then start moving eastward again, in apparent contradiction of the perfect uniform, circular motion posited by Aristotle. Arabic astronomers during the early Middle Ages, using sophisticated instruments such as astrolabes to map the heavens, continued to note these anomalies. One way of reducing the number of apparent imperfections in the model was to change certain basic assumptions, specifically the geocentric one. Putting the sun at the center of the imagined cosmos allowed Copernicus to maintain another assumption of far greater importance to him: the dualistic separation of the imperfect terrestrial realm and the perfect celestial realm, an assumption upon which rested the theological and moral structure of his Christian faith.

A saint, Ambrose Bierce remarked, is a sinner "revised and edited." If we mute the moral implication, the same might be said of scientists. Science has commonly been depicted as a hermetic undertaking, sealed from the impurities of the external world, and scientists such as Copernicus as disciplined, dispassionate, super-rational servants of truth. As it turns out, Copernicus's relaxed attitude toward data collection and rigorous computation rather disqualifies him as poster child for the now vaunted "scientific method." Likewise, revelations of Galileo Galilei's all-too-human qualities have cast a pall over his image as scientific hero. One need not go so far as Arthur Koestler once did in condemning him as "wholly and frighten-ingly modern" (Koestler, p. 363) for his evi-dent combination of a leviathan-sized intellect with thoroughly petty moral sensibilities. It is nevertheless true that scholarship on this, as on many other scientist-subjects, has not generally redounded to the benefit of the heroic stereotype.

Thus we must also revise our notions about Sir Isaac Newton (1642–1727), who long as-sumed in popular imagination the form of a secular saint, a view of him originally fash-ioned by Alexander Pope (1688–1744), who composed the now famous epitaph, "Nature and Nature's laws lay hid in night: / God said, Let Newton be! and all was light." But, to paraphrase the economist John Maynard Keynes, Newton was perhaps not the first of the modern scientists but the last of the alchemists. Newton devoted extraordinary efforts to alchemy and to investigating re-ligous prophecy rather than to the proper empirical pursuits we expect of a patriarch of modern scientists. He was a tireless translator and expurgator of bilibical texts, devoting at least as much energy and apparently more passion to his exegesis of the book of Daniel than to his formulations about gravitational force or his analysis of the composite nature of light. Of Newton's ambitions toward science, the late Betty Jo Dobbs observed, "If he, Newton, could but demonstrate the laws of divine activity in nature, the Christ operating in and governing the microcosm, then he could demonstrate in an irrefutable fashion the existence and providential care of the Deity." Dobbs continues, saliently, that al-though this was "a grand goal," it was "hardly a modern one" (Dobbs, p. 642–643).

The real significance of contemporary scholarship is not, however, in the tarnishing of cherished idols but in the much needed challenge it presents to the idea of science as an sovereign undertaking, unperturbed in method and motive by anything but rational thought.

## All Too Human

By attending to external contexts and other influences of the nonrational sort, we begin to appreciate that neither in motive nor method were the early practitioners (and pa-triarchs) of science quite what we might imagine or want them to be. Where their con-cerns seem familiar to us as fellow human beings, they seem not quite becoming to our ideal of a scientist. More generally, though, their concerns were not our concerns, their reasons not our reasons, and their motiva-tions considerably more subtle and complex than the caricatures of them we may carry in our heads. In particular, the assumption that guided earlier historians of science, the as-sumption that ideas might be treated as if they arose from the intellect independently of circumstance, is now thoroughly suspect. The more closely scholars consider their sub-jects in human, social, and cultural context, the more obvious it appears not only that the scientist is not isolated from external influ-ence but that the values and beliefs (many of them quite alien to our own) of the world he inhabits generally prefigure his choice of subject, his methods of study, and possibly even his interpretations. By the middle of the twentieth century, indeed, many schol-ars—most notably the sociologist Robert Merton and the scientist-historian J. D. Bernal—had concluded that an approach that emphasized social and cultural context was the only sensible way to begin to make sense of why and how scientists—past and present—do what they do.

The unwieldy and ever-growing mass of scholarship on Charles Darwin, sometimes called Darwiniana, illustrates this approach. Darwin saved historians much of the trou-ble of tracing the internal, purely intellec-tual influences on his work. He himself pointed out the intellectual context pro-vided by Thomas Malthus and Charles Lyell, without which his observations would have lacked two essential premises that provided the fertile soil, as it were, in which his observations on the variation of species might grow into a theory of evolu-tion. Malthus had pointed out the basic dis-parity between the rate at which animals, including humans, can create the means for survival and the much faster rate at which animals tend to procreate, thus ensuring an unremitting competition for the means of survival. At about the same time, Lyell had compiled much geologic evidence that pointed to the previously unsuspected antiquity of the earth. These observations, like Darwin's own, had everything to do with the fact that these men lived in a rapidly growing country. The global population, about 1800 C.E., had surpassed the one-billion mark for the first time. Englishmen such as Malthus, Lyell, and Darwin could observe the effects of population growth in combination with other factors that intensified the struggle to survive Malthus could observe the manifold growth of dozens of cities as a consequence of the Industrial Revolution, growth that created glaring social and economic disparities and so intensified the struggle for survival. Lyell could observe the unanticipated evidence of the earth's past (fossils of extinct species; geological strata), exposed by digging up the earth for the purpose of mining and railroad building to satisfy the demands of an expanding population. Darwin's thoughts about the diversity of life on earth benefited from these and other observations. He was especially fortunate to be living in a country with the largest navy in the world, just at the time it was approaching its imperial apex. He could use such resources to move about the world, observing a range of phenomena few others ever had. As a scion of the gentry, he had obtained a good education, which prepared him to make effective use of his ideas and observations, and he had sufficient leisure to devote to the exploration and contemplation of nature.

To acknowledge that more is at work in the process of scientific discovery than the operations of disembodied minds, one does not have to embrace a wholly reductive explanation of the origin of ideas in terms of circumstance. For instance, in the light of what has been said about Darwin and his time, it should not surprise us too much that another Englishman, Alfred Russell Wallace, devised an evolutionary model at almost precisely the same moment as Darwin. Indeed, a half-dozen contemporaries arrived at similar hypotheses, though with fewer empirical facts to support their claims. It is excessive to conclude that scientific discovery is predetermined by circumstance, but, concomitantly, it is insufficient to treat ideas as if they arise out of nothing but the insight and inspirations of the individual. Though it is one of the least helpful of platitudes to say so, the truth is somewhere in the middle, somewhere between material *reductio ad absurdum* and the notion that ideas "bubble up" from nowhere. One might just say that what German intellectuals used to point to with vague metaphysical aplomb as the *Zeitgeist* (spirit of the time) has achieved some empirical substance.

Another way of thinking about scientific discovery is to consider what may seem, viewed from the old perspective that mind operates independently of circumstance, a mysterious phenomenon: the not-infrequent coincidental discovery of the same basic idea. "Coincidences" in the history of science yield their mysteries if we simply consider them in context. The simultaneous interest in the behavior of gases by James Clerk Maxwell and Ludwig Boltzmann is but a manifestation of the broader concern in the mid- to late nineteenth century with the production and storage of energy, revealed most conspicuously in that period in the increasing use of steam machinery such as locomotives. The superficially remarkable co-discovery of calculus in England, Germany, and elsewhere, or the various and again virtually simultaneous invention of different ways of describing quanta by Werner Heisenberg, Erwin Schrödinger, and others, or again the co-discovery of the concept of evolution through natural selection, all become less remarkable or mysterious when viewed in the context of broader social and cultural trends. Furthermore, just as a change in ideas has social causes, the *persistence* of ideas, whether true or false—the assumption of geocentricity, the assumptions about the age of the earth, about the existence of ether, about the fixity of species, about the speed of light—is likewise conditioned by more than the isolated, unimpeded flow of scientific thought. The discovery of new ideas and adjustments of our scientific models are hampered as well as hastened by the myriad influences brought to bear by the culture in which scientific practice is immersed.

## Of the West

But how did our heroic, decontextualized ideas about science and scientists ever arise in the first place? Just as historians, especially in recent generations, can take considerable credit for challenging our historical stereotypes, they must also bear some responsibility for having created and perpetuated those stereotypes. Historians and their scholarly efforts must themselves be contextualized in order to expose the strong current of subjectivity that shapes their explanations of the past. Understandably, historians themselves have been reluctant to undertake this task.

One way to begin such contextualization with respect to the history of science, is with a question: Why, only in Western Europe in the early nineteenth century, many millennia after human beings began discovering facts of nature and inventing innumerable techniques for its manipulation (such as metallurgy, agriculture, irrigation, animal husbandry, astronomy, geometry, navigation, calendars, compasses, astrolabes, printing, ceramics, medicine, horticulture, engineering, and so on and on), would there appear a concerted effort to give a special name and history to a purportedly new method for obtaining knowledge? There is a rhetorical part to this question. The answer to the question "why" is implied: to wit, in order to provide a legitimating legacy to what was deemed a new and important form of knowledge. Less rhetorical is the problem of whether those who deemed science to be something new were justified in doing so. To consider this problem, it is necessary to situate a few of the early efforts at a history of science in some context.

Whewell wrote his *History of the Inductive Sciences* (1837) in order to shape for English-speaking world's assumptions about the sciences as something that was uniquely reliable because its various branches were bound by a common method. The idea that a particular method united the various sciences, plural, in turn encouraged the notion that there is such a thing as "science," singular. As the title of his work indicates, Whewell

emphasized the method of inducing from empirical (observable) facts toward more general truths. In part he was following the model of a respected predecessor, Francis Bacon (1561–1626), who, in his *Great Instauration* and *New Atlantis,* envisioned a sort of technological utopia arising from the systematic gathering of facts about the world and the subsequent manipulation of those facts for human benefit. In this respect, one may also view Whewell's account of science, not as "history" in the contemporary academic sense of objective inquiry into the past, but as a protracted (extending over three large volumes) argument to prove the past efficacy of the scientific method and its usefulness for the present.

Whewell was not alone among Europeans of his generation in positing the emergence of a uniquely reliable method of knowing, one that seemed to warrant its own history. Auguste Comte (1797–1857) drew up a short history of science for the French-speaking world. Like Whewell, Comte could look to the inspiration of a fellow countryman: René Descartes (1596–1650), who, about the same time Bacon was envisioning his scientific utopia, composed his *Discourse on Method,* which argued that reliable knowledge about the world could indeed be attained. In contrast to the English Bacon, who emphasized observation and the dogged collection of facts, Descartes argued that the senses were essentially unreliable and pointed instead to our rational faculties as the best means to arrive at clear and precise knowledge. Bacon and Descartes have become, in a manner of speaking, the Romulus and Remus of modern science.

The monolithic idea of science with which we are familiar had to be invented. That idea, in which science is conceived as something new under the sun, as a distinct sort of knowing with a specific method for arriving at a uniquely reliable form of knowledge, appeared to warrant its own history. The belief in this warrant emerged simultaneously with the emergence of Western Europe and the United States as global powers.

In England the connection between the changing social context and new ideas about the nature and prospect of knowledge is especially apparent. By the early nineteenth century, when the first histories of science began to appear, England had become the world's unrivaled sea power. As a result, it could secure its growing hegemony over the world's economic resources. England's craftman-inventors had already devised new methods for vastly accelerated production of textiles, which legislation fostered by making it impossible for foreign nations to compete. Other Englishmen had devised new sources of power, such as the steam engine, which were soon applied to new systems of transportation. Vast natural reserves of raw materials, such as coal and iron, could now be put to new and profitable uses: the construction of railroads, of bridges, of steamships, and the innovation of stronger materials such as steel. The ongoing enclosures of agricultural lands had concentrated the English population into cities and put them in need of employment by the burgeoning industrial age of the late eighteenth and early nineteenth centuries. From the viewpoint of those in a position to exploit its considerable resources, natural and human, England was a fortunate island; the contingencies of nature, culture, and political economy had allowed it considerable prosperity, and the fact that it was an island shielded its prosperity from the imperial ambitions of its neighbors while encouraging it to develop the merchant and military navy necessary to pursue its own imperial ambitions on a global scale.

By the end of the nineteenth and beginning of the twentieth century, the prosperity that England had for a time monopolized was being shared—or, more precisely, fought over—by many other European and transplanted-European nations, most notably the United States. It is not surprising that an account of the past in which European scientists were the pathfinders of this new European prosperity became generally amenable.

The idea, broadly outlined in the work of Whewell, that science was the unique cultural endowment of European civilizations found formalization in the work of the founder of the history of science as an academic discipline, George Sarton (1884–1956). At one level Sarton speaks of science as the legacy of all human beings, and he appears to approach science past with universal generosity. However, it does not take an especially acute reading of his many works to see that for Sarton, as for his friend and colleague Werner Jaeger (author of a famous encomium to the brilliance of Greek antiquity, *Paideia*), although all humanity may share the legacy of science, some parts of humanity have more of a share in building it than others do. Sarton attends to many of the innumerable inventions and innovations that are part of the human past. He makes it plain, however, that no other culture, however innovative, has quite displayed the scientific genius of the ancient Greeks and their direct cultural descendants, the Europeans. Arabic culture, which has more than a modest claim to scientific discovery, is treated by Sarton primarily as translators and transmitters of ancient Greek culture, an interpretation that has become a mainstay of textbook accounts of modern science (note the work of David Lindberg, for instance). In this and other respects Sarton was the bridge between the Whewellian model of the history of science and its contemporary practice. Whewell remarked, for example, that he would be disappointed to discover that Arabic culture had made any significant contribution to science. Despite less apparent bigotry and an avowed determination to approach the history of science in a cosmopolitan spirit, Sarton, like Whewell, clearly favored "Western" civilizations, and Classical Greek culture specifically, as the essential point of origin of the modern scientific tradition.

Just as Whewell's conception of science took form in the crucible of emerging English imperialism, Sarton's took shape during the "new" imperialism (as some historians call it), which spanned the last quarter of the nineteenth century up to the onset of World War I in 1914. The deliberate and often forceful dissemination of Europeans and their beliefs and institutions overseas encouraged the idea that Europeans had a great deal in common with each other, relative to the various subjugated peoples. The notion that Europeans and descendants of European colonists are all part of a "Western" culture began in this period to take its now familiar form. The idea of the West as a distinct cultural entity (an "imagined community," to use Benedict Anderson's phrase), one with coherent cultural legacies, was propagated as well by the intellectual efforts of well-known historians such as Oswald Spengler and Arnold Toynbee, and by perhaps less well known figures such as A. J. Balfour and Benjamin Kidd. Hereafter, the development of the history of science largely parallels the emergence of the now familiar idea of the "West," an idea that, after having achieved its modern form during the "new" imperialism, was rekindled, especially in the United States, during the Cold War of the mid- to late twentieth century.

Nearly two centuries after Whewell, and a century after its inception under George Sarton, *Isis*—the leading journal of the history of science—divides its subject up categorically among the following headings: twentieth century; nineteenth century; eighteenth century; seventeenth century; Middle Ages and Renaissance; antiquity; and general. In contemporary issues, the vast majority of essays pertain to the period from the seventeenth century onward and focus, with only occasional departures, on topics pertaining directly or indirectly to European science and scientists. Where non-Western science is discussed, it is invariably discussed in terms of its approximation to the methods and achievements associated with Western science. Thus, recent articles in *Isis* reveal how the work of certain Soviet and Chinese scientists were hampered by the non-Western values and institutions of the states in which they worked.

The ethnocentric, specifically Western bias still informing the discipline is most conspicuously revealed in the emergence in the mid-twentieth century of one of its guiding concepts. The phrase "scientific revolution" was, remarkably, popularized only about a half-century ago by the English historian Herbert Butterfield. Although application of the word *revolution* to scientific subjects goes back centuries, only in the middle of the twentieth century did there emerge the notion of a distinct turning point or watershed in the human capacity to acquire reliable knowledge. The phrase "scientific revolution" refers to a turning or revolving away from previous forms of knowing to something more scientific, a turning moreover accomplished largely by dint of individual scientific genius (primarily Western), working in isolation from, or often in opposition to, an intellectually hostile environment. As a historical concept with which to conjure, the "scientific revolution" became especially important in the very middle of the protracted war that was the twentieth century (or, as historians sometimes measure time, the "short" twentieth century, 1914–1989). More precisely, it became part of the scholar's intellectual currency between the onset of our modern "Thirty Years' War" (1914–1945) and the tumultuous beginnings of the Cold War. And this provides a clue about why the concept become a significant part of the discourse of the history of science. While there is as yet no unanimity about why World War I was fought, by the end of that war and the beginning of World War II ideological rhetoric had caught up with events and begun to explain them. For many Americans, English, and French the war came to be explained as a struggle for universal freedom—individual and national. Though the emergence of the Soviet Union as an ally against Germany in World War II complicated ideological formulations, by the end of the war, rhetoric had again managed to overtake reality. The West—meaning, increasingly, Anglo-American culture—stood for laissez-faire economics, freedom of thought, and the unfettered pursuit of truth (scientific and moral), as opposed to the state-directed, variously dictatorial, fascist, or totalitarian states of the recently defeated Axis nations (Germany, Italy, Japan) and the soon-to-be-defeated Soviet Union.

Our current popular notions of "science" and "scientific revolution" grew up together in the middle of the twentieth century, paradoxically at the same instant scholars were beginning to make substantial progress in rehumanizing the demigods of science. One may observe the "massive entrenchment" (Durbin, p. 27) of the concept of the "scientific revolution" during the 1950s and 1960s—that is, during the period when the Cold War was very hot. The remark of one leading historian of science during this period is also revealing. Charles Gillespie observed vis à vis the spread of the purportedly Western institution of science, "The hard trial will begin when the instruments of power created [in] the West come fully into the hands of men not of the West. . . . And what will the day hold when China wields the bomb: and Egypt: Will Aurora light a rosy-fingered dawn out the East? Or will Nemesis?" (Durbin, p. 5). The ethnocentric trajectory of the history of science established by Whewell and perpetuated by Sarton continued with little deflection well into the waning twentieth century. The attitude of governmental funding agencies underscores the point. In the mid-1950s the National Science Foundation chose "to downplay sociology and establish a research program uniting only the history and philosophy of science" (Durbin, p. 41). Of the three disciplines, sociology would lead most directly to a contextualized approach to science, whereas the history and philosophy of science, as they were then and in some respects still commonly are practiced, fostered an internalist, decontextualized approach.

Even while following the basic pattern of modern historiography by paying closer attention to the contexts—cultural, economic, political, and so forth—that make scientific discovery possible, historians of science still conceive of science as a modern, Western phenomenon. Dobbs, shortly before her death, remarked on the "Whiggishness" of much contemporary history of science. She might have added that "Whig" is and always has been a rather euphemistic formulation. To say that something is "Whig history" is to say that it is history written as if the events of the past were but prelude and preparation for our present. It is a way of ratifying the institutions and viewpoints of the present by explaining how the past leads inevitably to "us." "Whig history" is, in other words, a euphemisn for that less favored word, propaganda.

See also **Science in History**

**BIBLIOGRAPHY**

Bernal, John Desmond. *The Social Function of Science.* London: Routledge, 1939.

Bynum, W. F., E. J. Browne, and Roy Porter, eds. *Dictionary of the History of Science.* Princeton, NJ: Princeton University Press, 1981.

Cohen, I. Bernard. *Revolution in Science.* Cambridge, MA: Harvard University Press, 1987.

Dobbs, Betty Jo. "Newton as Final Cause and First Mover." *Isis* 85.4 (December 1994): 633–643.

Durbin, Paul, ed. *A Guide to the Culture of Science, Technology, and Medicine.* New York: Free Press, 1980.

Kragh, Helge. *An Introduction to the Historiography of Science.* Cambridge, U.K.: Cambridge University Press, 1987.

Laudan, Rachel. "Histories of the Sciences and Their Uses: A Review to 1913." *History of Science* 31 (1993): 1–34.

Lindberg, David C. *The Beginnings of Western Science: European Scientific Tradition in Philosophical, Religious and Institutional Context, 600 B.C. to A.D. 1450.* Chicago: University of Chicago Press, 1992.

Merton, Robert K. "Science, Technology, and Society in Seventeenth-Century England." *Osiris* 4 (1938): 360–632.

Needham, Joseph, ed. *Science and Civilisation in China.* 7 vols. Cambridge, U.K.: Cambridge University Press, 1954–present.

Ross, Sydney. "Scientist: The Story of a Word." *Annals of Science* 18.2 (June 1962): 65–85.

Sarton, George. *An Introduction to the History of Science.* 3 vols. Baltimore: Carnegie Institution of Washington, 1927–1948.

Thorndike, Lynn. *A History of Magic and Experimental Science.* 8 vols. New York: Macmillan, 1923–1958.

Whewell, William. *History of the Inductive Sciences.* London: Frank Cass and Co., (1837) 1967.

*Peter John*

# HUMAN SUBJECTS IN MEDICAL EXPERIMENTS.

The need to protect individuals subjected to human experimentation gained worldwide attention after world War II. The Nüremberg trials—at which an American military tribunal adjudicated cases from 1946 to 1949 against physicians in the Nazi party accused of inhumane and unnecessary medical experimentation on humans—signaled the harm that can come to human subjects if science and medicine do not receive oversight. The process of producing national regulation in the United States, however, was tempered by a widespread institutionalized belief that American physicians and researchers would never allow or participate in the unethical research practices in which the Nazis had engaged. Consequently, responses to specific incidents in American research, which had received public attention, have propelled U.S. federal regulation.

In spite of sensational cases that have shown otherwise, the image of U.S. research as ethically sound and nonexploitative has been the dominant vision of American science. It is clear that American physicians and researchers had a set of normative ethics that dictated four accepted conditions for conducting research on humans: (1) prior experimentation on animals, (2) willingness to self-experiment or to experiment on one's own family, (3) therapeutic benefit and/or absence of injury, and (4) consent or absence of coercion of the subject. Using this perspective, researchers were able to argue for self-regulation and against external control of research during the decades after World War II and continue to mobilize this argument against further regulation today.

In contrast to this image of self-regulating scientific practices, many American researchers, particularly physicians, have a long history of involuntarily using marginalized groups for human subjects research. Experimentation has systematically occurred using slaves, prisoners, mentally institutionalized people, and orphans. Criticism of these practices has even come from within the ranks of researchers; in 1966, American physician Henry Beecher published a whistle-blowing article in the *New England Journal of Medicine*. The article, which cited research studies published in medical journals, served to highlight serious ethical breaches being made by some American medical researchers and to accuse the medical profession of complacency in regulating research.

Beecher's article catalyzed the scrutiny of medical research and led to mass media coverage of major research projects like the Tuskegee syphilis study on African American men; the Willowbrook State Hospital (NY) study, in which mentally institutionalized children were injected with a live hepatitis virus; and the panoply of invasive studies on inmates at Holmesburg Prison (PA). These cases and others spurred the U.S. Public Health Service to issue preliminary guidelines to protect human subjects in 1969 and to develop them further in 1971. These federal guidelines specified the requirement for "informed consent" from all human subjects participating in all Public Health Service-funded research and for institutional review of research protocol prior to study commencement. In addition to these correctives initiated by the U.S. Department of Health, Education, and Welfare (now, Health and Human Services), the U.S. Congress created ad hoc advisory committees and conducted congressional investigations and hearings during the early 1970s to examine the extent of the problem.

## The *Belmont Report*

The U.S. Congress passed the National Research Act of 1974 to establish the National Commission for the Protection of Human Subjects of Biomedical and Behavioral Research, which was charged with recommending ethical guidelines that could shape federal regulation of human subjects research. Although the National Commission produced seventeen reports regarding a variety of issues in human experimentation, its most important and famous recommendations were published in 1979 in the *Belmont Report* (1979).

The *Belmont Report* identified three ethical principles to govern human subjects research: respect for persons, beneficence, and justice. The report then matched three applications to the respective ethical components: informed consent, risk/benefit assessment, and selection of research subjects. Defining "respect for persons" in terms of individuals as autonomous agents, the National Commission intended informed consent procedures to maintain and reinforce individuals' autonomy through the three ingredients of information, comprehension, and voluntariness. The second ethical principle of beneficence is described in two somewhat conflicting parts: "do no harm" and maximize possible benefits while minimizing possible harms. In the report, risk/benefit assessment was offered as the best way to ensure benefi-cence. The National Commission stressed the importance of measuring the risk of the re-search to the individual compared with the benefit of the research to the individual and, more broadly, to society. Finally, the third ethical principle identified by the National Commission is that of justice. They defined it in terms of "fairness of distribution." Con-cerned with the distribution of the benefits and the burdens of medical research, the National Commission discussed the need for fairness (but not equality). This principle and its application created a "right" to participate in medical research and were closely linked with respect for persons. Justice was not understood in terms of groups or a burden of participation; rather, it was understood that no one who could benefit from a research pro-tocol should be denied access to participation.

## Institutional Review Boards (IRBs) and the Common Rule

The task of transforming the recommenda-tions framed in the *Belmont Report* into fed-eral regulation fell to the Public Health Service. The 1981 regulation became known as the Common Rule (45-CFR-46) because it was designed for intramural and extramural research funded by the Department of Health, Education, and Welfare, but it was applied to human subjects research in all of the government's departments and agencies. In brief, the regulation requires a more vig-orous system of institutional review than was then in place.

According to the Common Rule, an insti-tutional review board (IRB) must consist of five members, who cannot all be of the same sex or profession. At least one member must represent specific scientific concerns, and at least one member must represent specific nonscientific concerns. One member of the IRB must not be affiliated with or related to someone affiliated with the institution for which he or she serves on the IRB. Finally, IRB members must not vote on their own re-search protocols or research in which they have a vested interest.

Outlined in the Common Rule, the most im-portant functions of IRBs are to ensure that risks are minimized and reasonable for the an-ticipated benefits, subjects are selected equi-tably, informed consent is sought and documented, and data monitoring is per-formed for safety. In order to operationalize in-formed consent for IRBs and investigators, the Common Rule specified that informed con-sent must include statements about (1) the re-search's purposes; (2) the procedures involved in the study; (3) a description of "reasonably foreseeable" risks; (4) disclosure of alternative nonexperimental procedures or treatments; (5) assurance of confidentiality; (6) any com-pensation that will be given to the subject should they be harmed during the study (however, compensation is not required); and (7) a statement that participation in the study is voluntary, the subject will not be penalized for not participating, and the subject has the right to end participation at any time. The regulation also specifies that informed consent must be presented in language understand-able to subjects and that it does not and should not waive subjects' rights.

Although the Common Rule applies to all federally funded research, the regulation is relatively weak for behavioral and social sci-ences because it creates a variety of systematic exemptions for institutional review and often dictates expedited review of those protocols. The strength of the *Belmont Report* and the Common Rule was concentrated in human subject research in the biomedical sciences.

## Vulnerable Populations

The Common Rule, with the recommenda-tions outlined in the *Belmont Report*, identifies three specific groups as vulnerable and in need of further explicit regulation: (1) preg-nant women and human fetuses, (2) prisoners, and (3) children. The first group is character-ized as a vulnerable population as a result of the thalidomide disaster, which received attention throughout the United States and Europe in 1962. Thalidomide was a drug that had been prescribed for nausea and insom-nia in pregnant women in the early 1960s. It soon became apparent that women who took thalidomide during the first trimester of their pregnancies gave birth to severely deformed infants. Because thalidomide poses no threat to women who are not pregnant, this tragedy highlighted the need to categorize pregnant women and fetuses separately in biomedical research and in the approval of new drugs. The Common Rule specifies that studies can only be done on pregnant women and fetuses when potential risks have been assessed through studies performed on pregnant animals and nonpregnant women. The regulation allows research that exposes a fetus to risk only if the study will have benefit to the woman or fetus. There is also an explicit clause that states that it is not permitted to provide any inducements, financial or otherwise, to women to terminate their pregnancies.

Prisoners were also identified as a vulnerable population owing to historical events. Following the Second World War and until the late 1970s, prisoners became an important population for biomedical research. Investigators argued that prisoners were an ideal population because they live in a controlled and controllable environment, so that prison experiments could take on laboratory conditions. However, prisoners had been grossly mistreated in many experiments and had been coerced into participating by the promise of better living conditions while in prison and shortened prison sentences. The Common Rule recognizes that incarceration may limit prisoners' ability to give voluntary consent. As a way to counteract this threat, the regulation limits the type of research that can be performed on prison populations to studies that are explicitly about prisoners, that investigate illnesses that are prevalent in prisoners, or that improve prisoners' health or well-being. Much of the federal regulation aims to respect the rights of prisoners as citizens and to eliminate the assumption that prisoners could pay their debt to society by participating in research studies.

The final group identified as vulnerable by the Common Rule consists of children. The regulation specified for protecting children has two main parts. The first element is that assent from children and written permission from parents is necessary for children to take part in research. The second part of the regulation restricts research on children who are wards of the state to studies related to their status as wards or to studies in which the majority of children included are not wards. This latter provision is a corrective for studies like those done on children at Willowbrook.

Many other groups have been discussed as potentially vulnerable in the context of human subjects research. Mentally disabled individuals may have a reduced capacity to give informed consent, yet, at this time, there is no explicit policy for including or excluding this group from research protocols. Military personnel and students have also been flagged as potentially vulnerable populations because, like prisoners, it is not clear that they are able to give truly voluntary consent. Severely ill populations may also be vulnerable because there is the possibility of creating a therapeutic misconception when there is little chance for benefit to patients, yet they consent believing that their health or quality of life will improve. One final group that can also potentially be characterized as vulnerable is the economically disadvantaged. This group may be compelled to participate in studies in order to receive financial compensation or access to health care.

## Cross-National Comparison

In 1964, the World Medical Association issued its *Declaration of Helsinki*, in which international guidelines for human subjects research were detailed. Physicians wrote the declaration as a means of creating and supporting principles of research ethics that transcended national boundaries. The hope was that the declaration would prevent research atrocities from happening in the future through a system of accountability, but it had no particular legal status. Nevertheless, it has served as a model for many industrialized countries as they struggled with creating regulation.

The United States was the first government to regulate human subjects research, and many other industrialized nations were influenced by the U.S. policy. Although the content of the guidelines—specifically, the use of research review committees (e.g., IRBs)—is quite similar among industrialized countries in North America, Europe, and Asia, the only country other than the United States that gave the role of oversight and development of regulation to the government, rather than some other organization, was New Zealand, where the public's attention had been directed to research misconduct. In the case of New Zealand, one study conducted from 1966 to the mid-1970s consisted of leaving cervical cancer untreated to observe the natural history of the disease. Investigations shocked New Zealanders when it revealed that many women participating in the study died unnecessarily.

Some commonalities in regulation in industrialized countries are based on the tasks or composition of the review committees. Most national guidelines' only incentive for submitting research protocols to be reviewed is the threat of losing research support for noncompliance. The United States, Australia, and Denmark are most serious about trying to create disincentives for not following the guidelines. Another common feature of review committees is that there is little monitoring after the initial review of the research protocol. Denmark is the only country to continue review by having spot-checks for on-going projects. In addition, review in most countries is generally decentralized and located in individual institutes. New Zealand and Denmark have a slightly different system, in which there are regional committees, rather than institutional ones. Finally, most nations require lay members to sit on the review committees, but these members are often in the minority. In New Zealand and Denmark, lay members need to account for half the composition of the review committee, and some need to be members of the groups being studied.

By contrast, developing countries in Africa, Asia, and Latin America have little to no regulation governing human subjects research, and many multinational pharmaceutical companies are using populations in developing countries as a means of escaping some of the restrictions imposed by industrialized countries. The Declaration of Helsinki is meant to give international guidelines in these contexts, but there is no monitoring to ensure compliance. Further investigation needs to address human subjects research in developing countries, particularly in the light of trends in globalization.

## Current Climate and Agitation for Change

The human subjects research landscape in the United States has shifted significantly since the Common Rule was enacted in 1981. One example of this change is manifest in the effects that advocacy groups have had on research practices and federal policy. The women's movement impressed upon the National Institutes of Health (NIH) the necessity of including women in clinical trials and resulted in policy requiring studies to include women or justify their absence from all medical research. Also, the AIDS movement succeeded in changing the conditions for access to AIDS research and in altering the power relations for access to research design decisions. These advocacy groups and others have had profound impacts on increasing the rights of human subjects.

The second type of change that is an important context for understanding the dynamics of human subjects research is increasing privatization and globalization of research sites. Within medical research, privatization involves a shift away from a reliance on academic centers as the primary producers of scientific knowledge to a market model characterized by outsourcing as the key principle to the distribution of labor. An example of privatization can be found in recruitment for medical research. Traditionally, recruitment for medical studies has relied upon the use of physicians' personal pools of patients. With the advent of privatization, monetary inducements, direct advertising, and centralized databases listing eligible subjects have become commonplace. Another example is the appearance of for-profit IRBs that are being used by academic as well as private institutions. Globalization is affecting medical research through the use of multisite studies and the selection of sites in developing nations.

A final example of change in the U.S. human subjects research is at the level of federal oversight of research sites, such as universities and hospitals. Since the early 1990s, the government has fully or partially suspended research in over 40 research organizations. The most well-known cases are the full suspensions of human subjects studies at Duke University in 1999 and at Johns Hopkins University in 2001. Hopkins's suspension came after a federal investigation following the death of a healthy 24-year-old woman participating in an asthma study. Following closely behind the death of a 19-year-old man who died in a gene-therapy experiment at the University of Pennsylvania in 1999, this Hopkins death signaled the need for government intervention to provide better protection of research subjects.

Although the Common Rule has been updated and revised since it was issued in 1981, little has actually changed in terms of its jurisdiction or scope. As a result, policymakers have been aware of the need to revise the regulation for almost a decade. The response by policymakers has been to appoint ad hoc committees and to create offices within the Department of Health and Human Services to study the ethics of human subjects research and to provide recommendations for new policy and regulation. Some updates to the regulations have been made regarding subjects' privacy and financial conflicts of interest for researchers and institutions.

These committees and reports have all recognized the need for change in the current system of human subjects research protection. From the Institute of Medicine's 2002 recommendation to create ethics review boards (ERBs) that complement IRBs, which are often more concerned with the scientific merit of the study than with debating ethics, to the Secretary's Advisory Committee on Human Research Protections Committee debating stem cell research policy, there are many facets to the national discussion about what needs to be done to better protect human subjects. Although many groups are poised to enact new regulation, there has been little movement toward actual change.

It is expected that the U.S. government will make amendments to the Common Rule in the light of the changing situation with human subjects research; it remains unclear, however, when these changes will occur and what they will be.

## BIBLIOGRAPHY

Annas, G. J., and M. A. Grodin, eds. *The Nazi Doctors and the N‚remberg Code: Human Rights in Human Experimentation.* New York: Oxford University Press, 1995. Details the N‚remberg code and its effects on systems of ethics and human subjects regulation.

Epstein, S. *Impure Science: AIDS, Activism, and the Politics of Knowledge.* Berkeley, CA: University of California Press, 1996. This is an excellent source for examining the impact of advocacy groups on medical research.

Faden, R. R., and T. L. Beauchamp. *A History and Theory of Informed Consent.* New York: Oxford University Press, 1986. This work is an extremely thorough and detailed history of U.S. federal policy regarding human subjects research.

Hornblum, A. M. *Acres of Skin: Human Experiments at Holmesburg Prison.* New York: Routledge, 1998. This book examines the range of medical studies that were conducted at Holmesburg Prison in Philadelphia, Pennsylvania. It is written in a very accessible style.

Institute of Medicine *Responsible Research: A Systems Approach to Protecting Research Participants.* Washington, D.C.: National Academies Press, 2002. This report contains the types of recommendations that are currently being discussed in U.S. policy circles.

Jones, J. H. *Bad Blood: The Tuskegee Syphilis Experiment.* New York: Free Press, 1981. One of the best works that describes the Tuskegee study.

Lederer, S. E. *Subjected to Science: Human Experimentation in American before the Second World War.* Baltimore, MD: Johns Hopkins University Press, 1995. This book is important for providing historical information about the effectiveness of self-regulation by physicians in human subjects experimentation before World War II.

McNeill, P. M. *The Ethics and Politics of Human Experimentation.* New York: Cambridge University Press, 1993. Excellent source for descriptions about industrialized nations regulations and systems of institutional review of human subjects research.

Rothman, D. J. *Strangers at the Bedside: A History of How Law and Bioethics Transformed Medical Decision Making*. New York: Basic Books, 1991. This is another book that describes the normative ethics of physicians during the twentieth century. It also contains in depth discussions about Henry Beecher as a whistleblower.

Rothman, D. J., and S. M. Rothman. *The Willowbrook Wars*. New York: Harper Collins, 1984. Description of the medical experiments conducted on mentally disabled children institutionalized at Willowbrook State Hospital in New York State.

Vanderpool, H. Y., ed. *The Ethics of Research Involving Human Subjects: Facing the 21st Century*. Frederick, MD: University Publishing Group, 1966. Good sample of essays detailing the range of discussions within scholarship on the ethics of human subjects research.

*Jill A. Fisher*

# HUMANITIES AND SCIENCE.

The sciences and the humanities, and the more specific disciplines within them, each produce knowledge using different practices. Over the course of the twentieth century, movements bringing the practices of the humanities to bear on the sciences have gained momentum. These movements are as diverse as the rhetoric of science, which analyzes modes of argumentation in scientific work; cultural studies of science, which inquires about how science and culture interact; the philosophy of science, which addresses epistemology and how we come to know truth through science; and the history of science, which develops narratives about major scientific events. All these combinations seek to make science understandable according to a humanistic discipline's terms. These interdisciplinary movements interrogate science about how scientific knowledge is produced and distributed. Production and distribution are frequently interrelated. This interrelation is addressed by one of the most recent interactions between the science and humanities. Science, technology, and society

(STS) studies include scholars in both the sciences and the humanities, in order to establish a socially conscientious perspective on scientific knowledge production and distribution. Before the sciences and humanities existed in relative harmony, however, they clashed on a number of historic occasions. This essay describes two such occasions. The first is the "Two Cultures" lecture delivered by Sir Charles P. Snow in 1959, a lecture that garnered a great deal of rancorous commentary. The second instance is the no less inflammatory Sokal hoax of 1996, when a leading cultural studies journal unwittingly published an article intended to be a parody. The article, written by a physicist, used caricatured versions of theoretical terms from the humanities to culturally analyze a similarly unrigorous version of physics. How science produces knowledge, and to what end, has in recent history frequently resolved into the question of whether science is constructed or essential; that is, whether it is only the product of social interactions and institutional interests, or whether it progressively reveals bedrock reality. The examples that follow demonstrate that even though the conflict between constructed versus essential has been considered a major debate in science-humanities interaction, describing the conflict exclusively in this way is a bit misleading. These examples explore some of the political and economic stakes of disciplinary formation, in both the sciences and the humanities. Humanities programs that address sciences can do so because science is embedded in culture; that is, the production and distribution of scientific knowledge seems so stable and enduring and truth-creating *because* it is held in place culturally, not in spite of its cultural basis. With this point in mind, when interdisciplinary movements identify science as "socially constructed," such a designation cannot be a way to dismiss humanities' intervention in the sciences, as disciplinary purists take it to be. Before turning to our two illustrative occasions, let us take a (necessarily broad) overview of how a divide developed between the sciences and humanities in the first place, a historical divide that informs their contemporary interaction.

## Disciplinary Emergence and Separation

As just suggested, many current academic interactions between the sciences and humanities take place within humanities' disciplinary terms, even though science seems to have a more pervasive presence in culture beyond the academy. For example, much interdisciplinary work that has contributed to a cultural understanding of science comes from the humanities' political work on race and gender. Such a perspective asks that scientific knowledge be accountable to these basic cultural categories; indeed, one of the most prolific sites of interaction between the sciences and humanities has been feminist science studies. Conversely, there is not nearly as much work done on how the conceptual tools of the sciences might be aimed at the humanities (though the social sciences work to make humanistic interests more scientific; and there are other instances of technology-mediated art and authorship, for example). This question of accountability has erupted from the humanities, rather than the sciences, for historical rather than natural reasons. The humanities have historically been charged with arbitrating culture, albeit in culture's varying manifestations, from "high" culture, to a more widespread ideological understanding of culture. This history is reflected in the disciplines' very names. The humanities are concerned with what seems peculiarly human—the beautiful, the ethical, the cultural, the reflective—whereas the sciences ostensibly are concerned with objective knowledge that is specific to the world regardless of cultural activity.

Reinstatements and challenges to disciplinary identity are connected to how knowledge has been institutionalized, and who has access to it. What we now know as the academic disciplines emerged over the late nineteenth and early twentieth centuries. Science's economic and disciplinary importance is tied to cultural history. For example, it was during this period that, in the Western industrialized world amid the swelling ranks of the middle and upper classes, people were becoming increasingly reliant upon science and technology, nationally,

as with the advent of modern technological warfare during World War I, and personally, with advances in medicine, transportation, communications, and so on. Following with these developments in nation building and private life, over the course of the twentieth century, education was becoming more accessible to more people, and its use value was likewise incorporated into everyday life. The G.I. bill, which allowed many men and women to go to college who would not otherwise have been able to do so, was introduced after World War II, and both responded to and helped create the fast-growing, and increasingly economically important, middle class. With the influx of new college students, standardized education requirements had to be established so that college degrees would mean roughly the same thing among people. Standardizations happened in the sciences and in the humanities, but for different reasons: in the sciences standardization controlled what would count as legitimate science, whereas standardization in the humanities maintained what was understood to be an education befitting enlightened citizens in a democracy.

By contrast to the sciences' usefulness for national defense and for modern ways of life, humanities departments were important to education because they offered literacy, writing skills, knowledge of history and culture—less "useful" benefits than sciences, to be sure, but nevertheless seen as crucial components of education. The humanities' offerings to postsecondary education espoused individual expression, associated with democracy, and somewhat highbrow cultural values historically associated with education. Deeply held cultural values and the university system developed reciprocally. Colleges and universities codified these values in research and education by creating modes of regulation, legitimization, and expertise throughout all academic disciplines. The sciences were codified as useful and the humanities as expressive. In short, academic disciplines are the products of the educational system, and, in concert with cultural values, produced by historical events.

These modes of regulation and legitimation that are the academic disciplines are bundles of practices. That is, they do not occur naturally, but emerge among social interactions. Another way to put this is to note that a scholar (in either the sciences or the humanities) who, for example, writes an article according to disciplinary conventions, is not only writing that article because of the disciplinary conventions; that scholar is at the same time maintaining or changing the conventions, by reinstating or challenging them. That is to say, disciplines are so named because they are the result of guiding and guided—disciplining—actions: actions of tenure, publishing, hiring and firing, admittance to programs, modes of evaluation, and the like.

This point about practices foregrounds that in the examples of interaction between the sciences and humanities such as those that follow, it is crucial to keep in mind that neither the disciplines proper, nor interdisciplinary movements are natural categories according into which new knowledge just plops into its given place. Indeed we might even claim that the opposite is true: that the knowledge we make is determined by existent disciplines and the possibilities they allow or foreclose. Accordingly, despite their self-representations, neither the sciences nor humanities developed according to a progressive, forward-moving purpose, with one unified goal or agenda. Rather, both the sciences and the humanities have developed according to multiple, social patterns of culturally sanctioned legitimation, patterns produced and reproduced through practices.

## Charles P. Snow's Two Cultures Lecture

Not all interactions between the sciences and the humanities became fraught in the way the following cases did. Revisiting these tensions offers historical perspective on disciplinary values. The first example here is the Two Cultures lecture, delivered by Sir Charles P. Snow at Oxford University as the annual Rede lecture in 1959. This event is frequently referred to as a kind of originary moment for contemporary interaction between the sciences and the humanities.

The lecture points out a rift between scientists and "literary intellectuals" (Snow's term). Snow's overarching point was that when operating together, the sciences and humanities offer a kind of wholeness of knowledge that is both progressive and ethical. To Snow, the existent rift between the disciplines meant that the industrial and scientific revolutions, with their progress unchecked by the humanities' ethics, were leaving most of the globe's population in the dust of poverty and overpopulation. In this 1959 lecture, Snow observed that modernity, with encroaching industry and warfare supported by the advancement of science, was setting in motion a global situation in which the rich were getting richer and the poor were getting poorer. The Two Cultures lecture claimed that the widening global economic rift was the direct result of the academic rift, and that the cultural distribution of scientific and humanistic knowledge should be accountable to each other. To Snow, humanities scholars positioned themselves as the ultimate arbiters of culture because of their monopoly on criticism of literature and the arts. The scientific revolution, as Snow saw it, was capable of alleviating global poverty, if only there was more communication between the sciences and the humanities, with the humanities softening the sciences, and the sciences making the humanities more useful.

Snow's main evidence for the widening rift between scientists and literary intellectuals was that neither knows the other's most basic tenets. For example, Snow points out that no literary intellectual is familiar with the Second Law of Thermodynamics, and similarly, few scientists have even heard of the popular author Charles Dickens, and those that have are likely to see him as an esoteric and difficult-to-understand writer. As a result of increasing specialization in the academy, evidenced by these inabilities to converse, scientists and literary intellectuals could no longer understand each other or the value of each other's work. Snow was optimistic that a world-redeeming ethics can be achieved by equal emphasis on different kinds of knowledge. Science could only move forward, his lecture implied, and forward could only mean easing the world's ills. The humanities were needed to temper the unrelenting usefulness of science.

The Two Cultures lecture became best known for the conflict it provoked, rather than for the content of the lecture itself. Snow's most vociferous detractor was literary critic Frank Raymond Leavis, whose criticisms often took the form of attacks on Snow's character. Leavis's main point is that the work of literary scholars is important in its own right. According to Leavis, the work of literary scholars had no business with the practical (and therefore lower) concerns of science. Leavis's criticisms, however, while among the most vitriolic and high profile, were not atypical. The very suggestion that the disciplines had something to offer each other seemed an attack on the deeply held values that had kept the disciplines separate within the logics of modernity.

Such values included the cultural authority of the humanities even while the sciences gained economic predominance. Leavis and others took the humanities to be unassailable—even natural—monuments of collected knowledge. The suggestion that they could be brought to bear on the sciences threatened this edifice (though Leavis's defensiveness arguably made the edifice seem more precarious than it would have otherwise) and called attention to the fact that it was "only" an edifice, rather than a natural formation, and therefore kept in place by social networks and practices. Counter to Leavis's intuition was the insight that social networks and practices are exactly what make a discipline so unassailable, and indeed it was practices that were ever more firmly embedding science into culture and making science impervious to a humanistic perspective. As a proponent of high modernism, and in the intellectual lineage of Matthew Arnold, who, in 1873 in an introduction to his book *Literature and Dogma*, famously defined culture as "the best of what is thought and said," Leavis's main stake in the questions Snow raised was that literary culture is not maintained by, and therefore cannot change, its practices. At that point in the century, the humanities were seen as more firmly planted than the sciences. In spite of Snow's main thesis about global poverty, the Two

Cultures lecture seemed only to provoke the sensibilities of staunch defenders of literature's stable, disciplined identity.

## The Sokal Hoax

By contrast, the publishing fiasco known as the Sokal hoax shows the sciences to be more firmly planted than the humanities. This more recent (1996) clash between the sciences and the humanities is no less vitriol-ridden than the Two Cultures conflict. The so-called hoax is named for Alan Sokal, a New York University physicist. The fiasco began when Sokal submitted a satirical article, written about physics from an extreme constructivist perspective, to *Social Text*, a leading journal in cultural studies. *Social Text* published the article, entitled "Transgressing the Boundaries: Toward a Transformative Hermeneutics of Quantum Gravity," as part of their special issue on what at the time was called the *science wars*. Shortly after the article's publication, Sokal published a somewhat smug revelation in *Lingua Franca* that he had written "Transgressing the Boundaries" using catchphrases and concepts that seemed to be trendy in cultural studies at the time, but essentially said nothing. According to Sokal, cultural studies, especially insofar as they had claimed to make inroads into the conceptual basis of science, had no logical cohesion, made dubious arguments, and generally lacked intellectual—that is, disciplinary—rigor. On Sokal's terms, this lack of rigor was reflected in the fact that the journal *Social Text* did not submit his article to physicists or to cultural studies scholars for review, but instead published it based on the evaluation of its editorial board, which comprised intellectuals in the humanities. He had submitted the article, which was absurd from the perspective of a physicist, as something of an experiment. Sokal's hypothesis was that the political left (with which he identified) in the humanities had sacrificed rigorous disciplinarity for overblown claims that came from a brand of radicalism more self-indulgent than politically purposeful.

Depending on perspective, Sokal's actions may or may not have proved a point about cultural studies, but what most strongly emerged from the responses to the hoax (from *Social Text's* editors, from academics in the sciences and humanities, and popular journalists), was the degree of difference between science and humanities scholars' respective practices of knowledge production. Ever since Snow first made his observations, science's infiltration of national and private life had continued, accruing ever more economic warrant. Meanwhile, "culture" had changed from Leavis's naturalized view, to include much more widely dispersed practices and ideologies, even as the humanities maintained their position as cultural arbiter, developing ever more sophisticated theoretical vocabulary for analyzing an expanding conception of culture. At the time of the Sokal hoax (and even today), scientific knowledge was seen as real, "useful," and distributable; meanwhile, especially from Sokal's perspective, the humanities claimed their importance to democracy in alienating jargon. Even science's most esoteric productions were seen as more crucial to the lives of U.S. citizens than those of the humanities. In a word, science seemed more *democratic* than did the humanities with their jargon and insular specialties.

Ultimately, the Sokal hoax battle in the so-called science wars of the 1990s proved to be the last gasp of the wars' sensationalistic antagonism, and helped shift the territory of the constructivist versus essentialist conflict. Humanities departments, and in this case, especially the discipline of cultural studies, had the conceptual vocabulary with which to leverage the claim that science is wholly constructed and rife with politically repressive logics. In the mid-1990s, cultural studies, and the nascent field of science studies, addressed these political issues by pointing out that science is contingent upon social networks of knowledge production. This in turn raised the stakes of the essentialist position, which held that science is truth, unadulterated by social position or bias. In other words, the "wars" leading up to the Sokal hoax focused on the ontological status of science, or what science is.

It has since been observed that the science wars were perhaps unproductively melodramatic, based on each side working with a somewhat cartoonish representation of the other. Sokal's article was an example of this overblown conception of the constructivist position. Much like Leavis's objections to Snow almost 50 years before, the clamor surrounding the Sokal hoax ultimately seemed more the result of scholars' guarding their own territory's epistemic privilege than engaging the specific terms of the debate. However, the Sokal hoax did shift the debate over science away from what science *is*, to what science *does*.

As a result of the Sokal hoax, the question emerged—still unresolved, but more promising—of how the sciences and humanities might be more responsible to each other within the academy and to the wider culture. Interdisciplinary work on globalization has begun to address Snow's point about the global economy's investment in the academic disciplines in, and scientists increasingly engage, for example, feminist concerns presented originally in the humanities, such as medicine's casting women in strictly reproductive roles. Some humanities scholars, such as Donna Haraway and N. Katherine Hayles, draw on their own scientific training to bring to the humanities different kinds of epistemological questions than the humanities have traditionally addressed. The emergent goal for the humanities is thus to invent new practices beyond simply taking refuge in their historically privileged cultural position. The emergent goal for the sciences is to invent practices that temper usefulness with social ethics, rather than continuing blindly with their economic warrant. The perspective each brings to the other suggests practices of interdisciplinarity that allow new types of knowledge to be produced and circulated.

*See also* **Humanities and Technology**

**BIBLIOGRAPHY**

Aronowitz, Stanley. *The Knowledge Factory: Dismantling the Corporate University and Creating True Higher Learning.* Boston: Beacon Press, 2000. A historical and polemical look at the impact of industry on colleges and universities.

Editors of *Lingua Franca. The Sokal Hoax: The Sham That Shook the Academy.* Lincoln, Nebraska: University of Nebraska Press, 2000.

The original text of Sokal's article and admission, *Social Text*'s editors' response, and responses to the scandal from journalists and academics in the United States and abroad.

Foucault, Michel. *The Order of Things: An Archaeology of the Human Sciences.* New York: Vintage, 1994. An important and intensive book on the history of forms of knowledge, beginning with the early modern period.

Haraway, Donna. *Simians, Cyborgs and Women: The Reinvention of Nature.* New York: Routledge, 1991. One of the key texts in feminist science studies, offering a powerful analysis of science's hold on culture and the political effects of this hold.

Hayles, N. Katherine. *How We Became Posthuman: Virtual Bodies in Cybernetics, Literature and Informatics.* Chicago: University of Chicago Press, 1999. A text that is central to both science studies and contemporary theory in the humanities, discussing analogies between literary and scientific formations.

Kuhn, Thomas. *The Structure of Scientific Revolutions.* 3rd ed. Chicago: University of Chicago Press, 1996. Originally appearing in 1962, this was the first book to suggest a model of scientific developments based on changing modes of legitimacy, rather than sheer progress. It is still referred to today.

Latour, Bruno. *Pandora's Hope: Essays on the Reality of Science Studies.* Cambridge, MA: Harvard University Press, 1999. An account of the science wars, and an description of the pertinence of science studies, as applied to historical examples.

Martin, Emily. *Flexible Bodies: Tracking Immunity in American Culture from the Days of Polio to the Age of AIDS.* Boston: Beacon Press, 1994. The anthropologist Martin traces the cultural implications of a medically defined object-the immune system.

Shapin, Steven, and Simon Schaffer. *Leviathan and the Air-Pump: Hobbes, Boyle, and the Experimental Life.* Princeton, NJ: Princeton University Press, 1985. A history of how scientific knowledge and empirical verification as we now know it came to be. Through accounts of Thomas Hobbes and Robert Boyle, this book demonstrates that epistemology and politics are intertwined.

Snow, C. P. *The Two Cultures.* Cambridge, U.K.: Oxford University Press, 1998. The text of Snow's original lecture, with an introduction by Stefan Collini that contextualizes the lecture and its aftermath in larger movements in culture and the academy.

Watkins, Evan. *Work Time: English Departments and the Circulation of Cultural Value.* Stanford, CA: Stanford University Press, 1989. A trenchant institutional analysis of English departments' position as cultural arbiters.

*Elizabeth Mazzolini*

# HUMANITIES AND TECHNOLOGY.

It is tempting to claim that the study of the humanities in the university has undergone a massive sea change in the past decade as a result of the large-scale implementation of digital information processing and retrieval. However, such an assertion is ahistorical, because it tends to perpetuate precisely the set of assumptions regularly interrogated by those scholars in the humanities who study technology, technological change, and the production of knowledge. Without doubt, the computer revolution has profoundly affected the study of the humanities, but that is only part of the story of the relation between the two.

Thinking about technology and the humanities leads to replacing the conjunction in favor of a preposition: technology *in* the humanities, technology *of* the humanities, and even technology *by* the humanities, which is just to say that the issues are multiple and densely interwoven. In beginning to unravel these relations a bit, it is possible to draw provisionally two different areas of inquiry: the idea of technology as an object of humanistic inquiry, and the role of technology in the practice of the humanities.

The term *technology* in its current usage is of recent vintage. In centuries past other words such as "applied science," "craftsmanship," or, much earlier, "techne" and "mechanic(k)s" stood in (partially) for what has come to be regarded today as "the scientific study of the practical or industrial arts" (*Oxford English Dictionary*). Technology began to be conceptualized in its present form almost as a mirror of what is currently called the humanities. They both have their source in the reorganization of the European university in the nineteenth century, with its strong division between the traditional university and the polytechnic schools. Such academic distinctions produced the "and" in "technology and the humanities": a seemingly impenetrable barrier where study in the humanities meant ignoring technology. Technological objects were precisely that: mere things, not worthy of study alongside the philosophically rigorous and socially fundamental concerns of humanistic inquiry. Conversely, a student in a polytechnic school might spend part of his or her day reading the classics, or at least practicing written communication, but the major portion of that student's time would be spent manipulating devices and testing objects.

Conceptually, the divide between learning technological practice and studying the humanities widened further through the simple but fundamental distinction in linguistics between the word and the thing. Such an obvious and clear principle (perhaps detailed most forcefully by Ferdinand de Saussure) split the objects of humanistic inquiry (texts) from their material circumstances (books) and the physical objects those texts named. A brief look at the scholarly apparatus that traditionally accompanies a text such as Henry David Thoreau's *Walden* provides a clear illustration. Notes offered to support the reader's understanding usually explain his references to such writers as Ovid, Shakespeare, Darwin, or Confucius, but only rarely are his discussions of the proper way to split a shingle, dry lumber, or attach a handle to an ax similarly annotated. These latter practices can be every bit as complicated as a Shakespearean sonnet, but, given their material (technological) circumstances, they are usually not regarded as nearly as important for the reader as the precise context of a literary reference.

## Technology as an Object of Humanistic Inquiry

In the latter half of the twentieth century, a number of humanities scholars turned their attention to the figure of technology in literature and other cultural artifacts. This is not to say that "the machine" had not been acknowledged as an omnipresent figure in literary texts for years, only to claim that scholars began to interrogate such machines as something more than literary tropes. In part this was an acknowledgment of the influence rationalized industrial production exerted in the first part of that century: a set of practices that profoundly affected philosophy, aesthetics, and literary production. Karl Marx's critique of industrial capitalism provided the vocabulary and perspective for much of the early work on representations of technology in literature and philosophy, perhaps best exemplified in the work of Raymond Williams and E. P. Thompson. The latter's "Time, Work-Discipline, and Industrial Capitalism" remains a classic example of a scholar taking seriously the material working of a device (in this case, the clock) as fundamental for both historical and literary understanding.

More often, though, humanists discussing technology evoke something like a zero-sum game: The more massively deployed a technology becomes, the more the humanity of humanism tends to diminish. Once again, Marx's critique lays the groundwork for the alarm sounded by a range of thinkers regarding the large-scale instantiation of rationalized technological practice and the alienation and degradation of the human. One need only recall the work of the Frankfurt School, particularly Herbert Marcuse's *One Dimensional Man,* or look to Antonio Gramsci's critique of Fordism in *The Prison Notebooks* to see the potency of this alarm. Such concerns, of course, are not the sole province of Marxist critics. Michel Foucault's harrowing description of the Western world as carceral dystopia in *Discipline and Punish* is a vision of a world where technologically inspired rationalized practices have led to the "Death of Man." Also, Martin Heidegger's frequently cited but seldom-understood classic *The Question Concerning Technology* is usually enlisted as support for a world view in which technology and technological practices are regarded with extreme suspicion by those who value a traditional conception of the human and the humanities.

In the past few decades, these stark visions have been tempered (though not wholly displaced) by intense and ongoing scholarly engagement with technological practice coupled with a redefinition of the scope of humanistic inquiry. To put it simply, objects—technological objects—have become the subject of research in the humanities. Areas of important inquiry include, for example, research in medical technology, feminism and technology studies, ecocriticism, and dwelling studies, as well as certain elements of cognitive science. This research is in part the result of university funding that has begun to stress interdisciplinary activity. In addition, societies associated with traditional disciplines in the humanities, such as literature, history, and philosophy, have begun to focus more specifically on technology. Cultural studies (particularly material cultural studies inflected by anthropology) also provides a theoretical framework for scholars to engage technology as a subject of inquiry. Perhaps most important was the "turn" in science studies from discussions of science as textually embodied and consequently destabilized by the ambiguities of language to the idea of science as practice or, to take a title from the work of Bruno Latour, to the study of "science in action." As a result, many researchers began to look seriously at the material circumstances of scientific practice, the use of instruments, and the sites of scientific knowledge production. One of the central texts in this last regard remains Steven Shapin and Simon Schaffer's monumental *Leviathan and the Air Pump*, a study that links political philosophy with science studies through the examination of a specific technological device, Robert Boyle's vacuum pump. Another significant figure in this shift of focus in science studies to technology is Latour, whose *Aramis, or the Love of Technology* (a study of a rail transportation system designed for the suburbs of Paris but never built) successfully draws together technological reports, political texts, philosophy, sociology, and literature (the framework of the book is a detective novel).

The net effect of this new-found interest in technological practices is an increasingly sophisticated understanding of the role technology plays in the production of society and our understandings of the human, as well as

a more complex notion of what constitutes humanistic inquiry. A text that draws together these threads in poignant fashion is N. Katherine Hayles's *How We Became Posthuman*. There Hayles begins to articulate the recent history of humanism (or the move into posthumanism) as the result of complex scientific and technological practices that demand a reformulation of the very idea of the human. What is important to recognize is that this move, unlike earlier technological-deterministic jeremiads, is not fraught with the same anxieties. Instead, Hayles is participating in a much broader rearticulation of the human and the humanities based precisely on the understandings developed by scholars who finally have turned to study technology itself.

## Technology and the Practice of the Humanities

Hayles's *How We Became Posthuman* has as its primary focus an examination of the history of digital information technologies, and so it leads directly into the second part of this discussion: the use of information technology in the humanities. Although questioning the efficacy of representational technologies has a long history in the humanities (think of all of the prologues to volumes of poetry addressing the often ill-clad verses being sent out into the cold world), serious inquiry into the emergence of the book and other forms of print technology is a fairly recent phenomenon. Elizabeth Eisenstein's massive study *The Printing Press as an Agent of Change* in many ways serves as the model and instigator for much research in this area, as does Marshall McLuhan's pivotal call to pay as much attention to the medium as to the message.

The twentieth century saw the rapid deployment of a broad range of new representational technologies including radio, phonographs, film, widely available amateur photography, television, telephones, and digital information technologies and networks. At the same time, there emerged a larger and increasingly literate populace that demanded cheap pulp fiction as well as the classics. Other not-so-obvious but nevertheless important information technologies in the early twentieth century included telegraphs and typewriters. This efflorescence of devices prompted in the late twentieth century an equally efflorescent scholarship on the cultural impact of those technologies. What has come to be regarded as an important precursor is Walter Benjamin's classic essay "The Work of Art in the Age of Mechanical Reproduction." Benjamin, also of the Frankfurt school, turns a somewhat nostalgic eye toward the "aura" lost in a world now flooded by the products of a new representational technology: photography. Nostalgia aside, this complicated meditation provides a vocabulary and perspective for critics to begin to understand the role any representational technology has in the production of the knowledge prized by humanities scholars. For example, Friedrich Kittler's study of reading habits in *Discourse Networks 1800/1900* successfully integrates representational technologies with reading practices, particularly as they relate to the consumption of literature. In addition, his study *Gramophone, Film, Typewriter* is a complex analysis of the broad cultural impact of each of the respective devices in the title, while at the same time he develops a nuanced interpretation of how each of those technologies figure in specific cultural productions (for example, the impact of the Malling-Hansen typewriter on the later work of Friedrich Nietzsche). Such analysis creates interesting possibilities for understanding the work of other early adopters of information technologies, such as the American poet William Carlos Williams, who often composed on the typewriter.

Still, the bulk of recent work on media technologies and the humanities is focused on networked new media. Interestingly enough, it takes on much the same dire tone as midcentury angst regarding the dehumanizing qualities of rationalized technological practices. According to many scholars, current media technologies are at least in part responsible for the death of literature and the ruin of the university. Without doubt, new media technologies are transforming the way the humanities are studied in the academy, particularly in terms of content delivery, research strategies, and pedagogical practice. In addition, the very product of humanistic research is being either repackaged (online versions of traditional print journals) or reconfigured (dynamic multimedia projects and Weblogs). In part, the dire tone results from the transformation of the professional rules of humanities practice and what counts as knowledge. The terms of the latter question were articulated in Jean-François Lyotard's *Postmodern Condition*, a book that helped define the zeitgeist of the late twentieth century. Lyotard's work is usually cited in efforts to establish the exact nature of postmodernism, but of more pressing importance is his reconfiguration of knowledge in light of new information technologies.

It is possible to see the result of this reconfiguration most vividly in the practice of teaching composition. Widely available word processors make ongoing revision of writing assignments less painful (and consequently more frequent), and Internet resources provide an abundance of reference material for those same papers. Perhaps more fundamental, though, is the shift in the kind of knowledge produced by such exercises. At midcentury in the United States, a typical essay would doubtless be the product of the patient and careful analysis of a limited number of texts, a practice fostered by the rapid expansion of the university after World War II and the development of the New Criticism at least in part to relieve the strain on the primary information technology of the era: the library. Today, writing projects ideally gather diverse materials into a logical and well-defined space; structure and organization are often prized over reflection and analysis. It is easy to view with dismay the students' increasing inability to perform close readings, but at the same time, it is impossible to deny that the traditional essay is rapidly being replaced by a different form of knowledge production.

With an increasingly sophisticated understanding of the function of technology in representation, and at the same time with increasingly powerful representational technologies at their disposal, some humanities scholars feel a sense of crisis, while others are beginning to experience the exhilaration of the immense potential of these new tools. The questions remain simple: How do these technologies facilitate traditional humanities scholarship? And how are the practices of humanities scholarship being reconfigured by these technologies? Unfortunately the answers are not that simple, though it is possible to gesture toward a few trends. A broad range of networked sites have been established to facilitate traditional humanities scholarship: Obscure texts are made available online, visual images are reincorporated into literary texts (as in the Blake archive), vast amounts of supplementary materials that had been largely unavailable are now sometimes easily had, and traditional publications can be supplemented by online versions. Put simply, when one overlooks the glut of ill-defined data pullulating on the Internet, networked digital technologies generally make traditional humanities research and publication easier.

However, it is naïve to assume that traditional humanities practice will not in some way be transformed by any innovation in representational technologies—a lesson taught by the history of the book as easily as by a history of the World Wide Web. Clearly, current information technologies are changing how the humanities are practiced, but it is important not to fall into the trap of "technological determinism"—the idea that new technologies will *inevitably* lead toward specific changes, which is why it is important to understand broader social transformations taking place in the information age (a history carefully detailed by Alan Liu in *The Laws of Cool*). The role of the humanities in the university is changing, not simply because of the adoption of network technologies but also because information technologies working within rapidly evolving business environments have changed the way the much of the world works, and the university is currently being reorganized along corporate lines—a shift every bit as significant as the nineteenth-century reorganization. Put simply, networked computers are not in themselves changing the labor of humanities scholars, but the broader organization of the "information society" is putting considerable

pressure on the academy to follow those business practices, including downsizing, innovative content delivery, team building, total quality management and assessment, and an increased emphasis on a model of production as opposed to the traditional model of quiet contemplation.

Although many view such prospects with dismay, it is important to realize that technology and the humanities have never been completely stable terms. Their mutual interactions over time have redefined both their relationship with each other and, more importantly, their own identities. The rise of the information age has brought with it considerable re-evaluation of a broad range of traditional practices and introduced new discourses on the "death of the human," which would necessarily bring with it the death of the humanities and, for many, the death of the university. However, study in the history and culture of technology and the humanities shows how such intimations of mortality are misplaced. In *The Question Concerning Technology* Heidegger articulates the common philosophical roots of *techne* and *poiesis*—seeing both as a making, a bringing-forth, and a revealing. An expanded version of his lesson is that technology and the humanities share a common heritage, they are not easily disentangled, and they are constantly re-examining what counts as a made object and, at the same time, redefining the human who is doing the making.

*See also* **Humanities and Science**

## BIBLIOGRAPHY

Benjamin, Walter. "The Work of Art in the Age of Mechanical Reproduction." In *Illuminations,* Translated by Harry Zohn. Edited and with introduction by Hannah Arendt. New York: Schocken Books, 1968.

Eisenstein, Elizabeth L. *The Printing Press as an Agent of Change.* New York: Cambridge University Press, 1979.

Foucault, Michel. *Discipline and Punish: The Birth of the Prison.* Translated by Alan Sheridan. New York: Vintage, 1977.

Gramsci, Antonio. *Selections from the Prison Notebooks.* Translated by Quintin Hoare and Geoffrey Nowell Smith. New York: International Press, 1971.

Hayles, N. Katherine. *How We Became Posthuman: Virtual Bodies in Cybernetics, Literature, and Informatics*. Chicago: University of Chicago Press, 1999.

Heidegger, Martin. *The Question Concerning Technology, and Other Essays* (1954). Translated by William Lovitt. New York: Harper and Row, 1977.

Kittler, Friedrich A. *Discourse Networks 1800/1900*. Translated by Michael Metteer with Chris Cullens. Stanford, CA: Stanford University Press, 1990.

Kittler, Friedrich A. *Gramophone, Film, Typewriter*. Translated by Michael Wutz and Geoffrey Winthrop Young. Stanford, CA: Stanford University Press, 1997.

Latour, Bruno. *Science in Action*. Cambridge, MA: Harvard University Press, 1987.

Latour, Bruno. *Aramis, or, The Love of Technology*. Translated by Catherine Porter. Cambridge, MA: Harvard University Press, 1996.

Liu, Alan. *The Laws of Cool: Knowledge Work and the Culture of Information*. Chicago: University of Chicago Press, 2004.

Lyotard, Jean François. *The Postmodern Condition: A Report on Knowledge*. Translated by Geoff Bennington and Brian Massumi. Minneapolis: University of Minnesota Press, 1984.

Marcuse, Herbert. *One-Dimensional Man: Studies in the Ideology of Advanced Industrial Society*. Boston: Beacon Press, 1964.

Marx, Leo. *The Machine in the Garden*. New York: Oxford University Press, 1964.

McLuhan, Marshall. *Understanding Media: The Extensions of Man* (1964). Cambridge, MA: Harvard University Press, 1994.

Saussure, Ferdinand de. *Course in General Linguistics* (1916). Translated by Wade Baskin. New York: The Philosophical Library, 1959.

Shapin, Steven, and Simon Schaffer. *Leviathan and the Air-Pump: Hobbes, Boyle, and the Experimental Life*. Princeton, NJ: Princeton University Press, 1985.

Thompson, E. P. "Time, Work-Discipline, and Industrial Capitalism." *Past and Present* 38 (1967): 56–97.

*T. Hugh Crawford*

# I

## INDIGENOUS KNOWLEDGE.

What is evoked in people's minds when they have the opportunity to take a look, right in the field, at the magnificent pyramids in South, Central, and North America? What thoughts pass through Westerners' minds when they arrive as tourists to the architectural marvels of Machu Picchu, Peru; the magnificent Betatakin and Keet Seel cliff dwellings of the Hisatsinom (Anasazi), Navajo Nation; or the vast and visibly complex water irrigation and road systems of the Incas? What can be learned regarding the impressive raised-bed cultivation systems used for millennia by the Indigenous Peoples of the Americas as well as in Africa and Asia, which are reportedly highly effective for soil preservation, irrigation, drainage, and control of frost and plant diseases? Is there nothing to be learned from the Aztecs, who had 10,000 hectares of *chinanpas* capable of feeding 100,000 people?

Though Westerners are often overwhelmed at the sight of such exquisite monuments of the Indigenous Peoples, they sometimes wrongly conclude that only the ancient civilizations excelled, lightly dismissing at the same time any capability or merit on the part of their direct descendants. Such observers do not realize that Native Americans of today continue to practice agriculture, medicine, architecture, and other technologies essentially in the same creative way as their ancestors did. Sometimes Westerners have such serious doubts about the Native Americans' skills and ingenuity that they prefer to believe that super-intelligent extraterrestrial beings came down to earth and built those marvels. The above-mentioned achievements are clearly not the result of

magical forces but, rather, the result of the inquisitive minds of objective and pragmatic people, capable of accurately grasping the nature of things. Is this body of knowledge held by the Indigenous Peoples comparable in some ways to Western science?

Can the poorest, most illiterate, and most dispossessed people of the world generate knowledge, technology, and specialized skills good enough to solve human problems? Can an indigenous society from the Amazon, which the Western imagination views as naked and in need of everything, have an organized set of structured knowledge about nature and life that could be profitably passed on to future generations? Are they capable of generating something that could be equated to Western science? When we think of science and knowledge, we are used to expecting scientific achievements to come almost exclusively from Western, literate, wealthy nations.

In this paper we take the point of view of the Indigenous Peoples in order to answer these questions. This means an approach based on experiences, history, and pragmatism rather than unending analysis and theorization. Furthermore, we focus on the medical knowledge of the Indigenous Peoples of the Americas as one example of the achievements of the world's indigenous peoples.

### The Indigenous Peoples' Medical Contributions to the World

When we are nervously seated in the waiting room in the dentist's office, wondering about the pain and discomfort we might experience, it is amazing to realize, once in the dentist's examination room, that an injection of a small quantity of the local anesthetic *lidocaine* in our mouth can make the experience much easier! For centuries millions of people worldwide have suffered the debilitating symptoms of malaria, and many of them have experienced the curative virtues of *quinine*, recognized as essential by physicians all over the world. *Tubocurarine*, to take another example, is a strong muscle relaxant used widely in abdominal surgery. In our gratitude and admiration for these medical accomplishment, we may unthinkingly give all the credit to Western biomedicine and, regrettably, forget completely to pay proper thanks to the fathers of this knowledge, who generously allowed humanity to benefit from these medicines: the Native American healers or doctors. The following discussion details how they are the source of these formidable developments of Western biomedicine and how they have contributed to the improvement of health worldwide.

*Quinine.* Quinine is the most important alkaloid found in the bark of the *quinaquina*, as it is known in Quichua, a tree native to the Andes Mountains in South America. *Quinaquina* (*Cinchona* sp.), an evergreen tree 10 to 15 meters tall with glossy leaves and fragrant pink or yellow flowers, contains a mixture of more than twenty alkaloids, the most important of which are quinine, quinidine, cinchonidine, and cinchonine. Although quinine can be synthesized, the procedure is complex and costly. For this reason, quinine and the other alkaloids are still obtained entirely from natural sources. Of the thirty-eight species of *quinaquina*, four species are economically valuable for the production of quinine: *C. calisaya*, *C. ledgeriana*, *C. officinalis*, and *C. succirubra*. For nearly four centuries quinine has been used as a Western biomedicine compound to fight malaria, the world's most devastating human parasitic infection, caused by the protozoan *Plasmodium falciparum*. Although treatment modalities for malaria are diverse, quinine remains an essential drug. It is currently used to treat cloroquine- and multidrug-resistant malaria, which is highly prevalent in South America, Africa, and Southeast Asia. According to the World Health Organization, in 1994 malaria was afflicting more that 500 million people and causing 1.7 to 2.5 million deaths each year. The Indigenous People of the Andes knew the *quinaquina* and its main medicinal properties long before the arrival of Europeans to South America at the end of the fifteenth century.

The story of how Europeans came to know the tree and later integrated it in the Western pharmacopoeia is truly captivating. It is said that in 1630 an Inca doctor known in Spanish as Pedro Leyva, who lived in what is now Loja in the southern highlands of Ecuador, revealed the medicinal uses and properties of *quinaquina* to the Jesuit missionary Juan López or, in other versions, to the Spanish Governor of Loja, Juan López de Cañizares. (Some stories have Leyva himself discovering the effects of *quinaquina* accidentally. In fact, missionaries had noticed its use by Native Americans as early as 1572.) One of the most famous legends behind the introduction of *quinaquina* in Europe is the story of Doña Ana de Osorio, countess of Chinchón and wife of the viceroy of Peru. In 1638 the countess contracted malaria; the fevers were slowly killing her. The Jesuit López or Governor López gave her the *quinaquina* that Leyva had revealed to him, and the Inca medicine cured her promptly and completely. Impressed by the virtues of the plant, she collected it and gave it to others who eventually brought it to Europe. Other historians believe that the philosopher Cardinal Juan de Lugo, also of the Jesuit order, was responsible for promoting the tree bark in Europe. This is why *quinaquina*'s many names include "Jesuit's powder" and "Cardinal's bark." Finally, in 1737 the Swedish botanist Carl Linnaeus named the tree after the Chinchón countess, and the name *quinaquina* was eclipsed in Western lore in favor of *cinchona*.

As early as 1640 cinchona began to be used to treat fevers all over Europe. The popularity and consumption of the plant increased dramatically over the following years. The virtues of the plant were constantly reaffirmed. By the end of the seventeenth century, Robert Talbor, a British apothecary, successfully cured King Charles II of malaria. It is probably these accounts that rendered cinchona an officially recognized medicine in 1677, when it was included in the London Pharmacopoeia as *cortex peruanus* ("Peruvian bark"). Later, in 1820, the French scientists Pierre-Joseph Pelletier and Joseph-Bienaim Caventou isolated the main alkaloid of cinchona, which they named *quinine*.

The demand for cinchona rose so dramatically that, by the mid-nineteenth century, wild cinchona trees were being exploited to extinction. Charles Ledger, an early bioprospector, sent a few cinchona seeds to Europe. The Dutch became interested in the tree and initiated a plantation of cinchona trees in their colony on the island of Java in Indonesia. Subsequently, and for nearly a hundred years, they controlled nine-tenths of the world's quinine supply. No profits were shared with the Andean peoples, the intellectual owners of the splendid *quinaquina*.

During World War II, after the Dutch lost control of Java, the U.S. government immediately arranged mechanisms to obtain *quinaquina* bark from the Andes. Under a Quinine Agreement between Ecuador and the United States, the American Office of Economic Warfare supplied the administrative direction, funds, and technical resources for a massive enterprise to collect *quinaquina* bark. The Ecuadorian Cinchona Mission, a joint U.S.-Ecuador institution, was soon established with a staff of more that a hundred technicians and bureaucrats, and thousands of Quichua plants experts and laborers. It is estimated that nearly six million kilograms of dried bark were collected.

A recurrent theme in the foregoing stories and facts is the Westerners' unwillingness to recognize the contribution and intellectual ownership of the Quichuas of the Andes. Linnaeus, a botanist, thought it appropriate to identify the tree using the name of the domain of the Spanish countess, Chinchón. The Missouri Botanical Garden in St. Louis published, in 1930, a proceeding of the celebration of the three-hundredth anniversary of the "first recognized use" of cinchona! Apparently, Westerners were unwilling or unable to realize that the Indigenous Peoples of the Andes had long since recognized the tree and its most important medical uses. The European colonizers and scientists chose to deny the Native Americans' contribution and preferred to portray themselves as the discoverers of the great treatment for malaria. Had Native Americans not had the essential knowledge about *quinaquina* and shared this knowledge, however, Western biomedicine would never have identified the tree, associated it with a devastating disease, or isolated from it an active substance, quinine, for the treatment of malaria.

*Cocaine and Local Anesthetics.*   Local anesthetics such as the lidocaine mentioned at the beginning of this section are due to another significant medical contribution from the Indigenous Peoples of South America to the world. The coca shrub, *Erythroxylon coca*, is a small tree native to the mountainous tropical regions of the Andes. For centuries, if not millennia, the chewing of coca leaves has occupied a central role in the lives of the Indigenous Peoples of northwestern South America. There is abundant material evidence of coca use as far back as 3000 B.C. Many figurines of coca chewers have been found in the coastal region of Ecuador, where the Valdivia people lived. The Carchi people of the northern highlands of Ecuador left many chewers' ceramics, including ash containers today called *iscupurus*. Mummies almost 2000 years old from the Nazca region of Peru customarily carry bags of coca leaves, now called *chuspas*, around their necks. Coca was so present and important that a wonderful Inca legend tells us that *Cocamama* (Mother Coca) was a beautiful goddess full of grace, health, and joy. All is not charm, and Cocamama, to the distress of many, was a promiscuous woman who was cut in half by her many lovers. Having no other means of survival, she managed to grow into the first coca plant. Coca leaves were consumed daily by the Indigenous Peoples of the Andes as both a food and a medicine. Coca was used to gain physical strength and endurance, to decrease fatigue, and to control pains such as toothaches and headaches. With the arrival of the Spaniards to South America, the coca plant entered the scene of the Western pharmacopoeia.

In the early years of contact, the Spaniards dismissed the claims made by the locals about the coca leaves. Soon enough, the Catholic clergy and colonial officials began to consider the practice of chewing coca leaves inappropriate. In 1551 the Bishop of Cuzco banned coca use, under pain of death by burning alive, believing that it was an evil agent and a distraction to the Church's evangelization work. Eighteen years later, however, King Philip II of Spain, wishing to protect the life and productivity of the enslaved Indigenous laborers, decreed coca not evil. As often occurred with indigenous natural resources in colonial times, coca then became one of the most important sources of revenue for the Catholic Church and the colonizers. Ironically, in the following years, the church distributed coca to their enslaved Quichuas in order to improve productivity in mines and crop fields. Jesuits are claimed to have said, "The Devil's coca will now be used to assist God's work."

By the mid-nineteenth century, coca was intensely studied and became an extremely popular product in Europe. The young German scientist Albert Niemann at Göttingen University first isolated cocaine in 1860. Then Karl Koller, a Vienna ophthalmologist, found cocaine to be very useful as an eye, nose, and throat anesthetic. By 1884 cocaine was quickly integrated and used widely by the medical community as a local anesthetic. Coca may have been introduced to Dr. Koller by Sigmund Freud, who studied coca enthusiastically and was a great advocate of its medicinal properties. He believed that coca could be used for depression and other mental conditions, and he used it to relieve its own social anxiety. Later, coca was present everywhere. The French druggist Angelo Mariani's mixture of wine and coca, called Mariani's Coca or Dr. Mariani's French Tonic, became a symbol of a miracle drink and medicine all over Europe and the United States. Even Pope Leo XIII sponsored Mariani's coca wine! It was during this wave of coca enthusiasm that in 1886 John S. Pemberton

invented Coca-Cola, which contained an extract of coca leaves (it still does, but the cocaine has been removed). By the end of the nineteenth century the U.S. pharmaceutical company Parke-Davis marketed cocaine as cigarettes, powder, a drink called Coca Cordial, and even a cocaine mixture that could be injected intravenously; the needle was included. Regrettably a history of misuse and abuse of coca followed, and cocaine addiction put a restraint to all the past fervor.

At the beginning of the twentieth century, concerns were raised about the toxicity and addictive properties of cocaine. In 1905, the work of Alfred Einhorn, who was searching for an alternative drug to cocaine, resulted in the synthesis of procaine, a prototypical local anesthetic. Procaine has the advantage of being less toxic than cocaine and nonaddictive. Since then, other closely related compounds inspired by the structure of cocaine have been synthesized. Among the most used in surgery and dentistry are procaine, lidocaine, bupivacaine, and tetracaine.

This collision of two worlds, the Western and that of the Indigenous Peoples of the Americas, exemplifies how traditional medical knowledge appropriation occurs. The persistent preference for recognizing a Westerner as a "discoverer" for simply bringing a plant to the Western public arena reflects the colonialist view that Indigenous people are nearly nonhuman, incapable of being owners or generators of knowledge, not entitled to respect for it. Even such a prestigious and sober textbook as Goodman and Gilman's *The Pharmacological Basis of Therapeutics* affirms that the anesthetic property of coca was "serendipitously discovered" in the late nineteenth century. It cannot be overemphasized that the Quichuas of the Andes had already been using coca for centuries to control pain!

*Curare.* In contrast to quinine and coca, curare did not attract the attention of the European masses. It remained a mysterious plant known only by medical experts. Curare is a poisonous resin used by the Indigenous Peoples of the Amazon River basin to hunt.

The preparation of curare is in the expertise of knowledgeable healers and hunters. The curare solution is coated onto small hunting arrows, which are shot with a blowgun to effectively paralyze animals. The most used species are *Chondrodendron tomentosum* and *Strychnos toxifera*. The poisonous nature of curare is due to the biological actions of its alkaloids: tubocurarine and the toxiferines. Curare acts as a neuromuscular junction—blocking agent, producing paralysis of skeletal muscles. The toxin causes death by paralysis of respiratory muscles.

Walter Raleigh is believed to have taken samples of curare to Europe in the late sixteenth century. Without the Native American source, he could never have noticed or come into contact with curare, made from a few of the thousands of plant species of the Amazon. Later, Charles Marie de la Condamine introduced curare to the French Academy, and Alexander von Humboldt, in 1805, stimulated the botanical investigation of the plant. The interest in curare re-emerged when doctors successfully used purified curare to treat patients with tetanus and spastic neurological disorders. When Richard Gill brought curare to the United States in 1940, the true clinical and surgical applications began. In 1942, Harold Griffith and Enid Johnson from McGill University, in Montreal, Canada, used curare, which until then had been considered as only a poison, to produce muscle relaxation. Tubocurarine is used today as a complement to general anesthetics in abdominal surgery. The severe muscle relaxation it produces, which would otherwise require dangerously large amounts of anesthetics, is beneficial in the operating room because it enables the surgeon to access and manipulate the internal organs. The use of smaller quantities of general anesthetics renders surgical operations safer, decreases anesthetic-related complications during surgery, and promotes quicker postanesthesia recovery. Without the original contribution of the Indigenous People of the Amazon, a substance some have dismissed as "a poison prepared by the savages of South America," Western biomedicine could hardly have achieved this breakthrough in surgery.

## From Colonialism to Equitable Partnership?

The contribution of the Indigenous Peoples of the Americas to the world has been vast. It has touched almost every single aspect of modern life, including food, land, medicine, architecture, mathematics, federative politics, education, and ecology, to name a few. *Quinaquina*, coca, and curare are only a small fraction of what Native Americans know and have shared. According to John Borchardt, the hundred Native American nations have altogether contributed 220 indigenous drugs to the pharmacopoeia of the United States of America. In a wonderful book edited by Nestor Foster and Linda Cordell, *Chilies to Chocolate: Food the Americas Gave the World*, we can find 128 species of plants and crops that the Indigenous Peoples of the Americas shared with the world, including maize, potatoes, tomatoes, cacao (chocolate), chili peppers, avocado, beans, rubber, peanut, pecan, and vanilla. Undoubtedly, the indigenous knowledge and science has been practical and extremely useful for humanity.

Because the Indigenous Peoples are dominated and dispossessed by stronger nations in America, and colonialist practices against them persist today, rarely do they benefit materially from the use of their knowledge. They are unable to exercise cultural ownership or independently defend their intellectual properties and heritage. Despite all the abuses of the past and present, however, a period of recognition of the Indigenous Peoples, of their culture, territorial and human rights, and of their valuable science appears to be beginning.

Indigenous Peoples themselves are strongly voicing their concerns about the appropriation and use of traditional knowledge. The Indigenous Peoples of the Americas, Africa, Asia, Australia, Europe, and the Pacific, united at the World Conference of Indigenous Peoples on Territory, Environment, and Development (May 25–30, 1992), endorsed the Kari-Oca Declaration and the

Indigenous Peoples Earth Charter, which states:

> our collective responsibility to carry our indigenous minds and voices into the future, that traditional knowledge of herbs and plants must be protected and passed onto future generations, and the usurping of traditional medicines and knowledge from indigenous peoples should be considered a crime against peoples.

Governments and international organizations are responding, sporadically, in positive terms. It is extremely encouraging to see that the Government of Peru passed Bill 27811 in August 2002 to protect Indigenous Peoples' rights and ownership over their ancestral knowledge. According to the bill, companies seeking to use Peruvian Indigenous knowledge must allocate no less than 10 percent of their profits to the Indigenous People's Development Fund of Peru. Panama legislature also passed Law 20 on June 26, 2000, addressing the same issue. Another Latin American country, Brazil, did the same with the enactment of Presidential Provisional Measure 2.186-16 in August 2001. In New Zealand this trend is in advanced stages of implementation concerning Maori knowledge; changes have been approved to trademarks legislation and to the Patents Act. Another exceptional development occurred with the approval by the South African parliament of the Traditional Health Practitioners Act in August 2004 to regulate the work of traditional healers. This national initiative was expected to affect an estimated 200,000 traditional healers, and the interest and enthusiasm was expected to continue for the next several years. The World Intellectual Property Organization (WIPO), through its Intergovernmental Committee on Intellectual Property and Genetic Resources, Traditional Knowledge and Folklore, proposed mechanisms to promote the recognition of traditional knowledge within the patent system and "reduce the practical likelihood that patents will be allowed that incorrectly claim inventions that make use of traditional knowledge and genetic resources" (WIPO).

The new appreciation of the Indigenous Peoples' culture and knowledge is reaching the Western scientific community as well. In the summer of 1999 the International Council for Science (ICSU) and the United Nations Educational, Scientific, and Cultural Organization (UNESCO) adopted the Declaration on Science and the Use of Scientific Knowledge. The declaration acknowledges that traditional and local knowledge systems are "dynamic expressions of perceiving and understanding the world that can make, and historically have made, a valuable contribution to science and technology" (UNESCO).

In America it took more than five centuries to go from the belief that Indigenous Peoples were savages and soul-less creatures, to the acknowledgment that they are peoples or nations with rights to be respected. Fortunately, we are at the dawn of a process of recognition and appreciation. The Indigenous Peoples' knowledge protection is being legislated, and early considerations on benefits sharing are proposed. Worldwide respect, protection, and recognition of the Indigenous Peoples' knowledge and rights could result in a more just world, the end of an era of colonialism, and a more equitable global human society.

See also **Grassroots Science; Science in History: Latin America; Technology in History: Latin America**

BIBLIOGRAPHY

Borchardt, John K. "Native American Drug Therapy: United States and Canada." *Drug News Perspectives* 3 (2003): 187–191.

Brown, Michael F. *Who Owns Native Culture?* Cambridge, MA: Harvard University Press, 2004. Up-to-date information and balanced views about what is called "cultural theft" of the Indigenous Peoples' heritage, including arts, traditional medicine, and sacred ceremonies. Current issues on cultural ownership, group privacy, intellectual property rights are presented.

Foster, Nelson, and Linda S. Cordell. *Chilies to Chocolate: Food the Americas Gave the World.*

Tucson: University of Arizona Press, 1992. An excellent source of information about the contributions of the Indigenous Peoples of the Americas. One of the very few books that keeps a respectful view and unbiased appreciation of the valuable Indigenous knowledge.

Harding, Sandra G. *Is Science Multicultural? Postcolonialisms, Feminisms, and Epistemologies.* Bloomington: Indiana University Press,1998. Excellent perspective on culturally dependent production of knowledge and science. Harding's discussion on culture as both toolbox and prison for science and technology is delightful.

McIntyre, Archibald Ross. *Curare: Its History, Nature, and Clinical Use.* Chicago: University of Chicago Press, 1947. A must for readers interested in the historical and clinical aspects of curare.

Ortiz-Crespo, Fernando. *Cinchona: Source of Quinine.* http://www.cuencanet.com/ortiz/cinchona.htm. Contains a book-length treatise (in Spanish, published posthumously) on the history of the use of *quinaquina* since the Spanish conquest, as well as excerpts in English, including translations of the earliest missionaries' accounts.

Perrine, Daniel M. *The Chemistry of Mind-Altering Drugs: History, Pharmacology, and Cultural Context.* Washington, DC: American Chemical Society, 1996. A very interesting historical account of psychotropic drugs along with sensitive cultural considerations. Succinct information on cocaine is provided.

Posey, Darrell Addison, and Kristiana Plenderleith. *Indigenous Knowledge and Ethics: a Darrell Posey Reader.* New York: Routledge, 2004.

Rainey, F. 1946. "Quinine Hunters in Ecuador." *National Geographic Magazine* (1946): 341–363.

UNESCO. "Declaration on Science and the Use of Scientific Knowledge." In *Science for the Twenty-first Century. World Conference on Science, Budapest, Hungary, 26 June–1st July 1999.* Paris: UNESCO, 1999. http://www.unesco.org/science/wcs/eng/declaration_e.htm.

Varey, Simon, Rafael Chabran, and Dora B. Weiner. *Searching for the Secrets of Nature: The Life and Works of Dr. Francisco Hernández.* Stanford, CA: Stanford University Press, 2001. The Nahua medical and botanical knowledge are described in a very respectful manner by Francisco Hernández, a Spanish physician sent by King Philip II to Central America to conduct a botanical exploration.

WIPO. *Recognition of Traditional Knowledge within the Patent System.* Geneva: World Intellectual Property Organization, 2004. http://www.wipo.int/edocs/mdocs/tk/en/wipo_grtkf_ic_7/wipo_grtkf_ic_7_8.pdf.

Mario Incayawar (Maldonado)

# INFORMATION TECHNOLOGY.

We are currently witnessing the birth of a new information infrastructure as powerful in its own right as the printing press. This new medium is helping us express knowledge in new ways and connect with each other across time and space in ways previously unimaginable, and it is having a huge social, economic, and political impact. Rather than in the proclaimed age of the Internet, however, we are continuing to live, as we have been living since the early nineteenth century, in the age of the database: Our ability to use information technology to marshal and analyze data has continued to rise in areas of government, science, and the economy.

## Building an Information Infrastructure

There are many models for information infrastructures. The Internet itself can be cut up conceptually a number of different ways. There is over time and between models a distribution of properties between hardware, software, and people. Thus, one can get two computers "talking" to each other by running a cable between them or by dedicating a given physical circuit (hardware solutions) or by creating a "virtual circuit" (software solution) that runs over multiple different physical circuits, with each small piece of a message possibly taking a different route. One can also (and this is still the fastest way of getting terabits of data between two cities) put a disk on a truck and drive it over. Each kind of circuit is made up of a different stable configuration of wires, bits, and people; but they are all (as far as the infrastructure itself is concerned) interchangeable.

Infrastructural development and maintenance requires work, a relatively stable technology, and communication. The work side is frequently overlooked. Consider the claim in the 1920s that with the advent of microfiche, the end of the book was nigh. Everyone would have a personal library; we would no longer need to waste vast amounts of natural resources on producing paper; all the largest library would need would be a few rooms and a set of microfiche readers. It was a possible vision, and one should not dismiss it out of hand just because it did not happen. Although anyone who has used a microfiche reader will attest that it is a most uncomfortable experience, it could well have been made considerably more comfortable if the same resources had gone into the failure as into the successful technology. The microfiche dream, however, like the universal digital library, runs up against the problem that someone has to sit there and do the necessary photography/scanning, and that takes a huge amount of time and resources. It is easy enough to develop a potentially revolutionary technology; it is much harder to implement it.

Further, one needs a relatively stable technology. If one thinks of some of the great infrastructural technologies that have been introduced (gas, electric, sewer, and so forth), one can see that once the infrastructure is put into place, it tends to have a long life. Electrical wiring from before World War II is still in use in many households; the sewer lines in major cities have been in the ground so long that in many cases there is no good map of them. The Internet is stable virtually rather than physically, through the mediation of a set of relatively stable protocols that enable it, as a whole, to survive any changes in the individual networks that it connects (for this reason, it can be called an internetwork technology rather than a network technology). However, there is nothing to guarantee the stability of vast datasets. At the turn of the twentieth century, Paul Otlet developed a scheme for a universal library that would work by providing automatic electromechanical access to extremely well-catalogued microfiches, which could link to each other. All the world's knowledge would be put onto these fiches. Otlet's vision presaged today's hypertext and World Wide Web. He made huge strides in developing this system (though he had the person power to achieve only a minuscule fraction of his goal). Yet, within forty years, with the development of computer memory, his whole enterprise was effectively doomed to languish as it does today in boxes in a basement—why retrieve information electromechanically using levers and gears when you can call it up at the speed of light from a computer? Much the same can be said of Vannevar Bush's never-realized but inspirational vision of the Memex, an electromechanical precursor to the computer workstation. Large databases from the early days of the computer revolution are now completely lost. The punched card, a technology whose first major use in the United States was to deal with the massive datasets of the 1890 census and that dominated information storage and handling for some seventy years, is hardly ever encountered today (except during elections). Closer to home, the "electronic medical record" has been announced every few years since the 1960s, and yet it has not been globally attained. Changes in database architecture, storage capabilities of computers, and ingrained organizational practices have rendered it a chimera. The development of stable standards, together with due attention being paid to backward compatibility, provides an fix to these problems—in principle. It can all unravel very easily, though. The bottom line is that no storage medium is permanent. Compact discs, for example, will not last anywhere near as long as books printed on acid-free paper. Our emergent information infrastructure, therefore, will require a continued maintenance effort to keep text accessible and usable as it passes from one storage medium to another and is analyzed by one generation of database technology for transmittal to the next.

## IT and Politics

The "open source" movement is archetypal of the new information infrastructure. This movement has a long history as IT trends go, reaching back almost to the origin of the Internet, proclaiming the democratic and liberatory value of freely sharing software code.

The Internet, indeed, was strung together on the basis of a set of software standards that, at least among the researchers engaged in the project, were freely distributed. The open source movement has been seen as running counter to the dominance of large centralized industries; the argument goes that it puts power over the media back into the hands of the people in a way that might truly transform capitalist society. This promise of cyberdemocracy is integrally social, political, and technical. While there is much talk of an "information revolution," there is not enough talk of the ways in which people and communities are constituted by the new infrastructure.

There has been much hope expressed that the new information infrastructure will provide the potential for a narrowing of the knowledge gap between the developed and the developing world. Thus, an effective global digital library would allow Third-World researchers access to the latest journals. Distributed computing environments (such as the GRID, being developed in the United States) would permit supercomputer-grade access to computing to scientists throughout the world. The example of the use of cell phone technology to provide a jump in technology in countries without land phone lines has opened the possibility of great leaps being made into the information future. As powerful as these visions are, they need to be tempered with some real concerns. The first is that although everyone who *has* access to an information infrastructure like the Internet may be an equal citizen as in ancient Athens, those without access are left further and further out of the picture (even more so than the slaves and women of Athens). Further, access is never really equal; the fastest connections and computers that can run the latest software tend to be concentrated in the First World. This point is frequently forgotten by those who hail the end of the "digital divide"; they forget that this divide is in itself a moving target. Third, governments in the developing world have indicated real doubts about the usefulness of opening their data resources out onto the Internet. Just as in the nineteenth century, free

trade was advocated by developed countries with most to gain because they had organizations in place ready to take advantage of emerging possibilities. so in our age the greatest advocates of the free and open exchange of information are developed countries with robust computing infrastructures. Some in developing countries see this as a second wave of colonialism: The first wave pillaged material resources, and the second will pillage information. All of these concerns can be met through the development of careful information policies. There is a continuing urgent need to develop such policies.

## International Technoscience and Information Technology

International electronic communication holds out the apparent promise of breaking down a First World/Third World divide in science. Remote manipulation of scientific equipment, as in the University of Michigan's Upper Atmospheric Research Collaboratory (UARC) project, enables scientists on the Internet to manipulate devices in places like the Arctic Circle without having to go there. The possibility of attending international conferences virtually is also being held out. And if universities succeed in wresting control over scientific publications from the huge publishing houses (a very open question), then access to the latest scientific articles may become possible and affordable for a researcher in the Australian outback. At the same time, there are forces working to reinforce the traditional center/periphery divide in science internationally. Even with the move to open up access to scientific publications and equipment, there is no guarantee that the "invisible colleges"—the relationship structures that operate informally to determine who gets invited to which conference and so forth—will change; indeed, the evidence seems to be to the contrary. Further, at the current state of technological development there is a signficant gap in information access between different regions of any given country or between different parts of the world. Consider the analogy of the telephone. In principle, anyone can phone anywhere in the world; in practice, some regions have more or less reliable phone services, which may or may not include access to digital resources over phone lines.

We can go beyond the continuing digital divide, however, to consider the possibility of mounting very large-scale scientific data collection efforts. Such efforts are central to the social sciences and to the sciences of ecology and biodiversity. With the development of handheld computing devices, it is becoming possible for a semiskilled scientific worker with a minimum of training to go into the field and bring back significant results. Thus in Costa Rica, the ongoing attempt to catalog botanical species richness is being carried out largely by "parataxonomists" who are provided with enough skills in using interactive keys (which help in plant recognition) to carry out their work almost as effectively as a fully trained systematist. Computer-assisted workers, together with the deployment of remote sensing devices whose inputs can be processed automatically, hold out the possibility of scaling up the processes of scientific research so that they are truly global in scale and scope.

## Distributed Collective Work

Collaborative work is central to the new knowledge economy. Traditionally, scientific breakthroughs have been associated with particular laboratories such as the Cavendish Laboratory in Cambridge, England. Such laboratories have always been a site for collaboration such as the exchange of ideas with visiting scholars, the holding of conferences, and the training of graduate students. However, it is impossible nowadays to imagine managing a large-scale scientific project without including far-flung collaborators, particularly if one is seeking to develop a truly global scientific culture.

Although technoscientific work is inherently collaborative, management structures in universities and industry still tend to support the heroic myth of the individual researcher. Many scientists turn away from collaborative, interdisciplinary work—precisely the kind of work that is most needed in order to develop policies for sustainable life—

because they are risking their careers if they publish outside of their own field. There is significant institutional inertia, whereby an old model of science is being applied to a brave new world.

This is coming out most clearly in the area of scientific publications. First in the field of physics and then in a number of other scientific disciplines, we are witnessing the spread of electronic preprints and electronic journals. Traditional academic journals run by huge publishing conglomerates cannot turn around papers quickly enough to meet the needs of scientists working in cutting-edge fields. Throughout the past two centuries, there has been a relatively stable configuration whereby journal articles have become the central medium for the dissemination and exchange of scientific ideas. Now there is no reason in principle why a scientist should not publish his or her findings directly to the Web. As in many sectors of the new information economy, the development of the new publication medium is leading to a reconsideration of just what kind of value the large publishing houses add to journal production. As more and more journals go online, they are being forced to go beyond their tradtional service of providing distribution networks and find ways of bringing their material onto the Web. It seems likely that all major scientific journals will soon be accessible on the Web, even though the economics of such distribution is not yet fully worked out (possibilities include paying for each paper downloaded, or buying an institutional subscription for a whole journal). A more far-reaching implication is that the journal article may no longer function as the unit of currency within the research community. Very large-scale collaborative databases, for example, such as the human genome databank, are a new kind of product that is made possible by the development of the Web.

The policy implications are clear. Great attention must be paid to the social and organizational setting of technoscientific work in order to take full advantage of the possibilities for faster research and publication cycles. There is a well-known paradox about the development of computing, known as the productivity paradox, according to which the introduction of computers into a workplace tends to lead to a lowering of productivity in the short term (about 20 to 20 years). Paul David and others have argued that what is happening here is that we are still using the old ways of working and trying to adapt them to the production of electronic text. A new academic field, social informatics, has grown up precisely with the goal of exploring social and organizational aspects of the new knowledge economy.

## Choices and Consequences

The choices that we are making now about the new information infrastructure are *irreversible*. The infrastructure is *performative* (in that it shapes the forms that technoscience will take) and it is *diffuse* (there is no central control). There is currently widespread belief in technical fixes for inherently social, organizational, and philosophical problems, such as curing the ills of incompatible datasets through developing metadata standards. Further, there is a disjunction between the policy and the informatics discourses. The emerging information infrastructure is extremely powerful; it is hoped that with good bricolage we can make it just and effective as well.

*See also* **Computers; Internet, The; Technology and Society**

### BIBLIOGRAPHY

Bowker, Geoffrey C. and Susan Leigh Star. *Sorting Things Out: Classification and Its Consequences.* Cambridge, MA: MIT Press, 1999.

Brown, John Seely, and Paul Duguid. *The Social Life of Information.* Boston: Harvard Business School Press, 2000.

David, Paul A. *Computer and Dynamo: The Modern Productivity Paradox in a Not-Too-Distant Mirror.* Center for Economic Policy Research No. 172. Stanford, CA: Stanford University, 1989.

Edwards, Paul N. *The Closed World: Computers and the Politics of Discourse in Cold War America.* Cambridge, MA: MIT Press, 1996.

Indiana University School of Library and Information Sciences. *Social Informatics Home Page.* 2003. http://www.slis.indiana.edu/SI.

Star, Susan Leigh, and Karen Ruhleder. "Steps toward an Ecology of Infrastructure: Design and Access for Large Information Spaces." *Information Systems Research* 7.1 (1996): 111–134.

Suchman, Lucy A. *Plans and Situated Actions: The Problem of Human-Machine Communication.* Cambridge, U.K.: Cambridge University Press, 1987.

Yates, JoAnne. *Control through Communication: The Rise of System in American Management.* Baltimore, Johns Hopkins University Press, 1989.

*Geoffrey C. Bowker*

# INSTITUTIONAL RELATIONS IN SCIENCE.

Like the family, law, and religion, science is a social institution. Social institutions create the fabric of any society, providing a persistent set of practices that solve problems and make life predictable.

Science is a special institution because it offers a creative way of knowing, and therefore interacting with the world, and because it is performed by highly trained personnel who govern the standards for entry, quality, and production of scientific artifacts that are reflected in scholarly journals, books, patents, and expert opinion. The institutional relations of science turn on how its practitioners interact with institutions that are *outside* of science but that are also significant *within* the society.

Contemporary society is largely a taken-for-granted product of twentieth-century scientific ingenuity. (Distinguishing "science" from "technology" is another matter, which will not be explored here.) In percentage terms, almost all scientists who ever practiced—in universities, corporate laboratories, or government facilities—are alive today (NSB, 2000). They continue to shape the social institution of science and its relations, increasingly complex and controversial, with other institutions that serve society's need to be informed, orderly, adaptive, and productive. Just as science has solved problems through the application of knowledge, so has it created new problems.

## Science as Problem-Solving and -Creating

At the interstices of institutions, science's dual role as problem-solver and creator is readily seen. We are all "stakeholders" in the issues that stretch the social fabric by challenging what should be done based on what we know. Today, what we know is derived from, or validated by, science. However, fundamental questions of when does life begin and when is one "dead" rely on clinical and biological interpretations that clash with religious belief, legal precedent, and free will. The stuff of scientific knowledge—fact, theory, experiment, cause, and probability— does not translate well in a world ruled by other methodologies that must determine guilt, liability, and policy. Science's "objectivity" becomes mere "value judgment" once the scientist leaves his or her institutional nest to confront other experts and ordinary citizens guided by *other* values, methods, objectives, and expectations.

The potential for conflict through a confrontation of values is vast in a democracy. Not only do citizens hold contrary views, but scientists may also disagree with one another owing to their knowledge of a highly specialized problem. This ensures institutional relations that are often, not surprisingly, tense. The point of contention is not just a matter of what is known, but how should that knowledge be acted upon. Who is likely to benefit and who may be put at risk?

To answer such questions, science is increasingly drawn into conversations with two of the more powerful social institutions—government and the media. Owing to its long history of support for scientific research and personnel, the federal government is a key actor in the allocation of public funding for enhancing the public good. One need only consider the transformational character of telecommunications in the last 15 years, creating e-commerce booms and busts; shrinking the world; and altering daily routines at work, home, and play (NSB, 2000).

Likewise, the role of the media for the two generations since Watergate can hardly be underestimated. Where once print dominated, there is now nonstop, global TV and Internet coverage, replete with a continuum of viewpoints. Whether unfiltered by eyewitness immediacy, "spun" by pundits with declared or undeclared political affiliations, or more cautiously analytical as the passage of time may allow, science as news competes with other institutional domains for coverage. This is both good news and bad.

The attention, shrouded perhaps in suspicion and skepticism (Goodell, 1985), helps to demystify what science is and what scientists do. While humanizing the profession, such attention can also deflate and expose the dark side of science—practitioners with diverse motives and human frailties. In short, what is "front page news" (Chubin, 1993) may capture a reality but not a statistically representative one.

And then there are the "gray areas," like the issue of stem cell lines. In 2001, a U.S. presidential advisory panel set limits on stem cell research because it destroys human embryos, which some regard as human lives, in the process. Even the director of the National Institutes of Health, himself a presidential appointee, has called for a lifting of the restriction. Recently, a medical ethics panel formed by Johns Hopkins University and composed of life scientists, philosophers, ethicists, and lawyers from the United States. and Europe has found that the 78 human embryonic stem cell lines now eligible for use in federally financed research will not serve all groups and are not safe for transplantation (Brainard, 2003).

## Cultural Lag: Trust Versus Understanding

The transition from research to human trials to human therapy illustrates what sociologists call "cultural lag." Scientific capability has outpaced society's capacity to regulate some of the very practices that can help to combat disease. Clinical trials of therapies based on stem cells may be only a few years away. What will determine their use goes far beyond what scientists can do in the laboratory. Such "forbidden knowledge" is endemic to any society that supports probing the unknown by highly specialized researchers who can earn a handsome livelihood doing so. Clearly, other institutions and perspectives, the *context* as well as the *content*, matter.

All social institutions—politics, media, business, law—have suffered from a diminution of public trust. Science has fared better than most. Yet the citizenry is largely science illiterate—consistently trusting (according to more than 20 years of public opinion polling), but not comprehending (NSB, 2002). That is a fragile state not readily changed by "popularizers" of science (the most illustrious of whom are deceased), such as Margaret Mead, Carl Sagan, Lewis Thomas, and Stephen Jay Gould, who illuminate institutional values and rituals while bringing public visibility and celebrity (and sometimes, peer jealousy) to their particular scientific passions. This approach differs significantly from the public images of science conveyed through most of the twentieth century (LaFollette, 1990).

Entering the twenty-first century, Neal Lane (1999), former director of the National Science Foundation and then presidential science advisor, spoke about the need for "civic scientists" committed to advancing appreciation for science among citizens in communities, the parents of children with the responsibility to vote. By listening as well as speaking with such audiences, scientists "tithe" to society—and nurture relations between science and other institutions.

## BIBLIOGRAPHY

Brainard, Jeffrey. "Low Number of Federally Approved Stem-Cell Lines Raises Ethical and Safety Concerns, Panel Says." *The Chronicle of Higher Education*, November 11, 2003. http://chronicle.com/daily/2003/11/2003111101n.htm.

Chubin, Daryl E. "Front-page Science: Positive Effects of Negative Images?" *BioScience* 43 (1993): 334–336.

Goodell, Rae. "Problems with the Press: Who's Responsible?" *BioScience* 35 (1985): 151–157.

LaFollette, Marcel. *Making Science Our Own: Public Images of Science 1920–1955*. Chicago, IL: University of Chicago Press, 1990.

Lane, Neal. "The Civic Scientist and Science Policy." In *AAAS Science and Technology Yearbook 1999*, edited by A.H. Teich, 237–242.

Washington, DC: American Association for the Advancement of Science, 1999.

National Science Board. "Science and Technology Policy: Past and Prologue. A Companion to *Science and Engineering Indicators: 2000*." Arlington, VA: National Science Foundation, 2000.

National Science Board. *Science and Engineering Indicators*: 2002. Arlington, VA: National Science Foundation, 2002.

*Daryl E. Chubin*

# INTERNET, THE.

Determining the impact that Internet technologies have on society is a daunting task. Years may pass before a technology finds its best use; for instance, it took more than twenty years before radio transformed entertainment by ushering in the broadcast era; radio was initially seen as a navigational tool for ships. And new technologies are sometimes considered threats to the dominant values of the time. From Carolyn Marvin's history of electricity, we learn that the telephone was seen as a threat to the bourgeois family. Because of it, contacts between the family and those outside it were suddenly much more difficult to supervise and "irregular courtships" threatened the control of parents over their children. "Somewhere between the expansive intentions of entrepreneurs," writes Marvin, "and the practiced exclusivity of familiar social codes, the telephone and other new media introduced a permeable boundary at the vital center of class and family, where innovative experiments could take place in all social relations, from crime to courtship" (p. 108).

In addition to being misjudged or feared, new technologies have also bred great optimism. Peter Daly reminds us that "as the first technology to link distant peoples, the telegraph was hailed as the silver bullet panacea of the Victorian era" (p. 23). Subsequent technologies all came laden with the hope that they would prove the solution to society's most difficult conflicts, and early cyber enthusiasts were no different. They tended to see the Internet as ushering in a new era of intense freedom in which the need for government would vanish and a new self-governing age would arise independent of the status quo power structure. One of the earliest proponents of this idea was Howard Rheingold, who believed that the Internet had great political power because it allowed people to ". . . challenge the existing political hierarchy's monopoly on powerful communications media, and perhaps thus revitalize citizen-based democracy" (p. 14). Nicholas Negroponte, another early and well-known writer on the impact of the Internet, predicted, "Without question the role of the nation-state will change dramatically and there will be no more room for nationalism than there is for smallpox" (p. 238). Lawrence Lessig describes the attitudes of his students circa 1995 with regard to the Internet: ". . . the students seemed drunk with what James Boyle would later call the 'libertarian gotcha'; no government could survive without the Internet's riches, yet no government could control what went on there. . . . Cyberspace, the story went, could *only* be free. Freedom was its nature" (p. 5) .

In a very important book on this topic, Pippa Norris (2001) summed up the range of opinion holders as follows: cyber optimists, cyber pessimists, and cyber skeptics. Cyber optimists see the Internet as allowing for greater citizen mobilization and empowerment because of the ease and comparatively low cost of organizing on the net. To the optimists, the Internet will increase democracy not just in those countries that are already democratic but it will become a tool for organizing globally around issues such as human rights and international poverty. Kevin A. Hill and John E. Hughes sum up the enthusiasm of the optimists as follows: "for utopian visionaries, the promise of nearly unlimited information delivered to your monitor in mere moments is the promise of a better democracy" (p. 2). To the cyber pessimists, the digital divide—the fact that most of America's poor and most of the world's poor lack access to the Internet—threatens to exacerbate the inequalities already present within the United States and around the world. And finally, cyber skeptics believe that technologies tend to adapt to so-

ciety and not vice versa. Michael Margolis and David Resnick (2000) sum up the skeptics' view as follows: "cyberspace is neither a mass breeding ground for liberated virtual communitarians nor a launching pad for electronic storm troopers bent on stamping out free expression and dissent" (p. 2).

So now that the Internet is in its "teenage" years, who is right? Cyber optimists can point to the surprising and successful campaign to ban landmines, which was organized largely via the Internet from a kitchen in Vermont; to the successful mobilization of antiglobalization protesters at rallies around the world; to the role of the Internet in fostering dissent in a closed society, such as China's; and to the successful campaign of Governor Jesse Ventura of Minnesota, who ran the first Internet campaign as a political independent and beat the two major political parties. Also, in spite of the fact that he lost the race for the Democratic presidential nomination in 2004, Governor Howard Dean's Internet campaign has thoroughly transformed how political campaigns are run. By allowing millions of people to make small contributions, thereby breaking the monopoly of "big money" over political candidates, the Internet may become a force for equality in democracy.

Cyber pessimists can point to the continuing existence of a digital divide in the United States and around the world to buttress their belief that the Internet is simply reinforcing rampant inequality. Although the number of low-income Americans online nearly doubled between 1998 and 2000, according to the *Digital Economy Fact Book*, the number of Americans online whose incomes are under $15,000 per year is only 12 percent, whereas the number of Americans online with incomes over $75,000 per year is 77 percent (p. 13). The digital divide is even more dramatic where the absence of such factors as basic literacy, electricity, and telephone lines means that billions of people around the globe are largely unaffected by the Information Revolution.

And cyber skeptics might well argue that the Internet, while making it easier to organize for some, has yet to show significant effect. After all, the landmine campaign did result in an international treaty, but the United States, the most powerful country in the world, has yet to sign it despite the fact that the campaign was organized in the United States. Antiglobalization protestors have appeared magically around the globe, but they have had little impact on the major trade agreements that are at the root of economic globalization. China put down the Falun Gong movement in spite of the fact that they used the Internet to organize and to surprise the authorities on numerous occasions. Jesse Ventura failed to revolutionize Minnesota politics, instead resigning after only one, fairly mediocre term.

Pippa Norris (2001) points out that classic theories of technological diffusion "have commonly followed an S-(sigmoid)-shaped pattern" (p. 30). This involves a slow rate of initial adoption followed by rapid adoption, and then falling prices as a result of rapid adoption. The cyber optimists point to this model to show that the gap between the rich and poor will soon close. Cyber pessimists, however, emphasize that inequalities that existed before the introduction of the Internet will continue to exist and be exacerbated. This is another way of saying that the rich will get richer and the poor will get poorer.

The actual research on this point offers support for both camps. The rapid spread and relatively low cost of technological diffusion in the United States—including the Internet—may be unique to the United States. As of 2001, one-third of all Internet users were still American; this is a fraction that has shrunk but that is still significant. Moreover, in America and in the rest of the world, Internet use correlates highly with other forms of communications media. Norris (2001) shows that countries rich in communications use the Internet, whereas ". . . people living in poor countries such as Burkina Faso, Yemen, and Vietnam were largely cut off from all forms of info-tech, including traditional mass media like radios and newspapers as well as modern ones such as mobile phones and personal computers" (p. 53).

But arguments about the extent and consequences of the digital divide are likely to be less important if the cyber skeptics are right and technologies adapt to society—not the other way around. Among those who use the Internet, are there significant changes in community, in citizen participation, in democratic process and in governance or has the Internet simply come to reflect the state of affairs that existed before it became as widely used?

There is no doubt, for instance, that the Internet has allowed for the creation of all sorts of "communities" of interest that could never have happened without it. For instance, people who have rare diseases can find each other in cyber space and share experiences and medical advice and people who like pop culture's Spice Girls can communicate with each other across continents. In an important essay on this topic, William A. Galston points out four aspects of the classic definition of community: limited membership, shared norms, affective ties, and a sense of mutual obligation. Leaving a virtual community is a lot less costly than leaving a physical community; thus, virtual communities tend to proliferate narrower and narrower groups of like-minded individuals, sacrificing perhaps, the ability to deliberate and compromise with people unlike ourselves. And there is little evidence that people in online communities develop shared norms, affective ties, or a sense of mutual obligation toward each other, which are key skills needed to live together peacefully in physical communities. Galston, a definite cyber skeptic, argues that the Internet could well lead to a form of "cyber-balkanization," in which communities of like-minded people reinforce each other and never learn the civic virtues of compromise and mutual respect— an outcome that could be very bad for civic and political life. Canadian political parties have been especially active in using various types of technology to facilitate communication and decision making among party members and yet, as Bill Cross concludes after studying the use of these new technologies, ". . . teledemocracy does not easily lend itself to collective decision making," thus confirming Galston's conclusion that an important part of community—the ability to come to common ground—is not necessarily facilitated by the new technology.

Similarly, there is little evidence that the Internet has ushered in a new era of widespread citizen participation. The title of Pippa Norris's (2002) article on the Internet and American elections from 1992 to 2000 says it all: "Revolution, What Revolution?"

Drawing on data from the Pew Research Center for the People and the Press that spans nearly a decade, she concludes that the Internet has not managed to bring new people into the political process, rather ". . . net political activists were already among the most motivated, informed, and interested in the electorate. In this sense, during recent political campaigns the net has been essentially preaching to the choir" (p. 76).

Nevertheless, after 1998 politicians took to the Internet with a vengeance, with some using it to augment their traditional campaign and some using it in the hopes that the Internet will provide them with a strategic advantage that others do not have. The following table shows the percentage increases in the use of Web sites by major party candidates for campaigning.

**Table 1. Website Usage by Candidates**

|                | 1998 | 2000 | 2002 |
|----------------|------|------|------|
| U.S. House     | 35%  | 66%  | 79%  |
| U.S. Senate    | 72%  | 91%  | 90%  |
| U.S. Governors | 95%  | 95%  | 99%  |

In parliamentary democracies, the growth in the use of the Internet on behalf of political parties has also been impressive. Norris (2001) tells us that in Western Europe there are ". . . over two dozen parties online per country, followed closely by Scandinavia" (p. 153).

There is no doubt that first radio and then television radically changed American politics. In fact, ever since the 1960 election, when candidate John F. Kennedy, clean shaven, handsome, and calm, bested candidate Richard M. Nixon, on television but not on radio, television has reigned supreme as *the* transformative technology in American politics. Since then, people have been waiting for the Internet to have an equivalently explosive effect on elections. What is going on in these candidate and party Web sites? Are they changing the nature of the elections in those countries? Is the interactive potential of the Internet being put into play to increase political dialogue between citizen and candidate or party leader? So far there is little evidence that that is the case. In a previous article, Norris said, "In 1998 most Internet campaigns looked like and functioned as electronic brochures and by the 2000 election cycle not much had changed" (2002, p. 89). A similar finding comes from democracies in other parts of the world. Writing about New Zealand's political parties online, Juliet Roper concludes that "the format of party political Web sites has so far failed to use the full range of available technology." Pieter W. Tops, Gerrit Voerman, and Marcel Boogers, writing about the use of the Internet in Dutch elections conclude that ". . . the World Wide Web is used mainly as a new medium to advise the electorate about candidates and party manifestos. Any opportunities toward interaction between candidates or the party headquarters can be seen as an attempt to improve vertical communication" (p. 98). And Karl Lofgren, writing about the use of the Internet in Danish political parties, refers to the production of "nice-looking 'shop-windows' for the parties to present mainly standard party information to the electorate and citizens" (p. 68).

Nevertheless, there are signs that political candidates are beginning to catch on to the power of the Internet. Governor Howard Dean, running for the 2004 Democratic nomination for president, used an Internet service called *Meet Up* to create a substantial network of activists around the country. His Internet savvy resulted in a burst of political contributions and catapulted him from long shot to serious candidate. The power of the Internet will most likely be felt especially in the American presidential nomination process; for instance, the rapidity of Internet information dissemination could mean that candidates who do well in one primary contest can capitalize on their victory by raising money and gathering supporters in time for the next primary—something that long-shot candidate John McCain was able to do in the 2000 election contest.

The debacle that occurred in the state of Florida in the 2000 presidential election led many to look toward the Internet and Internet voting as the solution to America's outdated election machinery. Proponents of Internet voting believe that its convenience will increase participation, especially among young people, and increase administrative efficiency and reduce mistakes. Opponents of Internet voting worry about the digital divide and the integrity of the ballot in an Internet system. Central to all truly democratic systems is the secrecy of the ballot, yet opponents fear that tight security features, such as the use of PINs or digital signatures, offer a way for unscrupulous officials to track people's votes. In addition, a wide array of computer fraud techniques such as "site jacking" and the use of "Trojan horses" to change a vote prior to transmission open up a whole new world of voter fraud.

Nevertheless, Internet elections have been held in a wide variety of places—Belgium, the Netherlands, Brazil, South America, England and the state of Arizona—and an increasing number of private elections have been held over the Internet. It is likely that, as states and countries upgrade their election technology, they will move away from machines and toward Internet voting at polling places. However, Internet voting from home, or remote internet voting systems are still some years away for the security reasons mentioned above. A study by the Internet Policy Institute (2001) concluded that "remote internet voting systems pose significant risk to the integrity of the voting process and should not be fielded for use in public elections until substantial technical and social science issues are addressed" (p. 34).

Finally, the Internet has been seen as a way to improve democracy by improving the quality and quantity of interactions between citizens and their governments. The first stages of electronic government resembled the first stages of the Internet in political campaigns. Governments put information on the World Wide Web, often without regard to its utility to the citizens. And then, in 1997, Vice President Al Gore and his National Performance Review published *Access America*, a document that outlined a digital future for government. It called for "... service improvements that will affect all Americans. It doesn't just propose electronic services, it calls for new ways to bring electronic options to all who want them, including those in underserved and rural areas" (p. 3).

Since the publication of that report, governments the world over have increased their use of the Web. A United Nations studypublished in 2002 found that 89 percent of the world's governments used the Internet in some capacity. But a much smaller portion of the world's countries—only 17 out of the 190 nations—actually allow citizens to interact with the government online and to complete transactions such as paying taxes online. Worries about privacy and security are holding up full electronic government just as they are holding up remote Internet voting, but these are slowly being overcome.

In addition to convenience and efficiency, is the use of the Internet in government contributing to better policy making? This is hard to assess, but it is possible that the Internet will allow average citizens to be heard in ways that were never possible before. For instance, the regulations that implement statutes are an integral part of the American legal system. And recognizing that, the law allows for a public comment period on regulations. But practically what this always meant was that a citizen had to hire a lawyer who lived in Washington D.C. to go to the reading room of the agency and look through the thousands of pages of public comments. As Cary Coglianese points out in a 2003 study, "the 'public' that participates in the rulemaking process is actually a very narrow slice of the entire citizenry." The vast majority of comments filed on regulations are submitted by corporations and industry groups.

Some departments of the U.S. government have embraced an initiative called *E-Rulemaking*. Coglianese reports that there is now a governmentwide portal, Regulations.gov, "to help citizens locate and submit electronic comments on any proposed regulation by any agency" (p. 6). In the 1990s, the U.S. Department of Agriculture set up a site where citizens could comment on the regulations on organic foods—a fairly controversial topic. The agency eventually received more than 250,000 comments.

The Canadian government's Web site allows citizens to write comments to ministers about policy. The hope and expectation of e-government is that it will promote more dialogue between citizens and their leaders and thus contribute to a better democracy.

*See also* **Computers; Technology and Society**

**BIBLIOGRAPHY**
Coglianese, Cary. "The Internet and Public Participation in Rulemaking." Regulatory Policy Program, Cambridge, MA: John F. Kennedy School of Government, 2003.
Cross, Bill. "Teledemoracy: Canadian Political Parties Listening to Their Constituents." In *Digital Democracy: Politics and Policy in the Wired World*, edited by Les Pal and Cynthia Alexander. Toronto: Oxford University Press, 1998: 132–145.
Daly, Peter H. "IT's Place in US History: Information Technology as a Shaper of Society." Cambridge, MA: Program on Information Resources Policy, 2002.
Eisenach, Jeffrey, Thomas Lenard, and Stephen McGonegal. *The Digital Economy Fact Book*, Third Edition, Washington, DC: The Progress and Freedom Foundation, 2001.
Galston, William A. "The Impact of the Internet on Civic Life," In *Governance.com: Democracy in the Information Age*. Washington, DC: Brookings Institution, 2002.
Gore, Al. *Access America: Reengineering Through Information Technology*, Washington, DC: Government Printing Office, 1997.
Hill, Kevin A., and John E. Hughes. *Cyberpolitics: Citizen Activism in the Age of the Internet*, Oxford, U.K.: Rowman & Littlefield Publishers, Inc., 1998.
Internet Policy Institute, *Report of the National Workshop on Internet Voting: Issues and Research Agenda*, College Park: University of Maryland, Freedom Forum and Internet Policy Institute, 2001.
Lessig, Lawrence. *Code and Other Laws of Cyberspace*. New York: Basic Books, 1999.
Lofgren, Karl, "Danish political parties and new technology, Interactive parties or new shop windows?" In *Democratic Governance and New Technology: Technologically mediated innovations in political practice in Western Europe*, edited by Jens Hoff, Ivan Horrocks, and Pieter Tops. London: Routledge, 2000.
Margolis, Michael, and Resnick, David. *Politics as Usual: The Cyberspace Revolution*. Thousand Oaks, CA: Sage Publications Inc., 2000.
Marvin, Carolyn. *When Old Technologies Were New: Thinking about Electric Communication in the Late Nineteenth Century*. New York: Oxford University Press, 1988.
Norris, Pippa. *Digital Divide: Civic Engagement, Information Poverty, and the Internet Worldwide*, Cambridge, U.K.: Cambridge University Press, 2001.
Norris, Pippa. "Revolution, What Revolution? The Internet and U.S. Elections, 1992-2000." in *Governance.com: Democracy in the Information Age*, Washington, DC: Brookings Institution, 2002.
Rheingold, Howard. *The Virtual Community: Homesteading on the Electronic Frontier*. Reading, MA: Addison-Wesley, 1993.
Roper, Juliet. "New Zealand Political Parties Online: The World Wide Web as a Tool for Democratization or for Political Marketing." In *The Politics of Cyberspace*, edited by Chris Toulouse and Timothy W. Luke, New York: Routledge, 2000.
Tops, Pieter W., Gerrit Voerman, and Marcel Boogers. "Political Websites during the 1998 Parliamentary Elections in the Netherlands." In *Democratic Governance and New Technology: Technologically Mediated Innovations in Political Practice in Western Europe*, edited by Jens Hoff, Ivan Horrocks, and Pieter Tops. London: Routledge, 2000.
United Nations Division for Public Economics and Public Administration and American Society for Public Administration. *Benchmarking E-government: A Global Perspective*. New York: United Nations, 2002.

*Elaine Kamarck*

# INVENTION IN HISTORY AND CULTURE.

The term *invention* refers to the activities by which individuals create new devices or processes that serve human needs and wishes. To create a device or process, an inventor must often investigate phenomena in nature. Sometimes an inventor need only observe nature closely to discover what will work, but in other cases, insights come only by experiment or ingenious manipulation. Because nature does not readily yield its secrets, one could say that an inventor "negotiates" with nature.

Yet invention is not simply building a device; an inventor must also connect his or her invention with society. In some situations, needs are well known and society readily takes up an invention. Since railroads in the mid-nineteenth century needed stronger rails and armies wanted stronger cannons, there was a ready demand for Henry Bessemer's new steel process in 1856. In other situations, though, there is no preexisting need and an inventor must convince society of an invention's value. For example, when Alexander Graham Bell invented the telephone in 1876, he found few people willing to buy it; indeed, it took the Bell Telephone Company decades to convince Americans that every home should have a telephone. Bell and his successor company not only had to invent the telephone, but also a marketing strategy that reflected the interests of users. In this sense, inventors "negotiate" with society.

What makes invention interesting, then, is that inventors stand astride the natural and social worlds. On the one hand, inventors must be willing to engage nature, to find out what will work; on the other hand, they must also interact with society, exchanging their inventions for money, fame, or resources. Thomas Edison was remarkable because he both perfected devices in the laboratory and convinced investors and the public to take up his inventions. Playing the heroic genius for the newspapers, Edison provided a myth of technological creativity that served as an antidote for the impersonal organizations that came to dominate American culture in the early twentieth century.

In negotiating with both nature and society, inventors are influenced by many factors. Their efforts are shaped by the resources available and the help they get from patrons and assistants. Inventors may utilize the skill of craftsmen as well as the theories of scientists. In connecting their creations with society, inventors must take into account the dynamics of the marketplace, the structure of business organizations, and the ways in which the state promotes and regulates tech-

nology. Most broadly, inventors must grapple with how a culture perceives technological change: is invention welcome or viewed with suspicion?

## Invention versus Technological Creativity

Archaeologists have found that the evolution of *Homo sapiens* is deeply entwined with technological creativity. Thousands of years ago, as early humans acquired hands with opposable thumbs, larger brains, and language, they learned how to use fire, plant crops, domesticate animals, make pottery, and work copper. Although these prehistoric developments are often called *inventions*, it is not accurate to do so; they were simply acts of technological creativity. Invention, instead, is a specific form of technological creativity in which individuals explicitly link a technological artifact with a need or wish. For these major prehistoric developments, we have no information about whether they were the product of an individual mind or a group effort. To our knowledge, the creators of these developments did not have to consciously connect their creations with needs or wishes.

As one form of technological creativity, invention has thrived in some societies and not in others. Although it is tempting to assume that all societies value technological change, this is not always the case. Since technological change may alter the social order, some societies do not encourage technological creativity. For example, despite contributions to science, politics, and art, the ancient Greeks were suspicious of change and not interested in new technology that might alter the social order.

Invention, then, is a historical phenomenon, occurring under certain conditions. First, the society must perceive technological change as conferring an advantage to someone in society—whether it be the rulers, investors, or a powerful class. And second, technological creativity must be seen as the product of an individual mind. Both imperial China (200 B.C.E. to 1700 C.E.) and early Islamic society (600 to 1400 C.E.) embraced technological change, and these societies introduced papermaking, printing, gunpowder, compass, and the la-teen sail. However, neither society viewed technology as the product of individuals or celebrated inventors as heroes.

## Invention in Renaissance Europe

It was in Europe and America that technologi-cal creativity came to be understood as inven-tion. The first technologists to call themselves inventors appeared in Renaissance Italy in the fourteenth century. As Italian city-states found themselves in competition and unable to gain an advantage over each other, several turned to technology. Whereas some cities improved their military technology, others sought inno-vation in manufacture. Offering to provide new technology, artists such as Leonardo da Vinci and Francesco di Giorgio created the role of the inventor. Like artists, inventors claimed that their ability to create new technology was based on personal knowledge and a flash of inspiration—the Eureka moment. Utilizing their artistic training, Renaissance inventors often sketched ideas for new machines. Their ideas could be fanciful, ranging from an un-dersea diver sketched by Jacopo Mariano to Leonardo's flying machines. They also skill-fully combined components (such as the wedge, screw, lever, pulley, and gears) to cre-ate clocks, sawmills, or a weight-driven spit for turning roasting meat.

In the notebooks of Leonardo and his con-temporaries, we see the importance of visual representation for invention. Renaissance in-ventors sketched both nature and machines since drawing helped them to observe and understand. During the Renaissance, artists and inventors established the ideas that na-ture could be improved by the human mind and that technology could contribute to so-cial progress.

## Invention and the Industrial Revolution

From 1500 to 1800, technological change be-came increasingly important as European states pursued exploration, military con-quest, trade, and manufacturing. These activ-ities depended on better ships, instruments, and weapons that in turn stimulated the development of new sources of power (coal and the steam engine), better materials (glass and iron) and new ways of organizing labor (factories). To encourage individuals to de-velop new machines and processes, govern-ments began issuing patents that gave inventors exclusive ownership of their cre-ations. The first patent was awarded by Flo-rence in 1421.

To promote technology, different European states pursued various strategies. In France, the strategy was to establish strong national institutions. To consolidate his power, Louis XIV sponsored royal industries in textiles and porcelain as well as a nationwide system of roads and canals. To design this trans-portation network, the French established in 1675 a special organization of engineers, the *Corps des Ingenieurs du Genie Militaire*, and the first engineering school, the *École Na-tionale des Ponts et Chaussées*, in 1747.

In contrast, the British saw invention as the prerogative of the individual, who should be permitted to develop and own new machines. British society would grow wealthy, argued the economist Adam Smith, if numerous people pursued their individual economic destinies. Much of the British Industrial Revolution was based on countless small changes in the design and manufacture of goods, yet a few inven-tors—Thomas Newcomen, James Watt, Richard Trevithick, and George Stephenson—concentrated on major developments, such as the steam engine and the railway. Well aware of the importance of linking inventions to prevalent beliefs, Watt's business partner, Matthew Boulton, would dramatically tell vis-itors to their factory that "we sell here, sir, what all the world wants: power."

Just as creative technologists in Britain called themselves inventors, so ambitious Americans did the same. As early as 1641, American inventors petitioned colonial legis-latures for patents. By the end of the Ameri-can Revolution, the British industrialization was well underway, and the American founding fathers appreciated the importance of stimulating invention. When they framed the constitution, one of the powers given to the federal government was to issue patents.

Spurred by first-hand experience of using machines in trade and farming, Americans readily invented. Recognizing the importance of agriculture for the new republic, Americans developed machines for harvesting or processing crops; Oliver Evans introduced an automated flour mill in 1790, Eli Whitney patented his cotton gin in 1794, and Cyrus McCormick demonstrated his mechanical reaper in 1831. Meanwhile, Robert Fulton (the steamboat) and Samuel F. B. Morse (the telegraph) contributed the transportation and communications technology needed to sustain an expanding nation.

Even though inventors contributed to the rise of industry in antebellum America, their efforts were circumscribed. Firms were generally small partnerships and lacked substantial capital. Most industries were marked by sharp price competition, which forced businessmen to avoid the long-term investment needed to improve technology. Although businessmen were willing to purchase patents, they kept inventors at arm's length. After discovering how to vulcanize rubber in 1838, Charles Goodyear spent another five years and $50,000 perfecting and patenting his process. Goodyear eventually sold his patent, but he died in 1860 with debts of nearly $200,000. Although Americans equated invention with progress, antebellum inventors found it hard to negotiate links between their creations and commercial opportunities.

## The Golden Age of Heroic Invention

In the 1870s, circumstances in the telegraph and electrical industries created a "Golden Age" for inventors. Notably, it was not the heroic origins of Morse's invention that made the telegraph industry a hotbed of inventors, but rather the appearance of the Western Union Telegraph Company. In the 1850s, Morse's telegraph was promoted by numerous small companies, but it soon became clear that the telegraph would flourish only if a single system connected cities across America. By absorbing its competitors and building the first transcontinental line, Western Union created a nationwide system in 1867.

But no sooner had Western Union achieved national dominance than it had to fight off critics and rival networks. As a monopoly, Western Union was seen as a threat to American democracy. Critics worried that Western Union controlled the flow of news and stock prices, and that it might use this power to ruin individual businessmen and manipulate the stock market. At the same time, another challenge came from Wall Street. Western Union had expanded by erecting lines along railroads and placing offices in train stations. This meant, however, that as new railroads were built, financiers could create their own telegraph networks and attempt to gain control of Western Union. Jay Gould pursued this strategy and eventually captured Western Union in 1881.

In response to these threats, Western Union employed price competition, political lobbying, and hostile takeovers, but invention also became essential. To remain dominant, Western Union needed better equipment. Likewise, critics and financiers realized that inventions might be used to penetrate the industry. In response, dozens of ambitious men experimented with new devices. Typical was Bell, who started inventing after learning that Western Union paid Joseph Stearns $25,000 for a duplex telegraph. Over 400 individuals patented telegraph devices between 1865 and 1880, but the most successful were two telegraph equipment manufacturers, Edison and Elisha Gray. Edison designed systems for both Western Union and Gould, and he used this patronage to build an "invention factory" at Menlo Park, New Jersey, in 1876.

At Menlo Park, Edison turned out spectacular inventions—an improved telephone, the phonograph, and an incandescent lighting system. Because Edison generated voluminous records, historians have explored his method of invention. At the start of a project, Edison reviewed the relevant science, existing technology, and potential markets. Inspired by the scientist Michael Faraday, Edison recorded his ideas in notebooks. Within his laboratory, he brought together machinists and scientists who built models and conducted experiments. And Edison was an effective research manager, identifying new lines of investigation and skillfully deploying his team. Edison's success at Menlo

Park prompted other inventors, including Nikola Tesla, Edward Weston, and Reginald Fessenden, to build their own laboratories in the 1880s. Even today, American inventors frequently invoke Edison as their role model.

## Inventors and the Corporation

But Edison and Menlo Park were soon eclipsed by other individuals and institutions. Because manufacturing and marketing new technology often required millions in capital and the creation of large organizations, business leaders concluded that they could not risk leaving invention outside the firm. Consequently, as electric light and power took shape in the 1880s and 1890s, inventors found themselves in new roles in relation to business.

Some inventors sought to maintain their creative independence. After working for Edison, Frank J. Sprague struck out on his own, designing direct current motors for streetcars. While he started several companies to develop his inventions, Sprague inevitably sold them to larger firms such as General Electric. Likewise, Tesla sold his alternating current motor patents to George Westinghouse in 1888 and used his new wealth to investigate radio waves. Unguided by knowledgeable patrons, Tesla fashioned a fantastic system for broadcasting power around the world. However, Guglielmo Marconi beat him out by developing wireless telegraphy for the lucrative market of ship-to-shore communication. With creative independence came the risk of failing to connect one's inventions with capital or markets.

In contrast to Tesla, Elihu Thomson understood the need to be connected to business and chose to work within the corporation. A chemistry teacher, Thomson began inventing arc lighting equipment in 1878 with Edwin J. Houston. Realizing that he knew little about manufacturing and marketing, Thomson sought out entrepreneurs who could help him. After two unsuccessful partnerships, Thomson found the right backer in Charles A. Coffin, a shoe manufacturer, who successfully raised capital and marketed electrical equipment. With Coffin's support, Thomson thrived as an inventor, and in 1892 the Thomson-Houston Company absorbed Edison General Electric to form the General Electric (GE) Company.

Thomson demonstrated to GE that invention was essential. But by the late 1890s, GE management realized that the size of their company (in terms of capital and organizational complexity) was such that they should no longer rely solely on individuals like Thomson. Unlike Western Union, which had let Edison move out to Menlo Park, GE felt that it could protect itself only by creating, in 1900, the first U.S. research and development (R&D) laboratory staffed with scientists. Over the next decade, AT&T, DuPont, Eastman Kodak, and Corning Glass also established labs.

But why did these corporations invest in a *scientific* laboratory? One reason was the growing supply of scientists. In the 1870s, few American universities offered advanced science degrees, but thanks to private philanthropy and federal land grants to state colleges, American universities by 1900 graduated hundreds of scientists annually. But there were also cultural reasons for choosing scientists over inventors. Although inventors do work methodically, they generally legitimate themselves by claiming that they possess unique personal knowledge. By claiming that invention occurs in a Eureka moment, inventors suggest that their work is discontinuous and unpredictable. Given this rhetorical stance, inventors were not appealing to managers trying to protect huge corporations. Yes, a genius like Edison can do great work, but should one bet the company on him?

Instead, as managers rationalized business in the early 1900s, they turned to scientists. If science is essentially about predicting the behavior of nature, then applying science to industrial problems should be predictable as well. Moreover, by taking a team approach and breaking down complex problems into routine experiments, scientists assured managers that they would get results. By promising to make invention predictable, scientists appealed to managers seeking to protect corporations from uncertainty.

## Invention in the Twentieth Century

Consequently, American companies in the twentieth century have assumed that R&D

labs—not inventors—will provide new technology. Several factors have reinforced this trend. The federal government has periodically investigated how corporations such as Standard Oil, AT&T, IBM, and Microsoft may have used technology to monopolize markets. Faced with the risk of antitrust litigation, many companies developed new products in-house to avoid the appearance of collusion. During both World Wars and the Cold War, the federal government invested heavily in scientific research, and scientists helped create the atomic bomb, jet airplane, radar, and computers. This overall trend is illustrated in patent ownership; in 2001, corporations and governments owned 86.9 percent of the patents granted and individuals only 13.1 percent.

Yet while DuPont promises "better living through chemistry," inventors have continued to revolutionize daily life. Electronic television was invented in the 1930s by Philo T. Farnsworth. In the 1960s, the integrated circuit was invented simultaneously by Jack Kilby and Robert Noyce, both of whom relied on hands-on knowledge. To start Apple Computer, Steve Jobs and Steve Wozniak invented in their garage. Similar stories could be told about the Wright brothers and the airplane or Gordon Gould and the laser.

To be sure, these inventions only succeeded through links to business. Kilby worked for Texas Instruments; Jobs and Wozniak sought advice from engineers at Hewlett-Packard. Farnsworth struggled for years only to see RCA capture the television market. Not only did these inventors make their devices work, they also struggled to persuade people to invest in their creations; they had to link the natural and social worlds.

Since the 1980s there has been growing interest in teaching invention as part of a general reform of science education. Early in this movement, the U.S. Patent Office sponsored Project XL to help teachers develop invention curricula, and it maintains a Web site for children. At the college level, the National Collegiate Invention and Innovation Alliance sponsors competitions for student inventors. With bequests from the inventor Jerome Lemelson, the Smithsonian Institution and

MIT support centers to study invention. Complementing the many biographies of inventors, STS scholars now investigate invention as both a social and cognitive process. Hoping to inspire more people to invent, books and Web sites celebrate minority, black, and women inventors. All these activities indicate that invention remains vital to American culture.

*See also* **Invention, Inventors, and Design**

**BIBLIOGRAPHY**

Berg, Maxine. *The Age of Manufactures, 1700–1820: Industry, Innovation, and Work in Britain.* 2nd ed. London: Routledge, 1994. A good general introduction to the British Industrial Revolution.

Bijker, Wiebe E. *Of Bicycles, Bakelite, and Bulbs: Toward a Theory of Sociotechnical Change.* Cambridge: MIT Press, 1995. Examines invention as a social process.

Brown, David E. *Inventing Modern America: From the Microwave to the Mouse.* Cambridge: MIT Press, 2002. Provides capsule biographies of major American inventors in the twentieth century.

Carlson, W. Bernard. *Innovation as a Social Process: Elihu Thomson and the Rise of General Electric, 1870–1900.* New York: Cambridge University Press, 1991.

Carlson, W. Bernard. *Technology in World History.* 7 vols. New York: Oxford University Press, 2005.

Carlson, W. Bernard, and Michael E. Gorman. "The Cognitive Process of Invention: Bell, Edison, and the Telephone." In *The Inventive Mind: Creativity in Technology*, edited by D. Perkins and R. Weber. New York: Oxford University Press, 1992: 48–79.

Crouch, Tom. *The Bishop's Boys: A Life of Wilbur and Orville Wright.* New York: W. W. Norton, 1989.

Ferguson, Eugene S. *Engineering and the Mind's Eye.* Cambridge: MIT Press, 1992. Argues that visual representation is essential to technological creativity.

Green, Peter. *Alexander to Actium.* Berkeley: University of California Press, 1990: 467–479. Includes an excellent chapter on science and technology in ancient Greece.

Hindle, Brooke. *Emulation and Invention.* New York: W. W. Norton, 1981. Discusses the inventive careers of Robert Fulton and Samuel F. B. Morse.

Hughes, Thomas P. *American Genesis: A Century of Invention and Technological Enthusiasm, 1870–1970.* New York: Viking-Penguin, 1989.

Jakab, Peter L. *Visions of a Flying Machine: The Wright Brothers and the Process of Invention.* Washington, DC: Smithsonian Institution Press, 1990.

Jenkins, Reese V., Robert Rosenberg, and Paul B. Israel, eds. *The Papers of Thomas A. Edison.* Baltimore: Johns Hopkins University Press, 1989. Five volumes to date.

Pretzer, William S., ed. *Working and Inventing: Thomas Edison and the Menlo Park Experience.* Dearborn: Henry Ford Museum and Greenfield Village, 1989.

Reich, Leonard S. *The Making of American Industrial Research: Science and Business at GE and Bell, 1876–1926.* New York: Cambridge University Press, 1985.

Reid, T. R. *The Chip: How Two Americans Invented the Microchip and Launched a Revolution.* New York: Simon and Schuster, 1984.

Riordan, Michael, and Lillian Hoddeson. *Crystal Fire: The Birth of the Information Age.* New York: W. W. Norton, 1997. Scientific history of the invention of the transistor.

Schwartz, Evan I. *The Last Lone Inventor: A Tale of Genius, Deceit, and the Birth of Television.* New York: HarperCollins, 2002. Biography of Philo T. Farnsworth.

Stanley, Autumn. *Mothers and Daughters of Invention: Notes for a Revised History of Technology.* New Brunswick: Rutgers University Press, 1995.

United States Patent and Trademark Office. "Patenting by Organizations, 2001." Technology Assessment and Forecast Report, 2002.

Weber, Robert J. *Forks, Phonographs, and Hot Air Balloons: A Field Guide to Inventive Thinking.* New York: Oxford University Press, 1992. Looks at invention as a cognitive process.

Wise, George. "A New Role for Professional Scientists in Industry: Industrial Research at General Electric, 1900–1916." *Technology and Culture* 21 (1980): 408–429.

W. Bernard Carlson

# INVENTION, INVENTORS, AND DESIGN

This entry includes three subentries

Overview

Design and Society

The Social Role of Inventors

## Overview

"Everybody complains, but inventors fix," said Jacob Rabinow (1910–1999), an inventor whose 230 patents involved a lot of "fixing." Inventors such as Rabinow create new technological devices, materials, processes, or systems; Rabinow's include the computer disk and text scanner. Although inventions influence social change, those who create these inventions are less obvious today. The typical nineteenth-century inventor was a self-employed person, a "lone inventor." Lone or not, these inventors stood out. In the twenty-first century, on the other hand, the typical inventor is more likely to be an invisible member of a corporate team. When an independent inventor comes up with a new device, we speak of "invention." When a corporate team does it, we call it "research and development" (R&D). The word *development* is particularly important. For instance, although the dry photocopying process was invented by one person, Chester Carlson (1906–1968), the Xerox 914, the first practical machine to put the process to work, was the product of a large corporate team that toiled over many years. So inventors may work from their basement or in the heart of a corporate R&D team. Independent inventors like Carlson are still important, but today about four times as many American patents are issued to corporations as to independent inventors.

Corporate and noncorporate inventors have much in common. They often display their abilities as children. Their mechanical ingenuity is seen through a special interest in how things work, by taking them apart, putting them back together, or modifying them in some way. This "Mr./Ms. Fixit" quality is of course also shown by future scientists and mechanics. Another key quality that separates the inventor is the desire to make things better. Howard Head (1914–1991), key inventor of the metal ski, developed it because he was unhappy with the way that wooden skis worked. It took him thirty-nine separate tries, but in the end Head changed the whole ski industry. Head's experience also suggests the third important quality: endurance.

While inventors have common abilities, however, they also differ on key dimensions. Some inventors barely finish high school, yet many professional inventors get college-level training in science and engineering, and some have higher degrees. This training helps them avoid the pitfalls that snare those without this education. Some inventors invent in several fields, while others work only in one. Inventors also range from those with a single patent to those (such as Rabinow) with hundreds of patents. A single patent, however, such as Samuel Morse's telegraph or the Wright brothers' airplane, may mark a major change in the technology available to society.

Inventions may begin with sudden insight. Edwin Land (1909–1991), inventor of the Polaroid camera, said that the idea for instant film development came to him in a single flash of insight. But inventions may also start through a carefully thought-out research program, as did the Sidewinder missile, whose features emerged gradually in the mind of U.S. Navy scientist William McLean (1914–1976). In 1949, after much research, McLean came to the conclusion that the Navy would need a low-cost heat-seeking missile. He put together a team at the China Lake Naval Ordnance Test Station, and they began working on the concept. By 1956 the missile was released to the Fleet. After many modifications, it is still in service today.

After an invention is conceived, the inventor (often with help) then must develop it into something workable. The road to workability may be long and hard. Head found that his ski took exceptional effort to bring to fruition. As mentioned, Head's metal ski required thirty-nine prototypes before one was found that ski instructors could not break. Head said, "Each time one of them broke, something inside me snapped with it" (Landrum, p. 159). When Head invented an oversize tennis racket (the "Prince"), it was refused a patent three times by the U.S. Patent Office. Head, however, persisted, finally demonstrating the racket's superior power in rigorous tests, and got his patent.

Even after the inventor has gotten the prototype to work, commercial success is not guaranteed. Philosopher Ralph Waldo Emerson claimed that if you invent a better mousetrap, "the world will make a beaten path to your door" (quoted in Yule, p. 138). Yet many inventors find the world indifferent to their inventions. Going into the marketplace, moreover, may bring on competition or patent infringement. Only one out of six patented inventions makes its inventor substantial profits. The reasons are many: lack of knowledge, lack of experience, lack of funds, or inability to solve key problems. Often the inventor has invented the wrong thing or is simply not good at business. Many inventors do not reach the end of the long road to success, but those who do, like Rabinow and Head, keep others inventing.

**BIBLIOGRAPHY**

Brown, Kenneth A. *Inventors at Work.* Redmond, WA: Tempus, 1988.

Hughes, Thomas Parke. *American Genesis: A Century of Invention and Technological Enthusiasm.* New York: Viking, 1989.

Landrum, Gene. *Profiles of Genius.* Buffalo, NY: Prometheus, 1993.

Rabinow, Jacob. *Inventing for Fun and Profit.* San Francisco: San Francisco Press, 1990.

Westrum, Ron. *Sidewinder: Creative Missile Development at China Lake.* Annapolis, MD: Naval Institute Press, 1999.

Yule, Sarah S. B., and Mary Keene. *Borrowings.* San Francisco: Dodge, 1889.

*Ron Westrum*

# Design and Society

*Designing* a technology is the activity of planning and producing a desired product, process, or system. In the standard linear model of technological innovation, in which each step follows sequentially from the preceding one, design is usually taken to be the starting point or the first step. According to this model, a technology such as a new computer is first designed (or invented), then prototyped, tested, manufactured, marketed, sold and finally used.

This simple linear model is now rejected by most scholars who study technological innovation. Often these different aspects of the process occur coextensively. For instance, the creation of a potential market for a new cell phone happens at the very same time that the requisite chips are being designed. Sometimes things happen in reverse order, with designers responding to market demand. Perhaps the most radical challenge to the linear model has come from the recognition that users of technology—the *last* step in the linear model—are often in the driver's seat and play an active role in designing technologies. For example, the first designs of the sticky note, the inline roller skate, the wind surfer, and the mountain bicycle were all produced by people who had no particular technological expertise but who were users and saw a need for something new. Sometimes users adapt current technologies to quite unexpected purposes, to which manufactures respond. For instance, farmers in the United States in the 1930s converted their Ford Model T automobiles into tractors by adding large rear wheels and tires. Henry Ford eventually responded to the farmers' inventions by developing his own Fordson tractor.

## The Design Process

Much technological design today is based upon existing knowledge and ideas. For example, civil engineers use their expertise and experience to design new bridges for highways. They are familiar with what kinds of of concrete and steel to use, which types of structures will bear loads and which will not, and what safety standards have to be met. It is extremely unlikely that the motorists who use the bridges will be able to come up with successful designs themselves. Areas of design that depend upon pre-existing engineering knowledge are increasingly automated, with computer-aided design (CAD) systems and knowledge-based engineering becoming indispensable in designing, say, a new airfoil for a modern passenger jet. With the growth of computer-aided manufacturing (CAM), design is today often integrated with other elements of the production process.

Some designers work alone, whereas others work in teams that can draw upon a wealth of different sorts of disciplines and experiences. New interdisciplinary design programs are common on campuses today. Within any one industry this pattern may change. For instance, the first commercial electronic music synthesizer, known as the Moog Modular System, developed in 1964, was largely the work of the inventor himself, the engineer Robert Moog, in collaboration with a single musician. By the time Moog's factory came out with the less expensive, more compact Minimoog synthesizer in 1970, his firm employed a small team of engineers and in-house musicians. Parts of the design were outsourced. During the 1980s and 1990s the large Japanese manufacturers of digital synthesizers, such as Roland and Yamaha, employed vast in-house teams of designers, marketers, and engineers, who would work together. Today, with globalization and better computer networking, the design process has once more become disaggregated. For instance, a synthesizer designer can sit at a computer in London and download designs to a coordinating office in Japan, which can then test-market a new product on musicians in Los Angeles.

Despite the growing complexity of design and its becoming embedded in manufacture, there is always room for contingency. Some designs come about almost by accident. One such example is the pitch wheel on keyboard synthesizers. A pitch wheel looks somewhat like a larger version of the volume control on a pocket radio or computer CD-ROM drive and is usually set to the left of the keyboard. By rolling the pitch wheel forward or backward, a keyboardist can "bend" the pitch of a note as a guitarist does by stretching the string out of line. This innovation happened by chance one day at the Moog synthesizer works in 1970, when an engineer happened to talk with the company buyer, who happened to have been a machinist earlier in life. What had been a rather clumsy rotary knob became transformed into a beautifully machined and balanced vertically mounted wheel with a click stop at normal concert pitch position. It was at the time one of many routine decisions in the design of the path-breaking Minimoog analog synthesizer, but it became a design feature that has lasted through generation after generation of digital synthesizers.

As a reminder of how complicated the design of a modern technological device in a global society can be, just think for a moment about one tiny aspect of the modern automobile: its use of automated voices to deliver safety warnings, driving directions, and the like, which is increasingly found on high-end automobiles. These automated voices, generated by small voice synthesizers contained in a chip, have to be designed. What works best? In a study of BMW drivers in Germany a soothing woman's voice was tried, but it was found that male drivers ignored the voice. Also, what is the most effective way to phrase a warning? Should the voice reprimand the driver ("Be careful! Slow down!"), or should it take the more neutral stance of pointing out to the driver that "road conditions are dangerous"? The latter warning turned out to be more effective. Such design issues have to be resolved and the best solutions found individually for different specific models of cars in different national contexts. The voice design in this case is outsourced to communications experts. A team based at Stanford University currently designs many of these voices for the world's leading auto makers.

Product design has increasingly had an aesthetic quality to it and, as a result, is often carried out in art schools. For instance, one of the leading design schools in London is housed within the Royal College of Art. Graphic artists and computer scientists play prominent roles in "design houses," where the design of anything from a company Web site or logo to a new product can be undertaken. A good reminder of the importance of the "look" of a machine is the story of the first successful photocopying machine. Chester Carlson invented the technique of xerography in 1937, but it was only in 1959, when the ugly-looking, dirty, and complicated photographic apparatus, inking mechanism, and rollers were housed within a clean white box—the legendary Xerox 914—that the photocopier as a product took off.

## Constraints on Design

All designs, whether stemming from adaptation of existing knowledge and technology or from the production of novel ideas and devices, have to work within constraints. These include ergonomic constraints (as in the design of the cockpit of a fighter jet to make the most efficient use of the human body shape), constraints imposed by the available materials, environmental constraints, and important legal and safety constraints. In the United States, for example, safety has been the field of a long-running battle between powerful corporate interests and public-interest groups. Some automobiles such as the Chevrolet Corvair, which Ralph Nader famously declared "unsafe at any speed" in 1965, were discontinued as a result, and others were redesigned to meet public concerns over safety. Whole industries have been curtailed in the United States because of such safety concerns, such as nuclear power, which has flourished within other cultural and legal contexts, as in France. The economics of design, whether the final price of a commercial product or the cost of a large-scale engineering project, are obviously crucial and are clearly governed by the wider societal distribution of resources, the working of markets, and so on. Corporate images and particular traditions built up within companies can be important constraints on design. For instance, a corporation such as Apple Computer will have a particular style. Apple users expect a certain look and feel to their computers and other products such as iPods. A device that does not look sufficiently "cool" to please demanding customers may sink without trace despite being technically adequate.

Quality and reliability are also key issues for designers and there are many trade-offs to be made. For example, expensive German laundry machines will last longer and wash clothes to a higher standard than a mass commodity washer made for the American market. Sometimes a device will become more successful when a company with a hard-won aura of quality lowers the bar to make a more affordable version. For example, Polaroid has found that its cheap i-Zone pocket camera, which produces low-quality, stamplike images, enjoys a huge success among teenage girls, who use the pictures as stickers.

## The Interaction of Design and Society

It should be clear now that society is integrally connected with almost every aspect of design. Most obviously, much technological design is produced directly in response to societal needs. Road engineers are commissioned to build a new section of highway that is needed to solve our transport problems. New medical technologies and new pharmaceutical products are designed to alleviate sickness and illness and to help the disabled. Genetically modified organisms are designed to get rid of troublesome pests or boost agricultural production. Military technologies meet societal needs for defense and fighting wars. Societal "need," however, is a tricky concept. Needs change according to what is actually possible. High-schoolers today tell their parents they need cell phones—a product that, when the parents were in high school, no student needed, because only a few wealthy users, if any, had them. Many technological needs are shaped by the advertising industry, which has grown relentlessly along with mass consumption. Also, many technologies are part of wider systems, and interdependencies develop between the different parts. For instance, a need for electric street lighting is possible only when there is a system of power distribution. Designs can also produce unintended consequences and dependencies. For example, when Henry Ford designed the Model T, he had no idea that, half a century later, automobile use in the United States would be connected with geopolitical concerns and America's dependency on oil from the Middle East. Some new technologies, such as genetically modified organisms, are resisted because of the fear of unintended consequences such as new herbicide-resistant "superweeds."

Design as an activity is a social process carried out within a particular social context. Engineers, artists, and technologists constitute professional groups who have their group norms, ways of training, procedures, standards, and so on. The knowledge, techniques, and practices they develop have a history and are nurtured within certain institutional contexts. Different approaches to design and engineering may be embedded within different national traditions, such as French engineers' early proclivity for highly mathematical methods in contrast to American engineers' more empiricist "trial and error" methods. These two different national styles of engineering led to the different designs of the first suspension bridges built in France and the United States in the early nineteenth century.

The politics of technology can be integrally involved with design, as the philosopher Langdon Winner has argued. Nuclear power reactors entail the design of systems of surveillance and control to prevent nuclear material falling into the wrong hands. The introduction of certain types of machines has led to job losses (as among skilled machinists after numerically controlled machine tools were introduced in the 1980s) and the strengthening of management control over workers. National politics can play a role, too, especially with prestigious high-technology projects. The Concorde supersonic airliner, which cost British and French taxpayers enormous amounts of money, was developed for the sake of national prestige and the need to show cooperation between the two countries in post–World War II Europe.

## Can Society Shape Design?

It is clear that technologies have an impact on societies in all sorts of ways, but can the very design of technologies be shaped or influenced by society? This is a topic that the sociology and history of technology has recently addressed by tracing the different meanings of technologies for different social groups. Different meanings can be attributed to the same technology; no technology carries its meaning forever; and meaning is always governed by the use to which a technology is put. For example, it might seem that a fighter airplane has a fixed meaning as a fast and highly maneuverable vehicle for delivering destruction from the air. During the first Gulf War in 1991, however, a new, more benign meaning appeared for these aircraft. Captured Iraqi Soviet-era MiG fighters were strapped down and their engines run at high power to blow out oil well fires with the exhaust. Conversely, the tragic events of September 11, 2001, demonstrated a new meaning for passenger jets as weapons of mass terror rather than modes of civilian transport.

This way of thinking about technology can help explain how new technologies develop. For example, in the transition from the high-wheeled "ordinary" bicycles of the Victorian era (the so-called "penny-farthings") to the modern safety bicycle it is useful to think about what this technology meant to different social groups at the time. The high-wheeled bicycles were used mainly for sport in parks by "young men of means and verve" (quoted in Bijker, Hughes, and Pinch) who used them to show off to each other and their lady friends. On the other hand, women and elderly men experienced the high-wheeled bicycle as unsafe because of the risk of falling off the saddle onto one's head (called "doing a header"—a real risk given the uneven nature of the mud tracks that passed for roads). Thus, there were two dominant meanings of the early bicycle: the bike as expression of machismo, primarily for sport, and the bike as excessively dangerous form of transportation. The transition to the safety bicycle can be understood only by following how the newer designs responded either to the sporting need (leading to larger-than-ever front wheels) or to the safety concerns (which eventually led to smaller and more evenly sized wheels).

Contestations over the meanings of technologies often appear in history. The battle over the best personal computer design between Apple and its rivals was in part a battle over the meaning we attach to our personal computers. These moments of contestation can be important guides as to how the wider society or culture can shape a technology. Such a contestation occurred over the first designs of electronic music synthesizers. Moog's synthesizer used a monophonic keyboard as its main way of controlling sound. At the same time that Moog developed his

synthesizer, a rival synthesizer was coming into use, invented by the musician and engineer Don Buchla. Working in the context of the sixties avant-garde milieu of Haight-Ashbury in San Francisco, Buchla conceived that the new source of electronic sound needed a new form of control, not one taken from an earlier technology, the organ. He designed synthesizers with touch-plate controllers that did not resemble conventional keyboards.

Buchla's synthesizer designs shared the meaning of many avant-garde and experimental composers, who thought that the new electronic music medium would be a way of making new sorts of music. Moog, on the other hand, found that his users included many commercial musicians, who wanted to use the synthesizer to play more conventional forms of music and to use it for new purposes, such as making "sound logos" (the few moments of electronic sound that sells a product, such as the "fizz" of Coca-Cola being poured) for the fast-developing advertising industry. The familiar keyboard with its white and black keys arranged in octaves served these groups better. Moog designed his modules around the "volt per octave" standard, which became the de facto industry standard for analog synthesizers. The meaning of the synthesizer as a keyboard instrument on which to play a wide range of different sorts of music, including conventional keyboard music, came to predominate, and Buchla's synthesizers were left only with a niche market.

In terms of design and society, the path taken by Moog is instructive, because Moog learned from the wider society and also shaped that society. Early on, Moog personally installed many of his synthesizers in studios and was able to watch how musicians used them and adapted his designs to their needs. Later he built a studio in his factory, employed house musicians, and offered free studio use to customers so that he could experiment and again learn from his users. He also started to shape the market, encouraging the new hobby of electronic music by running one of the first magazines devoted to it from his factory. Moog's salesmen played a crucial role as well by identifying and recruiting new users among young rock musicians, who wanted to solo with an instrument to match the sonic power and virtuosity of the electric guitar. By offering sound charts, which enabled musicians to set the synthesizer controls quickly on stage for producing various instrument sounds in live performance, and by setting up sales networks in retail music stores, Moog's marketers further reinforced the meaning of the synthesizer as a keyboard instrument.

The computer keyboard is another instructive example of how society becomes entwined with design. The keyboard layout in general use today, named QWERTY after the first six upper left letter keys, was inherited from the typewriter. The QWERTY keyboard was not the only design used on early typewriters; it was eventually accepted only after Underwood, the leading typewriter manufacturer, set up special schools to train typists. QWERTY is known not to be the most efficient design in terms of speed of typing or in terms of reducing repetitive strain injuries. In the early days of typing, speed was less of a premium than it is now, and it is even claimed that QWERTY was developed to slow down the typist intentionally so as to prevent type hammers from colliding and jamming. Because most users today are familiar with QWERTY, it is almost impossible now to change the technology even though theoretically better designs are available. A particular way of interacting with the technology is embedded within its design, playing, as it were, the role of a "script." Users follow the script because that is what they are familiar with. With new technologies, designers have to pay attention to the user script, or "configure the user" in a particular way. This is a crucial issue in the field of human-computer interaction, where it is important to understand the limitations on computer design stemming from the preferences and competencies of users. Sophisticated programming instructions that make too many demands on users may not be as successful as simple user interfaces that better meet users' abilities. How users will actually use a technology is so important for designers these days that some companies now employ anthropologists and sociologists to investigate this aspect of design.

## Gender and Design

Last, an important area in terms of design and society is gender. Technologies often assume a particular type of user, and some technologies are more likely to be marketed to women than to men and vice versa. Appliances used for domestic chores, such as dishwashers, washing machines, and stoves, are known in department stores as "white goods" (often having a functional white enamel finish) and are sold and displayed mainly to women in a different part of the store from where the more "sexy," exciting, and functionally differentiated entertainment appliances ("brown goods," often having a wood-grained or otherwise dark-colored finish), such as VCRs, DVD and CD players, and TVs, are displayed and sold mainly to men. That technologies are gendered is obvious if we think about cooking. Most women in the United States still have primary responsibility for operating the kitchen stove, but when it comes to the barbecue, the American man comes into his element. Having shown little previous interest in cooking, he suddenly becomes an expert on sauces, marinade techniques, and the best way to cook a steak! The outdoor, rugged nature of barbecue cooking (it is rare to see a white barbecue set) adds to masculine gender stereotypes and the notion that, when it really comes down to it, the male can still be the provider of the essential food for the family.

Gender is often explicitly incorporated into designs of technological devices. A striking example is found in the design of razors. Gillette's Sensor razors for women and men use the same blade cartridge, but the women's razor has a rounded body very different from the traditional straight T shape of the men's.

## Technological Design as Symbol and Substance

Technological design is inescapably part of society. Technologies can reflect the aspirations, dreams, and hopes of societies. They are symbolic. When technologies fail spectacularly, such as in the space shuttle *Challenger* and *Columbia* disasters, we witness not only the tragic death of a few people but also the death of a dream and a challenge to the identity of America as a "can do" nation. But technological design is also often routine and unnoticed. Indeed, the shuttle disasters pointed to the crucial design of small, unglamorous components such as O-rings, tiles, and polystyrene blocks. It is also often the design of the little things in life that we cherish most and make a real difference, whether freezer bags, sticky notes, or those pitch wheels on synthesizers.

**BIBLIOGRAPHY**

Bijker, Wiebe E. *Of Bicycles, Bakelites and Bulbs*. Cambridge, MA: MIT Press, 1995.

Bijker, Wiebe E., Thomas P. Hughes, and Trevor J. Pinch. eds. *The Social Construction of Technological Systems*. Cambridge, MA: MIT Press, 1987.

Cockburn, Cynthia, and Susan Ormrod. *Gender and Technology in the Making*. Thousand Oaks, CA: Sage, 1993.

Collins, Harry, and Trevor Pinch. *The Golem at Large: What You Should Know about Technology*. Cambridge, UK: Cambridge University Press, 1998.

Hughes, Thomas P. *Networks of Power*. Baltimore: Johns Hopkins University Press, 1983.

Nye, David. *American Technological Sublime*. Cambridge, MA: MIT Press, 1994.

Oudshoorn, Nelly, and Trevor Pinch, eds. *How Users Matter*. Cambridge, MA: MIT Press, 2003.

Petroski, Henry. *To Engineer Is Human: The Role of Failure in Successful Design*, reprint ed. New York: Vintage Books, 1992.

Pinch, Trevor, and Frank Trocco. *Analog Days: The Invention and Impact of the Moog Synthesizer*. Cambridge, MA: Harvard University Press, 2002.

Vaughan, Diane. *The Challenger Launch Decision*. Chicago: University of Chicago Press, 1996.

Winner, Langdon. *The Whale and the Reactor*. Chicago: University of Chicago Press, 1986.

*Trevor Pinch*

# The Social Role of Inventors

Innovation is the process by which new products and processes enter society. One of the forces shaping society is change in the physical equipment that people use to carry out their activities. These changes in the physical equipment come about through the activities of inventors and other developers of inventions. Inventors create new tools, and developers perfect the tools, turning them into socially useful products. Commercial producers provide large numbers of the products for society. Once produced in quantity, the technologies offer new powers, new efficiencies, and potentially, new social arrangements. New social arrangements, on the other hand, in turn provoke the search for and development of these new tools.

An important example is the steam engine, developed in England at the beginning of the eighteenth century to pump water out of mines. When James Watt (1736–1819) improved it through use of a separate condenser, the steam engine became a powerful prime mover, able to supply energy to mines, factories, ships, and locomotives. Commercial use of Watt's engine led to extensive changes in the way things were powered and produced. The railroads themselves led to extensive changes in the social fabric, quickly moving people and products over long distances.

The form of the steam engine was directed by the purposes of its human developers. Demand for a better prime mover from the mining industry helped direct Watt's attention to the costly and inefficient Newcomen engine. Watt's invention in turn led to formation of a commercial firm with entrepreneur Matthew Boulton (1728–1809) that developed and sold the engines. Intense demand for the efficient Watt engines led to industrial competition, patent infringement, and attempts by others to improve upon Watt's designs. The social matrix in which Watt's engine was embedded shaped not only its social effects but also future development of the technology.

There are several things to note about the steam engine. First, the invention was conceived and developed by James Watt, originally an independent inventor. Second, production of a commercial version required the involvement of a business firm. The act of invention seldom affects society directly. The social effects come as the invention is produced and distributed to society in a process directed by people we could call the invention's sponsors. Third, acceptance of an invention is hardly automatic. Boulton and Watt brought their products forward because there was a brisk demand for steam engines in mines. Often, however, demand must be created, especially if the idea is radically new to the users. This was true of the Xerox machine, whose developer (then named the Haloid Corporation) had to create a market for its product. Haloid's marketing department originally had thought there would be little demand for the product. Fortunately its engineers knew better and went boldly ahead, creating its first machines in 1950 and its world-beating 914 in 1960.

This article will now consider each of these three issues in more detail: the act of invention, the idea's development and production, and its reception by its future users.

## Invention, Design, and "R&D"

Inventions, by definition, are created by inventors. But what exactly is an inventor? Some inventors work alone, but today most "inventors" are in fact part of a corporation. About 80 percent of American patents are issued to inventors working in a corporate setting.

A good example of such development is the digital computer. The roots of the computer go back to the brilliant English scientist and inventor Charles Babbage (1791–1871) in the 1830s, who conceived complex mechanical computers. Babbage's ambitious designs proved difficult to build during his own lifetime. Further major development had to wait until Vannevar Bush (1890–1974) developed the Differential Analyzer at the Massachusetts Institute of Technology in the 1920s. This was the first effective analog computer. During World War II the first working digital computers were built in Britain, Germany, and the United States. One team at Harvard University was sponsored by International Business Machines and built an electro-mechanical machine called the Mark I. Another team, led by Presper Eckert (1919–1995) and John Mauchly (1907–1980) at the University of Pennsylvania, built a fully electronic machine, the Electronic Numerical Integrator And Computer (ENIAC). ENIAC filled a large room, had 17,468 vacuum tubes, and required 200 kilowatts of power and a highly skilled team to operate it. Similar teams built computers in Germany (Konrad Zuse's mechanical Z3) and England (Colossus). Although an individual could conceive of a computer (and many did), building one took help. This is true of many complex inventions, and is one of the reasons that today most inventions are developed by corporate teams.

The process of corporate invention is called "Research and Development," or "R&D" for short. Most of the effort (about 90 percent) is focused on development rather than research, because a corporation needs to focus on the immediately practical. Thus, although there is a lot of invention ("research"), it is the "development" that absorbs the lion's share of money and people. Most of the corporation's efforts are concentrated on getting products ready to sell. In fact, most corporate inventive activity is focused on making better versions of current products rather than totally new ones.

Inventions also evolve over time. The first computers seem very crude compared with today's sophisticated versions, which fit into the palm of one's hand and have internal memories that are enormous compared with ENIAC's. This continuous evolution is a product of further work that is still inventive, though less striking than the creation of the original invention. Refinement of the invention is usually called "engineering" or "design" work. Revision may seem less daunting, but it can require more intense effort than the invention itself. Thomas Alva Edison (1847–1931) did not invent the incandescent light, he simply created a better one. His efforts involved a staff of about twenty men and a worldwide search for a filament material. In the end he found success, not only in the light itself but also in the electrical system that powered it and then transformed society. Edison often continued to refine his own products long after their invention.

## Commercial Production

Once invention has taken place, we tend to assume that commercial production is automatic. This is not so. Getting something into production can be tough. According to Charles F. Kettering (1876–1958), himself a master inventor and head of General Motors' R & D efforts for many years, "the world's greatest endurance contest is getting an invention into a factory" (Boyd, p. 134). Very serious problems were encountered, for instance, in producing the first razor blades, the first diesel engines, the first photocopiers, and the little yellow adhesive notes that 3M calls Post-It™.

The effort involved, however, is not the only reason for considering production. The producer and other sponsors give the commercial version its distinctive shape and capability. The social effects the product influences are a function of its commercial versions, not that of the original invention. This impact can be demonstrated in many ways. If a television set costs half as much, many more people can afford to buy one. If a guided missile has twice the range, it may hit countries much further away. The original birth control pill was three times as strong as it had to be; reducing the dosage meant a pill with lesser side effects. When numerically controlled (NC) machine tools were first developed in the United States, they were more expensive than they had to be, because military requirements dominated the original forms. Leaded gasoline produced pollution that nonleaded gasoline does not. Examples could be multiplied indefinitely.

Most R&D, in turn, is aimed at upgrading current products, while a small amount is focused on totally new concepts. Consider the use of microwaves to heat food, a process first pioneered by inventor Percy Spencer at the Raytheon Corporation. Raytheon's appliance experts developed this basic idea into a commercial product, the large Radarange. Starting in 1947, the corporation sold the original versions of this large and expensive (several thousand dollars) product to hotels, restaurants, and railroads. It was a modest success. Then, under its subsidiary Amana, in 1967, Raytheon began selling a smaller version of the Radarange that would fit on a kitchen countertop. But an even bigger market was opened when American and Japan-ese firms revised the still large microwave downward into something the size of a breadbox, and the mass market "microwave" was created. By 1975 more microwave ovens were being sold than gas ovens.

In short, the inventor often delivers the invention to a commercial firm and not directly to society. The commercial firm then revises the invention and markets the product to society. Only when the inventor owns the firm making the invention does the inventor directly control the form and features of his invention.

Sometimes, too, an invention will not "get through" the commercial system because producing it will not yield a big enough profit. Introducing a new prescription drug in the American market costs something like $800 million. Thus an "orphan drug" situation may be created, in which making a drug would not produce large enough profits to justify such an investment. This situation is a great problem for the introduction of new drugs, just as the prevalence of lawsuits tends to limit, say, commercial interest in new birth control products. In other countries, with different market structures and government regulations, the dynamics of drug innovation may be quite different. A socialist country can have very different patterns in this regard from a capitalist one. Socialist countries such as France use more prizes and government support to stimulate invention than their capitalist counterparts do. By contrast, corporate profits appear to be the main incentive for Americans to invent.

## Resisting and Accepting

Society at large does not respond passively to inventions. Potential users may actively seek inventions or violently resist them. The reason is, of course, that acceptance of an innovation changes the social fabric. For centuries workmen resisted and destroyed new machines that changed the way that cloth and clothing were produced. They feared (with good reason) that their jobs might be displaced by these machines. The manufacturers, of course, sought the machines for just that reason; by paying fewer workers, they could make greater profits. Resistance can also take place because new procedures are unfamiliar and force changes to old habits. When the Linotype machine was introduced into commercial printing, some older printers committed suicide rather than learn the new methods. In history most new medical inventions were met by determined resistance.

An interesting case of social reaction to inventions is the Segway Human Transporter. Developed by inventor Dean Kamen (b. 1951), the Segway looks like an oversized manual lawnmower held upright. Using gyroscopes to maintain its upright position, the Segway can go long distances while holding a single individual. While Kamen was perfecting the device, publicity was used to create intense public interest. When the Segways were actually released to the market, however, sales were quite modest. The machines worked well, but the infrastructure to support them was simply not present. Think of the large number of social institutions we have to support the car, ranging from dealerships to gas stations to superhighways. Segways require special parking and electrical recharging facilities, and they must be supplemented by faster and longer-range transportation devices, such as cars, buses, or subways. Whereas tinkering with current products goes on constantly, infrastructure changes are a different matter, since they require changes in many social institutions. Segways represent such a major change.

Once a technology is created, its deployment in society channels further innovation. Once automobiles were accepted, the social institutions to support them had to be created. The emerging infrastructure of automobility led to stifling of alternative approaches, as interurban railroad tracks were ripped out and highways encouraged urban sprawl. The very social agencies necessary to create automobiles resisted change that might take society in different directions. Society became committed to the automobile. Segways may become part of the picture in time, but supporting them will require major commitments, adjust-

ments very much like those used to support the automobile.

Although products in a modern economy may stabilize from time to time because of government standards or monopoly producers, typically they are in constant change. Users force changes in technology by insisting on more features, more reliability, and lower prices. This creates a pressure for innovation that fuels research and development. Usually this pressure leads to constant change in products. There is always work for inventors.

## The Interplay of Invention and Society

The processes leading to social change in the devices and systems we use is not linear. When inventors invent, society does not simply react. Rather, there is a constant interplay between social demand leading to innovation and innovation leading to social adaptation. Furthermore, current invention is channeled by the technology and the power structure of society already in existence. Although we tend to see inventors as the drivers of social change, there is in fact an enormous variety of occupations associated with creating, changing, assimilating, and marketing inventions. Businesspeople respond to social demands, but through that response they shape the forms that products take. Firms organize teams to create more products and attempt to channel social demands to the products they make. There is a never-ending battle to determine the direction that society will take. Inventors play an important role in social change, but their activities meet resistance as well as support from other social forces.

**BIBLIOGRAPHY**

Boyd, Thomas A. *Professional Amateur: The Biography of Charles Franklin Kettering.* New York: Dutton, 1957.

Gausewitz, Richard. *Patent Pending: Today's Inventors and Their Inventions.* Long Beach, CA: Alston Publishing, 1983.

Israel, Paul. *Edison: A Life of Invention.* New York: John Wiley and Sons, 1998.

Kelly, Patrick, and Melvin Kranzberg, eds. *Technical Innovation: A Critical Review of Current*

*Knowledge.* San Francisco: San Francisco Press, 1978.

Kemper, Steve. *Code Name Ginger: The Story behind Segway and Dean Kamen's Quest to Begin a New World.* Boston: Harvard Business School, 2003.

Latour, Bruno. *Science in Action: How to Follow Scientists and Engineers through Society.* Cambridge, MA: Harvard University Press, 1987.

Pacey, Arnold. *Technology in World Civilization.* Cambridge, MA: MIT Press, 1990.

Rogers, Everett. *Diffusion of Innovations*, 4th ed. New York: Free Press, 1995.

*Ron Westrum*

# K

## KNOWLEDGE CONSTRUCTION.

Scholars in the social studies of science have developed a range of different approaches to the understanding of science. An interest in the *process* of *knowledge production* emerged in the late 1970s, just a few years after the new sociology of scientific knowledge (SSK) had taken off. This production-oriented approach shares with SSK the understanding that the content of (natural) science is accessible by way of empirical sociological analysis and should be subjected to it. Both schools of thought are convinced that science is not to be investigated merely as a social institution (in the tradition of Merton), that science's epistemic core is a matter of investigation in its own right and should not be abandoned to the philosophy of science. In their perspectives on science's epistemic core, however, the two approaches are complementary. Whereas SSK primarily focuses on the social causes of the scientists' convictions and knowledge-beliefs—that is, it centers on *science as knowledge*—the constructivist approach turns its attention to the constructive elements of scientific production—that is, it considers *science as practice.*

Interest in the process of knowledge production raises a number of important issues and has led to an improved understanding of science as a practical accomplishment. Scientific practice is firmly embedded in local environments; this observation brings the privileged sites of knowledge production into view: the scientific laboratories. The desire to analyze closely how knowledge is created calls for a specific methodological handle: an ethnographic approach, as chosen by "laboratory studies." These studies iden-

tify strategies that scientists employ in their work, strategies that are as diverse as forms of object configuration, discursive fact production, devices of representation, and social mechanisms of consensus formation. Since the first laboratory studies were published in the 1980s, the science-as-practice approach has proven particularly productive, and its concepts and methods have been extended to the investigation of new domains.

### Focus: Laboratory and Reconfigured Objects

An interest in how knowledge is produced (or constructed) brings the locales of such activities into view. Since the nineteenth century the laboratory has been developed as the paramount site of knowledge production in modern science. Although scientific knowledge is produced at other sites as well, the laboratory has come to symbolize and substantiate the power of science. This power can be linked to the rearrangement of the natural order that is effective within its boundaries: In the laboratory the phenomena of investigation are removed from their natural context, they become reshaped, or "reconfigured" in Karin Knorr Cetina's phrase, to the benefit of scientists. The shaping process "improves" upon the natural order in several ways: In the laboratory, objects or phenomena do not need to be dealt with *where* they occur, *when* they occur, and *as* they occur in nature. In a field science such as meteorology, for example, researchers need to take their equipment to remote places and wait for ephemeral phenomena to manifest themselves, but in laboratory settings scientists attempt to make things happen at will. In addition, lab objects differ in important aspects from their counterparts in natural environments. In contrast to field mice, for example, laboratory mice provide standardized research "material" that can be duplicated at low cost and is amenable to a full sequence of experiments. These examples suggest that modern laboratories rely on a highly complex technical culture that involves costly apparatus, elaborate instrumentation, and sophisticated technical procedures. Yet, the reconfiguration of the object world is only one characteristic element of the laboratory order. In addition, the social becomes reconfigured as well and is aligned with the specific requirements of the object world in the lab. For example, collaborations are forged to confront the object world optimally, with form and size of collaborations differing widely across fields, from the small-sized teams that cooperate in many life-science labs to today's particle physics collaborations, which may involve more than 2000 scientists. In this article the different construction processes involved in laboratory work are explored further, but the question of how such processes can be investigated empirically will first be raised.

## Approach: Ethnography and "Laboratory Studies"

The early analysts of scientific practice ventured into the research sites of fields as diverse as molecular biology, colloid chemistry, astronomy, or particle physics by employing what has since become termed the "laboratory studies" approach as exemplified by the work of Karin Knorr Cetina, Bruno Latour, Michael Lynch, Sharon Traweek, and Steve Woolgar. Laboratory studies borrow their approach and methodology from cultural anthropology: Analysts proceed in close analogy with ethnographers who study foreign cultures. Indeed, in comparison, scientific cultures are considered just as outlandish, their goings-on and practices equally worthy of scrutiny and in need of explanation. The main procedure for obtaining insight into a foreign culture is "participant observation," in which the ethnographer participates in the life of the culture and its locales (here, a laboratory) for a substantial period of time, observing the daily work and routine activities of its members, listening in on shop talk, interacting with the lab members in a range of informal and formal ways (such as interviews). When the observer is interested in how the scientists construe facts, participant observation has important advantages over other approaches that rely on text analysis or interviews alone. In contrast to an analysis of scientific publications, participant observation allows the analyst to get a grip also on the fleeting character of fact making, on what happens before a result is solidified and becomes accepted as valid within a scientific community. In contrast to approaches that rely uniquely on the scientists' oral accounts, participant observation allows the observer to uncover also those dimensions of scientific work that are taken for granted. In their everyday activities, scientists rely to a considerable extent on implicit knowledge, embodied practice, and practical skills that they hardly ever explicate. An ethnographic approach to laboratory culture, combined with ethnomethodology and discourse analysis, allows one to disentangle the intricate social machinery of fact making. In order to grasp such understanding, ethnographers need to reach a close understanding of the technicalities of scientific work in combination with the social relations that govern lab life. As ethnographers, laboratory students are required to learn the idioms of the observed field and spend a considerable amount of time, possibly as much as several years, in its environment. Yet, simultaneously, they need to keep their distance from the field, for at least two reasons. First, familiarity with the object of study needs to be bracketed to avoid "going native" (as ethnographers put it), which would entail a diminished attentiveness to the field's cultural practices as they unnoticeably become taken for granted by the observer. Second, social students of science should not let themselves be intimidated by the high status of science in our societies. Instead, scientific practice is to be investigated as irreverently as the practice of other "tribes" (or cultures) is.

## Results: How Facts Are Constructed

What, then, do close and ongoing on-site observations of "science in action" (in Latour's phrase) bring about? Perhaps the foremost observation concerns the *local nature* of scientific practice. Laboratory work exploits the contingencies of local contexts with respect to the equipment and research facilities at hand, the interactional circumstances, the conventions embodied in laboratories, the combined expertise gathered in a research team, and the organizational setting in which it is embedded. When exploiting local resources, scientists improvise and draw on a wide and imaginative repertoire of tentative solutions: Scientists "tinker" when they try to make things work. In this context, tinkering (or "bricolage") has no negative connotation; rather, it presents a characteristic feature of the scientists' activity. As a consequence, the research problems are locally constituted in much the same way as are the research objects, the tools, and the manner in which all of these are handled and become assembled.

Tinkering and improvisation make scientific practice appear messy and disorderly to the ethnographer at first sight, and this observation stands in contrast to the image of science as a straightforward, linear, and rational pursuit of solutions to predefined problems, as it is depicted in the scientists' research papers. This leads to another important observation: Scientists do not disclose the open, contingent, and negotiated character of practical work in their accounts, which are condensed and purified versions of what goes on in the laboratory. In the process of constructing knowledge, scientists "produce order" (in the words of Latour and Woolgar) and conceal the messy traces of their work. This implies that science does not merely represent reality as it is "out there"; scientific work is constructive, and it employs a whole construction machinery. What later appears as a natural phenomenon or as unproblematic data is the result of a complex production and selection process. Thus, terms such as "fabrication" or "manufacturing" which are sometimes used and the notion of "construction" refer not only to the scientists' active configuration of their research objects (as discussed earlier) but also to the process by which scientists make sense of their observations and transform data and statements into "facts." In

this perspective, the laboratory appears as a system of *fact construction*. Because laboratory studies have emphasized the central importance of various construction mechanisms for the production of knowledge, they have become closely associated with the concept of "social constructivism."

As a system of fact construction, the laboratory is a repository of a whole set of competences, practices, tools, and resources. In the manipulation of research apparatus and objects the scientists heavily draw on *embodied skills*. They need such skills to track down errors of a malfunctioning apparatus, breed cell cultures, effectively use a confocal laser scanning microscope, or interpret the ensuing images. Embodied practice remains implicit; it is a tacit asset that can be explicated only with difficulty. Novices learn laboratory skills by doing and by watching and cooperating with their experienced peers; manuals are of little help. Laboratory scholars were the first to emphasize the central importance of hands-on experience in scientific practice, which sociological and philosophical analysts had long considered a mainly theoretical activity.

Yet, not all competences reside primarily in the body. In addition to being a manipulative and physical occupation, scientific practice is also an interpretive, representational, and literary activity. Laboratory scholars have investigated the corresponding procedures and strategies of the scientists' daily routines.

Data and other outcomes and products of scientific practice are rarely—some would say never—unambiguous, complete, definite and univocal. They retain a high degree of interpretative flexibility. Eliminating ambiguity relies on negotiation and techniques of persuasion. In fact, not only the data and their interpretation are subject to negotiation; so are a myriad of other elements that intervene in the making of knowledge: the quality of research material, the applicability of a procedure to a particular problem, the reliability of a colleague's claims, the need to invest resources in a costly apparatus. This explains the abundance and intensity of oral (and, to a lesser extent, written) communication that forms an integral component of the scientists' work. In the laboratory, talk is part of the action and is not necessarily descriptive of the work in progress. Embedded in dense non-verbal communication, such as gestures or scribbles on the chalkboard, talk is inextricably tied to a specific situation and is infused with multiple references to it ("indexical expressions" in Harold Garfinkel's terms). This makes it difficult to understand its meaning outside of the context from which it originated. Agreement between scientists, then, is a result of situated procedures. By using methods from conversation and discourse analysis, students of laboratories have analyzed how agreement comes about. For example, they have shown how visual displays (such as autoradiographs) become "dissected" by talk and, as a result, lose their opaque nature.

The example of how to read autoradiographs points to yet another central element of scientific practice: the production and interpretation of images and graphic displays. The visualizations with which scientists work do not simply portray nature as it is; they are the result of a multilevel process of production, translation, and transformation that involves their active manipulation as well as interpretation. Images and displays serve, for example, to condense an extensive set of data graphically or to visualize and bring out new patterns in the data. The intricate visualization practice may be conceived as a "transformation of rats and chemicals into paper" (in Latour's words) that not only fosters understanding of research problems and results but also assists scientists in communicating their results across local contexts and in convincing their colleagues of the work's importance and validity. Graphic displays and other paperwork have the advantage that they travel easily: they are mobile and yet keep their characteristic features, they can be reproduced at low cost, and they can be recombined and compared with displays of other origins.

Besides images and graphic displays, the paperwork produced in the laboratory also takes the form of written documents of all sorts. Scientists casually scribble notes as part of the thought and interaction process; they report and document experimental, conceptual, and computational activities in laboratory notebooks, work out preliminary results on paper, and make them public in preprints. Finally, scientists stake their knowledge claims in *research articles*. The texts produced differ in their purposes and in the audiences that they target. Consider the difference between the casual notes in which a scientist records the daily ongoing activities and results for her own memory and that of close collaborators, on the one hand, and her research papers, which target the entire research community, on the other. A close reading of the latter reveals that both their author and the actual production process underlying their knowledge claims have been rendered invisible. This is achieved by avoiding words and expressions that point to the author as individual, to the local situatedness of research practice, or to its temporal structure and chronology. As a result, scientific discourse (and practice, more generally) appears to hide itself in articles. Objectivity effects derive from rhetorical procedures that constitute a central element in the transformation of statements into solidified facts. As a result, scientists seem to be simply "reporting natural facts" and the constructed nature of knowledge disappears from view.

## Outlook: Extending Laboratory Studies

The early laboratory analysts pioneered the opening up of the black box of natural science to social science analysis. They demonstrated that a sociological investigation of scientific practice is feasible, and they brought back from their laboratory visits interesting and, at times, provocative observations regarding the construction of scientific facts. At this first stage of the constructivist approach to knowledge, the conclusion that nothing epistemically special is happening in the scientific laboratory came as a surprise and raised wide attention. In contrast to the characterization of science as a purely rational pursuit, this approach emphasized the mundane and routine features of scientific practice.

Since then, laboratory analysts have been analyzed the general mechanisms of fact construction in more detail. Concurrently, an interest in scientific practice and laboratory culture has become widely shared across areas of science and technology scholarship—for example, by historians of scientific demonstration and experimentation, by symbolic interactionists interested in the social worlds of science, by cultural analysts of knowledge production in different cultural contexts, and by feminist scholars who uncover the gendered construction of knowledge. In this process, the constructivist approach to knowledge has been extended to new fields and topics of investigation. A small and necessarily biased selection of these topics will be considered in the following paragraphs.

As a theoretical notion, the laboratory is associated with the reconfiguration of natural and social orders that characterizes modern science. Although this conception resonates directly with a typical laboratory science such as molecular biology, it can also be fruitfully applied to other scientific fields and sites of knowledge production. Consider the field sciences, which have to make do with different kinds of local contingencies that frame their research practice, such as the need to manage heterogeneous sets of practices, sites, and actors. Moreover, laboratory and field are not as neatly separated as one might expect: Modern field sciences combine field measurements with laboratory work, while different forms of "labscapes" (in Robert Kohler's phrase) draw nature into the lab or bring the lab to the field. The lab-field border is managed and negotiated differently in different sciences. The clinical setting in modern biomedicine also constitutes a kind of field, and an extended body of sociological and historical literature has begun to address the processes of mutual constitution between the laboratory and clinical practice.

The idea of object reconfiguration in the laboratory can be applied as well to the alternate object worlds produced through computer simulation, although perhaps in a somewhat more metaphorical sense. Computer simulation allows for the constitution of digital laboratories in which the phenom-enon under investigation (be it a physical process, the design of a measuring apparatus, or the temporal evolution of climate) is amenable to reconfiguration and manipulation of great flexibility. In this case, when manufacturing facts, scientists are required to negotiate the different ontological orders and epistemic features between the simulated and the material object worlds. Extending the laboratory notion to include such simulated environments raises questions about the limitations of configuring objects in laboratory settings and about the possibilities of transferring laboratory results (back) to "real world" settings.

The central significance of local idiosyncrasies in the construction of facts comes under scrutiny when one considers the increasing importance of translocal computer-supported communication and cooperation in science. Virtual communication spaces appear to supplement and, occasionally, to replace traditional sites of knowledge production. As yet, it is an open question whether this development implies that the local situatedness of scientific practice loses its prime importance or needs to be reconsidered. The steady increase in collaborations that involve scientists and machinery across different sites, institutions, or national cultures calls for a fresh take on scientific practice and knowledge production. For example, negotiation processes might assume different forms and outcomes when occurring in face-to-face situations than when they take place by e-mail. Widening the angle of observation to address knowledge production in projects of international collaboration, on the other hand, allows one to explore also the diversity of national knowledge cultures, such as the distinctions between high-energy physics cultures in the United States and in Japan.

The constructivist approach to knowledge has produced an impressive number of empirically sound and detailed investigations of scientific practice in a wide array of scientific specialties. This approach has allowed its proponents to identify the general features of fact construction that are shared across contemporary sciences. Recently the diversity of scientific cultures and the particulars of fact construction mechanisms have moved to the center of attention. Empirical research has shown that specific features emerge more readily when one compares different cases. An important challenge for future investigations lies therefore in the direct comparison of different knowledge cultures with respect to their "epistemic machineries" (in Knorr Cetina's phrase)—that is, their machineries of knowledge production. This challenge becomes even more pronounced when one considers sites of knowledge production outside the traditional boundaries of academic science and research. The question then arises whether the concept of "laboratory" (and associated notions) as promoted by the science-as-practice approach can be productively applied to such other sites of knowledge production as well. Or will new conceptual frameworks be needed if we want to come to grips with knowledge practices in areas as diverse as currency markets, insurance companies, environmental agencies, patient organizations, or service industries?

## BIBLIOGRAPHY

Collins, Harry M. *Changing Order: Replication and Induction in Scientific Practice.* London: Sage, 1985. A convincing demonstration that replication in science is problematic.

Knorr Cetina, Karin. *The Manufacture of Knowledge: An Essay on the Constructivist and Contextual Nature of Science.* Oxford: Pergamon Press, 1981. Together with Latour and Woolgar's *Laboratory Life, The Manufacture of Knowledge* is today considered a classic of the science-as-practice approach.

Knorr Cetina, Karin. *Epistemic Cultures: How the Sciences Make Knowledge.* Cambridge, MA: Harvard University Press, 1999. Exceptional in its design in comparing two sciences (particle physics and molecular biology) with respect to their knowledge-making machineries.

Knorr Cetina, Karin, and Michael Mulkay, eds. *Science Observed: Perspectives on the Social Study of Science.* London, Beverly Hills, CA, and New Delhi, India: Sage, 1983. A collection with articles from (almost all) authors of early laboratory studies.

Kuklick, Henrika, and Robert E. Kohler, eds. *Science in the Field* (special issue), *Osiris* 11 (1996). A collection of articles drawing attention to characteristic features of the, as yet insufficiently studied, field sciences.

Latour, Bruno. *Science in Action: How to Follow Scientists and Engineers through Society.* Cambridge, MA: Harvard University Press, 1987. Based on a laboratory study, the book outlines the actor-network approach to studying science in action.

Latour, Bruno, and Steve Woolgar. *Laboratory Life: The Construction of Scientific Facts,* 2nd ed. Princeton, NJ: Princeton University Press, 1986. As one of the first book-length accounts of a laboratory study (the first edition was published in 1979), this volume presents a good introduction.

Lynch, Michael. *Scientific Practice and Ordinary Action. Ethnomethodology and Social Studies of Science.* Cambridge, U.K.: Cambridge University Press, 1993. A thoughtful analysis and reconstruction of both ethnomethodology and sociology of science, as well as of their complex interweaving.

Lynch, Michael, and Steve Woolgar, eds. *Representation in Scientific Practice.* Cambridge, MA: MIT Press, 1990. A valuable contribution to efforts to put scientific representations and visualizations on the agenda of science and technology studies.

Pickering, Andrew, ed. *Science as Practice and Culture.* Chicago and London: University of Chicago Press, 1992. A collection of papers providing ample visibility to the science-as-practice approach in its different varieties and the corresponding theoretical frameworks.

Schatzki, Theodore R., Karin Knorr Cetina, and Eike von Savigny, eds. *The Practice Turn in Contemporary Theory.* London and New York: Routledge, 2001. A volume juxtaposing contributions from practice thinkers across sociology, philosophy, and science and technology studies, drawing out similarities and distinctions between different practice approaches.

Sismondo, Sergio. *Science Without Myth: On Constructions, Reality, and Social Knowledge.* Albany: State University of New York Press, 1996. A volume exploring, among other topics, various concepts and metaphors of "social construction" as employed in recent science and technology studies.

Traweek, Sharon. *Beamtimes and Lifetimes: The World of High Energy Physicists.* Cambridge, MA: Harvard University Press, 1988. A widely read account on the world of particle physics research, with its places, machines, and communities, in anthropological perspective.

*Martina Merz*

# L

## LAW AND SCIENCE.

The study of science and law explores conflicts that arise when the law acts both as a consumer and a producer of science. Both types of conflicts are discussed in turn.

### The Law as Consumer of Science

Scientific research provides the law with important information for decision making. Science helps juries determine whether hair found at a crime scene matches that of the suspect, and it helps regulators quantify the extent to which pollution from a factory impairs the health of children in neighboring communities. However, because scientific information carries so much weight in legal decision making (leading in extreme cases to a criminal conviction or bankruptcy), the law has not been content simply to take the scientists' research at face value. Both the courts and the regulatory agencies employ multiple checks on the quality of science and expert analysis to ensure that the information is reliable. In fact, in its role as a consumer of science, the law can sometimes place intrusive demands on science and scientists, seeking answers to questions that lie beyond the realm of science and imposing quality checks that differ substantially from the controls the scientific community imposes on itself.

*Courts.* The courts are recipients of a great deal of scientific information, much of which is provided as part of the adversaries' efforts to present the cold, hard "facts" in support of their case. The assimilation of scientific knowledge by the courts typically proceeds through three stages, each of which presents opportunities for errors. First, the scientific research available to resolve a dispute in the judicial process is limited to the information

presented by the parties. Judges and juries are generally barred from conducting their own research or consulting independent scientists, except under tightly controlled circumstances. Indeed, in most states, judges and juries are not even allowed to ask questions of the parties' experts or take notes on their testimony. These restrictions on the incorporation of science into judicial proceedings present a potentially significant source of error. Lawyers representing the parties often lack the scientific expertise to identify the best experts and, more importantly, may lack the incentives to select mainstream experts that present testimony at the middle of the "bell curve" of scientific opinion. Instead, adversaries tend to select experts with more extreme views when this testimony will provide better support for their client's case.

During the second stage, the parties can seek to exclude or censor the scientific testimony of their opponents by proving that the challenged science is not reliable. Even qualified experts can be excluded from testifying if their testimony is not based on research capable of being replicated or tested. To determine the merits of these challenges, the judge might conduct an evidentiary hearing and rule on the challenged evidence separately. Courts screen scientific evidence before it reaches the jury for several reasons. First, excluding unhelpful expert testimony is believed to protect the jury from confusion, especially if the expert has impressive credentials and testifies on esoteric issues, but the testimony is unsupported by evidence. Excluding complex but unhelpful evidence can also speed up the adjudication of cases. Finally, excluding unreliable testimony provides attorneys with some, albeit weak, incentives to hire legitimate experts, although the attorneys might still be inclined to select experts who present only one perspective, rather than offering mainstream theories or methods.

Despite compelling reasons for judges to serve as gatekeepers over the quality of scientific evidence, courts have had a difficult time establishing a general test for determining whether scientific evidence and testimony is reliable. For much of the twentieth century, the federal courts relied on the *Frye* test for admitting scientific evidence, which considered only whether the testimony was "generally accepted" by the scientific community (*Frye* v. *United States*, 293 F. 1013, 1014 (D.C. Cir. 1923)). In applying the *Frye* test, however, the courts confronted problems in determining who the relevant scientific community might be, how to determine when results are "generally accepted," and whether the test would inappropriately exclude path-breaking science. As a result, in 1993 the Supreme Court revised the test in *Daubert* v. *Merrell Dow Pharmaceuticals, Inc,* 509 U.S. 579, holding that scientific testimony was reliable if the scientific methods used by the expert to support his or her conclusions were capable of being tested and replicated. Over the past decade, the trial courts have struggled with this new test because it places considerable responsibility on judges to evaluate whether methods are truly scientific.

The third stage of science assimilation into the courts entrusts the jurors and judge with weighing all of the admitted information, scientific and otherwise, under the applicable legal standard ("beyond a reasonable doubt" for a crime or "more probable than not" for a civil dispute). This stage grants great deference to the jury's weight-of-the-evidence decisions. Typically, jury verdicts are reversed only if the judge presiding over the case finds the verdict to be "clearly erroneous" in light of the evidence.

The assessment of whether the courts do a competent job of processing scientific information is mixed. Concerns about whether the *Daubert* test is truly the appropriate one for screening scientific evidence (in the second stage), or whether it is specific enough to be used in a competent and consistent way, have multiplied over the years. These concerns are reinforced by evidence that the courts' rulings on the reliability of challenged scientific evidence are erratic. Some judges, for example, have excluded expert testimony because they did not believe that the expert's opinion was supported by the weight of the underlying evidence, a determination that some argue should be left to the jury. Conversely, other judges routinely admit forensic evidence into criminal trials, such as fingerprint testimony, without confronting the deficiencies in the scientific methodology underlying the testimony.

Although surprisingly few empirical studies of juries' decision making have been performed, there is also sharp criticism of juries with regard to their competence to assess and use complex scientific information (the third stage). Some of the most criticized verdicts, especially in mass litigation concerning the morning sickness pill bendectin and silicone-gel breast implants, have been in conflict with scientific opinion on causation. Yet some commentators argue that these aberrant verdicts result not so much from the jury's lack of scientific competence but rather from the jury's inclination to "nullify" scientific evidence when, for example, a defendant is especially culpable or a prosecutor/policy is unfair. Although this theory of jury nullification is controversial, it does provide an alternative explanation for verdicts that are inconsistent with the weight of scientific evidence.

Regardless of how well juries and judges assimilate scientific information, it is evident from the description of the judicial process that current adversarial processes are a decidedly poor way to capture the best that science has to offer judicial decision makers. Yet one must keep in mind that the primary function of the courts is to resolve individual disputes. It is far more important to the functioning of the courts to do justice through an evenly balanced debate controlled by the opposing parties and formal rules of process than to provide an open forum for the most capable scientists and cutting-edge research to weigh in on these case-specific disputes.

*Regulatory Agencies and Legislatures.* The function of legislatures and agencies is different from that of the courts, because these political bodies are charged with solving larger and less clearly defined social problems. The legitimacy of the political branches thus depends in significant part on their ability to make effective use of the available scientific knowledge. As a consequence, legislatures and agencies place a higher premium on scientific comprehensiveness and accuracy in resolving social issues than do the courts. Rather than establishing formal rules for when science can be introduced into the decision-making process, legislatures and agencies consider any expert guidance and scientific research that informs their decisions. As a result, science is continually assimilated by the political branches: Expert staffers in both agencies and legislatures conduct their own research, call upon experts, identify studies in the literature, and otherwise attempt to ensure that scientific information is identified and used to the greatest extent possible.

The same three stages governing the assimilation of scientific information into the courts—(a) the identification of relevant science, (b) the exclusion of unreliable science, and (c) the weighing of the available scientific information—look quite different when the political branches use science. The agencies and legislatures employ far fewer formal checks and restrictions than the courts in their identification and screening of available evidence during the first two stages of the process, but they employ more intrusive oversight processes during the third stage, which is concerned with how these political bodies use the available scientific information in their final decisions. Concerns about hidden political biases in the agencies in particular led the U.S. Congress to establish numerous mechanisms to oversee agency use of science. For example, persons who object to the agencies' use of the available scientific information are allowed not only to submit comments to the administrative record underlying an agency action, but also to challenge the agency's action in court, although they must prove that the regulations are "ar-

bitrary and capricious." Congress, and in some cases the agencies, have created prominent science advisory boards to provide an independent evaluation of the scientific research used in regulation. Controversial studies used for regulation have sometimes been subject to expensive reanalysis or verification, and elaborate external peer review processes can also be employed. Use of scientific information by agencies and legislatures is also vulnerable to political attack, particularly when science of bad quality is used or obvious scientific errors are made. On more than one occasion Congress has held oversight hearings to investigate potential problems with an agency's use of science in the decision-making processes. Members of Congress also draw the media's attention to scientific errors committed by their congressional opponents, thereby producing strong incentives for lawmakers to use science competently in their legislative proposals.

It is thus not surprising that, according to the science-policy literature, Congress and the agencies do a relatively competent job of using science to support their decisions. Several book-length studies of regulatory science reveal that the science used by agencies is usually of good quality, despite occasional allegations by affected interests that the agencies are scientifically incompetent or that they use science to achieve political ends. A closer look, in fact, reveals that at their core most of these allegations reflect disagreements about underlying policy and value choices regarding how much evidence is enough to support expensive protective standards rather than disagreements about how well the agency used the available science.

Despite evidence that the political branches tend to find and use science of high quality, the literature does suggest that Congress and the agencies do a much poorer job delineating the point at which the science supporting a decision leaves off and the policy making begins. In part because science provides a convenient way of masking controversial policy decisions, legislatures and agencies tend to "scientificate" policy, making policy decisions appear as if they have or can be resolved exclusively by science. For example, rather than stating that an expensive air pollutant standard is set conservatively to err on the side of public health when the science is uncertain, the agency tends to state much more obliquely that the quantitative level was based on the "expert judgment of the administrator of the agency." The literature in science policy, as well as numerous laws and regulations, provides a number of examples of decisions that appear to be based on science or are supposed to be resolved with science, even when the tools of science are unable to answer the questions definitively.

*Future Decisions.*   Science creates substantial headaches for lawmakers and courts when the scientific issues or discoveries do not fit neatly within existing precedents or legal models. Technologies relating to abortion and *in vitro* fertilization, for example, are forcing the courts to confront new conceptions of personhood, motherhood, and reproductive rights. Likewise, developments in measuring and testing low levels of chemicals in the environment for adverse effects have forced regulators to revisit regulatory standards governing these risks. Genetic research has raised numerous important questions about genetic privacy, forcing courts and legislatures to grapple with the conflicts between this privacy and the utility of DNA technologies in facilitating the identification of criminals or providing medical screening.

*Spillover Effects on Scientists Themselves.*   As the law looks increasingly to scientists for answers, it also can intrude on the scientific process itself, particularly for scientists whose research is used to inform social decision making. The scientists, for example, may be confronted with deliberate efforts by affected parties to halt their research or tarnish their reputation simply because the discoveries affect these parties in unwelcome ways. Harassing subpoenas requesting the delivery of all laboratory research records in ongoing court cases and burdensome data-sharing requests through public-record statutes have been used by interested parties on both sides of an issue to distract or even intimidate academic or government scientists whose research has potentially adverse implications for their interests.

## The Law as Producer and Regulator of Science

The law not only consumes information; it also regulates and encourages the production of scientific information.

*Incentives for the Production of Science.*   The law encourages the production of scientific information in large part by providing scientists and research sponsors with a property interest in their discoveries. This property interest, for example a patent, produces a strong incentive for research in disciplines that promises discoveries with commercial value, such as biomedical research. In resolving patent claims, regulators often struggle to determine whether a scientific discovery is sufficiently novel or useful to be entitled to a protectable property interest. Developments in genetic research, for example, require regulators to decide whether the codes for valuable gene fragments, or the creation of new living organisms, can be patented, giving the discoverer the right to exclude and charge others for the material or information. Because legal decisions regarding intellectual property rights have significant implications for the sharing and use of these scientific discoveries, the scientific community has an important stake in how the law resolves issues relating to which scientific discoveries are ultimately protectable as private property.

Less important, but still worthy of note, the law also requires many manufacturers to undertake prescribed safety research on their products, especially for drugs and pesticides, as a condition of selling the products in the United States. Although this mandatory testing is conducted under rigid protocols, it is nevertheless a significant source of scientific activity, especially in the private sector. It is not clear how much of this research is useful to the broader scientific community, however, because a considerable amount of the research is never published, and it can be subject to trade-secret protection.

*Restrictions Governing Federally Funded Research.* A substantial portion of scientific research is federally funded. Researchers engaged in federally funded research do so under contracts that grant the federal government the ability to regulate features of the research, and as a result this federally funded research often comes with strings attached. One type of restriction derives from social considerations. For example, the federal government requires institutions receiving federal funding to establish oversight panels to ensure the ethical treatment of human subjects in all research. Federal law also prohibits the use of federal funds to conduct certain research on embryonic stem cells or human cloning. Some have argued that these prohibitions will put the United States at a competitive disadvantage because biomedical research, particularly on stem cells, promises important new discoveries and is publicly financed in other countries, such as Great Britain.

Federal standards also require that federally funded research meet minimal standards for quality. For example, federally funded researchers may not engage in "scientific misconduct." The data underlying federally funded research must also be available to the general public, who can request the information through the Freedom of Information Act or related public-record statutes.

By contrast, some restrictions on federally funded research, such as those on the ability of researchers to patent discoveries made with federal funds, have been lifted in recent years. This, in turn, has created its own ripple effect. Historically, scientists who made discoveries in the course of federally funded research were not allowed to claim intellectual property rights to their discoveries. In 1980, this restriction was eliminated by the Bayh-Dole Act, and federally funded researchers may now patent their discoveries and inventions. This new privilege has been the source of considerable criticism. The ability of federally funded researchers to obtain patents impairs open communication within the scientific community, with academic scientists in some fields imposing licensing fees on other scientists who seek to use their research or delaying publication or dissemination of their results until a patent is obtained. Giving researchers the ability to patent federally funded research has also caused some universities to provide greater support to departments that can provide an income stream to the university, such as chemistry departments, as opposed to other departments where profitable patents are unlikely.

*Prohibitions against Certain Types of Scientific Research.* Finally, there are legal restrictions that apply more broadly to all scientific disciplines, regardless of funding. Human cloning is illegal in a number of countries, although it is not prohibited for private research in most states of the United States. Unethical research on human subjects is also prohibited by an international treaty. In the United States, if research presents national security risks of grave danger, the government might intervene and preclude publication and dissemination of research findings. (These controls are still more aggressive when applied to federally funded research, and they can sometimes even restrict the nationality of those who are allowed to work in certain laboratories.) Practical enforcement of these unilateral restrictions is difficult, but the prohibitions exist nevertheless.

It should be noted, however, that because of the sanctity of free speech in the United States, exemplified by the First Amendment, significant prohibitions on research occur only in rare cases where the research is perceived to present extraordinary social risks or where research protocols are so repugnant to social morals that a legal prohibition is necessary.

*See also* **Law and Technology**

### BIBLIOGRAPHY

American Association for the Advancement of Science. *Science and Technology Policy Yearbook.* Washington, DC: published annually from 1991.

Angell, Marcia. *Science on Trial: The Clash of Medical Evidence and the Law in the Breast Implant Case.* New York: W. W. Norton, 1997.

Faigman, David L. *Legal Alchemy: The Use and Misuse of Science in the Law.* New York: St. Martin's, 1999.

Goldberg, Steven. *Culture Clash: Law and Science in America.* New York: New York University Press, 1994.

Huber, Peter W. *Galileo's Revenge: Junk Science in the Courtroom.* New York: Basic Books, 1993.

Jasanoff, Sheila. *The Fifth Branch: Science Advisers as Policymakers.* Cambridge, MA: Harvard University Press, 1990.

Jasanoff, Sheila. *Science at the Bar: Law, Science, and Technology in America.* Cambridge, 1995

Jasanoff, Sheila, Gerald E. Markle, James C. Peterson, and Trevor Pinch, eds. *Handbook of Science and Technology Studies.* Thousand Oaks, CA: Sage Publications, 1995.

Krimsky, Sheldon. *Science in the Private Interest: Has the Lure of Profits Corrupted the Virtue of Biomedical Research?* Lanham, MD: Rowman & Littlefield, 2003.

Nelkin, Dorothy. *Science as Intellectual Property : Who Controls Research?* New York: MacMillan, 1984.

Nelkin, Dorothy. *Dangerous Diagnostics: The Social Power of Biological Information.* New York: Basic Books, 1989.

Sarewitz, Daniel. *Frontiers of Illusion: Science, Technology, and the Politics of Progress.* Philadelphia: Temple University Press, 1996.

Sutton, Victoria. *Law and Science: Cases and Materials.* Chapel Hill, NC: Carolinas Academic Press, 2001.

*Wendy E. Wagner*

# LAW AND TECHNOLOGY.

Both law and technology have been essential to the evolution of society. From the farm-machinery and property-law reforms that allowed the agrarian revolution to flourish to the information technologies and intellectual property laws that allow the modern information and biotechnological revolutions to advance, both law and technology have served critical social roles. This essay examines two areas in which the law and technology interact in ways that influence each other's impacts on society: information technology (IT) and biotechnology.

## Information Technology, Law, and Society

Information technologies are remarkable for the rapid pace at which they have advanced following World War II, as well as for their pervasive presence in our personal lives and social institutions. Although many developments in IT have shaped, and been shaped by, the law, two areas stand out as particularly noteworthy: the legal and technological protection of information (information that is *owned*), and issues of informational privacy and communication (information that is *personal*).

*Information, Law, and Ownership.* The Digital Millennium Copyright Act of 1998 (DMCA) and the continuing legal actions the law has spawned are excellent examples of the intricate interaction between legal and technological forces. From movies to massive databases, information flows easily from one system to another as a digital stream of data. One powerful legal mechanism that has been developed to protect information is *copyright.* Copyright law protects fixed expressions of ideas—whether in print, on tape, in pictures, in binary code, or in other forms—and limits other people's uses of that work, so that the creators of the work may reap the rewards of their efforts. The ability to copy information easily and to move it across borders, between computer systems, and among people has resulted in an increasingly lucrative type of crime: *information piracy*—the theft of intellectual property that rightfully belongs to others.

Information piracy is not new: books have been illegally copied, or pirated, for centuries. The advent of digitized information, including books, music or movies, has, however, made it possible to make and distribute flawless copies at minimal expense, resulting in substantial lost profits to copyright holders. In 1988, the United States Congress enacted the DMCA in response to concerns about increased digital piracy, particularly in the music and motion picture industries.

In seeking to protect copyright holders and prevent piracy, a very wide-reaching provision was built into the DMCA—a provision that, in its very structure, serves as an excellent example of the interaction between the law that protects the information and the technology that determines how the information is "written." Section 1201 of the DMCA forbids people from making or trafficking in technology that would enable users of such technology to override or circumvent other technologies that are designed to control access to, or copying of, the copyrighted work. Thus, the law is shaped by the technologies available. That a legal provision such as Section 1201 can even exist is a direct result of the ability to "write" or "fix" the information in a digital form, and to incorporate into that "writing" technologies that control or prevent access or copying of that information. In this case, however, the availability of these copy- and access-control technologies, and in particular their direct incorporation into the law, has raised concerns.

One might reasonably believe that it is good to incorporate copyright protections into technologies, and to make it illegal to circumvent them, in order to prevent unauthorized uses. A problem arises, however, when the copy and access protections incorporated in the technology and protected by the DMCA prevent authorized uses, such as those uses falling under the doctrine of "fair use." Copyright existed long before the development of digital information systems, and a body of law developed surrounding allowable, or "fair," uses of copyrighted materials. The doctrine of fair use is what allows persons to photocopy an article or book chapter to use in research, or to "rip" a CD track to listen to on an MP3 player, both without seeking the prior consent of the copyright holder of the work. In each case a copy is made, and in each case a specific technology makes the copy (photocopier or computer), and with each technology there is risk of misuse, or piracy. Yet society has supported such uses in order to serve other worthy goals, such as scholarly research and personal expression. Critics of the DMCA say the built-in access and copy-control technologies restrict or eliminate legitimate, well-established fair uses of the copyrighted work.

DMCA critics further contend that by legally prohibiting the circumvention of technologies that would allow these fair uses, the DMCA favors the rights of the copyright holders over the rights of the public at the expense of the public good. Both the holders of copyrights and the advocates for fair use have legitimate stakes in this process, and their interests are clearly in tension. Although DMCA may have brought this tension over copyright to the fore at the end of the twentieth century, it seems likely that finding an acceptable balance among the many interests involved will reach well into the twenty-first.

*Information, Communication, and Privacy.* Imagine conducting all your business and personal correspondence on postcards. Imagine that instead of using telephones, you used megaphones to talk with your friends across a crowded street. Would you change what you write or say? Although these are intentionally extreme examples, the advent of email raised the question of just how private people can expect their online communication to be.

Among the widely pervasive developments within the world of IT are those that facilitate personal communication. Email has become for many people the primary means for personal or business correspondence. Yet email is a notoriously insecure system for critical communications. It is a fairly simple process to "eavesdrop" on a particular person's email communications, or to monitor mass communications for certain words or phrases (for example, for the name of a rival company in a corporate email system's "traffic").

In 1991 Phil Zimmerman wrote an encryption program called "Pretty Good Privacy" (PGP). PGP and similar encryption programs use an algorithm that encodes a readable message into an unreadable one that can be decrypted only with a "public key," a string of numbers and letters that the encryption engine uses to "unlock" the message. Zimmerman was concerned about the ease with which email could be scanned, without detection and with very little labor. PGP would allow the average user to encrypt email with a strong enough algorithm that it would be secure enough to deter even fairly motivated eavesdroppers. The program, distributed as shareware, was uploaded to the Internet (not by Zimmerman) and was thus available to the public in the United States and around the world. It might seem that Zimmerman was simply solving a problem with email, but the law took a different view. Early in 1993, the U.S. government began a criminal investigation against Zimmerman for exporting munitions in violation of the federal regulations on international arms trafficking. The law at the time considered strong encryption software to be munitions, or war materiel.

The government's view was rooted in the fact that encryption had traditionally been the purview of the military agencies, and it had been primarily used for secure communications of sensitive information. Therefore, views of encryption technologies tended to be dominated by concern for their potential to facilitate illegal activities by preventing the military defense agencies or law enforcement officials from monitoring communications.

With the widespread use of email, however, the use and utility of encryption needed to be rethought. People were sending business and personal information over a system that was easily interceptable. It became clear to policy makers that if people and businesses were going to trust and maximize the utility of email, they would have to feel confident that their privacy was ensured. Phil Zimmerman's case became a cause célèbre among privacy activists, civil libertarians, and prominent computer programmers, users, and scientists. In 1996 the government dropped its case against Zimmerman, but it would be another three years before federal regulations were changed to allow more liberal export of such technologies. Strong encryption technologies are now routinely built into many computer applications, including web browsers, word processors, and email applications.

The reconsideration of the nature of encryption technologies by the executive branch and the changing of the federal regulations governing their export are illustrative of how public perceptions and expectations can influence the law's approach to a certain technology. When encryption was used primarily by the military, the law treated encryption tools as military weapons technology for the purposes of import and export.

As individual citizens and businesses began to use a form of global communication that was insecure, however, encryption was increasingly seen as a tool that could be used to fulfill the public expectation that the personal expression of ideas (whether in commerce or in intellectual or interpersonal discourse) should be private and not subject to interception. In addition, the expectations for "electronic commerce" were substantial, and the government was expected to help facilitate, not hinder, the development of the "new economy." Ensuring the privacy of online communications was seen as critical to building consumer, and thus investor, confidence. The law changed its perspective on encryption not because such technologies no longer carried a risk of misuse, but in response to a growing societal need and the perception regarding the utility of such technology.

## Biotechnology and the Law

Biotechnology is the focus of considerable public attention, primarily for the dramatic changes it promises for human health, agriculture, and the environment. Along the way, its intersection with the law has brought about significant changes not only in how technology enables the law to perform its essential social functions, but also in how the law makes it possible for technology to achieve its promise. The end result, of course, is not always free of tension, as reflected in the two examples that follow: DNA (deoxyribonucleic acid) technology in the criminal justice system and the effects of patent law on biotechnology research and development.

*DNA Technology and the Law.* Various technologies have played a critical role in enabling the law to carry out its investigatory, judicial, and enforcement functions. The core mandate of the criminal legal system—judging guilt or innocence—is heavily influenced by the presence or absence of evidence linking a person to a crime, and nowhere has that influence been felt more than with DNA testing, sometimes referred to as "DNA fingerprinting."

DNA is unique to each individual (except for identical twins) and is found in every cell in the body. DNA testing initially emerged as a tool for assessing blood diseases. It entered the legal sphere in 1984, when British geneticist Alex Jeffreys realized that segments of DNA were "genetic markers" and were therefore as unique to each individual as a fingerprint. The accuracy and reliability of DNA testing has made it a powerful force in legal proceedings. DNA testing has not only helped link persons to crime scenes and thus obtain convictions, but it has also exonerated suspects, resulting in more than 150 postconviction releases of persons wrongly incarcerated, more than a dozen of whom were on death row. Although concerns have been raised about the calculations and testing methods used in interpreting DNA evidence, the technique has survived those challenges and is entrenched in legal systems throughout the world.

Despite these successes, the process of collecting and storing DNA samples in criminal databanks has raised a new set of challenges for the law. Initially in the United States, DNA databanks were limited to evidence gathered from persons convicted of sexual offenses. Soon, however, collection was extended to other violent crimes, and in recent years it has been expanded in some states to cover all felonies and certain misdemeanors. At least two states now allow for DNA to be collected from persons merely arrested for certain violent crimes. These new applications have raised concerns about potential abuses, especially since many states have not adopted clear policies on the disposal of samples or on the range of uses to which they may be put. Because DNA can be used to assess a person's risk for certain diseases or behavioral disorders, these criminal databases pose a potential threat to personal privacy. Hence, while DNA technology has helped achieve legal justice in ways that most would approve, it has raised additional questions to which the law will have to respond.

*Patents and Biotechnologies.*  Another area where biotechnology and the law have felt the power of each other's influence is in the context of patent protection, which grants limited monopoly rights to creators of new work. It is probably fair to state that the U.S. biotechnology industry has its origins in two events that occurred in 1980. One was the Supreme Court decision in *Diamond* v. *Chakrabarty* (447 U.S. 303 (1980)), allowing a genetically engineered microorganism to be patented. Since then, patents have been granted on a wide range of living organisms that have become integral to research in biotechnology. The other major 1980 event was enactment of the Bayh-Dole Act (P.L. 96-517), which gave academic institutions the authority to seek patents on their federally funded discoveries. This change in the law made the university research environment more attractive to industry, and it encouraged academic researchers and their institutions to seek commercial opportunities for their research. So, in the first instance, a court decision had a profound effect on the evolution of biotechnology in the United States by opening up new avenues of patentable subject matter, while in the second instance, a new law provided incentives for pursuing research with commercial potential, thereby increasing the importance of patent protection.

The major justification for patents is that the incentives and rewards they provide to inventors result in products that enhance society's quality of life. Absent such protection, some claim, private sector entities would not be willing to assume the high-cost risks associated with developing cutting-edge technology. Furthermore, in return for patent protection, one must disclose the essential details of the invention, thus enabling others to build on that original work and produce new discoveries. Questions have been raised, however, about whether recent developments in the granting of patents truly facilitate public access to technological progress or impede it.

*Patents and Research Tools.*  In considering the effects of patent protection on the development of biotechnologies, one needs to focus on both "upstream" and "downstream" technologies. "Upstream" technologies comprise research tools that enable researchers and companies to develop "downstream" technologies or products. Research tools may include such things as cell lines, gene sequences, genetically modified organisms, and laboratory equipment, access to which can affect further technological development. To the extent that patent owners lessen access to those tools—whether by restricting who is able to use them, charging high fees for their use, or demanding excessive royalty payments from products to which their tools contributed—they can adversely affect technological progress, including the development of diagnostic and therapeutic products.

The decision in *Madey* v. *Duke University*, 307 F.3d 1351 (Fed. Cir. 2002) has brought access to research tools to the forefront of debates over patenting. There is evidence that infringement of research-tool patents has been common, especially in universities, with such infringement justified on the basis of a "research exemption." Academic researchers claim that, because they are conducting research to further basic knowledge and not competing directly with patent holders, they should be free to use the tools without permission, based on legal precedent dating back to 1813. Industry patent holders affirm that such infringement has been widely tolerated, with the hope that the work of academic researchers on specific research tools would ultimately increase the value of the original patent. The decision of the court, however, has effectively eliminated the exemption for university researchers and throws into question how past practices will be affected. This is not a trivial matter, given the critical role that research tools play in advancing technology, and it is a good example of how the law could affect the future progress of technological development.

*Patents and Downstream Products.* As noted, patents are intended to ease the transition of new technologies into the marketplace to benefit the larger society. One obstacle to realizing those benefits is the proliferation of concurrent fragments of intellectual property rights by many different owners because of the increasing use of patents. If downstream developers must negotiate agreements and pay royalties to multiple upstream patent holders prior to commercializing a product, the work may be abandoned or slowed down.

Restrictions associated with diagnostic medical tests, especially in genetic testing, are also limiting access to benefits. One example is Canavan disease, a fatal central nervous system disorder. Patient groups that assisted in the research on the disease by donating tissue samples and raising funding for research found themselves without ready access to the diagnostic test after a hospital obtained a patent on the test. Access has been limited by both the cost of the test and the small number of laboratories authorized to conduct and evaluate the tests. Canavan disease is not the only example, as laboratories throughout the United States have ceased performing various genetic tests because of constraints placed on them by patent holders.

There is no doubt that patent protection has spurred innovation in biotechnology, but the nature and use of such protection has raised concerns that the pendulum has swung too far in the direction of limiting access to critical information. The challenge lies in finding the right balance between fostering discoveries through open access to basic scientific information and promoting discoveries that are more likely to emerge under the protection of patent rights.

## A Dynamic Relationship

As the examples of IT and biotechnology illustrate, the relationship of law and technology is a dynamic one, with each helping to shape the other over time. Advances in both IT and biotechnology have challenged traditional legal concepts and practices, forcing the law to adapt in order to respond to public expectations. The two types of technologies have also been embraced by the law to help it perform its critical social functions. It is hard to imagine a modern society in which the fate of technology is not influenced by the permissive or restrictive forces of the law. Equally unimaginable is a modern world in which the law is able to function effectively without the power that technology affords it. A paramount task for any society is to keep a watchful eye on both the law and technology to ensure as much as possible that the relationship between the two, while dynamic and evolving, is guided by the need to advance important social values.

*See also* **Law and Science**

## BIBLIOGRAPHY

Agre, Philip E., and Rotenberg, Marc, eds. *Technology and Privacy: The New Landscape.* Cambridge, MA: MIT Press, 1998. Contributors examine privacy and technology issues from a variety of frameworks and perspectives.

Bayh-Dole Act of 1980, Pub. L. No. 96-517, 94 Stat. 3015 (1980), as amended by Pub. L. No. 98-620, (codified as amended at 35 U.S.C. 200 et seq.).

Cohen, Wesley M., and Stephen A. Merrill, eds. *Patents in the Knowledge-Based Economy.* Washington, DC: The National Academies Press, 2003. Includes a series of chapters on the effects of post-1980 legislation and litigation on the U.S. patent system. Several chapters focus on information technology and biotechnology.

*Diamond* v. *Chakrabarty,* 447 U.S. 303 (1980).

Digital Millennium Copyright Act (DMCA), 112 Stat. 2860 (1998).

Eisenberg, Rebecca. "Patent Swords and Shields." *Science* 299 (February 14, 2003): 1018–1019.

European Community Directive on the Legal Protection of Databases, Directive 96/9/EC of the European Parliament, 1996 O.J. (L 78).

European Community Directive on the Privacy of Personal Data, Directive 95/46/EC of the European Parliament, 1995 O.J. (L 281).

Heller, Michael A., and Rebecca S. Eisenberg. "Can Patents Deter Innovation? The Anticommons in Biomedical Research." *Science* 280 (May 1, 1998): 698–701.

*Madey* v. *Duke University,* 307 F.3d 1351 (Fed. Cir. 2002).

Mertz, Jon F. "Disease Gene Patents: Overcoming Unethical Constraints on Clinical Laboratory Medicine." *Clinical Chemistry* 45 (1999): 324–330.

Nimmer, Melville B. (author 1963–85), and Nimmer, David (revision author 1986 to present). *Nimmer on Copyright.* New York: LexisNexis, continually updated. A classic legal reference book addressing issues of copyright.

Rudin, Nora, and Keith Inman. *An Introduction to Forensic DNA Analysis,* 2nd ed. Boca Raton, FL: CRC Press, 2001.

Schneier, Bruce. *Secrets and Lies: Digital Security in a Networked World.* New York: John Wiley & Sons, 2000. The author examines digital security issues with a broad perspective.

U.S. Copyright Office, Library of Congress. *DMCA Section 104 Report.* Washington, DC: Author, 2001. This report is required under the provisions of the DMCA; it provides solid analysis of some aspects of the DMCA.

*Mark S. Frankel and Brent Garland*

# LOGIC.

*See* **MATHEMATICS AND LOGIC.**

# M

## MATHEMATICS AND LOGIC.

Viewing mathematics and logic as social constructions can change our fundamental understanding of numbers, reasoning, and abstract thought in general. The discussion will consist of the following steps: First, our common-sense understanding, or lack of understanding, will be described. Second, a standard account of number and logic will be introduced and rejected. Third, a more sociologically congenial line of approach will be explored. Finally, some historical examples will be introduced. It is important to deal with matters of general orientation before looking at historical cases, because their significance depends on the interpretive framework in which they are viewed.

### Common-Sense Perplexity

If someone says he had three eggs for breakfast, the numerical aspect of the claim usually poses no special problems. If we were asked what the number three is, or what the word *three* stands for, on the other hand, confidence would probably give way to perplexity. If we said, "The number three stands for three things," we would then face the question, "Which three?" If we answered, "Any three will serve to give us the idea," our interlocutor might observe that each of us had her own set of ideas, so are there as many "number threes" as people with ideas of them?

Similar perplexities arise about logic and reasoning. We often speak of being "compelled" to accept a conclusion. If you accept *these* premises (say, "A is true" and "A implies B"), then you must admit *this* ("B is true"). Arguments of this form are called *modus ponens*. But what sort of compulsion is

involved here? Most of us are at a loss to explain it.

This lack of a ready account of number and logical compulsion is not confined to laypersons. Impressive mathematical and logical sophistication can be, and frequently is, accompanied by equal difficulty in explaining such matters.

### Platonism

These embarrassments have been elevated to the status of "philosophical problems," and a tradition of appropriately lofty answers has emerged to shed light on them. Since this tradition can be traced back to a philosopher from ancient Greece called Plato (*ca.* 428–348 B.C.), it is known as Platonism. The answers go like this: Numbers are "abstract objects" existing outside space and time, and mathematics records timeless truths about them. The number three is an object distinct from all others and distinct from the purely psychological ideas that we may have about it. Nevertheless we can come to know this "abstract world" and its intrinsic relations (for example, that the number six can be divided by the number two). The eminent Cambridge mathematician G. H. Hardy (1877–1947) expressed it like this: "I believe that mathematical reality lies outside us, that our function is to discover or observe it, and that the theorems which we prove . . . are simply our notes of our observations" (pp. 123–124). Not all supporters of Platonism would talk of our "observing" this reality. The logician Gottlob Frege (1848–1925) spoke of our intellect "fastening onto" mathematical truths. Little seems to depend on the metaphor. What matters to adherents of this tradition is the message that we discover, rather than invent, this realm of truths.

## Rule Following

The philosopher Ludwig Wittgenstein (1889–1951) had the idea of encapsulating these problems in a single example, that of rule following. Suppose we are following the simple arithmetical rule that generates the sequence of numbers that starts with 2 and whose next term is obtained by adding 2 to the previous term, thus: 2, 4, 6, 8, and so on. When we get to 1000, we must next say 1002. Here we encounter the logical "must," and, in fact, all the problems associated with mathematics and logic that have been mentioned are present in this example.

Platonists have their own picture of what is happening. The rule itself determines what is to happen if it is followed correctly. That we must say "1002" after "1000" is already laid down "implicitly" as soon as the rule is formulated. The right answer already exists in advance of our saying it. Platonists explain rule following by invoking the abstract world as an archetype that we follow. We trace out, as it were, what has already been faintly written, so that we know in advance what we must write down.

Wittgenstein rejected this as a myth. His objection is the closest thing we have to a decisive refutation of Platonism. He asked the question, "If I know in advance, what use is that to me when I come to give my answer?" His point was that our problem is to know what to write after we have written 1000. It is no use telling us that already, "in heaven" as it were, 1002 has already been written. We have to be able to get a glimpse of that fact and know that what we are glimpsing is the correct answer that we must repeat. But if we know enough about the archetype to be able to recognize it as the right answer, then we already know the very thing that the archetype was introduced to explain. The answer is question begging. It presupposes the skill it was meant to explain. Platonism, in whatever guise, is therefore empty and offers no more than high-sounding pseudoexplanations.

## The Humean Approach

Fortunately other intellectual traditions offer something better, such as the empiricism of the eighteenth-century historian David Hume (1711–1776), who focused on the real world and the psychosocial characteristics of the persons who inhabit it. Hume's goal was to do for the understanding of human nature and society what Robert Boyle and Isaac Newton had done for physics. In the introduction to his *Treatise of Human Nature* (1739), he said:

> 'Tis evident, that all the sciences have a relation, greater or less, to human nature; and that however wide any of them may seem to run from it, they still return back by one passage or another. Even *Mathematics, Natural Philosophy, and Natural Religion*, are in some measure dependent on the science of MAN; since they lie under the cognizance of men, and are judged of by their powers and faculties. (p. xix)

Hume went on to wonder what changes and improvements there might be in these fields if we were really acquainted with the character of human thought "and cou'd explain the nature of the ideas we employ, and of the operations we perform in our reasonings" (p. xix). Let us follow some of these suggestions.

## Mill and Wittgenstein

If, as empiricists say, all knowledge comes from experience, then the truths of arithmetic must depend on empirical truths. The addition 2 + 2 = 4 must be based on a simple law of nature about how, say, apples behave as they are sorted and counted. This was the view of John Stuart Mill (1806–1873), who followed Hume's approach. Mill wanted to turn deduction into induction. This view was and is widely scorned and nowhere more vehemently than in the pages of Frege's *Foundations of Arithmetic* (1884). Frege, whose approach was followed and refined by Bertrand Russell (1872–1970), sought to give mathematics a foundation in the simple, self-evident truths of symbolic logic. Where Mill wanted to prove 2 + 2 = 4 by counting apples, Frege and Russell engaged in lengthy exercises in symbolic logic. However, John L. Mackie (1917–1981) argued that, appearances to the contrary, these are really no advance on "apple" proofs. All that has happened is that symbols replace apples. If we could not already see that grouping two symbols with two more symbols enabled us to count four symbols, the "proof" could not be carried out. Once again, something that was meant to explain and justify our knowledge actually presupposes it.

Nevertheless there is something wrong with the simple empiricist view. Arithmetical truths are not treated like inductive truths and, therefore, in a sense are not inductive truths. The reason is that an alleged counterexample would never be allowed to refute them. Rather, it would be taken as a case where "something happened" (for example, new apples had been introduced, or created and destroyed, in the course of the calculation). Wittgenstein therefore added to the empiricist approach the idea that certain generalizations are given a special status. They become fixed standards of what is to count as "nothing happening" and hence as a baseline for the identification of causal processes. In other words, the inviolability of $2 + 2 = 4$ derives from its being held fast by convention.

If arithmetic seems to offer mysterious truths about a mysterious realm, this is because the social processes that give it its special status are not seen for what they are. Wittgenstein's thinking here followed a similar path to that of the sociologist Émile Durkheim (1858–1917), who made the transfigured and misunderstood character of social reality the basis of his entire discussion of religious experience and philosophical error. Common-sense perplexity about strange "things" called numbers, their peculiar ontology or mode of being, is just the failure to appreciate the role of the social processes behind our talk of number.

## Deductive Validity

The idea of a social convention can also shed light on logical compulsion. The compulsion in question comes from the obligation to avoid lapsing into invalid argumentation. By definition, a valid inference form is one that will never lead from true premises to false conclusions. *Modus ponens*, introduced above, is a typical candidate for this status.

How do we prove it will never lead us astray? To say it has never let us down in the past would be an inductive argument, not a conclusive demonstration. But if we look for a deductive proof, we run into the danger of circularity. It would beg the question to use *modus ponens* to prove *modus ponens*. Susan Haack has shown that the problem is both general and insurmountable.

Even if we cannot prove the validity of *modus ponens*, it certainly looks compelling. How could simple argument form fail to be valid? In fact, the cases where it can lead from true premises to false conclusions are familiar, but they are kept out of sight by being called "paradoxes." Consider the familiar paradox of the heap. You begin with a heap of sand. Now if you have a heap of sand and you remove one grain, you still have a heap. Nevertheless, if you repeat the process enough times, you finish up without a heap. This can be seen as a sequence of steps, each in the form of a *modus ponens* argument. The sequence begins with true premises and ends with a false conclusion: that one or zero grains is a heap. *Modus ponens* has led from truth to falsity. The standard response has been to claim that the premises are unacceptable because they are "vague" (as if logic didn't apply to anything vague). What is really happening is that the "valid" character of the argument form is sustained by dismissing counterevidence. The compulsion derives from a policy. As Wittgenstein said, we are compelled, but only in the same way as we are by the laws of the land and ordinary social and moral conventions.

## Analytic Validity

It is important to stress that the claim here is not that validity can be conjured up by arbitrary definitions of words or stipulations of rules. It is sometimes said that the validity of an argument derives from the meanings of the key, logical words or "constants," such as *and, or, if . . . then*, and so forth. Validity derives from conforming to the rules that specify the workings of these constants. This is called the theory of *analytical validity*. It is wrong and has been decisively refuted by the logician Arthur Prior (1960). He pointed out that if this were the source of validity, then introducing a new, nonstandard logical constant, and adhering to the rules that constitute its definition, should suffice to generate a new range of valid arguments. He duly invented rules for a "new" logical connective that sanctioned the deduction of any proposition from any other. If the theory were right, this would be a validity-generating procedure, but, of course, it completely violates all our logical intuitions. It is no use insisting that such definitions and rules can generate validity, provided that they are first "suitably" selected. That is just the point: Validity does not come from them but from something else to which they, in their turn, must duly conform. What is this something? Platonists have their answer ready. Empiricists and conventionalists can offer a less mystical, sociological account.

## Historical Examples

One of the virtues of a sociological approach is that it gives a new and deeper significance to studies in the history of mathematics. For a Platonist, historical study merely reveals the circumstances surrounding the act of knowing: It touches on the context of discovery but not that of justification. When the socially constructed character of the knowledge itself is appreciated, history reveals material that is constitutive of both the knower and the known and touches on both discovery and justification. The richness and depth of historical studies preclude anything more than a brief glimpse at some representative examples of this kind of work.

One fascinating historical theme has been the changing character of rigor and its bearing on the relation between "pure" and "applied" mathematics. Traditional historiography tells a story of progressive improvement in rigor. Thus, the infinitesimal calculus began experimentally and intuitively around Newton's time and only later came to be given a rigorous formulation as part of a general trend carried through in the nineteenth and twentieth centuries. (It was this trend that Frege and Russell sought to take to its ultimate conclusion.) Not all mathematicians looked with favor on this development or the narra-

tive of progress that typically accompanies and justifies it. Thus, Felix Klein (1849–1925), one of the twentieth century's greatest mathematicians, resisted the trend to formalization and purity. His stance was endorsed by applied mathematicians such as Theodor von Kármán (1881–1963), who carried it from Göttingen to the United States. Originally these positions were implicated in the struggles between traditional German universities and the newer *Technische Hochschulen*. Later, in the United States, they found expression in the changing fortunes of pure and applied mathematics initiated by World War II.

An example that shows the intimate connections between technical judgements about rigor and sociological variables is provided by the so-called "crisis" in the foundations of mathematics in the years after World War I. This was provoked by the movement called "intuitionism," led by L. J. Brouwer (1881–1966). Intuitionists rejected nonconstructive existence proofs. This type of proof shows that a certain mathematical entity exists (for example, a least upper boundary to the size of a quantity) but does not exhibit the entity itself (for example, by enabling the mathematician actually to write it down). Nonconstructive proofs depend on what is called the *law of the excluded middle*, which says that any given proposition is either true or false. This means that if the negation of a proposition can be shown to be false, then the proposition must be true. Much classical mathematics rests on inferences of this type. Would it not be more rigorous to insist that existence claims should be backed by an actual construction? Brouwer said yes; David Hilbert (1862–1943) said no. Historically, the credibility of the two positions seems to have depended on broad cultural factors about the general state of crisis in central Europe after World War I and, more locally, on the struggle for academic leadership and the competition for professorial chairs at Göttingen (see the works by Paul Forman, by Constance Reid, and by David Rowe and John McCleary).

The history of geometry has also provided a rich source of examples that show the contingent connections between mathematical judgments and social interests. For example, Joan Richards has shown that the study of non-Euclidean geometry had a specific meaning in Victorian England: It meant that the decision as to the real geometry of space was a matter for experiment. What had been a paradigm of *a priori* knowledge about reality was suddenly demoted. This was bad news for other claimants to *a priori* knowledge, such as the church. Mathematicians who wanted to dampen down the war between science and religion set about reinterpreting non-Euclidean geometry as nothing but the geometry of shapes stretched on three-dimensional surfaces, as on a sphere. The burst of interest in projective geometry was no accident given the social meaning of the exercise.

Sociological studies of more recent trends in mathematics, such as Donald MacKenzie, have naturally addressed the role of computer methods. Some mathematical conjectures, such as the four-color theorem, have long defied satisfactory proof but now, thanks to the computer, can receive an answer. (The four-color theorem says that no more than four colors are needed to color a map in which no two adjacent areas can be of the same color.) What, however, is the status of procedures that consist of millions of steps and hence defy any overview by human intellectual processes? Whether these really constitute proofs, or whether the meaning of the word is undergoing change, is a matter of judgment on which opinions may vary.

Even this brief glance at a few historical cases brings out the significance of the general principle that emerged in the earlier discussion. There are decisions and choices to be made and conventions to be created and sustained. The manipulation of symbols must correspond with and sustain the meaning that is accorded to them. Without this machinery of interpretation and taken-for-granted routine, mathematical and logical processes cannot operate, and as long as the machinery's nature and role are not understood, the processes will stay shrouded in mystery. The perplexing character of number and inference will then encourage obscuran-

tism rather than a cool look at human cognition. In reality, mathematics and logic cannot be divorced from the judgments, choices, and conventions of human practitioners. They stand, in Hume's words, "under the cognizance of men" and they are "judged of by their powers and faculties." As such they are the proper object of sociological curiosity.

## BIBLIOGRAPHY

Durkheim, Émile. *The Elementary Forms of the Religious Life.* Translated by J. W. Swain. New York: Free Press, 1915. A valuable theoretical resource for thinking about far more than religion.

Forman, Paul. "Weimar Culture, Causality, and Quantum Theory, 1918–1927: Adaptation by German Physicists and Mathematicians to a Hostile Intellectual Environment." *Historical Studies in the Physical Sciences* 3 (1971): 1–115. Controversial but very suggestive about mathematics.

Frege, Gottlob. *The Foundations of Arithmetic.* Translated by J. L. Austin. Oxford: Blackwell, 1959. Brilliant, dyspeptic Platonism.

Grabiner, Judith V. "Is Mathematical Truth Time Dependent?" *American Mathematical Monthly* 8 (1974): 354–365. Identifies the shifting role of mathematicians from court savants to teachers as playing a significant part in the growth of rigor.

Haack, Susan. "The Justification of Deduction." *Mind* 85 (1976): 112–119. (Puts deduction and induction on a level as equally unjustifiable.)

Hardy, Godfrey Harold. *A Mathematician's Apology.* Cambridge, U.K.: Cambridge University Press, 1940.

Hume, David. *A Treatise of Human Nature* (1739). Edited by L. A. Selby-Bigge. Oxford: Clarendon Press, 1960. Book III contains the first, and perhaps still the best, analysis of convention.

Kármán, Theodor von. "Some Remarks on Mathematics from the Engineer's Viewpoint." In *Collected Works of Theodore von Kármán,* vol 4. London: Butterworths Scientific Publications, 1956: 1–6.

MacKenzie, Donald. *Mechanizing Proof. Computing, Risk, and Trust.* Cambridge, MA: MIT Press, 2001. Impressive demonstration of the power of a sociological analysis.

Mackie, John Leslie. "Proof." *Proceedings of the Aristotelean Society,* sup. vol. 40 (1966): 23–38. It seems that Mill was right all along.

Mehrtens, Herbert, Henk Bos, and Ivo Schneider, eds. *Social History of Nineteenth Century Mathematics.* Boston: Birkhauser, 1981.

Prior, Arthur N. "The Runabout Inference Ticket." *Analysis* 21 (1960): 38–39.

Reid, Constance. *Hilbert.* New York: Springer-Verlag, 1996.

Richards, Joan. *Mathematical Visions. The Pursuit of Geometry in Victorian England.* London: Academic Press, 1988. A richly documented case-study.

Rowe, David E., and John McCleary, eds. *The History of Modern Mathematics*, Vol. 2, *Institutions and Applications.* Boston: Academic Press, 1989.

Sainsbury, R. M. *Paradoxes.* Cambridge, U.K.: Cambridge University Press, 1988. Chapter 2 gives a penetrating account of the paradox of the heap.

Wittgenstein, Ludwig. *Philosophical Investigations.* Translated by G. E. M. Anscombe. Oxford: Blackwell, 1958.

Wittgenstein, Ludwig. *Remarks on the Foundations of Mathematics*, 3rd ed. Translated by G. E. M. Anscombe. Oxford: Blackwell, 1978.

*David Bloor*

# MEDIA AND MEDICINE.

Media and medicine intersect in a number of ways, from the longstanding interests they share as powerful modern industries to the current proliferation of dazzling medical imaging technologies and feature-packed Internet sites. Medicine provides media with reliably popular content and expertise, while media provide medicine with modern communication systems to deliver its messages. Of the many dimensions of this relationship worthy of discussion, I will focus on three: (1) medicine-media partnerships in mass media production; (2) media representations of physicians, health, and disease; and (3) mass media in health communication.

## Medicine-Media Partnerships in Mass Media Production

The first movie theaters in the United States opened in 1905. Within two decades, more than one thousand films had been made on health and medicine, touching even such controversial topics as abortion, birth control, venereal disease, childbirth, euthanasia,

and eugenics. This early coincidence of interests had significant impact on both medicine and the film industry and on other medicine-media partnerships to come.

An example is the institutional coalitions forged to produce films on venereal disease (VD), later referred to as sexually transmitted diseases (STDs), during World War I. The positive reception in the United States of the 1915 theatrical film *Damaged Goods* suggested that the country might be ready at last to break the "conspiracy of silence" surrounding the VD epidemic. Certainly the military was ready: The prevalence of syphilis and gonorrhea among American troops as war approached threatened to compromise the nation's efficiency. Screened at a number of army and navy training camps, *Damaged Goods*—the story of a young man who marries and has a child before he learns he has syphilis—seemed to offer a promising new format for the education of military personnel. A key feature was the physician's prominent role in presenting the facts of syphilis to the young man—and to the audience. Medicine's recognized authority provided a scientific frame for the information and legitimated the shocking scene of real patients suffering the physical and mental ravages of tertiary syphilis (film reviewers were deeply impressed by these "repulsive yet actual results" of syphilis). Partnering with medical and media experts, the military added the production of training films to their existing arsenal of educational resources. Along with print materials, posters, slide shows, radio broadcasts, and lectures, the films would provide lessons on everything from basic hygiene to the recognition, prevention, and treatment of venereal disease.

Germ theory's growing impact had set the stage for mass media attention to medicine, both by advancing physicians' understanding and treatment of infectious disease and by providing thrilling microscopic images of the culprits. Many progressive physicians and public health leaders now supported a modern scientific approach to health education with its straightforward, nonmoralistic focus on "fighting infection." Ideally, of course, troops would be "physically fit and morally upright," but advocates of science favored secular messages to troops that were frank and graphic, provided unambiguous instructions for VD prevention (whether through abstinence, condoms, or chemical prophylaxis), and officially imposed penalties for violations.

As shown, however, in clashes that would recur in World War II and again in the HIV/AIDS epidemic, the decision to codify disease prevention into official media productions raised ideological, technical, and policy questions. What lessons should be communicated, and through which media? If troops were to be taught to use condoms and chemical prophylaxis, would explicit demonstrations be filmed? What if the films fell into the wrong hands? Would the single-minded strategy of "fighting infection" encourage—indeed, officially condone—immoral behavior? Should not self-control, moral character, and sexual continence—"fighting sin"—be modeled as well? Yet could these traits be made suitably masculine for men in combat? And should the federal government be in this "health propaganda" enterprise at all? As historian Allan M. Brandt so vividly demonstrates, the early campaigns of World War I rather uneasily tried to fight sin and infection simultaneously, the two approaches united only in the effort to eradicate prostitution and red-light districts from the vicinity of the training camps.

In the end, the combined power of science, a debilitating epidemic disease, and a great war moved the films toward medicine—that is, toward explicit language, images, and behavioral rather than moral goals. Clean living continued to be advocated, but with condom and chemical prophylaxis as its backup; moreover, troops were put on notice that those who contracted VD despite training could be left behind, docked of their pay, or court-martialed. Industrial-scale production yielded hundreds of films and related materials. To strengthen the "mental inoculation" effect of these media messages, patriotism and the war framed every production and invested every message with symbolic power: "Syphilis is the sniper, the saboteur behind the line." More practically, patriotism and war enabled the films to show explicit sexual images—male genitals, for example—without public censure.

Censure, and censorship, came after the war. President Woodrow Wilson had barely finished commending his returning troops on their dual victory, over both Germany and the "unseen enemy" of venereal disease, when the moral backlash began. Some films, including the highly successful training film *Fit To Fight* (*Fit To Win* in its civilian incarnation), were banned under the federal Comstock statute as obscene. But others, like the 1915 eugenics film *The Black Stork*, were banned in response to unintended viewer revulsion. One reviewer wrote that the *The Black Stork* was "as pleasant to look at as a running sore"; VD films were now characterized as a "succession of horrors," "realism gone mad," and "pathologically educational" (Pernick). These "unduly distressing" films, despite their recognized social and educational value, were subjected to "aesthetic censorship" and banned. By the 1920s, for example, films documenting the successful rehabilitation of troops wounded in combat, which had been regularly shown as a tribute to the vets' courage and heroism during the war, had been deemed "unduly distressing" and been banned.

Some physicians joined in the backlash against frank media presentations of medicine. Having tried unsuccessfully to distinguish "scientific" from "sensational" films, they concluded that explicit images of disease did little credit to the profession. While some physicians viewed any popular media treatment of medicine unseemly and unprofessional by definition, the American Medical Association (AMA) seemed more concerned that the wartime medicine-military-media production partnerships invaded medicine's professional sphere and smacked of "socialized medicine." A broader objection was the mass media airing of knowledge, topics, and practices that could engender public anxiety, revulsion, or calls for external regulation. Media practitioners, writes medical historian Martin J. Pernick, were likewise feeling the backlash: "While doctors divided over the medical value of the movies, the fledgling film industry was also bitterly split over the appropriateness of medical topics for the screen" (Pernick, p. 121). The upshot of these actions and reactions was that U.S. films were divided into categories. Commercial Hollywood studio films were now to be dedicated to the provision of entertainment only and were categorized as "theatrical." All other films, including those containing—and this applied to many health and medical films—"disgusting, unpleasant, though not necessarily evil subjects," were placed in the catch-all category "non-theatrical." Interestingly, as Pernick points out, this decision by the censors of the day inadvertently fostered the development of new film genres including familiar categories like educational, industrial, exploitation, documentary, and propaganda films. Health and medical films could potentially fit in any of these categories.

When World War II arrived, no one was surprised to find the VD problem still a problem, and the training film enterprise was reinvigorated. Training films were made on virtually every health topic and were shown constantly. Though exact figures are not available, one careful count for the month of June 1943 identified 478 training films in circulation and at least 100,000 screenings. Although these films incorporated much of the World War I "blitz against prostitution" ideology as well as the "anti-infection" versus "anti-vice" debate, they also show the creative influence of the many Hollywood actors, film-makers, and studios joining the war effort. John Ford and Darryl F. Zanuck's 1942 *Sex Hygiene* was the first training film of the war and, as film scholar Robert Eberwein writes, "the most famous and most viewed of all the training films." Although the script was quite complex, with flashbacks and other cinematic techniques, the final cut was left to the military. Seeing the completed film, Ford said, "I looked at it and threw up" (Eberwein, p. 66). Others included *Health Is a Victory, Plain Facts, Pick-Up, Easy to Get, Three Cadets*, and *VD Control: The Story of D.E. 733*. While these films continue the assault on prostitution (as the *American Journal of Public Health* put it, "prostitution is an Axis partner"), the message now was that *all* women were potential VD carriers (Eberwein, p. 80). The titles of *Pick-Up* and *Easy to Get* are meant to apply equally to syphilis and gonorrhea and to the women who carry them. In *D.E. 733*, the doctor uses a female anatomical model to show the men how a seemingly "nice girl" can be "filthy and dirty" inside.

A quite different World War II medical film is John Huston's extraordinary documentary *Let There Be Light*, which chronicles an experimental postwar program for the diagnosis and treatment of psychiatric disorders caused by the stress of combat (what we would now call "post-traumatic stress disorder"). The experiment is portrayed as successful, but despite its upbeat conclusion, its graphic portrayal of "men under stress" and the visual spectacle of their psychological damage (one man sobs uncontrollably, another is experiencing "hysterical blindness") were judged too shocking and demoralizing to be shown anywhere, and the film (like a number of other World War II documentaries) was shelved until recently.

By the 1950s, then, the medical film was represented by three very different kinds of projects. First were popular Hollywood films about doctors, medicine, and health, such as *Arrowsmith* and *Doctor Kildare*, discussed in the next section (and see the works by Richard Malmsheimer and Joseph Turow in the bibliography). Second were the institutional health and medical films just described. Third were the thousands of more specialized nontheatrical films produced throughout the century. Many of these are catalogued and described in Adolf Nichtenhauser's still unpublished manuscript *History of Motion Pictures in Medicine* (see also the analyses by feminist media scholars in the collection edited by me, Lisa Cartwright, and Constance Penley). What Nichtenhauser makes clear is the critical importance of technical knowledge of both medicine and cinema for high-quality medical filmmaking. Films that capture genuine medical insights and translate them into cinematic language are the exception rather than the rule. This challenge was perfectly understood by famed Chicago obstetrician and gynecologist Joseph B. DeLee, whom Nichtenhauser discusses at length. Best known outside his field, perhaps, as an early proponent of "medicalized childbirth" (routine use of techniques such as forceps delivery and Caesarean section), DeLee was also a pioneer of medical filmmaking. His 1933–1934 paper "Sound Motion Pictures in Obstetrics" lists the ingredients of a high-quality medical teaching film: Especially desirable is a cameraman trained in surgery. After that, materials, preparation, expert staff, props, director, script supervisor, rehearsals, sets, lighting, sound, titles, worksheets—all must be perfectly cinematic. "In short," he concludes, "one has to provide a bit of Hollywood in the operating room."

Since 1970, the breakthroughs in medicine and media technologies have produced numerous collaborative projects—far too many to attempt to summarize. Websites of the AMA, the American Association of Public Health, National Institutes of Health, and National Library of Medicine provide a small sampling of current work. Archives, where historical materials continue to come to light, attest to the value of film and other media products in better understanding the history not only of medical films but also of medicine and of film.

## Media Representations of Physicians, Health, and Disease

Since the early twentieth century, physicians have been common protagonists of popular novels, radio shows, and dramatic television series. In the 1950s alone, doctors appeared as characters in more than 400 commercial theatrical films. Many of these works are classic: *The Citadel, Arrowsmith, Not as a Stranger, Dr. Ehrlich's Magic Bullet, Men in White, The Story of Louis Pasteur, L'Enfant Sauvage (The Wild Child)*, the *Dr Kildare* movies, *The Doctor*, and *Coma*. Television series include *Medic; Ben Casey; General Hospital; Marcus Welby, MD; M*A*S*H*, and more recently *ER; Chicago Hope; St. Elsewhere; Dr. Quinn, Medicine Woman; Scrubs*; and *Medical Investigation*.

Most of these represent what British media scholar Anne Karpf identifies as "the medical frame," one of four broad categories of mass media coverage that translate material and data about health and medicine into stories that can be recognized and accessed by viewers. The notion of a "frame" that organizes media representations in familiar ways and makes them come to seem "natural" is based on relatively recent research that examines mass media's role in constructing—rather than simply reporting or reflecting—social reality.

The "medical frame" is the prototypical translation of medicine's orchestrated self-image—scientific, sovereign, and infallible—into narrative conventions very familiar to modern audiences: The setting is the gleaming hospital, and the physician is the star; in a world of white coats, stethoscopes, x-rays, and MRIs, he (until the 1970s, physicians, especially as shown in these productions, were almost always male) works tirelessly to conquer disease, wrestles with challenging cases and puzzling data, and succeeds through diagnostic brilliance, skillful hands, miraculous new drugs, and ever more dazzling instruments and technologies. The sentiments are exemplified by such dialogue as is put into the mouth of Paul Ehrlich (1854–1915) in *Dr. Ehrlich's Magic Bullet* (1940): "If I can only unlock the mystery of immunity, I will have the key to so much of the world's suffering." Medicine alone has the power to name the patient's reality and place it within the larger system of health and disease.

But *Coma*, a 1978 film based on physician-author Robin Cook's first blockbuster medical thriller, offers a darker version of the medical frame. The protagonist is a woman physician; as a surgical resident in a large Boston teaching hospital, she would have only recently graduated from medical school, for the profession in the United States began to admit women in significant numbers only in 1970. She was certainly a novelty for the medical film, both as the competent female physician she represents at the beginning and the action-hero/sleuth that she becomes as the plot grows more suspenseful and sinister. A key scene, in which she peels off her pantyhose so that she can climb into the hospital's ventilation system to identify the murder method, represents an anatomical dissection of medicine and the exposure of its inner workings. The patriarchal and systemic nature of evil and its profit motive are likewise exposed.

This concept inches toward what Karpf calls the "systemic" or "environmental frame," which sees disease produced by modern society and sustained through political structures and institutional priorities—industrialization, modernization, development, capitalism, colonialism, postcolonialism. More common to documentary film, this frame is exemplified by the film *Alice—A Fight for Life*. An acclaimed 1982 British documentary (revised and shown in the United States in 1984 as part of the PBS series *Nova* as "Asbestos: A Lethal Legacy"), it tells the story of a Yorkshire woman named Alice who is dying of mesothelioma. Alice's story unfolds in the larger social context of asbestos mining, production, and corporate cover-up of growing evidence of asbestos-related diseases. Yet it is structured as exposure of cover-up or conspiracy of corporate bad guys rather than routine existence of powerful special interests which are responsible behind the scenes for establishing the basic rules of the system. Several specific characteristics of the medical and health show on U.S. television follow from its individualistic focus. As summarized by Turow, these include the following: A significant gap exists between the structure of medicine and health care and the portrayal of that structure on television; drugs and machines dominate care rather than psychosocial issues; treatment is usually hospital-based; medical care is overwhelmingly appropriate, nonpolitical, and an unlimited resource; "politics" only enters in the form of moral and ethical questions faced by individual doctors (such as the "right to die"); and the real structural changes taking place in health care will be politically entrenched before they are represented on television.

## Mass Media in Health Communication

In 1981 media scholar George Gerbner reported in the *New England Journal of Medi-cine* that soap operas were the leading source of health information for many Americans. Physicians reeled. Despite the enthusiasm of individual doctors such as DeLee and Cook for "bits of Hollywood" and the profession's grudging acceptance of public health communications during national emergencies, physicians for much of the twentieth century distrusted and disdained the mass media, cultivating the public image of a noble and elite profession insulated from popular media, especially déclassé entertainment media like soap operas. Subsequent decades, however, have brought significant change. After the AMA was charged in the early 1980s with restraint of trade, physicians began advertising their services, pharmaceutical companies began advertising prescription drugs directly to consumers on radio and television, and cost and pricing of health and medical services and supplies became part of public debate. Portrayals of doctors and medicine in popular culture loosened up, and real physicians performed surgery on television, including extensive cosmetic surgeries on such shows as *Extreme Makeover*. Such behavior would earlier have been unthinkable.

At the beginning of the twenty-first century, television is regularly cited as a leading source of health and medical information—not just TV news coverage but the entire spectrum of programs, outlets, genres, and content areas. Indeed, health and medical media coverage have become so vast that comprehensive monitoring, let alone control, is impossible. As Anne Karpf wrote in 1988, "Media's long-standing interest in health and medicine has swelled in recent years into an obsession." The AMA has thus made media communication a priority: Not only are individual physicians free to speak with the media; it is their responsibility to do so. In short, physicians have chosen to join the media revolution rather than fight. Gerbner's report on soap operas was perhaps a harbinger of this change.

Although soap operas, or daytime serials, have traditionally been held in low cultural regard by social, intellectual, and policy elites, the soap genre illustrates the specific "cultural work" that a given media genre can do. Soaps carry out unique educational tasks unavailable to scientific journals, network news coverage, and official mass media public health campaigns. The U.S. network soap *General Hospital*, for example, successfully introduced a continuing HIV/AIDS storyline in 1995 by linking it to popular adolescent characters and weaving it into other ongoing storylines. While combining education with entertainment is not an established tradition of commercial U.S. media, a growing number of health and medical organizations now formally recognize the power of entertainment media to reinforce and popularize official health messages. The Centers for Disease Control and Prevention (CDC) participates in an annual "Soap Summit" that makes awards to high-quality health programming on entertainment shows, while the Websites of the AMA, the American Public Health Association, and the Henry J. Kaiser Family Foundation track and evaluate media coverage of health and medicine and provide links to related resources.

Other countries are demonstrating the use of the serial or soap opera format as part of comprehensive media communication initiatives for social change, as demonstrated by Phyllis Piotrow and collaborators and by Arvind Singhal and Everett Rogers. The growing field of education-entertainment media (also known as "edu-tainment media," "pro-social health communication," and "social marketing") has historical roots in programs like *The Archers*, a long-running postwar BBC-Radio serial designed to familiarize rural audiences with innovative agricultural techniques. A prototypical education-entertainment serial today is South Africa's *Soul City*, a popular, prize-winning enterprise now in its second decade. Like other successful edu-tainment achievements, *Soul City* is a collaborative public-private effort with a systematic consultation process involving not only medical, media, creative, and education specialists but a wide array of stakeholders from all sectors of South African society, in-

cluding health and media consumers across the economic spectrum. The heart of the enterprise is the serial drama *Soul City*, broadcast regularly on television and radio in all eleven official languages of South Africa; each season the show takes up a different health theme (infant nutrition, HIV/AIDS, domestic violence). To reinforce and diversify the central health themes and messages, *Soul City* produces and distributes printed and electronic materials; creates spinoff programming; works with schools and community organizations; sponsors special programs, live events, and other outreach activities; and maintains a comprehensive Website. *Soul City* now serves as a prototype for similar efforts in other African nations.

*Soul City* is a creative effort to meet the pressing challenges for health communication. The global HIV/AIDS epidemic will continue to demand the full-scale deployment of media resources and innovative health communication media strategies. While these need to go beyond individual behavioral change, condoms remain virtually unadvertised on U.S. network television (although they are advertised on some radio stations)—especially in the context of ubiquitous and quite explicit television ads for Viagra and other drug treatments for male impotence.

Emergency health communication is a growing global priority, brought home by the unexpected emerging infectious diseases and bioterrorist threats of the late twentieth century. Despite federal assurances of readiness for any new crisis after the World Trade Center and Pentagon attacks of September 11, the mailing of anthrax spores to U.S. journalists produced a nationwide "anthrax panic." When health departments and hospitals complained that they got better information from Cable News Network (CNN) than from CDC, the health communication crisis was evident. This will undoubtedly be a top priority in health communication in the years ahead.

## BIBLIOGRAPHY

DeLee, Joseph B. "Sound Motion Pictures in Obstetrics." *Journal of the Biological Photographic Association* 2 (1933–1934): 60.

Eberwein, Robert. *Sex Ed: Film, Videos, and the Framework of Desire*. New Brunswick, NJ: Rutgers University Press, 1999.

Karpf, Anne. *Doctoring the Media: The Reporting of Health and Medicine*. London: Routledge, 1988.

Malmsheimer, Richard. *"Doctors Only": The Evolving Image of the American Physician*. New York: Greenwood Press, 1988.

Nichtenhauser, Adolf. *History of Motion Pictures in Medicine*. Unpublished Manuscript, History of Medicine Division, National Library of Medicine. Manuscript Special Collections 380. ca. 1950.

Pernick, Martin J. *The Black Stork: Eugenics and the Death of "Defective" Babies in American Medicine and Motion Pictures since 1915*. New York: Oxford University Press, 1999.

Piotrow, Phyllis Tilson, D. Lawrence Kincaid, Jose G. Rimon II, and Ward Rinehart. *Health Communication: Lessons from Family Planning and Reproductive Health*. Westport, CN: Praeger, 1997.

Singhal, Arvind, and Everett M. Rogers. *Entertainment Education: A Communication Strategy for Social Change*. Mahwah, NJ: Lawrence Erlbaum Associates, 1999.

Treichler, Paula A., Lisa Cartwright, and Constance Penley, eds. *The Visible Woman: Imaging Technologies, Gender, and Science*. New York: New York University Press, 1998.

Turow, Joseph. *Playing Doctor: Television, Storytelling, and Medical Power*. New York: Oxford University Press, 1989.

*Paula A. Treichler*

# MEDIA AND SCIENCE.

By incorporating information and images of science into news and entertainment, the mass media provide a critical link between society and the scientific enterprise. Journalists report new research findings while television documentaries and popular science books summarize the state of knowledge on topics as diverse as anthropology and astronomy. Movies and novels transform scientists into fictional heroes or villains and weave scientific terminology and concepts into plots about romance, crime, or politics.

Science communicated to mass audiences in such formats resembles, but is not identical to, the science communicated among specialists in textbooks and journals, differing in purpose and authorship as well as level of detail. Mass-media or popularized science seeks to inform and entertain its audiences rather than to educate them. Media professionals shape the content of popular science with sensitivity to market forces and audience tastes rather than necessarily to scientific standards.

## Science Journalism

Although the particular content of science news reflects national values and interests within each country, such as attitudes toward public education and economic development, science journalism worldwide shares certain characteristics and challenges, primarily because the scientific community's professional communications, interactions, and import are global in nature. An international press corps routinely covers major scientific events and controversies.

Science news includes stories about the results and conclusions of research studies; research processes, instrumentation, and ethics; the implications, applications, and uses of scientific knowledge; and the research system's organizations, institutions, funding, politics, regulation, and international cooperation. In addition, news about other topics may include scientific information and explanations or statements from scientists (discussions about the reliability of DNA testing in crime or disaster stories, for example).

The type of news story that typifies contemporary science reporting evolved over the last hundred years, as both science and journalism became more professionalized and scientists' accomplishments became increasingly newsworthy. Before then, science news reports contained little criticism or interpretation. Newspapers reprinted prominent experts' lectures in their entirety and published accounts of inventions, astronomical events, and mysterious natural phenomena with minimal interpretation. By the 1900s, however, concerns about public health prompted increased press attention to medical research, and such inventions as radio and the airplane generated interest in the scientific principles underlying their design. Excitement over Marie Curie's research on radioactivity and Albert Einstein's theory of relativity, the use of chemical weapons during World War I, and controversies such as the 1925 conviction of John T. Scopes for teaching about evolution in a Tennessee classroom all helped to bring science onto the front page.

As the scientific community became more concerned about its public image, groups such as the American Chemical Society (ACS) established news offices to facilitate the flow of information to the media and to shape news coverage overall. Beginning in 1921, Science Service, a nonprofit news syndicate underwritten by a prominent publisher but managed by scientists, supplied news copy and photographs to journalists and helped to produce radio interviews with scientists.

Through the 1920s, few journalists specialized in science, but newspapers began to assign journalists to the science beat. Formation of the National Association of Science Writers (NASW) in 1934 in the United States signaled an important step toward professionalization, followed by creation of the Association of British Science Writers (1947) and eventually the International Science Writers Association (1967), European Union of Science Journalists' Association (1971), and World Federation of Science Journalists (2002). Other specialty organizations unite print and broadcast communicators interested in environment, medicine, or space topics.

World War II brought a new level of censorship to science journalism when governments attempted to prohibit publication of information on certain topics, most notably those associated with atomic-bomb research. Most journalists cooperated voluntarily, but Cold War clashes over how nuclear energy should be discussed in the press exacerbated tensions between government and the media. These conflicts foreshadowed debates over secrecy and openness that have since surfaced in media coverage of such fields as biotechnology and computer security.

The 1957 launch of the Soviet earth satellite *Sputnik* and subsequent expansion of American and USSR space programs added another dimension to science coverage, linking it more closely to national political issues. Likewise, the development of a vibrant environmental movement in the 1960s drew journalists' attention to scientists' roles in measuring pollution and suggesting remediation approaches. Along with coverage of the tremendous strides being made in biology and medicine, and continued concern about nuclear proliferation and biological terrorism, the range of science news coverage has expanded steadily in all media. During the 1970s a number of major newspapers established special science sections or features; many of these, such as *The New York Times* weekly "Science Times" section, continue today.

Science grabs the most attention in the news when it is relevant to a major event, crime, disaster, or military conflict, or whenever the scientific community engages in acrimonious debate or becomes entangled in controversy. The 1979 accident at the Three Mile Island (TMI) nuclear power plant in Pennsylvania and the explosion of the space shuttle *Challenger* prompted many science journalists to question the reliability of official channels for information. After the TMI crisis, a nonprofit advocacy group, Scientists' Institute for Public Information, established the Media Resource Service, which from 1980 until 1995 brokered connections between experts and media representatives and helped to establish standards for such interactions.

The selection of scientific sources and interview subjects continues to challenge journalists in all media. Some scientists who appear frequently in the news achieve prominence for reasons other than their professional accomplishments. These "visible scientists" interact comfortably with journalists and media producers; they are at ease in front of microphones and cameras, giving hundreds of interviews, perhaps even hosting television programs. Margaret Mead, Linus Pauling, Carl Sagan, Jane Goodall, Paul Ehrlich, and others have become celebrities with worldwide name recognition.

One continuing debate among science reporters involves the attempt by scientific journals to withhold information about new work (that is, to embargo it) until peers can review results and the authors have received professional credit. Journal embargo policies began to emerge in the 1960s, concurrent with the media's rising social influence, when two prominent physics editors declared that they would reject any manuscripts describing work already discussed in the general press, arguing that premature publicity encouraged sensationalism. Although most contentious in biology and medicine, where the preeminent rationale is to protect patients from trying unevaluated treatments or drugs, embargo policies are used throughout science, sometimes as a way for the publications and organizations to manipulate news coverage. Reporters who refuse to honor embargoes may find their future access to a journal or laboratory restricted.

## Nonfiction Books, Magazines, and Documentaries

Science books directed at mass audiences, although not routinely subjected to the same rigorous technical review as journals, provide valuable sources of information for readers of all ages and all backgrounds. Such books as *The Atomic Age Opens!*, quickly published in August 1945, helped to explain atomic energy to the world; Rachel Carson's *Silent Spring* (1962) outlined the scientific evidence for unanticipated adverse impacts of pesticides. Witty and literate scientists such as Isaac Asimov, Lewis Thomas, Philip Morrison, E. O. Wilson, and Stephen Jay Gould have charmed and enlightened readers for decades. From Paul de Kruif's *Microbe Hunters* (1926) to Stephen Hawking's *A Brief History of Time* (1988), science books have also regularly appeared on best-seller lists.

Magazines have been another important channel for dissemination of information, generally seeking to promote the cause of science rather than criticize it. Periodicals such as *Popular Science Monthly*, founded in 1872, initially borrowed authority from close connection to the scientific community and relied on contributions from scientists. The early content of *Scientific American*, which has been published continuously since 1845, reflected nineteenth-century interest in invention and technology, but its circulation soared after 1948 when new owners broadened the scope to include more political issues, such as arms control and environmental policy, alongside well-written explanations of theoretical work in physics, chemistry, and mathematics. *Scientific American* is now published in fifteen foreign-language editions, such as *Le Scienze* (Italian), *Swiat Nauki* (Polish), and *Kexue Zazhishe* (Chinese). Norway's premier popular science magazine, *Naturen*, has been in publication since 1884. Germany's first popular science magazine, *Umschau in Wissenschaft und Technik* flourished from 1897 to 1984. *National Geographic Magazine*, published by the National Geographic Society, has, since 1888, offered glimpses of exotic places through the eyes of archaeologists, anthropologists, biologists, and explorers. In the 1970s a flurry of new science magazines were created in many countries, with perspectives and political affiliations ranging from conservative to radical.

## Radio, Film Documentary, and Television

In the early days of broadcasting, radio offered a comfortable platform for scientific discourse and public education, leading many scientific organizations to offer radio lecture series in the 1920s. Commercial broadcasters, however, soon reconfigured their science programming to please audiences who were more easily captivated by drama, comedy, and music, and groups such as the American Medical Association (AMA) and American Association for the Advancement of Science (AAAS) began to adopt entertainment formats for their informational programming. Science Service's *Adventures in Science*, broadcast on CBS from 1938 to 1957, infused drama into its brief descriptions of the latest discoveries in physics, chemistry, and biology. *The Human Adventure*, produced by the University of Chicago from 1939 to 1946, used orchestras and Hollywood actors to re-create great moments in science. The Smithsonian Institution's radio program, *The World Is Yours* (1936–1942), included segments featuring research by its scientists. AAAS produced various series, such as *Science in the News* (1936–1940), in the belief that listening to scientists might encourage public confidence in scientific reasoning. Radio lectures, such as General Electric's *Excursions in Science* (1936–1949), were often reprinted and distributed to listeners upon request. Government agencies, dairy councils, pharmaceutical companies, and even insurance firms produced public health and agricultural series and thousands of print supplements.

In the decades following World War II, radio played a brief but important role in the discussion of the promise or peril of nuclear energy through such documentaries as *The Sunny Side of the Atom* and *The Quick and the Dead*. ACS, AAAS, and other scientific organizations have continued to subsidize prerecorded segments to be inserted within radio news broadcasts; *Science Update*, AAAS's award-winning series, has been on the air since 1987. Astronomy groups and universities regularly produce radio alerts about upcoming astronomical events, such as the University of Texas series *Star Date*. Many national networks, such as the Canadian Broadcasting Company, maintain science units or offices. The Australian Broadcasting Company has carried Robyn Williams's *The Science Show* and related programming since 1975.

Radio allowed the public to listen to great scientists; film helped to expand the visualization of their work. Many documentaries have been created for dual purposes, as both popular entertainment and public education, and others incorporate film or video originally created for research purposes. Among the most popular, however, are fantasies disguised as stories about nature, such as the films produced by Walt Disney Studios, by oceanographer and explorer Jacques-Yves Cousteau, and by the National Geographic Society. These productions explore the connection of humans to the natural environment, revealing nature through science-based conclusions about habitats, climates, and geography. The first Disney full-length nature film, *Seal Island* (1948), was followed by such popular films as *The Living Desert* (1953), which were also shown on the *Disneyland* television show. Nature films have been international successes in part because they require little narration and can be easily dubbed. Cousteau's first feature films, such as *Le Monde du Silence* (1956), won praise for their cinematography; his series of U.S. television specials, *The Undersea World of Jacques Cousteau*, which premiered in 1968, brought him international fame.

The British Broadcasting Corporation (BBC) has long been regarded as the foremost producer of television science. Two weekly series, *Tomorrow's World* (begun in 1965) and *Horizon* (begun in 1964), have been broadcast for over forty years, products of a BBC science unit established in 1963. By 1974, about 3 percent of all BBC programming consisted of science, in addition to segments included within regular news and public affairs programs. The first television science programs, such as the BBC's *Inventor's Club* (1948) or *The Johns Hopkins Science Review* (1948–1955) in the United States, emphasized live demonstrations or lectures by experts; audiences were expected to sit passively and absorb the material provided. Later programs such as *Watch Mr. Wizard* (1951–1965; 1971–1972) were directed at children or else presented science as a narrative account of progress through research, occasionally enlivened with special effects and animation.

Two miniseries of the 1970s—*The Ascent of Man* and *Cosmos*—set new standards for science broadcasting through innovative use of computer-generated graphics and spectacular special effects, such as taking viewers "inside" a living cell or "outside" the solar system. In each series, a charismatic host ensured success. These extravagant, expensive projects required the patronage of multiple underwriters, and they whetted audiences' appetites for glamorized science. The success of *The Ascent of Man*, a joint production of BBC and Time-Life Films starring Jacob Bronowski, first broadcast in England in 1972 and in the United States in 1975, demonstrated that audiences wanted more than technical explanations and were interested in the moral and social implications of science. *Cosmos,* hosted by the astronomer Carl Sagan, provided an exhilarating, witty perspective on astronomy and astrophysics. Twenty years after its 1980 premiere and several years after Sagan's death, the programs were still being rerun on the Discovery Channel and were available in video worldwide.

Another product of that time was the public television series *NOVA,* which imitated *Horizon*'s formats and tone and has dominated American television's representation of science since 1972, just as CBC's *The Nature of Things* (broadcast since 1960 and hosted by geneticist David Suzuki since 1975) has led Canadian efforts. Australian Broadcasting established a Science Unit in 1964; its premier series, *Quantum,* was on the air for 16 years, replaced in 2001 with *Catalyst.*

Although physics, astronomy, archaeology, and chemistry topics may be found throughout these documentary series and specials, the majority of their content involves human and animal biology. Nature programming, exemplified by the programs of British naturalist David Attenborough (*Life on Earth* and *The Life of Birds,* for example), has been consistently popular, starting with Marlin Perkins's *Zooparade* (1949–1957) and *Wild Kingdom* (1963–1988) and continuing with the cable channel Animal Planet.

### Fiction

Novels, television dramas, and Hollywood movies, which do not pretend to be educational, present fictionalized science and scientists with varying levels of accuracy. Science's potential reigns supreme, rivaled perhaps only by religion and spirituality; in the movies, science is a power for both good and evil, apt to be exploited by governments, corporations, demagogues, or criminals.

Respect for science varies widely in entertainment plots. Sometimes, scientific knowledge is blamed indirectly for a problem— radiation from fictional atomic tests, for exam-

ple, turns ordinary insects into monsters—yet few novels or movies reject science's ability to improve life if given the chance.

When fictional scientists are protagonists, they are likely to be constructed as romantic heroes, absent-minded inventors, "mad" evil villains, or eccentric reformers battling unresponsive bureaucracies. In the years after 1945, they began to be shown as auxiliary to the action, advising politicians (as in the character Dr. Strangelove) or inventing defenses against runaway viruses, monsters, and comets that others would implement.

Until the late twentieth century, the standard wardrobe for all types of scientist characters has been a white lab coat, although fictional laboratories now tend to feature more relaxed dress codes, include more women and minorities, and emphasize teamwork rather than suspicious isolation and independence. Dramatizations continue to romanticize scientists' intelligence, stamina, and motives, as if all scientists were brilliant, industrious, and unconcerned for personal gain or political power. Modern novels such as Michael Crichton's *The Andromeda Strain* (1971) and *Jurassic Park* (1993), which have been successfully translated into movies, explore complex themes in microbiology and paleontology, rejecting the older stereotypes of a villainous "mad" scientist like Frankenstein in favor of characters who perceive science as a tool for social good.

The subgenre of "science fiction," a term applied to technically plausible or scientifically derived fantasies that transport viewers forward or backward in time, has projected generally positive images of science since the nineteenth-century novels of H. G. Wells and Jules Verne. Mass-market science fiction matured as a genre in the 1920s with the founding of Hugo Gernsback's magazine *Amazing Stories* (1926) and flourished with Hollywood space movies and television's innovative aliens and adventurers. Although anthology series such as Rod Serling's *Twilight Zone* and space dramas such as *Star Trek* have engaged in obvious speculation, they acquire much of their presumed authenticity through association with real space explorations, aided by the extensive news coverage of NASA missions.

Science fiction dramas often contain important subtexts about the social or ethical implications of technological progress, and yet they may also convey contradictory messages about science's role in social progress. Plots routinely center on ingenious inventions and discoveries even as travel to outer space is portrayed as necessary to escape unchecked pollution, nuclear war, or other problems created by an irresponsible technological society. Throughout radio, television, and the movies, fictional representations help to shape public attitudes toward science and its appropriate role in society, but they may also introduce ambiguity when juxtaposed in the media with news and documentary presentations on the same topics and issues.

*See also* **Media and Technology**

### BIBLIOGRAPHY

Attenborough, David. *Life on Air: Memoirs of a Broadcaster.* Princeton, NJ: Princeton University Press, 2002.

Burnham, John C. *How Superstition Won and Science Lost: Popularizing Science and Health in the United States.* New Brunswick, NJ: Rutgers University Press, 1987.

Foust, James C. "E. W. Scripps and the Science Service." *Journalism History* 21 (1995): 58–64.

Friedman, Sharon, Sharon Dunwoody, and Carol L. Rogers, eds. *Communicating Uncertainty: Media Coverage of New and Controversial Science.* Mahwah, NJ: Lawrence Erlbaum Associates, 1999.

Goodell, Rae. *The Visible Scientists.* Boston: Little, Brown, 1977.

Kiernan, Vincent J. "Ingelfinger, Embargoes, and Other Controls on the Dissemination of Science News." *Science Communication* 18 (1997): 297–319.

LaFollette, Marcel C. *Making Science Our Own: Public Images of Science, 1910–1955.* Chicago: University of Chicago Press, 1990.

Lambourne, Robert, Michael Shallis, and Michael Shortland. *Close Encounters?: Science and Science Fiction.* Bristol, U.K.: Adam Hilger, 1990.

Mitman, Gregg. *Reel Nature: America's Romance with Wildlife on Film.* Cambridge, MA: Harvard University Press, 1999.

Nelkin, Dorothy. *Selling Science.* New York: W. H. Freeman, 1987.

Penley, Constance. *NASA/TREK: Popular Science and Sex in America.* London: Verso, 1997.

Silverstone, Roger. *Framing Science: The Making of a BBC Documentary.* London: British Film Institute Publishing, 1985.

Sobchack, Vivian Carol. *Screening Space: The American Science Fiction Film,* 2nd enlarged ed. New Brunswick, NJ: Rutgers University Press, 1998.

*Marcel C. LaFollette*

## MEDIA AND TECHNOLOGY.

Academic literature contains a long history of claims about the transformative power of communication technology. Optimists imagine that communication technologies will create the kinds of social bonds that lead to universal understanding, world peace, informed citizens, and new realms of individual freedom. Pessimists worry that the capital-intensive nature of communication technologies systematically distorts communication to promote the agenda of the social interests that have a financial stake in the technology. As a society, we have already had more than 150 years of experience with instantaneous communication technologies and it is clear that these technologies do shift the terrain upon which culture develops. New patterns, scales, and intensities of experience emerge from every major technological advance; the development of the Internet as a mass medium has generated a new high-water mark for claims about the social significance of technology. Modern communications have drastically altered the ordinary terms of experience and consciousness, the ordinary structures of interest and feeling, and the normal sense of being alive or of having a social relation. The Internet studies scholar Steve Jones's work titled *Virtual Culture* provides clear evidence of the distinctive power of communication technology to generate cultural innovations.

Recognition of the connection between communication and society is long standing. As John Dewey emphasized, society exists not only *by* transmission, *by* communication, but it may also be said to exist *in* transmission, *in* communication. This definition exploits the ancient identity and common roots of the terms *commonness, communion, community,* and *communication.* Such a ritual view of communication is directed not toward the proliferation of information but toward the maintenance of society in time, and not the act of imparting information, but the representation of shared beliefs. Communication is the symbolic process in which reality is produced, maintained, repaired, and transformed. To study communication is to examine the actual social process by which significant symbolic forms are created, apprehended, and used. As the human adventure enters a new millennium, media culture continues to be a central organizing force in the economy, politics, culture, and everyday life. How do changes in communication technology influence what we can concretely create and apprehend? The starting point of the analyst too often determines how this question is answered, leading to an almost unbridgeable chasm between camps forecasting utopian and dystopian cultural outcomes.

## Media as Spectacle

One important movement in cultural thought projects the development of advanced communication technologies to envision a dystopian future. Many scholars fear that the spread of the Internet, television, and other media may actually contribute to a decline in civic involvement, with people failing to participate in the democratic process as they retreat into private worlds of saturated infotainment. This critical perspective focuses attention on the power of the media to shape a culture's understanding of itself. Douglas Kellner uses the term *spectacle* to highlight the cultural power of the media: "media spectacles are those phenomena of media culture that embody contemporary society's basic values, serve to initiate individuals into its way of life, and dramatize its controversies and struggles, as well as its modes of conflict resolution" (p. 2). Media culture drives the economy, disseminating the advertising and images of high-consumption

lifestyles that help to reproduce the consumer society. Media culture also provides models for everyday life that replicate high-consumption ideals and personalities that sell consumers commodity pleasures, solutions to their problems, new technologies, and novel forms of identity.

In the past decades, spectacle culture has evolved significantly. Every form of culture and more and more spheres of social life are permeated by the logic of the spectacle. Movies are bigger and more dazzling than ever. Systems with five hundred or more cable television channels proliferate endlessly with movies, talking heads, infomercials, home shopping channels, and specialty niches aimed at every imaginable demographic segment that offers the possibility of commercial profit. These channels do more than sell consumer goods. Media culture provides fashion and lifestyle models for emulation and promote a celebrity culture that provides idols and role models. Media culture excels in creating megaspectacles of sporting events, world conflicts entertainment, and media events, such as the O. J. Simpson trial and the death of Princess Diana. In doing so, media culture is more important than ever in serving as a force of socialization, providing models of masculinity and femininity, socially approved and disapproved behavior, style, and fashion. Kellner argues that media culture arbitrates social and political issues—deciding what is real, important, and vital. Construction of media spectacle in every realm of culture is one of the defining characteristics of contemporary culture and society.

If the Internet-based economy deploys spectacle as a means of promotion, reproduction, and the circulation and selling of commodities, then media culture itself becomes an autonomous force proliferating ever more technologically sophisticated spectacles to seize audiences and increase the media's power and profit. Kellner worries that the power of the media spectacle to shape experience and everyday life becomes a tool of pacification and depoliticization. On the receiving end of the spectacle is the spectator, a reactive viewer and consumer of a social system predicated on submission, conformity, passivity, leisure, and consumption. This deep alienation from the human potential for creativity and imagination undercuts the possibility for self-activity and collective practice.

This critique is a restatement of the worries of Theodor Adorno and others of the Frankfurt School, whose writings, beginning during the early twentieth century, diagnosed a society entirely shaped by cultural industries given over to the needs of economic interests: capitalism is victorious, the autonomy of individuals is radically reduced, the capacity for critical thought is minimal, there is no real space for a public sphere in an era of transnational media conglomerates and a pervasive culture of advertising. The pronouncements of this era must have been overstated, for unless there had been at least some cultural space open to human freedom it would not have been possible for a new generation of communication technologies to further erode the human condition in the early years of this new century.

## Media as Superculture

If Kellner and other scholars imagine the future as a nightmare so deep that the sleeper can never awaken, other media analysts project a future so stunningly bright that the resulting blindness itself goes unrecognized. During the early 1990s, a range of futurists extending from countercultural engineers on the left to business leaders on the right confidently predicted an era of limitless information, universal connection, and unbounded individual freedom. In books, academic addresses, government testimony, and the business press, thought leaders such as Microsoft CEO Bill Gates, MIT Media Lab head Nicholas Negroponte, and Time Warner CEO Gerald Levin argued that digital information technologies would enable a "friction-free capitalism" that would revolutionize not simply the economy but human culture in all its manifestations.

Even after the Internet bubble collapsed, more-modest claims about the liberating power of the Internet have retained widespread currency. The scale, scope, and volume of Internet communication experienced tremendous growth throughout the 1990s; at the beginning of the twenty-first century, about 500 million Internet users extend across the globe. Spatially distant persons can join together via electronic networks of various kinds to develop, reinforce, and cultivate cultural contacts and communities. The complex functionality of communication technology interacts with the open-ended nature of contemporary cultural resources, creating opportunities for more varied niche identities and lifestyle groupings. Symbolic, communicational, and cultural complexities inevitably contribute to the increased fragmentation, acceleration, and personalization of life experience. Instead of being troubled by these developments, James Lull argues that these transformations usher in a new age of what he terms *superculture*. Unlike previous eras when values and ways of life were mainly tied to local contexts and influences, cultural forms today circulate far more widely and are used in strikingly imaginative ways. Lull is enthusiastic about the general direction augured by the technological transformations: "creativity and hybridity have always been at the heart of cultural construction and embody some of the best human tendencies" (p. 161).

The resulting struggles over culture and identity on a global scale have become core issues for academics, politicians, and citizens in all countries. Informational capitalism creates historically new forms of social interaction that embody and provoke a multitude of contradictory tendencies. Some see inherent technical characteristics of the Internet as enlarging the arena of cultural freedom, believing that the Internet is widely used in extremely creative, even revolutionary ways that defy supervision and control, even democratizing some routine global communications.

Our models of communication create what we disingenuously pretend they merely describe. Communication scholarship is inherently reflexive. When we model communication, we not only describe particular behavior; we create a particular corner of culture that determines, in part, the kind of communicative world we inhabit. Stories about technology play a distinctive role in our understanding of ourselves and our common history. Technology, the hardest of material artifacts, is thoroughly cultural from the outset: it is an expression of and creation of the very outlooks and aspirations we pretend it demonstrates. Technologically determined effects derive from a broad set of assumptions in which what is technological is a configuration of materials that affect other materials, and the relationship between the technology and human beings is external. Because our culture is so complexly and completely intertwined with mediated pathways of communication, it is meaningless to draw a line marking technology on one side of an analytic division and culture on the other side. Internet technology in particular dematerializes communication and in many of its aspects transforms the subject position of the individual who engages within it. Thus, the Internet resists the basic conditions for asking the question of the effects of the technology. As Mark Poster observes, "the only way to define the technological effects of the Internet is to build the Internet, to set in place a series of relationships which constitute an electronic geography" (p. 205).

Technological determinism peaks with every new major technological innovation, but the relationship between culture and technology is more complex than the utopian and dystopian prophets imagine. How, then, does one understand the cultural consequences of new communication technologies? How do changes in forms of communication technology affect the constructions placed on experience? How does such technology change the forms of community in which experience is apprehended and expressed? Given the distortions inherent in monopoly-directed technological development, economic approaches either ask the wrong questions or supply the wrong answers in thinking about the connection between technological development and social progress. The relationship between technological development and consequential forms of social life seems significant, but

what approach is most useful for making sense of the connections between structure and culture, and the character and quality of community and democracy?

## Media as Social Ecology

The study of media, technology, and culture calls for an approach able to unpack complex intellectual questions that cross traditional disciplinary boundaries. One of the most innovative and important movements is the perspective of *social ecology* developed by the Chicago School. This approach emerged when the deep-rooted philosophy of pragmatism developed by John Dewey and George Herbert Mead informed the practical questions and methodological concerns of a community of sociologists that included Robert Park, W. I. Thomas, Ernest W. Burgess, and Robert McKenzie. Borrowing the concept of ecology from biology, the Chicagoans applied it to society. Social ecology was identified as the study of human beings in relation to the phenomena of space and distance. In practice this meant the study of interrelationships among groups of human beings and how these were influenced by natural geographical boundaries. Social ecology is powerful because of its focus on interdependency, interaction, and the new types of communities and social bonds created when advanced technologies enlarge the scale of social interaction.

The Chicagoans argued that the shift from a rural to an urban society had weakened familial and community ties, leading to the kind of atomization of society that Émile Durkheim had typified as organic solidarity. At the same time, the Chicagoans recognized that change was inevitable. Unlike their predecessors, the social pathologists, they did not hark back to the past. Instead they saw that the industrial city irreversibly transformed social organization; the problems of the city could not be solved by abolishing the city, but rather could be solved only through a process of social reform. The city is not merely a physical mechanism or an artificial construction. It is involved in the vital processes of the people who compose it; it is a product of nature and, particularly, of human nature. This starting point leads to an emphasis on the individual's constant negotiation of social reality.

Like the dramatic social changes posed by the industrial revolution that motivated the development of the Chicago School, advanced information technologies and intense flows of ideas, culture, data, and capital also reconfigure natural geographic boundaries, although on a much larger scale. Globalization requires that we no longer focus on a single metropolis, but instead on the patterns and flows of a more tightly bound global order. Information flows—via the Internet, corporate databases, or advanced media—are key conduits that constitute the consequential flows of a new global social, cultural, political, and economic era.

In an age when new media technologies are ever expanding, the concept of community has evolved from that of a local geographic scale to one of a global scale. As Internet technologies expand, communication and relationship building are possible beyond the boundaries of cities, states, and countries. And just as the Internet facilitates global communication with ease, it has allowed for new forms of communication within smaller geographic communities through the creation of online community networks.

It is a direct progression to go from talking about personal communities linked by phones, planes, and cars to talking about virtual communities linked by computer-enabled communication media such as electronic mail and video conferencing. After all, a computer network is a social network when it connects people and organizations. Although hopes and fears are exaggerated, computer-mediated communication will affect work and community. After decades of empirical study of community, sociologist Barry Wellman sees dramatic changes in social pattern, but these changes are not the result of an external technological force acting on society. Technologies that support desired social patterns become more tightly woven into the social fabric. Wellman describes a move away from densely knit, tightly bound work groups in offices and factories to more loosely coupled organizations, with shifting roles, collaborations, and reporting structures.

The changes Wellman detects at the micro level of human interaction alter patterns at the global level as well. Anthropologist Arjun Appadurai's framework for exploring the relationships among five dimensions of global cultural flows relies upon *scapes*, which are perspectival constructs (Appadurai, 1990). These situated landscapes are navigated by agents who both experience and constitute larger formations as imagined worlds. In extending the perspective of the Chicago School of Sociology to grapple with the complexities of globalization, it is still necessary to combine a scientific objectivity and sensitivity to the lived experience of informants in order to map an object of study, or a terrain of relations. However, understanding the terrain of relations in a world of intense global flows, as Appadurai demonstrates, requires a more complex kind of mapping. Metropoles have never been autonomous spaces, but rather contemporary locations fully shot through with and integrated into global flows.

In the public spheres of many societies, there is concern that policy debates occurring around world trade, copyright, environment, science, and technology set the stage for life-and-death decisions for ordinary farmers, vendors, slum dwellers, merchants, and urban populations. And running through these debates is the concern that the discourses of expertise that are setting the rules for global transactions have left ordinary people outside and behind. One positive force that encourages an emancipatory politics of globalization is the role of the imagination in social life. It is possible to detect the beginnings of social forms without the predatory mobility of unregulated capital. Such social forms have barely been named by current social science, and even when named their dynamic qualities are frequently lost. Thus terms such as *international civil society* do not entirely capture the mobility and malleability of those creative forms of social life that are local transit points for mobile forms of civic and social life.

One task of a newly alert social science is to name and analyze these mobile civil forms and to rethink the meaning of research styles and networks appropriate to this mobility. Anthropologists Michael Fischer and George Marcus call for "a text that takes as its subject not a concentrated group of people in a community, affected in one way or another by political economic forces, but 'the system' itself—the political and economic processes, spanning different locales" (p.91). Such an effort requires a fractured and discontinuous plane of movement and discovery among sites as one maps an object of study and needs to posit logics of relationship, translation, and association among these sites. There is no longer an outside to modernity and capitalism because everyone on the planet is affected, even if only indirectly, by complex, interrelated, and shifting flows of trade, media, and culture. This influence seems bound to increase as we move into the next millennium. Multisited ethnographies aim to provide a new understanding of a world without traditional walls, where the traditional categories of media production, cultural consumption, and knowledge production are called into question. These approaches foreground discourse and the self-conscious, self-reflective use of language, focusing attention on the novel ways in which facts and authority are mobilized through legal, scientific, technological, or economic forms of representation.

Multisited ethnography as a research method has a strong affinity with the study of technologies, especially communication technologies. Interactive media contexts, which are designed to cross existing geographic boundaries and create new forms of sociality in emerging "communities of interest," are perhaps the most fruitful areas in which to deploy such theories developed in the subdiscipline. Processes of meaning and communication are often understood to be constitutive of political and economic interests. An extensive literature exploring information technologies as cultural contexts has begun to emerge. Prompted by the work of such scholars as Arturo Escobar, who framed cyberculture in terms of concerns with the cultural constructions and reconstructions on which new technologies are based and which they in turn help to shape, a host of communication scholars, sociologists, anthropologists and cultural critics are analyzing how specialized discourses, interfaces, new environments, speed, multiple identities, and brand environments characterize technologies as forms of life and contribute to an ongoing conversation examining communication technology as a distinctive cultural practice.

*See also* **Media and Science**

**BIBLIOGRAPHY**
Appadurai, Arjun. "Disjuncture and Difference in the Global Cultural Economy." *Public Culture* 2:2 (Spring 1990): 1–24.
Carey, James W. *Communication as Culture*. New York: Unwin Hyman, 1989.
Dewey, John. *The Public and Its Problems*. New York: Holt, 1927.
Escobar, Arturo. "Welcome to Cyberia." *Contemporary Anthropology* 35. (June 1994): 211–231.
Jones, Steve. "The Internet and Its Social Landscape." In *Virtual Culture: Identity and Communication in Cyberspace*, edited by S. Jones. London: Sage, 1997: 7–35.
Kellner, Douglas. *Media Spectacle*. New York: Routledge, 2003.
Lull, James. *Culture in the Communication Age*. London: Routledge, 2001.
Marcus, George, and Michael Fischer. *Anthropology as Cultural Critique: An Experimental Moment in the Human Sciences*. Chicago: University of Chicago Press, 1986.
Poster, Mark. "Cyberdemocracy: Internet and the Public Sphere." In *Internet Culture*, edited by D. Porter. New York: Routledge, 1996: 201–218.
Wellman, Barry. "From Little Boxes to Loosely Bounded Networks: The Privatization and Domestication of Community." In *Sociology for the Twenty-First Century Continuities and Cutting Edges*, edited by J. L. Abu-Lughod. Chicago: University of Chicago Press, 1999: 94–114.

*John Monberg*

# MEDICAL COMMUNITY, THE.

This entry includes two subentries

　　Communication with the Public
　　Health Communication

The current structure of the medical community was largely shaped by scientific breakthroughs that began to appear with increasing regularity during the nineteenth century. Better technology and more tools led to more accurate diagnoses, which improved treatment. This increased public confidence, which translated into a growing demand for health care services and thus required an increasing supply and widening range of facilities and health care skills. The need to help people pay for care spawned a whole new sector of health care system-related organizations and occupations that were devoted to formulating payment arrangements and plans. Plans were introduced by both private sector organizations (such as insurance companies) and public sector organizations (such as government agencies). As things progressed, many of the organizational or occupational groups began hiring policy analysts to monitor, evaluate, and develop proposals designed to reform health care arrangements with an eye toward advancing a particular set of priorities.

Two parallel trends, apparent in all industrialized countries, contributed to this expansion—the aging of the population and the increase in the prevalence of chronic illness. The pharmaceutical industry responded to the increasing demand for drugs to treat an expanding number of people living with chronic illnesses. The growing level of demand in combination with increasing numbers of persons seeking care resulted in the steady rise in the percentage of the national budget going to health products and services.

In short, the delivery of medical care has evolved into a vast medical-industrial complex. The result is that an attempt to change any part of these arrangements, most notably cutting costs, is likely to have a negative impact on some category of participants even as it benefits other participants—and the attempt is sure to be resisted.

It is interesting to consider how an enterprise of this size and scope evolved given that it is based on many participants' contributing to a process that was not only unplanned, but was also unforeseeable, largely because scientific discoveries are not predictable. Part of the explanation is that the growing level of confidence in and demand for health care services has been providing the incentives for generating a vast array of new products and services in response to the enormous profit potential. This is why there is a consensus about the importance of keeping this sector strong, matched by a high level of disagreement about the best way to go about achieving that objective. This is not to dismiss the fact that many in this enterprise are motivated by the desire to help those in need of medical attention. Let us consider how all the parties that are now part of the medical community got there and how they get along.

## The Medical Profession

Medicine as a field of endeavor achieved great gains over the twentieth century. Society recognized that the rigor of medical training meant that doctors had an extensive base of scientific and applied knowledge at their command. The extent to which medicine actively promoted this view of itself in order to gain social advantages and rewards, including prestige and income, became the topic of debate during the latter half of the twentieth century. The concept of medical *professionalism* is at the core of this debate.

Doctors set themselves apart by claiming and achieving recognition as "professionals." Their success in this effort is generally attributed to two core identifying characteristics: (1) prolonged training in a specialized branch of knowledge and (2) a service orientation. The social rewards flowing from this achievement, which increased significantly during the first half of the twentieth century, are tangible, receiving a generous income, and symbolic, gaining prestige. The process was so successful that it provided a model for other occupational groups hoping to achieve similar results.

## Specialized Knowledge and Trust

Doctors were eager to distinguish themselves from shopkeepers in the business of selling goods and services to customers. Although shopkeepers may not be more motivated to take advantage of their neighbors than anyone else in society, trust is a critical factor in the doctor-patient relationship. Patients are in a particularly vulnerable position in their interactions with doctors. Accordingly, doctors have been interested in finding ways to assure patients that the treatments provided were both efficacious and in the patients' best interests, rather than for the monetary gain of doctors. Doctors were also interested in distinguishing themselves from others offering curative services. To name only a few of their competitors, there were hydropaths, homeopaths, naturopaths, chiropractors, and osteopaths. It was not until the end of the nineteenth century that *allopathic medicine* gained ascendancy. Several of the other forms of practice survived, and newer ones have been added—acupuncture, for example—and are now known as *alternative*, and more recently as *complementary*, forms of medical practice. Osteopathy is now considered mainstream medicine.

Allopathic physicians were able to present evidence regarding the efficacy of the cures they promoted by predicting outcomes with and without treatment. They gained control over the development of the medical body of knowledge and its application by providing empirical evidence, espousing the peer review of outcomes, and committing themselves to sharing the knowledge gained through experience. Recognizing that meeting these objectives required collective effort, they formed professional associations.

## Professionalism and the Professional Association

The significance of the emergence of a national professional association in fostering the professionalization process becomes clear when we consider the responsibilities it assumed. The American Medical Association (AMA), established in 1847, became the forum for developing a code of ethics, disseminating knowledge through professional meetings and publications, promoting improvements in medical education, and making a public commitment to monitor the performance of fellow physicians. The association was instrumental in establishing professional journals designed to communicate new discoveries and promote successful therapies. Physicians in other countries established associations as well: the British Medical Association was established in 1832 and the German Medical Association not until 1931.

## Professionalism and Medical Education

An important objective in medicine's effort to professionalize focused on medical education. Although medical education became linked to universities very early (around 1200 in Italy), it did not become firmly grounded in science until the last century. Medical education and training in Britain preceding the twentieth century is a case in point. While physicians were receiving a broad university education with heavy emphasis on philosophy in the most prestigious British universities, surgeons were undergoing vocational training as members of the barber-surgeon guild. (A remnant bit of evidence of this connection remains in the symbol used by barbers then and now to identify a barber shop—the barber pole, a bloodstained staff grasped by patients while undergoing bloodletting.)

In the United States, the vast majority of doctors received their medical training in apprenticeship arrangements. Established practitioners earned a portion of their income by accepting fees for training apprentices. Concern regarding the value of this form of training led elite members of the profession to invite the Carnegie Foundation, known for its commitment to education, to look into the matter.

The person invited by the Carnegie Foundation to carry out an investigation of the quality of medical education in the United States was Abraham Flexner. The project was initiated in 1906 and culminated in the Flexner Report (1910). The standard of comparison Flexner used was the program of study offered at Johns Hopkins Medical School. The effect of the report was the failure of a large number of medical schools and the upgrading of all other schools. There were 175 schools when Flexner began his review and 107 a decade later.

The long-term effect of the Flexner Report was to ensure that medical education in the United States would emphasize training in the basic sciences. By the second half of the twentieth century, the focus on scientific medicine, which was so well received earlier in the century, led to the criticism that medicine is more interested in treating body parts than the whole person. The emphasis on basic science, empirical research, and a more scholarly approach to translating such findings into practice had one other important effect: it helped to advance medical specialization.

## Specialization

The first specialty to emerge in the United States, in 1916, was ophthalmology, the treatment of eye diseases. By 1944, there were twenty specialty groups offering certification. Fearing that specialization would produce a schism in the ranks, the AMA announced that it would not permit any new specialties to become established after that year. That resulted in a proliferation of subspecialties (numbering approximately 120 by the end of the twentieth century). The AMA was particularly concerned about the negative impact that specialization might have on the status of general practitioners.

General practitioners did become dissatisfied. They responded by creating a specialty of their own in 1971, referred to as *family practice* and requiring *residency* training lasting three years. Prior to this time all physicians completed a one-year *internship* before entering into practice. Specialists went on to residency training for at least two more years. The effect was that all physicians practicing in the United States would now be required to complete a minimum of three years of residency training. Many complete several additional years in subspecialty fields of medical practice.

Two parallel developments, which began to take shape over the first half of the twentieth century, contributed to the rising level of dissatisfaction, first among general practitioners and later among the majority of doctors. One development was the dramatic change in hospitals; the other development

was the emergence of the "third-party payer," namely, medical insurance.

## The Hospital

The history of hospitals, like the history of medicine, can be traced to the Greek and Roman eras. Hospitals were generally operated under the auspices of generous patrons, typically wealthy women. It is not clear how efficacious they were. More is known about medieval European hospitals. These were places to which only the destitute and abandoned went as a last resort; they functioned as "poorhouses" and "death houses." Preceding the first decade of the twentieth century, patients had only a fifty-fifty chance of surviving hospitalization in the United States.

Hospitals began to change as surgeons became increasingly more dependent on hospitals. They needed more equipment and more staff to perform surgery than they could accommodate in their offices. Surgical success rates were also improving, so much so that middle-class patients were becoming less reluctant to go to a hospital for care. Surgeons were in a position to bring this new wave of middle-class, paying patients to the hospital of their choice. That persuaded hospitals to respond to the surgeons' demands regarding improvements.

Hospital administrators, who formed their own association in 1899, renamed it the *American Hospital Association* in 1908. In 1952, they joined together with doctors to form an accrediting organization mandated to carry out formal reviews, now known as the *Joint Commission on Accreditation of Healthcare Organizations* (JCAHO).

## Hospitals as Training Facilities for Emergent Occupations

The increasing rate of surgery performed in hospitals made clear the need for better-trained hospital workers. Hospitals took on the responsibility for training a wide range of health care personnel. Many health occupations (now often referred to as allied health) grew out of on-the-job training organized by doctors who needed assistants to perform particular tasks. As the tasks evolved and the training became more systematic, training moved into colleges, often as two-year programs at first, and many expanding into four-year programs later. Hospitals continue to follow the same pattern as new, highly sophisticated pieces of equipment come into wide use.

## The Special Case of Nursing

Hospitals began training nurses first. Completion of a three-year training program resulted in a diploma and registered nurse (RN) designation. By the middle of the century, nursing leaders were arguing that hospitals were taking advantage of the low-cost labor provided by nursing trainees. That plus the general interest in the value of professionalization, for which medicine served as an excellent model, caused nursing leaders to argue that university-based education was essential.

Universities developed bachelor's degree programs first and went on to create master's and doctoral programs shortly thereafter. By the 1960s, nurses could receive a registered nurse certification by completing one of three different types of programs: a three-year diploma program, a four-year bachelor's degree, or a two-year associate's degree. The two-year programs were developed during the 1960s in an effort to expand the supply of nurses.

This did not resolve the basic problem, namely, the never-ending shortage of nurses, which is periodically severe enough to close whole wings of hospitals. Hospitals have experimented with a variety of approaches to address the problem, mostly by dividing up nursing tasks and then training other, usually less-skilled workers to perform them.

None of the solutions that have been tried so far has succeeded in overcoming the nursing shortage. At the same time, the problem is becoming more severe because patients admitted to the hospital are now more seriously ill than was true in the past. The result is that fewer nurses are now available to deal with more severely ill patients. More troubling is the fact that a growing body of research shows that the quality of nursing care in the hospital affects patient outcomes, a fact that has become apparent because hospital care is being more closely monitored in an effort to control costs. That

leads us to the topic of health insurance and third-party payers.

## Health Insurance

Being unable to pay for medical care was not a problem during the first half of the twentieth century. When patients did begin seeking care more regularly, doctors would sometimes treat them even if they could not pay. But it was impossible for hospitals to do that. They simply could not afford to keep their doors open and hire staff if patients did not pay their hospital bills. The problem became especially acute during the Great Depression. This is when Baylor Hospital, in Dallas, Texas, devised a plan whereby people would pay 50 cents a month for twenty-one days of hospital coverage. The plan was so successful that it quickly spread across the country, eventually becoming known as the Blue Cross and Blue Shield systems. The features of the plan are worth noting. Blue Cross paid whatever the hospital charged. The rate was adjusted based on *utilization*, that is, the cost of treating people over the past year. Everyone paid the same amount, *a community rate.* Blue Cross was chartered as a not-for-profit organization. Blue Shield, which paid for the doctor's portion of the bill, evolved a little later based on the same principles.

For-profit insurance arrangements developed as well but did not become well established until the 1940s. The shift was dramatic. Less than 10 percent of the population was covered by health insurance at the start of World War II. By the end of the war, 50 percent were covered. That happened because wages were frozen, so benefits were the main incentive that employers could offer to attract employees during a period during which men were at war and only a limited number of women were prepared to go to work.

Over the following decades, health insurance became a standard benefit offered by large companies. However, not everyone was fortunate enough to receive this benefit, such as those working for small companies, retirees, and others not connected to the labor market, most notably, poor women and their children. Recognition of the plight of the last two categories led to the Great Society programs of the 1960s, when Medicare and Medicaid came into existence (legislated in 1965 and fully operational by 1967). Medicare is the program for anyone over age 65 and for certain categories of disabled persons under 65. Medicaid was originally a program for poor women and their children. (By the end of the twentieth century, Medicaid was spending more on the elderly, blind, and disabled than on poor women and their children.) The development of these two public health insurance programs has had a profound effect on the health care delivery system in general and on the medical community in particular.

The American Medical Association was vehemently opposed to the introduction of the two public health insurance programs. The AMA argued that this was socialized medicine and would lead to dictatorial bureaucratic rules that would interfere with the doctor-patient relationship. That did not happen, at least not in the way opponents predicted.

What did happen in the short run is that doctors' incomes began to rise significantly. This is not surprising considering that the programs allowed large numbers of people to seek medical care who had not done so previously because they could not afford it. The programs guaranteed reimbursement for treating the new influx of patients. Because the flow of patients did not recede after an initial upswing as predicted, costs continued to rise. The unanticipated cost of the programs caused planners to find ways of reducing costs without cutting services.

## The Golden Age of Medicine

Before turning to the changes brought about by the rising cost of medical care, let us review medicine's achievements over the first half of the twentieth century. It succeeded in asserting control over the development of medical science by locating medical education in the university. It gained control over the application of medical science via licensure, which gave medicine the exclusive right to do two things—prescribe and perform surgical procedures. Doctors succeeded

in gaining society's respect for their accomplishments, which society was willing to reward handsomely in both prestige and income. Not surprisingly, they look back on the post–World War II period as the Golden Age of Medicine. That golden age began to decline by the end of the 1960s.

The same features that worked to inspire public confidence in medicine also worked—for better or worse—to restrict the entry of competing occupational groups into the field. The battle over the right to treat patients between the medical profession and a variety of competitors has not abated. A number of occupational groups have lobbied for the right to prescribe medications, among them, nurse practitioners and clinical psychologists. Other groups have lobbied for the right to do more extensive procedures and to be reimbursed by third-party payers. Chiropractors succeeded in having their treatments reimbursed. Others have had less success in overturning legal restrictions and getting third-party reimbursement.

Medicine focused on improving the *quality of care* throughout the first half of the twentieth century. It succeeded. Society did not become concerned about two other objectives that have come to define the challenges facing health care delivery systems until the decade of the 1960s. These are access to health care services for all members of the society and cost containment. All countries confront the challenge of achieving all three—quality, access, and cost containment—at the same time.

## Access to Care and Cost Containment

Health insurance was originally designed to pay for treatment once the patient was sick, meaning that preventive care had to be paid for "out of pocket." Therefore, people did not seek care until the problem was already more serious and more expensive to treat. Health maintenance organization (HMO) legislation passed in 1974 promised to address this. The idea was that prepayment for all the care a person might need would allow patients to seek care at the earliest, least expensive stage of illness. Organized medicine was opposed to the prepaid aspect but found it hard to argue against because it was presented as "health maintenance."

The most successful prepaid practice arrangement in existence then (and now) was Kaiser Permanente of California. It served as a model. Two features were especially important: its nonprofit status and the fact that affiliated physicians were salaried and worked on a full-time basis. That became known as the *staff model.* Newer HMOs began negotiating with groups of physicians—organization to organization—signing up group practices under the same contractual understandings about reimbursements, referral to specialists and hospitals, and even treatment modalities. That became known as the *group model. Mixed model* arrangements came next, that is, contracts including a share of profits, bonuses for cutting costs, and so on.

It did not take long for insurance companies to recognize the potential for profit. They entered the market during the late 1970s, enrolling a steadily increasing number of persons by offering lower rates. They did this by directing their efforts to large companies with younger, healthier, white-collar work forces. Competition was fierce. New HMOs started up and others failed, all in a very short space of time.

As parent organizations sponsoring HMOs, the major insurance companies were ready to accept lower profits, even a certain amount of lost income, in order to stay in the race over the long run and beat out the competition. They also provided the resources to integrate inpatient (hospital-based) and outpatient (office-based) care in order to gain greater control over the entire course of treatment. The result was that HMOs evolved into *managed care organizations* (MCOs). The predictions were that only a handful of MCOs would be around when the dust settled. At the time of this writing, these predictions are very close to being fulfilled.

It is worth noting that the for-profit MCOs label the percentage of operating funds spent on patient care as the "medical loss ratio." Using incentive arrangements designed to cut operating costs is considered good business practice. Critics say that the concept captures exactly what is wrong with health care arrangements that focus on profit.

The growth of managed care has had some negative consequences: People working in sectors that carry a high risk of injury or illness began finding insurance harder to get and certainly more expensive when it is available. This is more problematic when the economy falters and companies need to reduce expenses. Cutting the costs associated with health insurance for employees becomes a major consideration. The result is that approximately 15 to 18 percent of people in the United States have been uninsured over the last two decades of the twentieth century. An even greater number are uninsured for part of the year.

This is in contrast to the situation in virtually all other industrialized countries, which provide *universal coverage,* that is, health insurance for everyone in the country. Most have some form of national health insurance. England is the notable exception. It established the National Health Service following World War II, and it operates all health care institutions and employs all health care personnel. Accordingly, England is the only major industrialized country that operates a health care system that can be identified as socialized medicine, even though it is generally not identified in this way. The U.S. health care system is also unique among industrialized nations. It is the only country that leans so heavily on *market mechanisms* and the private sector. Although the government spends a considerably greater portion of the national budget on health care than any other industrialized country, the government portion is still smaller than in any other industrialized country. Critics of health care arrangements in the United States argue that the lack of health insurance is a major part of the explanation for the fact that life expectancy is lower in the United States than it is in many industrialized countries.

## The Market Approach to Health Care Delivery

The consequences of the market approach to organizing the health care delivery system in the United States have been significant and largely unanticipated, and remained far from resolved at the beginning of the twenty-first century. Consider the level of control that MCOs have at their command. Managed care organizations have enrolled or effectively captured about 80 percent of patients. This permits them to require enrollees to seek care only from participating doctors and hospitals, which has forced doctors to participate in managed care arrangements if they are to have enough patients to stay in practice.

The entry of MCOs has also had an immense effect on the medical community. The predominant form of medical practice that evolved during the first half of the twentieth century was a fee-for-service reimbursement arrangement and solo practice. Some doctors worked in partnerships of two or three. General practitioners' offices were scattered throughout the community instead of being located in hospitals or professional office buildings, as has become the norm by the end of the twentieth century in most urbanized areas. Patients were well acquainted with doctors who maintained offices in the community. Patients shared information about the treatment they received, which had the effect of inspiring confidence and trust in doctors. Doctors developed mutually gratifying relationships with their patients.

As increasing numbers of doctors began to specialize and move away from the community, they became strangers. Trust and confidence declined; malpractice increased. The growth of for-profit medicine did not cause this to happen, but it did exacerbate these trends. Managed care organizations portrayed doctors as motivated primarily by money and not to be trusted. They promised to protect patients from greedy doctors. However, when the MCOs went on to restrict patients' options and deny treatment recommended by doctors, patients decided that the MCOs were even more greedy. Confidence in the whole enterprise declined. The fact that Americans have a long history of distrusting government makes the situation difficult to fix. Public opinion surveys consistently find that Americans want changes but are split on whom to trust to

achieve change—government or the private sector.

The decline in the level of confidence in the health sector is paralleled by an increasing reliance on alternative medical practitioners. The irony is that this comes at a time when impressive medical advances are being reported almost daily. Although negative outcomes related to the use of alternative medications are reported far less frequently, they include serious injury and death. One interpretation of the shift toward alternative medical practitioners is that patients want more personal attention. Patients are more interested in improving the quality of their lives than in prolonging life under adverse conditions. However, patients are not in a position to bring about change because they have no organization to represent them and to articulate their demands.

Doctors are also casting about for mechanisms to have their voices heard and their demands addressed. They recognize that they need powerful organizations to do this but disagree about whether they should rely on a corporate model or an alternative. They face a very different agenda from the one that drove their successful pursuit of professionalism at the beginning of the twentieth century. By law, doctors are still considered to be independent contractors, very much like small businessmen who sell products and services in their shops. As such, they are not permitted to join together to set prices, which would be anticompetitive, causing restraint of trade and a violation of antitrust law. Insurance companies, the parent organizations of the MCOs, are not prohibited from setting the price they will pay doctors. The American Medical Association has identified this as an "uneven playing field" and is lobbying to have legislation passed to exempt doctors from antitrust legislation. Legislators are reluctant to do this, fearing that it might lead to other groups' demanding similar legislation and thus create a new set of complications that legislators would have to resolve.

Although doctors may not bargain collectively, they may negotiate over fees and conditions of work with MCOs on an individual basis or do so as members of a corporation that contracts with MCOs on an organization-to-organization basis. Managed care organizations have no interest in negotiating with scores of individuals. Physicians have responded by creating *independent practice organizations* (IPOs), which collectively represent individual members in contracts with MCOs. This allows individual practitioners to have working arrangements with several MCOs at the same time. The advantage is that they have a large pool of patients. The disadvantage is that the amount of paperwork increases considerably because each MCO uses different forms and has different restrictions.

Doctors who enter into a contract with a single managed care organization, and who accept a fixed salary and a set of working conditions, confront a somewhat different problem related to collective bargaining. They say that their position is more that of an employee rather than that of an independent contractor. Managed care organizations counter with the argument that doctors oversee the work of nurses, technicians, and clerks, making them supervisors, who are also not permitted to join collectively and work in opposition to the interests of the organization that employs them. These arguments are not a matter of idle speculation. They take place in court. The results are neither consistent nor predictable. Some courts find in favor of doctors who are interested in organizing collectively based on the claim that they have virtually no managerial responsibility or authority and should be allowed to organize to protect their interests. In other cases, the court finds the argument presented by the managed care companies more convincing, generally claiming that doctors' primary interest is economic gain and that giving in to their demands would result in an unacceptable increase the cost of care.

The argument is about doctors' right to unionize. There are three major physicians unions in the United States that are steadily increasing their membership. Giving doctors the right to bargain collectively could have significant social implications. The steps that doctors took to achieve professionalism during the twentieth century provided a powerful model for other occupational groups to follow, and it is possible that medicine's success in gaining the right to bargain collectively could have a comparable effect over the twenty-first century.

## The Future of the Medical Community

The shape of the medical community can be expected to change. How it will look in the future will depend on both the course of scientific discovery and the prevailing sociopolitical value system that interprets the significance of those discoveries. For example, genetic research appears to be full of promise at present. However, it is not clear how society will react to breakthroughs that allow geneticists to add intelligence or eliminate genes linked to certain diseases before knowing what if anything else those genes affect. Since there is no reason to expect demand for health care to decline, will new occupational groups and new technologies develop to deal with decisions about the distribution of scarce health resources? Or will investment in research continue in the hope that it will eventually lead to less costly treatments? It is clear that life expectancy is increasing but that it is not necessarily illness free. Is it possible to refocus attention to providing services that increase quality of life? One category of researchers now argue that medical care that focuses on the individual does not improve society's health nearly as much as measures that focus on the health of the whole population. These as well as many other issues are being debated. It is possible that we do not know the right questions to ask to prepare for the future; more certain is that fact that we cannot know how they will be resolved.

*See also* **Biomedical Technologies; Health Care, Access to; Medical Profession; Medicine and Society**

### BIBLIOGRAPHY

Anderson, Gerard, Uwe Reinhardt, Peter Hussey, and Varduhi Petrosyan. "It's the Prices, Stupid: Why the United States Is So Different from Other Countries." *Health Affairs* 22 (May/June 2003): 89–105.

Blendon, Robert, and John Benson."Americans' Views on Health Policy: A Fifty-Year Historical Perspective." *Health Affairs* 20 (March/April 2001): 33–46.

Flexner, Abraham. *Medical Education in the United States and Canada*. New York: Carnegie Foundation for the Advancement of Teaching, 1910.

Goode, William J. "Encroachment, Charlatanism, and the Emerging Profession: Psychiatry, Sociology and Medicine." *American Sociological Review* 25 (December 1960): 902–914.

Institute of Medicine. *Insuring America's Health: Principles and Recommendations*. Washington DC: National Academies Press, 2004.

Law, Sylvia. *Blue Cross: What Went Wrong?* New Haven: Yale University Press, 1974.

Needleman, Jack, Peter Buerhaus, Mattke Soeren, Maureen Stewart, and Katya Zelevinsky. "Nurse Staffing Levels and the Quality of Care in Hospitals." *New England Journal of Medicine* 22 (May 30, 2002): 1715–1722.

Starr, Paul. *The Social Transformation of American Medicine*. New York: Basic Books, 1982.

Stevens, Rosemary. *In Sickness and in Wealth: American Hospitals in the Twentieth Century*. New York: Basic Books, 1989.

Woolhandler, Steffie, and David Himmelstein. "Paying for Health Insurance—And Not Getting It." *Health Affairs* 21 (July/August 2002): 88–98.

*Grace Budrys*

# Communication with the Public

Popular health advocacy preceded the U.S. Civil War, when it was delivered through public lectures, books, almanacs, pamphlets, and the occasional specialized magazine, such as the *Water-Cure Journal* in 1845. Heavy advertising of commercial products, such as medical devices and proprietary compounds, emerged as the major source of public health information used by the U.S. population by the mid-nineteenth century. Manufacturers of so-called proprietary, or patent, medicines were among the first to market their products nationally. Total advertising expenditures for proprietary medicines soon exceeded those of all other products combined.

Health reform movements emerged in a morally charged context to suppress "vice" and "vicious habits," such as alcohol, tobacco, and narcotics consumption, as well as prostitution. The Women's Christian Temperance Union campaigned for state laws requiring public-school instruction in health, hygiene, physiology, and the dangers of alcohol and tobacco consumption; these laws were adopted in all states by the turn of the twentieth century. Citizens' organizations such as the New York Society for the Suppression of Vice, headed by Anthony Comstock, aided in the passage of the Comstock Laws of 1873, which regulated "obscene, lewd, and lascivious" materials, as well as information "intended for the prevention of conception or the procuring of abortion," including books, pamphlets, advertisements, or notices of any kind. Arrested under these laws in 1878, the infamous abortionist Ann Lohman, who had amassed a fortune under the name "Madame Restell," spent nearly $60,000 annually advertising her services. It was not unusual for nineteenth-century advertising budgets to exceed $100,000 per year, and some reached the million-dollar mark.

Morally based, religious reformers tended to dominate nineteenth-century medical communication. Not until the late nineteenth century did health communication to the public become the domain of physicians and lay persons associated with the emerging popular enthusiasm for personal health, physical fitness, and hygiene from which sprang social pressures for cleanliness. The so-called sanitary movement became the twentieth century public health movement.

## Public Health and Mass Media in the Early Twentieth Century

The expansion of the public health infrastructure led to the 1912 consolidation of the United States Public Health Service (USPHS), which conducted public health campaigns, such as those against venereal disease (VD) during World War I in cooperation with state health departments. The mass media campaign against VD involved significant congressional funding through the Army Appropriations Act (1918) for sex education campaigns and a national system of VD clinics, which quietly transmuted into birth control clinics in the 1940s.

Turning to mass media, the USPHS began radio broadcasting in 1921. Short newsreels on health topics produced by studios such as Twentieth Century Fox were shown in movie houses as early as 1912. Broadcast media sought to control information by banning such words as *syphilis* from the airwaves. In 1934, CBS prevented New York state health commissioner and member of President Franklin Delano Roosevelt's Science Advisory Board, Thomas Parran, M.D., from going on air because the latter refused to delete the word *syphilis* from his script. In his later capacity as surgeon general (1936–1948), Parran campaigned to "put the words *syphilis* and *gonorrhea* into the language" through the print media, which was far more receptive than the broadcast media in the mid-1930s.

The USPHS sought to reach the self-organized African-American health movement, one of the cornerstones of early civil rights activity. The Rosenwald Fund, a private philanthropy devoted to the health and welfare of African Americans, partnered with USPHS in 1930 to begin a project on the epidemiology and treatment of syphilis that would become the notorious Tuskegee Institute study. Outreach efforts included mobile syphilis units that worked in conjunction with religious programs. In 1932, the USPHS opened the Office of Negro Health Work, which served as a clearinghouse to promote National Negro Health Week educational programming. Headed by Roscoe C. Brown, an African-American dentist originally hired by the USPHS Division of Venereal Disease to provide health education to blacks, the Office of Negro Health Work broadened its mandate beyond sex education to "disease prevention" and general fitness. The grassroots Negro Health Movement housed in that office conducted community campaigns until 1950, losing its institutional footing when the USPHS "integrated" itself and terminated attention to "separate needs."

The 1930s witnessed an influx of lay popularizers and journalists into the health field, as well as the emergence of professional organizations of health educators. Life insurance companies developed public health and safety campaigns to prevent unnecessary death. The glorification of wartime medical research led to increased optimism about authoritative cures and so-called magic bullets due to successes with contagious and infectious diseases and the expansion of the research infrastructure. The post-World War II reliance on marketing techniques drawn from behavioral science placed an emphasis on individual lifestyle rather than public health. The USPHS launched a sophisticated advertising campaign against tobacco smoking that accompanied the nationally publicized 1964 release of the highly influential *Surgeon General's Report on Smoking and Health*. Television advertisement of cigarettes was banned in 1970 in an amendment of the Federal Cigarette Labeling and Advertising Act of 1965.

## The Impact of Social Movements on Health Communication

Official efforts have been underway since the 1970s to strengthen health and consumer education. The National Consumer Health Information and Health Promotion Act (1976) was intended to support research and demonstration projects on the increased application of health knowledge and to secure the cooperation of the communications industry in ensuring access and accuracy. Consumers' concerns about accuracy are often voiced in relation to advertising claims. The road to direct-to-consumer advertising was paved by the Kefauver-Harris Act of 1962, which amended the Federal Food, Drug, and Cosmetic Act to specify that medical advertisements must contain the following: (1) statements that are neither false nor misleading, (2) a "fair balance" of risks and benefits, (3) "material facts," and (4) a brief summary including all risks. After the first wave of direct-to-consumer advertisements in the early 1980s, consumers and manufacturers requested further study of their effects. In response, the FDA imposed a moratorium, which was lifted in 1985. However, the FDA did not publish draft guidelines until 1997. Although there is no formal approval process for pharmaceutical commercials or Web sites, the FDA sends out notices to sponsors who violate the terms of the 1962 amendments regardless of whether the violation involves understatement of risk or overstatement of effectiveness.

The social movements of the 1960s can be credited with the expansion of consumer rights and patients' rights, and the immense growth in consumer-oriented interest groups in the nonprofit and nongovernmental sectors. Consumer interest in and use of health information—as well as the activity of seeking preventive information-varies by gender. Women constitute the primary audience for health information, perhaps due to their continued assignment of the gendered role of primary caregiver within families. Not surprisingly, the women's health movement took an especially proactive role in preparing and distributing its own medical information. The most notable example was the Boston Women's Health Book Collective, originally comprising twelve members of the women's liberation movement. Finding that health information, whether delivered publicly or privately, was often biased or inaccurate, they produced a book that has gone through multiple editions to sell more than 4 million copies worldwide: *Our Bodies, Ourselves*, first published in 1973. They formed the Women's Health Information Center in 1980 and supported the founding of the National Women's Health Network—the first national women's health advocacy membership organization to lobby for greater attention and resources for women's health. The network established a Black Women's Health Project in Atlanta, Georgia, in 1981. Founded in 1983 by health activist Byllye Y. Avery, the National Black Women's Health Project developed with the mission of effectively communicating health information to African-American women.

Rising popular interest in science, medicine, and health led to significant expansion of coverage by the lay press from the mid-twentieth century on. Relations between the lay press and the scientific press are often depicted as oppositional. There has been an empirically documented increase in the portrayal of scientific and medical issues as controversies by the lay press. Regular crises in medical broadcasting often arise when there is disagreement among clinicians and researchers. However, the effects of public controversy are not confined to those of diffusion within the lay press. Popular press coverage amplifies the impact within core scientific journals. Studies that receive attention in the lay press receive markedly more citations in the scientific press for up to a decade following publication. This suggests that the popular press may serve as a filtering mechanism for scientists as well as the public.

## An Attentive Public and the Effort to Ensure the Accuracy of Health Information

Concerns about accuracy and validity have long haunted the public's reliance on commercial media for medical information. Studies conducted in 1970 showed that up to 5 percent of commercial television broadcasting carried misleading or inaccurate health information. Mass media has been especially susceptible to commercial influence but is an important source of health information. Some studies now show that television is the primary source of health information for the majority of U.S. adults, with physicians, mass-circulation magazines, and specialized magazines sharing second place, and the Internet trailing far behind. Even where medical information is accurate, its content is often limited to a few sources (chiefly the *Journal of the American Medical Association*, the *New England Journal of Medicine*, and *Science*). Content analysis reveals that the media concentrate on medical technology and pharmacological "fixes," operate with an extremely limited conception of medical expertise, and are fascinated with a few conditions. Mention of preventive measures, environmental influences on disease, or occupational factors is extremely rare.

Increased electronic communication capabilities have changed medical communications disseminated to the public in several ways. Defined as the use of electronic telecommunications technologies to deliver and support health care where distance separates the participants, telemedicine applications are expanding rapidly after a period of decline during the late 1980s and early 1990s, when transmission costs were high. Foreshadowed as early as the 1920s by imagined "radio doctors," the telemedicine of the 1950s and 1960s was initially confined to rural populations, transmission of radiological images, and urban emergency telemetry (which remain among its most common uses). The sheer scale of the health care system, the imperative to reduce costs, and the development of interactive technologies have made telemedicine attractive despite the obstacles to its implementation. As a form of health communication, however, it offers the possibility of individual health care delivery from a distance.

Effects include widespread availability of material that did not undergo peer review, difficulties with archiving, and concerning regarding the privacy and confidentiality of medical records. The pace, style, and informality of argumentation on the Web may make "agonistic forums" more common in ways that pressure the accuracy and validity of the available information. The "cold fusion" incident is often cited as one in which unreliable, conflicting, and contradictory information was released early via electronic media, only to be overtaken by older and more reliable communicative means. Studies conducted in the mid-1990s showed that only two percent of U.S. adults utilized the Internet to obtain health information, but since those studies there has been an increase in health sites, which currently top two thousand.

Public dissemination of the results of scientific and medical research raises the question of audience receptiveness to health communication. Trend data from national assessments of "biomedical literacy" reveal only modest growth in adult comprehension of scientific vocabulary and the nature of scientific inquiry since the late 1980s. Levels of biomedical literacy correlate with general education levels, suggesting that public health campaigns may increase knowledge gaps due to individuals' differential capacity not only to consume but also to make use of information they receive and to apply it to their own health situations. Education levels also affect how individuals rank the seriousness of particular diseases. More educated individuals tend to make up the "attentive public" for health policy, and thus legislation and research have often been geared more toward the needs of the already affluent. Pharmaceutical advertising of so-called lifestyle drugs, marketed largely to middle- and upper-class individuals, illustrates this last point.

"Publication by press conference" is frowned upon within the scientific community. Attempting to counter this phenomenon, the medical press has developed three regulatory mechanisms: peer review, the Ingelfinger rule, and the news embargo. First, the peer review process to which manuscripts are subjected was developed to ensure the validity and accuracy of published findings. Second, originality is preserved by not publishing an article whose substance has previously been reported, a mechanism dubbed the *Ingelfinger rule,* after an editor for the *New England Journal of Medicine,* Franz J. Ingelfinger. Despite these brakes on untimely, premature, or redundant publication, interactions between the media and medical and scientific researchers can sometimes lead to what the American Medical Association called a regrettable "rush to judgment," which occurred during the maternal crack-cocaine scare of the late 1980s. Third, medical research news embargoes were pioneered by Morris Fishbein, former president of the American Medical Association and editor of its prestigious journal from 1924 to 1949. Embargoes specify the date and time before which findings from medical or scientific research may not be released to the public. Usually the date and time specified for the release of information to the public is the evening before publication of the journal or concurrent press briefings. However, articles and related materials are made available to physicians and journalists before that date and time. Proponents of news embargo protocols such as those used by the *Journal of the American Medical Association* defend them on grounds that they enable journalists to prepare higher-quality investigative articles by leveling the playing field in terms of deadlines, and they protect physicians from being caught off-guard in the case of findings that have implications for clinical practice. A controversy ensued in July 2002 when a *Detroit Free Press* reporter allegedly breached an embargo by posting a story on the premature termination of a clinical trial involving hormone replacement therapy for postmenopausal women. Although the women who had participated in the trial under the rubric of the National Women's Health Initiative received letters detailing its early termination, and a press conference had been scheduled, the story broke despite the efforts of the principal investigator and journal editor, who cautioned the reporter as to the sensitivity of the findings and the necessity for observing the embargo in order to avert public panic. Although medical embargoes are sometimes overused, it is clear that they have a place in the guaranteeing the orderly release of accurate medical information.

Medical embargoes represent temporary measures. Other factors, such as the social stigma of certain conditions or lack of clarity about symptoms, combine to prevent health information from reaching the public. Not until the U.S. Surgeon General's 1988 campaign was there a national-scale effort to educate the public about HIV/AIDS transmission or risk reduction practices, because of the Reagan Administration's reluctance to address the issue. In HIV/AIDS campaigns, the lay press has had a huge impact in terms of generating beliefs about disease transmission, affected population, and necessary behavioral changes.

## BIBLIOGRAPHY

Boston Women's Health Book Collective. *Our Bodies, Ourselves.* New York: Simon and Schuster, 1973.

Brandt, Allan M. *No Magic Bullet: A Social History of Venereal Disease in the United States since 1880.* Oxford: Oxford University Press, 1985.

Burnham, John C. *How Superstition Won and Science Lost: Popularizing Science and Health in the United States.* New Brunswick, NJ: Rutgers University Press, 1987.

Fellman, Anita Clair, and Michael Fellman. *Making Sense of Self: Medical Advice Literature in Late Nineteenth-Century America.* Philadelphia: University of Pennsylvania Press, 1981.

Field, Marilyn J., ed. *Telemedicine: A Guide to Assessing Telecommunications in Health Care.* Washington, DC: National Academy Press, 1996.

Fontanarosa, Phil B., and Annette Flanagan. "Policy Regarding Release of Information to the Public." *Journal of the American Medical Association.* 284 (2000): 2929–2931.

Gwyn, Richard. *Communicating Health and Illness.* London: Sage Publications, 2002.

Johnson, Timothy. "Medicine and the Media." *New England Journal of Medicine* 339.2 (1998): 87–92.

Karpf, Anne. *Doctoring the Media: The Reporting of Health and Medicine.* London: Routledge, 1988.

McCann, Carol R. *Birth Control Politics in the United States, 1916–1945.* Ithaca, NY: Cornell University Press, 1994.

Miller, Jon D., and Linda G. Kimmel. *Biomedical Communications: Purposes, Audiences, and Strategies.* New York: Academic Press, 2001.

Mullan, Fitzhugh. *Plagues and Politics: The Story of the United State Public Health Service.* New York: Basic Books, 1989.

Nathanson, Constance A. "Disease Prevention as Social Change: Toward a Theory of Public Health." *Population and Development Review* 22.4 (1996): 609–637.

Phillips, David P., E. J. Kanter, B. Bednarczyk, and P. L. Tastad. "Importance of the Lay Press in the Transmission of Medical Knowledge to the Scientific Community." *New England Journal of Medicine* 325.16 (1991): 1180–1183.

Signorielli, Nancy. *Mass Media Images and Impact on Health: A Sourcebook.* Westport, CT: Greenwood Press, 1993.

Smith, Susan L. *Sick and Tired of Being Sick and Tired: Black Women's Health Activism in America, 1890–1950.* Philadelphia: University of Pennsylvania Press, 1995.

*Nancy D. Campbell*

# Health Communication

Health communication consists of a wide variety of activities, both purposive and unintended, that inform and influence decisions that affect individual and the public's health. Health communication can take many forms. It is quite similar to and overlaps with health education and health promotion. Distinctions between health education, promotion, and communication are difficult and imprecise, but Table 1 provides working definitions of these three fields. Health communication is a hybrid discipline that draws from principles and research developed from the practice of marketing, public relations, journalism, and communication as well as clinical medicine and public health. Health communication is the study and use of both mediated and person-to-person messages processed at multiple ecological levels, focused on health-related influences and outcomes.

One of the central tenets of health communication is that the meaning of messages depends on many factors, including those at individual, interpersonal, contextual, environmental, and societal levels. Further, the meanings present in any communications may consist of both intended and unintended meanings. The practice of health communication contributes to health promotion and disease prevention in many ways that affect individual quality of life as well as the health of communities and populations. In order to illustrate the many domains, perspectives, and activities that constitute health communication, this entry uses the ecological levels of analysis (that is, individual, organizational, community, and society) as an organizing framework.

## Individual

There are at least four ways health communication research has informed health behavior at the individual level. First, individuals turn to their close friends, trusted associates, and even casual acquaintances when making health-related decisions such as what to eat, whether to exercise, what care to seek, and specifically whom to seek it from. Individuals have predispositions formed from myriad influences, but their day-to-day decision making is often shaped by their interactions with others who share the values, beliefs, and information that influence behavior. Interpersonal communication is critical for finding a provider for conditions ranging from primary care to locating complementary and alternative therapists and many other services.

Interpersonal communication in health care settings is critical for disseminating information concerning risk behavior, diagnoses, and treatment strategies. Communication with health care providers has been studied con-

**Table 1. Distinctions between Health Education, Health Promotion, and Health Communication**

|  | Education | Promotion | Communication |
|---|---|---|---|
| Unit of analysis | Individual | Community | Society Population |
| Intervention of choice | Train providers | Empower Communities | Change Population |
| Medium of choice | Interpersonal counseling | Training and community organization | Mass Media |
| Common settings | Hospitals, clinics | Underserved communities rural towns | Countries Regions Metropolitan areas |
| Role of communication | Improve quality of care | Empower individuals and groups | Inform population Persuade population |

siderably, with many factors identified that influence patient-provider interaction. Studies have shown that providers, particularly medical doctors, vary considerably in the degree to which they listen to their patients. Physician communication behaviors are important because they are closely linked with patient satisfaction, adherence to treatment, comprehension of health-related information and overall health status. Doctors can be trained to enhance their communication skills with their patients, and this improved interaction has been associated with greater patient compliance to treatment recommendations. Thus, the training of health professionals and consumers in effective communication skills has many direct and indirect benefits to the well-being of patients and their loved ones, such as affecting the quality of health care delivery and health care outcomes. Similarly, a number of studies have been conducted to test whether providers can act as agents of behavior change. For example, brief interventions have had some success in prompting providers to persuade their clients to change their substance-using behaviors.

## Community

Health communication is used at the community level to support health promotion programs in numerous ways. First, health communication principles are used to design media materials such as posters, flyers, and newsletters that enable local public health agencies to communicate with their populations. Second, many community-based health promotion programs use lay providers to increase access to care. These lay providers receive training in health communication, and the programs use frameworks and principles developed partly from health communication experiences. Third, health communication plays a role in the development and formation of community coalitions to promote health. This communication occurs among coalition members and between the coalition and its constituents.

Community-level interventions have benefited from health communication's focus on audience research. Developing and implementing community-level programs requires dialogue between community residents and leaders, many of whom may be distrustful of outsiders who purport to have their interest at heart. The process of identifying community assets and creating community empowerment requires dialogue and communication, the end result of which is more sustainable and more effective programs.

## Health Care Settings

Characteristics and patterns of communication within health care settings (hospitals, HMOs, clinics) among providers and staff influence the type of care patients receive. Physician-to-physician communication influences the choice of therapy and determines the diffusion of medical therapies/ technologies. These organizations also use communication to inform their patients about health education activities as well as other services. Further, health care practitioners can use telecommunication technologies to provide diagnostic and counseling services to mitigate the barriers imposed by geography, space, and time. These telemedicine activities can alter the traditional environment of the doctor-patient encounter and potentially change relationships between health care providers and health care organizations. Besides creating a new level of interaction among health care professionals, new communication technologies increase patient access to health information, support, services, and personnel. Many patients now turn to the Internet as their source for medical information.

Health communication principles are at work when organizational development teams identify opinion leaders or construct workgroups within organizations to manage the change process. Opinion leaders or champions can be identified via network-analysis techniques in which the flow of communication and influence is traced through the organization. Communication networks form the backbone of most organizations, and using these networks to accelerate change and make it more responsive to employee needs has proven advantageous.

## Societal

Communication at the societal level—more so than other levels—is both purposive and non-purposive. Purposive communication consists of public information campaigns designed to create awareness, change attitudes, and encourage the adoption of healthy behaviors. These campaigns may take the form of Public Service Announcements (PSAs), such as a campaign to prevent sexually transmitted infections, but may also use other media formats and distribution channels. Nonpurposive health communication can occur through news reports, such as Magic Johnson's announcement of his newly discovered HIV status, which had an unprecedented impact on calls to AIDS hotlines. Nonpurposive health communication also occurs in the form of popular culture expressions on radio, television, and film. For example, studies have shown that characters in the most popular films will frequently exhibit significantly more use of alcohol, tobacco, and other drugs than is actually consumed in society. Further, the consumption of these substances is rarely accompanied by any portrayal of the consequences of such use. Thus, popular entertainment programs may tend to promote deleterious behaviors.

Recently, health communication research has directed attention to entertainment-education (EE) strategies to promote healthy behaviors. EE is the deliberate use of entertainment formats such as street theater, soap operas, films, and music video to promote pro-social ideas and practices. EE has been used extensively in developing countries to promote family planning and other health behaviors and is being used in the United States with attempts to incorporate pro-social messages on entertainment television. For example, the Harvard Alcohol Project advocated including scenes and dialogue promoting use of designated drivers on popular TV programs.

## Integration

The challenge facing health communication is finding ways to integrate behavior change programs across the different ecological levels. Because behaviors are shaped by communications at each level, the influences can be complex and sometimes contradictory. Although a health care provider may tell patients they should eat less and exercise more, the barrage of TV commercials advertising fatty foods and the presence of fast food restaurants in one's neighborhood make following a healthy regimen difficult.

Further, scholars studying health behavior often specialize in studies focused at one of these levels; experts in interpersonal communication do not study mass media and vice versa. Hence, models of behavior change have often omitted factors and communication effects at different ecological levels. Studying communication effects at different levels requires experimental methodologies that explicitly account for these differences. For public health practitioners, interventions then need to be created at multiple levels of analysis, often with considerable community involvement. Such interventions take time and effort to create.

To address this challenge, health communication (and other) scholars have created interventions that incorporate diverse communication channels. Prominent examples of this approach include large-scale cardiovascular disease programs that combine mass media channels (such as television, radio, newspapers, mass distributed print media) with interpersonal communication methods (such as group training classes for smoking cessation and aerobic exercise) to influence knowledge, attitudes, and risk behaviors related to heart disease. Mass media channels are typically employed at the onset of these campaigns to increase public awareness of the need for behavior change, and interpersonal channels are used later to present reinforcing materials and persuade audiences to engage in recommended risk-reduction behaviors. The inclusion of both mass media and interpersonal communications channels was expected to be associated with greater changes in health behavior than reliance upon a single channel of influence. Thus, health communication approaches are often combined with educational, community development, and empowerment strategies to address the many determinants of health.

In addition to using multiple channels of influence, successful health communication interventions seek various levels of change. For example, such an intervention might seek the modification of individual behaviors as well as shifts in policy and environmental norms that support and contribute to behavior change at the community level. There may be different target populations of interest as well as multiple intervention strategies to address each program objective. At a broader community level, statewide tobacco-control programs have developed highly sophisticated countermarketing campaigns to foster policy change and support a variety of community prevention and cessation programs. Campaign messages in Florida's tobacco-control program, for example, have provided youth with important information about tobacco use while also empowering them to join the statewide youth antitobacco group as a means of diffusing campaign messages interpersonally and garnering support for tobacco control policy.

In an effort to focus on the outcomes and effectiveness of health communication interventions, these programs have explicitly incorporated program evaluation and research within stages of intervention design and implementation. Intermediate outcomes, such as changes in knowledge and attitudes, have helped make these interventions more likely to achieve their ultimate behavior-change outcomes. Health communication scholars and programs will continue to face the challenge of integrating theory, research, and practice as they attempt to create both individual and societal change.

## Importance of Research

The public health field has been engaged in efforts to promote health-related behavior using communication principles for some time. These efforts include promotion of hygiene behaviors, safety and accident prevention, substance abuse prevention, adoption of healthy lifestyles and eating habits, family planning and contraceptive use, and many others. The knowledge and application of health communication principles has greatly enhanced public health practice. At the same time, public health application has helped pushed the communication field to be more relevant and improve its theoretical power. While health communication activities alone cannot eliminate the prevailing social conditions related to poor health, the feedback between research and practice will aid individuals and communities in identifying needs and taking appropriate actions to better their health and well-being.

The health communication approach to health promotion and disease prevention is marked by four overarching principles. First the process is dialogic, consisting of sharing ideas and reaching mutual understanding. Second, the process is continuous, in that continued dialogue and refinement are encouraged. Third, the process entails understanding the meanings embedded within communication acts. Fourth, the process is conducted within a scientific framework, including formative research so that ideas and decisions are based on evidence rather than intuition. These principles guide the creation and dissemination of behavior change programs and can help minimize the unintended negative effects of programs.

## BIBLIOGRAPHY

Brodie, M., U. Foehr, V. Rideout, N. Baer, C. Miller, R. Flournoy, and D. Altman. "Communicating Health Information through the Entertainment Media: A Study of the Television Drama *ER* Lends Support to the Notion That Americans Pick Up Information While Being Entertained." *Health Affairs* 20.1 (2001): 192–199.

Cline, R. J. "At the Intersection of Micro and Macro: Opportunities and Challenges for Physician-Patient Communication Research." *Patient Education & Counseling* 50.1 (2003): 13–16.

Earp, J. A., E. Eng, M. S. O'Malley, M. Altpeter, G. Rauscher, L. Mayne, H. F. Mathews, K. S. Lynch, and B. Qaqish. "Increasing Use of Mammography Among Older, Rural African American Women: Results from a Community Trial." *American Journal of Public Health* 92.4 (2002): 646–654.

Finnegan, John R., and K. Vishwanath. "Communication Theory and Health Behavior Change: The Media Studies Framework." In *Health Behavior and Health Education: Theory, Research, and Practice*, 3d ed., edited by K. Glanz, B. Rimer, and F. Marcus Lewis. San Francisco: Wiley & Sons, 2002: 361–388.

Glanz, Karen, Barbara K. Rimer, and Frances Marcus Lewis, eds. *Health Behavior and Health Education: Theory, Research, and Practice*, 3d ed. San Francisco: Wiley & Sons, 2002.

Glik, D., G. Nowak, T. Valente, K. Sapsis, and C. Martin. "Youth Performing Arts Programs for Health Promotion—Education for HIV/AIDS Prevention and Health Promotion." *Journal of Health Communication* 7.1 (2002): 39–57.

Green, Lawrence W., and Marshall W. Kreuter. *Health Promotion Planning: An Education and Environmental Approach*, 3d ed. Mountain View, CA: Mayfield, 1999.

Institute of Medicine, Committee on Assuring the Health of the Public in the 21st Century. *Future of the Public's Health in the 21st Century*. Washington, DC: National Academies Press, 2003.

Kreps, G. L. "Consumer/Provider Communication Research: A Personal Plea to Address Issues of Ecological Validity, Relational Development, Message Diversity, and Situational Constraints." *Journal of Health Psychology* 6.5 (2001): 597–601.

Maccoby, N., J. W. Farquhar, P. Wood, and J. Alexander. "Reducing the Risk of Cardiovascular Disease." *Journal of Community Health* 3 (1977): 100–114.

Maibach, Edward, and Roxanne Louiselle Parrott, eds. *Designing Health Messages: Approaches from Communication Theory and Public Health Practice*. Newbury Park, CA: Sage Publications, 1995.

McLeroy, L., D. Bibeau, A. Steckler, and K. Glanz. "An Ecological Perspective on Health Promotion Programs." *Health Education Quarterly* 15 (1988): 351–377.

Niederdeppe, J., M. C. Farrelly, and M. L. Haviland. "Confirming 'Truth': More Evidence of a Successful Tobacco Countermarketing Campaign in Florida." *American Journal of Public Health* 94.2 (2004): 255–257.

Ong, L. M. L, J. C. de Haes, A. M. Hoos, and F. B. Lammes. "Doctor-Patient Communication: A Review of the Literature." *Social Science & Medicine* 40.7 (1995): 903–918.

Perednia, D., and A. Allen. "Telemedicine Technology and Clinical Applications." *Journal of the American Medical Association* 273.6 (1995): 483–488.

Piotrow, Phyllis Tilson, D. Lawrence Kincaid, Jose G. Rimon II, and Ward Rinehart. *Health Communication: Family Planning Communication: Lessons for Public Health*. Westport, CT: Praeger, 1997.

Rice, Ronald, and Charles Atkin, eds. *Public Communication Campaigns*, 3d ed. Thousand Oaks, CA: Sage Publications, 2001.

Rogers, Everett, and Arvind Singhal, eds. *Entertainment-Education: A Communication Strategy for Social Change*. Mahwah, NJ: Lawrence Erlbaum, 1999.

Roter, Debra L., and Judith Hall. *Doctors Talking to Patients/Patients Talking to Doctors: Improving Communication in Medical Visits*. Westport, CT: Auburn House, 1992.

Siegel, M., and L. Biener. "The Impact of an Antismoking Media Campaign on Progression to Established Smoking: Results of a Longitudinal Youth Study." *American Journal of Public Health* 90 (2000): 380–386.

Snyder, Leslie B. "How Effective are Mediated Health Campaigns?" In *Public Communication Campaigns*, 3d ed., edited by R. Rice and C. Atkin. Thousand Oaks, CA: Sage Publications, 2001: 181–190.

U.S. Department of Health and Human Services. *Healthy People 2010*, 2nd ed. (with *Understanding and Improving Health and Objectives for Improving Health*), 2 vols. Washington, DC: Author, 2000.

Valente, Thomas W. "Evaluating Communication Campaigns." In *Public Communication Campaigns*, 3d ed., edited by R. Rice and C. Atkin. Thousand Oaks, CA: Sage Publications, 2001: 105–124.

Valente, Thomas W. *Evaluating Health Promotion Programs*. New York: Oxford, 2002.

Valente, Thomas W., and Darleen V. Schuster. "The Public Health Perspective for Communicating Environmental Issue." In *New Tools for Environmental Protection: Education, Information, and Voluntary Measures*, edited by T. Dietz and P. Stern. Washington, DC: National Academies Press, 2002: 105–124.

Wallack, L., and L. Dorfman. "Media Advocacy: A Strategy for Advancing Policy and Promoting Health." *Health Education Quarterly* 23.3 (1996): 293–317.

*Thomas W. Valente and Darleen V. Schuster*

# MEDICAL PROFESSION.

This article provides an overview of the development of the profession of medicine in the United States over the last two centuries, including the struggle to create an organized medical profession. In addition, we examine the impact of the medical profession on the provision of health care by public health institutions and hospitals. The second half of the article discusses a number of issues that emerged in the second half of the twentieth century and which continue to challenge the profession.

## The Emergence of Organized Medicine

Doctors have not always enjoyed the status, prestige, and economic rewards that they receive in today's world. During the eighteenth century, medicine was among the least desirable trades for an individual to pursue. There were many reasons for the low social standing of physicians. First, doctors were not in high demand because caring for the sick was seen as the responsibility of female family members, midwives, or lay practitioners. In addition, if one were inclined to seek care from a physician, the time involved required traveling long distances and sacrificing a day's labor. Finally, physicians had very few proven therapeutic techniques available to them.

In the eighteenth and early nineteenth centuries, the practice of medicine did not require a degree, license, or any certification. The majority of doctors learned their trade through apprenticeships and received a certificate of proficiency, usually in general medicine. The first American school of medicine opened in 1765 at the University of Pennsylvania, and by the early1800s medical schools began developing throughout the United States. Many of these schools were "for-profit" entities and provided training of questionable quality to their students. Although all schools at the time had requirements for a medical degree, many of them overlooked or loosely enforced these standards.

Throughout the nineteenth century, many different theories of healing and practice existed. Advocates of each of these theories organized into medical sects, the largest of which was the allopaths, or regulars as they were called because they had received a formal, or "regular," education from a medical school. They practiced mostly the heroic methods of treatment, such as bloodletting (bleeding a patient to remove impurities), purging (administration of substances to produce vomiting or diarrhea), and later surgery, along with other highly invasive procedures that often caused more harm than healing. Other sects, including homeopaths, eclectics, and Thomsonians, relied on less invasive and more naturalistic treatment methods. The differences in practice among the sects often led to disagreements, competition, and hostility.

The lack of standardization in medical training and practice was a source of major frustration for practitioners who sought to increase the status and legitimacy of the medical profession, and it provided the impetus for the founding of the American Medical Association (AMA) in 1847. Many medical practitioners saw the AMA, which was dominated by regulars, as trying to prohibit all "non-regulars" from practicing medicine. Nevertheless, the profession began to coalesce because of two pivotal events: first, the 1875 requirement by the Michigan state legislature to incorporate a division of homeopathy into the University of Michigan State Medical School, which brought students of disparate medical philosophies into the same basic science classrooms; and, second, advances in science and technology that brought about effective forms of diagnoses and treatment that were widely adopted by practitioners of various backgrounds.

Although variations in practice and training were declining by the beginning of the twentieth century, it was not until 1901 that the medical profession became unified and organized under the AMA. This was largely the outcome of the AMA's decision to rewrite its constitution, stipulating that national association membership would be granted only to physicians who were members of both their county and state medical associations. As a result, the AMA became organized at the state and district level and membership increased from 8,000 members in 1900 to 70,000 in 1910. This dramatic rise in membership changed the history of the medical profession. The profession was finally unified and organized, and the requirement for paying membership dues provided a source of income for the AMA. The dues were used to fund the AMA's regulative activities—namely, seeking government support and developing a standardized medical education.

The implementation of a standardized medical education and the passing of strict licensing and practice laws developed concurrently. Although the first law in the United States requiring a license to practice medicine was passed in 1763 in Norwich, Connecticut, it did not include proficiency standards or penalties for those who practiced without a license. In 1877, Illinois passed the first state medical practice law that granted a state board of medical examiners the authority to prohibit the granting of licenses to graduates of substandard medical schools. Similar laws were subsequently adopted by a number of other state legislatures. The legitimacy of the new laws and their effectiveness in controlling medical practice was established in 1889 in the case of *Dent* v. *West Virginia* (129 U.S. 114). Frank Dent was an eclectic physician who was fined and convicted by the state of West Virginia for practicing medicine without having either a state license or a diploma from an accredited medical school. The case went to the Supreme Court, where West Virginia's decision was upheld. This decision sent a message to the public, as well as to other nonlicensed practicing doctors, that states and the Supreme Court would uphold and enforce medical practice laws.

After the AMA had successfully organized as a profession and had obtained support from state legislatures, they directed their attention to developing a standardized medical curriculum. In 1904, the AMA formed the Council on Medical Education, which developed and set the following minimum requirements for practicing medicine: having a high school diploma, completing four years of medical study, and passing a state licensing exam approved by a state board of medical

examiners. These new educational and regulatory requirements resulted in the closing of over eighty substandard medical schools. Further improvements in the quality of medical training resulted from the publication of the "Flexner Report" in 1910, which was an in-depth study of the nation's medical schools that revealed varied and inadequate training programs. Today the requirements to practice have evolved to include four years of undergraduate school, four years of medical school, three to eight years of internship and residency, and the passing of a state license exam.

## Public Health

As the profession of medicine was developing and influencing the delivery of care to individual patients, the field of public health was also emerging and having an influence on the health of communities. The first state board of public health was created in 1855 and was followed by the founding of the American Public Health Association in 1872. In addition, the first bacteriological laboratory was developed under the auspices of the Marine Health Service in 1887; this first endeavor in public health research eventually evolved into the National Institutes of Health (NIH) in 1948. Public health services fall under the federal umbrella of the Department of Health and Human Services.

The first public health officials, originally known as sanitarians, focused their attention on improving the physical environment by removing human and animal waste from communities. These wastes were believed to produce foul air, or miasmas, that the public believed caused diseases such as cholera. Although incorrect, the miasma theory of disease causation inspired a massive sanitation movement, leading communities to clean their streets and to dispose of their garbage and waste away from public areas, thus reducing the occurrence of waste-borne diseases.

A major shift occurred in the conceptualization of disease causation in the 1860s and 1870s, when scientists such as Louis Pasteur and Robert Koch discovered bacteriological organisms and were able to link the presence of bacteria to specific diseases. This bacteriological revolution led in the 1880s to a new notion of disease causation known as germ theory. Subsequently, sulfonamide drugs and later penicillin, accidentally discovered by Alexander Fleming in 1929, were used to treat bacterial infections. Germ theory motivated public health officials to take a new approach to disease control, emphasizing personal hygiene—that is, brushing teeth, washing hands, and keeping a clean house. Public health officials also urged the public to see a doctor regularly, which significantly increased the status and business of practitioners.

Throughout the twentieth century, tension existed between public health institutions and the profession of medicine when it came to determining which institution would be responsible for the delivery of health education, treatment, and services. Public health officials favored the use of dispensaries, or drug distribution centers, as the vehicle for providing health care interventions in the early to mid-twentieth century. Although dispensaries were a convenient mechanism for the delivery of care to growing city populations, doctors felt that health departments were overstepping their boundaries by not working under the authority of physicians and the AMA. By drawing on their political clout in state legislatures in the 1920s, the AMA was able to suppress the growth of the public dispensaries, which eventually disappeared altogether. In the 1930s, neighborhood health centers emerged to fill the void left by dispensaries. However, they too were unable to compete with the AMA in securing adequate state and federal funding and eventually went out of existence. It was not until the public health reforms of the 1960s that community health centers, which qualified for federal support, came into existence. These clinics provided an essential source of health care for the uninsured. Although public health departments continue to play a role in the provision of health care to individuals—for example, by providing immunizations—in the early twenty-first century they play a more prominent role in promoting the overall health of communities by focusing primarily on health education and disease prevention activities. This focus is especially prominent in the promotion of lifestyle behaviors, such as smoking cessation, increase of physical activity, and prevention of sexually transmitted diseases.

## Hospitals

Both the medical profession and public health were enhanced by the development of the modern-day hospital. Prior to the discovery of antisepsis in 1847 by Ignaz Semmelweis and its innovative use by Joseph Lister to disinfect surgical instruments, the hospital was considered by many as a place to die or a place for those who did not have a home or family to provide care. In addition, before the development of various medical tools, there was little reason or need to move a sick person from home to a hospital. As medical technology advanced and new machines were invented, such as the X-ray machine, developed accidentally in 1895 by Wilhelm Conrad Roentgen, doctors needed a place where they could see patients and have access to the new specialized equipment. At the same time the industrial revolution and the influx of immigrants brought about an increase in the number of people living independently within rapidly growing cities who needed the services of hospitals when they were sick. The combination of these factors, along with interested investors, brought about the development of the modern hospital.

## Doctor-Patient Relationship

As the profession of medicine evolved in response to sociocultural and technological changes, so did the relationship of doctors to their patients. In the eighteenth century the patient was the authority on his or her illness and was responsible for providing details regarding symptoms to the physician, who lacked sophisticated diagnostic tools. The doctor-patient relationship began to change with the introduction of new medical technologies such as the stethoscope, invented in 1916, which allowed the doctor to gather information independent of the patient's input. These new technologies changed the balance of power and decision making in favor of the physician who was trained to diagnose and treat diseases. In the twenty-first century complex, sophisticated treatments and technologies continue to be developed that create even greater imbalances of power in doctor-patient communications and decision making. In response to this challenge, medical school curricula and practice have begun to emphasize patient-centered approaches to communication in order to involve the patient more in decisions about their health.

## Cost of Health Care

One of the ongoing challenges in the late twentieth and early twenty-first centuries has been the rising cost of health care and the increase in the number of uninsured individuals. Historically, there have been four methods of payment for health care: first, "out of pocket" payments for health care at the time the service is received; second, direct private insurance; third, employer-based insurance programs; and fourth, government-funded insurance plans.

The concept of health insurance grew in popularity throughout the twentieth century. For example, by 1940 Blue Cross, which was established in 1929, had over 6 million subscribers. At the same time, the poor and elderly, who did not qualify for employer-based insurance programs, were left without insurance and had to pay out of pocket for their own care, which few could afford to do. To address the lack of equal and accessible health care for the poor and elderly, the federal government in 1965 enacted Medicaid, which provides health insurance to the poor, and Medicare, which provides health insurance for the elderly.

The debate over health care coverage and cost continues today. In 2004, the United States was the only industrialized country besides South Africa that does not have a universal health care system. As a result, 45 million Americans are left without a guarantee that they can access health care. In addition, the cost of health care has continued to rise-the percentage of the gross national product (GNP) spent on health care increased from 4.5 percent in the early twentieth century to 7.3 percent in the late 1970s to 14 percent in the early twenty-first century, the highest of any industrialized country.

## Diversity among Medical Practitioners

Another challenge that the profession of medicine faced in the late twentieth century was the lack of diversity among its members. The movement for women's rights brought attention to the male-dominated nature of the profession and called for greater recruitment and acceptance of women into medical schools. Although the acceptance of women initially faced resistance, medical school admissions had become balanced between the sexes by the early twenty-first century. Similarly, the civil rights movement of the 1960s called attention to the low numbers of African Americans enrolled in medical schools. Changes in enrollment by race have been slow. In 2004 there were about 1,200 African American students enrolled in America's top twenty-six medical schools, or an average of 8 percent of the total incoming classes in these schools. Although greater numbers of women and minorities have entered the profession, the struggle for race and gender equality in pay and position is ongoing.

## Changes in the Organization of Practice

Over the last half of the twentieth century, a number of changes occurred in the organization of medical practice. Traditionally, physicians who provide care through private or group practices were considered the backbone of the AMA. A new type of medical practice that emerged in the late twentieth century is the large provider organization that employs physicians—a hospital or a health maintenance organization (HMO). This option often involves less professional autonomy and greater restrictions on practice in the form of regularity norms, such as diagnostic group rates, yearly budgets set by the HMO, Professional Standard Review organizations, and a strict administrative hierarchy. Another form of medical practice is academic medicine—that is, undertaking teaching and research at a university medical school and seeing patients at an affiliated teaching hospital.

Physicians today also have to negotiate their position in relation to the growing number of paraprofessionals and alternative health practitioners. From 1950 to 1970 the number of people working in the health care field increased from 1.2 to 3.9 million. The number of paraprofessionals has steadily been increasing over the last half-century, and many states have passed legislation expanding the professional scopes of non-physician clinicians

The trend toward medical specialization has been another ongoing concern as the numbers of physicians leaving general practice has increased. This trend has had a disproportionate impact on rural communities. In the 1930s, for example, less than 25 percent of doctors identified as being specialists, but in 1966 69 percent considered themselves specialists. Increasing specialization of medical practice has also led to a decline in membership of the AMA because most specialists join their own professional associations instead. Today AMA membership includes less than half of all practicing doctors.

### Research and Technology

The agenda for medical research has developed in response to number of considerations, such as the major causes of morbidity and mortality at any given time and basic advances in scientific knowledge. For example, at the beginning of the twentieth century, the major causes of morbidity and mortality were infectious diseases. This motivated a search for the development of vaccines for the common childhood diseases. As infectious diseases were brought under control, the focus of medical research in the late twentieth and early twenty-first centuries has been on developing new technologies and treatments for chronic diseases such as cancer, heart disease, obesity, and HIV/AIDS. In addition, advances in basic scientific knowledge—such as the discovery of DNA by Wilkins, Crick, and Watson in 1953—have brought about a "genetic revolution," inspiring the Human Genome Project—the largest worldwide study seeking to map each chromosome to inherited traits and body functions. This knowledge has in turn ushered in a new era in the search for cures and treatments for diseases.

### A Look Ahead

The profession of medicine will continue to change and evolve in response to external events: the emergence of new and reemerging diseases, advances in basic science and technology, and changes in the social, cultural, and political environment.

**BIBLIOGRAPHY**

American Association of Medical Colleges. Available online at http://www.aamc.org.

Starr, Paul E. *The Social Transformation of American Medicine: The Rise of a Sovereign Profession and the Making of a Vast Industry.* New York: Basic Books, 1982.

Tesh, Sylvia Noble. *Hidden Arguments: Political Ideology and Disease Prevention Policy.* New Brunswick, NJ: Rutgers University Press, 1988.

U.S. Department of Labor, Bureau of Labor Statistics. Physicians and Surgeons. Available online through http://www.bls.gov.

*Margaret Kerr and Jeanette Trauth*

# MEDICAL TECHNOLOGIES.

Medical technologies are the most visible and familiar expression of the power of biomedicine in modern life. They have become a central part of human experiences including birth, family relations, work life, aging, and death, no matter if one is ill, disabled, or healthy. All societies make instruments for evaluating the body's ability to function physically and cognitively, and all fashion tools for rehabilitation, research, and healing. What differs is the way that concepts of disease and wellness, scientific theories, and technical developments interact within particular economic, cultural, historical and political environments to produce technological systems that persist while others are discarded or subordinated. Although complex technologies are associated with Western, allopathic medicine found in more industrialized societies, a variety of technological systems can coexist in any society.

For purposes of this review, *medical technologies* will be defined as the various kinds of devices, instruments, and procedures used for diagnostic, therapeutic, rehabilitative, preventive, or experimental purposes in medicine, including the associated practices and systems in which they exist. Medical technologies include communication, computational, mechanical, chemical-pharmaceutical, and biological elements and increasingly are hybrids of several of these components. Medical technologies can replace or change bodily function, or even extend some functions far beyond what the average body could perform. Apparatuses can be used to extract information with which to establish a body's biological and social status, monitor it over time and circumstances, and report the findings to various types of experts across widespread networks. From this activity, large databases can be created with which to define health and illness, reformulate categories of normal and abnormal, make judgments about individuals and populations, provide predictors of risk, and then plan future services and technologies. In this way, assumptions about deservedness, capability, and behaviors are built into both the technologies and the interpretation of the data they produce. The dramatic transformations in life, labor, and governance have drawn many researchers to medical technology as an endlessly fascinating and fertile ground for exploring questions about science, technology, and society.

Technology has been variably positioned in the study of health and medicine. Some social scientists study technology as a part of the material culture of a society. In this view, artifacts are regarded as simply one of several by-products of social processes. Priority is placed on narratives, social interactions around the artifact, and adaptations or innovation processes thought to direct the form that artifacts take. Such social determinist views are counterbalanced by equally narrow views suggesting that technologies stand outside of society, inevitably determining human actions. Such studies follow the trajectory of a particular device or suggest an imperative to adopt complex technologies to solve health and social problems, spiraling costs, and care out of control. In this view, tools are seen to exert powerful impacts "upon" society.

All of these studies are significant for their contributions to exposing ethical dilemmas, inequities caused by poor allocation of goods and services, unintended consequences, and the changes in relationships between patients and physicians and other caregivers. Studies of the medical-industrial complex dating from the 1970s also drew attention to the influence of power and profits in the development of certain technologies and their consequences.

Yet technologies operate differently under varying circumstances and participants interacting with them. Devices do not simply succeed because of their inherent characteristics or fail because of social problems. Students of science, technology, and society have more recently reframed questions to examine technological *systems*, rather than specific techniques or tools, because this broadens the view to include the social dynamics of power relationships, organizational structures and networks, and global political-economic implications in addition to daily work practices and arrangements of instruments and techniques. By placing medical technologies in contexts, researchers show how cultural assumptions are embedded in discourses and practices around artifacts, as well as in the design of the tool and its enabling systems.

This review will focus on three directions in work on medical technology, broadly grouped as knowledge produced by interactions with technologies; work and social relations, and the relationship of technologies and the body.

## Medical Technology and the Production of Knowledge

The rise of scientific medicine centered the biosciences in matters of life and labor and transformed notions of governance over societies. Medical technologies designed to detect, diagnose, and monitor variability became a chief mode of collecting and aggregating information about individual bodies and populations. As such, technologies have become integral to creating understandings of the processes of disease and well-being, in terms of the kinds of information they yield, the way cultural assumptions are built into their design, and the way interventions may change the way certain conditions come to be seen as a problem or disorder. The data produced then influence how people live their lives and manage disorders as well as how others (such as advisory experts, insurance companies, and policy makers) assign responsibility, establish policy, invest capital, and channel resources. The products of technology frame the way we think about risk, understand human relations (kinship, biological compatibility, individual and communal rights), and make plans for the future.

Prior to the advent of modern technologies, information was gleaned from patients' narratives about their complaints and from outward bodily signs. Physicians used their senses to detect differences in sound, temperature, and appearance. Auscultation and pulse taking were employed to diagnose physical and emotional irregularities. Skull formation, body proportions, and skin color were used as signs of internal well-being as well as indicators of intellectual and moral character. To increase the reliability lacking in such human judgments, self-regulating instruments (such as blood pressure detectors) and machines with which to peer inside the body (such as X-rays) were developed. Quantitative measurements, graphs demonstrating relationships, and visual representations of "what's in there" were seen as objective measures of function and status. The relative role of individuals' reported experiences, descriptions, and histories diminished, transforming the relationship of patient and physician to one mediated by technologies.

Yet simply having more information does not always help. In fact, the superabundance of data often creates even more ambiguity. For example, the visual records produced by X-rays, ultrasound, magnetic resonance, and radioisotope-detecting scanners are often understood to be accurate representations of reality, even though the images produced may not closely resemble structures and physiology, may be reconstructed mathematically, and require significant interpretation by skilled workers. To compare results across patients, protocols and rules for collecting and interpreting data are required to standardize the process. However, these are based on sometimes arbitrary criteria, and the interpretation of data can change according to the context in which they are produced as they are circulated, edited, and juxtaposed with certain narratives.

To illustrate, positron emission tomography (PET) uses color and light patterns to reveal metabolic functions (blood flow, chemical activity) in the brain, whereas radiological images show structures (tumors, malformations). Brain function is redefined using computationally derived visual data rather than observation, patient history or direct measurement. When attempts are made to connect such abstract representations to behavior and cognition, difficulties arise because of the flexibility of interpretation and because behaviors change under varying circumstances outside the PET lab. Nevertheless, PET images exert a powerful effect in the sense that they affect categories of sane or insane, normal or abnormal, as Joseph Dumit demonstrated in his ethnography of brain imaging technology.

The importance of gaining agreement on the value and reliability of a technique for clinical decision making is seen in Edward Yoxen's history of the use of ultrasound for medical purposes. The original idea of adapting military uses of sound waves for therapeutic purposes (such as tumor destruction) was abandoned in favor of generating two-dimensional images based on differences in sound wave absorption to identify structures in fluid-filled areas. A variety of technical and institutional reasons were at play, but without consensus between diverse groups of medical and technical professionals about the way that images could reliably be represented, the technology may not have become stabilized enough to become a routine diagnostic tool.

Ultrasound indeed became a routine procedure for pregnancy, ostensibly to look for fetal abnormalities. However, data has shown that its use makes little difference in outcomes. Instead, the ultrasound image of the fetus became a cultural icon as much as a medical image, often thought to create bonding and emotional response to the fetus, as well as establishing a separate identity of the fetus as a person. The role of technologically produced images produced through endoscopy and ultrasound became central in the context of public debates over abortion in the United States, as Rosalind Petchesky showed. In fact, anti-abortion activists lobbied to equip all abortion-providing clinics with sonography equipment, in the hope that women would see the unborn fetus and choose not to abort.

Among the earliest to analyze the way interactions of social, political, and economic conditions coconstruct disease categories and the introduction of particular technologies was Alonzo Plough. Policy debates about the high cost of chronic illness and how to provide treatments equitably were intensifying during the time kidney dialysis was being established. The complex pathophysiology of kidney failure extends to many other organ systems, but the availability of dialysis to replace the function of renal tissue narrowed the focus to one set of technical solutions. The result was a large influx of federal funding for dialysis and, later, transplantation as the preferred way of dealing with long-term costs over other care alternatives. In the process, concepts of efficiency and cost-benefit were incorporated into a newly created category of "end stage renal disease" (ESRD).

Assumptions about the body, social relations, and mechanisms of power can create or expand definitions of disorders and the kinds of interventions deemed necessary. Detection of the presence or absence of genes or proteins or biochemical changes in individuals can suggest a calculable probability of getting a disease but cannot say with certainty what will actually occur. Such tests may identify individuals and groups who enter a different classification: the "worried well" or those who are "at risk" of developing disease although they do not manifest symptoms. Treatments may not be available, but individuals in such newly created categories are encouraged to consume products, utilize further diagnostic tests, and change their lifestyles as a result of the information. One result may be a new form of social stigma.

In all of these examples, technologies affect subjectivity. Reproductive technologies, genetic interventions, and diagnostic predictions serve to identify who is a "mother" or progenitor, a "patient," or a "consumer" and who is "normal" or "at risk."

## Medical Technology and Work

Clearly, technology is more than just the tools a physician employs. New arrangements of power, knowledge, tools, and work evolved together as the practice of medicine became more science-based. Historians observe that the shift from obtaining information directly from patients (through the senses) to indirect inferences (through instruments) results in a loss of tacit skills for physicians. Indeed, as practitioners came to insist on machine confirmations of their observations, a different type of analytical work skill was required. New professions and organizations evolved around the changing kinds of care.

At the same time, the development of complex, cumbersome, and costly tools for diagnosis and treatment necessitated sharing among a number of patients and practitioners in central locations. The hospital thus became established as the site for both providing care and learning. Sites of operation, organizational structures, and even architecture can change, as seen when large-scale equipment formerly used for research enters community hospital settings or when some functions move out into outpatient settings. The use of telemetry and information technologies creates a distributed network of abstract information, such that knowledge is not sequestered within bounded communities of expertise.

The history of certain diagnostic tools reveals how technologies can dramatically change the division of labor in medicine, sparking the creation of new classes of workers and specialties. As technologies moved into hospital settings, some physicians wanted to identify with them as a matter of prestige and thus kept control over operations and interpretations. Other forms of work were then turned over to new groups of skilled workers. Radically new equipment designs or concepts can also disrupt existing work routines, social relations, and rituals. Computerized tomography, for example, not only transformed hospital physical structures and health care capital expenditures but changed the relation among pathologists, radiologists, and technicians. Devices enabling minimally invasive procedures fitted well within the marketing and economic climate of the 1980s, which prioritized less costly delivery of care, disseminated into ambulatory care facilities.

The classic work of Anselm Strauss and coworkers examined the ways technology transforms both routine tasks and urgent care. Care involves making difficult choices about *how* to apply technology and *for which patients*—choices by necessity often made by those at the level of operating the equipment rather than purely by physicians. Much of medical work involves articulating patients' symptoms with established routines, standardized protocols, and infrastructure constraints (such as payment schemes, work schedules, and policies about access to certain technologies). Multiple elements are thus modified and aligned to make things work within technological systems at hand. As Marc Berg showed in his study of computerized decision support techniques, this articulation work can also reformulate medical criteria and limit therapeutic options. In such cases, it is often the manufacturer's sales representative, not a "medical" expert, who forms the link between knowledge and making technology usable.

The role of diverse kinds of expertise has been a productive new area of study. New voices of authority and knowledge are emerging that problematize what was previously taken to be a dichotomy of "medical" versus "lay" expertise. For example, legal, actuarial, and marketing perspectives participate in the construction of both technologies and the consuming subjects who will use them. For-profit radiological imaging centers now sell access to high-tech diagnostics without a physician's order, and direct marketing of genetic and diagnostic test kits to the public shifts the roles of providers, regulators, and patients, placing responsibility for health on consumers. Importantly, these studies connect the close-in view of work around technologies with the broader political, economic, and social environments in which they occur.

Many studies suggest that technology and its associated expert elites determine cultural meanings and social or political choices. Rayna Rapp provides a corrective. Her research on women undergoing amniocentesis illustrates that individuals make decisions by placing outcomes within the context of their own family and work lives, traditions, and personal beliefs rather than on test findings alone. Such studies of the diverse ways in which individuals weave scientific knowledge into their own concepts of well-being contribute to the understanding of how dominant concepts are incorporated or resisted as they circulate through society. Although all forms of expertise are not equivalent, a number of new studies are examining social movements and activism as phenomena that directly influence medical knowledge, the direction of research, and the availability of certain therapies.

## Medical Technology and the Body

Humans have long fashioned substitutes for missing or damaged body parts and tools to extend bodily capabilities. Prosthetics have primarily been made for sensory, mobility, or structural purposes, but whereas earlier devices "filled in" for a missing part, many now mimic function. Examples include artificial limbs and joints, eyeglasses and artificial retinas, hearing aids and cochlear implants. More recently, attempts have been made to create substitutes for vital organs with complex metabolic functions, such as kidneys and hearts. In many cases the augmenting part can be designed to perform far better than a human part could, possibly raising the bar for expectations of bodily functioning and even longevity—at least for those who have access to the technologies and can afford them.

Body part substitution involves multiple kinds of technologies—chemical, biological, mechanical, electrical, and, increasingly, computer-based. In most cases, these are made of synthetic, externally applied materials and are relatively inert, although living or dead human and animal sources have been used, as in organ or bone marrow transplants. Emerging techniques are more intimately integrated into multiple bodily processes; self-assembling molecules, genes and stem cells that stimulate the body's own repair mechanisms, and interactive biological-electromechanical devices (such as neural cortex control devices or myoelectric arms) are examples. In such cases, the assumed categories of "the natural" or biological and "the cultural" or technological become difficult to tease apart.

For many, such techniques further complicate meanings of the body and the interrelatedness of parts to the whole person. The questions of what constitutes the human and how such minglings of mechanical, electronic, and biological components may affect identity are key to a growing number of investigations of the relationship between bodies, technology, and society. Renée C. Fox and Judith Swazey, pioneers in the study of transplantation and artificial organs, note that the very notion of interchangeable "spare parts" is based on concepts of the body as mechanistic and fragmentary. The idea of using complex technologies to continually repair specific aging or malfunctioning parts prompts Fox and Swazey to ask what constitutes success, when other bodily systems will continue to fail.

The design and development of replaceable parts is based upon beliefs in the ability of technology to transcend nature and liberate the body to go beyond its frailties. Inherent in this epistemology are assumptions about valued physical characteristics, bodily sufficiency or deficiency, and how the body can be made to perform within a range of "normal" or "enhanced" capabilities.

Marilyn Strathern observes that the ability to augment or transform biology occurs within the context of consumer desire and contemporary medical marketplaces, in which the relationship of bodies and technologies conforms to cultural expectations of what the body should be and how it should perform. Assisted reproductive technologies (ART) and preimplantation genetic diagnosis, for example, arose in a context of consumer desire for specific kinds of children on the part of those who can afford the high procedure costs. At the same time, with the technical ability to derive an embryo from the ovary of a fetus, to clone organisms, to enable a postmenopausal woman to bear a child, or to enable a woman to give birth to a genetically unrelated child, the biological and genealogical "facts of life" are redefined, as Sarah Franklin put it. Medical and lay interpretations of women's biology, genetic relatedness, and narratives of ART experiences create new social and kinship relations.

Technology plays a key role in sustaining biological processes, which in turn means that the boundaries between the organic and life and death have become malleable. For many, these changes raise further questions about the conditions under which certain kinds of interventions can and should be made. Despite the myth of technology's ability to resurrect or sustain life, most technologies do not reverse degenerative or dying processes. At best they are a temporary measure, allowing time for legal or other pragmatic decisions to be made and social separation to occur. Technology is often blamed for unnecessarily extending suffering, for disrupting death rituals, and for using expensive, scarce resources on futile cases. Nevertheless, the application of life-extension technologies has become a compulsory protocol for the care of many patients. Cardiopulmonary resuscitation (CPR), for example, has become a mandatory procedure, even though in most cases the patients cannot be revived. Paramedics are not allowed to proclaim patients dead in the field; rather, they are required to initiate revival efforts and transport patients to a hospital, where a physician can either continue efforts or proclaim death. CPR, like other technologies, thus exists within medical systems in which concerns about legal liability and professional roles coexist with cultural concerns about the finality of death and its malleability.

The advent of respirators, parenteral nutrition, dialysis, and other resuscitative technologies created ambiguous states of life and death. The ability to control biological processes externally and suspend the dying process artificially has caused enormous uncertainties about personhood in the absence of interactive consciousness and bodily control and has cast doubts on existing social and medical understandings of the end of life. New medical-legal classifications such as "persistent vegetative state" and "brain death" were created within a continuum of life and death to deal with the ambiguity and make it possible to take such actions as removing organs from bodies that had been designated legally dead (but biologically living) for transplant into other bodies.

Practices and meanings around the body at death differ among societies, however, and the use of technological interventions is historically situated. For example, the struggle Japan has had with both the definition of brain death, as practiced by other societies, and the ethics of transferring organs from the dead to the living is often blamed on "traditional aversion to technology." Margaret Lock counters such essentialized depictions, pointing out that the individualism, utilitarianism, and rationalism that undergird organ donation are distinctly Western values. I have showed how similar debates were played out in Germany against the multiple histories of euthanasia and medical experimentation on vulnerable people under National Socialism and of the political-legal concept of bodies belonging to the state under the former socialist East Germany.

## Medical Technology and Society

These examples demonstrate the powerful effect medical technologies have on life, health, work, and social relations. Studying medical technological systems in their entirety shows that it is not the innovation of new tools alone that creates such transformations, nor can technologies be considered as autonomous agents. Rather, political-economic changes; shifts in cultural views of health, longevity, and disorder; and new and old forms of knowledge and power interact to produce certain technologies to solve both medical and social problems.

*See also* **Medicine and Society**

### BIBLIOGRAPHY

Barley, Stephen. "The Social Construction of a Machine: Ritual, Superstition, Magical Thinking and Other Pragmatic Responses to Running a CT Scanner." In *Biomedicine Examined*, edited by Margaret Lock and Deborah Gordon, 497–540. Dordrecht, The Netherlands: Kluwer, 1988.

Berg, Marc. *Rationalizing Medical Work: Decision Support Techniques and Medical Problems*. Cambridge, MA: MIT Press, 1997.

Blume, Stuart. *Insight and Industry: On the Dynamics of Technological Change in Medicine*. Cambridge, MA: MIT Press, 1992.

Cussins, Charis. "Producing Reproduction: Techniques of Normalization and Naturalization in Infertility Clinics." In *Reproducing Reproduction: Kinship, Power and Technological Innovation*, edited by Sarah Franklin and He-

lena Ragoné, 66–101. Philadelphia: University of Pennsylvania Press, 1998.

Dumit, Joseph. *Picturing Personhood: Brain Scans and Biomedical Identity*. Princeton, NJ: Princeton University Press, 2003.

Fox, Renée C., and Judith Swazey. *Spare Parts: Organ Replacement in American Society*. New York: Oxford University Press, 1992.

Franklin, Sarah. *Embodied Progress: A Cultural Account of Assisted Conception*. New York: Routledge, 1997.

Hartouni, Valerie. *Cultural Conceptions: On Reproductive Technologies and the Remaking of Life*. Minneapolis: University of Minnesota Press, 1997.

Hogle, Linda F. *Recovering the Nation's Body: Cultural Memory, Medicine, and the Politics of Redemption*. New Brunswick, NJ: Rutgers University Press, 1999.

Howell, Joel. *Technology in the Hospital: Transforming Patient Care in the Early Twentieth Century*. Baltimore: Johns Hopkins University, 1995.

Koenig, Barbara. "The Technological Imperative in Medical Practice: the Social Creation of a 'Routine' Treatment." In *Biomedicine Examined*, edited by Margaret Lock and Deborah Gordon, 465–497. Dordrecht, The Netherlands: Kluwer, 1988.

Lock, Margaret. *Twice Dead: Organ Transplants and the Reinvention of Death*. California Series in Public Anthropology, vol. 1. Berkeley: University of California Press, 2001.

Nelkin, Dorothy, and Laurence Tancredi. *Dangerous Diagnostics: The Social Power of Biological Information*. New York: Basic Books, 1989.

Petchesky, Rosalind P. "Fetal Images: The Power of Visual Culture in the Politics of Reproduction." *Feminist Studies* 13.2 (1987): 263–292.

Plough, Alonzo. *Borrowed Time: Artificial Organs and the Politics of Extending Lives*. Philadelphia: Temple University Press, 1986.

Rapp, Rayna. *Testing Women, Testing the Fetus: The Social Impact of Amniocentesis in America*. New York: Routledge, 1999.

Reiser, Stanley Joel. *Medicine and the Reign of Technology*. Cambridge, U.K.: Cambridge University Press, 1978.

Strathern, Marilyn. *Reproducing the Future: Anthropology, Kinship, and the New Reproductive Technologies*. New York: Routledge, 1992.

Strauss, Anselm, Shizuko Fagerhaugh, Barbara Suczek, and Carolyn Wiener. *Social Organization of Medical Work*. Chicago: University of Chicago Press, 1985.

Timmermans, Stefan. *Sudden Death and the Myth of CPR*. Philadelphia: Temple University Press, 1999.

Yoxen, Edward. "Seeing with Sound: A Study of the Development of Medical Images." In *The Social Construction of Technological Systems: New Directions in the Sociology and History of Technology*, edited by Wiebe Bijker, Thomas Hughes, and Trevor Pinch, 281–303. Cambridge, MA: MIT Press, 1987.

Linda F. Hogle

# MEDICAL VALUES AND ETHICS.

Over the course of two centuries, medicine has gained such great professional stature that it has become a prototypical profession. Licensing laws coupled with a standardized medical curriculum, control over hospitals and clinics, the rise of biomedical sciences, and avoidance of governmental regulatory control allowed medical practitioners to set the criteria for their own work, supported by a specialized set of private and governmental institutions and sustained by an ideology of expertise and service. As a consequence of far-reaching medical hegemony, healing practices with rich traditions, such as homeopathy and acupuncture, have been defined as alternative, whereas osteopathy and several patient activist movements have largely been co-opted by mainstream medicine. Over the course of the twentieth century, medicine boldly extended its reach into the social control of deviant behavior, leading to an increasing medicalization of a number of problems—previously outside the jurisdiction of medicine—to redefine them as individual pathologies. Further jurisdictional expansions in the last two decades of the twentieth century have been characterized as the biomedicalization of society, to draw attention to the medical manipulation of life itself from customized genetic pharmaceutical treatments to large randomized, double-blind clinical trials. Medicine has morphed into the Biomedical Technoservice Complex Inc., in which health rather than an absence of disease has become an insatiable desire to be quelled by consumption.

In light of the stunning manifestations of new medical power, it is easy to overlook that health care is ultimately about people's suffering and dying. Furthermore, much, of the anxiety and suffering might be iatrogenerated if one takes a broad definition of iatrogenesis to include not just medicine, but also the health care system in all its aspects. Indeed, the expansion of medical knowledge and power has given rise to a host of contradictions, complexities, and ambiguities as a result of power sharing, conflicting jurisdictions between medical subspecialties, commodification, opposing value systems, care inequities, objectification, health care priorities, and escalating health care costs with inevitable rationing of medical care. When these issues impinge on medical care, the health care field has incorporated a new specialist: the bioethicist. And although bioethicists play a central role in justifying the implementation of new medical technologies, social scientists have both critiqued the limitations of a purely bioethical approach and pointed out multiple "ethical" problems of advanced medical technologies that remain unrecognized by the emerging bioethical profession.

## The Rise of Bioethics: Professionalization

American bioethics represents an intellectual and professional movement oriented toward addressing the ethical quandaries associated with biomedical advances in modern medicine related to advances in biology (e.g., the sequencing of DNA, human genetic engineering) and the growth in sophisticated, and sometimes heroic, medical technologies (e.g., neonatal and adult intensive care units, organ transplantation, life support technology, reproductive technologies, dialysis treatment, stem cell research). The field of bioethics emerged in the 1960s in response to advances in biomedical knowledge and technology, as well as to larger societal concerns about the authority of modern medicine to claim expertise in guiding the application of such advances. Three central societal concerns about modern medicine led to the emergence and development of bioethics. First, the medical profession's exclusive claim to authority and autonomy in decision making related to patient care came into question. Before the 1960s, "medical ethics" in medical decision making had been a process internal to medicine, a private matter between physician and patient, with the physician possessing ultimate decision-making power. Until abortion on demand was legalized through *Roe* v. *Wade*, for example, physicians and internal hospital committees operated as moral and medical gatekeepers of abortion services for women under the often-subjective rubric of "medical necessity," the only mechanism for obtaining abortion during its period of criminalization. With the advent of advanced medical technologies, the growth in biological knowledge, and a broader social climate of declining trust in institutional authority, the profession's claim to exclusive expertise related to medical decision making, the idea that physicians "knew best" what benefited patients, was newly understood as a paternalistic practice. Second, the treatment and safety of human subjects in experimental research were other societal concerns. In the wake of public knowledge about unethical research practices such as the Tuskegee study, a federally funded research project that intentionally failed to provide medical care to African-American men with syphilis, the need for establishing protection for human subjects from harm in biomedical research was a priority. Third, concerns about inequity and "justice" in health care played a role in the emergence of bioethics. With a lack of national insurance coverage and unequal access to care in the United States, social disparities in morbidity and mortality, potential inequalities in applying and rationing new technologies, and the growing corporatization of medical care, social critics sought a way to address injustices inherent in the U.S. health care system.

Bioethics began as a marginal academic field external to the medical profession and resided primarily within the academic disciplines of philosophy and specialized centers for bioethics research. Bioethics has since established itself as a rationalized, bureaucratized profession institutionalized within academic medicine and government and has been fully integrated into the medical education curriculum. That the field of bioethics now represents a consolidated profession in the United States is evidenced by academic chairs and degree programs in bioethics; the establishment of fifty centers for bioethics research within academic medicine; an enormous professional literature on the subject in bioethics journals; and the establishment of formal and informal regulatory committees to guide, regulate, and resolve ethical issues in medicine. Institutional Review Boards to ensure the protection of human subjects in biomedical research at facilities receiving federal funding were established through federal legislation; Institutional Ethics Committees are now required in hospitals to resolve ethical dilemmas; and a new category of bioethics professional, the "clinical ethics consultant," has been developed to assist in ethical conflicts in medical settings. At the level of federal government, since 1978 bioethics has been institutionalized in ongoing presidential commissions and councils dedicated to studying ethical problems related to advances in biomedical knowledge and technology, and providing policy recommendations.

The success of bioethics in establishing itself as a new profession within medicine was based not merely on its concern with social values, but more importantly on its claim to unique expertise in analyzing and resolving ethical dilemmas of modern medicine. The basis of this exclusive claim to authority in ethical matters was the moral methodology of "principlism," the combination of four philosophical principles intended to resolve bioethical dilemmas in medicine. These principles include autonomy; beneficence, the duty to contribute to a person's welfare, especially improving health; nonmaleficence, to refrain from harm; and justice, including fairness in the distribution of health care. Principlism has been promoted by traditional bioethicists as an analytical methodology capable of solving ethical problems across the board in medicine and has retained its cultural status as the central theoretical basis of American bioethics.

## Critiques of Bioethics

Bioethics, as a profession and philosophical methodology, has recently been the subject of critique among scholars of the field. Most routinely, bioethics has been criticized for three shortcomings: a lack of attention to social and cultural factors related to ethical conflicts in medicine; a tendency to uphold, rather than critically question, the authority of organized medicine; and the use of principlism as a sound and effective methodology for resolving medical-ethical dilemmas.

Scholars have critiqued a principle-based bioethics for its lack of attention to, and wariness of, social and cultural factors related to ethical dilemmas in medicine. Such factors, including the politics of health care; structural changes in health care, including the effect of managed care on relationships between providers and patients; the social and cultural context of organized medicine; social factors that contribute to the need for medical attention; and doctor-patient relationships in which ethical dilemmas unfold, are left largely untouched by bioethicists.

Further, bioethics is critiqued not only for bypassing the social and cultural context of ethical dilemmas themselves, but also for not recognizing that the bioethics enterprise itself is historically situated and socially constructed. Bioethics is largely taken for granted as a natural development in the sometimes-troublesome progress of modern medicine, a natural outgrowth of scientific advancement. The crucial drawback of this "science as progress" perspective is that bioethics consequently lacks the knowledge that the notion of "moral values" itself emerged from social and historical circumstances, not from formal philosophical analysis. In his historical account of the rationalization of public bioethical debate about human genetic engineering, John Evans establishes that the current character of bioethics as formal rationality is the result of intense professional competition for jurisdiction over ethical problems in the public sphere, not the outcome of a natural progression of philosophical thought. In particular, the necessity of government advisory commissions to secure a universal working knowledge from which to form policy recommendations largely shaped the establishment of current bioethics.

In addition, critics of bioethics have used the term *social bioethics*, the ethics of actual practice, to bring to light the socially constructed nature of moral and ethical dilemmas and solutions in daily medical practice. Rather than investigating how bioethical dilemmas should be normatively resolved, social scientists have investigated how the daily professional practice of patient-care provider dilemmas defined as "ethical" emerge and are resolved in practice. Renee Anspach, for example, studied life and death decision making in academic neonatal intensive care units. She found that ethical conflicts emerged only in situations in which staff and parents differed about the course of treatment. Ethical conflicts are further avoided when medical staff work out differences of opinion in closed sessions and present a united front to the parents. Parents' opinions only matter when they want to continue treatment after staff has determined that further treatment is futile. Even then, staff further diffused dissent by employing preemptive strategies, persuasion, and psychologizing the anger of the parents. Similarly, Daniel Chambliss noted that "ethical" problems involving nurses tend to recur in hospitals when power discrepancies persist involving nurses who provide hands-on care for patients, but do not play a dominant role in treatment decision making.

Bioethics has also been criticized for its tendency to uphold, rather than question, the authority and practices of organized medicine. Though bioethics began its professional journey as vocal criticism of medicine, scholars have suggested that bioethics lost its critical edge in its relationship to medicine over time; bioethics has focused on the micro aspects of medical practice related to ethics and medical dilemmas, but has left untouched larger problems associated with the institutional structure, power relationships, and organization of medicine. As a result, the problem of ethics in medicine has largely been framed by bioethicists as conflicting values in relationships between providers and patients, not between organized medicine and the provision of care. Where bioethics has developed into a profession dedicated to easing the relationships between patients and providers related to ethical issues, critics view it as a mediator between medicine and the public. Instead of explicitly siding with patients and the lay public, bioethics has largely functioned to justify both medical treatment and nontreatment. Comparing the introduction of solid organ transplantation techniques in the United States and Japan, for example, Margaret Lock has shown how Japanese society gradually built a public consensus about organ donation while in the United States such fundamental decisions were facilitated in commissions and presented as a *fait accompli* to the public.

Lastly, critics have questioned the legitimacy and effectiveness of principlism for the resolution of medical-ethical conflicts. Critics of bioethics' use of principlism reject the notion that it is possible to agree upon a universal set of values to assess ethical dilemmas in medicine. It is not clear that principlism, as a set of abstract concepts, is practically useful as a guide for daily medical practice or public debates about bioethics. And although principlism claims to give equal weight to each principle, there has been an overemphasis on autonomy as the ideal end in bioethics. The principles of beneficence, nonmaleficence, and justice are less easy to resolve. With beneficence, it is difficult to determine whether beneficial ends are met, and for whom: the individual, the family, or society? It is also difficult to ensure that nonmaleficence in ethical dilemmas is possible as numerous technologies and treatments involve some iatrogenic illness or potential harm. Lastly, as the U.S. health care system is not organized around universal health care coverage, the issue of justice is nearly impossible to resolve. Although autonomy is the main

principle employed for resolving medical-ethical dilemmas, it, too, is controversial. In conducting social bioethics research, social scientists have determined that autonomy is not necessarily achieved in medical practice, even as it is institutionalized through policies and ethical stances in medicine. The case of death in the United States is instructive for this finding. In Western societies, death is closely managed by medical professionals into diverse and partially contradictory cultural ideals including the "natural" death in intensive care units where interventions are gradually removed to mimic a slow, but not lingering death trajectory; the palliative hospice death; and, still tentatively, the physician-assisted suicide. To enhance autonomy in these dying situations, Congress passed the Patient Self-Determination Act in 1990 to mandate that patients in health care agencies receiving federal funding be given information about legal rights regarding living wills and durable powers of attorney for health care at admission. Most observers, however, have concluded that this legislation and living wills in general have not succeeded in empowering patient decision making.

In contemporary U.S. society, bioethics has thus taken the lead to justify advanced medical technologies and the structural problems that their implementation entails. Social scientists have not only critically analyzed the legitimizing functions of bioethicists, but have also addressed multiple ontological and political changes of advanced medical technologies ignored by bioethicists.

## Science and Technology Studies and the Ethics of Biomedicine

One contested area of advanced medical technologies that received much attention from social scientists is reproductive technologies, particularly infertility treatments. Whereas from a clinical bioethical perspective, the dominant issue in infertility treatment is whether patients are fully informed about the treatments they assent to, feminist researchers have analyzed the dynamic relationship between clinical innovation and the changing experience of reproduction. They have challenged the gendered culture and inequalities of science and have offered an exploration of the political economy of the rapid introduction of *in vitro* fertilization (IVF) under the guise of "procreative choice" and have also critically examined the political wrangling over the status of embryos, eggs, and sperm. Women's ability to freeze eggs and delay pregnancy might offset some of the inequalities of the occupational market, but still puts the burden of reproduction on women, rather than on men, even though 30 to 40% of infertility has been linked to men. Feminists have shown that the ideology of reproductive choice obscures the commodification of infertility, racial and social access inequities, and the anguish of the treatment. Whereas women might actively submit to the objectification inherent to IVF technologies, IVF has a relatively low success rate and might compromise resulting babies' health. *In vitro* fertilization further raises questions about the ownership of embryos and the right of children to trace biological donors. Genetic diagnostic applications enter these discussions because these technologies provide more accurate information about susceptibility to genetic diseases, necessitating a critical examination of the overt and hidden messages of genetic counselors about disability and risk factors. The uncoupling of biology with conventional relations of reproduction has also instigated research on the biological basis of parenting and kinship, especially in parental configurations that differ from the heterosexual mold and the biological nature of life itself.

Furthermore, governmental bodies (with the help of bioethicists) in Europe and North America have attempted to regulate the sale of embryos, germ line genetic alteration, cloning, and the creation of embryos and stem cells for research purposes to ensure control over the impact of new reproductive technologies. At stake is whether these technologies revolutionize reproduction and kinship or whether they anchor traditional reproductive gender roles, or even, when these techniques are used for sex selection in developing countries, whether they herald a new era of eugenics. The potential of reproductive cloning of humans, has raised issues about genetic reductionism of social identity and the further commodification of body parts on a cellular

level. Stem cell research is governed differently depending on cell origin (from embryo or adult) and purported use (therapeutic or reproductive). These developments promise important ontological changes in the reproductive chronology of generations; the uniqueness of the self; the temporality of embodiment; relationships between donors and recipients (which is already complicated in "traditional" organ donation); risks of disability and mutations; and ownership, both financial and legal (i.e., with patents), of cell lines and genetic materials that can be mined for decades. Social scientists have questioned the moral basis, rather than equity or economic grounds, of governmental regulation.

Social scientists studying eugenics, abortion, euthanasia, genetic engineering and cloning, drug trials, biomedical information systems, the body parts markets, and reproductive technologies have thus cast a much wider critical net than have bioethicists in problematizing the ontological, and political-economical changes that these medical developments entail. For now, the result of social scientific analysis has resulted mostly in indirect awareness raising and documenting both the medical hegemony and the many forms of resistance rather than direct clinical decision making of bioethicists. While some social scientists have joined the ranks of bioethicists, others have become increasingly critical of the professional agenda and moral authority of bioethics itself.

*See also* **Bioengineering; Clones and Cloning; Human Subjects in Medical Experiments; Tuskegee Project, The**

**BIBLIOGRAPHY**

Anspach, Renee. *Deciding Who Lives: Fateful Choices in the Intensive Care Nursery.* Berkeley, CA: University of California Press, 1993.

Beauchamp, T. L., and J. F. Childress. *Principles of Biomedical Ethics.* Oxford: Oxford University Press, 1979.

Bosk, Charles L. "Professional Ethicists Available: Logical, Secular, Friendly." *Daedalus* 128 (1999): 47–68.

Chambliss, Daniel F. *Beyond Caring: Hospitals, Nurses, and the Social Organization of Ethics.* Chicago: University of Chicago Press, 1996.

Clarke, Adele E., Janet K. Shim, Laura Mamo, Jennifer R. Fosket, and Jennifer R. Fishman. "Biomedicalization: Technoscientific Transformations of Health, Illness, and U.S. Biomedicine." *American Sociological Review* 68 (2003): 161–194.

Conrad, Peter, and Joseph W. Schneider. *Deviance and Medicalization: From Badness to Sickness.* St. Louis: Mosby, 1980.

Cussins, Charis. "Ontological Choreography: Agency for Women in an Infertility Clinic." In *Differences in Medicine: Unraveling Practices, Techniques and Bodies,* edited by Marc Berg and Annemarie Mol. Durham: Duke University Press, 1998: 166–201.

Evans, John. *Playing God? Human Genetic Engineering and the Rationalization of Public Bioethical Debate.* Chicago: University of Chicago Press, 2002.

Farquhar, D. *The Other Machine: Discourse and Reproductive Technologies.* London: Routledge, 1996.

Fox, Renee C. "The Evolution of American Bioethics: A Sociological Perspective." In *Social Science Perspectives on Medical Ethics,* edited by George Weisz. Dordrecht: Kluwer Academic Publishers, 1990: 201–217.

Fox, Renee C. "Is Medical Education Asking Too Much of Bioethics?" *Daedalus* 128 (1999): 1–26.

Fox, Renee C., and Judith P. Swazey. *Spare Parts: Organ Replacement in American Society.* Oxford: Oxford University Press, 1992.

Franklin, Sarah. *Animals and Modern Cultures: A Sociology of Human-Animal Relations in Modernity.* London: Sage, 1999.

Franklin, Sarah, and Jeanette Edwards. *Technologies of Procreation: Kinship in the Age of Assisted Conception.* London: Routledge, 1999.

Haraway, Donna J. *Simians, Cyborgs, and Women: The Reinvention of Nature.* New York: Routledge, 1991.

Light, Donald W., and Glenn McGee. "On the Social Embeddedness of Bioethics." In *Bioethics and Society: Constructing the Ethical Enterprise,* edited by Raymond DeVries and Janardan Subedi. New Jersey: Prentice Hall, 1998: 1–15.

Lock, Margaret. "Contesting the Natural in Japan: Moral Dilemmas and Technologies of Dying." *Culture, Medicine, and Psychiatry* 19 (1995): 1–38.

McNeil, Maureen, Ian Varcoe, and Steven Yearley, ed. *The New Medical Technologies.* New York: St. Martin's Press, 1990.

Mohr, James C. *Abortion in America: The Origins and Evolution of National Policy, 1800–1900.* New York: Oxford University Press, 1978.

Parens, Eric, and Adrienne Asch. "The Disability Rights Critique of Prenatal Genetic Testing: Reflections and Recommendations." *Hastings Center Report* Special supplement, 1999.

Rosenberg, Charles E. "Meanings, Policies, and Medicine: On the Bioethical Enterprise and History." *Daedalus* 128 (1999): 27–46.

Rosenberg, Charles E. "What Is Disease? In Memory of Owsei Temkin." *Bulletin of the History of Medicine* 77 (2003): 491–505.

Seymour, Jane. *Critical Moments: Death and Dying in Intensive Care.* Buckingham: Open University Press, 2001.

Starr, Paul. *The Social Transformation of Medicine.* New York: Basic Books, 1982.

Teno, J., J. Lynn, N. Wenger, R. S. Phillips, D. P. Murphy, A. F. Connors Jr., N. Desbiens, W. Fulkerson, P. Bellamy, and W. A. Knaus. "Advance Directives for Seriously Ill Hospitalized Patients: Effectiveness with the Patient Self-Determination Act and the SUPPORT Intervention. SUPPORT Investigators. Study to Understand Prognoses and Preferences for Outcomes and Risks of Treatment." *Journal of the American Geriatrics Society* 45 (1997): 500–507.

Timmermans, Stefan. *Sudden Death and the Myth of CPR.* Philadelphia, PA: Temple University Press, 1999.

*Emily S. Kolker and Stefan Timmermans*

# MEDICINE, SCIENTIFIC.

Talcott Parsons noted that "modern medicine is organized about the application of scientific knowledge to the problems of illness and health" (p. 432). Indeed, ever since Enlightenment's critical thinking opened up the knowledge base of Ancient Greek healing practices, leading figures have attempted to steer medicine toward a true scientific practice and science has become the hallmark of contemporary health care delivery and its knowledge base. In premodern medicine, usually dated until the early nineteenth century, orthodox medical practitioners had a limited therapeutic repertoire, relying on a handful of medicines—*materia medica*—with radical and obvious effects to adjust the body's internal equilibrium. Disease was viewed as a fluid, highly idiosyncratic entity specific to the constitution of the individual in relationship to his or her environment and the physician needed to typify a patient's humoral constitution in order to treat. Both patient and practitioner understood that the healer's task was to regulate, restore, and console to relieve anxiety and pain. As satirized by the French playwright Molière, health care providers relied as much on impression management techniques as on therapeutics to bind paying patients in the volatile early modern market.

In the middle of the nineteenth century, under influence of the French school of pathology with its emphasis on direct clinical observations, diseases were seen as distinct clinical entities, with a characteristic etiology, course, and symptomatology. Physicians and surgeons discovered localized pathologies postmortem and began to evaluate the effectiveness of therapeutic techniques statistically as well as to construct statistical profiles of diseases. Leading practitioners initially renounced therapeutics altogether, but gradually new insights and emerged from morbid anatomy. The idiosyncrasy of healing moved from patients to diseases itself, leading to both a physiological reductionism and universalism of knowledge in medicine in which the pathological and physical signs of disease prevailed over symptoms.

The hospital, earlier treated as a charitable poorhouse, became viewed as an accessible treasure trove of clinical material. These "materials" were further investigated in the emerging laboratories, which would not only thrust experimental medicine forward into cell biology but also change medical education, particularly in Germany, but also later, when students returned, also in the United States, particularly at Johns Hopkins and the Rockefeller Institute for Medical Research in New York. In one of the greatest scientific breakthroughs of that century, Louis Pasteur and Robert Koch established the "germ theory"—the notion that disease is caused by invasion of the body by microscopic living organisms. The new bacteriology not only led to vaccines for scourges such as anthrax, but also played a dominant role in tropical medicine. In the late nineteenth century and twentieth century, exchanges between life sciences, spin-offs from established disciplines, and chance generated a steady stream of innovations in immunology, nutrition research, endocrinology, neurophysiology, psychiatric disorders, surgery, and ultimately genetics with the announcement of the double-helical structure of DNA by Francis Crick and James Watson in 1953.

## The Golden Age of Medicine

While laboratories revolutionized therapeutics and the knowledge base of medicine, physicians also consolidated their professional status via educational reforms and state-sponsored licensing arrangements over the course of the nineteenth century. The American Medical Association lobbied successfully to fight third-party, including regulatory, encroachment on the patient-doctor relationship. After physicians gained control over large public hospitals from lay patrons or religious orders, they established smaller specialist hospitals with well-equipped surgical theaters and laboratories. Favorable legislation led to a building boom, and hospitals emerged as the power centers of modern medicine. Nursing and other allied professions developed their own professional agendas within the institutional frame of the hospital.

As part of post–World War II optimism initially generated by the development of penicillin and sulfa drugs in the early 1940s, the federal government stimulated research funding by establishing the National Institutes of Health to support basic scientists in biochemistry and physiology. Physicians who went to medical school in the 1920s and 1930s and practiced in the post–World War II era witnessed the social and scientific transformation of medicine and reaped its benefits. The leaders of that generation combined residency with laboratory work and specialized in a medical subdiscipline. Practicing physicians became applied scientists, engaging in experimental medicine to treat chronic diseases, such as cancer and pulmonary and cardiovascular diseases, which had begun to make up the leading causes of death. They saw the medical knowledge base grow in leaps and bounds and had to continuously adapt their clinical practice as new technological innovations became available. Patients were generally trusting, and physicians could actually offer curative treatment.

Still, the summit of medical power also contained tensions of problems to come at the end of the century, and, similar to the successes of medicine, they also originated in the increasing biomedicalization of medicine. Donald Light sums up the intertwined dilemmas: "To wistfully remember 'the Golden Age of Doctoring' is to forget that it was also the age of gold, the age of unjustified large variations of hospitalization and surgery caused by autonomy and lack of accountable standards, the age of large portions of tests, prescriptions, operations, and hospitalizations judged to be unnecessary by clinical researchers, the age of medicalizing social problems, the age of provider-structured insurance that paid for almost any mistake or poor investment anyone happened to make, and the age of corporations moving in to reap the no-lose profits of such a world by exploiting the profession on its own terms"(p. 202) The expansion of medical care led to great inequities in access to health care in the United States, with more than 40 million people lacking access to regular primary care at the beginning of the twenty-first century. In addition, the health budget keeps spiraling, increasing from 4 to 13 percent of GNP over the last 50 years. Medical technologies and a moral imperative to treat when technologies are available have been singled out as the origin of contemporary inequities.

## Biomedicalization and Evidence-Based Medicine

Further changes in the jurisdiction of medicine under pressure of technological and scientific advances in the last two decades of the twentieth century have been characterized as the era of "biomedicalization" to draw attention to the reconfiguration of health care knowledge production, provision, distribution, and information management centered on life itself. Information gathered from computerized record keeping coupled with biotechnical innovations and biostatistical manipulations led to a shift in previously immutable frontiers of medicine (such as giving birth one decade or more after onset of menopause, recovery after complete heart failure, and genetic design of life). As a consequence, health itself rather than disease has become a target of interventions: health has become a personal moral responsibility to be fulfilled through improved access to knowledge, self-surveillance, prevention, risk assessment, and the consumption of biomedical goods. When health becomes part of ongoing self-transformation, biomedicine generates its own medicalized stratification centered around disease constituencies and risk factors.

One of the most prominent recent manifestations of the rationalization of the health care field is the turn toward evidence-based medicine, a movement to infuse medical decision making with clinical research. The concept of evidence-based medicine was coined by a group of clinical epidemiologists frustrated by the perceived inability of care providers to consistently apply scientifically validated research, resulting in great variation in clinical care delivery and quality.

In an evidence-based practice, health care providers are expected to tailor diagnostic and therapeutic patient care based on a critical assessment of the biomedical literature. This literature is assessed and ranked according to epidemiological principles with a preference for randomized clinical trials and distilled in ready-to-use clinical practice guidelines. Professional organizations offer evidence-based medicine as a service to their members, third-party governmental organizations and insurance agencies have begun to develop their own evidence-based guidelines and databases, and the principles have spilled over to allied professionals and even governance. Humanist critics have decried the Taylorist standardization implied in evidence-based medicine, arguing that clinical-practice guidelines enforce a "cookbook" model of practicing medicine. Critics drawing from a political economy perspective focused on the cost-benefit aspects of evidence-based medicine. They viewed the rise of evidence-based medicine as another rationing tool to insert economical cost-cutting principles into the health care field.

Implementation and outcome studies, however, show that both criticisms are premature. Because they are presented as educational tools, the guidelines rarely manage to modify professional behavior. Social scientists have pointed out that the implementation of evidence-based medicine requires the active submission of the care provider to protocols and guidelines, but without enforcement mechanisms such submission remains voluntarily and ineffective. Even the incorporation of practice guidelines in reimbursement schemes has a diminished rate of return and generated a backlash against managed care. The health care provider's professional autonomy thus largely prevails.

At the same time, evidence-based medicine has succeeded in redefining the knowledge base of medicine, indicating the triumph of clinical epidemiology over pathological anatomy in the twenty-first century. In the current era, the clinical gaze lingers no longer in the tissues of corpses but flourishes in the tight linkage between the biomolecular-genetic laboratory and the large randomized, double-blind clinical trials. The biotech and pharmaceutical industry set the agenda of medical knowledge production, not only by searching for genetic markers and molecular proteins to design customized treatments but by sponsoring the large population-based clinical trials that form the basic material for evidence-based medicine. Clinical trials research vested in new bioinformatics and biostatistics continuously redefines both the content of health interventions (e.g., by defining "unnecessary" treatments) and the scientific criteria of medical knowledge production by standardizing drug evaluation. Epidemiological principles have thus penetrated the taken-for-granted infrastructure of health care delivery and knowledge production. This closed knowledge loop dominates the knowledge base (with several forms of "alternative and complementary" medicine excluded from knowledge production and a further "biologization" of diseases) and raises serious equity questions when women, children, and ethnic minorities have been systematically excluded from clinical trials. Evidence-based medicine further indicates the bureaucratization of disease, where diseases gain specificity through technological engagement and where risk factors such as cholesterol levels turn into proto-diseases to be managed in a risk society.

Although contested by some historians of epidemiology, the rise of scientific medicine drawing from the information, biological, physical, chemical, and neurological sciences is credited for increasing life expectancy in Western societies. Infant mortality that was estimated to vary from 15 to 30 percent in the second half of the nineteenth century, for example, has been reduced to less than 1 percent, in part through better nutrition and housing, a higher standard of living, multiple advances in pediatric knowledge, public health campaigns, and a lower fertility rate. In the wake of evidence-based medicine, further promises of control await, with visions of individualized gene therapies, xenotransplantation, electronic patient records, telemedicine, innovative reproductive technologies, and therapeutic cloning. At the same time, the old foes of mortality and pain have never been completely repressed, patients complain about the objectification of contemporary health care, and health care providers lament the far-reaching rationalization of health care.

## BIBLIOGRAPHY

Beck, Ulrich. *Risk Society: Towards a New Modernity.* London, Newbury Park, CA: Sage Publications, 1992.

Clarke, Adele E., Janet K. Shim, Laura Mamo, Jennifer R. Fosket, and Jennifer R. Fishman. "Biomedicalization: Technoscientific Transformations of Health, Illness, and U.S. Biomedicine." *American Sociological Review* 68 (2003): 161–194.

Epstein, Steven. "Bodily Differences and Collective Identities: The Politics of Gender and Race in Biomedical Research in the United

States." *Body & Society*, 10.2–3 (June–September 2004): 183–203.

Evidence-Based Medicine Working Group. "Evidence-Based Medicine: A New Approach to Teaching the Practice of Medicine." *JAMA* 268 (1992): 2420–2425.

Freidson, Eliot. *Professionalism Reborn: Theory, Prophecy, and Policy.* Chicago: University of Chicago Press, 1994.

Hess, David. *Can Bacteria Cause Cancer? Alternative Medicine Confronts Big Science.* New York: New York University Press, 1998.

Kaufman, Sharon R. *The Healer's Tale: Transforming Medicine and Culture.* Madison, WI: University of Wisconsin Press, 1993.

Light, Donald W. "The Medical Profession and Organizational Change: From Professional Dominance to Countervailing Power." In *Handbook of Medical Sociology*, edited by Chloe E. Bird, Peter Conrad, and Allen M. Fremont. Upper Saddle River, New Jersey: Prentice Hall, 2000: 201–217.

McKeown, Thomas. *The Role of Medicine: Dream, Mirage, or Nemesis?* London: Nuffield Provincial Hospitals Trust, 1976.

Meckel, Richard A. *Save the Babies: American Public Health Reform and the Prevention of Infant Mortality 1850–1929.* Baltimore: Johns Hopkins University Press, 1990.

Parsons, Talcott. *The Social System.* Glencoe, IL: The Free Press, 1951.

Rosenberg, Charles. *Explaining Epidemics and Other Studies in the History of Medicine.* Cambridge: Cambridge University Press, 1992.

Rosenberg, Charles E. "The Tyrany of Diagnosis: Specific Entities and Individual Experience." *The Milbank Quarterly* 80 (2002): 237–260.

Starr, Paul. *The Social Transformation of Medicine.* New York: Basic Books, 1982.

Timmermans, Stefan, and Marc Berg. *The Gold Standard: The Challenge of Evidence-Based Medicine and Standardization in Health Care.* Philadelphia, PA: Temple University Press, 2003.

*Stefan Timmermans*

# METRICS OF SCIENCE.

As scientific activity has grown over the centuries, the activity of observing scientific growth has also grown. Measurement of scientific activity is an area of inquiry in its own right (called *scientometrics*), but it has also been a craft encouraged by governments as an information base for policy. What is measured about science has evolved as theories about the institutions of science and technology have evolved. Measures of science and technology have been used to test the pulse of national innovation systems, to measure effectiveness at the program and institution levels, and to evaluate individuals.

This essay will trace the emergence and development of the most common forms of science metrics, along with their relationship to theory on the one hand and policy on the other.

## Roots in Information Science

Scientometrics was born in the observations scientists made about regular patterns in the literature they worked with. In the 1920s, for example, Alfred J. Lotka, a chemist, demographer, ecologist, and mathematician, discovered that the number of authors making $n$ contributions is about $1/n^a$ of those making one contribution, where $a$ is often nearly 2 (Lotka's law). In 1934, after studying a bibliography of geophysics covering over 300 journals, Samuel C. Bradford noted that the journals tended to fall into three groups, each with about one-third of all articles: (1) a core of a few journals, (2) a second zone, with more journals, and (3) a third zone, with the bulk of journals (the Bradford distribution). George Kingsley Zipf, a linguistics professor at Harvard in the 1930s, found that a similar distribution characterizes words: The quantity under study is inversely proportional to the rank (Zipf's law).

Such occasional observations eventually crystallized into a field of study. A word for the quantitative study of scientific information production was first coined by Vassily Vassilievich Nalimov (1910–1997) in Russian as *naukometriya*, translated into English as *scientometrics*. Nalimov was a librarian working at the All-Union Institute for Scientific and Technical Information in Moscow in the 1950s. Drawing on earlier generalizations scattered across the sciences, he focused on mathematical regularities in the scientific literature. Over the last few decades, following Nalimov's lead, library science evolved into *information science*. An information scientist does not specialize in archiving printed materials but rather in understanding the structure of information and developing tools for accessing it. Information science and scientometrics have had strong reciprocal influences.

A central tool in both fields is the *Science Citation Index* (SCI), first published in the early 1960s. The SCI is based in concept on an indexing tool in law, *Shepherd's Citation Index*, which indexes recent legal cases by their references to older ones. Since more recent publications cite older ones in their footnotes, one can find the recent literature that is related to an older document by checking what has cited it. The concept is simple, but providing the citation index tool in the 1960s required computing capacity that was just developing. The company that produces the SCI, the Institute for Scientific Information, was founded by Eugene Garfield, counted among the distinguished information scientists of the late twentieth century. His collected *Essays of an Information Scientist* chronicle the early uses of both publications and citations to study how science works.

Information science continues to contribute concepts and models for use in scientometrics. For example, Loet Leydesdorff's work on the contributions of communication theory in scientometrics applies information theory to the study of science as a self-organizing system, exploring such areas as probabilistic entropy measures for studying complex developments in networks.

## Links to the Sociology of Science

The sociology of science was founded as a subfield in the late 1950s, and, as early as the 1960s, work at the intersection of information science and sociology of science was generating insights into the social organization of knowledge production. Derek de Solla Price, a historian of science with a quantitative bent, led the way. Price's fascination with measuring science began in Singapore, where he was a junior professor of physics. While the college library was under construction, he was storing copies of the *Philosophical Transactions of the Royal Society*, ordered from England, as they arrived, each volume in one pile. Price noticed after a while that if one imagined a line drawn along the top edge of the piles, it traced an exponential curve. This observation led him to a lifelong study of exponential growth in science.

One of his observations was that exponential growth led to differentiation among knowledge communities. A human being can keep up with only a finite amount of new scientific information. Scientists therefore are led to specialize, choosing a scope for their knowledge that represents a manageable amount to keep up with. This is the literature of their "specialty areas," and it is produced by their most relevant colleagues, who form "invisible colleges." Price's observations on this phenomenon in his book *Little Science, Big Science* and in his seminal article on networks of scientific papers published in *Science*, stimulated a spurt of interest in specialization in science. Several other books picked up the theme. Henry Menard, an earth scientist, noted the tendency for disciplines to subdivision in his *Science, Growth and Change*. Diana Crane linked the publication base of specialties to measures of social interaction in her *Invisible Colleges*.

Eugene Garfield's new information tool also provided an impetus to specialty studies. Henry Small, a historian of science, and Belver Griffith, an information scientist, developed a technique for grouping published articles using the citations to them in the more recent literature, a technique called "co-citation clustering." The results captured very well Price's concept of invisible colleges and showed these areas forming and merging as local areas of order in the vast sea of scientific knowledge. The technique led to an upsurge of interest in mapping change in scientific knowledge at a detailed, subdisciplinary level, either at one point in time or the changes over time.

As the computational tools developed, a second form of mapping the fine structure of scientific knowledge was developed: co-word analysis. Again, it was brought into scientometrics by way of a sociological theory, this time actor-network theory. Developed by Michel Callon and his colleagues in Paris, co-word analysis makes maps using any words associated with any publications (words in titles or abstracts, controlled-vocabulary keywords, and so forth, from articles, reports, or other written products), using the co-occurrence of the words in the written products to create the structure. Actor-network theory posited that power in science, as in other social realms, derives from network relationships. Co-word maps provided a way of displaying networks and analyzing their dynamics.

The sharply skewed distribution of citations to scientific papers indicated quite different levels of attention being paid to them by the research community. Some observers therefore took citations to be an indicator of quality. Studies of stratification processes in science adopted publication and citation analysis as tools at an early stage of their development. Scientists objected from the beginning to the notion that high numbers of citations were indicators of quality. Citation context analysis (the examination of the text surrounding the footnote when it appears in the citing article) dismissed some of the myths but added to the skepticism in other ways. Empirical work demonstrated that only a small percent of citations are negative; that they are concept markers, used to bring important ideas from the literature into the text; and that the experimental results connected to a theory were more likely to be cited than the theory itself, a difference clearly not related to any absolute quality scale.

## Links to the Economics of Research and Development

Patents are another form of publication associated with science and technology. The word *patent* means "open," and in fact, filing a patent is a way of making the knowledge it embodies public, in exchange for exclusive rights to develop and market a technology for a set period of time. Patents are taken as partial indicators of inventive activity, since not all inventions needed to be patented in order to be protected in a commercial sense.

Common alternatives are copyrights and trade secrets.

In the early 1980s, the U.S. National Science Foundation supported the development of patent citation databases, that is, databases that included citations from patent applications both to other patents and to the scientific literature. These data gave a new tool for studying the emergence of technologies, the relationships among them, and the relationships among the different institutions that both produce patents and receive patent citations. Once centralized in a single contractor because of the data processing capability required, this data source and approach are now accessible to any researcher who can download data and analyze them on a desktop computer.

Again, the new source of data intersected with active questions in a field of research—this time, the economics of research and development. Like the sociology of science, this area appears to be a subfield of a discipline but actually involves scholars from many disciplinary backgrounds. This field focuses on how industry translates research and development into new products and processes, and thus contributes to economic growth. The analysis of patent data has nicely illustrated several observations in this field.

Patents and patent citation data have also been useful in tracking the growing relationship over the 1980s and 1990s between university research and industrial use. A large majority of papers cited by U.S. industry patents, for example, report on publicly supported research conducted at academic and government institutions; only about a quarter are authored by industrial scientists.

Another striking finding was the local connectedness of technological knowledge. Keith Pavitt had been arguing that research knowledge is a public good only in theory. In practice, research knowledge is likely to be used close to home because of the tacit components necessary to use it effectively. Patent citations nicely demonstrated that both the scientific and technology sources firms draw on in creating new products and processes are likely to have come from an institution relatively close to home. Patent citation studies demonstrated that there was a strong national component in citation linkage, with citations within countries happening twice as frequently as would be expected by chance. Furthermore, in the 1990s, U.S. patents were increasingly citing U.S. research.

The 1990s saw growing attention in the economics of R&D to the network nature of these local innovation systems, through the concept of cluster. Analysis of patent data has also been helpful in representing and exploring this concept.

## National Science Indicators

Long before literature-based metrics of science developed, the advent of large-scale government support for science brought several other policy-relevant measures into play. During World War II, scientists were an unexpectedly valuable resource, creating the weapon that eventually ended the hostilities: the atomic bomb. In the 1950s, then, the governments of many industrialized countries started registries of their scientific and engineering personnel, so that they would be able to locate them if another emergency arose. These registries became the beginning of national human resource statistics, which rapidly developed into regular surveys.

Likewise, data on government funding itself began to accumulate in the immediate postwar period. In the United States, for example, the Office of Naval Research had been formed to continue the relationship between university research and the Department of Defense that had been established during the war. The National Institutes of Health were set up to fund more health-related biological research. And in the early 1950s, the National Science Foundation finally went into operation. These new research ventures, along with the existing research funding through the U.S. Department of Agriculture, began to pour money into university research. With its congressional mandate to collect information relevant to national policy, NSF began to collect data on that funding. The support grew very fast, with another upward turn as the National Aeronautics and Space Administration (NASA) was added to the set of major funding agencies after 1958.

By the mid-1960s the National Science Board decided to launch its now-familiar report, *Science Indicators*. The Science Board is NSF's governing body, but it also has broader responsibilities for advice on science policy in the United States. The *Indicators* project was launched with ambitious goals of finding ways to evaluate federal policy efforts towards science and technology and making informed choices. It has grown into a widely used and widely imitated encyclopedic compilation of statistics on science and engineering in the United States. At the same time the U.S. effort was starting, the OECD (Organization for Economic Cooperation and Development) was beginning to gather R&D statistics and was developing its *Frascati Manual*, the authoritative source for definitions of terms used in science and technology statistics.

The *Science Indicators* project and the *Frascati Manual* both adopted terms and categories that reflected the linear model of innovation. The statistics were organized into inputs, outputs, and impacts, but inputs have dominated in amount of data presented. Inputs included the counts of scientists and engineers that had started with the national registries, and all the funding data. Outputs were publications and citations, adopted from the contemporaneous developed of analysis based in the then-new *Science Citation Index*. International comparisons on both input and output indicators became commonplace in the volume, including comparison on perhaps the most commonly-used policy-relevant science and technology statistic: GERD (Gross Domestic Expenditure on Research and Development) as a percent of Gross Domestic Product.

Over the years, the *Science Indicators* report has grown. It includes much more data on education for careers in science and engineering than it used to (including test results that indicate how well young people in various countries are mastering basic math and science skills). Patent data and patent citation data have been added. Collaboration indicators, constructed from the SCI-based publication data set, have tracked the broad-based and steady growth in the number of collaborators per article as well as the institutional diversity of authors. The volume is studded with special reports on quantitative studies of particular aspects of science and engineering. But most of the innovations that go beyond the original input indicators have been temporary, and input numbers remain the mainstay of the system. Impact data have come and gone.

The idea of indicators volumes has been popular, however, both in OECD countries that collaborate on statistical standards through that organization and in other countries, such as China. The European Union has produced a trans-Europe report, with a more analytic slant than the U.S. version, brought in by economists of R&D.

## Performance Measurement

Evaluation is one of the basic social processes of the scientific community. *Ex ante* evaluation has traditionally been done qualitatively through peer review. In *ex post* evaluation, however, peer review is often supplemented with quantitative performance measures based in scientometrics.

For individuals, quantity of publication has long been taken as a sign of the contribution of new knowledge to a field. Promotion processes in both university and government laboratories use publication rates as one factor in the evaluation, and some expect high impact, as indicated in citation counts, as well. Since these numbers are used by evaluators who work in the same field as the person being evaluated, they can be seen in the context of expectations for the field.

Using publication and citation counts across fields, however, carries more complications. Anthony F. J. Van Raan and his colleagues have done extensive comparisons of laboratories in the Dutch university system, normalizing citation counts by field using average citation rates for journals. Doing publication and citation profiles of research institutions has become common practice in evaluation in many parts of the world, although it is not particularly suitable for institutions where publications are not an expected output or where researchers publish in many local journals not covered by the SCI.

Many funding programs have been required over the last decade to provide performance indicators, and publications commonly appear among the output indicators, along with students trained. Patent data have been used at regional and national level as indicators of inventive activity to help evaluate innovation policies.

Collaboration measures such as co-authorship of papers or patents are also sometimes used for evaluation. For example, one goal of the European Union's research programs is increasing collaboration across Europe, so its evaluation office has commissioned publication-based studies of those patterns. Funding programs aimed at encouraging innovation networks also use literature-based collaboration indicators for *ex post* program-level evaluation. Many other forms of data are also used in evaluation, with surveys and case studies predominating.

## Summary and Conclusions

Every human activity presents measurable aspects, but the measurable aspects of scientific activity have received particularly active attention, in part because science itself relies so heavily on quantification to understand the world. The basic metrics of science are people, funding, publications, and patents. Publications and patents provide the additional tools of authorship and citations, which feed into increasingly sophisticated analytic techniques, especially based on network analysis. The spread of computing power over the last few decades has put most of these forms of data and analytic techniques into the reach of any research team, and they are widely used. Scientometrics as a field contributes to and draws from evolving theories of how science and technology operate as well as the world of research and innovation policy.

## BIBLIOGRAPHY

Callon, Michel, Jean-Pierre Courtial, and Françoise Laville. "Co-word Analysis as a Tool for Describing the Network of Interactions between Basic and Technological

Research: The Case of Polymer Chemistry." *Scientometrics* 22.1 (1991): 155–205.

Cozzens, Susan E. "The Knowledge Pool: Measurement Challenges in Evaluating Fundamental Research Programs." *Evaluation and Program Planning* 20.1 (1997): 77–89.

European Commission. *Third European Report on Science and Technology Indicators: Towards a Knowledge-Based Economy.* Brussels: European Commission, 2003.

Garfield, Eugene. *Essays of an Information Scientist.* Philadelphia: ISI Press, 1977.

Godin, Benoît. 2003. "The Emergence of Science and Technology Indicators: Why Did Governments Supplement Statistics with Indicators?" *Research Policy* 32 (2003): 679–691.

Griffith, Belver C., Henry G. Small, Judith A. Stonehill, and Sandra Dey. "The Structure of Scientific Literatures. II: Toward a Macro and Microstructure for Science." *Science Studies* 4 (1974): 339–365.

Leydesdorff, Loet. *A Sociological Theory of Communication. The Self-Organization of the Knowledge-Based Society.* Parkland, FL: Universal Publishers, 2000.

Moed, Henk F., Wolfgang Glänzel, and Ulrich Schmoch, eds.. *Handbook of Quantitative Science and Technology Research: The Use of Publication and Patent Statistics in Studies of S&T Systems.* Dordrecht, The Netherlands: Kluwer Academic Publishers, 2004.

Narin, Francis, Kimberly S. Hamilton, and Dominic Olivastro. "The Increasing Linkage between U.S. Technology and Public Science," *Research Policy* 26.3 (1997): 317–330.

National Science Board. *Science and Engineering Indicators.* Washington, DC: National Science Board, 2004.

Price, Derek J. de Solla. *Little Science, Big Science.* New York: Columbia University Press, 1963.

Small, Henry G., and Belver C. Griffith. "The Structure of Scientific Literatures. I: Identifying and Graphing Specialties." *Science Studies* 4 (1974): 17–40.

*Susan E. Cozzens*

# MILITARY, THE, AND TECHNOLOGY.

The relationship between technology and war is particularly intimate. Humans need weapons to be efficient killers. It is hard to imagine a battle without at least stones and clubs—crude technologies, but effective at close quarters. Some scholars postulate that the demand for better war weapons drove the development of technologies such as metallurgy and shipbuilding and also led to the invention of bureaucracy to manage war. Lewis Mumford (1970) argued that ancient armies were the first machines, bringing masses of tools and complicated processes together in regimented action. Perhaps cities, irrigation systems, or large rituals were actually the first machines, but the evidence is compelling that even during ritual wars before the founding of cities (civilization), the cult of the weapon was a powerful technological impetus and that the better weapons (including protective armor) were often decisive in victory. For all this, ritual and even ancient war were basically conservative technologically. The blades of weapons name the slow transitions—Stone Age to Bronze Age to Iron Age to age of steel.

The rise of city states and their organization of battle transformed war from semi-religious skirmish and kidnap to the most horrible political instrument humans have invented. Despite sophisticated early theories about war such as the unsurpassed Chinese collection *The Art of War*, ascribed to Sun Tzu, most wars were about brute strength in numbers, well-made weapons, and the skill of certain elite aristocratic troops such as chariots and cavalry. Later, the lethal value of slingers, archers, and heavy infantry, often citizens, was crucial. In any event, almost all battles ended with butchery, including sieges.

Butchery is a precise way of describing war, as the historian John Keegan revealed in his masterful 1976 history of the experience of soldiers: *The Face of Battle.* Cut down by the blades on chariot wheels, stabbed and trampled by phalanx or horse, shattered by explosion, ripped by gunfire, seared by napalm—war is always about mutilating and destroying human bodies.

This explains the persistence of magical and emotional thinking around war. Honor, *élan*, fate, courage, and divine intervention are important figures in both ritual and ancient war, all the more so since technological change was slow and intermittent. The armies of ancient Egypt could have fought pretty equally, in technological terms, with anyone in the next 2500 years. In the 1500s, however, reason and experiment were fostered, and soon they were turned to the ballistics and navigation problems of war. War began to be rationalized, it became "modern," and 500 years later industrialized slaughter (as in the assembly-line deaths of the Western Front, submarines, the Holocaust, and strategic bombing) had reached epic proportions. War was still emotional, but the emotion was sublimated through an instrumentalist rationality into constant improvement in military technology.

## Modern War

In many respects modern war not only was made possible by the technological breakthroughs of the Renaissance and after; in turn it largely shaped the European nation-state system that politically structures the world today. The influential John Nef even argued that human technological progress was almost completely driven by war (1963), but most historians argue that the relationship is more complex. Sometimes war or the fear of war produces new technologies; sometimes it adapts old ones. New technical developments sometimes lead to military applications and even new doctrines; sometimes not. Often the ideology about the place of technology in war can precede the changes in war itself.

In the 1500s the Italian political operator and philosopher Niccolò Machiavelli called for the application of rationality to make war more politically useful through total battle. This is exactly what came to pass. Aristocracy gave way to meritocracy, and the pace of weapons development accelerated remarkably. The historian Geoffrey Parker (1988) and others rightly ascribe the spread of European dominance around the world to military innovation, although it was often not the weapons but the way they were utilized that was decisive.

As modern war developed and matured, so did the modern nation-state, industrial society and science, and European and North American colonialism. The modern world system coevolved with modern war. Military innovation was institutionalized, culminating during World War II in a plethora of technological innovations including radar, sonar, computers, missiles, jet airplanes, and nuclear bombs.

Whole strategies were built on new technologies (as in the case of blitzkrieg) or were influenced by them. In particular, the breaking of the Japanese and German codes using new systems, including computing machines, was decisive.

Modern war, however, fell victim to its own success. The nightmare of World War I did not merely end the Romanov and Ottoman empires in Russia and the Middle East, the near consensus in Western culture that war was natural was also shattered. At the heart of this new view that war was not inevitable was the argument that, whatever its value had been in the past, contemporary war had been so transformed by technology that heroism and any other martial virtues were now rendered impossible or mute. For many, the technological efficiency of it—its deadliness beyond human scale—meant that war was now politically and morally unacceptable.

This view has continued to gain strength over the intervening decades, as military technologies have far surpassed those of World War I. After the Vietnam War it came to be held by a surprising number of veterans, officers, and enlisted men as well as military historians and experts.

But another stance toward technology and war has also thrived under the rapid changes of the twentieth century: a love for the technology. A professor of American literature, Bruce Franklin (1988), has described how, over the years, many U.S. policy makers have been attracted to superweapons as a solution to the problem of war. Robert Fulton, Thomas Edison, and other great inventors were deeply involved in military developments, seeking technological solutions to the problems of battle. Even though President Lincoln had to intervene personally to get the Sharps repeating rifle adapted during the Civil War, over time the U.S. military became more and more proactive technologically, fostering the combat development of trains, balloons, telegraph, machine guns, accurate artillery, steel battleships, and airplanes.

After World War I, the majority of all military leaders agreed that new technology was profoundly important. Reflecting on how machine guns and artillery cut down a generation of European manhood, generals and politicians alike vowed never to be "technologically surprised" again. War became technophilic; in love with any new and potential weapons.

World War I introduced tanks, planes, radios, trucks, and submarines to total war. These technologies proved to be incredibly important in World War II, along with such new "wizard weapons" as radar, missiles, and computers. Finally, at the height of World War II, weapons of mass destruction (WMD) were developed, destroying modern war's core logic. Total war was no longer possible. Some weapons, at least in some forms, became for most people too horrible to use.

The Cold War that followed was replete with paradoxes. Its stability depended on mutually assured destruction (MAD), the promise that a superpower would use these horrible weapons in retaliation. It apparently was enough to keep conflict to bloody but "limited" wars such as Korea and Vietnam. But these were strange wars of stalemate and even defeat for the formerly dominant European and North American powers. War was no longer modern.

## Postmodern War

So what should war be called now? There is no real agreement. Over fifty terms have been used to relabel war since the end of World War II. Among the most interesting are "permanent war," "pure war," "perfect war," "postmodern war," "high-technology war," "technological war," "technowar," "cyberwar," "computer war," "high modern war," "hypermodern war," "third-wave war," "net war," "information war," "info war," "iwar," "hyper-real war," "neocortical warfare," "sixth-generation war," "fourth-epoch war," and "nonheroic war." Some of these are official terms for war, but most are from war theory. What links them all is that technology is central to their definitions. It has become impossible to think of war without the relentless perpetual changes in military technology.

Historians of war have noted that for the last six hundred years there have been a series of war-changing technological jumps, termed "revolutions in military affairs" (RMA). They include the infantry revolution (fourteenth century), the artillery revolution (fifteenth century), the revolution of sail and shot at sea (fourteenth to seventeenth century, a slow one), the fortress revolution (sixteenth century), the gunpowder revolution (sixteenth and seventeenth centuries), the Napoleonic Revolution (eighteenth century), the land war revolution (nineteenth century), the naval revolution (nineteenth and twentieth centuries), and then a series of revolutions during and between the World Wars (mechanization, aviation, information) and the nuclear and electronics revolutions afterward.

The earlier technological revolutions were separated by large blocks of time, and often took place over centuries. Since the start of World War I in 1914, RMAs have come often and suddenly. It makes more sense now to talk of a permanent revolution in military affairs. Innovation is constant, especially in the realm of information technology and such specialities as remote weapons and battlespace informatics.

To produce these new generations of weapons and other military technology, the industrial economies were militarized in constant preparation for apocalyptic and more limited conflicts. In both the West and the Communist countries the military parts of the economy were favored, such that President Dwight Eisenhower even warned of the dangers posed by the "military-industrial complex" to American democracy.

This system has helped shape the world economy. The first transistors, integrated circuits, and computers were all developed in the favorable climate of massive U.S. military purchases and cost-plus contracts. The Internet itself began as a military research project to improve communication between military and civilian researchers working on Department of Defense contracts and to test the usability of a distributed nonhierarchical network. Such networks were deemed necessary to control nuclear weapons in the case of total war, when large parts of any command and control system were expected to be obliterated.

Fortunately, the command and control system for nuclear conflict was not tested in war. But some analysts credit the technological arms race of the Cold War, along with other factors, for eventually forcing the Soviet Union into collapse.

As reason and technology justified the MAD nuclear policy, it was assumed that they could also solve the problems of limited war. The real test was Vietnam. Financial support by the United States for France's attempt to reassert its colonial domination of Indochina turned into direct military intervention to prevent a Communist victory. Confident of its technological superiority, the United States set out to wage the "perfect war" (Gibson 1986). Lip service was paid to winning the hearts and minds of the Vietnamese, but faith was put into "vertical envelopment" (helicopter attacks), the electronic battlefield, precision bombing (jets and drones), superior firepower, and "McNamara's Wall," a high-tech death zone intended to stop the flow of weapons along the Ho Chi Minh Trail between North and South Vietnam, named after the Secretary of Defense, Robert Strange McNamara.

A Harvard Business School professor with an expertise in bargaining theory, McNamara joined the U.S. Army Air Corps strategic bombing campaign during World War II to do systems analysis (also called operations analysis and operations research), calculating casualty rates, pilot tours of duty, and plane replacements, and solving other problems. After the war, he marketed himself and a team of such experts to Ford Motor Company, where they "rationalized" its procedures. Later, when President Kennedy appointed McNamara as Secretary of Defense, most of his old team came with him to the Pentagon.

As in business, the theory of systems analysis is that institutions today are combinations of humans and machines and that the flows of information, material, and energy between them can be closely controlled, usually through information technology. Operations can also be successfully managed this way. However, it turned out that systems and technologies could not bring the United States victory in Vietnam. In reality, the theories of asymmetric war seem to apply in the postmodern era. Because they used appropriate technology and kept the political aspects (popular support, especially in Vietnam and the United States) of the war at the center, the North Vietnamese outlasted the United States.

Even though the Vietnam War was lost, the commitment of the U.S. military to systems and to new technologies as a way of dominating the battlespace has, if anything, grown stronger. It is still producing technological innovation in every aspect of the military. Logistics, uniforms, training, weapons, and weapon platforms are all in never-ending cycles of improvement. Everything possible is being done to improve both the automation of weapons and the efficiency of human-machine systems.

Information technology represents a particularly important part of this dynamic. In World War II it was generally accepted that "force of fire," the amount of metal and explosives you could hurl at your enemy, was the most important military advantage. But now many of the most forceful weapons cannot even be used. So there has been a growing emphasis on collecting information and using it. Officially, information is now the main force multiplier for NATO, and this is a doctrine that is accepted in Russia, China, and Israel as well.

The importance of information processing (collecting, evaluating, using) had become so evident that in the 1990s there was a flurry of wild claims of a special kind of information (or info or cyber or net) war. But such pure information war is a myth. In reality war continues to be about killing and the terror it evokes.

## War in the Twenty-first Century

Technology will certainly continue to shape war profoundly. It is erasing old distinctions between nonstate and state combatants, because one does not need a government now to kill thousands. Terror and war also are bleeding together; human-weapon suicide bombers, nuclear deterrence, and cruise missiles all involve terror. The WMDs that ended modern war are particularly important. Continued improvements in biotechnology, materials science, rocketry, and other areas means that the proliferation of nuclear, biological, and chemical weapons is probably inevitable. This has already contributed to fundamental changes in the international system, including regional alliances, international legal frameworks, and the concept of preemptive war. Constraining WMDs will be very difficult.

The history of limiting military technological developments is not encouraging. In the Middle Ages there were attempts by the Catholic Church to outlaw the use of crossbows against Christian enemies (infidel or heretical enemies not being expected to have any reason to respect the ban), but it did not last. More successful was samurai Japan, which limited the use of firearms for over 300 years (1543–1879). Both of these cases involved attempts to stabilize a cultural order threatened by new military technologies, and Japan was an exception in its isolation. In general, military innovation has been difficult to suppress because, in a multipolar world, it only takes one adopter to force everyone to the next level—hence, arms races.

Sometimes such competitions are disguised, such as the race to put a man on the moon. It is clear now, especially in light of the intercontinental missile defense system (Strategic Defense Initiative, or SDI) of President Reagan and the limited missile defense system of President George W. Bush, that the overall goal of the Soviet, U.S., European, Chinese, and Israeli space programs has been to use, or even dominate, space militarily. Space assets for surveillance and communications have become increasingly important in war. Plans for space-based weapons platforms are in development, and in the United

States, for example, the Space Forces are expected to become the latest branch of the military.

The debates about SDI did have a very important unforeseen effect. Many scientists and engineers came forward with their opinions about the unworkability of the designs for the gigantic, automatic system to stop a Soviet missile attack. While there were problems with the physics and general engineering, it was its computer problems that were particularly notable. SDI would have been the largest computer system in history, but even its simulation would have required the largest computer system ever. In debates on SDI's technical aspects it became clear that there were severe limits on what military systems dependent on information technology could accomplish. These limits are based on work such as Gödel's incompleteness theorem (all formal systems are either limited or have paradoxes) and the Church/Turing thesis that demonstrated the same limits for an infinite computer.

When one puts this into the context of war itself, that complex, violent, unfathomable experience that involves not just triumphing over nature but over a human opponent, it becomes clear that managing war is impossible. Battle systems that purport to dominate war without casualties (for at least one side) will ever fail, unless the opponent is incredibly weak.

This is a key feature of the strange geography of the twenty-first century international system. Supposedly there is one superpower, the United States, but, because it is so dependent on high-technology approaches to combat that severely limit U.S. casualties, the United States is actually quite limited in its potential military actions, just at a time when it seems more threatened then ever because of the threat of nuclear, biological, or chemical attacks, conventional military weapons, or just hijacked civilian high technology. So the United States has made a massive commitment to new weapons, continual innovation, and carefully chosen interventions.

The Cold War system seems to be continuing with somewhat new alignments and with much less stability than the Soviet-

West stalemate. Meanwhile, the military establishments of all the industrial powers, from China to NATO, rush the latest technologies and their associated doctrines into action while researchers explore the military potential of dozens of new areas, including nanotechnology, biotechnology, lasers, small nuclear munitions, anti-gravity systems, sound, noxious smells, intimate human-machine weapon systems (cyborgs), intelligent autonomous weapons, and psychological operations. But the history of technology and war strongly suggests that any technological solutions will be short-lived. Political problems, in the long run, are not solved by new weapons, and the problems of the twenty-first century cannot be solved by war. In fact, the high-technology war/terror that has developed may be the most important problem of all.

*See also* **Biological Terrorism**

## BIBLIOGRAPHY

Brodie, Bernard, and Fawn Brodie. *From Crossbow to H-Bomb*. Bloomington: Indiana University Press, 1973. A fine, readable account of the long history of technology and war.

Dyer, Gwynne. *War*. New York: Crown Publications, 1985. Graphically rich and historically clear history of war with due emphasis on the growing importance of technology. Linked to a fine television documentary available on video.

Edwards, Paul. *The Closed World: Computers and the Politics of Discourse in Cold War America*. Cambridge, MA: MIT Press, 1996. Charts in detail the origins of military computerization and the influence it has had on American science, especially psychology, and culture.

Ekstein, Modris. *Rites of Spring: The Great War and the Birth of the Modern Age*. Boston: Houghton Mifflin, 1989. A beautiful and horrific description of how the technologically driven carnage of the Western Front transformed twentieth-century culture.

Ellis, John. *The Social History of the Machine Gun*. New York: Pantheon Books, 1973. Gripping account of the macabre dance between machine gun technology, modern war (including colonial conquest and class conflict), and Western civilization.

Franklin, Bruce. *War Stars: The Superweapon and the American Imagination*. Oxford: Oxford University Press, 1988.

Gibson, James. *The Perfect War: Technowar in Vietnam*. Boston: Atlantic Monthly Press, 1986.

Gray, Chris Hables. *Postmodern War: The New Politics of Conflict*. New York/London: Guilford Press, 1997. An extended analysis of contemporary war with an emphasis on how technology has shaped its major aspects.

Keegan, John. *The Face of Battle*. London: Penguin, 1976. This book revolutionized the history of war by focusing on the actual experience of combat.

Keegan, John. *The Mask of Command*. New York: Viking, 1987. Argues convincingly that changes in war technology have fundamentally changed the nature of military leadership.

Krepinevich, Andrew. "Cavalry to Computer: The Pattern of Military Revolutions." *National Interest* (Fall 1994). Reprinted in *Foreign Policy* (Winter, 1998): 82–83.

Machiavelli, Niccolò. *The Art of War* (1520). Translated by Ellis Farnsworth. De Capo Press, 1990.

Mumford, Lewis. *The Myth of the Machine*. Vol. 1: *Technics and Human Development*. New York: Harcourt, Brace & World, 1967.

Mumford, Lewis. *The Myth of the Machine*. Vol. 2: *The Pentagon of Power*. New York: Harcourt Brace Jovanovich, 1970.

Nef, John. *War and Human Progress*. New York: W. W. Norton, 1963.

Parker, Geoffrey. *The Military Revolution: Military Innovation and the Rise of the West, 1500–1800*. Cambridge, U.K.: Cambridge University Press, 1988.

Scarry, Elaine. *The Body in Pain*. Oxford: Oxford University Press, 1985.

Sherry, Michael. *The Rise of American Air Power: The Creation of Armageddon*. New Haven, CN: Yale University Press, 1987. Sobering analysis of the Allied strategic bombing campaign of World War II.

Sun Tzu. *The Art of War*. Translated by S. B. Griffith. Oxford: Oxford University Press, 1962.

Van Creveld, Martin. *Technology and War: From 2000 B.C. to the Present*. New York: Free Press, 1989.

Zulaika, Joseba, and William A. Douglas. *Terror and Taboo: The Follies, Labels, and Faces of Terrorism*. New York: Routledge, 1997. Examines the many different definitions of terrorism and the politics behind them.

*Chris Hables Gray*

# MISCONDUCT, SCIENTIFIC.

Scientific *misconduct* is a term that entered public discourse in the early 1980s through the federal regulatory process. The precise scope and meaning of the term have been changing over time, the result of a protracted (and ongoing) discussion about the nature of deviations from accepted practices of scientific research and the appropriate response to such behavior. This entry will explain the term, placing it in historical and regulatory context. It will also summarize the principal findings of social research concerned with the prevalence and causes of scientific misconduct and indicate likely future directions.

## Research Misconduct Defined

On December 6, 2000, the President's Office of Science and Technology Policy (OSTP) released its official definition of *misconduct*, which affects all federally-funded research:

> Research misconduct is defined as fabrication, falsification, or plagiarism in proposing, performing, or reviewing research, or in reporting research results.

- *Fabrication* is making up data or results and recording or reporting them.
- *Falsification* is manipulating research materials, equipment, or processes, or changing or omitting data or results such that the research is not accurately represented in the research record.
- *Plagiarism* is the appropriation of another person's ideas, processes, results, or words without giving appropriate credit.
- Research misconduct does not include honest error or differences of opinion.

In order to conclude that research misconduct has occurred, it is necessary to show that:

- There be a significant departure from accepted practices of the relevant research community; and

- The misconduct be committed intentionally, or knowingly, or recklessly; and
- The allegation be proven by a preponderance of evidence.

Note the definition excludes "honest errors" and "differences of opinion" and must include "significant departure from accepted practices" in science. This definition is the outcome of a long and contentious effort by various committees convened by prestigious organizations, including the National Institutes of Health (NIH), the National Science Foundation (NSF), the National Academy of Sciences (NAS), and the Institute of Medicine. It unifies disparate conceptions of the phenomenon produced by those committees and resolves more than a decade of sometimes heated controversy.

One reason for the dispute is that in scientific inquiry there is a thin and indistinct line between a brilliant conjecture and a misleading conclusion, between an honest error and a deceptive fabrication, between an educated judgment of data quality and a falsifying act of trimming or concealment. In consequence, definitions of what constitutes scientific misconduct, mechanisms for detection and accountability, and methods of inquiry and remedy have been contested and changing. Since the early 1980s federal government officials and university administrators have shifted from viewing science as a self-regulating entity to viewing it as an organizational function subject to oversight and regulation. Rising regulatory interest in research misconduct has been accompanied by increased scholarly attention to its extent, causes, and consequences. Scientists, in turn, have exerted counterpressure, arguing that autonomy to innovate and freedom to err are essential for creative inquiry. The dispute continues and will probably endure.

## History

In the past twenty years scientific misconduct has attracted much notoriety, accompanied by congressional scrutiny and agency oversight, but historical evidence suggests that the underlying behaviors have existed for centuries. *Reflections on the Decline of Sci-*

*ence in England* (1830) by Charles Babbage offers the first systematic treatment of scientific misconduct, characterizing various types of misconduct as data "forging," "trimming," and "cooking." But the phenomenon itself precedes Babbage's treatment by centuries, and according to current standards the behavior of such as Ptolemy, Isaac Newton, Friedrich August Kekulé, Galileo Galilei, and Louis Pasteur would attract considerable scrutiny.

Gregor Mendel's plant genetics experiments are one of the more closely studied historical cases. Statistical estimates based on a subset of Mendel's experiments show that observations as close to expectations as those Mendel reports would occur in only 1 out of 30,000 replications. Experimental reconstructions of his work also find that observed data were uncomfortably close to expectations, lending support to the contention that the data were adjusted to suit expectations. Mendel's biographer suggests that he may have destroyed his own notebooks before his death in order to protect his work from future scrutiny. Of course, such judgments apply contemporary standards of conduct to a case that is more than a century old, and one may question the fairness and value of such "presentist" assessments.

At its inception in the seventeenth century, science rested upon gentlemen's trust of one another, augmented by the nascent institution of peer review. Changes in the profession, in the organization of scientific work, and in the coupling of science to other social purposes (government—particularly defense and health—and business), however, have combined to alter the vocation of science and shake the foundations of trust. Such changes figure into explanations of misconduct, discussed subsequently.

## Rates of Scientific Misconduct

Estimating the prevalence of scientific misconduct is difficult. There are two main sources of information about scientific misconduct: organizational reporting (from funding agencies, universities, and research institutes) and survey data (collected directly from researchers themselves). Each gives a distinctly different impression of the prevalence of misconduct. The Office of Research Integrity (ORI), under the aegis of the Office of Public Health and Science of the U.S. Department of Health and Human Services (DHHS), compiles allegations and cases of misconduct in DHHS-funded projects. In 2002, 99 institutions received 163 allegations of misconduct and opened 83 new cases of misconduct activities, resulting in 67 inquiries and 31 investigations. The number of cases reported to ORI had increased in each of the past four years. For the purpose of calculating a rate of scientific misconduct, if the number of cases or the number of allegations is to be placed in the numerator, it is not at all clear what number to place in the denominator. Should it be the number of research projects? The number of scientists? The number of publications? Dividing 100 or 200 instances of misconduct by any such denominator would yield a very small fraction, and we probably cannot estimate the true rate of misconduct within one or two orders of magnitude.

In contrast, questionnaire surveys that ask scientists whether they have observed any of a variety of behaviors that qualify as misconduct suggest much higher rates. Melissa Anderson and her colleagues surveyed doctoral students to ask about a variety of research practices perceived as misconduct. They found that the average doctoral student had observed (or had other direct evidence of) 2 to 5 cases of misconduct by a faculty member or other graduate student, with 53 percent of respondents stating they "probably" or "definitely" could not report suspected misconduct by faculty without anticipating retaliation, and 77 percent said their departments are "not very" or "not at all" active in preparing students to cope with ethical issues such as misconduct. There is also substantial likelihood of university-industry liaisons that might pose conflicts of interest or commitment.

There are significant problems with any method of estimating incidents of misconduct. For organizational statistics there is the problem of underreporting, confounded by differences in definitions and delays in completing investigations; as a result of the latter, the number of incidents in the current year may not be known for three years or more, when cases are closed. Self-report surveys have problems of their own, including clarity of definitions, response bias (in both directions), and multiple reporting of a single incident (if the question is, "Have you ever observed someone . . . ?"). Perhaps more chillingly, scientists systematically resist efforts to gather better data. A large-sample national survey proposed by ORI and a smaller-scale survey planned by researchers funded by ORI met with resistance from scientific societies and from the Office of Management and Budget, respectively.

## The Political Context of Scientific Misconduct

Whereas many scientists and their professional societies argue that the peer review process is adequate for detecting and correcting misconduct, others insist that science should be overseen by external agencies at the federal and university levels. The shift toward external institutional regulation of scientific misconduct began with congressional hearings leading to the 1985 Health Research Extension Act, which regulated the Public Health Service (PHS). In 1986 PHS released federal regulations for defining and handling cases of misconduct (42 CFR 50.101), followed in 1987 by the National Science Foundation's definition (45 CFR 689.1). Both definitions of misconduct originally went beyond fabrication, falsification, and plagiarism, and the PHS regulation still includes "other practices that seriously deviate from those that are commonly accepted within the scientific community for proposing, conducting, or reporting research." The NSF regulation, however, dropped the "other serious deviation from accepted practices" clause in 2002.

Although many minor infractions of misconduct might escape detection, several high-profile cases have received disproportionate media coverage, attracting attention from government officials and the general public. Some examples include the Stanley Pons and Martin Fleischmann cold fusion case of 1989; the Gallo HIV blood test patent dispute, which ended in 1995; and the Bell Laboratories superconductivity data fabrication case, concluded in 2002. But the David Baltimore case is perhaps the best known of all and has had the greatest lasting impact.

The actors in this tragedy were engaging: a Nobel laureate, a powerful chair of a House of Representatives subcommittee, two immigrant scientists, and a pair of fraudbusters without portfolio. The themes were timeless and all too human: pride, ambition, hubris, and a shattering application of power. What began in 1988 when a postdoc experienced difficulties in replicating results produced by her laboratory director led to investigations by two universities, the NIH, and the Secret Service; hearings on Capitol Hill, spiced with heated exchanges between Baltimore and Chairman John Dingell; national news coverage, often on page one; a published retraction; some tainted reputations; universities divided; and a long and detailed book documenting it all. In 1996, after years of controversy, Baltimore and Thereza Imanishi-Kari were exonerated by a federal panel that rejected a misconduct ruling by the ORI, but that quiet outcome belies the emotional and reputational damage endured by the principals and the institutional damage incurred by academic science in the U.S. The Baltimore case, more for its prominence, duration, and vitriol than for the importance of its science or the veracity of its allegations, marks a shift from internal self-regulation of science to a commitment to external regulation.

## Explanations of Scientific Misconduct

The fundamental question of why misconduct occurs in science has been extraordinarily difficult to answer. Within the science studies literature four primary explanations for scientific misconduct have been advanced: individual psychopathology, anomie, alienation, and social control. Each is addressed in turn.

*Individual Psychopathology.* Individual psychopathology, which contends that personality flaws cause scientists to cheat in their work, is the most intuitive and least likely explanation for scientific misconduct. Popular among practicing scientists, this explanation overlooks the larger social context in which science is practiced: All deviant acts are claimed to result from the personality defects or imperfect socialization of particular individuals. Ironically, the personality traits that seem to characterize successful scientists (imagination, boldness, creativity, and single-mindedness, among others) are also cited among the character flaws that lead to misconduct. According to this explanation of misconduct, the indicated remedy is simply to remove the bad apple and leave the barrel alone.

*Anomie.*   *Anomie* offers a well-established explanation for deviance. In classical social theory, a social group is anomic when rapid social change causes mechanisms of social control to become outmoded. One result of this state of moral deregulation is deviance. Technological innovations, rampant specialization, and changing organizational contexts, as well as new scientific goals and the changing relationships between performance and reward, ensure that much of modern science is characterized by just such rapid social change.

According to a more social psychological treatment of anomie, deviance results when certain members of society lack the legitimate means to achieve a socially desirable end. Under such conditions individuals may reject socially endorsed means of achieving such goals and instead use innovative means (deviance) to achieve desired ends. Applied to the scientific community, this position would argue that a high premium is placed on originality, but the means to achieve this goal (intellect, mentorship, training, assistance, resources) are differentially distributed among the population, resulting in anomic circumstances. Scientists finding themselves in such circumstances may resort to innovative but inappropriate means to produce results, publish papers, and receive recognition: deviance.

Alternatively, anomie may be an extreme form of sociological ambivalence, arising in uncertain circumstances where tensions arise about which of two or more contradictory principles or values should be honored. The result is normlessness, which results in scientific misconduct. Changes in the social organization of science, accompanied by changes in values and behavioral guidelines, and the presence of contrasting norms of scientific practice, may give rise to this form of anomie and result in scientific misconduct.

While anomic theories of misconduct represent an improvement over theories of individual psychopathology, they too are problematic. One challenge is the difficulty involved in investigating anomie empirically: Anomic conditions might characterize a scientific field or specialty, yet only a few scientists would be susceptible to their influence. What accounts for differential susceptibility? Also, Robert Merton's social psychological formulation implies that misconduct would be more common among scientists with fewer intellectual resources, but available evidence suggests quite the opposite: Scientists at prestigious institutions are well represented among those accused of misconduct.

*Alienation.*   *Alienation* refers to the separation of workers from the products of their labor, their selves, or other workers, as well as to the feelings of estrangement and malaise resulting from this situation. Although popular conceptions of science view it as a deeply engaging and nonalienating career, scientists seldom own the means of producing their work and usually have assigned to their employer the rights to their intellectual property (through technology transfer offices and university patent rights). When research becomes highly specialized, finely divided, or strongly mediated by instrumentation, connections between daily tasks and larger scientific goals may become blurred. Similarly, when products of research become esoteric or are collectively produced, so responsibility and credit are diluted; when the interests of working group members diverge because of unequal access to desired resources (such as autonomy, prestige, and job security), feelings of alienation and scientific misconduct may ensue. As early as 1918 Max Weber delineated the alienating circumstances of contemporary academic research, and recent evidence suggests that similar conditions prevail today.

As an explanation of misconduct, alienation has problems similar to those of anomic explanations. The structural condition of alienation is pervasive, but misconduct is comparatively rare. What additional factors explain susceptibility? Alienation is both a structural condition of employment and a sentiment or affective response to that condition, but it is not clear which should be given priority in the explanation, and the two conditions might not coincide.

*Social Control.* A final perspective has been developed to explain misconduct by examining the conditions under which social controls are applied to scientific work. Rather than examining the act of misconduct, this perspective focuses on public attention and policy concerns related to misconduct as well as on how and why certain behaviors come to be defined as misconduct in the first place. This theory posits that, under certain social conditions, increased attempts at social control of science will ensue, resulting in a greater likelihood that misconduct will be searched for and discovered. Rather than assuming scientific misconduct to be increasing, it argues that attempts to impose oversight and control are on the rise. Daniel Kevles, for example, argues that attention to misconduct in science arose at a time of increasing national concern for all forms of wrongdoing, and notes that there was no similar rise in misconduct cases or concerns in other advanced industrial nations. There are other reasons why attempts over social control of science may be increasing, resulting in increased attention to scientific misconduct. First, rising public expenditures for science, particularly "big science," accompanied by greater public visibility and higher performance expectations, make science a more public enterprise, which invites increased monitoring. Second, the ability of science to manipulate and control nature makes it a valuable resource attractive to powerful social groups, so it will be monitored and managed closely by those who may benefit from its usage. Third, by virtue of its unparalleled ability to lend legitimacy to decision making, science has become a key source of intellectual authority and prestige. Fourth, science has become a modern religion, surrounded by such a strong moral environment that scientific misconduct provokes moral outrage. With such high premiums placed on the purity of science, control and monitoring are likely to follow. Fifth and most importantly, in areas of defense, medicine, economics, legal disputes, policy, and general international prestige, science has become closely intertwined with politics.

Taken all together, such increased connections between science and other powerful social institutions would predictably yield greater attempts at scrutiny and control. The social control argument also holds that, in addition to an increased frequency in misconduct findings as a result of increased visibility and accountability, the strong connectivity between science and other professions and organizations can actually foster misconduct. Groups involved in the scientific enterprise will find it in their interest to make sure scientific products are received in a timely and cost-efficient manner. Intermingling of scientists with these groups may also have consequences for the career trajectories, roles, and principles guiding science.

Although misconduct will continue to challenge scientists and policy makers, it is important to keep in mind that most scientists conduct research with integrity and honesty. At the same time, additional investigation of the phenomenon is needed. There have been virtually no empirical studies as to the causes of misconduct. Comprehensive statistics reporting rates of allegations, findings of misconduct, and change over time are not available. Although the creation of a federal standard surely helps in identifying the problem, enforcing integrity in research remains a challenge. As research teams grow in size and the trend toward international collaboration continues, we can anticipate the need for an international definition of misconduct.

**BIBLIOGRAPHY**

Anderson, Melissa S., Karen Seashore Louis, and Jason Earle. "Disciplinary and Departmental Effects on Observations of Faculty and Graduate Student Misconduct."*Journal of Higher Education* 65.3 (1994): 331–351.

Anderson, Melissa S. "Uncovering the Covert: Research on Academic Misconduct." In *Perspectives on Scholarly Misconduct in the Sciences*, edited by John M. Braxton. Columbus: Ohio State University Press, 1999.

Blumenthal, D., Nancyann Causino, Eric Campbell, and Karen Seashore Louis. "Relationships between Academic Institutions and Industry in the Life Sciences: An Industry Survey." *New England Journal of Medicine* 334 (1996): 368–373.

Broad, William, and Nicolas Wade. *Betrayers of the Truth.* New York: Simon and Schuster, 1982.

Chubin, Daryl, and Edward J. Hackett. *Peerless Science: Peer Review and U.S. Science Policy.* Albany: State University of New York Press, 1990.

Dunn, Leslie Clarence. *A Short History of Genetics.* New York: McGraw-Hill, 1965.

Francis, Sybil. "Developing a Federal Policy on Research Misconduct." *Science and Engineering Ethics* 5.2 (1999): 261–272.

Geison, Gerald. *The Private Science of Louis Pasteur.* Princeton, NJ: Princeton University Press, 1996.

Hackett, Edward. "Science as a Vocation in the 1990's." *Journal of Higher Education* 61 (1990): 241–279.

Iltis, Hugo. *Life of Mendel* (1932). New York: Hafner, 1960.

Kevles, Daniel J. *The Baltimore Case: A Trial of Politics, Science, and Character.* New York: W. W. Norton, 1998.

Merton, Robert K. "Science, Technology and Society in 17th Century England." *Osiris* 4 (1938): 360–632.

Office of Inspector General. *Key Regulations.* Arlington, VA: National Science Foundation. http://www.oig.nsf.gov/misconscieng.htm. This page links to the pre-2002 and current NSF definitions of misconduct.

Office of Research Integrity. *Policies/Regs/Statutes.* Rockville, MD: U.S. Department of Health and Human Services, Office of Research Integrity, 2003. http://ori.dhhs.gov/html/misconduct/regulation_subpart_a.asp. This is the PHS definition.

Office of Science and Technology Policy. *Federal Policy on Research Misconduct.* Washington, DC: Executive Office of the President, 2000. http://www.ostp.gov/html/001207_3.html.

Piegorsch, Walter W. "Fisher's Contributions to Genetics and Heredity, with Special Empha-

sis on the Gregor Mendel Controversy." *Biometrics* 46 (1990): 915–924.

Shapin, Steven. *A Social History of Truth: Civility and Science in Seventeenth-Century England.* Chicago: University of Chicago Press, 1994.

Weber, Max. "Science as a Vocation" (1918). In *From Max Weber: Essays in Sociology* (1946), edited by Hans Gerth and C. Wright Mills. New York: Oxford University Press, 1958.

Zuckerman, Harriet. "The Sociology of Science." In *Handbook of Sociology*, edited by Neil Smelser, 511–574. Beverly Hills, CA: Sage Publications, 1988.

*Edward J. Hackett, David Conz, and John Parker*

# MUSEUMS.

This article seeks to answer, first, how historical transformations have changed the ways science is being presented in museums, and whose knowledge enters the institutions. Second, it examines how contemporary scientific knowledge is transformed into objects, which can be exhibited publicly. Finally, it looks at what kind of public science is incorporated into contemporary exhibitions and in which ways.

## Please Touch! About (Old) Museums and (New) Exhibition Spaces

With *Inside the White Cube: The Ideology of the Gallery Space*, the artist and art theoretician Brian O'Doherty stimulated a debate about exhibition strategies and the complex relationship between economic and aesthetic questions as well as the role of the gallery space. His work, dating from 1976, triggered a widespread discussion of how artists design and construct their work in relation to the museum, the displayed object, and the visitor.

Three of his main arguments for reflecting on science in museums and exhibitions are still valid today. O'Doherty describes the ideal gallery space as the archetypical image of the twentieth century. Accordingly, it has to be white, clean, and neutral. This strongly contrasts with the notions of earlier centuries, when the panels were hung very close to each other, one piece of art above, beneath, or next to the other. It was the frame of the picture that created and marked a distinct boundary between the panels. In the twenti-eth century, the single object was presented on its own, without any reference to any kind of context. The ideal observer was the perceiving eye itself. The whiteness of the walls helped to keep away any kind of inter-ference between the visitor's eye and the piece of art itself.

O'Doherty could not have written his es-says had he not observed a shift of standards. The neutrality, however, was questioned by people having to do with gallery space: artists, mediators, curators, critics. The objects had lost their ascribed virginity; the content began to transgress the frame. Furthermore, some artists unexpectedly started to leave the ascribed rooms and occupy the entire gallery space for their performances.

The notion that the meaning of a work is shaped by its contextual preconditions and that the material, as well as the institutional, frame enables the interpretation of what one sees can be applied to the analysis of the evo-lution of science in museums. That insight can be seen as a guide through this article. Namely, important changes can be observed with regard to exhibiting scientific knowl-edge since the 1970s as well, representing one of several transformations in the long history of demonstrating science in public arenas. First, a larger number of museums are concerned with scientific issues and peo-ple involved in mediating them. Since the 1970s, specialized curators for science and technology, scientific communication ex-perts, and, increasingly, scientists themselves are responsible for the creation of exhibi-tions. Second, new ways of arranging scien-tific knowledge in exhibitions arose, pointing to new preferences and understandings. Fi-nally, new and different attention was given to the visitors, in an effort to attract as di-verse and broad a public as possible.

## Shifting the Frames and Framing the Contents: A Brief History of Science in Museums

The important role natural scientific objects, mechanical machines, electronic tools, and semipublic experiments have played in the history of science stands in contrast to the contemporary assertion of newness to the ex-hibition of scientific knowledge. In fact, the recently opened museums and science cen-ters in general follow a rich tradition dedi-cated to the display of nature's and humanity's wonders and monsters. During the seventeenth and eighteenth centuries, a colorful set of traders, adventurers, botanists, and aristocrats brought home diverse and previously unknown things. They were keen to store the treasures they had found on their trips around the world in so-called cabinets of curiosities.

In the nineteenth century, university col-lections and natural history museums incor-porated such cabinets of curiosities. Yet they increasingly arranged the objects according to the emerging classification systems of the newly developing natural sciences along dis-ciplinary boundaries. Thus, new ways and means of collecting and display came into existence. The Science Museum in London (founded 1857), the Deutsches Museum in Munich (founded 1903), the French Palais de la Découverte in Paris (first exhibition: 1937), and the American Museum of Natural His-tory (founded 1869) in New York were a few renowned examples of this development.

In so-called workshop-museums, crafts-men introduced their newly invented ma-chines. Technical artists were in charge of designing exhibits. Lectures provided infor-mation on the advancement of science in a time when neither the educational nor the scientific system was fully institutionalized. Visitors were asked to participate not only as observers, but also as witnesses to newly cre-ated experiments. Like the world fairs, these museums related to the industrial develop-ments after the mid-nineteenth century. They were prominent examples of the national en-deavors to stimulate technological progress by displaying the country's expertise and ca-pabilities.

During the twentieth century it became in-creasingly difficult to collect and to exhibit the whole range of the scientific and technological inventions being made. The expansion of the scientific system complicated the organization of exhibitions according to such formerly dom-inant criteria as nature, culture, and technology.

Scientific institutions and museums increasingly drifted apart. For the post–World War II period it is possible to observe a trend toward avoiding the presentation of systematic scientific collections, and we meet new approaches in the conceptualization of exhibitions. San Francisco's Exploratorium: The Museum of Science, Art and Human Perception, founded in 1969 by the noted physicist Frank Oppenheimer, was one of the first science centers that prominently declared the educational character of interactive museum-displays.

The foundation of the Exploratorium was paralleled by the introduction of new educational concepts in schools and cultural institutions in the 1970s. The following decades brought about a skyrocketing of science centers all over Europe and North America, complementing the range of the traditional natural science and technology museums. In the belly of whale-like buildings, on board huge sailing-boat-constructions, and in postmodern housings, science and technology were brought to a public as large as possible. The observation that most museums of natural history and anthropology in the early twenty-first century seem to follow the call for interactivity confirms the statement of cultural theoretician Mieke Bal that museums of the nineteenth century were flagships of colonialism whereas museums of the twentieth century seem to follow a grand educational project.

## Scientific Objects and Displayed Things: Linking Knowledge and Materiality

Step by step, old systematic and classifying modes of representing scientific knowledge, along with their benefits, have been replaced by novel media and techniques used to provide new images and varieties of objects with different meanings. But how is it possible to understand the different ways and increased efforts to demonstrate scientific work and results publicly and as interactively as possible? Even for the traditional natural history and science museums, representations do not stand only for themselves; they are also embedded in social networks and cultural meanings.

Cultural identification and social involvement are important factors in modern exhibition strategies. As already mentioned, interactivity has been increasingly used involve visitors. Thousands of hands-on exhibits wait for children, youths, and adults to be experienced physically and actively. Visitors are asked to discover the magic of science, to experience new wonders, to rediscover the marvels of technology and research. Furthermore, guided tours, labels with a great deal of information, highly differentiating brochures, and precise instructions introduce the visitor to the complexities of a modern scientific world.

Among the most prominent issues on display in the early twenty-first century are new technological trends and problems and modern scientific research topics, as well as canonical scientific knowledge. Climate change, gene therapy, the appearance of DNA structures, and the future possibilities of media and information technologies can be found in many exhibitions in various countries. Visitors are confronted with contemporary public debates and issues on political agendas. Research practices and experimental cultures as well as research results and breakthroughs are made accessible in a museum setting, and the exhibitions serve as mediators for scientific institutions.

The displays in modern exhibitions of science differ in their origins from the objects in the older museums, which were incidentally found or purposefully acquired on long adventurous trips and assembled in a systematic composition to represent nature. In contemporary science museums systematic collections have taken a back seat. Many of the thematically arranged exhibits are constructed in museum-owned workshops, although some items are donated from contemporary sites of natural scientific knowledge or practice. Some pieces are purchased from research laboratories and cannibalized from outdated machines, tools, and instruments. Henrietta Lidchi writes that such objects hold a hybrid identity, belonging to a scientific as well as a cultural referential system. This holds true also where remnants of old natural history collections are transformed into contemporary science exhibitions and assigned to their niche as curiosities.

Several theories help explain the shifting role of objects in multiple contexts. Bruno Latour, for example, argues that a single scientific object does not stand for itself but circulates as an immutable mobile in a system of references. In his understanding, scientific objects are structured through scientists, but they also play a role by themselves in the endless procedures from which scientific knowledge evolves. A globe, for example, marks geographic regions of the earth and allows people to learn about them. It might also stand in a museum, serving historians to recognize shifts of national boundaries over time and opening a space for interpretations of worldwide traveling technologies and territorial conquests and battles. A display of an authentic mammoth tooth might be measured by zoologists, but at the same time it fills a gap in a historical chain, being classified by evolutionary biologists and touched by school children. Such chains of references, of course, vary according to time, place, and social context. They play different roles in the heterogeneous collections of contemporary scientific displays in the science museums found all over Europe and the United States. If a global enterprise opens up an information center and releases the newest research on solar technology, the enterprise is distributing not only basic knowledge, but also a brand it wants to mediate. When a research group installs an interactive exhibit one time in a foyer of a research institute and another time at an art fair, the reactions of the public will find their way back into scientific papers and expert reports according to the site of the exhibit.

Some theoretical approaches move a step further. Systems of representations produce meaning by displaying objects. Observers such as Mieke Bal and Ian Hacking stress the production of new meaning by classification and display and therefore point to the diversity of references and their differing impacts on society. Bal points out that objects, in their mere being, resemble the purest form of objectivity—making a statement as such. Hacking stands prominently for the assumption that representing must be understood as intervening. Objects of scientific knowledge do not just stand for something; they manipulate the ways the world can be seen, too. Certainly, the displays serve as signifiers, be it for a word or a material object, a scientific law, a typical experiment, or scientific achievements and contemporary technological problems in general. Yet they stand for something more, something eventually absent, something scientifically linked, which obviously cannot be shown and is perhaps not yet named. Being present in the exhibition and shown in the museum, a representation creates new understandings of the so-far unknown. Thus, the objects refer beyond the epistemic content to different places in a social system, to various contexts and institutions—not necessarily only scientific ones. Sharon Macdonald and Gordon Fyfe argue that museums are part of their social contexts and that the contents also reflect natural and social orders. Hence, exhibits still work as technologies of classification and accompany the social changes of the scientific system, but they do it in a new way.

## The Visitors Within: New Concepts and Involvements

People involved with exhibiting scientific knowledge struggle with the problem of having to increase the numbers of visitations. In addition to economic pressures, they are faced with political demands that science produce usable results. With the expansion of the educational system, questions of authorship certainly have become crucial as a larger number of scientists and professionals (curators, educators, designers, and so on) constantly criss-cross the boundaries between museums and academic institutions. In many cases, museums take on a role complementing the educational system. They work with novel strategies to involve a highly diverse public and have become important sites of formal and informal learning.

Since the 1960s, Western democracies have experienced increased political pressure to improve the integration of the public into various institutions. This tendency has reached the academic world as well. Although popularization—the public dissemination of scientific accounts, as described by

Stephen Hilgartner—implies a notion of simplification still disliked by many scientists, national governments and science policy organizations insist that scientists should communicate with the public, and scientific institutions try to react in order to maintain their legitimacy and credibility. Public understanding of science programs dates back to much earlier, obviously—the founding of a Committee on Public Understanding of Science by the American Association for the Advancement of Science in 1958 being only one example—but the efforts to go public gained new momentum. In the mid-1980s the British Royal Society issued a report called "Public Understanding of Science." John Durant and others described a gap between science and society in a 1989 article, which became later very well known, in *Nature*. Since then, government agencies, private and public foundations, and academic institutions themselves have strengthened their efforts to bridge this gap, but of course they have faced the danger of reproducing it at the same time.

Critics addressing the "public understanding of science" programs subsequently pointed to the one-sidedness of the information, which was thought to flow usually in a linear one-way street from the informed scientific specialist to the uninformed public. Brian Wynne postulated that this view had its source in a "deficit model" based on a predominance of expert knowledge. Bruce Lewenstein argued that in order to have scientifically literate citizens, it is necessary to inform them about the social forces at individual and societal levels of knowledge production and the role of process in the genesis of scientific facts. Of course, scientific knowledge mediated by museums is confronted with this criticism. Accepting a museum's difficulties in changing that structure, curators, program designers, and pedagogues have nevertheless tried to solve the problem with interactive exhibits. The declared aim was to actively involve the visitors. The public should be able to experience science as researchers do in academic institutions and even participate in scientific experiments. Through the exhibition design, laypeople should be transformed into scientists—even

if only for the restricted amount of time a museum visit would last.

The hands-on exhibits that marked the beginning of interactive exhibition in the 1970s have long since obtained little sisters and brothers. Today, "Jump-On", "Body-In", and "Try-Out" exhibits, along with even theatrical performances, are commonplace in science centers and museums. But what has happened to the scientific information originally on stage? The strength of interactivity, the focus on hands-on exhibits, and the incorporation of knowledge through new means led to the assumption that a cognitive exploration of scientific thinking and the display of scientific collections would no longer suffice.

The action-oriented forms for presenting scientific knowledge refer to and reflect the changing institutions producing that knowledge. They indicate the new roles dedicated to the public—participation is the key word. Furthermore, they point to a novel epistemic status of the material objects within these institutions. For example, in the 1980s a small European particle physics institute sought to convince the lay visitor of the need and importance of detector research by displaying a fully functional detector inside a Plexiglas container. The detector would even be explained by the scientist if a visitor met him by chance in the hallway. A guest of the same institute who visits its affiliated science center twenty or more years later can be an observer of a historical transformation: one can now step into a "Cosmic House," a huge installation, to learn about the same experiments. Dozens of small lights and flashes, emanating from laser lamps, enable visitors to "detect," and therefore to "see," the cosmic rays with their own eyes. Guests might be amazed by the accomplishments of detector research while visualizing cosmic rays, but they cannot see the detector anymore. It has been incorporated in the ceiling or the floor construction. From a historical standpoint, the visitor witnesses a change from being an outside observer of a scientific object toward experiencing the bodily involvement of someone usually uninvolved.

Looked from a theoretical perspective, Lidchi's description of ethnographic museums also holds for science exhibitions: Museological praxis decodes a text and turns something unfamiliar into something that can be easily understood. Of course, the object is simultaneously encoded—veiled again for exhibiting purposes, as in the example of the visualized background radiation by the installation Cosmic House. Yet an interactive exhibit exceeds the reach of a readable text. Science as something complex and hardly accessible can be brought closer to the public by using familiar instruments. And even if the audience's knowledge about detector research had not grown after the visit, the guests were able to link everyday associations with what science might be about. Following Laurent Thévenot, the French sociologist who talks about pragmatic regimes of engagement with the world, one can classify interactive exhibits like the one just described as embedding themselves and the public in a regime of familiarity. This is primarily a way to govern an individual's path through the world while orienting that individual toward some kind of good and trying to articulate an approach to reality. Through such pedagogical concepts, visitors enter the realm of science with a somatic-emotional attitude rather than an intellectual cognitive attitude.

## Who Speaks in the Name of Science?

Although natural history and science museums originally set out to conserve and display extensive collections, the science centers and museum foundations set up in the last thirty years of the twentieth century seem to rely less on representation. Whereas in the cabinets of curiosities the wonders of nature seemed to inscribe themselves into the text of the exhibition, a whole set of actors and institutions are now involved in designing interactive science centers where a heterogeneity of objects and people can meet.

Science has reached a point at which it can no longer be perceived as representing and communicating a single set of epistemic truths, a homogeneous group of actors, and a standardized institution. We have to think of the manifold actors involved in the production of those effects and the wide range of topics covered by scientific exhibitions. This leads us to ask—apart from the scientific knowledge—who the authorities are. Who is speaking in the name of science when nature has stopped talk-

ing? The museum is not simply a place the scientific institutions use for spreading the fruits of their knowledge and bringing them closer to the public. Museum theoreticians, designers, curators, and mediators are only the most visible of the professional experts in this game. The numerous national and international governmental agencies involved might be given as just another example. Andrew Barry suggests the concept of a "technological society," referring to a "specific set of attitudes towards the political present" and certain kinds of problems raised by technological developments. Different social areas are preoccupied with these problems, under a diverse set of obligations. In this connection one has to understand the contemporary differentiation processes that have reached the academic professions as well. Negotiations are going on behind the museum scenes among scientists, academic managers, museum curators, and others concerning what to show and in which temporal and spatial order. These negotiations might lead to the problematic positioning of old beside new exhibits and texts beside things under the key notion of reflexivity. The hybrid forms in which scientific knowledge in exhibitions is now presented may provide a way to understand that science does not hold all answers anymore, but is one of society's projects to reflect on burning questions of our times.

How then can science in museums ultimately bridge the gap and add to a better understanding of contemporary developments? Museums, of course, still play an important role as symbols of communities, in the way Benedict Anderson saw them. Yet, nationality and history seem strangely hidden in the new science centers, although there exists no European or American Science Center so far. The exhibits unite people in their experiences with contemporary evolutions in science and technology. But concerning their heterogeneity and hybridity, it seems clear that, even as symbols for communities, the exhibition sites are contested. Although the knowledge and with it the material objects transferred to the museum are to a large degree devoid of the institutional context they originate from, the museum tries to serve as a neutral space, where the knowledge and objects can be reconnected to a changing world.

The objects and the educational programs around them do not stand for the scientific and technological system per se but as its little, decontextualized helpers. They allow the public a recontextualization in a familiar manner. Obviously, it is not possible that people understand science as cultural practice and institutional setting completely and fully, but under the paradigm of hands-on exhibits they are meant to experience social closeness to an increasingly differentiated scientific system. In a science center the visitors are re-educated with regard to the standards they know from everyday life. They are asked to touch, to experiment, to taste, to smell, and to feel. Through interactivity they are, furthermore, involved in a system of evaluations. Going home they are able to apply and reproduce their capabilities to appreciate and to accommodate to new technologies, and possibly to evaluate and to challenge the changed role of science in an evolving society as well.

*See also* **Public Understanding of Science; Public Understanding of Technology**

## BIBLIOGRAPHY

Anderson, Benedict. *Imagined Communities. Reflections on the Origin and Spread of Nationalism.* New York: Verso, 1991.

Bal, Mieke. *Double Exposures: The Subject of Cultural Analysis.* London: Routledge, 1996. This is an analysis of the relations between things and stories.

Barry, Andrew. *Political Machines: Governing a Technological Society.* New York: Athlone Press, 2001. This theoretical work on science and technology in society includes aspects of the social role of interactivity in museums.

Bennett, Tony. *The Birth of the Museum.* London: Routledge, 1995. This theory and history of the modern museum depicts its role in reforming social routines and individual behaviors.

Durant, John, ed. *Museums and the Public Understanding of Science.* London: Science Museum in association with the Committee on the Public Understanding of Science, 1998. This collection of essays by various museum professionals and theoreticians deals with several problems related to exhibition making and contemporary exhibition techniques and strategies, such as the role exhibits play for education and their effects on the visitors.

Durant, John, Geoffrey A. Evans, and Geoffrey P. Thomas. "The Public Understanding of Science." *Nature* 340.6 (1989): 11–14. This crucial article stimulated the debate on the gap between science and the public and introduced the "public understanding of science" discussion.

Farmelo, Graham, and Janet Carding, eds. *Here and Now: Contemporary Science and Technology in Museums and Science Centres.* London: Science Museum, 1997.

Hacking, Ian. *Representing and Intervening: Introductory Topics in the Philosophy of Natural Science.* New York: Cambridge University Press, 1995.

Haraway, Donna. *Primate Visions: Gender, Race, and Nature in the World of Modern Science.* New York: Routledge, 1989.

Hilgartner, Stephen. "The Dominant View of Popularization: Conceptual Problems, Political Uses." *Social Studies of Science* 20.3 (1990): 519–539. The author criticizes the construction of a demarcation line between genuine and popularized knowledge, terms often used by scientists in public debates as political arguments.

Irwin, Alan, and Brian Wynne. *Misunderstanding Science? The Public Reconstruction of Science and Technology.* Cambridge, UK: Cambridge University Press, 1996. The book contributes to the debate on "public understanding of science" from the perspective that not only is the public in need of scientific knowledge, but also needed is a closer look at what the public already knows and what it really says and a more critical analysis of the structure of scientific institutions themselves.

Latour, Bruno. *Pandora's Hope. Essays on the Reality of Science Studies.* Cambridge, MA: Harvard University Press, 1999. Case studies by one of the representatives of the actor-network theory elucidate how things, people, and activities are creating scientific knowledge and how it is embedded in a social world.

Lewenstein, Bruce V. "What Kind of 'Public Understanding of Science' Programs Best Serve a Democracy?" In *Science Studies: Probing the Dynamics of Scientific Knowledge,* edited by Sabine Maasen and Matthias Winterhager. Munich: Transcript Verlag, 2001: 235–255.

Lidchi, Henrietta. "The Poetics and the Politics of Exhibiting Other Cultures." In *Representation: Cultural Representations and Signifying Practices,* edited by Stuart Hall. London: Sage Publications, 1999: 151–168, 184–208.

Macdonald, Sharon. *Behind the Scenes at the Science Museum*. New York: Oxford, 2002. This is a rich and detailed ethnography of the making of an exhibition in a science museum.

Macdonald, Sharon, and Gordon Fyfe, eds. *Theorizing Museums: Representing Identity and Diversity in a Changing World*. London: Blackwell, 1996. This volume comprises anthropological, sociological, and cultural studies on how spaces and times, differences and identities, and, finally, classifications and practices influence contemporary museum sites.

O'Doherty, Brian. *Inside the White Cube: The Ideology of the Gallery Space*, expanded ed. Berkeley: University of California Press, 2000. With this groundbreaking work O'Doherty launched a debate in the art world on how the content of an exhibition and the visitors' experience is structured by the institutional frame surrounding it.

Thévenot, Laurent. "Pragmatic Regimes Governing the Engagement with the World." In *The Practice Turn in Contemporary Theory*, edited by Theodore Schatzki, Karin Knorr Cetina, and Eike von Savigny. London: Routledge, and New York, 2001: 56–73. This moral and political sociology discusses how people evaluate the realities they live in and how they are able to act in accordance with their judgments.

Wynne, Brian. "Knowledges in Context." *Science, Technology and Human Values* 16.1 (1991): 111–121.

*Priska Gisler*

# N

## NATURE, CONCEPTS OF.

The phrases "nature," "natural," and "the natural world" are so commonly used today that many of us remain unaware of their multiple meanings and their historical complexity as concepts. Similar terms in other languages and cultures, such as the Japanese phrase *namo no shizen*, translated as "nature in the raw" or the Chinese term for nature, *ziran*, suggesting "spontaneity" (Bruun and Kalland, pp. 246 and 175) further increase the difficulty of fully comprehending these concepts. The English word *nature* comes from the Latin word *natura*, which literally means "birth" but far more often means the inherent quality or character of a person or thing or of the world as a whole. The oldest (thirteenth-century) use of the English word refers likewise to the inherent or nascent power of a thing. During the fourteenth century the term *nature* had come to refer also to the creative, regulative process in the world, personified as a goddess *Natura*. By the seventeenth century, however, when the world came to be understood by European philosophers as a matter of mechanism, the concept of "nature" became deactivated and depersonified—turning from a living process or being that activates and informs things and relationships in the world into the material world itself.

The ways the term *nature* is used to refer to a process or agent of change in the world, or alternatively, to the world as a whole, can have profound political, economic, and moral ramifications. For example, Lynn White, a historian of technology writing in 1967, traced what he called "the historic roots of our ecologic crisis" to Western people's assuming that nature was not an agent in its own right but was made by God to serve the interests of human beings. This assumption was derived from the account of creation offered in the Book of Genesis, which informed both European thought and practices during the Middle Ages. The triumph of Christianity over pagan religions overturned the animist-based understanding of nature that had prevailed in Europe and replaced it with two key ideas: first, that nature was not imbued with spirit; and second, that nature was subject to God's command that man subdue the earth and maintain dominion over it.

Nature is never referred to as an independent, causal entity in Genesis. Nature as a whole is addressed only in references to "the earth," with its waters, its air, its land, its dust and clay and living beings. None of these aspects of the earth, however, are inherently "natural" in the sense of being self-generated. All of them are explicitly created by God. Spirit resides at first exclusively in a transcendent Creator, who then breathes it into the first man. In cultures with animistic beliefs, people respect and value nature or natural objects, beings, or places precisely because they are inhabited by or invested with spirit(s). Lacking any spirit of its own, though, nature in the Judeo-Christian view deserves respect and value only insofar as it either reflects the nature of its creator or meets the needs of mankind.

In her book *The Death of Nature* (1980) the environmental historian Carolyn Merchant claims that despite challenges from Christian ideas, the previous pagan and Greek concepts of an organic Nature were integrated tightly enough into the social, economic, and political fabric of medieval culture to withstand the assault from Christianity for many more centuries. Up to the end of the Renaissance, Merchant suggests, the concepts of nature prevailing in Europe were still mostly derived from a combination of Greek and local pagan ideas, such as the belief that many significant features of nature were inhabited or protected by major and minor divinities. Among these spirits, Merchant claims, Mother Earth (who may have derived from the ancient Greek goddess Demeter) came to have a particular significance, checking human tendencies to exploit and disrupt the physical environment by representing nature as the inviolable body of a sacred mother-figure.

Another concept, derived from Plato's *Timaeus* but developed more fully by the Neoplatonists, distinguished between ordinary nature and a "higher nature." True nature, in this view, is a matter of pure, abstract, and mathematical form and can be apprehended only by reason. The world that we experience sensually is merely an imperfect reflection of that true nature. Belief in the ultimate significance of such a higher, mathematically formal nature focused the attention of late Renaissance natural philosophers away from the irregularities of ordinary experience and onto the carefully controlled perception of rational, mathematical, and mechanical patterns that characterize nature when it is reduced to its most abstract qualities such as matter and motion.

In the late Renaissance this belief in the mathematical character of a higher "nature" could easily be reconciled with the Judeo-Christian account of nature found in Genesis, particularly in the first chapter. There, God performs a series of grand, abstract acts of division and multiplication, with nature as the result. Creation begins with a sequence of geometrical bifurcations as God separates light from darkness, day from night, sky from earth, sea from land. This is followed by an orderly creation of life forms. Through successive acts of separation, naming, and replication, God invests a formless void with form, place, meaning, function, and temporality. God then goes on to create man—both male and female, "in our image, after our likeness" (Genesis 1:26). Male and female

"man" are symmetrical like the rest of creation, but are distinguished as the only beings in that world said to have been made in the divine image. God then distinguishes them further by giving them "dominion" over the earth and all the living things that move. The world is thus established as an ordered hierarchy, with mankind dominating nature from a position intermediate to the earth and the heavens, the material and the spiritual realms.

Merchant argues that the old notion of the earth as a sacrosanct female divinity weakened under the combined pressure of Neoplatonic and Judeo-Christian ideas about nature and of new technologies centered on farming, mining, and building. People engaged in the activities of delving into and reshaping the physical earth steadily overcame pagan strictures against violating the female earth's nurturing body. The newly synthesized late-Renaissance concept of nature as a strictly material world opened the modern path to human domination: making close observations and experiments to systematically uncover the hidden regularities and relationships informing natural phenomena, and using technology, guided by this knowledge, to physically reshape and control the earth. Following that path has brought us, she claims, to our current state of environmental crisis.

At about the same time that such a mechanistic approach to nature was developing, however, the Western colonial enterprise was leading to a renewed hope of returning to the state of grace and harmony with nature that had marked mankind's experience of Eden before the fall into sin and exile. This notion of nature was informed partly by the first chapter of Genesis but primarily by the second and third chapters. In this intimate account of Creation, the Lord makes man from earth with His own hands and blows the spirit of life into his nostrils. After choosing a well-watered place and planting a garden filled with trees and fruit, the Lord tenderly settles His human creation into this Eden. The relationship between the first man and nature, as represented by the garden, is to be reciprocal—the garden will feed and delight the man, and the man will "cultivate and care for" the garden (Genesis 2: 15).

The first man, Adam, begins to live with the animals, the birds, and the first woman, Eve, in a state of innocence, within an abundant garden that provides food, water, shade, and pleasure. When the human beings transgress God's command, however, both their innocence and state of harmony with nature and that of all their descendants are ended. God curses the earth from which Adam was made, and condemns both Adam and Eve to a condition of struggle and at least partial enmity with respect to the rest of creation. They are cast out of Eden and into a dry and barren wilderness. Adam's care and cultivation of the garden, which has provided them with such easy abundance, is exchanged for a toilsome, sweaty battle against thorns, thistles, and dust to compel the earth to bear enough plants to sustain human life until death returns them, too, to dust, and this is to be the relationship of human beings to nature ever after.

Steeped in knowledge of the Bible, and armed with the New Testament promise of eventual redemption from Adam's original sin and subsequent fate, European colonists moved into a strangely familiar world that appeared to be underpeopled—in part as a result of the European diseases that devastated the indigenous peoples. Because of its beauty and signs of abundance, its relative freedoms from the close restrictions of European social, political, and economic structures and habits, and its parklike, open woods that were nevertheless in proximity to an apparently trackless wilderness, the explorers and colonists in New England immediately perceived their environment as a newly accessible Garden of Eden. The New World promised to fulfill the prophecy of Isaiah 51:3 that God would comfort the people of Zion and "make her deserts like Eden, her wilderness like the garden of the Lord."

As the historian William Cronon points out in his seminal work *Changes in the Land,* all too many European colonists, often Spring arrivals with only the limited experience of the lush "strawberry time" to inform them, assumed that the New World represented the opportunity to bask in the leisurely enjoyment of nature's paradisiacal abundance while barely lifting a hand to pick up the fruit, nuts, and game with which they expected to be perpetually surrounded. Even when New England's cold, disease, and regular periods of seasonal scarcity eliminated individuals who eschewed hard work, and repeatedly decimated the ill-prepared colonies, the lure of Eden relocated to the New World persisted in the American imagination and helped carry the colonists continually westward in search of an easier, more abundant life than the one they were perpetually attempting to leave behind them.

General faith in the possibilities of enjoying the pleasures of a natural Eden in this lifetime tended to clash in the American experience with a more Puritan faith that hard work and abstinence from natural pleasures marked those predestined for true Paradise. Despite such conflicts, both approaches contributed to the rapid exploration and exploitation of the American landscape, converting much of it into a domesticated, Western pastoral vision of nature as a carefully cultivated garden. The pastoral vision was reinforced in the late seventeenth and eighteenth centuries, both in America and in England, by an Enlightenment appreciation of God as a beneficent, intelligent "watchmaker" who had designed everything in nature specifically to provide for the needs of mankind. Human beings, by following the natural laws of physics and society that God had built into an orderly universe, could apply their own divinely instilled intelligence to the task of successfully remaking the world into a recreation of Eden as an ideal state of harmony, abundance, and providential order.

Such a tidy, domesticated vision of ideal relations between human beings and a human-centered natural world did not satisfy the imagination of everyone, however. For some, the true value of nature lay not in its capacity to accommodate mankind's needs but in its capacity to inspire men and women to contemplate power, beauty, and therefore the Divine, on a scale that was clearly beyond human power to imitate or control. This nineteenth-century Romantic vision of nature in terms of the sublime resonated, not with the second and third chapters of Genesis but with the Book of Job, with its majestic passages in chapters 38 through 41, where God thunders out such questions to Job as

> Have you ever in your lifetime
> Commanded the morning
> And shown the dawn its place
> for taking hold of the ends of the earth?

and

> Have you entered into the sources of
> the sea,
> Or walked about in the depths of the
> abyss?

The implications of the Transcendental Romantics' appreciation of the sublime challenged the continued domestication of the American West. Advocates of this Romantic vision came to believe that nature was most valuable where it was least habitable, and they began to argue for the importance of preserving wilderness as a place apart from human society.

Their arguments for preserving nature by restricting human interference with it were bolstered by observations made by both scientists and humanists about the state of the natural and human environment by the middle of the twentieth century. When twentieth-century technologies were combined with a rapid acceleration in population growth and Western domination of most of the world's natural resources, increasingly powerful Western efforts to control and shape the natural world into a productive human paradise generated unintended, large-scale side effects, such as soil erosion and decline in fertility, air and water pollution, excessive deforestation, environmental toxicity, and the unexpected decline and extinction of increasing numbers of benign species, together with the increasing resistance to extermination of plant, animal, fungal, and bacterial species identified as pests or diseases.

If one driving force behind Western cultural efforts either to subdue and exploit the earth outright or to convert the wilderness into a man-made Eden was indeed Western religious thought, environmental problems stemming from the overzealous domination and exploitation of nature became inescapable when Western religious attitudes were merged with fully developed capitalism and the power of industrial technologies. Numerous scholars since the late 1960s have therefore concluded that the solution to current problems lies partly in preserving pristine parts of the natural world from human intervention altogether, and partly in learning how to interact with nature in less destructive ways by following the examples set by cultures with religious traditions that are not so human-centered or mechanistic with respect to nature and with economic systems that were different from modern Western industrial capitalism. Prehistoric cave dwellers, premodern pagan Europeans, pre-Columbian as well as modern Native Americans, Zen Buddhists, Hindus, and contemporary Third World animists, among others, have all been considered at one time or another in the last 35 years to offer positive alternatives. Each of these peoples, it has been hoped, must have a more respectful attitude toward nature than Westerners do and so could reasonably be expected to act in ways that are less destructive of their environment.

Although preliminary works have frequently proposed ways in which people of these various cultural traditions can serve as models of harmonious interactions with the environment, based on their fundamental attitudes of human fear, respect, oneness, love, or aesthetic appreciation for nature, natural spirits, or the environment in which they live, more detailed studies of the concepts of and practices concerning nature in each of these cultures have shown that environmental practices are not always in keeping with religious or philosophical ideals and that human populations have consistently been able to justify altering and make use of their environments in ways that serve short-term human interests.

For example: Javanese animist traditions that identify certain parts of the forest and particular trees as sacred because of the spirits that inhabit them have been shown nevertheless to provide ways to appease the spirits or transfer the responsibility for violating spiritual sanctions. For centuries this has allowed valued trees and even whole forests to be cut down and used or sold, even in the absence of Western economic pressures. As forests have shrunk substantially and steadily over the last 250 years, the spiritual strictures protecting them have been increasingly modified and weakened to accommodate the cutting.

Oral histories, rituals and folk-tales, along with artworks, suggest that both modern and pre-Columbian Native Americans, though with much cultural variation, generally see nature as an organic community. Many groups respect elements of the landscape itself, as well as various plants and animals, as entities inhabited with spirit and significance at least as much as humans are. However, Native Americans also made regular use of fire to clear forests of undergrowth and unwanted trees, to improve hunting conditions, and sometimes to clear sites for growing crops. Even though the fires improved the habitat and so increased numbers for some species of plants and animals, they also made it impossible for other indigenous species to survive.

In another example, Japanese culture is rightly acclaimed for the high value it places on a sensitive and articulate appreciation of natural beauty, and the Japanese, too, have beliefs that natural places are imbued with spirits. They are also among the most technically proficient people in the world when it comes to controlling, shaping, and exploiting nature, both agriculturally and aesthetically. As Arne Kalland points out in a recent article on "Culture in Japanese Nature," the Japanese reserve their highest praise for those aspects of nature that are most tightly framed or controlled by culture, such as the gardens and bonsai trees so carefully designed to evoke an idea of mountain wilderness or the cherry blossom and full moons that are admired *en masse* in prescribed ways from traditional viewpoints at highly ritualized viewing parties. Meanwhile, the modern Japanese fishing industry's impact on oceanic fisheries has been overwhelming, and the Japanese government has joined with Scandinavian and Northern Native American groups to pressure world regulatory bodies to allow large-scale whaling operations to resume, despite other severe environmental pressures on the existing cetacean populations.

Even attempts to look to the Neolithic for a more pristine relationship between humans and nature have found little encouragement in the findings of recent scholarship. As the population biologist Jared Diamond argues, the prehistoric humans who created such sensually compelling images of animals may also have contributed in a major way to the relatively sudden extinction of the megafauna in prehistoric Europe and North America through the practice of highly effective, yet wasteful hunting techniques.

Despite contemporary Western concepts of pristine nature as something apart from human beings—something that can be admired and protected from the outside—the existence of such nature, at least on this planet, is itself a cultural fiction.This fiction is rooted in the Western notions of ideal forms as well as in the image of Eden as it was before God installed the first man there, or as it has been since the Fall—inaccessible, yet somehow waiting for our return. For as long as the human genus has existed, however, human beings have been part of the fabric of nature, and therefore so has human culture and technology. However much we may construct specific concepts of nature, we also, in a sense, "practice" nature into being along with ourselves through all our actions and effective beliefs. What we need, especially in Western society, to be most aware of, is that our relationship with nature is reciprocal and encompassing. We cannot remove ourselves from nature, nor can nature remove itself, or be removed, from us.

## BIBLIOGRAPHY

Barbour, Ian G. *Western Man and Environmental Ethics*. Reading, MA: Addison-Wesley, 1973. A useful and forward-thinking collection of essays on ethical dimensions of ecology, culture, and environmental thinking, including a reprint of Lynn White's seminal piece,

"The Historical Roots of Our Ecologic Crisis," with responses from Lewis W. Moncrief and René Dubos plus a follow-up by White.

Boas, George. "Nature." In Philip P. Wiener, ed. *Dictionary of the History of Ideas*, Vol III, 346–351. New York: Scribners, 1973. A thoughtful brief essay on Western concepts of nature from the Greeks through the Romantics.

Bruun, Ole, and Arne Kalland, eds. *Asian Perspectives of Nature: A Critical Approach*. Copenhagen: Nordic Institute of Asian Studies, 1995. A recent collection of scholarly studies of nature in a variety of Asian cultures and places, most of which offer specific evidence to challenge the validity of both a general "Asian" perspective on nature and the idea that Asian cultures are universally more effective at protecting the environment than Western cultures are. Of particular significance for the present essay are the articles by Peter Boomgaard, on "Sacred Trees and Haunted Forests in Indonesia, Particularly Java, Nineteenth and Twentieth Centuries," pp. 48–62; Arne Kalland, "Culture in Japanese Nature," pp. 243–257; and Ole Bruun and Arne Kalland, "Images of Nature: An Introduction to the Study of Man-Environment Relations in Asia," pp. 124.

Cronon, William. *Changes in the Land: Indians, Colonists, and the Ecology of New England*. New York: Hill and Wang, 1983. A foundational monograph in ecological history that has been particularly influential in comparing the efficacy of Native Americans favorably with that of the European colonists in manipulating the natural environment of New England. The work drew attention to the previously unconsidered ecological consequences of the European takeover of North America.

Cronon, William, ed. *Uncommon Ground: Rethinking the Human Place in Nature*. New York: W. W. Norton, 1996. A collection of contemporary, interlinked essays on the cultural construction of nature that proposes new ways of thinking about contemporary and past human connections to nature. The pieces by Candace Slater, Kenneth Olwig, and Carolyn Merchant on the theme of Eden as a guiding moral construction, along with Cronon's own introduction and the pieces by Michael Barbour, Jeffrey Ellis, and James Proctor on contested terrains, are particularly thought-provoking.

Diamond, Jared. *The Third Chimpanzee: The Evolution and Future of the Human Animal*. New York: Perennial, 1992. A highly readable perspective on human physical and cultural evolution, stressing human interrelationships to the environment.

Finch, Robert, and John Elder, eds. *The Norton Book of Nature Writing*. New York: W. W. Norton, 1990. An accessible and useful collection of excerpts from literary and scientific primary source writings on nature.

Herron, John P., and Andrew G. Kirk, eds. *Human/Nature: Biology, Culture, and Environmental History*. Albuquerque: University of New Mexico Press, 1999. A provocative collection of contemporary essays based on a conference on environmental thinking about human connections with nature.

Hoad, T. F., ed. *The Concise Oxford Dictionary of English Etymology*. Oxford: Oxford University Press, 1986.

Leiss, William. *The Domination of Nature*. New York: George Braziller, 1972. A foundational study in the history of Western thought about nature, particularly strong in the premodern periods.

Lewis, Charlton T., and Charles Short, eds. *A Latin Dictionary*. Oxford: Clarendon Press, 1879. Searchable at http://www.perseus.tufts.edu/cgi-bin/resolveform?lang=Latin.

Merchant, Carolyn. *The Death of Nature: Women, Ecology and the Scientific Revolution*, reprint ed. San Francisco: HarperSanFrancisco, 1990. A important, work on the cultural history of environmental thought and practice in the early modern period, drawing on a range of literary and philosophical sources to support its claims about the demise of an organic, feminized view of nature under the pressure of both technological change and newly powerful concepts of a mechanized universe.

Nash, Roderick. *Wilderness and the American Mind*, revised ed. New Haven, CT: Yale University Press, 1973. A classic account of the history of the concept of wilderness in American literature, philosophy, and politics.

Thomas, Keith. *Man and the Natural World: A History of the Modern Sensibility*. New York: Pantheon Books, 1983. A penetrating history of English thought, attitudes, and practice with respect to nature, drawing on a wide range of cultural, political, and literary sources.

Torrance, Robert M. *Encompassing Nature: A Source Book [on] Nature and Culture from Ancient Times to the Modern World.* Washington, DC: Counterpoint Press, 1998. An ambitious, rich and useful attempt at a comprehensive, cross-cultural sourcebook of excerpts from writings on nature across the last two and one-half millennia.

White, Lynn. "The Historic Roots of Our Ecologic Crisis" (1967). In *Western Man and Environmental Ethics,* edited by Ian G. Barbour. Reading, MA: Addison-Wesley, 1973. A ground-breaking, frequently reprinted essay that has continued to provoke thought and controversy since it was first presented as a lecture by White in 1966.

*Linda M. Strauss*

# NETWORKS AND LARGE-SCALE TECHNOLOGICAL SYSTEMS.

Large-scale systems and networks, such as electricity production and distribution, railways, and telephone systems, exhibit three core properties. First, they are *heterogeneous;* that is, they are composed of elements that are different in kind. This is important because, although it is obvious that a technical system includes technological components (computers, control centers, power stations), such components operate properly only in a context that also includes people, knowledge, social institutions, and naturally occurring processes. To understand large-scale systems, all of these have to be explored in addition to the technologies themselves. Second, they exhibit *emergent properties;* that is, they have effects that depend on the patterned interaction between their different components. For instance, an electricity supply system only works if its different parts (the inputs of those who work in it, its hardware, its software, and a host of arrangements legal, organizational, and economic) are carefully orchestrated. It is obvious, but needs to be stated, that the individual components of large-scale systems do not have these properties outside their context within the system. Third, it follows that the elements that make up the system are them-

selves *shaped,* and indeed arguably *created,* by the system network of relations within which they are located. People can work properly within a technical system only if they are trained, paid, motivated, and provided with an appropriate environment. An example such as Three Mile Island reveals that the operators of a nuclear power station are effective components of the system only if they have the necessary information and control systems in place. The same, however, is also the case for technical components of a system, which need (as in the case of a nuclear power plant) to be replaced, moderated, maintained, cooled, and operated if they are to work properly. In short, the attributes of individual elements in a system depend on how they interact with the other elements.

The following account starts by exploring these themes for the classic system technology of electricity generation. It then considers how system and network thinking can be applied to large-scale networks that are less obviously technological or "systemic." Next it explores the control of large-scale systems and considers questions of risk and hazard in relation to the configuration of systems. Finally, it considers the possibility that large-scale systems and networks work in less "systemic" ways and, indeed, depend on their success because they are imprecise, fluid, or only partially coherent.

## Networks of Power

There have been large-scale systems in existence since the invention of agriculture and trading. Certainly some kind of "system" is necessary for the kind of political, military, technical, and economic organization implied by the ancient civilizations that arose independently in the Middle East, China, and in Central and Southern America. It is also difficult to understand the European imperialist expansion from the fifteenth to the nineteenth century without reflecting on the interconnected organization of administration, military power, economics, technology, printing, and communications, as well as the diseases dispersed by that imperialism.

However, the exemplary cases of large technical systems developed in the nineteenth century, with the growth of industrialization, urbanization, and populations. These include the railroad; telegraph and telephone systems; and such public utilities as sewers, gas, and electricity.

The term *networks of power* was popularized by the historian Thomas P. Hughes in his magisterial study of the growth of electricity generation and distribution in the Western industrialized world between 1880 and 1930. In the 1870s, there was no public electricity supply. By the 1930s (and decades earlier in most European and U.S. cities), there were large-scale electricity utilities supplying millions of domestic and commercial customers. How was this extraordinary achievement possible? Hughes answers this question by attending to the activities of a number of important innovators, including inventor and entrepreneur Thomas Edison, who was convinced that an appropriate electricity system could compete with the dominant gas mode of lighting. He started his investigation in 1876 by seeking to devise an incandescent light bulb filament of the appropriate resistance. This involved scientific and technical experiments by a team of assistants. At the same time, however, this technical work was closely linked to economics. Indeed, in his notebooks, the technical and the economic are combined together. The price of copper is juxtaposed with losses due to resistance, the thickness of wire, and calculations about the distances for transmitting electricity economically at different voltages and currents. In 1878, he achieved the principles of a workable system. Electricity would be distributed underground from a central power station at 100 V. Incandescent lights would be wired in parallel. Consumption would be metered, and the system would be able to light houses more cheaply than gas. More capital was raised, and there followed a period of intense negotiation and lobbying (the gas industry was very powerful and resisted Edison's plans). But in 1882 the Pearl Street generating plant on Manhattan was completed, and the first "network of power" came online.

Hughes also describes the work of other innovators, such as Samuel Insull, who became president of Chicago Edison in 1892 at a time when this company was one of twenty small power companies in that city. Under Insull's leadership over a thirty-year period, the company turned itself into the monopoly supplier of electricity in the city using a combination of economic, technical, and political techniques. First, rival companies were bought up. Second, they were integrated and supplied from large power stations using high-voltage AC current, which could be efficiently transmitted over longer distances. The voltage was stepped down at substations and in many cases converted to DC for local supply. Third, as the network grew it became possible to manage the load across the system in order to maximize efficiency and minimize the generation capacity needed to meet that load. Demand was diversified. Transport, factories, and domestic consumption peaked at different times of day, so it was most efficient to supply all of these within a single system instead of depending on one or two. At the same time, the job of the load dispatcher was created. The load dispatcher was the person in control of the system who collected and collated real-time information about the changing demands in different parts of the network and routed power accordingly. The result was a high degree of efficiency as more capacity of the power stations was used for more of the time.

Hughes's work is crucial to the understanding of large technical systems, and it illustrates the three system features mentioned at the beginning of this article: their heterogeneity, emergence, and the way in which they shape their components. Hughes also argues that important innovators such as Edison were "strong holistic conceptualizers." They thought in system terms, refused to confine their thinking to the standard pigeonholes, and imagined and created "seamless webs." Edison, Insull, and others mixed up the technical, the scientific, the financial, the political, and the

organizational in a process that has sometimes been called *heterogeneous engineering*.
The result was the creation of networks or
systems with emergent properties that
shaped their own components—for instance,
as with the load dispatcher who depends on
collated information to undertake his essential system task.

## Large-Scale Systems

Large technical systems and networks of
power have became increasingly important
over the last 200 years, and at the beginning
of the twenty-first century it is commonplace
that much of the world is caught up, enmeshed, and indeed made up of, large technical systems and networks. The container
revolution; the track-and-trace technologies
used by manufacturers, distributors, and re-
tailers; air traffic control systems; the complex choreography of technologies used in,
for instance, patient records or blood donations and transfusions; military technologies
and systems of all kinds; satellite navigation
systems; Internet protocols; connectivities
and routers; the mass media; and many parts
of experimental natural science; not to mention the design and operation of complex installations, such as aircraft, power stations,
ships, and industrial plants of all kinds—all
of these are examples of networks or large
technical systems.

If heterogeneous connectivity is commonplace, so too is the use of metaphors of connectivity in the rich industrialized world.
Terms such as *network, Internet, command,
communicate, control, joined up thinking, human
engineering* point to the importance of connectivity and heterogeneity in contemporary
ways of thinking about society and technology. Unsurprisingly, therefore, network and
system metaphors have been used to understand large-scale organizations whose raison
d'être has directly to do with technology.

Consider, for instance, the work of nineteenth-century medical innovator Louis Pasteur. The sociologist of science Bruno Latour
explores the way in which a network of relations between farms in France and Pasteur's
laboratory in Paris was engineered. French

farmers were losing cattle to anthrax, and
most veterinarians interpreted this in terms
of miasmatic theories of disease. Pasteur was
convinced that the disease was caused by a
microbe. But how to show this? How to protect cattle against the condition? And how to
convince farmers that they should follow his
advice?

The answer is that it took a heterogeneous
organizational effort. Pasteur collected soil
samples from afflicted farms, and took them
to his laboratory. There, in artificial circumstances, and removed from competition, the
anthrax bacillus was isolated, cultured, and
shown to infect cattle. But how to weaken it
in order to make a vaccine? The answer was
to subject it to chemical and physical stress.
There was much trial and error, but in the end
a weakened strain was isolated that conferred protection against the disease without
causing sickness. But how to convince French
farmers? The answer was set up a large-scale
experiment on a farm at Pouilly-le-Fort.
There the conditions that had been created in
the laboratory—vaccination, vaccinated animals, and unvaccinated animals—were recreated, and the farmers and veterinarians were
enrolled into the system.

This is another example of "heterogeneous
engineering." By the time the network was in
place, farmers and farming practices, veterinarians, bacteria, farm animals, the reputation of miasmatic theory, official statistics on
cattle death, the importance of Pasteur's laboratory, and Pasteur's own scientific reputation had *all* been assembled together and
modified. As with networks of electrical
power, the economic, the political, the scientific, the technical, and the natural worlds
are combined and reworked in an emergent
and creative system. The root processes—
heterogeneity, emergence, and the shaping or
creating of the components of the system—
are the same. Indeed, Latour presses the logic
one stage further by arguing that Pasteur
himself can be understood as a product of
this system. He too is shaped by his circumstances, circumstances that include his training, but also, for instance, the presence of
statistics that revealed anthrax to be a

scourge for French farming. Pasteur is better understood, then, not as an individual genius but as an effect of the system that produces his laboratory and his reputation.

## System Configurations

In the two cases discussed above, network and system patterns work to create centers. Pasteur's laboratory becomes what the sociologist of technology Michel Callon has called an "obligatory point of passage." Farmers who wished to protect their cattle needed to buy his vaccines and turn their farms into small-scale versions of his laboratory by following very specific protocols. In short, the laboratory was an obligatory point of passage. An analogous argument can be made for networks of power: in practice, if customers wished to use electric light in Manhattan, then they needed to detour via Edison's company.

But the patterns and configurations of systems are important in other ways. For instance, Charles Perrow argues that complex and tightly coupled technical systems are very vulnerable to breakdown. *Complex systems* are those with many branches and feedback loops, so any failure is likely to ramify in unexpected ways and produce widespread disruption. By contrast, breakdowns in linear systems have more predictable and limited consequences. *Tightly coupled* systems are those in which events are transmitted through the system too fast to be controlled. In such systems, breakdowns ramify uncontrollably. Examples of complex and tightly coupled systems include personal computers and nuclear power stations. Personal computers hang up because software interacts rapidly and in unpredictable ways: intervention is impossible. More seriously, Three Mile Island and Chernobyl failed because nuclear reactors are both complex and tightly coupled. Once one problem arose, it ramified through the system with more or less catastrophic consequences. Perrow suggests that it is dangerous to depend on technologies that are complex and tightly coupled, if the consequence of their failure is catastrophic. This is because it is only a matter of time before they go wrong and succumb to what he calls a "normal accident."

This is a logic that can be extended beyond large technical systems. It is plausible, for instance, to suggest that aspects of world agricultural organization are both tightly coupled and complex: the explanation fits epizootics such as foot-and-mouth disease outbreaks in northern countries. It is also possible to argue that global warming counts as a "normal accident" because, despite the fact that processes of atmospheric change are slow, it nevertheless seems that they are moving too fast for effective political action.

## Optimism, Pessimism, Invisible Work, and Politics

Large-scale systems and networks are difficult to treat neutrally or matter-of-factly. Much commentary falls either into hype (as with much talk about information and communication technologies [ICTs] and genomics) or into pessimism (where technologies are seen as inherently dehumanizing or hazardous). There is a secular change here. Most people in northern countries were probably optimistic about large technical systems at least up until the 1950s. However, with the creation of atomic weapons and increase in concerns about the environment, pessimism has grown, and many are now ambivalent about such systems.

This polarity relates to another, which has to do with the effectiveness and desirability of centralized control. The electricity systems of Edison and his successors worked best if they were designed and managed centrally. Many contemporary technologies, such as air traffic control, are similar. However, there are other technologies—the decentered character of routing in the Internet is an example—that seem to work well, indeed perhaps best, if they are partially decentered. But this is complicated because it is likely that in practice most large technical systems are less coordinated and coherent than either their advocates or their critics tend to think. Thus, feminist writers on technoscience have frequently observed that technical systems depend on so-called invisible work by underprivileged people that is necessary to the system but invisible to, and sometimes prohibited by, those in positions of authority. The practice of the "work to rule" used by employees in industrial disputes reveals the logic. It is probably impossible to specify all the details of a task within a complex technical system. This means that if employees work to rule, the system breaks down. Alternatively, the rules may specify tasks that have to be partially ignored if the system is to work. For instance, prescribed safety procedures often impede smooth system functioning, although this is poorly recognized by those who seek to control or regulate systems. Many accident investigations find that employees have broken rules, but do not recognize that rule breaking is also an integral part of safe operation.

If large technical systems work because they are not perfectly organized and coordinated in practice, this suggests the need to find ways of appreciating the importance of partial coordination without at the same time necessarily treating this as a defect. Recent studies in the discipline of science, technology, and society have attempted to do this by elaborating such notions as "multiplicity." The argument is that technical systems are patched together locally and, in practice, sometimes but only sometimes achieve relative coherence and consistency. Then it is argued that they are held together *both* by inconsistencies and noncoherences (which are necessary for the reasons mentioned above) *and* by practices that insist upon and reproduce their coherence and consistency (for otherwise they would, indeed, simply fall apart). Finally, and counterintuitively, it is argued that this implies that technical systems and projects are *both* singular and coherent on the one hand, and multiple and noncoherent on the other.

This noncoherence has a variety of implications. For instance, it is consistent with the insight of feminist technoscience writer Donna Haraway, who argues that it is never possible to obtain an overview of any system: if they are noncoherent, then so too our knowledge of them will also be patchy, partial, situated, and sometimes noncoherent. This suggests in turn that the control hopes of the system builders

need to be abandoned or modified. It also suggests that it is important to explore the politics and the practices of more fluid technical systems that reflect local circumstances and practices. Flexible approaches are now common in management practice, but there are alternative and less dominatory political possibilities. It is, for instance, at least sometimes possible to devise technical systems that do not require a center for command, communication, and control. Such systems are sometimes to be found in agriculture, and also in "appropriate technologies" as, for instance, in the case of a pump that is widely used to bring clean water to rural areas in Zimbabwe. The success of this technology in large part lies in the fact that it is adaptable and is widely variable in practice, and the firm that manufactures it demands neither uniformity nor royalty payments. This, then, is a technological system without a strong center, one that carries a political agenda and social implications that are quite unlike those of conventional large-scale systems. Most writers on technical systems agree that artifacts have politics, but the character of those politics only become clear if we both think about their uses and treat them as heterogeneous networks or systems rather than as isolated objects.

*See also* **Energy Systems**

**BIBLIOGRAPHY**

Callon, Michel. "Some Elements of a Sociology of Translation: Domestication of the Scallops and the Fishermen of Saint Brieuc Bay." *Power, Action and Belief: A New Sociology of Knowledge? Sociological Review Monograph,* 32, edited by John Law. London: Routledge and Kegan Paul, 1986: 196–233.

de Laet, Marianne, and Annemarie Mol. "The Zimbabwe Bush Pump: Mechanics of a Fluid Technology." *Social Studies of Science* 30.2 (2000): 225–263.

Diamond, Jared M. *Guns, Germs, and Steel: A Short History of Everybody for the Last 13,000 Years.* London: Jonathan Cape, 1997.

Haraway, Donna. *Simians, Cyborgs, and Women: The Reinvention of Nature.* London: Free Association Books, 1991.

Hughes, Thomas P. (1983), *Networks of Power: Electrification in Western Society, 1880–1930,* Baltimore: Johns Hopkins University Press.

Hughes, Thomas P. "The Seamless Web: Technology, Science Etcetera." *Social Studies of Science*, 16 (1986): 281–292.

Latour, Bruno. *The Pasteurization of France*, Cambridge MA: Harvard University Press, 1988.

Latour, Bruno. *Aramis, or the Love of Technology*, Cambridge, MA: MIT Press, 1996.

Law, John. "Technology and Heterogeneous Engineering: The Case of the Portuguese Expansion." In *The Social Construction of Technical Systems: New Directions in the Sociology and History of Technology*, edited by Wiebe E. Bijker, Thomas P. Hughes, and Trevor Pinch. Cambridge, MA: MIT Press, 1987: 111–134.

Law, John. *Aircraft Stories: Decentering the Object in Technoscience*, Durham, NC: Duke University Press, 2002.

Law, John. "Disaster in Agriculture, or Foot and Mouth Mobilities." *Environment and Planning A*, (2005).

Law, John, and Annemarie Mol. "Local Entanglements or Utopian Moves: An Inquiry into Train Accidents." In *Utopia and Organization*, edited by Martin Parker. Oxford: Blackwell, 2002: 82–105.

Perrow, Charles. *Normal Accidents: Living with High Risk Technologies*, Princeton, NJ: Princeton University Press, 1999.

Scott, James C. *Seeing Like a State: How Certain Schemes to Improve the Human Condition Have Failed*. New Haven, CN: Yale University Press, 1998.

Suchman, Lucy. *Plans and Situated Actions: The Problem of Human-Machine Communication.* Cambridge, U.K.: Cambridge University Press, 1987.

Wajcman, Judy. *Feminism Confronts Technology.* Cambridge, U.K.: Polity Press, 1991.

Winner, Langdon. "Do Artifacts Have Politics?" In *The Social Shaping of Technology*, edited by Donald MacKenzie and Judy Wajcman. Philadelphia: Open University Press, 1999: 28–40.

*John Law*

# P

## PAIN AND CULTURE.

The meanings of *pain* and *fatigue*, and the states that are considered painful or tiring, vary cross-culturally. Many pain researchers make note of anthropological accounts of exotic non-Western rituals, such as the "hook-hanging ritual" of rural India or the Sun Dance "self-torture" ceremonies of the North American Plains Indians. Scarification, tattooing, and piercing within Western subcultures similarly indicate the contextual character of pain, as the tolerability of pain is shaped by its infliction in meaningful personal rituals. The individuals' past history and the immediate cultural context shape the meaning and experience of pain, influencing pain tolerance and expression. Indeed, it may be that the pains today are so different in kind and meaning that even though the physiological bases of pain may be the same, the total experience is fundamentally different from pain in other historical contexts.

Expression of pain is one dimension of cultural variation. Stoicism, which can be taken to mean either "tolerance of pain" or "an ennobling understanding and acceptance of life as filled with suffering," is noted to be both a general feature of pre-Christian and Judeo-Christian Western philosophies. Stoicism in the sense of lack of expressiveness of pain, which is assumed to be a sign of high pain tolerance, is often associated with particular cultures, such as the British in the American imagination. Mark Zborowski found Jewish and Italian hospital patients more expressive of discomfort than Irish and "Old American" patients. While this research today might be regarded as stereotyping, it does give indications about how class and character are evaluated in terms of the expression of pain. The "Old American" patients were singularly reticent about expressing pain, and similarly disapproving of the expressions of pain in others. More recently, high expressivity, for example prevalent among a pool of Puerto Rican immigrants using a New England pain management clinic, was seen as inappropriate by the Anglo-American staff of the clinic, although in the Puerto Rican context this expressivity would have been unremarkable.

The degree of expressivity is often interpreted as a marker of class, gender, and character. For example, stoicism is noted as a feature of the established white middle class American repertoire of handling pain. Women are frequently stereotyped as having lower pain tolerance because they will express discomfort rather than suppress it. Reports of pain in women are often not taken seriously, despite the stereotype of low pain tolerance. The pain associated with childbirth signals a certifying rite of passage, and pain tolerance is especially valued in this context. Controversies over the use of anesthesia in childbirth illustrate the tension between a valorization of pain tolerance and the imperative that scientific medicine has to alleviate pain.

A less frequently considered assumption that varies cross-culturally is that pains are only individually experienced. In Western society it is an important assumption that pains cannot be shared. In other cultures, however, this is not taken to be true. For example, couvade, or "sympathetic labor," is the pain of childbirth experienced by a male partner or other relative of a woman delivering a child. It is often treated as a psychiatric problem and baffles Western researchers with their assumption about individualism.

With this view of pain as culturally situated, many pain researchers and clinicians carefully negotiate the variability of pain experience in treating patients. Apparently, however, the cultural specificity of contemporary pain researchers and clinicians (white, middle-class, English-speaking) often prevents application of these cross-cultural insights in practice and presents barriers to effective diagnosis and treatment. That is, anthropological knowledge has helped to shape a specific *theoretical* approach to pain research and management, but this development has not lead to a widespread development of *practices* effective in multiethnic clinical situations.

## Pain Research and Western Medicine

Roselyn Rey outlines a history of research and thinking on pain. Currently, the *nociceptive* theory of pain regards holds sway. In this model, pain is a functional expression of impending harm to the integrity of an organism. Pains coming from identifiable illness or injuries are called *organic* pains, as opposed to *psychogenic* pains, which do not have these bodily referents and seem to originate in the mind. Pain is now largely understood to be "in the brain" and not situated at the sites of injury or illness. Some argue for an entirely biological approach to pain that dissolves the distinction between organic and psychogenic pains. Ronald Melzack and Patrick Wall articulate what is the prevailing model of pain. The "gate control theory" of pain differentiates between the sensation threshold and the perception threshold. People can perceive a stimulus at the sensation threshold and perceive it to be painful or unpleasant at the perception threshold. The gate control theory is a cybernetic model of pain perception, interpretation, and behavior, and it is dominant in theory and clinical practice. Physiologists argue that the sensation threshold for humans is a relatively stable range of sensitivity to neurophysiological stimuli. Nonetheless, both levels of sensation are considered both culturally circumscribed and situationally constructed.

The situational characteristic of pain (for example, that it can be moderated by attention or distraction or exacerbated by anxiety) simultaneously provides pain clinicians with techniques for managing chronic pain and works toward configuring pain management as a matter of will. Behavioral strategies are built around alleged personality styles, in the hopes of improving the effectiveness of pain management techniques.

There are three general techniques for pain measurement. The first is based on subjects' reports of the intensity and character of their pains. The McGill pain questionnaire is the most widely used model and has been translated into many different languages. It allows patients to describe the location, severity on a scale of 1 to 10, and terms such as "hot" or "cold" and "sharp" or "dull" to describe their discomfort. A second pain measurement model describes the intensity of a pain in terms of the amount of an analgesic used to relieve it. The latter is not used frequently but has emerged out of research meant to test and standardize pain relievers. The third category of techniques uses technologies for visualization and measurement, which could be called "dolorometers," for the measurement of suffering. The newest among these is functional magnetic resonance imaging (fMRI), which tracks patterns of brain activity while a patient is experiencing pain. Other techniques include electromyography (EMG), which measures nerve activity through sensors placed on the skin.

Pain has its uses. Pain serves physicians as a diagnostic tool. Reports of pain are considered authentic when reports can establish causality in illness and injury. Physicians learn to use reports of discomfort as a way to develop diagnoses. In such cases, including those of "heterotopic" or "referred" pains distant from the affected part, physicians are interested in believing pain reports. Pain can serve to validate the helping professions. Pain, in this instance, legitimates the healing power of the physician. Specialists in pain and pain management, such as practitioners in chronic pain clinics, take an interest, as physicians do generally, in the treatment and authentication of pain. There are tensions in authentication processes, however, in part because physicians act as certifying authorities or gatekeepers to avenues of compensation and treatment, and in part because they are concerned with proper diagnostics of organic causes of pain. For example, researchers suggest that pain sufferers can use pain to gain attention. This attention is gained through manipulating interpersonal relations or through receiving certification from compensating institutions such as the state or an insurance firm. This is seen as an inauthentic use of pain, and it sits uneasily in a tacit nexus of a culturally endorsed stoicism and a professional mandate for the appropriate treatment of the individual.

Chronic pain is on the rise in Western societies, but chronic pain generally falls outside of the colloquial realm of credibility. The nociceptive model of pain breaks down with chronic pains, and researchers have either denied that chronic pains are "real," worked to revise the model, or positioned chronic pain in an ambiguous relationship to acute and organic pains. Isabelle Baszanger discusses two modes of understanding chronic pain: a clinical approach of patient-centered medicine, focusing on what seems to be a psychiatricmodel, and another approach, largely based on behavioral modifications. People reporting chronic pain are often accused of malingering, moral weakness, or mental illness such as depression, because most common notions and experiences of pain presume that it is temporary. Medical researchers and clinicians also frequently discredit chronic pain, although social groups and researchers have made significant gains in legitimating the claims of chronic pain sufferers.

In addition, there are multiple representations of where "the problem" in evaluating pain lies. Many books and other narratives place the blame squarely in the hands of physicians and "the medical establishment" for discrediting people's pain. Pain practitioners, particularly those in the professional pain management associations, prefer to present themselves as much more sympathetic, although still maintaining a subtle discourse that discredits chronic pain and problems of malingering and psychogenic pains. Physicans direct blame toward drug enforcement agencies, who wish to control the most powerful analgesics. Physicians who prescribe opioids for patients face higher degrees of scrutiny by law enforcement and state agencies. Problems of opioid diversion, and physician addiction, are real, of course, but perhaps mismanaged because of contradictory cultural messages wavering between "just say no" and "medicine can fix everything." Social movements of pain sufferers have engaged states for legislation, and medical boards for discussions of best practices, leading to attempts to require attention to pain reports for acute and chronic conditions, and, particularly, to provide palliative care at the end of life. Recent controversies in pain research also include the experiences children have with pain, and the associated problem of undermedication, as well as detailed technical discussions of the neurophysiological mechanisms of pain. For example, since pain has a large learned component, do very young children feel pain? If so, to what extent and with what consequences?

## Pain and Knowledge, Objectivity, and Subjectivity

The relevance of pain to philosophers and sociologists of knowledge falls along several dimensions. First of all, pain calls into question definitions of knowledge, objectivity, and subjectivity often taken for granted by both lay and expert audiences. Secondly, pain research demonstrates interesting processes in the development of technology and measurement. Thirdly, it illustrates differing issues of trust and authority around science. Finally, pain has interesting social and cultural dimensions, tied to recent social movement activity and also tied to scholarship discussing the sociocultural symptoms of contemporary society.

Pain and fatigue are generally perceived to be undesirable states. There are, of course, positive evaluations of these conditions, such as the "good pain" or "good tired" of athletic effort or the necessary pain of rites of passage. But the existence of these positive meanings to pain indicates that pain and fatigue are context-dependent and constructed. To consider pain *knowledge* about the state of the self is problematic when evaluated from formal definitions and criteria of scientific knowledge, such as reproducibility, objectivity, or intersubjective stability. Colloquial usage in Western cultures generally, although conditionally, acknowledges self-reports of pain and fatigue as reliable. Conventional Western interpretations of pain also situate it as internal to the individual. Subject-object dualism is foundational to theories of knowledge and much Western philosophy. Similarly, the dualism between mind and body also gets complicated by research on pain. Recognizing that pain is all in the brain places pressure on pain sufferers to be stoic or ignore their pain, since it is "all in their head." Although distraction, relaxation, and other mental techniques can help some sufferers, these do not necessarily completely relieve pain or help at all in some cases.

Elaine Scarry argues for the ineffability of pain, particularly its position in human experience *before* language. In her analysis, the infliction of pain in war and torture is the destruction of language and reality. Pain is, in this perspective, an ineffable, nondiscursive or nonrepresentable essence of bodily experience. This places pain in the realm of pure subjectivity and positions a pain sufferer as the sole author of her own pain experiences. Since objectivity and subjectivity are assumed to be completely opposed, this creates a problem for knowledge, or validation of pain by others.

The McGill pain questionnaire is the current standard for pain clinicians and researchers. It manifests high intersubjective validity and reliability. For example, the same patient interviewed by different people at different times will fairly reliably redescribe his pain in similar terms. The test is reliable across adminstrations, contexts, and interviewers, which indicates that it has high "objectivity" for the social scientists and practitioners who use it. It also demonstrates high validity in measuring similar constella-

tions of pain descriptors for a given ailment across patients. These are indicators of a kind of objectivity and the emergence of standardized languages and measures of intensity and characteristics. However, since the pain questionnaire is based on subjects' self-reports, it is, for some researchers, *by definition* subjective and thus unreliable. This has lead to the emergence of the technological measurements such as fMRI and EMG already mentioned.

Lorraine Daston (1992) has historicized contemporary usage of the term *objectivity*, noting that as this concept had developed in scientific and philosophical discourse since the seventeenth century, it has a tripartite meaning of methods, metaphysics, and morals. For social scientists and many pain clinicians, then, the "objectivity" of the McGill pain questionnaire rests on a matter of methods that are independent of any particular observer. For researchers interested in independent pain measurement with dolorometers, this is a problem of finding the "real" measurements of pain and is a matter of metaphysics. Given a subject-object split prevalent in the contemporary scientific discourse, and major philosophical orientations to it, one might *have* a pain, but one cannot *know* a pain. Both approaches to pain measurement, however, do engage the moral meaning of objectivity as that of being "unbiased." Social scientists, physicians, and engineers regard a reported pain as suspect if the sufferer and author of that pain may have something to gain from it.

The dolorometers—devices used for "objectively" evaluating pain—are designed to provide quantitative evaluations of pain character and severity and are attempts not only to make pain discursive but to make it quantitative, and not reliant upon the "subjectivity" of pain sufferers. Research is projected on technological interventions meant to verify pain and fatigue, transform knowledge and authority about pain from the sufferer to the device and thence to the scientific expert who uses the device. The measurement process becomes a process of objectification of both the pain and the pain sufferer. These interventions require a generalized, and problematic, transfer of authority over evaluating a personal state or experience from the subject of pain to an observer, most often a physician or nurse in the clinical setting or engineers and neuroscientists in research settings. In the transformation the subject-as-author of pain becomes the subject-as-object of measurement.

## Trust, Social Movements, Sociocultural Symptomatics

Pain is clearly an issue of trust and authority. If people suffering pain were always believable and telling the truth, and if pain had a common metric, or if pain were to be understood directly by others through strong empathy, there would be no need for externalized technologies to measure pain independently of people's accounts of it. The cultural context of pain experience is thus also a context that has dimensions of trust. Trust is undermined in contemporary contexts when someone is perceived as gaining benefits or attention through their pain experience. Chronic pain also sets up situations of distrust because of cultural expectations of pain as temporary and from unambiguous organic causes. Trust also flows in different ways along gender and cultural dimensions. Women's reports of pain experience are often not taken as seriously as men's, although men are expected to demonstrate higher pain tolerance with low expressivity. Pain reports by cultural groups that do not have high expectations of stoicism for themselves are often discredited by groups who do expect high degrees of tolerance. Experts, who have seen a wide range of pain patients even if themselves not pain sufferers, often do not take specific experiences of pain as significant for the individual because experts have seen others handle similar complaints.

These issues of trust, and related controversies in the diagnosis and management of pain, have led to the emergence of social movements by pain sufferers to gain more voice in pain research and treatment. While pain physicians have professional associations where they communicate ideas, pain patients and their families have developed similar groups, and challenged the expertise

of scientific authorities, for understanding the pain experience. These social movements engage discussions about access to appropriate medication, best practices for treatment, and disability rights. Chronic pain is associated with elusive diseases such as fibromyalgia, chronic fatigue syndrome, and multiple sclerosis, and pain groups have developed connections with researchers and clinicians to develop expertise and enhance the credibility of sufferers. Pain groups also connect with associations concerned with end-of-life issues, such as palliative care for relieving pain in terminal illness, hospice care and pain management, and euthanasia and doctor-assisted suicide discussions. In all of these groups, the rights of patients, and the ethics concerning the cessation of human suffering, are foremost issues.

Although pain and suffering have always been a feature of human life, there are some suggestions that chronic pain is rather new on the social landscape and needs to be explained. Anson Rabinbach (1992) noted that fatigue served as an ideological tool in debates about the limits of industrialization in the beginnings of modernity. For example, for Josefa Ioteyko, a Polish-born, Paris-trained physiologist of the late nineteenth century, "fatigue had a normative dimension, which, like pain, protects against sufferings inflicted by work and by society" (Rabinbach, p. 141). Similarly, Alfred Binet, in 1899, warned of widespread social decline with fatigue. The New York physician George Miller Beard in the 1860s described neurasthenia as a "disease of the unreliable subject," a disease of fatigue affecting well-situated young men and indicating a weakness of will. Its counterpart, hysteria, was the disease of similarly situated young women and marked as a perversion of will. More recently, Melzack has noted that hysteria, depression, hypochondria, and neurosis form a constellation of pathologies around chronic pain, the most problematic "kind" of pain in Western culture. Similarly, the problems of chronic pain and chronic fatigue are taken as markers of the malaise of postmodernity, as the signal illnesses of contemporary life. Synesthesia replaces neurasthenia, and the chronic displaces the acute in the transformations from the perils of an industrial society to those of a postindustrial or "risk" society.

In the West, colloquial usage generally although conditionally acknowledges self-reports of pain as reliable. The problem of the "subjectivity" of pain and fatigue rests on their "interiority," which renders reports of these states suspect in relation to conventionally held norms of "objectivity." This suspect character is based on the apparent incommunicability of pain and fatigue and the perception of the complete individuality and idiosyncrasy with which they are experienced. Although researchers have ameliorated the acute and chronic pains of many sufferers, much pain remains elusive. Similarly, fatigue is difficult to measure and has conflicting meanings. For example, the concept of fatigue is used to convey the limits of the human body. These limits are taken to indicate the limits of the development of modernity, limits not to be transgressed without dire consequences. But to some, pain and fatigue represent capacities of the body to be transcended in the name of and by way of technique, efficiency, and progress. Such complex and contradictory meanings of pain make it useful for studying knowledge and objectivity, and the body in culture and society.

## BIBLIOGRAPHY

Baszanger, Isabelle. *Inventing Pain Medicine: From the Laboratory to the Clinic.* New Brunswick, NJ: Rutgers University Press, 1998.

Bates, Maryann F. *Biocultural Dimesions of Chronic Pain: Implications for Treatment of Multi-Ethnic Populations.* Albany: State University of New York Press, 1996.

Daston, Lorraine. "Objectivity and the Escape from Perspective." *Social Studies of Science* 22 (1992): 597–618.

Delvecchio-Good, Mary Jo, Paul E. Brodwin, Byron J. Good, and Artur Kleinman, eds. *Pain as Human Experience: An Anthropological Perspective.* Berkeley, CA: University of California Press, 1994.

Hardcastle, Valerie Gray. *The Myth of Pain.* Cambridge, MA: MIT Press, 1999.

Helsinger, Elizabeth K., Robin Lauterbach Sheets, and William Veeder. *The Woman Question: Society and Literature in Britain and American, 1837–1883.* Volume 2: Social Issues, Chapter 2, "Science," 56–108. Chicago: University of Chicago Press, 1983.

Melzack, Ronald, and Patrick D. Wall. *The Challenge of Pain.* New York: Basic Books, 1983.

Morris, David B. *The Culture of Pain.* Berkeley, CA: University of California Press, 1991.

Rabinbach, Anson. *The Human Motor: Energy, Fatigue, and the Origins of Modernity.* Berkeley, CA: University of California Press, 1992.

Rey, Roselyn. *The History of Pain.* Cambridge, MA: Harvard University Press, 1995.

Scarry, Elaine. *The Body in Pain: The Making and Unmaking of the World.* New York: Oxford University Press, 1985.

Sheridan, Mary S. *Pain in America.* Tuscaloosa: University of Alabama Press, 1992.

Zborowski, Mark. *People in Pain.* San Francisco: Jossey-Bass, 1969.

*Jennifer L. Croissant*

# PATIENTS, MEDICAL COMMUNICATION WITH.

A common scenario of patient communications involves a person who feels pains in her chest and also more tired than usual. These forms of discomfort suggest to her that something is wrong with her health, so she visits a doctor. In the office, she describes her suffering and the doctor listens. Then he asks her about her history of such complaints, listens to her heart through a stethoscope, and conducts some tests. Considering all of this information, he makes a diagnosis that she is probably suffering from coronary artery disease and prescribes a drug to be taken daily.

Even in this truncated and stereotypical scenario, there is a complicated set of communication transactions to examine. (1) The body communicates ill health to a person through pain. (2) The person communicates, now as a patient, "dis-ease" to a doctor through words; next, these words—now representing symptoms, plus instruments and tests, as signs—communicate a differential diagnosis to the doctor. (3) The doctor in turn communicates a diagnosis, prognosis, and prescription to the patient. Each of these stages involves a number of assumptions, power relations, and variations that are explored in this article, attending in particular to the twenty-first-century U.S. context of managed care and direct-to-consumer advertising.

## Perceiving Illness

Talcott Parsons' early work on "illness behavior" introduces the idea that perceiving illness, or recognizing oneself as sick, is a prerequisite for seeking out healing. However, what is perceived as illness is historically, culturally, regionally, and individually variable. Medical anthropologists and sociologists have explored how identifying symptoms in the first place, such as noticing one's state of stress, is not a "natural" course of action; symptoms are not at all phenomenologically stable across persons and environments. A key mode of illness perception includes felt suffering, that is, experiences of pain or incapacity that interrupt one's daily activities and intrude as demands upon one's attention, often requiring that one turn to others for assistance. The seemingly intrinsic obviousness of this category of felt suffering has been investigated by anthropologists who have studied the different ways in which emotional, bodily, and social complaints are expressed cross-culturally. Common expressions, such as "My heart hurts" or "I am stressed," have many interpretations, and it is not at all clear how to settle whether and how there is a bodily reality here that is usefully separate from social and emotional ones. Of course, the different ways in which people are taught to experience themselves as ill are caught up in culturally inflected assumptions about the relationship between social categories, bodies, and pathology. For instance, Margaret Lock has explored the construction of menopause as a "deficiency disease" and the subsequent use of hormone replacement therapy in the United States—a process that many Japanese physicians find surprising and even disrespectful. These cultural variations inform the critical projects of those social scientists who explore the hegemony of biomedical explanations of illness, and the growing public capitulation to the idea that we are constantly "at risk" for one thing or another.

Indeed, phenomenological noticing, or consciously attending to, forms of suffering has become inseparable from forms of health worrying, such as about germs, risks, and asymptomatic conditions. Nancy Tomes has tracked the persistence of germ consciousness in consumer advertising throughout the twentieth century, pointing to the emergence of the notion of "selling health," and the crucial role of advertising campaigns in promoting public hygiene and practices of prevention. Dangers to health, popularized as deriving from pollution, modernity, and poor diet, were later produced as specific health risks and "risk factors" in the second half of the twentieth century. Large-scale epidemiologic studies such as the Framingham heart study and later clinical trials made headlines as news that redefined the reader's state of health, often by implying that what had formerly been considered normal was now to be understood as a living state of risk—or even as an ongoing state of suffering from a disease without knowing it.

Together, these uses of the mass media to shape public opinion about health set the stage for direct-to-consumer (DTC) advertising: the mass promotion of prescription-only pharmaceuticals through print, radio, and television. Direct-to-consumer advertising is currently legal only in the United States and New Zealand, but pharmaceutical company-sponsored illness awareness campaigns, public screenings, and clinical trials for marketing purposes are growing worldwide. Direct-to-consumer advertising primarily solicits persons as "patients-in-waiting," that is, those who do not know they need the pharmaceuticals until they encounter the advertising. This leads to a scenario that is very different from the opening example of patient communication:

> A person watches a TV and sees a commercial informing him that tiredness and changing eating habits may be signs of depression. Skeptical, but slightly disturbed and curious, he takes down a Website address given in the commercial, goes online, and finds a

checklist that he fills out. The checklist tells him that he may indeed be suffering from depression and should talk to his doctor. He researches the condition and the drug further and decides that perhaps Paxil will help him. Visiting his managed care office, he is seen by a nurse practitioner. He describes his symptoms following the template provided by the checklist. These symptoms accord with the checklist that the nurse uses, and he walks out with a three-month prescription for Paxil.

In this scenario, a mass communication about the possibility of a disorder suggests to a person that he is probably suffering from a disease without knowing it. What before had been perceived as everyday variation (like sleeping difficulties) are reframed as bona fide medical symptoms, of which he had previously been ignorant. Skeptical of advertising in general and medical information in particular, like many others, he uses the Internet to investigate the facts. By 2002, over 45 million Americans had gone online to help themselves or another person deal with a major illness. According to a Pew Institute Survey, 20 percent of these people had diagnosed or treated a medical condition on their own, without consulting a doctor. In such cases, the very perception of an illness becomes an interactive, social, and cognitive negotiation.

An increasing worry in the United States is that the ubiquity and power of DTC advertising persuades many individuals who do not necessarily need medications to seek them out. *Prevention Magazine*'s surveys of DTC have found that 32 percent of consumers have talked to their physicians about an advertised medicine because of an ad (61 million people), and 13 percent of consumers have talked to their physician about a health condition for the first time because of an ad (25 million people). Physicians surveyed by the FDA have said that DTC advertising had created anxiety in many of their patients and that, because of these ads, their patients have sought after unnecessary prescriptions. In other words, DTC advertising is actively cre-

ating patients as well as passively communicating with them.

Given this phenomenon, in recent years one of the primary tasks of the FDA has been to repeatedly draw the line between legitimate and illegitimate communication about diseases. It has to insist that companies do not attempt to grow the market for their products by blurring the distinction between real disorders and "variations of normal daily functioning." For example, a banned radio advertisement for the antidepressant Effexor began, "Hey you, listening to the radio . . . how're you feeling these days? Okay? Not bad? Come on, is that where you want to be?" Moreover, the pharmaceutical industry has relied not only on DTC advertising, but has also exploited the third-party credibility of public relations, including the use of advocacy groups and the news media. Advertising is much of the time viewed as the reinforcement of branded illness awareness, which has already been created through these other PR channels. This kind of diverse infiltration shows the need for social scientists to attend to how persons come to perceive themselves as ill, and how this changes along with the ongoing media field, as well as culture, biology, and environment.

## Talking with Doctors

Medical sociologists and anthropologists have used the distinction between *illness* and *disease* to mark the difference between lived experience and the medical definition of it. Arthur Kleinman has argued that, in practice, the doctors reformulate illness into the technical terms of their theories of diseases and disorders. This approach to patient communication emphasizes the asymmetrical distribution of power in doctor–patient relations, in which doctors control patients' access to health information and health care resources. Systemic discrimination along race, gender, and class lines is a constant concern here. Cross-cultural encounters especially are fraught with miscommunication because many doctors are not trained in understanding the many modes of somatic expression. The question here is how thoroughly doctors are able to pursue the possibilities inherent in phrases such as "my heart hurts." This approach thus requires attending to whether the doctor actually listens to the entirety of the lived experience of suffering or whether the doctor employs a preconceived set of categories and prejudices that pigeonhole patients into disease models. While simplistic, this opposition enables us to understand how poverty, abuse, and environmental problems can be medicalized and treated only as organic syndromes, thus perhaps effacing the real causes of distress.

Talking with doctors is more than a unidirectional transaction, however. Medical anthropologists have focused on the interpersonal dynamics between doctors and patients, showing how storytelling is a crucial (but often underanalyzed) part of living with illness or disability, and of obtaining medical treatment. The reduction of personal notions of illness to "folk" or "lay" beliefs is particularly problematic. As analyzed by Byron Good, "belief" has come to be an implicitly pejorative concept applied to "other" cultures, but it must be symmetrically applied to understand our own relationship to what we consider to be proper biomedical understandings of bodies, illness, and disease. Patients of all sorts necessarily impart unique, personal, and social narratives to the supposedly neutral biological facts of their suffering. Similarly, Emily Martin's ethnographic work among patients of various socioeconomic backgrounds examines differences in how they understand the immune system. Metaphor is a key conceptual device that binds together different social spheres— weaving together health, nationalism, militarism, and biomedicine—and affecting the kind of communication that takes place between doctors and patients.

Communication between doctors and patients is thus mediated by the interpretive frames that the doctor applies to the patient's speech. Good and others have explored how clinicians with different medical training can make very different interpretations of what ails the same patient. For instance, rather than hearing the speech as a direct expression of somatic complaints, therapies such as psychoanalysis provide a way of listening to speech itself as a symptom. Instead of treating the patient as a rational actor who describes what he or she feels in a straightforward manner, the doctor interprets what is said as revealing a deeper level of distress and meaning. In this case, it might be said that the patient is unconsciously communicating with the doctor. In fact, one of Sigmund Freud's descriptions of psychoanalytic technique included that analysts develop a "hovering" attention to the analysand—not focusing on anything in particular, but rather "receiving" the analysand's unconscious through their own unconscious. Psychoanalysis was incorporated into mainstream medicine from the 1930s to 1950s as "psychosomatic medicine," which emphasized attending carefully to the emotional relationship between the physician and patient as essential to a positive clinical outcome. The importance of this emphasis continues today as hospice, and in renewed attention to positive communication in placebo studies.

Instruments and tests are another means of getting behind a patient's words. A stethoscope, MRI scan, or blood test may be used to figure out what is really wrong with the patient. Here the patient's body appears to be communicating directly with the doctor. In the extreme case, the initial patient's complaint is reduced to a trigger that gets the patient into the office but nothing more. Whereas many doctors combine all of these approaches, historians of medicine have traced a trend in which doctors increasingly listen to their instruments rather than patients. In this trend, power again shifts in favor of the doctor over the patient, and the lived experience of suffering and the social context of illness are both denied in favor of an exclusively biological definition of disease.

Talking with doctors is significantly shaped by the institutional form of health care. Physicians earn their living from practicing medicine and the incentives contained in their method of payment can affect the nature of communication they have with their patients and the kinds of medical decisions they make. Managed care in the United States is a form of private health care characterized by the linkage of clinical, financial, and administrative responsibility for the care of a population. Before managed care, fee-for-service operations created a basic incentive structure for doctors to do more, whereas the payment capitation systems typical of health maintenance organizations (HMOs) basically encourage quick office visits and discourage referrals to specialists. In this context, other modes of health communication such as support groups and Internet sites have been directly supported as enabling patients to better manage their own health.

Thus, the steady rise of consumerism and patient empowerment is in constant negotiation with medical professionalism, expertise, and the domination of medical knowledge. As highlighted in the second scenario of patient communication, DTC advertising capitalizes on this situation by structuring its messages such that patients who take them up are directly provided with the same language of "symptoms" that the doctor uses to make a diagnosis. This scenario depends on the fact that most patients do not in fact communicate with specialists (cardiologists, psychiatrists, gastroenterologists); in managed care settings, they almost always talk only to primary care physicians, nurse practitioners, and physician assistants. In fact, these nonspecialists prescribe 80 percent or more of all antidepressants, cholesterol-lowering statins, and acid-reflux disease pills. These practitioners' education about these drugs is often conducted by the same companies that design the consumer campaigns, and thus the process of communication between patient and prescriber is often short-circuited in favor of rapid prescription writing. In this scenario, power over diagnosis shifts to managed care and pharmaceutical companies over both doctors and patients.

## Communicating Diagnosis and Treatment

Doctors are intensely aware of the power of their words in this process of diagnosis and prescription. Often there remains a set of treatment possibilities, and the doctor is required to offer the patient a choice between them. Doctors narrate these choices in such a way as to persuade the patient that one approach is better than another, a process that Cheryl Mattingly has usefully called *therapeutic emplotment*. Underlying this apparent manipulation is the fact that all communication between doctor and patient is already framed; the patient is always trying to see past what the doctor says to what she really means. In this manner, the psychoanalytic relationship is reprised on the side of the patient.

The act of communicating a diagnosis by doctor to patient is performative, changing the status of the patient into someone definitely suffering from a particular disease, enabling various forms of insurance coverage, and perhaps also socially labeling that patient. Prenatal and genetic screening, mental illness, cancer, and AIDS are all exemplary situations in which the social and emotional power of a diagnosis of disease or risk can profoundly reorder the patient's life. Ideally, doctors and other health care professionals are extensively trained in how to communicate with patients about these.

Communication about a number of illnesses has been improved only after intense patient activism. In addition to collective action around basic health research, people have come to form social movements around specific illness categories. Breast cancer and AIDS have been the most studied of these in part because of their visible success and the fact that they have provided templates for other social movements. As observed by Alberto Melucci, the very existence of a social movement around an issue functions as a message in society communicating that, in this case, the illness is a social problem. Patient movements provide a forum for communicating shared experiences of specific forms of suffering and for agitating changes in the manner in which the illnesses are treated in the health care system. Groups often produce educational Websites and pamphlets for doctors and for patients to aid both parties in communicating more effectively about the illness. For illnesses that remain socially controversial, such as attention deficit disorder, chronic fatigue syndrome, and autism, patient communication involves negotiating the costs and benefits of the diagnosis—weighing the value of access to resources with the social stigma attached to having the diagnosis.

Communication with patients frequently begins only with the diagnosis and prescription. Ongoing communication between patients and health care providers has many different faces. Rather than simply improving a patient's condition, medical treatment often induces a new medicalized lifestyle for the patient, one fraught with tensions among enjoyment, medications, side-effects, hopes, and fears. These days, *noncompliance* is the key term in contemporary biomedicine to name the tension between the lived cost of carrying through a prescribed regimen of treatment and patients' choices to seek out alternative treatment forms.

In managed-care environments, a prescription is often the goal of a doctor visit, and given the number of different health care providers a patient must deal with, the patient's relationship with her medication can become a surrogate for the relationship with her doctor. More and more, the stability of this relationship is mediated by the brands of such drugs. Indeed, brand loyalty has become a focus for the pharmaceutical industry to drive medication compliance. Loyalty is managed through direct company-to-patient communications via checklists, mailings, phone contacts, and e-mail, in addition to the "emotional" attachments that marketers work hard to generate through drug advertising.

The complex and varied contemporary environments of health care communication require new analytic tools on the part of the social scientist. Direct-to-consumer advertising has become a form of medical education, and in general we are witnessing a rise in health care consumerism in which patients are actively encouraged to be "information seekers" and to make choices among competing drug companies and insurance companies. Many patients no longer consider doctors as their first or even primary source of health care information. Alongside the growing diversity of information channels that communicate to people about their health are the changing ways in which people identify with each other, organize to share their experiences, and instigate social changes around

the identification and treatment of illness. Issues of stigma, discrimination, and health disparities remain serious, unsolved problems.

*See also* **Patients' Rights**

## BIBLIOGRAPHY

Aronowitz, Robert A. *Making Sense of Illness: Science, Society, and Disease.* Cambridge, U.K.: Cambridge University Press, 1998.

Blaxter, Mildred. *Health and Lifestyles.* New York: Routledge, 1990.

Epstein, Steven. *Impure Science: AIDS, Activism, and the Politics of Knowledge.* Berkeley, CA, University of California Press, 1996.

Fox, Susannah, and Lee Rainie. *The Online Health Care Revolution: How the Web Helps Americans Take Better Care of Themselves.* Washington, DC: The Pew Internet and American Life Project, 2000. http://www.pewinternet.org/PPF/r/26/report_display.asp.

Good, Byron, Henry Herrera, Mary-Jo DelVecchio Good, and James Cooper. "Reflexivity, Countertransference, and Clinical Ethnography: A Case from a Psychiatric Cultural Consultation Clinic." In *Physicians of Western Medicine*, edited by Robert A. Hahn and Atwood D. Gaines. Dordrecht, The Netherlands: D. Reidel Publishing Co., 1985: 193–221.

Good, Mary-Jo DelVecchio. *American Medicine: The Quest for Competence.* Berkeley, CA: University of California Press, 1995.

Kleinman, Arthur. *The Illness Narratives: Suffering, Healing and the Human Condition.* New York: Basic Books, 1988.

Kommers, Nathan, and Lee Rainie. *Use of the Internet at Major Life Moments.* Washington, DC: The Pew Internet and American Life Project, 2002. http://www.pewinternet.org/PPF/r/58/report_display.asp.

Lock, Margaret. *Encounters with Aging: Mythologies of Menopause in Japan and North America.* Berkeley, CA: University of California Press, 1993.

Martin, Emily. *Flexible Bodies: Tracking Immunity in American Culture from the Days of Polio to the Age of AIDS.* Boston: Beacon Press, 1994.

Mattingly, Cheryl. *Healing Dramas and Clinical Plots: The Narrative Structure of Experience.* Cambridge, U.K.: Cambridge University Press, 1998.

Melucci, Alberto. *Challenging Codes: Collective Action in the Information Age.* Cambridge, U.K.: Cambridge University Press, 1996.

Metzl, Jonathan M., and Joni Angel. "Assessing the Impact of SSRI Antidepressants on Popular Notions of Women's Depressive Illness." *Social Science and Medicine* 58.3 (February 2004): 577–584.

Mol, Annemarie, and John Law. "Embodied Action, Enacted Bodies: The Example of Hypoglycaemia." *Body and Society* 10.2-3 (June-September 2004): 43–62.

Nelkin, Dorothy. *Selling Science: How the Press Covers Science and Technology.* New York, W. H. Freeman, 1987.

Parsons, Talcott. "Illness and Role of the Physician: A Sociological Perspective." *American Journal of Orthopsychiatry.* 21 (1951): 452–460.

Rapp, Rayna. "Constructing Amniocentesis: Maternal and Medical Discourses." *Uncertain Terms: Negotiating Gender in American Culture,* edited by Faye Ginsburg and Anna Lowenhaupt Tsing. Boston: Beacon, 1990.

Slaughter, Ed, and Martha Schumacher. "Prevention's International Survey on Wellness and Consumer Reactions to DTC Advertising of Rx Drugs." *Prevention Magazine* (2000).

Starr, Paul. *The Social Transformation of American Medicine: The Rise of a Sovereign Profession and the Making of a Vast Industry.* New York: Basic Books, 1982.

Taylor, Verta. *Rock-A-By Baby: Feminism, Self Help, and Postpartum Depression.* New York: Routledge, 1996.

Tomes, Nancy. *The Gospel of Germs: Men, Women, and the Microbe in American Life.* Cambridge, MA: Harvard University Press, 1998.

U.S. Food and Drug Administration. "Attitudes and Behaviors Associated with Direct-to-Consumer (DTC) Promotion of Prescription Drugs: Main Survey Results." Center for Drug Evaluation and Research, Division of Drug Marketing, Advertising, and Communications, 1999. For other data, see http://www.fda.gov/cder/ddmac/ and http://www.fda.gov/cder/ddmac/Final%20Report/FRfinal111904.pdf.

Waitzkin, Howard. *The Second Sickness: Contradictions of Capitalist Health Care.* New York: Rowman and Littlefield, 2000.

*Joseph Dumit and Nathan Greenslit*

# PATIENTS' RIGHTS.

Science and technology, sometimes referred to as "technoscience," intermingle in the production of Western medical theories and procedures to treat a host of medical conditions. Western or "scientific" medicine, although located in the cosmopolitan context of technoscience, is also located within Western philosophical, legal, and sociocultural understandings. A hallmark of Western social and political philosophy is that individuals have rights; yet an examination of rights demonstrates that they are frequently debated. As scientific medicine grew, discussions of rights, who had them, and who could infringe upon them were paramount. Although politicians, physicians, patients, and local communities may disagree over what particular rights should be assured the citizen-patient, general concerns over rights continue to shape the face of medicine by determining what is and is not a reasonable treatment or procedure, defining who is and who is not an acceptable patient, and identifying the venues and circumstances in which medical information is gained, stored, and revealed.

This exploration of patients' rights in the context of scientific medicine begins with a short discussion of patients' rights with particular attention to the principles of autonomy and justice—both key concepts in the narratives of American identity. Next, the flexibility of patients' rights claims is demonstrated by examining some instances in which conflicts arise. Following these discussions, attention will finally be turned to cybermedicine, genetics, and privacy, all three of which stand as foundational issues in future discussions of patients' rights in an increasingly computerized medical context.

## Rights as a Philosophical Ideal and Social Conflict

Though different definitions of rights exist, in general they are defined in ways that present them as innate aspects of an individual that keep his or her body, mind, and things separate from the claims or harms of others; rights are entitlements that allow individuals to protect and express themselves. Though individuals may debate the question of what is the source from whom or which rights are given, granted, or declared, the notion that human beings have within them, and sanctioned from without, some entitlement to express their thoughts and protect or use their things (including their bodies) is a concept embraced by Western, scientific medicine.

Although many agree that rights exist, they may disagree over what actions to take when rights conflict or individuals are harmed. Multiple court cases dot the landscape in which rights have been acknowledged, determined, and, by some views, undermined. With each conflict new approaches to rights are considered, and often more conflicts ensue. The outcome is less that rights are settled upon and more that rights are continually renegotiated. In fact, some bioethicists and legal scholars (such as Lori Andrews and Pilar Ossario) argue that it is best to think of rights as existing within a "bundle." With some medical actions, some aspects of a patient's rights are infringed upon, but other aspects of the patient's rights remain unharmed. In this light it is less that a medical act is right or wrong than that it may be interpreted as more right than wrong in a specific context.

Once medical procedures are sanctioned for use by their developers and supporters, they are used on patient bodies. Though medical devices and treatments often transgress the patient body, ideally they do so with attention to the patient's rights. Only by acknowledging and safeguarding a patient's rights is medicine considered or presented as an acceptable social interaction. Theoretically, physicians should respect the views of their patients and not act to harm or assail them. The Hippocratic Oath and its obligation to "do no harm" is a clear reminder to physicians that patients entrust them with their bodies and rely upon the knowledge and assistance of the medically trained. Nonetheless, medical interactions take place in a context that is not equal between a healthy, medically knowledgeable provider and an ill, less medically knowledgeable patient. Although "equal rights" may exist, not all individuals are "equal"; lack of education or resources and differences in race or gender (to name a few examples) may put individuals at scientific or medical disadvantage.

The sociology of medicine and science describes the subtle and overt differences between practicing professionals with privileged or elite knowledge and those who lack such advantage. As medicine professionalized, power through elite knowledge was sought, sanctioned, and protected. Along with emergent medical specialties came the development of hospitals and other institutions to house patient bodies and professional knowledge. Patients were work objects upon whom insights were practiced, inscribed, and gained (not always to their advantage, as the first patients in such institutions were at times victims of disease spread through unhealthy and misunderstood medical practices). Even now, issues of power through knowledge still trouble the physician-patient relationship. The differences (and similarities) between patients and physicians form the context in which medical interactions including experimentation, diagnosis, and treatment develop. Within these medical contexts, patients' rights are theoretically protected and safeguarded. Yet the ideals of protecting patients' rights are not always upheld. Discussions of patients' rights necessarily juxtapose the theoretical ideals of protection and respect with the unfortunate circumstances of past harms, current transgressions, and fears of future impingements. Concerns over patients' rights are one of the few mechanisms sought to curtail the possible infringements arising from a physician's power over a patient.

## Rights in Conflict

Western philosophy and theology, in conjunction with physician experience, formed the basis of what is now Western bioethics (or biomedical ethics). Today bioethics is an increasingly multidisciplinary field. Although all theorists bring with them theoretical underpinnings and questions specific to their fields, it is nearly inescapable that all examinations of patients' rights must come to terms with the classic building blocks and ethical principles associated with bioethics. Although theorists debate which principles are most important or practical, the "Georgetown Mantra" of autonomy, beneficence, nonmaleficence, and justice is frequently cited and recognized as a useful and insightful grouping. Although all four principles are useful when examining patients' rights, autonomy and justice are particularly important when exploring the rise of scientific medicine, because they stand as cornerstones to the narrative of American identity. The emphasis on the individual and his or her civil rights stands American identity separate from many other Western, cosmopolitan nations and societies. American identity shaped scientific medicine through its concerns for patient autonomy, justice, and, by extension, privacy and confidentiality. Simply put, patients' rights ensure individuals a certain degree of freedom and respect within a medical context that can be limiting and impinging.

As Western science and medicine grew, some individuals expressed concerns over the informed consent of patients and the reasonability of medical investigations. These arguments and concerns were borne out in political and judicial interactions and often learned through public outcry. Some of these concerns stemmed from the Nuremburg trials following World War II and increasing public awareness of human experimentations committed by the Nazis. Although war crimes offer important sites for investigating patients' and human rights claims, other less obvious, but equally political, sites also exist and serve as case studies from which patients' rights continue to develop. For example, in the Tuskegee syphilis study (1932–1971), poor African-American men from Macon County, Alabama, were recruited for a study to determine the natural course of syphilis. The men were only observed and not offered penicillin to treat the disease (to be clear, when the study commenced, penicillin was not available, and the medications that were available were potentially harmful). The goal of the study was to learn about the progression of syphilis, not to assist a patient with syphilis. The study continued for decades (well after widespread availability of penicillin) until 1972, when a member of the Associated Press broke the story that the study was ongoing and that its subjects were not offered penicillin or other antibiotics. Relatively soon after the news article hit the press, the study ended and its subjects were awarded compensation in a settlement with the federal government. A number of criticisms may be leveled at the Tuskegee study, including shoddy research deign, lack of informed consent, and inability to act in the interests of the patient. Although some of these concerns are tempered when the case is viewed through a 1932 understanding of physician-patient relationships, it remains useful in that it highlights the importance of differences in assumptions and contexts separating physicians and patients, in this case, race and education (it is also important to note the length of the study and question why antibiotics were not offered once available). Although the study was designed to observe the natural course of syphilis, thereby not offering treatment options to subjects, the infringement of these men's rights to information about their participation in the study and its potential ill effects should be criticized. This and other case studies in the twentieth century, highlighting lack of consent and abuse of power, helped to lay the groundwork for critical examinations of patients' rights within medical experimentation and clinical interactions. The changing mood in the United States over concerns for individuals and protections of their civil rights affected scientific medicine by establishing a context in which policies and protocols that limited and guided research on human subjects were possible. For example, the 1979 Belmont Report, produced by the National Commission for the Protection of Human Subjects in Biomedical and Behavioral Research, remains a foundational document for today's institutional review boards (IRBs). The report stresses the need for respect for persons, benefice, and justice in the oversight of medical research with human subjects.

The Tuskegee study illustrates the circumstances in which an autonomous and aware patient's rights are infringed upon by the patient's community or physician. Not all patients' rights concerns are this clear. In the Tuskegee study rights are determined (such as the right to treatment or informed consent) and a patient is identified (that is, an African-American man with syphilis). A less clear debate, but one with no less implication for rights, exists between the rights of a woman carrying a pregnancy and the rights accorded her unborn, developing fetus. In some cases, claims of fetal-patient rights may force women to undergo treatment options or medical procedures against their wishes. Unlike rights conflicts in which one autonomous individual is positioned against another, in the case of the maternal-fetal relationship one "individual" is held within another. When the rights of the fetus-patient are seen as more important than the woman-carrier (the maternal environment), it is possible to argue against protecting her rights and, instead, to extend medical assistance to the fetal patient. The ability of scientific medicine to render the fetus in utero visible or identify it as a patient threatens the rights claims of some pregnant women. In these cases a woman's decisions and rights claims may stand in stark contrast to those of members of the community surrounding her, exacerbating a tenuous political situation. Simply put, when the rights of individual patients conflict, it is not always clear what should be done to resolve the situation. Autonomy and justice, although useful ideals, are not always possible for all parties involved.

## Rights in the Twenty-first Century: Cybermedicine, Genetics, and Privacy

As scientific medicine continues to develop, different issues concerning rights come to the fore. Perhaps the most significant concerns expressed for the future of medicine and patients' rights center around concerns of access to information and patient privacy. As computer networks grow and medical interactions (including the act of storing patient information) become increasingly computer-based, concern over who has the right to access and use the information grows. Similarly, as genetic screening and treatments develop, it is likely that some patients will wish to withhold information regarding their personal genotypes from their

physicians and others (such as insurance companies and employers). Both of these issues are important today, and with the continued growth of computer networks and their threats to patient and community privacy, these issues will likely take center stage in future rights debates and merit detailed consideration.

With increasingly powerful computer technology and its widespread use in public and private spaces, Western societies are changing. Although some greet these changes as potentials for leveling the playing fields between different individuals and groups, others are concerned about unbridled, or at least unmonitored, developments that may threaten our rights and potentially our existence. While some optimistically hope for utopian changes in technoscientific existence, others are wary of a slippery slope leading "civilized" societies into new eras of miniaturization, conformity, and danger. Although these two ends of the spectrum are built upon competing notions of the outcome of increasing computerization and the establishment of ever-powerful and centralizing cybernetworks, they share the undeniable position that computers, the networks that carry them, and the notions that keep them in existence are particularly powerful tools. These tools will continue to affect patients' rights and medical developments in the future.

Cybermedicine, or the use of computer information and cybernetics in medical procedures, makes medicine more flexible, less visible, and prone to security mishaps. Although cybermedicine ushers in numerous changes to patient experiences, two particularly important considerations for patients' rights debates are clear: First, computerized patient records may be prone to hacking and other potential misuse; and second, patients are increasingly able to find and share information online. Cybermedicine offers patients increased possibilities to assert their rights and wishes at the same time that it threatens them with potential abuse of their information for the advantage of others.

Proponents of cybermedicine argue that with computer networks (such as informational Websites on the Internet) and through computerized medical technologies, patients are better able to assert their rights to medical information and appropriate care practices. Through computer networks and the tools that encompass them, patients are better positioned as medical consumers. Patients with computer access are able to form virtual communities and seek out specialized help and services when needed. Patients with little known conditions or "orphan diseases" are now able to find assistance in a variety of guises, including online support groups, sympathetic physicians for second opinions, and vast amounts of medical information; for example, the National Organization for Rare Disorders and WebMD offer extensive Websites with medical resources for physicians and patients alike. Proponents of cybermedicine reason that patients' rights claims to information and second opinions are hollow if patients are unable to access the necessary information to make informed decisions and, ultimately, to offer informed consent to medical procedures. In these cases cybermedicine ushers in a new era in which patient desires for information and their ability to find sympathetic and informed caregivers increase dramatically. At the same time, geographically separated patients may offer one another important emotional support through electronic communication.

Although computer technology and information networks may be useful to patients who have access to them, individuals who gain access to patient records databases are able to read and share information about patients with or without their consent. As scientific medicine continues to develop within hospitals in which patients no longer see one physician but instead are seen by health care teams, the likelihood for patients having personal information revealed about them without their consent increases. Today, computers hold patient records and other forms of personal information. Such systems may be accessed by any number of medical and nonmedical individuals wanting this important, yet personal, information. Indeed, the rights of a health care team to access the information needed to do its job may threaten the

rights of patients to keep their medical information confidential. As computer networks continue to house medical information, and as medical teams continue to become flexible and interchangeable, patients' abilities to control who has access to their information may be reduced and some patients may fear entering medical interactions in which guarantees of medical confidentiality are nearly impossible to keep. As more diseases and genotypes become known, it is likely that individuals will have their records kept on a computer or computer network. If this is the case, computer fraud and safety enter into issues of patients' rights in ways never before imagined. Will patients have a right to keep their personal genotype and medical information private, and what might privacy suggest in a medical era noted for its increasing cybernetic linkages and networks? These emerging questions and concerns over patient privacy and access to medical information will continue to fuel patients' rights concerns for years to come.

Just as cybermedicine relies upon access to information, genetics, and the technologies to identify and alter them, rely upon using the information contained within them. With the mapping of the human genome and increasing public attention to the role that an individual's genes play in his or her overall health, it is not surprising that some debates and concerns surrounding the future of medicine stem from the potential roles that genetics and genetic treatments may suggest for individuals wishing to keep their genotypes private. Currently, genetic screening offers many individuals the chance to know whether they carry a gene (or genes) that cause or predispose to certain conditions such as Huntington's disease. Some patients embrace these technologies as empowering, allowing them to plan their lives in a fashion that accounts for potential future diagnoses. Nonetheless, much genetic information runs in families; genotypes are "handed down" over generations. To know one person's genotype is potentially to peek at another's, with or without the latter's consent. Continued developments in genetic screening and potential developments in miniature robotics and even nanotechnologies may allow patients to know more about their bodies, to know more about those around them, and, in theory, to create new treatment options, including medications designed for specific genotypes. As information becomes more exact and broader in scope (that is, as more characteristics and traits are identified), concerns for individual privacy over a person's genotype will increase, likely with beneficial and traumatic outcomes.

## Rights Today and Tomorrow

Articles and reports expressing concern over patients' rights continue to fill U.S. newspapers and magazines as private citizens and elected politicians debate the merits of a "right to health care" and a "Patient's Bill of Rights." These concerns reflect the Western technoscientific context, which accepts rights claims as an integral part of consuming and producing fair and just medical practices. As with past examinations of patients' rights, issues of autonomy and justice are particularly important to understanding medicine in an American context. Undoubtedly, these and other issues will continue to shape scientific medicine for decades to come. Patients' rights vary over time and reflect the social, cultural, and legal context in which they are examined; as medical procedures develop and medical interactions occur, different rights claims come to the fore. Continued questioning by physicians and patients alike will prove invaluable as rights are negotiated into the twenty-first century and beyond.

*See also* **Medical Community, The: Communication with the Public; Patients, Medical Communication with**

### BIBLIOGRAPHY

Andrews, Lori. "My Body, My Property." In *Taking Sides: Clashing Views on Controversial Bioethical Issues*, 6th ed., edited by Carol Levine, 280–287. Guilford, CN: Dushkin Publishing Group, 1995.

Arney, William Ray. *Power and Profession of Obstetrics*. Chicago: University of Chicago Press, 1982.

Cassell, Eric J. "The Principles of the Belmont Report Revisited: How Have Respect for Persons, Beneficence, and Justice Been Applied to Clinical Medicine?" *The Hastings Center Report* 30.4 (2000): 12–21.

Fuchs, Stephan. *The Professional Quest for Truth: A Social Theory of Science and Knowledge*. Albany: State University of New York Press, 1992.

Henderson, Gail E., Nancy M. P. King, Ronald Strauss, Sue E. Estroff, and Larry R. Churchill, eds. *The Social Medicine Reader*. Durham, NC: Duke University Press, 1997.

Joy, Bill. "Why the Future Doesn't Need Us." In *Technology and the Future*, edited by Albert H. Teich, 295–317, Belmont, CA: Wadsworth Publishing, 2003.

Levine, Carol, ed. *Taking Sides: Clashing Views on Controversial Bioethical Issues*, 6th ed. Guilford, CN: Dushkin Publishing Group, 1995.

Munson, Ronald. *Intervention and Reflection: Basic Issues in Medical Ethics*, 4th ed. Belmont, CA: Wadsworth Publishing, 1992.

Ossario, Pilar. "Legal and Ethical Issues in Biotechnology Patenting." In *A Companion to Genethics*, edited by Justine Burley and John Harris, 408–419. Malden, MA: Blackwell Publishers, 2002.

Pence, Gregory E. *Classic Cases in Medical Ethics: Accounts of the Cases That Have Shaped Medical Ethics, with Philosophical, Legal, and Historical Backgrounds*. New York: McGraw-Hill, 1990.

Pence, Gregory E. *Recreating Medicine: Ethical Issues at the Frontiers of Medicine*. New York: Rowman and Littlefield, 2000.

Siegler, Mark. "Confidentiality in Medicine: A Decrepit Concept." In *Taking Sides: Clashing Views on Controversial Bioethical Issues*, 6th ed., edited by Carol Levine, 148–152. Guilford, CN: Dushkin Publishing Group, 1995.

Starr, Paul. *The Social Transformation of American Medicine: The Rise of a Sovereign Profession and the Making of a Vast Industry*. New York: Basic Books, 1982.

Wertz, Richard, and Dorothy C. Wertz. *Lying-In: A History of Childbirth in America*, expanded ed. New Haven, CN: Yale University Press, 1989.

*Deborah Blizzard*

# PERSPECTIVES ON MEDICINE AND SOCIETY.

In 1800 a major metropolitan hospital could perform a few hundred surgeries a year. In 1924, the Mayo brothers performed 23,626 operations in their Minnesota clinic. In 1900, a hospital stay nearly always meant an accompanying infection, even if the patient presented with a compound fracture of an arm. Over one hundred years later, infection is not as ubiquitous, but if contracted, all the more deadly. Such statistics and features are complicated to evaluate. Medicine has become "more." There are more surgeries, more drugs, and more people working in health care industries; medical institutions have received more money, from paying customers as well as from the state. There has also been more resistance to institutional medicine, on the part of disease and by individuals as they engage or are forced to engage the medical "system." Infection is on the rebound as bacteria become resistant to the antibiotics of the twentieth century. The incidence of diseases like tuberculosis fell in the early twentieth century only to have dramatically risen again. There has also been more of a disparity between those with access to medicine and those without. The social organization of medicine itself has changed. Historian Roy Porter puts it this way: "Medicine used to be atomized, a jumble of patient-doctor transactions . . . what could be more different from today? Medicine has now turned into the proverbial Leviathan, comparable to the military machine or the civil service, and is in many cases no less business- and money-oriented than the great oligopolistic corporations" (p. 628). In addition to the organizational features of medicine and medical and surgical innovations, the social meanings of medicine change readily. Physicians have been seen as selfless missionaries on behalf of health in the model of Albert Schweitzer and as evil experimentalists such as the Nazi doctor Josef Mengele. Medicine has also been decried for the way in which it uses concepts like "normal" and "pathological" to describe a range of cultural practices. Globally, communities are divided by the ways in which certain diseases are seen as more worthy of attention and resources than others. If one were to do the accounting on medicine since the nineteenth century, then, there would be a mixed balance sheet. For each positive development, there seems to be a series of negatives that were not entirely anticipated. The

following article divides the medical history of the late nineteenth and twentieth centuries into the medical categories of investigation, diagnosis/prognosis, cure, and decorum.

## Investigation

During the past 200 years, the practice of medicine has increasingly drawn on scientific results and has incorporated a model of investigation broadly in alignment with scientific research trends. This has had implications for physicians who have had to make the transition from clinician to researcher, but also for patients. Once the object of scrutiny by trained medical eyes looking for symptoms and signs, patients have become the object of scientific scrutiny, examined by ever more precise instruments and put into large studies, spanning national research agendas. The developments discussed in the following paragraphs, then, are inventions and discoveries that have marked medicine's move from a clinical practice centered on *healing* to a scientific practice, also concerned with *knowing.*

*Wartime Innovations.* War has long driven technological innovations in medicine; the nineteenth-century Crimean War motivated the professionalization of nursing, thus changing the infrastructure of medical institutions. In the twentieth century, the many bloody wars provided theaters for developing medical and surgical skills as well as such drugs as antibiotics, anesthetics, blood products for transfusion, vaccination programs, and psychological treatments.

Vaccination tends to be heralded as the triumphant victor of the war years. The first large-scale vaccinations were given to the British and French soldiers in World War I. By the end of World War II, there were vaccines available for tetanus, diphtheria, yellow fever, poliomyelitis, measles, whooping cough, and hepatitis. Although effective, these vaccination programs were slow to become available worldwide, leaving underdeveloped parts of the world struggling with continued illness. Vaccination itself was also not without risks. An early disaster in 1902 in Mulkowal, India, fatally infected people with tetanus in contaminated plague vaccine. Vaccination also raised the specter of medical experimentation and worries about medical control of vaccines. The combined MMR (measles, mumps, rubella) vaccine in 2000 caused a controversy similar to one surrounding the syphilis vaccine in 1870. The worry in 1870 was that the vaccine could cause syphilis; in 2000, the MMR vaccine was worried to be associated with autism. So, even though vaccination has come a long way since the days of cowpox, the development of vaccines in the context of war has not ensured equity in their delivery or assuaged fears of risks that come along with vaccination.

*Antibiotics: The Wartime Spin-Off.*   A crucial shift in the medical understanding of disease and cure came with the discovery of antibiotics. Even though the first antibacterial action was noted by Louis Pasteur as early as 1877, it was Alexander Fleming, working on infection resistance during World War I, who identified penicillin. It was not initially thought to be a promising discovery because penicillin was difficult to manufacture in quantity. However, in 1940 Howard Florey kept alive a patient who was near death from septicemia for four days before running out. After approaching pharmaceutical companies in the United States and then in the United Kingdom, Florey tested his mass-produced penicillin on war wounds in North Africa to great success. Penicillin became a war "spin-off." The pneumonia fatality rate before penicillin's production hovered around 30 percent. After penicillin, it dropped to 6 percent. So successful was penicillin that in the 1950s it accounted for 10 percent of the U.S. pharmaceutical industry's revenues. Streptomycin and isoniazid followed, and tuberculosis was a disease that was predicted to be completely wiped out during the twentieth century. However, as soon as the new antibiotics were implemented, resistance quickly became an issue, both in wartime and peacetime.

Whereas pain had not exactly prevented surgery or medical intervention in the past, anesthesia was particularly valued on the battlefields, where amputation was common, and in the birthing room, where pain was all too evident. Queen Victoria took chloroform during her labor, and the battlefields of the twentieth century became excellent places for using and researching the effects of anesthesia. Although anesthesia was undoubtedly comforting for those undergoing surgery, it came with the concomitant danger of addiction. Nineteenth-century experiences with ether, opium, and chloroform encouraged Sigmund Freud's encomiums to cocaine as an anesthetic, which he mistakenly thought would not be addictive.

*Regulating Medical Research.*   World War II also offered an opportunity to reconsider medical experimentation and research in warring states. The Nüremberg Code gave ten explicit rules to govern medical research; chief among them was that the consent of a medical subject was essential before medical research could be carried out. Despite the clear messages of World War II and the Nuremberg Code, there is still substantial controversy over the use of prisoners of war as well as soldiers as "guinea pigs" for medical experimentation and vaccination. The First Persian Gulf War did little to remove lingering doubts about the role of experimentation in military medicine; evidence of a Gulf War syndrome is hotly contested, but it is speculated to result from cocktails of vaccines and antibiotic agents given to soldiers before and during deployment in the Middle East.

Ethical regulation, however, is only one form of regulating medical research. The late twentieth century has seen increasing attention to regulating medical research, and increasing frustration at the difficulties in attempting regulation. One central issue is making medical research more publicly accountable. As case studies of medical innovation reveal, there is a widening gap between medical research and public concerns and priorities. Some signs of this include the increased influence of private interests in regulating research, most notably in pharmaceutical research areas. Programs to facilitate citizen participation have been scaled down; there are few places where lay interests intersect with medical experts. Also, there is a worry that public information about medical research is either not available or not accessible. Even though the Internet has become a leading broker for medical information, there are still worries about who has access to this information and that the general quality of available information is low.

## Diagnostics

The holy grail of medical research is its ability to assist with the diagnosis of illness and the prescribing of the appropriate intervention. As such, the growth and development of diagnostic aids and tools has grown alongside basic and clinical medical research, sometimes bridging the two.

*Seeing the Inside.* One of medicine's longstanding limitations was the inability to diagnose internal illnesses. The body did not yield to palpation and perception alone, and using an autopsy to diagnose internal illness was not satisfactory to patient or physician. Early scopes, including the stethoscope (1816), the ophthalmoscope (1851), and the gastroscope (1868), allowed medical practitioners to see or, in the case of the stethoscope, hear, the heart, lungs, and intestines; the inside of the eye; and inside the stomach, respectively. But it was not until 1895, when Wilhelm Conrad Röntgen found "invisible rays," that pictures could be produced of the inside of the body. These "X" rays, when projected through the body with cathode rays and onto photographic plates, yielded radiographs that displayed bones, shattered bullets, the dense mangle of tumors, and the shadowy tubercular lung. The diagnostic potentials of X-rays seemed limitless, and even now, X-rays are regularly given for the same reasons. However, in the twentieth century, it was found that X-rays caused cancer and exposure should be limited. Supplementing X-rays are CAT (computerized axial tomography) scans, which also use X-rays but produce high-resolution 3-D images, and PET (positron emission tomography) scans, which present imagery based on radioactivity emitted from patients who have been administered radioactive glucose. In the latter scan, different areas absorb the glucose in different amounts; brain areas that have been dam-

aged and are inactive from stroke, for example, absorb little glucose and are revealed as a dark spot in the scans. The latest imaging technology to see inside the body is MRI (magnetic resonance imaging), which uses no ionizing radiation at all. Instead, it relies on the resonance of hydrogen atoms when exposed to energy from magnets. Images obtained via CAT, PET, and MRI allow patients to remain under relatively normal physiological conditions, while giving a clear picture of damaged and healthy regions. They also show soft-tissue injuries, which ordinary X-rays cannot reveal. Ultrasonography, or ultrasound, which was developed from naval technology, is applied in cases in which X-rays are impossible and CAT, PET, and MRI are either undesirable or unnecessary. This usefully applies to diagnosing fetal disorders in the womb. It is undeniable that these imaging technologies have added to the stock of medical theories helping to diagnose disease. However, increasing powers of diagnosis have not always worked alongside cure; this raises the problem of being able to diagnose a patient but not being able to offer appropriate treatment or cure.

*Smaller and Smaller.* Microscopy also allowed diagnosis at the level of the cells and even at more minute levels, complementing the structure and tissue level of diagnosis afforded by modern scanning and ultrasonography. By 1946, objects could be magnified 200,000 times, and in the 1970s, scanning electron microscopes revealed the world of the cell and cellular-level disease. Chronic diseases, such as diabetes and cancer, looked very different at the cellular level, and, coupled with new theories about genetics, the minute structures of the pancreas and other organs seemed promising locations where diagnoses could be made or confirmed. Biopsy could yield samples of a range of tissues, and, increasingly, medical education has included looking down microscopes alongside looking down throats.

*Genetic Diagnosis.* Other diagnostic breakthroughs were made possible by the burgeoning science of genetics. Alkaptonuria (characterized by dark urine and arthritis) was the first metabolic disease suggested to have a congenital basis in the literature in 1897. Other diseases were suspected to be congenital, and biochemistry combined with a knowledge of Mendelian laws helped to identify Down syndrome (trisomy 21) and sickle-cell anemia as genetic disorders. This has opened new research possibilities as well as fronts for intervention. It is now common to screen fetuses for genetic disorders before birth, with Down syndrome remaining a concern, but joined by myriad others. The Human Genome Project, nominally completed in 2000, has been touted as the first step at identifying and then diagnosing the full complement of human genetic disorders. However, there is considerable controversy over the results of the project. It remains to be seen whether identifying genes can guide medical intervention and prognosis, because gene expression and gene interaction are not yet fully known.

*Diagnosis Without a Doctor.* Technological innovation has meant that the clinic and the laboratory can enter domestic space with increasing ease. Home thermometers for monitoring fever as well as resting body temperatures are ubiquitous in the first world. Pregnancy tests were among the first home testing kits that supplanted the visit to the doctor's office or clinic. Diabetics routinely check their own glucose levels, people with high blood pressure can buy blood pressure cuffs to check their numbers at home, and home urine and blood tests are growing in availability and quality. Diagnostics that started out as research initiatives and were gradually outsourced from doctors to laboratories are now coming home. With greater access to diagnostic tools, patients are put in a position of having more and more information, some of it quite technical and produced by their own hand, at their disposal when seeking a medical prognosis and treatment.

## Cure

"An ounce of prevention is worth a pound of cure," goes the old adage. The emphasis on disease prevention instead of curing after the disease process has begun has been a hall-

mark of late twentieth-century medical practice. But the pound of cure has not entirely taken the back seat. Indeed, medicine has identified an ever-growing number of ailments that might be amenable to medical intervention. Studying those ailments closely has been paramount, with new research areas pushing medicine to adopt scientific methods and standards in order to measure the success of cures more accurately. This section deals with disease prevention, drug therapy, reproductive medicine, and the clinical trial.

*Prevention.* The twentieth century mass-produced vitamins, tried to perfect vaccines, and developed sciences of nutrition and public health in attempts to shore up preventive measures that would encourage and ensure health. These measures have attacked two of the leading causes of disease in the developed world in the twentieth century: heart disease and cancer. Advances in immunology also allowed medical intervention to be conceived of in a new way. Instead of a range of illnesses brought about by individual reactions to foreign bodies, an *immune system* united these reactions into an integrated theory of antibody response. At first, this meant better control over blood transfusion reactions because blood types and Rh (rhesus) factor could be considered for donors and recipients; a common cause of complications was thus avoided. This opened up a world of interventions into allergies and other immune system disorders. This first endeavor at understanding immunity as warding off foreign invaders (the notion itself owing much to war imagery) has, however, been complicated by the identification of autoimmune diseases, such as lupus, Crohn's disease, and notoriously, HIV/AIDS. Acquired immunodeficiency syndrome has done much to alert lay publics and researchers alike of the power of autoimmune disease. Most immunological treatments rely on the body's own immune system, the very target of the human immunodeficiency virus (HIV), thus rendering standard immunological techniques impotent in the face of the disease. The optimism of twentieth-century medicine that it was only a matter of time before diseases could be cured was dealt a considerable blow.

Indeed, the optimism of the mid-twentieth century was attacked from multiple directions. Epidemiologic and demographic studies suggested that the increasing life span enjoyed by patients in the West was due to public health measures, including cleaner water, public hygiene, and a rise in general standards of living and not by the new antibiotics or herculean interventionary strategies by physicians or technologically advanced hospitals. As evidence, epidemiologists produced data showing that the poor, women, and minority groups, whose living standards had not improved, still faced illness more often than the wealthy. This did little to slow the pace of the development of drugs, surgical techniques, and other interventions, but it did point out that the biomedical model of disease might be too limited to explain health disparities.

*Drug Therapy.* If prevention failed, one could still live with illness; drug therapies have made illness more bearable for those who have had access. In a countertrend to the growing pessimism about medical cures, optimism has replaced a rather conservative approach in the development of drugs. In the nineteenth century, French and German chemists had isolated many active ingredients from plants, including quinine, strychnine, emetine, caffeine, and acetylsalicylic acid (aspirin). These "natural" compounds were synthesized, and by the 1920s there were several synthetic antipyretics (fever reducers) and analgesics available to patients. However, in the early years of drug development, pharmaceutical companies went down a conservative path, most drugs being derived from natural sources. Salvarsan, a synthetic treatment of syphilis, was an exception, but, because it was known as *606* and was preceded by 605 unsuccessful attempts, it is easy to see why companies did not expect immediate success or recompense. The Bayer company was the first to break from the conservative path, and in 1932 it announced the use of Prontosil, a synthetic dye, for eliminating deadly streptococcal infections. The Pasteur Institute in Paris

discovered that Prontosil's mode of action was to form another compound, sulfanilamide, in the body. An English pharmaceutical company raced to patent their version, called *sulfapyridine,* in 1938. This pattern of discovery and patent was to define drug research and development for the rest of the century. Vitamins, hormones, insulin, and sulfa drugs were all available between the wars to treat deficiencies, reproductive disorders, diabetes, and infections. The conservative notion that drug development was expensive and framed by trial and error was replaced with an optimism that "the chemotherapy of infectious disease" was promising in clinical and economic terms.

*Research and Reproduction.* It was not until the 1960s that hormones became available in the form of a pill to control fertility. Initially developed to control fertility and population growth in the developing world (available in the 1950s), "the Pill" was later embraced by women in the West, but its use remains a source of concern as recurrent studies indicate links with cancer and circulatory disorders. Historians have noted that the Pill has left several legacies of medical research. First, the Pill reveals medical institutions' struggles with gender and sex; indeed, the Pill was initially greeted with criticism that intervention into natural reproduction was unethical. By contrast, advocates finally applauded medical research for putting the power of reproduction in the hands of those who bore the burdens of reproduction. Second, the Pill might also be seen as the first so-called "lifestyle drug" of the twentieth century.

Reproductive research, however, has meant more than the Pill and has pushed medical research into the media spotlight and encouraged political debate about medical research into reproduction. Medical researchers have been criticized for developing designer babies alongside the sometimes contradictory criticism that new reproductive technologies have not been successful enough at solving infertility problems. These contradictory pressures on medical research into reproduction have been framed in two rhetorics, one of hope and one of fear. On the one hand, "designer babies" is a signifier that merges the natural category of "baby" with the unnatural category of "design," conjuring a fearsome image of manufactured life. On the other hand, intervention into infertility conjures hope for couples who desire a child. These competing rhetorics have framed much of contemporary debate on reproductive technologies and show evidence of structuring other public debate on controversial medical research.

*The Clinical Trial.* From Hippocratic times there have been tensions between healing and knowing, and between healing someone in particular and the process of healing in general. A key feature of twentieth- and twenty-first-century medicine has been the rise of the clinical trial and evidence-based medicine, the latter of which is rhetorically positioned as the gold standard of medical evidence. The randomized, controlled clinical trial is designed to eliminate physician bias (and, critics maintain, judgment) by evaluating the efficacy of medical treatments, not by the perception of medical workers or the patient, but by comparative standards between treatments. These standards are assessed by administering to patients randomly assigned as control subjects, who are given either nothing or the best known current treatment, or as test subjects, who are given experimental treatments thought to be better than the best known treatment. The first randomized, controlled trial was designed to establish the effectiveness of the new drug streptomycin against tuberculosis in 1946 (although James Lind is sometimes credited with the first trial, in 1747, on the causes of scurvy). The streptomycin trial showed that streptomycin was more effective than traditional methods in halting tuberculosis. Since 1993, the Cochrane Collaboration has stepped in to analyze a bewildering array of clinical trials for drugs, surgery, and even medical behaviors, sometimes with conflicting or ambiguous results or divergent methodologies. The collection mediates research results for such parties as doctors, midwives, nurses, technicians, and lay audiences of patients. The need for such a research mediation service flags the transition of medical research from a specialized area of academic interest to the center of medical practice and lay interest. The rise of mediation organizations raises questions about who is most capable of evaluating medical research and, equally important, who is in the best position to communicate best practice in medical research.

Clinical trials have also produced tensions between disease models. Substances like drugs must first be tested on animals. Questions are raised as to whether animals are adequate models for human disease processes. If drugs are deemed successful in animal trials, they then are tested on large human populations. Here, questions are raised about statistical results and how they are applied to the varied experiences and presentations of human pathology. Clinical trials on cancer drugs underscore these unanswered questions, as does research on efficient therapies for AIDS, for which no animal model has proven adequate and life expectancies under standard therapies range dramatically.

## Decorum

Medical practitioners can boast that they have one of the oldest examples of a decorum, a written set of norms for social behavior and practice, dating to the late Hippocratic writers of the third century B.C.E. But, what constitutes appropriate behavior in medical contexts is not always so clear. In the past 100 years, propriety has changed drastically, in part as a response to technological developments in disease investigation, diagnosis, and cure. In the nineteenth century, the "silent doctor" reigned, and reserve in bedside manner, secrecy, and restraint defined medical interactions. The rise of the belief that patients should be involved in medical decision making has made some patients complain that they know too much about their conditions and that they are harassed by talk and information from all sides. The twentieth century has also seen the rise of knowledge intermediaries in medicine; nurses, midwives, technicians, translators, and caregivers come in seemingly infinite specialties and all treat patients with the codes of their profession. Insurance companies and health maintenance organizations step in to direct patient care and adjust costs; no longer do physicians send their bills at the end of the year as they might have done at the turn of the twentieth century.

The rise of clinical research has also markedly changed medical decorum. Clinical trials, as they collect data from large populations, are carried out with multiple centers of organization through research facilities, hospitals, and individual physicians. This necessitates elaborate communication networks that eschew the empathy of the classical doctor for the technical precision of the research scientist. Not that empathy does not have its place. The "new" biopsychosocial model of clinical practice emphasizes the role of medical practitioners' own expectations and feelings about the patient and illness; this model predicts that patients will respond better to medical practitioners who address their physical ailments alongside personal concerns and their social context. The conflicting modes of communication demanded by clinical trial research and the biopsychosocial model are not lost on clinicians who report that the pressure on them to communicate in multiple modes has never been greater.

Where will the patient be standing or sitting? The place of consultation has also radically changed. In the United States at the end of the twentieth century, an extreme example, only 2 percent of medical consultations happened in the home. Other systems, including those in the United Kingdom, Australia, and France, have encouraged general practitioners to make home visits, but the general trend is to consult patients in clinical or even research settings in major urban centers. And then there is the matter of whether or not patients in consultation should be sitting or standing. This, it seems, is a marker of how long consultations or visits should last. At the end of the twentieth century, the range of visit times is roughly from 3 to 15 minutes in most First-World nations. A study in the 1980s reflected that, although verbal exchange was central to the eventual diagnosis, patients were given an average of 18 seconds and, at the far extreme, 2 minutes, to tell their physician the reason for their visit. Tellingly, Julian Tudor Hart quotes a U.S. doctor as follows: "I have only one sit-down desk in three rooms, because when you sit down you double the office call length. You get someone sitting on their backside; they talk longer than when they're standing. . . . I'm here predominantly as a medical man, and if people are waiting I don't think they would appreciate me sitting here talking about something other than medicine" (p. 552). Medical decorum, then, has changed as medicine has changed from a set of practices of healing to a set of practices equally concerned with knowing.

## More of Medicine

Only an estimated 1 in 17 workers in the American health system is a physician. The rise of the medical technician and other knowledge intermediaries in medical work raises questions about who actually produces medical knowledge. Is the physician still at the center of medical research, diagnosis, and cure? In the West, at least, the physician has moved to the periphery to share the medical functions with other professionals. In many parts of the world, people have long relied on local experts for their medical care, and midwives and nurses provide the bulk of professional medical care. For the West, the twentieth century has brought about the medicalization of lives in new ways and even common sense is seen to have a medical basis. Infant care, interpersonal relationships, and communication styles have all been seen to be objects for medical scrutiny and expert opinion, leading some critics to suggest that everyday life has been medicalized and common sense has been scientized. And yet, medicine is equally a product of social forces, from war to social imperatives to have children. How the relationship between medicine and the social good is imagined in the next century is an open question and is related to the "more" that medicine has become. This opens yet other questions about whether more medical research is needed in particular areas, whether or not lay publics want more accountability from medicine, and not least, where more economic support will come from for future medical innovation.

## BIBLIOGRAPHY

Beckman, H. B., and R. M. Frankel. "The Effect of Physician Behaviour on the Collection of Data." *Annals of Internal Medicine* 101 (1984): 692–696.

Bloomfield, B. P., and T. Vurdubakis. "Disrupted Boundaries: New Reproductive Technologies and the Language of Anxiety and Expectation." *Social Studies of Science*, 23. (August 1995): 533–551.

Bourke, J. *An Intimate History of Killing: Face-to-Face Killing in Twentieth-Century History.* London: Granta, 1998.

Canguilhelm, G. *On the Normal and the Pathological.* Translated by C. R. Fawcett, ed., and R. S. Cohen). Dordrecht, The Netherlands: Reidel, 1978.

Cooter, R. "War and Medicine." In *Companion Encyclopedia of the History of Medicine,* edited by W. F. Bynum and Roy Porter, 1526–1563. London: Routledge, 1993.

Dutton, D. B. *Worse than the Disease: Pitfalls of Medical Progess.* New York: Cambridge University Press, 1988.

Gaudilliere, J. P. and Lowy, I., eds. *The Invisible Industrialist, Manufacturers and the Production of Scientific Knowledge.* London: Macmillan, 1998.

Hart, J. T. "Going to the Doctor" In *Medicine in the Twentieth Century,* edited by R. Cooter and J. Pickstone. London: Harwood, 2000.

Junod, S. W., and L. Marks "Women's Trials: The Approval of the First Oral Contraceptive Pill in the United States and Great Britain." *Journal of the History of Medicine.* (April) 2002: 117–160.

Kevles, B. *Naked to the Bone: Medical Imaging in the Twentieth Century.* New Brunswick: Rutgers University Press, 1997.

Marks, H. *The Progress of Experiment: Science and Therapeutic Reform in the United States 1900–1990.* Cambridge, UK: Cambridge University Press, 2000.

Montgomery, S. *The Scientific Voice.* London: The Guilford Press, 1996.

Porter, R. *The Greatest Benefit to Mankind: A Medical History of Humanity from Antiquity to the Present.* London: Harper Collins, 1997.

Porter, R. *Blood and Guts: A Short History of Medicine.* New York: W. W. Norton, 2003.

Tiles, M. "The Normal and the Pathological: The Concept of a Scientific Medicine.' *The British Journal for the Philosophy of Science.* 44.4 (December 1993): 729–743.

*Joan Leach*

# PERSPECTIVES ON TECHNOLOGY AND SOCIETY.

Although the problem of the relationship between technology and society is both highly significant and of long standing, recent research has proposed a series of important alternative conceptions of the problem. A wide variety of social scientists have been involved, among them, sociologists, anthropologists, political scientists, scholars in communications studies and in cultural studies, especially those working in the interdisciplinary area known as *science and technology studies* (STS) or sometimes known as *social studies of science and technology.* As discussed in this section, crucial to this recent research is the challenge to the way in which the problem was originally conceived. The upshot of this research movement is to suggest a new way of framing the problem. Instead of asking about the connections between technology and society, and following a research program aimed at specifying those connections in greater and more specific detail, recent research proposes that technology is not thought of as separate from society but, in the many particular senses to be articulated in the following paragraphs, as constitutive of it. The importance of this approach to technology and society is not just that it provides greater insights into the relationship between the two. More significantly, it provides an important way of challenging and rethinking key elements of philosophical positions associated with essentialism and objectivism in general.

To develop this argument, this section discusses four main models of, or perspectives on, the relationship between technology and society. The point is made that each of these perspectives parallels debates about the character—essential, objective, or otherwise—of science and scientific knowledge. The strengths and weaknesses of each perspective are outlined. The implications of these new perspectives are then considered using an empirical example of the "role" that politics and society play in the new technologies of airline passenger management.

## Technology Is Neutral

A well-known position holds that technology is essentially neutral, both with respect to its technical effects and to its social and political qualities. In other words, technology is thought of as having no necessary effects or consequences, other than those that transpire in virtue of its use and deployment. This means, for example, that organizations are more or less free to deploy technologies as they see fit and in ways which they regard as conducive to desired organizational outcomes. At the societal level, this implies that societies are similarly free to choose the technologies they wish to achieve desired outcomes. (Of course, the analogy immediately reveals a key weakness of the position: the assumption that "desired outcomes" are either unambiguous or are the result of consensual agreement within a society). The argument that the effects of technology depend (solely) on how they are used exactly parallels the debate about the social responsibility of scientists wherein some have taken the position that scientists themselves cannot be held responsible for the ways in which their science is deployed. From their point of view, the outcomes of scientific research are objective and made available to society at large on the basis that decisions about use fall within the provenance of politics, not science. Similarly, the position that technology is neutral envisages a process of design and development of technology that is largely divorced from difficult decisions about subsequent use and effects. The pro-neutral position further points to the virtues enjoyed by technological developers and innovators of designing and inventing while being unfettered by the "constraints" of taking possible adverse uses into account.

A celebrated example of controversy about technology in this perspective concerns weaponry and gun control. Those complaining about the increase in violence and armed assault in cities and (especially) in schools, point to guns as the root cause. By banning or severely restricting the sale of arms and handguns, they argue, we can reduce the adverse effects. They unequivocally locate the problem in the character of the technology. The pro-gun lobbyists respond that "guns don't kill people—people do." Thus, in line with the view that technology is neutral, the pro-gun lobbyists argue that it is the deployment of the technology, not the technology itself, that is to blame for the effects. Their argument draws on and feeds into ideas about freedom of choice and individual rights, that is, the notion that government should not interfere with citizens' freedoms to decide whether and how to deploy technology.

## Technology Determines Society

A second set of perspectives portrays technology at the center stage of societal change. The view here is that technology is (literally) the engine of significant developments in history and in the ways in which humans relate to each other and organize themselves. The most famous statement of this position is attributed to Karl Marx—despite disagreement among Marxist historians as to what precisely he intended—who said that "the hand mill gives you society with the feudal lord, the steam mill gives you society with the industrial capitalist."

There are, however, two main difficulties with this position of technological determinism. First, it is notoriously difficult to isolate the role of technology in any specific historical change. Can we say that mobile telephony has given rise to a more mobile lifestyle in societies where this form of electronic communication has been taken up, or was this a societal change that was about to happen in any case, and that merely coincided with electronics manufacturers' spotting the market opportunity? At best, we can say that the effects of technology are always entangled with changes that may or may not have been due to social, cultural, and historical changes. Second, the claim that technology determines the course of societal and historical change depends on being able to unambiguously specify the character of the technology that is effecting the change. In other words, it assumes an essentialist view of the technical nature and capacity of a technology. Against this assumption, many examples of the introduction of new technology emphasize how uncertain are the effects and hence the capacity of the technology. The characteristic of technologically related change seems to be that outcomes are counterintuitive. They frequently confound expectation, so that the "actual" capacity of technology to change is constantly being rewritten and reinterpreted in the light of events. Is the mobile phone the cause of new patterns of social behavior among young people? Well yes, but only because we now agree that, quite against original expectations, text messaging emerged as an unintended by-product of this technology. The fact that new technologies are frequently prone to spectacular and costly failure similarly exposes the processes whereby "the capacity" of the technology is the subject of continual renegotiation and dispute as experts and others debate what went wrong and what in fact was technologically possible. To overlook this feature of technologies in favor of a summary account that "technologies drive history" is to miss perhaps the most important element of dealings with technology.

## Technology as a Social and Political Construction

This perspective significantly counters the preceding two by emphasizing the social and political processes involved in the design, development, and use of new technologies. This perspective is based on the constructivist tenet that "it could have been otherwise." In other words, the technology could have turned out differently. It could have ended up with a different design, technical capacity, and putative effects. At the core here is the argument that design and development are not straightforward extrapolations from existing technical knowledge and options. Instead, the technology is to be understood as the contingent outcome of a series of social, political, and organizational factors.

However, this perspective encompasses a range of differing positions both with respect to the particular sources of contingency—for example, between the interests of social groups, the political bias of designers, or gender influences—and with respect to how far authors are willing to push the idea of contingency. As discussed herein, differences in the latter respect have an important bearing on the capacity of the analysis to confront questions concerning essentialism and determinism.

For authors such as Langdon Winner, the position that technology is neutral is "somnambulist." It overlooks the political dimensions of technology to such an extent that we uncritically accept what we are told about its use and deployment. Indeed, these political dimensions are inscribed in the technology in such a way that the technology seems neutral. So what we think are choices in the use of neutral instruments of technology are in fact responses to (and collusion with) politically impregnated technology. Winner's project is thus to draw attention to the political processes involved in the design, construction, and implementation of new technology. A famous example is his discussion of the bridges built by the architect Robert Moses on Long Island. Winner claims that the bridges were designed (consciously or unconsciously) to have a particular social effect. Poor people and blacks, who normally used public transit, were kept off the roads because the twelve-foot-tall city buses could not travel under the new overpasses. One consequence was to limit access to Jones's Beach, which was the architect's widely acclaimed public park.

For those advancing the social construction of technology, claims about the neutral or determining qualities of technology overlook the social and historical contingency of the ways in which the technology is shaped. Thus, in the classic example of the social shaping of bicycles, the form and function of the modern bicycle is shown to be the outcome of a series of processes affected by the interests of social and political groups. For example, the interests of the manufacturers of weapons and of sewing machines, the former having recently lost their market with the ending of a war, contributed to the first development of metallic wheel rims and frames. The particular interest in bicycles for sporting purposes as well as the emphasis on speed, led to larger and larger front wheels (the penny farthing). But controversy over the use and safety of these machines eventually led to new designs that were more in line with what we now recognize as a modern bicycle. Researchers in social construction of technology point out that had the arguments gone another way or had another social group emerged as powerful campaigners, we could have ended up with quite a different bicycle.

In arguments of both the social and the political constructions of technology, authors invoke the "it could be otherwise" clause as a way of exposing different factors involved in the construction. Notably, however, they both deploy specifically limited interpretations of this clause. For example, attention is drawn to political construction as a way of showing how the "actual" effects of a technology disadvantage some users (and empower some others). The nature and status of these effects are thereby largely taken as given, often through an appeal to a kind of common sense, such as that certain things can be done with nuclear power stations that cannot be done with a ballpoint pen. (In this respect, the work echoes the Marxist determinist position.) Research that highlights the social factors tends also to make partial use of the constructivist clause. The arguments are most comfortable when dealing with actual controversies between alternative design suggestions and proposals, rather than considering why only such a delimited set of options was thought possible in the first place. Why not square-wheeled bicycles for instance? By implication, the argument suggests that considerations of "efficiency" are similarly to be understood as the outcome of processes of social construction. Our contemporary perception of penny farthings as quaint and slightly ridiculous is a reflection of contingent criteria for what has come to count as "efficient." But the social constructivist perspectives tend not to go so far.

Avoiding the (apparently bizarre) question about square wheels has the unfortunate effect of overlooking the ways in which considerations of how what counts as "natural," "efficient," or "obvious" are themselves part and parcel of the ways in which participants apprehend different aspects of the problem as self-evident or problematic. The social construction work also proposes that the relevant social factors are most evident when technological failure opens inquiries into the construction process and whenever controversy rages, but the social construction work, questionably, suggests they become mute when parties reach a consensus about the most appropriate technical solution.

## Technology as Interpretive Action

This fourth perspective addresses these last issues by insisting on a more action-oriented, interpretivist approach, which is consistent with the view that ideas about technology and society are thoroughly interpenetrated. According to this perspective, the two great concepts of technology and society are not merely linked by the flow and influence of "factors" one upon the other, but mutually constitute one another. Just as, from the perspective of ethnomethodology, "society" is the constant "accomplishment" and "production" of its members' actions, so this perspective focuses on the ways that "technology" is "achieved," "rendered," and "constituted" as an unavoidable, essential and constitutive feature of the constant reproduction of social order.

A starting point for this perspective is the observation that technology is typically experienced as a matter of considerable ambivalence. By contrast with the grand claims for the ways in which technology is changing society, or how the Internet is significantly affecting our lives and so on, it is notable that ordinary everyday usage of technology often involves considerable inconsistency and uncertainty. We know, for example, that technology is both good and bad: a few mouse clicks will make available a cornucopia of knowledge or will spread race hatred and pornography. We love and hate the technology. The simple mastery of PowerPoint presentations can provide the basis for a slick performance (and still impress some in corners of academia), yet the experience is also one of engaging with a coercive technology, one that compels the author to reduce wisdom and discursive eloquence to little bullet points and that extends anxiety by making it possible to change things right up to the final moments before delivering the presentation. Finally, we know that the technology both works and does not work. It is wonderfully convenient to be able to send a colleague a long paper attached to an e-mail, yet it may well be that a server is down, the modem has stopped working again, or the phone connection is disrupted.

Relatedly, we know that the reception and use of technology is unevenly socially distributed. This is well illustrated by MacKenzie's "uncertainty trough," a schematic representation of the relation between the degree of uncertainty (along the y-axis) in a particular technological claim or product, and the social distance (along the x-axis) from the site of production of the claim or product. Based on the production and reception of measures of nuclear missile accuracy, MacKenzie demonstrates that this graphical relation takes the form of a U-curve. Nearest the site of production, there is high uncertainty. The manufacturers themselves hold a "realistic" view about the technology in question, whereas those further removed from the site of production—for example, military personnel and politicians—are much more confident in the technology. Yet further away from the site of production, uncertainty increases again, corresponding to people there being disinterested in, and largely alienated from, the institutions that are manufacturing and using the technology. Although initially developed for the case of measuring nuclear missile accuracy, the uncertainty trough has a useful applicability to technologies in general.

The move from the third to the fourth perspective is the most recent and perhaps the most controversial development in research on technology and society. In particular, it raises the question as to the point of invoking the "it could be otherwise" clause if not as a means of revealing the (actual, underlying?) social and political factors. The answer is that a more thorough "technology as interpretive action" perspective reestablishes the provocative power of this kind of work to urge a reappraisal of basic philosophical assumptions about essentialism and objectivism.

At the core of the difference between the third and fourth perspectives are two related but distinct senses of the idea of *interpretive flexibility*. As we have seen, a first sense of interpretive flexibility (characteristic of technology as social and political construction) stresses the relevance of social, political, cultural, and organizational aspects to the technical outcome, so that the nature and capacity of the technology is not the straightforward extrapolation from existing technical systems and solutions, but instead is the consequence of contingent circumstances. In an important sense, social, political, and other factors can be understood as *built into* the technology. This sense is summarized in the STS tradition by slogans such as "technology is congealed social relations"; "technology is politics by other means"; "technology is action at a distance"; "technology is society made durable" (and "software is society made malleable").

The second sense of interpretive flexibility—characteristic of technology as interpretive action and less common in the STS literature—is that the apprehension, use, and interpretation of technology is also subject to an in principle unknowability. The first sense implies that a consensus emerges from the passage of technical development such that all parties reach agreement on the nature and capacity of the technology that has evolved. The second sense argues for a more open-ended approach. It suggests that the business of interpretation and use extends beyond any putative moment of consensus. One implication is that we need to understand technology in use as a constant process of interpretation and understanding, or, if one allows the textual metaphor, as an ongoing accomplishment of reading the technology text. Substantial evidence from studies of new technologies supports this second sense of interpretive flexibility. For example, research into the supposed effects of Internet-based technologies shows the centrality of counterintuitive outcomes.

The connection between the two senses of interpretive flexibility is important for understanding the role of technology in governance. In the first sense, technology can be understood as having had certain social and political predilections "built in." For example, in Winner's account, the bridges designed by the architect Robert Moses are said to embody the racial bias of the architect because they prevent the passage of public transport. We are thus encouraged to "read off" the structures of governance from what we discover about the circumstances of the production and development of the technology. It follows that different technologies and different technological paths can afford and give rise to different forms of governance. In line with the general concern about the diminishing lack of clarity and structure as organizations move from hierarchical to network forms of organization, networked technological systems are proving more popular. Thus, Castells suggests it is a happy coincidence that Internet technologies appear just at the point in our history when institutions recognize the greater advantage of networked forms of organization.

One problem with this first sense of interpretive flexibility is that the reading of structural form is questionable. It is unclear, for example, whether and to what extent the bridges actually had the effects Winner ascribes to them. Rather, the very attribution of effect seems to be part and parcel of Winner's own determination of a putative underlying "racial bias." Although this is suggestive in unlocking and deconstructing the contention/assumption that technologies are neutral, and have fixed capacities, it merely shifts the weight of essentialist reasoning from the technical to the sociopolitical. In other words, the presumption is that whereas we cannot straightforwardly read off the *technical* characteristics of the technology, we are somehow capable of reading off the (actual) *social and/or political* characteristics that apply.

The second sense of interpretive flexibility instead insists on the deep indeterminacy of "readings" of the social and/or political from a technology. It encourages the heuristic starting point that all interpretations and uses of technology, including understandings of their effect and impact as well as their origins and their putatively inculcated politics, are inescapably contingent and are at best construable in terms of the radically local circumstances of their apprehension. Or, put in less-trenchant terms, readings of the governance implications of technology are occasioned. In a resonance with the well-known proposal from ethnomethodology that we take such interpretations and uses as topic rather than (just) as resource, this second sense of interpretive flexibility opens up a research program of analyzing how determinations (interpretations, uses) of governance structure are generated, made, and sustained, rather than simply taking them as a self-evident starting point. This means that a primary focus for analysis would be the ways in which technologies are "made to do" political work.

The key STS slogans that were introduced earlier now require important modification. They should read as follows: "technology is *an attempt to congeal* social relations"; "technology is *an effort to pursue* politics by other means"; "technology is *the pursuit of* action at a distance"; "technology is *the striving to make* society durable" (and "software is society made malleable").

This conceptualization of technology as interpretive action opens up three further important themes: (1) the nature and extent to which technology development can be understood as involving the configuration of its users; (2) a reappraisal of the rather static current models of the producer–consumer relationship (in which producers and consumers are respectively considered as the proponents and dependents on systems of governance); and (3) a reevaluation of the problems of affordances and postessentialist approaches to the "textuality" of technology (where "affordance" can be understood as a feature of the technology text temporarily and reflexively accomplished in the light of germane accountability relations).

## Governance and Accountability Relations in Technology

The final part of this entry illustrates some of the themes of the preceding section in relation to a specific example: an ethnographic study of the governance and accountability relations associated with the deployment of security technologies for controlling the flow of passengers at airports.

Renewed concerns about the nature and effectiveness of government have prompted renewed consideration of the framework, structures, and institutional forms that make government possible. A shift has been noted away from forms of governance based on bureaucratic authority to more complex, hybridized forms in which self-renewing networks play pivotal roles. This is governance without *government*. The move from hierarchy to network paradigms is said to bring intractable coordination and control problems. However, few of these recent discussions give serious attention to the role of technology in governance. In their focus at the level of institution and on the connection between the structural form and effectiveness of government, they are almost silent about technology. Which of our four perspectives on technology and society are best suited to redress this situation?

"Mundane" technologies—in the sense of widespread, common, and taken for granted—offer a special opportunity for exploring governance and social arrangements because they are potentially implicated in pervasive apprehensions and interpretations of governance and accountability relations. The promotion and use of pervasive, mundane technologies can be understood as a potential solution (whether intended or not) to public problems by the establishment and/or enactment of governance relations. This raises three main questions: (1) What are the relative value and utility of the concepts of governance and accountability as these derive from different social science perspectives; (2) how are ideas about the terms *mundane* and *public* experienced and consti-

tuted in practice, and how do practices stabilize around these terms; and (3) how are networks of governance involving various technoscientific entities assembled and maintained through accountability relations?

In order to address these questions, the empirical foci of the research were three interrelated case studies of mundane technologies at various stages of development—local authority recycling, traffic management, and airport security—each designed as multisite ethnographies. This involved following connections and links between individuals, groups, and agencies, and crossing conventional boundaries. For the airport security case, the research drew on observations, interviews (with passengers, passenger representatives, managers, and airport authority (AA) personnel), survey work, and the experience of having one's biometric identity recorded. A particular focus was on the various developments for controlling and managing passenger flow proposed by the AA, a privatized former U.K. government ministry that now makes huge profits for shareholders (£1 billion a year) and has contracts and/or owns airports in Britain, Europe, the Middle East, Australia, and the United States. The AA operates in a context of needing to enhance profits and shareholder dividends, and of sensitivities to security issues following September 11th, 2001, and the so-called war on terror.

Some of the more visible developments included the proposed use of radio frequency identification (RFID) tags in boarding cards (with the backing of private companies) and biometric identity systems (being pushed both by government and by private companies, which are tendering for the contracts to suggest and promote ways biometrics could be used). But the management of passenger flow also involved discussions about a wider range of technologies: different kinds of electronic equipment (scanners, databases, and so on) and the use of new and/or bigger signs to be posted around the airport buildings.

Part of the development effort involved the airport authority's finding out more information about passenger flow. This includes measuring passenger movements as flows of people, mapping passengers' ability to find their way, taking account of passenger representative groups' views on regular airport usage, and considering ways of getting passengers to shop more (because airport authorities make more revenue from shopping than from planes). It is clear that the possible technological options sit at a nexus between a tricky combination of security, way finding, and shopping. Each of these possibilities raised questions for AA management about which technology would work, would they work in combination, in what combination, and so on.

These materials suggest that the development of airport security technology involves a complex interplay of considerations and interests of different parties. It could be argued (in line with the first sense of interpretive flexibility) that the particular technical outcome, for example, the decision to use biometric security checks at a point in the flow of passengers when they have already passed through the shopping areas, is the result of the ascendancy of commercial interests over those of airline security. The results of the ethnography show, however, that this is at best a misleading simplification. Instead, and in line with the second sense of interpretive flexibility, the apprehension of the location and effect of the technology is itself an attempt at the reflexive enactment of the governance and accountability relations that are said to be in place. The technology is both the subject and site of this reflexive interpretation, on the one hand, and a discursive resource for attempting to close down competing interpretations, on the other. It is not so much that a putative mode of governance and specific regimes of accountability are concretized (congealed) in the technology. Nor are the "politics by other means" straightforwardly evident. Rather, "the technology" is best understood as itself an occasion and resource for organizing and deliberating upon a preferred set of responses "to the technology."

It is clear, overall, that many approaches in this area tend to conceptualize technology as an incidental player, as a mere appendage to systems of government and power. The "technology as neutral" perspective fits this approach. On the other hand, "technological determinism" attributes uninterrogated technical effects on governance. "Technology as a social and/or political construction" unlocks the black box but tends to shift definitive attributions to social and political factors. "Technology as interpretive action" seems to offer the best understanding of the ways in which science and technology are increasingly central to the formation and maintenance of systems of accountability and governance. Technology can be understood as a concerted effort to make governance durable. As already indicated, this framework needs careful handling, not least because of the temptation of simply "reading off" aspects of governance "from the technology." Governance should instead be understood as being actively enacted in and through the apprehension and use of the technology, leaving issues of interpretability, durability, and accountability always open ended.

Finally, then, one can see the significance of perspectives on technology and society that pursue the injunction that technology should be understood as interpretive action. The central ideas of ambivalence, the social distribution of certainty and configuring the user together reinforce the contention that "technology" and "society" are mutually constituting. Even beyond the suggestion that technology and society are co-constructed, this perspective insists on the ways in which technological discourse is an ongoing attempt to constitute society; and societal discourse is an ongoing attempt to constitute technology. The importance for research on technology and society is that both be understood as mutually provocative.

## BIBLIOGRAPHY

Bijker, Wiebe, Thomas Hughes, and Trevor Pinch, eds. *The Social Construction of Technological Systems*. Cambridge, MA: MIT Press, 1987.

Grint, Keith, and Steve Woolgar. *The Machine at Work: Technology, Work and Organization*. Cambridge, U.K.: Polity Press, 1997.

Latour, Bruno. *We Have Never Been Modern.* London: Harvester Wheatsheaf, 1993.

MacKenzie, Donald. *Inventing Accuracy: A Historical Sociology of Nuclear Missile Guidance.* Cambridge, MA: MIT Press, 1990.

Winner, Langdon. "Do Artefacts Have Politics?" In *The Social Shaping of Technology*, edited by D. MacKenzie and J. Wajcman. Cambridge, MA: MIT Press, 1999: 28–40.

Woolgar, Steve. "What Happened to Provocation in Science and Technology Studies?" *History and Technology* 20.4 (2004): 339–349.

*Steve Woolgar*

# PHARMACEUTICAL ANTHROPOLOGY.

Since the end of World War II, Western pharmaceuticals have gained popularity everywhere mainly because of the successful use of penicillin and its spectacular effect. They often represent a large part of health expenses in developing countries. In 1981, for instance, the people of Bangladesh spent nearly $75 million on Western pharmaceuticals. This was about 60 percent of all health expenses. However, 70 percent of these drugs were useless from a therapeutic point of view. As a result of lack of knowledge and inadequate control, Western pharmaceuticals are often misused and become a health hazard.

The idea that some medicines have an intrinsic power is widespread. Nevertheless, different cultures may have a distinct conception of the nature of that power. The notions of potency of a medicine and the expectations about its functioning are culturally defined and may vary greatly from one culture to another. Without the adequate accompanying knowledge, the borrowing of and self-medication with medicines from other cultures may be harmful. Western pharmaceuticals are often integrated not only into indigenous healing processes but into cultural belief systems as well. They are viewed through local concepts of healing and are often attributed special power and efficacy because they come from far away, arrive in modern packages, and are applied by nontraditional means, such as injections and capsules. They sometimes even become indigenized; they are used in a given

community as if they were authentic local products. In such cases, the effects of Western pharmaceuticals are described with traditional concepts of efficacy. In certain cultures, for example, diseases as well as medicines are classified as "cold" or "hot." A cold medicine is viewed as appropriate to treat a hot disease and vice versa. Indigenized medicines are used in culture-specific way, such as powdered and sprinkled on wounds or dissolved in herbal teas. They are sold in small neighborhood stores; they are given local names or, conversely, their names are given to traditional medicines. This process of indigenization is called *cultural reinterpretation.*

Cultural reinterpretation of Western medicines can be found in several countries. In the Philippines, Diatabs (loperamide hydrocloride) and Polymagma (attapulgite), two types of antidiarrheals, have been on the market there for decades and they are very popular. They are sold in small stores and are very well known. The Philippines consider them good for hardening and giving shape to stools. These are the same properties that they attribute to fruits such as star-apple and guava, with which they traditionally treat diarrhea.

In Cameroon, tetracycline capsules are easily available in markets, small shops, and from peddlers. This antibiotic is used in a widespread fashion, as it is believed to be efficient for the treatment of any disease. It has become so popular that it has been given the local name of *folkolo,* which means "wound healer," probably because of the common practice of sprinkling the content of the capsules into a wound.

In Brazil, the antibiotic *Terramicina* (Terramycin, oxytetracycline) is widely available at a low cost, and it has been on the market for many years. This medicine has also been indigenized. It is viewed as an intestinal stabilizer and its popular use is a single dose for intestinal ailments. It is also very appreciated for the treatment of wounds. People do not only take it orally to prevent infection, but, as the traditional method of herbal treatments dictates, they mix the contents of the capsule together with pork fat and apply it locally. This antibiotic is so popular that Brazilians have given the name of *terramicina de mato* ("herbal Terramycin") to a local herbal medicine used traditionally for the same purpose.

Traditional healers may also prescribe Western pharmaceuticals. In Ecuador, it was observed that while treating a man suffering from *llaqui*, a culturally bound syndrome similar to depression and anxiety, a Quichua healer prescribed him a liter of lemonade in which several aspirin tablets (around ten 300-mg tablets) were dissolved. The patient had to drink it at once, entirely. No studies have yet been done to verify whether it is a common practice. In Sri Lanka, Ayurvedic healers frequently prescribe Western pharmaceuticals. They say they do so because patients insist on receiving the most potent medicines.

Westerners as well borrow medicines from cultures all around the world without the proper traditional knowledge and subsequently misuse them. Kava-kava has been used traditionally by Pacific Islanders for centuries. They take it as a tranquilizing tea, in low concentration. In the last few years, it has become very popular in the United States and Canada. It can be said that it has been indigenized to the Western world. Because it is a plant, people view it as a natural and harmless tranquilizer. Suiting cultural preferences, the kava-kava is sold in capsules of increasing dosage. Misuse and overuse have led to several cases of hepatic intoxication, and recently the government of Canada has banned this product.

## BIBLIOGRAPHY

Silverman, Milton, Philip R. Lee, and Mia Lydecker. *Prescriptions for Death: The Drugging of the Third World.* Berkeley, CA: University of California Press, 1982.

Van der Geest, Sjaak, and Susan Reynolds Whyte, ed. *The Context of Medicines in Developing Countries: Studies in Pharmaceutical Anthropology.* Amsterdam, The Netherlands: Het Spinhuis, 1991.

*Lise Bouchard*

# PHYSICS AND ASTRONOMY.

Throughout much of the twentieth century, efforts to understand the nature of scientific knowledge—efforts undertaken by philosophers, historians, and some sociologists of science—have singled out physics and astronomy as paradigmatic scientific fields. They are paradigmatic in two senses: as ideal and as model. That is, physics and astronomy have been taken to represent scientific knowledge at its best and, at the same time, to set a standard for how science ought to be done. If one reads the literature in history and philosophy of science before 1980, one finds that many of the cases, examples, and historical breakthrough moments of science that are mentioned or explored come from either astronomy or physics. From the discussions about truth and falsification among the early twentieth-century philosophers of science such as Karl Popper and Rudolph Carnap, to Thomas Kuhn's demonstration of the alternating continuities and discontinuities in scientific practices, to the critical analysis of the explanatory powers of the laws of nature in Nancy Cartwright's work, physics and astronomy provide the illustrations—sometimes critiqued, often admired—of how scientific thinking is thought to work.

Obviously, physics and astronomy have contributed enormously to our understanding of the natural world, not only providing a common set of explanations but, thereby, shaping our material, social, and natural environment. How fast do things fall? How do galaxies form? What is the shape of the universe? When and how did the earth develop? How does matter hold together? What is going on in the sun? What explains the attraction and repulsion between inanimate bodies? What is matter and how does it behave? Increasingly, such questions are the subject of large-scale, intricate, experimental research programs that require hundreds of researchers and millions of dollars to complete. Many of them continue to be the subject of debate. Others have—at least for the time being—been answered to a such a degree of consensus that they set the agenda for further research or serve as working hypotheses. Yet others have been answered conclusively: They have been shown to work in applications varying from the flight of airplanes to the magnification of objects in telescopes and microscopes to magnetic resonance imaging of the body to the explosion of rockets and atomic bombs. By way of such applications and their consequences, physics and astronomy have an unmistakable impact on the way we live, on our social arrangements, and on notions of what it means to be human in the (post-)modern world; the iconic example is the atomic bomb. Their "sociality," in terms of their direct *effects* on society and culture, cannot be denied.

It is equally obvious that physics and astronomy, as human endeavors, are social in terms of how they work (for an early inventory of the various ways in which science is a social enterprise, see the work by Barry Barnes and David Edge). The sociality of their *practices* has been thoroughly documented in recent science studies. Not only is becoming a member in these areas of inquiry a matter of apprenticeship, initiation, and socialization, as Sharon Traweek has argued; Karin Knorr-Cetina, among others, has shown that the arrangements in which research and teaching are carried out—especially when large and complex collaborations are involved—bank on productive social networks, include complicated human resource issues, demand enormous public resources with an appended request for accountability, and require pertinent cultural sensibilities in all concerned. Robert Merton has pointed out that scientific conduct and the processes by which scientific facts are established conform to social rules, norms, and conventions, some of which, as Simon Shapin and Simon Schaffer have argued, have their origins in the civil society of the seventeenth century. Visible and invisible, institutionalized as well as implicit, connections between practitioners within and among fields are formative of the kinds of theories these practitioners construct, according to Sal Restivo. Andrew Pickering and Harry Collins have documented the learned, social behaviors and acquired intuitions that together make one a participant in the culture of these disciplinary fields: Knowing how to strike a balance between competition and collaboration, recognizing merit

and promise, cultivating connections, discerning between good and bad equations, and being able to decide at first sight whether a paper is solid or crackpot are just a few of these skills.

Physics and astronomy are interesting as social knowledge-making *practices* with *effects* on society, but examining these fields sociologically also has a broader cultural significance. If we want to illuminate the connections between science and society, as this volume attempts to do, it is not only the practices and effects of the physical sciences, but also their widespread treatment as *exemplary* sources of knowledge, that deserves our attention. This attitude toward physics and astronomy fuels, but also reflects, a common assurance that the thought developed in those sciences leads to generalizable, universal, objective, representative, and supreme knowledge that not only predicts and explains, in a "grand theory," the behavior of the materials that constitute the universe but that also puts these materials in their proper place. This assurance—and here is the significant twist—depends on a conception of physics and astronomy as independent of everything social and cultural, immune to human influences. This denies the embodied and material sociality that, as we now recognize, constitutes the practices of science, including physics and astronomy. This tendency, moreover, on the part of generations of analysts of science to make an asocial, disembodied physics and astronomy the model and the norm has produced two artifacts: It has made generalizability, universality, and objectivity the measure of good science and the "cosmology" of modern times, and it has brought forth a "mythology" suggesting that the disciplines of physics and astronomy, in fact, produce such science.

Those two notions have been interrogated by sociological and anthropological approaches adopted in science studies in the last two decades of the twentieth century, and the results of that inquiry are under review in the remainder of this article. Taking an anthropological view of science in culture, and building on the understanding of science as social practice, it problematizes precisely these two common assurances of and about physics and astronomy. First I discuss their "mythical" position at the pinnacle of knowledge; second, I address their mandate to establish the "cosmology"—the order of things in the known universe—that underwrites twenty-first century Western society.

## Physics and Astronomy as Model of Knowledge

As an ethnographic tool to make science, and their beliefs about science, "strange," I often ask my undergraduate science and engineering students to create a hierarchy of the disciplines. Invariably mathematics is at the top, immediately followed by physics and astronomy; the humanities end up at the bottom, and biology is often pejoratively called "a hum" (humanities-like). These 17-to-21-year-old aspiring scientists already "know" the (or an) order of things: Absolutely certain that math produces the "best " knowledge because of its "purity," they find physics a close second because it abstracts and because it promises to "explain everything." It also helps that physics is "really hard." Astronomy is empirically sound, observing the universe "as it is," and the humanities offer "just stories." In giving these answers my students reproduce—and reinforce—a common myth about the cultural place of astronomy and physics: In the hierarchy of intellectual understanding, according to cultural belief, these scientific disciplines are, and rightfully belong, at the top.

A myth is both legend and falsehood; usually, it is both at the same time. In this case, one might make visible the implicit metrics of value that underwrite the students' hierarchy; argue about what constitutes brilliance, intelligence, or difficulty; or dissect physical and astronomical theory as a story—an origin story—in its own right. One also has to acknowledge, however, that the legend is truth-making in some ways. A self-fulfilling prophecy, my students' belief in their hierarchy results in social practices that inform their choice of major, that determines the credibility of scientific fields and their associated experts, that lays out a fine-grained economy of respect, and that guides the

classroom behavior of both students and professors. The myth thus has a profound effect on learning, institutional orders, and political practices, at least at the level of the particular undergraduate institution in which we coexist. There is power in these rankings, and this power has effects.

But a myth is also "a traditional story about heroes or supernatural beings, often explaining the origins of natural phenomena or aspects of human behavior; . . . a character, story, theme, or object that embodies a particular idea or aspect of a culture; something or somebody whose existence is or was widely believed in, but who is fictitious" (*Encarta World English Dictionary*). Physics and astronomy textbooks treat the protagonists of their fields in just such ways: Physicists appear as disembodied brains, floating in space. Astronomers are considered "lone wolves": Heroes who deny their bodies, observing long, lonely nights on mountaintops or launching balloons carrying astronomical instruments at the South Pole, they suffer deprivation and hardship in order to retrieve knowledge. Whereas both the "standard model" in high-energy physics (the current theory of fundamental particles that constitute matter and the forces by which they interact) and launching an observatory on Antarctica require large numbers of cooperating bodies, materials, and funds, physicists and astronomers are persistently portrayed, in their brilliance, as supernatural beings who embody ideas but not the social conditions (infrastructures, collaborations, education, family, and funds) that make such ideas possible. Their existence is widely believed and thoroughly fictitious: Conceptualizing them as "lone wolves" or "brains floating in space" is a distortion of the realities of their existence, at best.

In an effort to understand this mythology—this belief in the disembodied and a-social character of physics and astronomy—I have asked physicists what is beautiful about physics, another way of asking what makes people believe in its quality. One of my informants (personal communication, December 2004) told me: "We like reduction. We live in a reductionist field, and we really, really like reduction." And a lot gets reduced; in the making of physical and astronomical theory, much of the work is about eliminating noise, finding a pure signal, recognizing interference, paring down data to a signal that signifies a difference from background clutter. Similarly, in the making of the history of physics and astronomy, it is often the social—the conditions that enable the production of theory—that is reduced. Purity, abstraction: The ability to reduce is a powerful, seductive tool. Such reductionist moves are not restricted to physics and astronomy, as Bruno Latour and Steve Woolgar demonstrated in their study of the construction of a biological fact. But in physics and astronomy they have an impact on our cherished ideas about the sciences at the core.

## Physics and Astronomy as Source of Order

The promised result of such reductions is a "Theory of Everything." Cosmology, a field that crosses astronomy and physics, is in search of a definitive theory that explains the origin and structure of the universe. String theory, currently the "sexiest" field in physics, aspires—in the words of Brian Greene, its well-known proponent—to a "framework with the capacity to explain every fundamental feature upon which the universe is constructed" (Greene, 1999). Greene explains further:

> [I]f we did finally have the unified theory, if we did finally have the deep laws of the universe in hand, that in a very real sense would also be a beginning. It would be the beginning of our quest to use that deep understanding to fully explore this universe, to fully understand black holes and stars and galaxies and even the big bang, to fully understand how things got to be the way they are. So in many ways, it would just be the start. A unified theory would put us at the doorstep of a vast universe of things that we could finally explore with precision. (PBS)

In other words, physics and astronomy position themselves as the producers of the

"cosmology," in the anthropological sense, of the late-twentieth/early-twenty-first-century Western world: the system that explains and gives meaning to the natural order of things.

Such a cosmology has general and generalizing implications. It pervades cultural sensibilities and makes claims of universality. It produces an origin story that positions humans and things in relation to each other; it articulates laws of nature that guide moral and ethical behavior and gives the protagonists of the story a sense of who they are. If physics and astronomy do not succeed in providing such guidance, they certainly aspire to create the framework in which natural and moral orders fall into place. And here is where the sociality of physics and astronomy takes yet another shape: Is not the desire for a cosmological origin story, a narrative that puts everything in its place, an expression of culture already doing its work?

But physical theories and cosmological origin stories cannot come out of the blue. They are not produced a-socially and outside of cultural contexts. In his monograph about the Laser Interferometer Gravitational Wave Observatory (LIGO), Harry Collins (2004) argues that the social materiality of the instruments and experiments is formative of their results. In discussing the book, one of Collins's protagonists portrays the interferometer as a one-time social collaboratory: "Collins was viewing a long-term, one-time experiment, which unfolded in front of his eyes, and which could not have been done anywhere else . . . and which cannot be replicated" (personal communication, December 2004). In saying that LIGO is idiosyncratic, bound to time, place, and a particular collective of participants, this scientist suggests that the science that the gravity wave detector will produce is idiosyncratic as well. That is not to say that the science is unreliable; its reliability depends on iteration upon iteration, observation after observation, matching after matching, until the experts are certain that the patterns in the data are significant and real. What my informant is telling me, however, is that the data coming from LIGO are a product, or an artifact of the instrument.

They are bound to the instrument. It is because this instrument was built that they are observable, in this particular form, at all.

What I have claimed in this article is the following: Physics and astronomy have widely been regarded as the exemplary sciences, at the pinnacle of knowledge and representing the scientific method in its purest form. Historical, philosophical, and, to some extent, sociological work on science has contributed to this common view of astronomy and physics, by building much of its analyses around examples from these scientific fields. From these analyses derives the impression that physical and astronomical knowledge is a representation of a natural reality, as it presents itself to the observing eye or by way of the experimenting hand. The methods of understanding deployed in these fields are, supposedly, neutral and objective and produce unbiased knowledge that is uncompromised by human intervention. As a result, the knowledge and methods produced and used in physics and astronomy are understood to provide a representation of the universe that is as accurate as it gets. In other words, because their methods are tested, tried, and objective, these sciences provide knowledge of the universe that describes the order of things accurately. This rendering hinges, crucially, on the absence of human intervention in the theoretical realm.

To say, in contrast, that physics and astronomy are social enterprises, connected in so many ways to the social and cultural worlds in which they flourish, has profound consequences. It not only points to the collaborative, collective, and consensus-seeking nature of the work in and of those areas of scientific practice; more importantly, it forces one to rethink the promises of objectivity, universality, representationality, and generalizability that were derived from the practices of astronomy and physics in the first place!

Without denying intellectual advances made in astronomy and physics, I have argued that the prominent place that these fields take up in the cultural world is not only, or simply, due to the intellectual merit of their accomplishments. It is also a product of the social transactions performed, and cultural formations arranged, in and around these highly visible, intensely material, and profoundly historical practices of inquiry. Such social transactions and cultural formations—and therefore the nature and effect of physics and astronomy as knowledge-making practices—are material, discursive, real, and imagined *all at the same time*. What counts is that they have particular and inescapable world-making effects. Although some might argue that physical and astronomical knowledge, by virtue of its merits, must naturally guide the way in which modernity envisions its universe, science (or knowledge) is a co-production of scientific practices and visions of the universe that are already inscribed in what modernity entails.

## BIBLIOGRAPHY

Barnes, Barry, and David Edge, eds. *Science in Context. Readings in the Sociology of Science.* Milton Keynes, U.K.: Open University Press, 1982.

Beck, Ulrich. *Risk Society: Towards a New Modernity.* London: Sage, 1992.

Buchwald, Jed Z. *Scientific Practice: Theories and Stories of Doing Physics.* Chicago: University of Chicago Press, 1995.

Carnap, Rudolph. *Philosophical Foundations of Physics: An Introduction to the Philosophy of Science.* Edited by Martin Gardner. New York: Basic Books, 1966.

Cartwright, Nancy. *How the Laws of Physics Lie.* Oxford: Clarendon Press, 1983.

Collins, Harry. *Changing Order. Replication and Induction in Scientific Practice.* Chicago: University of Chicago Press, 1992.

Collins, Harry. *Gravity's Shadow: The Search for Gravity Waves.* Chicago: University of Chicago Press, 2004.

Galison, Peter. *Image and Logic: A Material Culture of Microphysics.* Chicago: University of Chicago Press, 1997.

Greene, Brian. *The Elegant Universe.* New York: Norton, 1999.

Knorr-Cetina, Karin. *Epistemic Cultures. How the Sciences Make Knowledge.* Cambridge, MA: Harvard University Press, 1999.

Kuhn, Thomas. *The Structure of Scientific Revolutions* (1962). Chicago: University of Chicago Press, 1986.

Latour, Bruno, and Steve Woolgar. *Laboratory Life: The Construction of a Scientific Fact.* Princeton, NJ: Princeton University Press, 1986.

Merton, Robert K. *The Sociology of Science: Theoretical and Empirical Investigations.* Chicago: University of Chicago Press, 1973.

PBS. *The Elegant Universe.* Interview with Brian Greene. July 2003. http://www.pbs.org/wgbh/nova/elegant/greene.html.

Pickering, Andrew. *Constructing Quarks: A Social History of Particle Physics.* Chicago: University of Chicago Press, 1984.

Popper, Karl. *The Logic of Scientific Discovery.* New York: Basic Books, 1959.

Restivo, Sal. *The Social Relations of Physics, Mysticism, and Mathematics.* Dordrecht, The Netherlands: Reidel, 1983.

Shapin, Steven, and Simon Schaffer. *Leviathan and the Air-Pump: Hobbes, Boyle, and the Experimental Life.* Princeton, NJ: Princeton University Press, 1985.

Traweek, Sharon. *Beamtimes and Lifetimes: The World of High Energy Physicists.* Cambridge, MA: Harvard University Press, 1992.

*Marianne de Laet*

# POLITICAL ECONOMY OF HEALTH.

The political economy of health is a theoretical framework used to study health inequalities. It proposes that health disparities are determined by social structures and institutions that create, enforce, and perpetuate poverty and privilege. Rooted in (but not restricted to) classical Marxist historical materialism, political economists of health take a critical, historical approach to analyzing the social production, distribution, and treatment of health and disease. They analyze the relationships between health status and political-economic institutions throughout the world, with particular emphasis on the detrimental health effects created by capitalist relations of production and sustained by specific political-economic arrangements. The political-economic approach focuses on the social determinants of disease etiology and health inequality rather than on individualistic, biologistic, geographical, cultural, or psychological explanations of ill health.

## The Social Production of Disease

The political economy of health is an interdisciplinary framework used by anthropologists, economists, geographers, political scientists, sociologists, and experts in medicine and public health. Research undertaken in the field is conceptual and theoretical, as well as empirical and applied. Political economists use a variety of methodologies, including epidemiological, ethnographic, and historical research. They are committed to community-based, participatory methodologies and to *praxis*, defined as putting theory into action. Political economists of health often put their skills at the service of labor organizations, liberatory social justice movements, and community activists to develop strategies for health-enhancing social change.

Political economists of health traditionally begin their analyses with a focus on social class. Following the nineteenth-century political theorists Karl Marx and Friedrich Engels, political economists argue that ill health is often determined by one's relationship to the means of production, with a fundamental division between workers and owners. They point out that the logic of capitalist accumulation provides little incentive for owners to invest in healthy environments, safe workplaces, supportive social services, affordable housing, or high-quality medical care for workers if such investments will subtract from potential profits. Capitalism fosters disease by exploiting workers, minimizing health-promoting investments, maximizing short-term profits, and ignoring the long-term consequences on health and the environment. By promoting political institutions that support and enforce capitalist accumulation, the medical-industrial complex that grew up in the twentieth century has been able to shift the costs of disease to states and workers though mechanisms such as health financing, industry subsidies, and private medical care and insurance schemes. The discourses of biomedicine, meanwhile, obscure the origins of suffering and prevent people from understanding the sources of poverty and disease. The economic structures of capitalism, as well as the political and medical institutions that support and enforce them, are thus implicated in the social production and unequal distribution of disease and medical care.

Political economists of health often take a comparative cross-national perspective to analyze public health systems and medical institutions. They have been especially interested in the evolution of health and medical care under socialism. The purpose of such research is to identify the relationship between political regime, economic production, and health status and to frame and evaluate alternatives to biomedical capitalism. A comparative focus can also be used to analyze the effects of global domination by the Western model of biomedical capitalism. For example, political economists might investigate the effects of exporting Western models on medical education, the so-called brain drain that lures talented doctors and scientists from underdeveloped to developed countries, and health services and financing, as well as the effects of patent laws governing the distribution of pharmaceutical products and other medical devices. One strength of the political economic approach is its ability to identify structural causes by focusing on the historical, economic, and political processes that underlie particular epidemiological profiles and sociomedical challenges.

## Intellectual Antecedents

The political economy of health is sometimes labeled radical, but Merrill Singer, a critical medical anthropologist and community health activist, notes that political economy made eminent sense to nineteenth-century social theorists and physicians who demonstrated the unity of medicine and society. Intellectual ancestors of the political economy of health traced to mid-nineteenth-century Europe include the German pathologist-anthropologist-activist Rudolf Virchow (1821–1902), a social reformer and government critic. He is known by political economists for his studies of public hygiene, sewage disposal, housing conditions, and the safety of food supplies, which prompted him to encourage the government intervention to prevent disease and improve living conditions. His classic study of a typhus outbreak in Upper Silesia, Prussia, in 1848 concluded that poor social conditions were responsible for the epidemic.

Another prominent intellectual ancestor is socialist philosopher Friedrich Engels (1820–1895). Engels, who was Karl Marx's closest collaborator, wrote *The Condition of the Working Class in England,* in 1845. Focusing on the links between the Industrial Revolution, the capitalist mode of production, and ill health, Engels argued that disease was caused by the rapid growth of an alienated, urban-dwelling proletariat that lived in impoverished conditions characterized by a lack of sanitation, harrowing and hazardous working conditions, dependence on child labor, poor housing stock and overcrowding, inadequate nutrition, and poverty. Capitalism, he showed, might produce great wealth for some, but it also produced poverty and ill health for many.

The theoretical integration of medicine and social science was lost, in many parts of the Western world, in the late nineteenth and early twentieth centuries. The rise of scientific medicine and the germ theory of disease focused attention on biological (and eventually molecular and genetic) levels of analysis. In the process, a rift developed between medicine and sociomedical sciences (including social epidemiology). With increasing disciplinary specialization, the political economy of health was fragmented as it was integrated into sociology, anthropology, public health, social epidemiology, geography, political science, economics, and environmental studies. Despite social medicine movements in Britain and the United States in the 1930s and 1940s, the political economic perspective became diluted as parallel movements developed in subfields such as social epidemiology, ecosocial models, social medicine, community-oriented primary care, critical medical anthropology, and the political economy of biological anthropology, to name a few. Indeed, the political economy of health is sometimes incorrectly claimed as a descendant, rather than an ancestor, of these fields. Only in the late twentieth century, after biomedicine was divorced from the social sciences, has it been possible to consider the political economic perspective as "radical."

Although the political economy of health was regarded with suspicion or hostility in the United States and Europe during the first half of the twentieth century, it did continue to flourish in other parts of the world. The Latin American *medicina social* (social medicine) was exemplified by the work of the physician and future president of Chile, Salvador Allende, who was a direct intellectual descendant of Rudolf Virchow. In 1939, Allende wrote the influential *La Realidad Médico-Social Chilena* (The Chilean Medical-Social Reality), which identified workers' living conditions as the primary source of individual ill health. Allende's proposed solutions included "income redistribution, state regulation of food and clothing supplies, a national housing program, and industrial reforms to address occupational health problems" (Waitzkin et al., 2001, p. 1593). This social justice–oriented vision of state–citizen relations threatened the status quo in Chile. Allende, who had been elected president in 1970, was assassinated in a CIA-sponsored coup in 1973 after attempting a peaceful transition to socialism. Exiled Chilean physicians who fled the ensuing military dictatorship took medicina social to other countries, including Mexico and the United States, and inspired a generation of health activists. Research and activism under the banner of social medicine continue today under the banner of the *Asociación Latinoamericana de Medicina Social*, focusing on workers' health, critical epidemiology, and the impact of the biomedical-industrial complex. Proponents hold that good health and adequate living standards are basic human rights, which are the government's responsibility to guarantee. The history of Latin American social medicine has recently been brought to the attention of English-speaking readers through the efforts of Howard Waitzkin, Celia Iriart, Alfredo Estrada, and Silvia Lamadrid (2001).

Political economists have always been interested in comparative health and medical systems, and the relationship between political regime and health service provision. They have studied socialist systems, including Mao Zedong's system of "barefoot doctors" in China, which were cited as a source of inspiration for the primary health care movement adopted by the World Health Organization and UNICEF in 1978. The primary health care movement urged governments to focus on preventive rather than curative care, provide universal access to primary health care services, sponsor community participation in health, and train village health workers. Political economists of health have been equally interested in the Cuban revolution of 1959, which led to health sector reforms, universal access to preventive and curative care, and advances in medical education. Cuba's health indicators surpassed those of all other Latin American countries in the late twentieth century, providing an example of what was possible when a government put a high value on public health. Many political economists of health and international health activists worked in Nicaragua after the 1979 revolution. Though battered by a U.S.-sponsored counterrevolutionary war, Nicaragua was able to implement exciting and innovative approaches to health promotion and medical care.

## Contemporary Variations

There was a resurgence of interest in the political economy of health in the United States in the 1970s, spurred by the revitalization of Marxist scholarship and the growth of dependency and world systems theory, as well as civil rights, feminist, and antiwar activism. Dependency and world systems theorists were following the lead of the Economic Commission on Latin America and the work of Andre Gunder Frank and Emmanuel Wallerstein, who argued that modernization and economic growth would not result in improved standards of living for the residents of third-world countries. Development in "core" or "metropole" regions, they argued, was made possible by the extraction of goods and services from the "periphery" or "satellite" regions. Underdevelopment was not the result of backwardness or lack of integration into the world system, but of the expansion of global capitalism. Political economists writing in this vein argued that disease and malnutrition resulted from export-oriented agriculture and biomedical ideologies and practices that obscured the social causes of ill health.

In the 1980s, political economists of health engaged in theoretical debates, with some arguing that economic determinism could be as limiting as biological determinism. The theory of poststructuralism, associated with French philosopher and historian Michel Foucault and outlined in *The Birth of the Clinic*, brought attention to the concept of *biopower*. Emerging nation-states adopted the power to shape, define, construct, and control bodies, populations, and life and death, and to make the effects of that construction appear normal and natural. Poststructuralist perspectives emphasize the importance of locating one's knowledge within particular histories and subject positions; this insight, in turn, allows analysts to excavate the hidden histories and invisible motors that drive medical policies, ideologies, and practices. Foucault cited the creation of the hospital, the invention of vital statistics and other population-measuring techniques, and the controlling medical gaze as examples of the coercive instruments used by bureaucracies and states to exercise power. Political economists of health retain their attention to social class, for example, by criticizing epidemiological surveillance systems that do not collect information about social class. But in response to these theoretical debates, political economists have broadened their investigations to include the health effects of stratification due to gender, race and ethnicity, national origin, religion, and sexual orientation. An overemphasis on the deleterious effects of capitalism, they acknowledge, has sometimes prevented them from appreciating the steps people take to protest, oppose, and overthrow harmful policies. This realization has inspired many political economists of health to study health activism, resistance strategies, and emancipatory social movements.

The world economic recession of the early 1980s; the disintegration of the Soviet Union in 1989; and neoliberal reforms, such as free trade agreements, structural adjustment, and the consolidation of global capital caused governments worldwide to retrench from comprehensive primary health care programs and devote greater attention to cost cutting, privatization, and improved productivity in the health sector. Health is increasingly construed as a pathway to economic growth and development, as well as an untapped source of revenue for the private sector. In this context, political economists of health have become more vocal, pointing to increasing gaps between the haves (insured) and have-nots (uninsured), which in the United States is exacerbated by a failure of political will to promote universal health coverage and by conservative opposition to the notion that health is a right of citizenship. Political economists of health have renewed their passionate appeal to world leaders to regard health as a basic human right.

Among the most forceful advocates of a humanitarian commitment to global health is Paul Farmer, a physician-anthropologist and infectious-disease specialist who employs the term *structural violence* to refer to the systematic and institutionalized structures that deny food, shelter, and economic opportunity to poor people. Farmer's crusade calls attention to the increasing global gap between the rich and the poor, the corrupt and impoverished social vision of world leaders, and the shameful and wholly preventable disease burden and daily indignities that the poor are forced to bear as a consequence. He shifts attention from the sick themselves to those who produce sickness; one of his books is appropriately titled *Pathologies of Power: Health, Human Rights, and the New War on the Poor.*

Another response to globalization, privatization, and free trade has been the emergence of the field of political ecology, which grew out of geography and environmental studies. Political ecology focuses on the politicized environment, or what some have called *ecological Marxism*. Disease is related to global problems—sometimes called the *new world disorders*—that in turn are linked to deteriorating environments. Because global health is increasingly determined by access to finite environmental resources, such as forests, clean air, food, oil, oceans, and drinking water, political ecologists of health have analyzed occupational health hazards created by international trade agreements, such as NAFTA (North American Free Trade Agreement). They have examined systematic exposure to toxins and poisons, such as coal dust, mercury, tobacco, cocaine, heroin, and lead, as well as wide-scale industrial contamination such as what took place at Chernobyl and Bhopal. Political ecologists of health have looked at the links between land degradation, contamination of the food supply, widespread antibiotic resistance, the privatization of water, and the eruption of epidemic infectious disease. They point out the ill health effects when world leaders invest in war and militarization while cutting health budgets and allowing pollution, disease, and famine to persist.

Political economists of health argue that international donors and policy makers, such as the World Health Organization and the World Bank and International Monetary Fund, have appropriated the rhetoric of health promotion while ignoring the fact that poverty and inequality are the main determinants of disease. By focusing on technical rather than political or structural fixes, these international organizations propose to invest in health as a means to facilitate economic growth and productivity and to facilitate private-sector profits; in other words, to further the accumulation of capital. Private corporations may attempt to co-opt and profit from health improvements, for example, by tying international aid to government reforms designed to privatize the health sector. Political economists vehemently object to the notion that health improvements should be regarded as investments in "human capital" solely for the purpose of improving labor productivity and enhancing private profit. They prefer a more humanitarian approach; Hans Baer, Merrill Singer, and Ida Susser, for example, support "the creation of a democratic ecosocialist world system and the pursuit of health as a human right" (p. ix).

Contemporary political economists of health have taken diverse paths, but they share the conviction that disease is socially produced and caused by structural factors. They point out that the biomedical and technological advances and investments of the past fifty years have not succeeded in preventing disease, improving overall health, or contributing to a better quality of life for many people. Epidemiological polarization—the gap between rich/healthy and poor/unhealthy—has grown, they say, signaling an obvious failure of biomedical capitalism to live up to its promise. Political economists often use the metaphor of "refocusing upstream," that is, walking up the riverbank to see who (or what) is responsible for pushing in the drowning victims who are being pulled from the river and treated downstream. They study the effects of epidemiological polarization on embodied social distress, in which people suffer disproportionate levels of disease (such as diabetes or hypertension) based on their exposure to racism and wealth disparities. Epidemiological polarization can also be studied at the level of international health policies (such as pharmaceutical investments) that allocate medical resources differentially according to a country's ability to pay. Political economists of health reject the fragmentation that separates social/biological, health/disease, mind/body, environment/individual, nature/nurture, politics/economics, developed/underdeveloped, theoretical/applied, and technical/political. Some of the most outspoken and articulate proponents of the political economy of health in the early twenty-first century English-speaking world include Baer, Lesley Doyal, Ray Elling, Farmer, Elizabeth Fee, Nancy Krieger, Vicente Navarro, Randall Packard, Singer, and Waitzkin. All argue that improved global health will require dramatic social change and a renewed commitment by states to protect citizens and the environment from the exploitative effects of corporate greed and state-sponsored global capitalism. Political economists argue that this will not be achieved through medical advances, simple reforms, tinkering with health policy, or promoting the interests of the elite. It will require an adequate analysis of structural problems—what Nancy Krieger calls "a vision of social justice" (p. 671)—accompanied by the political will to create health improvements, reductions in poverty and inequality, and movement toward a more just and equitable society.

*See also* **Health Care, Access to**

**BIBLIOGRAPHY**
Baer, Hans A., Merrill Singer, and Ida Susser. *Medical Anthropology and the World System: A Critical Perspective.* Westport, CT: Bergin and Garvey, 1997.
Farmer, Paul. *Pathologies of Power: Health, Human Rights, and the New War on the Poor.* Berkeley, CA: University of California Press, 2003.
Krieger, Nancy. "Theories for Social Epidemiology in the 21st Century: An Ecosocial Perspective." *International Journal of Epidemiology* 30 (2001): 668–677.
Waitzkin, Howard. "Report of the WHO Commission on Macroeconomics and Health: A Summary and Critique." *The Lancet* 361 (February 8, 2003): 523–526.
Waitzkin, Howard, Celia Iriart, Alfredo Estrada, and Silvia Lamadrid. "Social Medicine Then and Now: Lessons from Latin America." *American Journal of Public Health* 91.10 (2001): 1592–1601.

*Lynn M. Morgan*

# POLITICS, SOCIAL JUSTICE, AND THE ENVIRONMENT.

To understand the intersection of politics, social justice, and the environment, it is important to define the terms. *Politics* can be described as the process through which collective power is generated, organized, and distributed in society. *The environment* is the sum of the conditions of the natural, material, and social surroundings. *Justice* concerns the societal negotiation of fairness and ethical behavior. *Social justice* involves a complex negotiation between equality of opportunity and equality of outcomes for various social groups. Social justice is sometimes defined as *distributive justice*, meaning a fair distribution of goods and services as well as fair distribution of the ability to participate in and influence political processes and decision making.

The intersection of politics, environment, and justice can be seen most clearly in the environmental justice movement. *Environmen-tal justice* (EJ) is a broad term encompassing all human needs and the quality of life. It refers not only to the condition of the natural environment (such as the prevalence of pollution) but also to the economic conditions, health status, human safety issues, housing and utility concerns, and the civil rights of the residents of a community. Thus, environmental inequities are not separable from economic inequities or political inequities. The EJ movement is a hybrid of two earlier, primarily homogenous, social movements:

1.  The environmental movement, which was typified by white, middle- and upper-class people voicing concern over endangered species and the loss of natural habitats
2.  The civil rights movement, which was led by people of color and focused on human rights and equal access and input into the political and institutional structures that govern them

The composition of the EJ movement is very heterogeneous, including many women's voices as well as those of minorities and people of all social classes. Typically these movements are local and coalesce around immediate issues of concern, dissolving once the issues have been resolved and reforming should new problems emerge. This makes EJ very different from its parent environmentalism and civil rights movements, because it is flexible and "biodegradable" and can consist of a diverse membership, depending on the problem. The types of issues and constituencies of these groups varies across the United States and internationally.

The EJ movement was first associated with African-American and Hispanic groups protesting the siting of polluting or unwanted facilities in their urban neighborhoods. The movement expanded to include rural groups of minority and poor citizens as well as cross-class and cross-race groups brought together by the transitory nature of pollution. It also includes working-class white neighborhoods, as in the case of Love Canal, a now infamous housing development built on a hazardous waste disposal site in New York. The concerned group was organized by women noticing unexplained illnesses in their children. Women of all races and classes have played a pivotal role in the EJ movement in the United States and abroad. Blue-collar workers and farm workers have also been involved in the struggle for a safe work environment. Janitorial service workers and the dry cleaning industry are examples of labor groups that have pushed for cleaner practices (Gottleib, 2001). There have also been alliances between unionized industrial plant workers and poor and minority residents fighting for cleaner industrial practices.

While in some poor and minority communities the EJ fight has been against the addition of unwanted materials and facilities in their communities, other groups, such as Native American communities, have also had to fight against unfair natural resource extraction and uses on their lands. That is not to say that they have not also had pollutants, such as nuclear waste, disposed of on their property, but they have additional EJ concerns as well. The problem is complex and ranges from the diversion of fresh water from Hispanic and Native lands to open-pit mining without proper compensation or consideration of subsequent water pollution. The EJ framework has also been used by indigenous communities to argue that they should be allowed to continue traditional practices of hunting, forestry, and other natural resource usage even though it violates the national environmental agenda. They argue that these laws prevent them from making a living and providing food for their families as well as curtailing traditional religious and cultural practices, endangering their ability to thrive.

Internationally, environmental injustices, particularly in developing nations, have been attributed to the systemic tendencies of globalization. Among the problems has been the shift of industrial waste to poorer nations because of their cheap land, low labor cost, and lax environmental regulation. Women and children have been unfairly disadvantaged by natural resource practices that leave them less able to provide food and water for their families. The "Chipko" protest movement, in rural South Asia, is an example of women protest-

ing nonsustainable uses of their forests from which they gather fuel and food. Poverty and lack of education also drive some of the nonsustainable practices of indigenous peoples. EJ in these regions needs to include enhancing the local economic infrastructure as well as the political empowerment of the people. Other factors such as education, gender equity, and proper utilities infrastructure also factor into the EJ framework on an international level.

*See also* **Rain Forests**

**BIBLIOGRAPHY**
Bullard, Robert D. *Dumping in Dixie: Race, Class, and Environmental Quality,* 3rd ed. Boulder, CO: Westview Press, 2000.
Gottlieb, Robert. *Environmentalism Unbound: Exploring New Pathways for Change.* Cambridge, MA: MIT Press, 2001.
Hofrichter, Richard, ed. *Toxic Struggles: The Theory and Practice of Environmental Justice.* Philadelphia: New Society Publishers, 1993.
Schwab, Jim. *Deeper Shades of Green.* San Francisco: Sierra Club Books, 1994. An examination of environmental justice movements across the United States. Foreword by Lois Marie Gibbs.
Shiva, Vandana. *Staying Alive: Women, Ecology and Development.* London: Zed Books, 1988. A lens into the EJ movement among women in India and other developing nations.
Szasz, Andrew. *Ecopopulism: Toxic Waste and the Movement for Environmental Justice.* Minneapolis: University of Minnesota Press, 1994.

*Barbara L. Allen*

# PROFESSIONAL RESPONSIBILITIES IN MEDICAL RESEARCH.

In the last decade of the twentieth century and in the first part of the twenty-first, there have been several high-profile accounts of research fraud reported in the media. Hardly a month passes without the reporting of misconduct in scientific or medical research involving prestigious academic institutions and respected researchers. The problem is not confined to North America; serious cases of research misconduct have been reported in other countries, such as Britain, Germany, Japan, and Denmark.

In the United States, two incidents generated a great deal of interest in the prevalence of misconduct in research. In September 2002, Jan Hendrik Schön, a physicist at Bell Laboratories (Bell Labs) in Murray Hill, New Jersey, was found by an independent investigative panel to have fabricated data sixteen times. Some of his research in nanotechnology, if true, had been considered Nobel Prize–worthy. (Researchers at Bell Labs had won six Nobel Prizes in physics since its founding in 1925.) Dr. Schön was subsequently dismissed for fabricating and manipulating data. Withdrawing all of his fraudulent research from the corpus of scientific knowledge however, has been a difficult and protracted task.

In the same year, Victor Ninov, a scientist at the Lawrence Berkeley National Laboratory in California, claimed that he had discovered a new element, number 118, to be added to the periodic table of elements, which would have constituted a huge scientific breakthrough. After an internal investigation found that Dr. Ninov had faked data supporting his alleged discovery, however, he was fired from the prestigious laboratory.

The medical research community in Germany is still reeling from investigations into the largest case of research misconduct in the country. In 1997 two cancer researchers—Friedhelm Herrmann and Marion Brach—were accused of publishing papers based on fabricated data. The alleged fraud took place when they both worked at the Max Delbruck Center for Molecular Medicine (MDC) in Berlin. In its final report, members of an independent task force set up to investigate the research scandal found that data in at least ninety-four papers had either definitely or very probably been manipulated.

Since 1998 the United Kingdom's General Medical Council has revoked the licenses of several doctors who were reported for scientific misconduct by the Association of British Pharmaceutical Industry. One of the more publicized cases was that of Dr. Geoffrey Fairhurst, who in 1996 was struck off the Council for forging consent forms of patients who were entered into a research study sponsored by a drug company. Dr. Fairhurst also happened to be the vice-chairman of the local medical ethics committee.

Do these reports indicate a rise in the incidence of research misconduct in recent years? This question has been the subject of debate for years. Interestingly, many of the publicized cases of research misconduct in the early twenty-first century were being brought to the attention of the public at a time when other cases of massive fraud in the wider world were being uncovered. The Enron and Arthur Andersen accounting scandal, as well as the plagiarism allegations leveled against such world-renowned scholars as Stephen Ambrose and Doris Kearns Goodwin, serve as prominent examples. This prompts the question: Is the apparent rise in the number of cases of fraud in medical research simply a part of a larger societal phenomenon? Answering this question would be easier if it were possible to ascertain whether there is indeed a significant rise in cases of fraud in science and medical research.

Unfortunately, it is almost impossible to determine conclusively the overall incidence of fraud in research in the United States and other countries. One of the main reasons for this is the paucity of data on reported and investigated misconduct cases. In the United States, it is still widely believed that research fraud is generally uncommon and that the number of reported cases remains relatively small compared to the amount of research being conducted at any given time.

The U.S. Office of Research Integrity (ORI), the federal body charged with investigating misconduct in research sponsored by the National Institutes of Health, received about 1,000 allegations of misconduct between 1993 and 1997. Out of this number, 150 were investigated and about 75 were found to be cases of misconduct or fraud. Further, the National Academy of Sciences panel on Scientific Responsibility and the Conduct of Research found in 1992 that "the number of confirmed cases of misconduct in science was low compared to the level of research activity" (National Academy of Sciences, p. 95).

Other schools of thought, however, believe that fraud in medical research is generally underreported and that much of the misconduct is not caught in time or at all. In its 2000 *Annual Report,* the ORI states that a three-year decline in research misconduct activity was reversed in 1999 as institutions reported a moderate increase in such activity in their annual reports on possible research misconduct filed with the ORI. A study published in 1993 in *American Scientist* found that between 6 and 9 percent of respondents said they were personally aware of results that had been plagiarized or fabricated within their faculties.

In addition, public concern about misconduct and fraud in research appears to be cyclical. The current wave of publicized incidents is certainly not without precedent.

## How Is Research Misconduct Defined?

One of the problems in the research arena is the lack of a clear, concise, and consistent definition of what constitutes misconduct in research. Terms such "misconduct in science," "research fraud," and "intellectual dishonesty in science" are rather amorphous and sometimes create legal confusion. In addition, the phrases do not make clear distinctions between various questionable activities in the wide spectrum of activities encompassed under the label of research.

Government, academic, and private research institutions use different definitions of fraud, as do countries, further compounding the definitional problem. A panel set up in 1989 by the U.S. National Academy of Sciences, National Academy of Engineering, and the Institute of Medicine attempted to define research misconduct by distinguishing three categories of activities:

a. Misconduct in Science, referring to fabrication, falsification, or plagiarism in proposing, performing, or reporting research. It does not include errors of judgment; errors in the recording, selection, or analysis of data; differences in opinions involving the interpretation of data; or misconduct unrelated to the research process.

b. Questionable Research Practices, referring to actions that violate traditional values of the research enterprise and that may be detrimental to the research process. Examples in this category include failing to maintain important data for a reasonable period, refusing to give peers reasonable access to unique research materials or data that support published results, and misrepresenting speculations as fact.

c. Other Misconduct, referring to actions that are not unique to the conduct of science, although they may occur in a laboratory or research environment. Such practices are subject to generally applicable legal and social penalties: they include negligence, vandalism, misuse of funds, and sexual harassment.

Because of the lack of consensus on the definition and on procedures to deal with research misconduct, the U.S. Congress and the Department of Health and Human Sciences set up a Commission on Research Integrity in 1995. One of the Commission's mandates was to refine the definition of research misconduct, which it attempted to do:

Research misconduct is significant misbehavior that improperly appropriates the intellectual property or contributions of others, that intentionally impedes the progress of research, or that risks corrupting the scientific record or compromising the integrity of scientific practices. Such behaviors are unethical and unacceptable in proposing, conducting, or reporting research, or in reviewing the proposals or research reports of others. Examples of research misconduct include, but are not limited to the following:

*Misappropriation:* An investigator or reviewer shall not intentionally or recklessly

a. Plagiarize, which shall be understood to mean the presentation of the documented words or ideas of another as his or her own, without attribution appropriate for the medium of presentation; or

b. Make use of any information in the breach of any duty of confidentiality associated with the review of any manuscripts or grant application.

*Interference:* An investigator or reviewer shall not intentionally and without authorization take or sequester or materially damage any research related property of another, including without limitation the apparatus, reagents, biological materials, writings, data, hardware, software, or any other substance or device used or produced in the conduct of research.

*Misrepresentation:* An investigator or reviewer shall not with intent to deceive, or in reckless disregard for the truth,

a. State or present a material of significant falsehood; or
b. Omit a fact so that what is stated or presented as a whole states or presents a material or significant falsehood.

(Commission on Research Integrity 1996, pp. 15–16)

Scientists met the Commission's report with much hostility because of the classification of certain practices as misconduct; also, the consequences of being accused of scientific misconduct were dire and in some cases, according to some scientists, unwarranted.

In its *Policy and Procedure for Inquiring into Allegations of Scientific Misconduct,* the British Medical Research Council includes the following in its description of misconduct: failure to follow established protocols if this failure results in unreasonable risk or harm to humans, other vertebrates, or the environment; and facilitating of misconduct in research by collusion in or concealment of action by others.

In addition, it is important to note that research misconduct can arise from acts of *commission* (as described so far) as well as acts of *omission* (see point b under misrepresentation in the Commission's report). According to the ORI, the omission of data is considered falsification when it misleads the reader about the results of the research. In addition, we would argue that deliberately choosing *not* to gather data on variables that are potentially damaging to the sponsor or investigator (rather than merely omitting to disclose negative data) is of great ethical con-

cern, particularly where such information has the potential of significantly altering the research study's outcomes. Perhaps the most important omissions are not "research misconduct" per se, but acts that result in a biased view of the underlying scientific or clinical reality by, for example, *not* conducting research on an unpatentable drug that might nonetheless be equal or superior to patented and more expensive drugs.

Despite these and other attempts to define acts that fall into the category of research misconduct, much discrepancy remains in the research world on exactly what types of conduct constitutes fraud.

## Why Does Research Misconduct Occur?

Several reasons have been offered to explain why researchers engage in misconduct. It has been argued that many researchers feel great pressure to be prolific in research. The relationship between publication rate, promotion, and tenure, Arnold Relman has argued, is so strong that it is probably a factor in research fraud. Breakthrough discoveries bring important recognition and prestige to researchers, and these set the stage for a race to be the first to announce and publish such discoveries.

It is also important to note that publication of research is important in obtaining and maintaining grant funding. Further, it has been observed that an increasing number of researchers are under pressure to publish findings that are favorable to their sponsors as more work is primarily sponsored by industry. This phenomenon brings the question of conflicts of interests to the fore.

A report entitled *The Responsible Conduct of Research in the Health Sciences,* published in 1989 by the National Research Council, pointed out that:

> Investigations of cases of scientific fraud suggest that various factors in the research environment may contribute to the occurrence of scientific misconduct even though they are not the direct causes of these occurrences. Examples include pressures to "publish or perish", and emphasis on competition and secrecy in research with

their peers and mentors. There is concern that not only ethics, but also the quality of scientific research in general may suffer in this environment. (p. 1)

In the realm of biomedical research, another reason offered to explain misconduct is the growing disconnect between researchers in pharmaceutical companies that sponsor research and the researchers carrying out the actual trials on human subjects. To improve efficiency in the clinical trials, particularly those involving large populations, pharmaceutical companies outsource clinical trial management to private companies. These companies, popularly known as contract research organizations (CROs), are run by people who are not necessarily loyal to pharmaceutical companies or research subjects, and who have direct financial interest in ensuring that the research is conducted within a set time frame, at minimum cost, and with the least administrative burden. The incentive for misconduct thus becomes much higher because of the competing interests and high stakes involved.

In addition, many researchers work under minimal supervision, regulation, and administrative oversight. Several years can pass before anyone notices irregularities or misconduct in the researchers' work.

All these factors can be said to contribute to or increase the motivation and opportunity for researchers to engage in fraud or misconduct in their work.

## Consequences of Research Misconduct

Science is still viewed by the public as a largely pure, rational, and objective discipline. By tampering even minimally with the truth, a scientist is seen as committing a sacrilegious act. The scientist is seen as having abused the special privilege of trust that society bestows on all scientists: to discover the truth using scientifically accepted methods, and to publish the honest results of such inquiry.

Fraudulent research not only adds false information to the organized body of scientific knowledge (which others rely upon and cite), but also increases the public's distrust of science and scientists in general. As a result, several countries and institutions are trying to prevent misconduct from occurring in the first place by removing the motivation and opportunity to commit fraud.

In clinical trials with mortality and morbidity endpoints, inaccurate or biased results can result in the release of potentially unsafe pharmaceutical products to the market; these products may consequently pose a serious danger to the health of consumers.

To deal with researchers who have been found guilty of misconduct amounting to fraud, various institutions and countries use different measures. The scientific research community prefers to self-regulate, and the manner in which a particular case is handled usually depends on such factors as the quality of the evidence, the gravity and nature of the misconduct, and the professional status of the researcher. Typical responses include dismissal of the scientist from his or her position, retraction of publications, and debarment from funding.

Another consequence of fraud or misconduct in research has been the re-examination of the role and responsibilities of coauthors in joint publications. In the Schön case, one of the coauthors was a highly respected, senior researcher. Although that coauthor was later exonerated from misconduct charges, the American Physical Society now holds senior authors to a high level of responsibility for the integrity of the research findings; it does recognize, however, that junior coauthors have limited responsibility. The Schön case certainly illustrates the fact that it is in the professional interest of coauthors to make sure that the data cited are authentic, because their names and reputations are also on the line.

In addition to these institutional responses, a new genre of lawsuits has begun to emerge in the United States. These suits target medical researchers in medical malpractice cases, with the statements of claim focusing on ethical misconduct, such as improper informed-consent procedures or undisclosed financial conflicts of interests.

## Efforts to Prevent, Detect, and Investigate Fraud in Research

Several efforts are being undertaken to prevent and deal with misconduct in research at the institutional level, within professional groups, and even at the national level. Many universities and research institutions are developing policies to deal with misconduct and to build curricula on research ethics.

Every research institution in the United States that receives federal grants is required to have a formal system in place to investigate allegations of research misconduct. In addition to these institutional systems, the federal government has set up a national body, the ORI, in the Office of Public Health and Science (OPHS), which is within the Office of the Secretary of Health and Human Services. The ORI is charged with promoting integrity in biomedical and behavioral research supported by the National Institutes of Health at about 4,000 institutions worldwide.

The ORI monitors institutional investigations of research misconduct and facilitates the responsible conduct of research through educational, preventive, and regulatory activities. It therefore acts as an oversight agency; the institutions themselves are required to investigate allegations of fraud. The ORI does have the power to recommend a finding of research misconduct and the imposition of administrative action to the Assistant Secretary for Health.

Other bodies in the United States that are responsible for ensuring research integrity include data safety monitoring boards (DSMBs), which are generally sponsored by the National Institutes of Health, other federal health agencies, as well as private research sponsors engaged in clinical research. DSMBs are set up to ensure the safety of human subjects enrolled in clinical trials and to guarantee the integrity of scientific data.

In addition, professional and academic institutions play an active role in trying to reconcile the various inconsistencies in the definition of research misconduct and investigative procedures. As previously mentioned, the National Academy of Sciences, the National Academy of Engineering, and the Institute of Medicine released a lengthy

report in 1992 that attempted to clarify the meaning of research misconduct.

Other countries plagued by allegations of fraud among their researchers have also set up administrative structures to prevent, detect, and deal with research misconduct. Germany has few guidelines to deal with research misconduct; however, its major university grant-giving body, the Deutsche Forschungsgemeinschaft (DFG) has set up procedures to investigate fraud. In addition, the Max Planck Society—an independent nonprofit body in Germany that maintains institutes dedicated to research in the biological, physical and natural sciences, medicine, the arts, and humanities—has set up a system to investigate research misconduct in its various institutes.

In the United Kingdom, the Medical Research Council, Biotechnological and Biological Sciences Research Council, and General Medical Council have established policies and procedures for dealing with misconduct in research. These organizations have, however, received much criticism for being ill-equipped to self-police. The Committee on Publication Ethics (COPE), a group of British medical journal editors, has developed a code of conduct aimed at preventing the publication of fabricated data. The International Committee of Medical Journal Editors has also published a revised statement on publication ethics that deals with authorship and accountability.

Norway, Denmark, Finland, Austria, and Australia also have national systems in place to deal with scientific misconduct. In Japan, a government advisory group, the Science Council of Japan, recommended in 2003 that allegations of research misconduct be investigated by third-party committees run by national ministries of scientific societies rather than universities and institutes.

## Looking beyond the Horizon

There is no question that the research landscape in the twenty-first century is very different from what it was in the mid-twentieth century. Financial relationships among industry, investigators, and academic institutions are increasingly complex and fraught with competing interests. The stakes are high and competition among researchers for grant funding is stiff. There is great pressure on researchers to publish or perish, as the saying goes. It must be remembered, however, that misconduct in scientific research is not a new phenomenon and that science is generally believed to be a self-correcting discipline.

Some commentators fault the peer review system that governs scientific publication for failing to detect fraud, fabrication, and plagiarism. This attribution, we believe, is mistaken. The peer review system of scientific communication is predicated on the assumption that the observations reported by scientists are, in fact, accurate reports of the observations they made. Peer review provides quality control and sophisticated critique of methodology and interpretation; it is a scientific, not a forensic activity. Although it is true that peer reviewers occasionally uncover scientific misconduct, especially when it is clumsy, it should be neither excoriated for failing to detect misconduct nor counted on as a reliable way to uncover misconduct.

Researchers and institutions must remain responsible for the integrity of their research. Misconduct comes at a very high price for the researcher, for the corpus of scientific knowledge, and for the public's trust in the research enterprise. It must be avoided, and if it occurs, it must be dealt with thoroughly in accordance with prudent policies. Perhaps one of the most important initiatives that institutions can develop for their researchers is a robust curriculum in research ethics. We would even go further, suggesting that such education should not start at the postdoctoral level, but that serious consideration of scientific integrity, including courses in research ethics for budding scientists, should be undertaken from the undergraduate through the graduate level.

Finally, as succinctly put by David Miller and Michael Hersen,

Governmental agencies in power, the administrators in academia, the leading researchers in the field, and the mentors of junior faculty must engender in these individuals the thirst for knowledge and the excitement of discovery. However, in so doing they must repeatedly underline and underscore that there are no shortcuts in science, for the most part the discover is slow, plodding, painstakingly, with the road to ultimate success paved with numerous stumbling blocks, dead ends and wrong turns. (p. 241)

*See also* **Medical Values and Ethics; Misconduct, Scientific**

### BIBLIOGRAPHY

Abbot, Alison. "Science Comes to Terms with the Lessons of Fraud." *Nature* 398 (March 4, 1999): 13–17.

Braunwald, Eugene. "Cardiology: The John Darsee Experience." In *Research Fraud in the Behavioral and Biomedical Sciences*, edited by D. Miller and M. Hersen. New York: John Wiley & Sons, 1992: 55–79.

Broad, William, and Nicholas Wade. *Betrayers of the Truth: Fraud and Deceit in Science*. Oxford: Oxford University Press, 1985.

Commission on Research Integrity. *Integrity and Misconduct in Research*. Washington, DC: Office of Research Integrity, 1995.

International Committee of Medical Journal Editors. "Sponsorship, Authorship and Accountability." *Medical Journal of Australia* 175 (2001): 294–296.

Jones, Ann Hudson, and Faith McLellan, eds. *Ethical Issues in Biomedical Publication*. Baltimore: Johns Hopkins University Press, 2000.

Medical Research Council Ethics Series. *MRC Policy and Procedure for Inquiring into Allegations of Scientific Misconduct*. MRC Ethics Series. London: Medical Research Council, 1997.

Miller, David, and Michael Hersen, eds. *Research Fraud in the Behavioral and Biomedical Sciences*. New York: John Wiley & Sons, 1992.

National Academy of Sciences. *On Being a Scientist: Responsible Conduct in Research*. Washington, DC: National Academies Press, 1989.

National Academy of Sciences, National Academy of Engineering, and the Institute of Medicine. *Ensuring the Integrity of the Research Process*. Responsible Science, vol. I. Washington, DC: National Academies Press, 1992.

National Research Council, Committee on the Responsible Conduct of Research. *The Responsible Conduct of Research in the Health*

*Sciences*. Washington, DC: National Academies Press, 1989.

Normile, Dennis. "Japan Seeks Answers to Rise in Misconduct." *Science* 301 (2003): 153.

Relman, Arnold. "Fraud in Science: Causes and Remedies." *Scientific American* 260 (April 1989): 126.

Swazey, Judith P., Melissa S. Anderson, and Karen Seashore Lewis. "Ethical Problems in Academic Research." *American Scientist* 81 (Nov.–Dec. 1993): 542–553.

Thelen, Mark, and Thomas M. Di Lorenzo. "Academic Pressures." In *Research Fraud in the Behavioral and Biomedical Sciences*, edited by D. Miller and M. Hersen. New York: John Wiley & Sons, 1992: 161–181.

*Angela Wasunna and Thomas Murray*

# PROFESSIONAL RESPONSIBILITIES IN SCIENTIFIC AND ENGINEERING RESEARCH.

Discussions of professional responsibilities in scientific and engineering research often proceed with discussions of the responsibilities of individual professionals. This "fast forward" ignores the fundamental truth that there are no individual professionals without professions, and professions are collective bodies whose identities require criteria that transcend those of the particular individuals in them. This entry begins with a classification scheme for professional responsibilities in scientific and engineering research. After laying out this scheme, it discusses collective agency and collective responsibility, in order to help explain how professions develop views about their collective responsibility, and the relationships between those views and those of individual professionals. It ends by providing specific examples in support of its position.

## Classifying Professional Responsibilities

It is possible to classify kinds of professional responsibilities and to identify distinctively professional and moral concerns. The classification scheme can be sketched along two axes, one enumerating relationships and the other, research-related activities. One commonly used classification scheme for relationships divides professional responsibilities into those between

- The individual professional and other professionals/the profession
- The individual professional and the employer or client
- The individual professional and third parties.

This division overlooks entirely the responsibilities of a profession itself. One reason for the oversight might be that this recognition depends on the recognition of collective agency and collective responsibility. Once this possibility is acknowledged, the potential parties to take into account multiply. Professional responsibilities involve

- The relationships between individual professionals and their own and other professional groups, as well as the relationships of individual professionals and their own professional groups to clients, employers, patrons (institutional and individual), and to students in their own professions and others
- The relationships of individual professionals and their own professional groups to third parties, including public and private organizations and individuals, and the citizenry or public writ large.

Along the other axis, that of research-related activities, the issues can be divided into two kinds: those arising in the conduct of science and engineering research and those arising with its influence in society. For example, issues concerning human subjects would fall into the former category, whereas issues concerning the consequences of research on intelligence would fall into the latter.

## Collective Agency

It is easy for people to assume that only individual human actors are responsible agents. On the other hand, people often hold organizations and nations responsible for the effects of what they do; and it is very difficult to see how that is the same thing as holding individuals in the organizations or nations responsible. Sometimes people think it perfectly acceptable to hold a body responsible after all those individuals who were directly involved have died. It is not unheard of for professional organizations to be sued and found guilty of conflict of interest, after the committee that issued the standard had been disbanded.

There are conceptual difficulties in according the status of agents to groups of various types, rather than exclusively to the individual human beings who populate the groups. However, more difficulties arise if the powerful effects of organizational and group action, which go far, far beyond individual human capacities, are not recognized.

A joint commitment cannot be understood as the conjunction of the commitments of the individuals involved. For example, when a fictional Maggie and Louise start a research project, they constitute "a body" or "a plural subject" that intends to do it. They are jointly committed to certain arrangements and behaviors; they are a "we," as in "We intend to spend Saturday working out the protocol." It would be strange, were Louise not to meet Maggie, as she had indicated she would, and in response to her rebuke say, "Well, it was only a personal commitment of mine." The joint commitment stands independently of Louise's personal action; the joint commitment has implications for the individual commitments in such cases. Indeed, when people make such commitments, they may find themselves bound to perform certain actions with, for, or to unnamed individuals; they may find themselves with obligations toward many.

These kinds of obligations are not moral obligations. People make agreements to do immoral or nonmoral things. They come to share immoral or nonmoral attitudes or beliefs. The feeling of obligation to one's fellows, despite their failings, is an emotion that few have never felt. Group members feel entitled to rebuke one another when some responsibility or other to that plural subject—to the "we" of it—is unfulfilled. In these circumstances, members believe that their rights (not necessarily moral rights) have been disregarded. The social rules—the rules of that social group, rules to which members should conform—have not been followed, and those who have disobeyed can and even should be punished. The rules are, to put it one way, "the way we do things."

Another phenomenon is associated with social groups—whether they are political or scientific, religious, or recreational. Groups as plural subjects hold beliefs; and their members react with surprise and hostility when one among them challenges those beliefs. This phenomenon seems to be universal to groups. Groups view a member's challenge as a betrayal to the joint commitment to uphold their collective beliefs. Knowledge of this response helps to explain fears of speaking up against collective or conventional wisdom. Collective beliefs are enablers, allowing groups to take much for granted and move forward; they are also unavoidable and powerful and potentially repressive social forces.

An organization, a group, a political party, a scientific discipline, a professional association, all may be plural subjects. As such, they exhibit behaviors that need to be characterized and classified as core group or social phenomena, not as the sums of the individual attitudes and actions of their members.

## Collective Responsibility

One implication of the previous discussion is that the normative notion of responsibility is different from, or perhaps broader or more inclusive than, the notion of moral responsibility. Tom can be responsible for covering his buddy's back while his buddy is fudging the data. As the dictionary definition indicates, one meaning of *responsibility* involves being held to account or render satisfaction—being answerable. Individuals hold other individuals responsible in this manner; a group holds its members answerable for those actions that are constitutive of and for the group.

Another implication is that there are at least two senses of collective responsibility. One sense is that of a number of agents, each of whom has a particular responsibility for the well-being of something. It can be the well-being of the group or of individual group members. Another sense is what the group as itself an individual entity—a plural subject—must assume. Thus, the group undertakes a coordinated set of actions that allows it to complete a research project or cheat in a course. Through those actions it can develop an archive of previous tests and results, papers, or data; sell material to clients; provide special assistance to members who fall behind; and the like. Individual members identify and assume personal responsibilities in recognition of the group's needs and goals; the group has expectations of its members. Individual members can hold the group accountable when problems arise: "We have a responsibility to use a wide variety of sources, so we develop a complete literature review." Or "We have a responsibility to use a wide variety of sources, so none of us falls under suspicion."

Moral responsibility includes both senses of collective responsibility. Unlike legal responsibility, professional moral responsibility is forward-looking and shared. Moral responsibility wishes to consider how to improve future behavior, not just evaluate past behavior, and recognizes that one person's responsibility, say to ensure an airplane's safety, does not preclude others from having the same responsibility. The emphasis on sharing takes notice of the collective nature of responsibility. However, this notion of sharing does not exclude each individual from having a different set of responsibilities from any other. Nor does it require a plural subject or joint commitment.

Responsible behavior also requires that agents be prepared to act appropriately when called upon; that is, agents must be able to respond appropriately to the unforeseen and even unforseeable. These behaviors require going beyond following predetermined rules. The rules must be evaluated and sometimes disregarded or even, perhaps, discarded in light of particular circumstances and the agents' goals.

In further explanation of the claim that responsibility is shared, forward-looking responsibility is differentiated; it is institutionally divided. That is, who is responsible and for

what will depend on the agent's role and the enterprise in which the agent is involved. The notion of institutions is quite broad. For instance, a congressional candidate who hires a pollster is participating in the social institution of elections. The health maintenance organization that hires a physician is participating in the social institution of health care. The sociologist and her graduate students, studying street corner gangs, are participating in the social institution of research. Responsibilities of participants in these institutions depend on what their distinctive practices and ends are. The particular behaviors that are required cannot be limited to predetermined rules, although the revisions that are allowed will be limited by the parameters of those practices and ends. Those involved are expected to be able to perform the required tasks and actions.

These institutions appear to fulfill the requirements for plural subjects. Their members have joint commitments of particular kinds; they share certain beliefs; they feel entitled to rebuke each other when obligations are not met. Engineers, scientists, physicians, lawyers, babysitters, parents, judges, and so on all have particular responsibilities as a function of their roles in their institutions. People have many roles and thus many kinds of collective responsibility. But besides asking about which rules to apply and when to apply or disregard them in these roles, people can also ask questions about the institutions themselves, the institutions' rules, and their roles in them.

So far, the characterizations of responsibility in the preceding paragraphs do not suffice to make behaviors, or institutions, moral. Odd as it sounds, Tom can function as a responsible cheater, if cheating can be called a social institution. It happens that the institution is often illegal and many would say immoral. Now, some people would say that Tom is participating in the institution of research or education, and those institutions frown on cheating. At least in part, that is true; it is also true that institutions may encompass groups with conflicting views about their members and those members' responsibilities.

Moral responsibility comes with the recognition that morality functions in a differentiated social enterprise reflecting on and evaluating the variety of ends of human life. Morality is itself a social institution with a distinctive range of concerns: concerns not for specific goods or purposes, but rather with the general good, with having us do what we ought. Human beings and institutions use the resources of the institution of morality in judging qualities of other social institutions. Humans are members of numerous institutions. In a membership capacity in an institution, they may act as plural subjects, taking the "we as a body" point of view about a matter of institutional significance. Or they may step out of that role and view the matter from a personal perspective or the perspective of another institution in which they are members, even the perspective of morality.

Cheater Tom can do that of his institution of cheating, or many others. He can ask moral questions of educational, political, economic, professional, scientific, cultural, religious, familial, tribal, or other institutions. He can puzzle about the boundaries between them. He can ask similar questions of his roles in them. He can turn to other members of his gang and raise these moral questions. They may be likely to respond from the plural subject point of view, rebuking Tom for raising matters they believe are settled. Then again, some or all of them may decide to change their priorities and operations. These deliberations can be distinctively moral, involving the relevant range of actors, considering a specific range of consequences, with the authority to revise or override the applicable rules.

## Issues of Professional Responsibility in Scientific and Engineering Research

People expect members of a profession to exhibit mastery and control of a basic body of abstract knowledge. Its members know that other members and persons outside of it have certain expectations. The members share numerous common beliefs and joint commitments. They recognize that "we do

things in certain ways." Their technical and social standards seem to suffice to classify professions as plural subjects or groups that exercise collective agency and joint commitments.

These technical and social standards also allow professions and their members to claim certain expertise as within their special competencies. By making these claims and having them attested to in certain ways, the professions and their members set boundaries. Thus, their roles are differentiated. Boundaries establish the arenas within which they can exercise certain privileges and from which they can exclude nonmembers.

Collectivities have powers beyond the individual, as well as responsibilities. For the professions, these include powers of self-regulation and social control. Professions in different countries and at different times assume different degrees of responsibility and have different degrees of autonomy for ensuring that their members meet professional standards of competence and ethics. The more responsibility they assume and the more conformity they create, the more their professions can be said to exercise a social control that is self-regulating. As research becomes more closely involved and associated with social purposes other than basic production of knowledge, more external scrutiny and oversight are likely. If increased scrutiny becomes counterproductive, it may be worthwhile to think about ways to insulate research from increased oversight, perhaps by changing the incentive systems that couple it to economic reward systems and strong hierarchical arrangements.

Beyond these basic competencies and privileges, people have developed further expectations of professions: that they exhibit the ideal of service. Many professions have, and more and more of them are developing, codes of ethics (for a collection of them, see Center for the Study of Ethics in the Professions). The professional societies debate ethical issues at their meetings. They include attention to ethical issues in their governance structures and publications. Ethics has become an increasingly visible characteristic, a standard part of professional life, marking professions as sites of institutionally divided moral responsibilities.

Doing things in certain ways and for certain purposes becomes definitive of correct behavior. However, these standards are sometimes contested, within a group and between the group and other groups or individuals. Different laboratories may have approaches with technical and ethical differences. Members of a lab or a professional group may perceive their individual professional responsibilities to have both technical and ethical components independent of the group's standards. Outsiders will also challenge the perceptions. Professional values will change and evolve; it will be important for professions to remain open to these possibilities.

Currently, attention to professional responsibility in science and engineering research concentrates more on issues that arise in the conduct of research than in its social influence. This is true for individual professionals and for professions as institutions, as well as for other organizations with responsibilities for research performance and oversight. It is certainly true that ethics in research conduct is an important topic, and not least because inattention could easily lead to societal mistrust of research and researchers, and more oversight. Issues of compliance and conscience are both involved in these circumstances. It is also true that professions and organizations responsible for their oversight, as distinctive agents, can easily recognize the centrality of such matters to their functioning, being less closely concerned with broader matters.

Various social institutions have assumed and been assigned responsibilities to foster ethics in research conduct. As could be expected, appraisals of their effectiveness are mixed. In the United States, federal agencies that support research have issued new requirements for organizational oversight of research ethics and human subjects protections, and for education concerning research ethics and human subjects protections. Professional associations have developed guidance for academic institutions on misconduct policy

and for their members on publication ethics. University associations have also developed policy guidance on misconduct; they as well as professional groups have distributed recommendations for good mentoring practice in graduate education.

Caroline Whitbeck characterizes the overall ethical concern as one for trustworthiness in research, and she cautions against narrowing the subject area to issues of research misconduct. The caution is to view the issue as one of conscience (from the perspective of the institution of morality), not just one of compliance (the institution of law). The focus should be on all aspects of education, training, and practice necessary to provide requisite assurance of trustworthiness. Issues arise in teaching and collegial relationships, publication activities, and research activities. Issues arise in research supervision—beyond fabrication, falsification, and plagiarism. Issues of laboratory safety, animal welfare, or environmental protection also need consideration, where appropriate. Conflicts of interest, issues of privacy and confidentiality in data handling, and fairness in competition are other examples.

Compared with issues in the conduct of research, the issues of science, engineering, and society receive less attention from scientific or engineering societies or governmental agencies with specific responsibilities to encourage, support, or undertake research. As plural subjects or collective agents, such organizations can concentrate more easily on activities in the conduct of research. These activities are more intrinsic to their functioning as the sorts of institutions they are. The responsibilities they assume and are asked to assume, when focused on research conduct, are distinctive. Even though the particular standards may be hotly disputed and controversial, the fact view that there must be standards with respect to these matters becomes recognized as part of the morally differentiated role of the institution. As time goes by, they become matters for compliance—not, or not just, conscience.

There are areas where issues in the conduct of research and issues of science, technology, and society overlap. In these areas, the morally differentiated role may include or come to include broader social and moral concerns. Human and animal subjects protections provide examples.

Another important question that falls into this overlap between conduct and societal effects involves methodological value judgments, which are an unavoidable part of research. Biases in such judgments, as can occur when relevant populations are overlooked for instance, are never acceptable. However, the results of data collection and interpretation can differ considerably as a function of scientifically acceptable differences in these judgments. Individuals and organizations bring different technical and ethical standards to bear in discussions about these matters, and struggle over whose standards will prevail. An example is alternative judgments about how to assess risk.

In many cases where questions about the consequences of research arise, research-related institutions and their members view the phenomena as less their responsibility. Rather, they view them as responsibilities that belong as much or more to others.

Kristin Shrader-Frechette argues for a principle of ethical objectivity in the actions of scientists and engineers when evaluating whether and how to undertake research or report its results. She believes that this principle requires individual scientists and engineers to make independent assessments of public rights and research consequences so as to provide greater protection for individual human beings and the environment. From the perspective of the institution of morality, individual scientists and engineers—like everyone else—can evaluate their actions and undertake to behave in morally responsible fashion. However, professional organizations and their members do not share particular technical and ethical standards that bear on these matters. Shrader-Frechette's position overlooks the force that groups as plural subjects provide, standing behind professional requirements. On the other hand, as these standards develop, more protective moral positions may become a distinctive part of these social institutions in the future.

Disciplines and professions are social institutions that exhibit distinctive roles incorporating technical and ethical standards. As collective agents, they recognize and hold their members to these standards. Negotiations about their acceptability and appropriateness go on within the professions, between them, and between professions and other social institutions. It is easier for professions to recognize and assimilate standards with regard to their distinctive research roles than it is with regard to broader social questions involving their research.

*See also* **Misconduct, Scientific**

## BIBLIOGRAPHY

Airaksinen, Timo. "Professional Ethics." In *Encyclopedia of Applied Ethics*, edited by Ruth Chadwick. San Diego, CA: Academic Press, 1998: 671–682.

Braxton, John M. *Perspectives on Scholarly Misconduct in the Sciences.* Columbus, OH: Ohio State University Press, 1999.

Center for the Study of Ethics in the Professions. *Codes of Ethics Online.* Chicago: Illinois Institute of Technology. http://ethics.iit.edu/codes/index.html.

Gieryn, Tom. "Boundaries of Science." In *Handbook of Science and Technology Studies*, edited by Sheila Jasanoff, Gerald E. Markle, James C. Peterson, and Trevor Pinch. London: Sage, 1995: 393–443.

Gilbert, Margaret. "Remarks on Collective Belief." In *Socializing Epistemology: The Social Dimensions of Knowledge*, edited by Frederick F. Schmitt. Lanham, MD: Rowman and Littlefield, 1994: 235–256.

Gilbert, Margaret. *Sociality and Responsibility: New Essays in Plural Subject Theory.* Lanham, MD: Rowman and Littlefield, 2000.

Ladd, John. "Collective and Individual Moral Responsibility in Engineering: Some Questions." *IEEE Technology and Society Magazine* 1.2 (June 1982): 3–10.

May, Larry. *The Morality of Groups: Collective Responsibility, Group-Based Harm, and Corporate Rights.* Notre Dame, IN: University of Notre Dame Press, 1987.

Richardson, Henry S. "Institutionally Divided Moral Responsibility." In *Responsibility*, edited by Ellen Frankel Paul, Fred D. Miller Jr., and Jeffrey Paul. Cambridge, U.K.: Cambridge University Press, 1999: 218–249.

Shrader-Frechette, Kristin. *Ethics of Scientific Research*. Lanham, MD: Rowman and Littlefield, 1994.

Whitbeck, Caroline. "Research Ethics." In *Encyclopedia of Applied Ethics*, edited by Ruth Chadwick. San Diego, CA: Academic Press, 1998: 835–843.

*Rachelle D. Hollander*

# PROFESSIONS, OCCUPATIONS, AND TECHNICAL ROLES IN THE SCIENTIFIC COMMUNITY.

Few practitioners of the social studies of natural science have taken the standpoint of the material conditions under which knowledge is produced, at the level of the larger social and economic forces or the division of labor at the workplace. Moreover, most have studied projects in academic or small, nonprofit settings from an epistemological perspective, that is, from the optic of how scientists come to agree that they have found a given object or have established certain relations. The impression of much of this work is that scientific research is an autonomous activity that is informed by a "culture" and is mediated only by their respective subjectivities, intellectual traditions, and training, and not principally by power. With a few exceptions, the environments in which the overwhelming majority of scientists work and the gradual merger of science and technology in the performance of nearly all scientific labor are bracketed when not completely ignored. Although the social study of science has yielded great insights into the nature of scientific work, this article seeks to make a contribution to bringing the stars down to earth. I have not ignored these insights but have transported them into the categories of labor process studies.

## Science and the Industrial Revolution

The great scientific achievements during the eighteenth century were made mostly by independently wealthy men, such as Robert Boyle, Benjamin Franklin, and Isaac Newton, for whom science was an avocation in that they did not earn a livelihood from these activities. They conducted their experiments in personally financed home-based laboratories. It was not until Michael Faraday—a person of financially modest means whose work on electromagnetism literally electrified the nineteenth century—was invited by the Royal Institution to become a resident and was thereby provided facilities to perform his research that science became institutionally bound and the scientist became, de facto, an employee. He was not paid a salary, but the Royal Institution provided him with room, board, a small stipend, and laboratory facilities to conduct his work.

But the Industrial Revolution, in which science played a vital role, changed the face of scientific work. By the late nineteenth century, science had become a vocation that required a specific credential for those who wished to practice it. Because corporations and the government required scientific workers in a wide variety of spheres of economic and social life, science education expanded in institutions of higher learning, advanced degrees were granted, and, in the United States, the laboratory became a ubiquitous facility within some land-grant state universities and a few elite private universities. The federal government employed scientists in agencies such as those tied to navigation, shipping, and agriculture. By the turn of the twentieth century, Charles Steinmetz, one of the more eminent scientists of his era, was invited from Germany by General Electric Corporation to establish a laboratory dedicated to both basic and applied research in electromagnetism and the development of the industrial means of production and consumer products. But GE was not singular in this innovation. The famed Bell Labs eventually hired thousands of scientists and engineers. Stimulated by military needs but also by the growth of mass medicine, by World War I the example of the pioneers was emulated by dozens of corporations, notably the developing chemical and pharmaceutical industries.

By World War II typical scientists were salaried employees. They worked either for industrial corporations, such as GE, ATT, U.S. Steel, GM, DuPont (whose early business was manufacturing gunpowder), Merck, and Union Carbide; in university-based laboratories, in which they were additionally obliged to teach; or they were employed directly by the federally owned and managed defense facilities, such as Los Alamos and Brookhaven, or in "independent" research institutes linked to large universities, such as the Scripps lab, where Jonas Salk and his colleagues made scientific discoveries that were the foundation for product development activities aimed at finding a cure for polio.

Because scientific discovery and its allied technical development were conducted chiefly by salaried employees, did the production of scientific knowledge become a labor process like any other? Is science produced according to the hierarchical organization of work entailed by the prevailing industrial division of labor? Do scientists confront the same issues of job security (and insecurity) and problems of promotion and income that are characteristic of other forms of qualified, credentialed intellectual labor as does, say, academic labor? Are the objects and processes of scientific discovery constrained by institutionally controlled methods of work, dominant intellectual paradigms, and managerial control at the workplace? The answer is yes, but it would not be accurate to reduce scientific labor to these boundaries. In rare instances, which every student of the history of science knows well, knowledge breakthroughs may occur when through the publication of experimental or theoretical reports an individual or research collaborative, forces a paradigm shift. Yet, as Thomas Kuhn has shown in his historical studies of the moments of transformation in physics, the condition of gaining status as legitimate scientific knowledge is the approval of the recognized authorities in the field, who are the gatekeepers of past knowledge as well as the putative shepherds of the new, a dual role that upon more than one occasion has delayed and even suppressed the broad acceptance of new scientific knowledge.

## The Scientific Labor Market

As in any other occupation, there is a scientific labor market. It divides into private industry— corporations that maintain their own research facilities, usually not in basic research but in research and development (the technoscience of transforming the results of basic and applied research to designing new products), and independent, for-profit research organizations that, increasingly, are the recipients of contracts from large corporations to perform specific tasks, usually in applied science or products. These research corporations typically hire scientists on a contract basis, provide few if any health or pension benefits, and lay their researchers off at the completion of the project. The scientific worker may be rehired if the company wins a new contract, but if not, the worker is back on the job market. Although well paid, these researchers are part of a growing temporary and contingent labor market that has afflicted almost every sector of American business.

The academic labor market for scientists is similarly divided. Only a minority are in research institutions, and, in the United States, these are the sites of most basic and applied research, although product development has become more common since the 1980s even in the most prestigious universities. But, except under unusual circumstances, almost all academic scientists have teaching as well as research duties. In order to keep an appointment in a research university, the scientist must obtain grants and write articles reporting research results. Most American colleges and universities are primarily teaching schools. These teaching institutions—third-tier public universities and community colleges, and all but a handful of private four-year liberal arts colleges—require scientists as well as other professors to shoulder heavy teaching loads, from which only grants and awards can permit the project participants to purchase course reductions.

It is becoming harder to get grants at all levels of the academic system. Most are awarded by agencies of the federal government. The young scientist who is not working as researcher under an established senior professor must spend much of his or her time writing grants. Twenty-five years ago, the leading federal granting agencies, especially the heavily funded Department of Defense, National Science Foundation, and National Institutes of Health, awarded funds for one in three grant applications. This was the era when money was relatively plentiful for the physical sciences. Now, even though funds for research in the biological sciences have increased over the past decade, the successful grant ratio is about one in ten. This means that grant writing is a major component of the research scientist's job description. The senior scientists may spend as little as one month performing this task. But the less-successful researchers may spend as many as six months trying to get a grant to do their own research, and some do not succeed at all. In either case, since, under recent federal science policy, funds are earmarked for "dedicated" research—which favors knowledge that enhances economic development and university–corporate partnerships—scientists must be prepared to work on projects that correspond with the priorities of the grantor and the companies with which they seek to collaborate.

## Scientific Work

The work of scientists now has a strong political and economic dimension. In the Cold War era, even though money for basic research came from the Department of Defense, it was relatively unfettered. Now the object of scientific labor is determined from outside the scientific community, although the funding process must be generated by the scientists themselves. Furthermore, led by MIT scientists who made an agreement with biotechnology companies in the early 1990s that gave the companies a share of their patents in return for contributing the research funds that had been withdrawn from public sources, university–corporate partnerships have flourished in the sciences, thus edging basic research to the margins. Moreover, as the embryonic stem cell controversy illustrates, the Bush administration imposed political and ideological criteria on what federal agencies are willing to fund, regardless of the peer review process or the eminence of the scientists. Scientific organizations have protested the politicization of science, but its practical consequence is that some scientific activities are proscribed or severely reduced.

The scientific workplace is organized according to the division of labor among intellectual, technical, and manual workers, as with other professions. Although the intellectual worker enjoys relatively more autonomy than the others, the director of the laboratory or the project has overall supervisory responsibility, and the scientist is ultimately subordinate and accountable to him or her. Research is directed by senior scientists, who function largely as managers and administrators. Many of the administrative tasks are somewhat distant from the actual process of scientific work but are vital to its success because the cost of conducting research has escalated exponentially since World War II, and the nature of the tasks involve five, ten, or a hundred workers at various levels of the occupational hierarchy. The director's duties include supervising and writing budgets for corporate or university approval; vetting and lobbying for large grants, which invariably require the director as well as the prospective principal investigator of a project to meet with officials at various federal agencies; and directing the research of team members who may be principal investigators for a segment of the project. The principal investigator, in turn, works with various staff members under her or him, such as full-time research scientists, graduate students, and technical support people.

What do practicing research scientists do every day? The first task is to observe, the second to measure. Each requires machines of various sorts. The contemporary research site is often a gaggle of technological devices. Almost all scientific work entails the interaction of the scientist with machines that mediate the relation of humans to nature. That is, modern science is nearly always technoscience, even for marine biologists, zoologists, and plant biologists, who, in the popular imagination, have direct interaction with ecosystems. As in any other science, they use instruments for observation and measurement, but these instruments vary according to the specific character of the investigation. After the research questions have been asked, and the division of tasks has been specified among researchers—an accomplishment that entails collective discussion, involving managers, scientists, and sometimes technical workers—the laboratory organizes itself as a work site marked by two distinct sections: workstations and offices. The workstations are typically equipped with computers, which, in both physical and biological sciences, function as devices of calculation and inscription. In the dominant biological field of genetics, one of the most important subfields is biophysics. The work of the biophysicist is done on the computer, with the assistance of programs that the researcher writes, often with the help of a computer programmer, which constitutes his "lab." But other biogeneticists still work with more extensive technologies and spend some of their day splicing genes through the use of various machines. These machines are sometimes standard and are purchased from scientific machinery firms. But, in all the so-called natural sciences, the scientist herself becomes a designer and even a mechanic: She makes a machine designed specifically for an experimental algorithm that she has adapted from traditional elements of scientific methods. Larger facilities may employ a toolmaker or machinist to make the device, but invariably the scientist is chiefly responsible for its design and might draw the blueprint herself. The machine enables the scientist to observe what she is looking for, but the form of the "object" is subject to the effects of the instruments. In these instances, the concept of "observation" is invariably a form of intervention. In basic and applied projects, time is also spent talking with colleagues. Sometimes this "talk" is conducted in formal settings during the workday: the project staff meeting, which usually excludes technical staff. These meetings are held to assess progress, share findings, and, often, to debate what has been seen and what has been measured. But there are informal settings for talk: the lunchroom or cafeteria, the rest room, or an "after-work" drink at a local bar. As Andrew Pickering and Sharon Traweek have observed in their studies of basic research in physics, these conversations are more than chitchat. The informal communication between researchers often clarifies the interpretation of the day's work or of the work so far. Through gestures as well as speech, the researchers may gain a kind of peer review of what they have done or, for that matter, what they have seen and calculated.

Scientists working on intense research projects have no fixed hours. The boundary between labor and leisure is ineluctably blurred. As Andrew Ross has shown in his study of a for-profit dot-com at the cutting edge of the media industry, the excitement of being part of the technological avantgarde and the financial incentives associated with innovation prompted many in this company to stay in the workplace for hours after the prescribed work day, to take materials home, and spend their evenings working. Many reported they dreamed about the content of the work and experienced life as centered in the work, rather than in families and other outside friends. The after-hours activities of the engaged research scientist resembles this model. Thus, the nature of scientific "work" resembles the prevailing division of labor, insofar as there are relations of domination and subordination, but also departs from the model considerably. The scientist often lives in an environment of high anxiety, in part because the pressure to "produce" is excruciating. Perhaps one of the central sources of this stress is the writing imperative, as described next.

The end product of the experimental process is, as Bruno Latour and Steve Woolgar have shown, to inscribe results in the form of an article published in (hopefully) a leading scientific journal. But throughout the day, the researcher takes time to record observations and perform provisional calculations, with the assistance of computer programs or by hand. But article writing is an intrinsic component of the job description. In private research institutes or centers, government agencies, and academic institutions, if the researcher hopes to be offered a permanent job, in universities to achieve tenure, or, if a contract worker, to be offered a tenure-track position, writing is the name of the game. Moreover, careers are made not by doing good experimental work alone, but by making public the results of one's efforts.

And a single successful publication is not enough. The aspirant must demonstrate continuous publications in order to be considered for promotion or even for permanent employment. As good jobs have become scarce in scientific fields because of the reorganization of the labor market, the competition has intensified for publication. Writers are often obliged to publish in second- or third-line journals, but not necessarily because their work is somehow inferior. The peer review process, which is supposed to assess the quality of the submission, is, like similar processes in the humanities and social sciences, subject to judgments that may be less than fair-minded. Because reviewers are overwhelmingly recruited from the mainstream of the discipline, new ideas—even if the research is conducted under the most rigorous methodological procedures—may encounter hostility from one or more readers. And in a crowded field, even one negative review might sink the submission. However, even if the submission is rated "revise and resubmit," the writer is obliged to spend weeks, perhaps months, in the rewrite, without a guarantee of publication.

Large industrial labs generally follow a fairly rigid hierarchical model. Scientists and engineers are subject to more specialized tasks with a much more elaborate division of labor. Experiments are more routine, and technological devices are usually more standardized. And, since the nature of the research is devoted to developing and perfecting new commercial products, the scientist derives most of his knowledge from work done elsewhere. Only on rare occasions does the work reveal new concepts. At the same time, the scientists may still be required to publish if they want to advance to senior status or to management. But the routines associated with the work of most scientists in these settings make them more subject to the division of labor of any industrial workshop, where the autonomy of the worker is highly restricted. In many company labs, for example, scientists and engineers who often work on the same floor and collaborate with each other must sign or punch in when they show up for work. At the same time, unlike the re-

search scientist, except for the key people, they do not take their work home.

Yet the workplace for a large industrial company, which for fifty years after World War II offered scientists job security and fairly good wages and benefits in return for subordination on the job, has become as unstable as the global economy. Lacking union protections, many older scientists have increasingly experienced permanent redundancy, either because the firm has outsourced work to research companies or because—as management contends—the scientists have not been able to keep up with new knowledge and techniques of investigation. Others have been hired in their place but, as in some major corporations, no longer as permanent employees but as contract workers. They may earn higher salaries but have few, if any benefits.

In contrast to the natural sciences, almost all of the work in the *social sciences* is performed in academic settings; when it occurs, collaborative work usually consists of two to four researchers, and, except for experimental and development psychology, the laboratory is not the main site of knowledge production. But showing how social science is produced is no less an urgent task. If we can free ourselves from the obsession that revealing the secrets of nature is somehow privileged, perhaps in the near future we can hope that students of social science will reflect on their own practice.

*Stanley Aronowitz*

# PSYCHOLOGICAL STUDY OF SCIENCE AND TECHNOLOGY

Although the inner relationships among science, technology, and human life are the subject of intense scrutiny within science and technology studies, research generally tends to focus on society, culture, economics, or politics. The world of the individual, however, and the psychological dimensions of science and technology remain largely neglected. There is, however, much interview-based research that echoes the voices of individuals, but analysis that addresses the subjective aspects of human practice and how people directly experience science and technology in their lives tends to be marginal and unsystematic. It is, therefore, not surprising that the conceptual and methodological vocabulary necessary for such a task remains underdeveloped and imprecise. This deficiency may stem from the widely held supposition that because humans are social beings, it is extremely problematic to differentiate the human from the world, the individual from society, people from technology, and so on. But, however unsatisfactory an abstract dichotomy of individual and society or people and technology may appear, we cannot overcome this dualism simply by abolishing the distinction altogether. Although social beings through and through, we are not identical to the social world we have created or to the relations it engenders, nor do they exclusively define us. We are capable of "self-relatedness" and of positioning ourselves toward the world: we can reflect, think through things, exert influence, or change things. It is this possibility of relating as subjects both to ourselves and to the world that opens up the perspective of psychological inquiry. This entry takes a close look at how the individual and society are interrelated and aims to show why the psychological dimension and the perspective of the subject need to be given greater weight in the study of science, technology, and society. The article will question why psychology itself has contributed so little to this area and will present an approach that seeks to adequately represent the individual side of human life in the study of science and technology.

## The Sociality of Human Subjectivity and the Need for a Psychology of Science and Technology

If we look at the psyche as a historical phenomenon and examine it from the perspective of its phylogenetic development, then we can understand it in its most fundamental form as an activity that mediates between the individual organism and the world. First signs of psychical life, for example, can already be observed in the amoeba. It is a living creature, which is capable of recognizing relevant aspects of its environment (light, temperature) and of moving around in accordance with its needs as an organism—it is able to mediate sensitively and actively between its inner and outer world. During the course of its phylogenetic development, this general capacity of the psyche becomes ever more complex and differentiated: The perceptive faculties become more precise as eyes, ears, and the sense of touch become more refined; the abilities to think and learn develop; sociability and the facility to communicate with fellow species members emerge; and emotionality, motivation, and the ability to manipulate and use tools develop. This enables the species to refine the dynamic play between inner needs and lacks (like hunger) and activities to overcome them, and to engage with and exert influence on the natural world (for example, the collective hunt for prey seen in some species, or the use of sticks to "fish" for termites in others).

When we consider the phylogenetic development of the psyche on the path to becoming human, we observe how the capacity to manipulate and utilize things develops into the ability to create and use tools—the beginnings of technology. We can observe a parallel process whereby the initial capacity to communicate phonetically within social groups develops to produce language and consciousness, which are prerequisites of science and knowledge, and the possibility of communal understanding. This is what makes us uniquely human: We are not simply living with others of the species in a direct relationship to a preexisting "natural" world; instead, we actively and consciously create our own *societal world*: collectively and in accordance with our needs and imaginings. It is this living of life through social and material structures that we (in contrast to other living creatures) have ourselves constructed that allows us to speak of societally mediated individual life: a life process mediated by structures of our own creation and that we can grasp, use, and change.

Thus, it makes no sense to try to understand the world of science and technology in abstraction from individuals and their inner relations to it. We create the world of science and technology as individuals from within specific linguistic and cultural contexts, communities, institutions, work structures, and so on—that is, in a societally mediated form. However, we are also directly confronted with this world: We live our lives in and through it; and, in the process, we change ourselves. Therefore, we as individuals are not involved in a one-sided relationship to the technoscientific world, being simply determined by and identical to it; the situation is rather that of a *two-sided relationship*, where we are both determined and determining. On the one hand, this relationship involves appropriating artifacts—using them and investing them with meaning and so on. But on the other hand, we also produce or coproduce things; and it is precisely this two-sided nature of human activity that constitutes the specificity of our subjectivity. As the human psyche is *the* prerequisite for this subjectivity and what enables us to create and inhabit the world of science and technology, and since we live our lives in societal form, mediated by the knowledge and artifacts we produce, it follows that the psychological dimension must be a central concern of science and technology studies.

The epistemology of a psychologically informed study of science and technology must reflect this twofold nature of human subjectivity. Most studies that try to include the perspective of the individual follow the traditional epistemology of the natural sciences and approach human practice objectively, from an external, third-person standpoint. Here, the individual is understood as either an object determined by artifacts or as a user who utilizes them, or as an information or experience resource, a constructor of meaning, and the like. But such analysis remains one-sided: Humans and their actions are reduced to specific functions, often leading to a reification of the individual subject. In contrast, a genuinely psychological analysis, which acknowledges the two-sided potential of human action, approaches human practice *subjectively*, analyzing from an inner, first-person perspective. Only thus is it possible to develop a comprehensive understanding of human subjectivity, and only thus can the richness of human experience and action be adequately repre-

sented. When we talk about a first-person perspective and an epistemology that takes the standpoint of the subject, we are not recommending an individualistic perspective, which delves into the internal, private lives of people (this is, after all, none of our business). We recommend instead a consideration of the direct experience and activity of individual subjects and the conflicts and contradictions they encounter as they engage with the scientific and technological world.

However, it is not enough just to explore people's subjective experience; this alone does not constitute a subjective conceptualization of human practice. Since human experience and action not only is mediated by the world but actually creates it, we also need to explore the meaning of this created world. During the course of our history, science and technology (and in fact our whole social world) have not simply been of general use to humankind; instead, they have always manifested themselves in contradictory structures and forms, embodying specific interests and ideologies, powers, and politics. Therefore, any adequate analysis must illuminate the objective meaning of the knowledge and artifacts produced, and question how their inherent action maxims in turn influence human experience and action.

A two-sided approach, therefore, would seem to offer the most promising route to a psychology of science and technology. It would take as its starting point the subjective problems and contradictions experienced by people in a technoscientific world, but would also explore the meaning of this world itself and its inherent contradictions. Thus, it would be in a position to comprehend the conflicts people face in daily life practice within their proper worldly context.

## The One-Dimensional Scientific Self-Understanding of Traditional Psychology

Within psychology, for over three decades now, there have been demands for a comprehensive engagement with science and technology; and however justified these demands may be, they have yet to be adequately answered (any exceptions tending merely to justify the rule). Although itself both a science and a field of study that directly contributes to the development and design of technological products, such as in the area of engineering psychology, the social debate about the human implications of science and technology seems to bypass psychology completely. Why is this the case? What lies at the root of psychology's reluctance to participate in the study of science and technology despite the widely perceived necessity of just such a program?

Ironically, the reasons stem from the traditional proximity of psychology to technology and to mechanistic ways of thinking, especially in regards to psychology's understanding of itself as a science. When we consider the emergence of psychology from under the wings of philosophy in the middle of the nineteenth century and its consolidation as a scientific discipline in its own right, we can observe how its main strands have grounded their scientific self-understanding and academic identity, not on a specific mode of understanding of their subject-matter, but rather on the basis of a theoretical language and methodology borrowed from the natural and technosciences. From the elemental psychology with its strong affinity to the elemental model found in chemistry (also being developed at the time); to Gestalt psychology, which was developed on the basis of Gestalt physics; from behaviorism with its physiological terminology to the computer language of cognitivism—all these various strands owe their categories and basic concepts to natural science. Attempts to forge a genuinely psychological theoretical language—for example, in William Stern's blueprint of a personalistic psychology, Klaus Holzkamp's foundation of a subject-scientific psychology, or Jerome Bruner's outline of a narrative psychology—have remained the exceptions.

This reliance on a scientific vocabulary taken from the natural and technosciences, including a statistically based experimental method, implies a reduction of psychological phenomena to purely cause-and-effect relationships. Thus, the various dimensions of human life are represented in the abstract form of numbers and variables, resulting in a very limited and deterministic perspective. On the one hand, the experiences and actions of human subjects are reduced to measurable and quantifiable "responses" and "outputs"; on the other hand, the diversity and complexity of the real world and its meaning for the subject are reduced to a measurable "stimulus" and "input" form. This mechanistic methodological conception underlies the scientific assumptions and self-understanding of traditional psychology. This illustrates to what extent academic psychology has reduced itself to a technology and to a technological form of thought. And in the process, it has rendered itself structurally incapable of articulating the two-sidedness of human subjectivity or of developing a deeper understanding of either the process of making or the repercussions of the technoscientific world.

Despite all the insights and empirical evidence from science and technology studies demonstrating the social and cultural constructedness of scientific knowledge, mainstream psychology still clings doggedly to a positivist epistemology. With the accompanying inability to imagine the connection between science, technology, and people other than through a one-dimensional deterministic framework, traditional approaches have little to offer. We can see here how scientific structures—in this case theoretical language and methodology—can systematically hinder the development of knowledge, and how crucial a fundamental rethinking of the traditional epistemology has become.

## Psychologies of Science and Technology

But beyond the universalistic technoscientific tradition, the history of psychology also includes epistemologies that articulate the particularities and the diversity of human subjectivity. From the psychoanalysis of Sigmund Freud to William Stern's critical personalism, from phenomenology to the cultural-historical school, from discursive psychology and social constructivism to the various critical psychologies, we can find a variety of approaches that seek to understand psychological phenomena within their proper natural, cultural, and societal context.

Within these traditions of thought, we can find comprehensive projects of human self-reflection, where the psychosocial and historical implications of both science and technology are actually thematized. All these approaches make the effort to bridge the traditional dichotomy between the natural and the social sciences/humanities as well as to conceptualize people, not as causally determined objects, but as socially situated, acting subjects. And in drawing on qualitative methods, discourse analysis, and ethnography, they have developed specific subject-psychological methods of empirical investigation that enable research to focus on people's problems rather than assuming the people themselves to be the problem.

Within these approaches, we can speak of two different perspectives that have emerged to explore the issue of science and technology. Firstly, we have those who tend to take the perspective of the subject: they examine people's experiences and thought processes, the way they construct meaning and identity, and what role repression and the unconscious play—all within the context of their dealings with science and technology. Historically, Freud was one of the first to analyze from the perspective of the subject and to proceed from there to examine the progress of science and technology critically. "Man has," according to Freud, "become a kind of prosthetic God. When he puts on all his auxiliary organs, he is truly magnificent; but those organs have not grown on to him and they still give him much trouble at times" (p. 91). Freud recognized that science and technology meant not only increasing freedom from natural constraints and a general easing of life, but also implied something rather different. He interpreted the march of culture and technology in the light of people's internal psychic structures and drives, believing that the darker and more destructive side of humanity also finds its expression here. The work of Sherry Turkle provides a further example of a psychological study of technology from the perspective of the subject. Turkle examines the intimate relationship between human subjects and computers. She understands computers as evocative objects, she describes how virtual digital worlds can generate new conceptions of the self, and she explores to what extent cyberspace could become an experimental field for new, less-restrictive forms of identity. Secondly, approaches have emerged that shift the focus onto science and technology themselves and attempt to analyze their structure and meaning for human life by questioning what action maxims they might represent for the human subject. Kurt Danziger is someone who has done important work in this area. He has furnished detailed evidence of how closely the roots of modern psychology and the development of psychological categories and methods are intertwined with sociohistorical and cultural conditions. Others analyze the position, power, and functioning of psychology in contemporary society and have discovered a form of governmentality in traditional psychological concepts and practices—a form of regulative knowledge and social technology primarily concerned with governing others and governing ourselves. Here, we can also find empirical studies that show how psychological concepts such as intelligence or learning disability tend to categorize and label people instead of addressing people's actual needs; leading to a process whereby the individual becomes acquired by the concept and branded as a problem case.

Both perspectives yield important insights about the ways in which subjects actually experience science and technology, the sociohistorical situatedness of academic psychology, and the extent to which the action maxims embodied in the scientific and technological structures are enabling or restricting for individuals. However, in both perspectives there remains a dichotomy between subject and structure, obscuring the inner relationship between the individual and the created world. One perspective concentrates on the richness and diversity of subjective experience; but how this experience relates to specific technoscientific conditions remains unclear. The other perspective analyzes the nature of technoscientific structures and how they implicitly order both social life and human identity; but experience and action, the concrete practices of the subjects dealing with the structures, are not included in the analysis. A comprehensive analysis would, therefore, need to integrate both perspectives to a two-sided approach. Starting from the perspective of the subject, it must not only place its experiences and interactions at the center of its inquiry, but also include the action maxims embodied in the technoscientific structures, thus revealing the inner relationship between the subjective experience/interaction and the world of science and technology.

As we stand at the beginning of the twenty-first century, the gap between technoscientific progress and a social consciousness of how it is affecting our communities and ourselves seems greater than ever. The inclusion of a psychological dimension to the study of science, technology, and society could help to redress this imbalance and open the way to a new type of scientific culture. This would be a culture that reflects critically on the consequences of what it produces and helps individuals and communities both to understand the ambivalences and contradictions they encounter and to explore their possibilities for influence and choice in today's technoscientific world.

## BIBLIOGRAPHY

Anders, Günther. *Die Antiquiertheit des Menschen. Band 2. Über die Zerstörung des Lebens im Zeitalter der dritten industriellen Revolution* [The Obsolescence of Human Beings: The Destruction of Life in the Age of the Third Industrial Revolution]. 3rd ed., Munich: C. H. Beck, 2002. One of the best philosophical critiques on how modern technologies are changing individual and social life; unfortunately, still not available in English.

Bruner, Jerome. *Acts of Meaning*. Cambridge, MA: Harvard University Press, 1990. Based on a precise critique of the information process paradigm of cognitive psychology, this study illuminates the centrality of narration and culture in the making of meaning.

Danziger, Kurt. "The Social Origins of Modern Psychology." In *Psychology in Social Context*, edited by Allan R. Buss. New York: Irvington, 1979: 27–45.

Danziger, Kurt. *Constructing the Subject: Historical Origins of Psychological Research*. Cambridge, U.K.: Cambridge University Press, 1990. A thoughtful social-historical analysis of the construction of psychological knowledge.

Foucault, Michel. "Technologies of the Self." In *Technologies of the Self*, edited by Luther H. Martin, Huck Gutman, and Patrick H. Hutton. Amherst, MA: University of Massachusetts Press, 1988: 16–49.

Freud, Sigmund. "Civilization and Its Discontents." In *The Standard Edition of the Complete Psychological Works of Sigmund Freud*, Volume XXI, translated under the general editorship of James Strachey. London: Hogarth Press, 1975: 55–145. A classic psychoanalytic interpretation of science, technology, and culture.

Gergen, Kenneth J. *The Saturated Self: Dilemmas of Identity in Contemporary Life*. New York: Basic Books, 2000. A powerful social constructionist account exploring the influence of twentieth-century technologies on human relationships and the self.

Harré, Rom, and Grant Gillett. *The Discursive Mind*. Thousand Oaks, CA: Sage, 1994. An excellent introduction to the social and discursive constitution of psychological processes.

Holzkamp, Klaus. *Grundlegung der Psychologie*. Frankfurt: Campus, 1983. A key work that lays the foundation of psychology as a science of the subject; explores in detail the historicity and sociality of human subjectivity; unfortunately, not available in English.

Louw, Johann. "Psychology, History, and Society." *South African Journal of Psychology* 32 (2002): 1–8.

McDermott, R. P. "The Acquisition of a Child by a Learning Disability." In *Understanding Practice: Perspectives on Activity and Context*, edited by Seth Chaiklin and Jean Lave. Cambridge, U.K.: Cambridge University Press, 1993: 269–305.

Rose, Nikolas. *Inventing Our Selves: Psychology, Power, and Personhood*. Cambridge, U.K.: Cambridge University Press, 1996. An important analysis, inspired by the work of Michel Foucault, on how the theories and practices of psychology are intrinsically linked with the rationalities and technologies of political power in contemporary liberal democracies.

Scarry, Elaine. *The Body in Pain: The Making and Unmaking of the World*. New York: Oxford University Press, 1985. A persuasive study of the relationships linking bodily pain, imagination, and the process of human creation.

Shadish, William R., and Steve Fuller, eds. *The Social Psychology of Science.* New York: Guilford Press, 1994. This volume offers contributions that attempt to develop a psychology of science on the basis of the positivistic epistemology of traditional psychology; and it reveals the limits of such a project.

Turkle, Sherry. *Life on the Screen: Identity in the Age of the Internet.* New York: Simon and Schuster, 1995. Sherry Turkle's work is a milestone in the psychological study of technology. In this book she explores the intimate relationships emerging between people and the new information technologies.

Vygotzky, Lew S. "Thinking and Speech." In *The Collected Works of L. S. Vygotsky,* Volume 1. Problems of General Psychology, edited by Rorbert W. Rieber and Aaron S. Carton, translated by Norris Minick. New York: Plenum Press, 1987: 37–285. One of the most influential texts of the cultural-historical school. First published in 1934, it remains an exciting account of how the sociality of human subjectivity is constructed through language and communication.

Walkerdine, Valerie. *The Mastery of Reason. Cognitive Development and the Production of Rationality.* London: Routledge, 1988. A profound post-structuralist analysis of school mathematics as a discursive practice and how subjectivity is constituted within this social discourse.

*Ernst Schraube*

# PUBLIC UNDERSTANDING OF SCIENCE.

The phrase "public understanding of science" now refers to at least four phenomena: (1) what the public understands of science as practiced by professional scientists, typically as measured by public opinion surveys; (2) public representations of science, especially as found in the mass media; (3) the appropriation of science by various constituencies, often with the aid of local mediating bodies such as "science shops"; and (4) the substantive public object (*res publica*) constituted by those with an interest in science—that is, both scientists and nonscientists—on the basis of which decisions about research focus and resource allocation are taken. Generally speaking, research into (1), (2), and (3) has provided the political impetus for the development of (4), which will be the main focus of what follows.

## The Origins and Development of the Concept

The phrase "public understanding of science" (or PUS) recurs periodically in the second half of the twentieth century to cover various Cold War–related activities. Most of these were initiated by scientists concerned that public support for science be based on scientifically sound grounds and not simply the word of politicians keen to use science as a smokescreen for politically dubious judgments to build weapons, exploit the natural environment, and so forth. However, PUS evolved from an exercise in social responsibility to a concerted field of inquiry only in 1985, with the publication of the Bodmer Report by the Royal Society of London. Bodmer tried to link widespread ignorance of basic scientific facts and theories to a decline in science enrollments and, more importantly, a decline in the United Kingdom's intellectual and economic stature on the world stage. At first it was presumed that ignorance of science reflected hostility, born of real or perceived threats that science brought to ordinary people's lives. However, this "ignorance" coexisted with an unprecedented amount of science popularization and generally favorable public attitudes toward science.

Somewhat surprisingly, relatively little subsequent research has directly challenged the Bodmer Report's core assumption that raising the levels of science literacy would enhance the nation's global standing. Instead, PUS research has tended to treat the allegations of mass scientific ignorance as symptomatic of more "local" forms of knowledge on which people normally and successfully rely. Scientists' failure to recognize those forms of knowledge—that come from, say, farmers' experience with their animals—were readily shown to be a source of resistance to science-led public policies. The specific British origin of PUS has anchored the development of PUS studies in two respects.

First, the United Kingdom is a relative latecomer to nationalized science policy. Until the establishment of national research funding councils in the 1960s, the amateur ethic had prevailed in British science, at least as ideology. Science was funded from ordinary university coffers, philanthropic trusts, and sometimes—though often reluctantly—industry. Traditionally lacking the political intimacy long present in Germany, France, Japan, and (after World War II) the United States, British scientists were accustomed to periodic campaigns for public support. Thus, PUS has a long pre-history in Faraday's experimental demonstrations and Huxley's public lectures at the Royal Institution in the nineteenth century, as well as in the more recent best-selling books by Richard Dawkins and Stephen Hawking. Indeed, the United Kingdom has the highest per capita readership of popular science books in the world. The urgency surrounding PUS in the late 1980s was traceable to Margaret Thatcher's fiscal conservatism and the winding down of the Cold War, which together threatened to roll back the institutional gains that British science had made over the previous quarter-century. With the demise of the Soviet Union and the worldwide ascendancy of neoliberal regimes, the rest of the world soon came to face a version of the United Kingdom's predicament.

The second anchoring effect is the United Kingdom's historic aversion to formal public institutions. This has led PUS researchers to stress the sheer voicing of nonscientific forms of knowledge at the expense of designing policy-making forums for integrating them with more scientific forms of knowledge. Seen idealistically, PUS has promoted a "secularization of science" that authorizes communities to adopt knowledge relevant to their specific needs and to register complaints when state-led scientific initiatives run roughshod over local interests. But seen cynically, PUS has enabled politicians to play off against different constituencies as they decide how best to win elections.

The symbiosis of PUS's idealistic and cynical visions has led to the funding of much sociological research that has proved internationally influential, especially under the leadership of Brian Wynne at Lancaster University. Nevertheless, the overall political effects have been ambivalent: To be sure, scientists are now routinely questioned about the applicability of their research—and the levels of risk they carry—in specific contexts. Moreover, public consultation has become a regular feature of science policy making. At the same time, however, the public itself rarely participates as decision makers. On the contrary, the sphere of discretion available to politicians has widened, as they become adept in sampling public opinion by techniques ranging from telephone surveys and Internet polls to focus groups and consensus conferences. Predictably, these often yield a variety of contradictory results, from which the savvy politician can then pick and mix as she pleases.

The idea that different types of input enable better decision making makes sense only if the methods used to acquire the data provide interestingly different accesses to the population and if there is some clear procedure for integrating these diverse inputs. Otherwise, one is simply left with a lot of noisy data open to indefinite interpretation. Bad social science then becomes a cover for "politics as usual." This problem is exacerbated in cases involving science and technology because it is unclear what it would mean to "represent" public opinion on these matters.

Science and technology tend to have a pervasive impact across all sectors of society, yet people generally do not have any fixed views about them. Not surprisingly, then, those groups with the clearest views—"stakeholders"—often end up defining the terms of representation. The solution to this problem is not cleverer techniques for surveying public opinion, but new conceptions of scientific citizenship and new policy-making institutions that enable the public to develop its views about science and technology. The historic genius of democratic regimes rests on institutions that render the political whole more, not less, than the sum of its citizen parts. In this spirit, this article focuses on the *consensus conference*, whose jury-like structure ensures that the public's views are not dominated by self-declared stakeholders.

## Platforms for Citizen Science: Consensus Conferences

Before the current wave of PUS research, the expression "citizen scientist" referred to an Enlightenment ideal exemplified by the U.S. founding fathers Benjamin Franklin and Thomas Jefferson, who were innovators in science and technology as well as politics. In both contexts, they were seen as contributors to a single world-historic development. In contrast, nowadays "citizen science" refers to at least two distinct senses of democratic knowledge production that are somewhat at odds with each other. The first and perhaps dominant sense clearly renounces the Enlightenment ideal altogether. Here "citizen science" signifies the credence given to a form of "local knowledge," regardless of its scientific credentials. This knowledge is desirable as both an expression of local identity and a methodological corrective to the more abstract and general forms of knowledge pursued by professional scientists. Much sociologically inspired PUS research is captured by this rubric. Thus, the citizen scientist might veto large-scale schemes that contradicted local experience.

However, "citizen science" may also aim to extend and deepen the ordinary democratic franchise by giving voice to people traditionally excluded from state policy decisions in which science plays an increasingly large role. Although this perspective acknowledges the alienation of scientific inquiry from the prerogatives of citizenship, it nevertheless points to a reintegration. Perhaps the most significant institutional innovation along these lines has been the consensus conference, popularly called "citizens juries," or, more grandiosely, "experiments in deliberative democracy."

The consensus conference has attracted a wide international constituency, ranging from community activists in the United States (such as Richard Sclove) to science museum directors in Europe (John Durant, for example) and science and technology studies scholars in Japan (such as Tadashi Kobayashi). Since 1985, over fifty consensus conferences have been held in fifteen countries on four continents. Topics have included cloning, genetically modified foods, gene therapy, genetics research more generally, the "informatization" of society, and nuclear waste disposal. In at least one country, Denmark, consensus conferences are regularly convened whenever the parliament is considering legislation relating to science and technology.

A consensus conference is usually convened for three to four days, though the time may be spread across consecutive weekends, given the citizen-jurors' other life commitments but also to encourage them to deliberate informally over the Internet. In addition, prior focus-group research may be needed to ensure that the consensus conference's remit is neither too broad nor too narrow. Normally a consensus conference has two phases. In the first, ten to fifteen members of the public are empowered to take testimony from various experts and interest groups. As in a trial, no prior collusion of witnesses is permitted to present the jury with a uniform account. Indeed, the citizen-jurors are encouraged to probe discrepancies in expert testimony. In the second phase, the jurors deliberate among themselves to arrive at policy guidelines for legislation governing the issue. The jurors themselves draft the guidelines, as in a constitutional convention. The results are then turned over to the elected legislature.

Ultimately it is more important that traditionally marginalized people participate in the decisions than that their views remain intact by the end of the process. Indeed, insofar as the consensus conference is a crucible for forging a more democratic society, it is expected that all parties will alter their views over the deliberations. It does not necessarily follow that they will reach a unanimous judgment—and in this respect "consensus conference" may be a misnomer. But this sense of citizen science implies that, even if a form of knowledge is "local" or "scientific," it is not intrinsically authoritative in political settings. Rather, the forum is itself a site for original knowledge production.

Experts have been consistently impressed by the seriousness with which consensus conference members have taken their assigned tasks, and especially the cogency of the resulting policy guidelines. To be sure, many citizen-jurors begin with strongly held views about, say, the appropriate uses of genetics research. However, by the end of the conference, they come to distinguish between what they would allow for themselves or their families and what they would allow others to do. They may still never personally seek gene therapy, but now they can see why others might want to do otherwise.

Recognizing the difference between private and public interests is crucial for science policy in a liberal society. There will always be people whose value commitments prevent them from supporting biomedical research or treatments that would, say, introduce an animal substance into a human body. The question is whether they would allow others to do otherwise. Consensus conferences are crucibles for forging a sense of permissible differences, from which the entire society may ultimately learn. Without this recognition, it is not clear how scientific innovations can acquire widespread democratic support. Every innovation begins life as a minority dissent that aims to change the norm. Yet, if everyone's voice is heard equally, then either the majority will drown out dissent or, if there is no clear majority, the least offensive, or "safest," policy will win. Consensus conferences are designed specifically to counter the tendency of democracies to associate every new idea with the attempt by one interest group to gain advantage over the rest.

A consensus conference should not be confused with a "town meeting" in which those with strong interests, both for and against a proposed policy, dominate the discussion and determine the outcome. Rather, as in ordinary jury selection, those empowered to decide policy are segregated from interested parties, whose only role is to offer testimony.

Although the citizen-jurors often have strong personal views, these will be diverse and not necessarily connected to their day jobs. Such are the conditions needed for collective deliberation.

The fact that consensus conferences have been tried in so many countries suggests that science is not quite yet a "public thing," a *res publica*. PUS campaigns have succeeded in drumming up interest in science without necessarily providing outlets for expressing and applying that interest. The polarized response of the United States and Europe to the introduction of genetically modified foods illustrates the problem. Without something like consensus conferences to focus collective thought and action, people will respond as either passive consumers or militant protesters. Neither group constitutes an adequate public engagement with science.

Despite their impressive results, consensus conferences have not received the whole-hearted support of either the natural or social scientific communities. Biomedical scientists, in particular, are often uncomfortable with the tendency of consensus conferences to blur the distinction between the conduct and application of scientific research. They argue that the public should determine how research is used, not how it is actually done. The argument is based both on scientists' fears for their own autonomy and on concerns about the state of lay knowledge. However, the line dividing research conduct and application is clearer in theory than in practice. When a consensus conference considers a matter such as gene therapy or genetically modified foods, many issues arise that straddle the divide: How will the treatments or products be tested before being made generally available? Which groups are most likely to be advantaged and disadvantaged from the development of these treatments or products? In addressing these questions, the citizen-jurors will of course take testimony from the relevant experts.

Social scientists raise a somewhat subtler problem. They argue that science has "multiple publics," each with its own set of interests. Moreover, the state of scientific research itself is subject to multiple interpretations and considerable uncertainty. A consensus conference cannot be expected to resolve all these differences into a single "public understanding of science." A historical perspective is useful in answering this objection. The advance of democracy has been always dogged by skepticism about the capacity of people with radically different viewpoints to flourish under a common framework of government. For such skeptics, a democracy can exist only among the like-minded, not those whose interests are in constant tension. Thus, sociologists who study science today tend to privilege "local" forms of knowledge cultivated by particular communities over attempts—say, by the state or industry—to impose a specifically "scientific" understanding of things. Consensus conferences would appear to provide yet another opportunity for just this sort of imposition to occur.

Ultimately, the answer to this objection is that the track record of legislatures and constitutions is much more positive than the skeptics make out. The consensus conference is simply the latest chapter in this history of democratic conflict management. Moreover, because science has become integral to everyone's life, now more than ever it needs to be incorporated into the ordinary mechanisms of government. Beyond merely dwelling on the "uncertainty" of scientific knowledge, consensus conferences can provide the basis for what legislatures do best, namely, conduct corrigible social experiments, only this time on science itself.

## Science Wars and Speaking for Science

Before the 1830s, when "scientist" began to refer to someone with specific professional qualifications, the pursuit of science had two countervailing qualities: On the one hand, it was expected only of those (few) with sufficient leisure; on the other, the results of science were accorded universal validity, even if most people remained ignorant or passive with regard to their production. This paradox persists today. Indeed, science appears to be sociologically unique as a form of knowledge—certainly in contrast with reli-

gion or even politics—in that most of its believers have little specific knowledge of what it is they believe. Thus, there is widespread cross-cultural agreement that Einstein was the premier intellect of the twentieth century, but how many in this consensus can correctly express Einstein's estimable insights? If anything, greater knowledge of science appears to be correlated with greater skepticism of science's goodness. Not surprisingly, to this day the most effective critics of science have come from within the scientific community, since there remains no widely institutionalized means of soliciting public input into the conduct of science. In this respect, science remains *for*—but not *by*—the people.

From Plato's *Republic* to Auguste Comte's "positivist polity," the tension between elitism and universality in PUS has inspired philosophers to imagine that a noncoercive social order could be built by establishing clear patterns of deference to the appropriate scientific authorities. However, as the checkered career of the "expert witness" in law illustrates, the ability of scientists to function with such authority depends on presenting a united front, with scientific agreement providing a basis for public agreement. Unfortunately, the requisite level of agreement among scientists is rarely achieved, even when defending science itself. Thus, once the end of the Cold War exposed scientists to a much more competitive funding environment, they responded by campaigning for science as if its multifarious activities could be justified as an extension of their particular research orientation.

For example, 1992 alone witnessed the publication of two distinguished works of science advocacy: *Dreams of a Final Theory* by Steven Weinberg, a theoretical physicist, and *The Unnatural Nature of Science* by Lewis Wolpert, an experimental embryologist. Not surprisingly, Weinberg portrayed science as aiming for "beautiful theories" whose internal coherence resists all attempts at falsification, while Wolpert stressed the falsification of fixed ideas as the only guarantee that one is actually doing science. The one tests his theories on instruments (particle accelerators) that have little clear application outside their immediate research context, while the other conducts experiments in a medical school where applications are in the forefront of the researcher's mind. Yet, despite their radically different accounts of contemporary science, both lay claim to the 2500-year-old legacy of Western rationalism and demonize the philosophers and especially sociologists of science who would call that legacy into question.

This last practice has triggered the so-called Science Wars, which came to a head in 1996, when a disgruntled U.S. physicist, Alan Sokal, managed to publish an article in a leading cultural studies journal that combined scientific nonsense and politically correct references. The "Sokal Hoax," as Sokal's old teacher Weinberg dubbed it, brought to mainstream public attention the contested nature of "speaking for science." Although sociologists had long known that science speaks in many voices, for scientists themselves this was a potentially damaging revelation, which (so they feared) would give policy makers an excuse to cut science budgets or, perhaps even worse, channel resources toward unorthodox, even "alternative," forms of inquiry that more directly satisfied both the spiritual and material needs of the populace. It quickly became clear that scientists found it difficult to countenance a politics of science in which scientists "represent" our knowledge of reality in the same sense that a politician might represent her constituency.

As the Science Wars have progressed, two diametrically opposed strategies have emerged to protect scientists from the larger sense of public accountability implied by sociologists' accounts. On the one hand, some scientists have agreed with Weinberg that scientists virtually *own* science. Not surprisingly, then, every so often—as in the Sokal Hoax—they must catch trespassers who try to extract surplus cultural value from science without having first undergone the relevant training. On the other hand, others have followed Wolpert, who portrays scientists as modest toilers whose competence does not extend beyond the confines of the laboratory. They envisage scientists delivering a fully mapped human genome on the public's doorstep, but then quickly moving on to their next research project without involving themselves in the political implications of what they have done.

Two issues are likely to dominate the PUS agenda for the foreseeable future. The first is the institutionalization of public participation in science policy decision making. One clear proposal is to make consensus conferences a regular and binding part of the policy process, perhaps akin to the primary and convention season that precedes a general election in the United States. The second issue involves the communicative process that joins scientists and their publics. As scientific research increasingly impinges on public concerns about health and safety, professional scientific bodies have called for the tightening of procedures by which scientists and science journalists report to policy makers and the public at large. At the same time, however, the spread of graduate programs in "science communication" points to the emergence of a new field that may well set its own professional guidelines for treating PUS as more than simply good public relations for science.

*See also* **Public Understanding of Technology**

## BIBLIOGRAPHY

Collins, Harry, and Trevor Pinch. *The Golem: What Everyone Needs to Know about Science.* Cambridge, UK: Cambridge University Press, 1993.

Fuller, Steve. *Science.* Minneapolis: University of Minnesota Press, 1997.

Fuller, Steve. *The Governance of Science.* Milton Keynes, UK: Open University Press, 2000.

Fuller, Steve, and James Collier. *Philosophy, Rhetoric, and the End of Knowledge.* 2d ed. Hillsdale, NJ: Lawrence Erlbaum, 2003.

Hess, David, *Science in the New Age.* Madison: University of Wisconsin Press, 1993.

Irwin, Alan. *Citizen Science.* London: Routledge, 1995.

Irwin, Alan, and Brian Wynne, eds. *Misunderstanding Science? The Public Reconstruction of Science and Technology.* Cambridge, UK: Cambridge University Press, 1996.

Joss, Simon, and John Durant, eds. *Public Participation in Science: The Role of Consensus Conferences in Europe.* London: Science Museum, 1995.

Ross, Andrew, ed. *Science Wars.* Durham, NC: Duke University Press, 1996.

Sclove, Richard. *Democracy and Technology*. New York: Guilford Press, 1995.

Sokal, Alan, and Jean Bricmont. *Fashionable Nonsense,* New York: Picador, 1998.

Weinberg, Steven. *Dreams of a Final Theory,* New York: Radius, 1992.

Wolpert, Lewis. *The Unnatural Nature of Science.* London: Faber and Faber, 1992.

*Steve Fuller*

# PUBLIC UNDERSTANDING OF TECHNOLOGY.

Public understanding of technology covers a spectrum that matches technologies and situations but takes only a few modal forms. It can be informed by a sense of wonder. It can be informed by a sense of oppression, alienation, or fear. Primarily, it is functional: public understanding is vastly and quietly engaged in *operative* understanding of the technologies of everyday life. This operative understanding reflects public engagement in the great and diverse daily accomplishment of the activities of everyday life—on "getting through the day" and of the cycle of individual, family, and public life in the material and social world.

Against the background of a vast and everyday labor of engagement with and operative understanding of a multiplicity of *normal technologies*, technology issues arise. Technology issues often concern *workplace technology* (for example, safety issues concerning X-ray machines formerly used in shoe stores to fit children's shoes; the potential "long term, low dose" problem with dental X-rays, which has led to increased shielding of patients during routine dental X-rays; or the growing incidence of hospital-induced super-infections that result from the Darwinian competition of pathogens with antibiotics in hospital settings). Technology issues also arise concerning *distant technologies*, such as new technologies that come into use in situations and locations that are not easily understood or even moderately accessible to the public. When an issue concerning technology comes to public attention, public understanding takes different forms in different sections of the public.

When technology is encountered in the form of an unusual event (for example, the development of penicillin or the polio vaccine, the moon landing, the Chernobyl disaster, or the pollution of the oceans and the rapid decline of stocks of fish) or in the form of social or technical issues relevant to everyday life (such as the ability to track hurricanes closely by satellite, global warming, the addition of new "security" protocols to air transport or entry to public buildings, or the globalization of jobs) it may be perceived as full of wonder. Or it may be perceived as oppressive, as alienating, or as fearsome. Since there is not a single public, several publics coalesce around problems or issues in an ongoing speech and action process that may or may not lead to a general consensus. There is often more than one form taken by public understanding, and different experiences, situations, and contingencies will inform public understanding of technology in different sections of the public.

## Public Understanding of Elite Promotion of Technology

A familiar "technology is wonderful" theme is often promoted by governments, corporations, and educators. Public understanding of technology may be encouraged by entertaining such visions. However, in the main, the public is both permissive and inherently skeptical. It permits elites to carry on technological programs, but it then waits for progress to happen before endorsing them. This response is intelligent, because such 1950s visions as "energy too cheap to meter," the "automatic kitchen," or a coming era of mass leisure by the year 2000 never came about. We did get virtually universal access to electricity at high prices, and we got semi-automatic coffee makers, but that is different from what was promised. Today, we are promised that "star wars" missile defense systems will protect us against foreign governments and that intensive mining of all personal data to permit "total information awareness" by government will protect us against terrorists. We are also told that globalization of both manufacturing and white-collar jobs is inherently good for the national economies and that genetic modifications to crops will not come loose from their intended applications to infiltrate the biosphere. Based on the record, from a public perspective the new set of promises for the social use of technology are as good as the earlier promises of things to come. Yet it is certain that, both in special situations and in odd moments of reflection, especially when a technology opens a new experience of capability for the individual (for example, an umbrella or a new personal computer) or of well-being for families, communities, and society (for example, successful medical treatments and pain relieving drugs), technologies are appreciated.

Technology can also be deadly oppressive, as when a member of civil society faces a blitzkrieg or loses civil rights to state surveillance. Technologies can be highly alienating in effects, as in the displacement of local business and craftspeople by "big box" suppliers whose low-wage staffs lack technical understanding of products. Technology can be frightening, especially when it seems out of control or provides a power that is operated without respect for individual or communal interests. All of the various modes of public understanding of technology (operative, wonder and appreciation, oppression, alienation, and fear) may be rational. Often, when the public expresses concern about a technology, it is because the public takes a wide sphere of interests and relationships into account.

## Public Understanding of Normal Technology

For the normal technologies we employ every day, we typically have a high degree of personal control within a very small set of relevant options. We drive to work, or we may go by subway or city bus. We light the gas or turn on or off the electric service. We prefer this kind of doorknob over another, but nearly all of them only open doors. A narrow ballpoint pen or a pen with a fat barrel may be easier to grasp, but it is still a ballpoint pen. We select a yellow highlighter or a different one. We step on an accelerator, but the auto stays within a three-dimensional

Newtonian frame, plus time. We use these technologies but pay little attention as we operate them, because the technical and social options offered by each everyday technology are routine and few.

We use normal technologies competently but superficially. If we stop to think about it, we may see a new relationship and tailor some aspect of a technology or its use in a new or more appropriate way. But we usually do not, that being one of the characteristics of most of our use of normal everyday technology. Socially, normal technologies become innate; we use them the way we use speech.

Intuitive technical understanding is gained through seasoned experience in practical use of a technology, whether it is an umbrella or a personal computer. That relatively unreflective technological intuition is possible stems from two characteristics of the technologies. First, technologies are based on the way the world works in a material sense—they are in correspondence with underlying order. Second, the technologies used by humans have been developed, adapted, and tailored for human use. These factors—the material base and the tailoring for human use—provide cues to underlying patterns that enable us to achieve a kind of generalized intuitive understanding of the everyday technologies of our epoch. This intuitive understanding reaches to the use of technologies, not to understanding how they work. The intuitive relationship extends our abilities and provides generalized competence in the use of the everyday technologies of our epoch. All normally constituted adults achieve high competence in everyday technologies: it becomes a generalized ability.

## Public Understanding of Work Technologies

The technologies of our work life are of another order. Often the technologies of work are expensive and restricted to work locations. For example, to know what it is to work on an automobile assembly line, one must work on the line; to know the technology of rocket or satellite guidance fully, one must work where it is possible to guide spacecraft. To fully understand the technology of a research lab, one must work in the lab with colleagues who embody a community of specialized knowledge. That is another characteristic of technology: To know it fully, you have to use it in its material context. One day you are a part of the public that appreciates auto making from the outside; two days later you may be a "left-hand seat bolter," repetitively applying a power tool to place bolts to hold the left hand seat in every car that comes down the line. However, even with a technology as seemingly simple as using a power bolter, it may take two to three weeks before the technology is fully understood and competently exercised. It may take three to five years of experience in different line positions in the plant, and several experiences with union and management incidents, before the full technology of the production line begins to be understood in its social and organizational context. Only small portions of technology can be learned in isolation. It is characteristic of modern society that we are not very interchangeable outside of the shops in which we gain technological experience.

## Public Understanding of Distant Technologies

Any technology of which we lack practical experience is distant to us. We may know of it, but we cannot control it and we do not know it in a practical sense. Sometimes we are simply aware of distant technologies by reading about them. For each of us, the multitudinous diversity of human technology is largely unknown and distant. It is true that we receive the products of a multitude of other areas of work life with which we are not familiar, and so it seems that the individual participates in the technology of an epoch when, in fact, we only know the technologies of our shops. We participate in the collective human technology of our epoch through products.

When a product is received, the work effort and the technologies employed are usually completely opaque to the individual whom the product touches, because the work effort and the technologies have been completely embodied. We may have not a clue to how the TV works or why the electric light comes on, even though we are adept at turning on the TV set and turning on the light switch. Most of us know no more about cattle than we meet in a hamburger or a milkshake. If we are typically unreflective about everyday technologies, we are even less aware of the embodied technologies that make up the physical surround of our constructed environment.

The typical way in which we come to some understanding of distant technologies is through the media. For example, educational TV may present a show on some area of work such as farm life or building motorcycles. We may be incidentally introduced to an unfamiliar work environment through entertainment. A TV series developing stories around a particular workplace (such as a police bureau or hospital emergency room) may incidentally introduce an understanding of much of the actual work life of the shop, including a representation of its range of available technologies and something of the internal orientations of workers (for example city police or emergency service technicians) toward the technologies.

Part of the story of the past century is that the official state-sponsored approach to technology was celebratory. For example, there are many technology films and publications in which the underlying message is that technology brings human progress. Although inventors today are not so much celebrated as tolerated (as "geeks"), throughout most of the twentieth century the public representation of the inventor was of a potentially amazing person working away in a lab. These portrayals often incorporated a patriotic or corporatist theme of social solidarity and were paid for by governments and corporations.

## Public Understanding of Ownership of Technology

In the largest sense, technologies are the property of the people and not of individuals, corporations, or particular countries. Legal systems, of course, enforce narrow ownership for a time that is a large proportion of a human lifespan. Law often restricts use of technology to ensure control and financial return to own-ers. Law threatens consumers for inventive or self-directed uses of technologies. For exam-ple, many tools come with the disclaimer that any use in a manner other than as described by the manufacturer is a violation of law. Sim-ilarly, corporate Websites often have a strong warning to the effect that use of the site for any use other than a highly restricted one will be prosecuted. The public is technically blocked from sharing a DVD purchased in one country with a friend in a different zone. VHS cassettes as well as DVDs are legally restricted so that they may be shown in private but not in social settings. Downloading music from the Inter-net is criminalized although the technology for this use of music, unanticipated by the handful of corporations that control access to music, is universally available and pervasively used. Similarly, in the short term, technologies that have a commercial or military application are highly restricted so as to provide a strate-gic advantage to one human organization or another. However, in the long run all organi-zations that attempt to restrict scientific and technical communication are porous—they can only slow, burden, or confuse communica-tion for a brief moment in human history. Human technology is a characteristic of the human species, what in the nineteenth century would have been called the "species being." Its essence is that it belongs to the whole.

## Inherently Regulatory

The public is aware that technology is often a means of social control, but it is probably not very aware that while technology is "techni-cal," it is also *inherently* regulatory and politi-cal. Often we are not fully aware of the political dimensions of technology even as we are regulated and kept to certain pathways. As Herbert Marcuse wrote in his critical analysis of what he termed "one-dimensional" society, the intersection of technology and the political is exemplified by the on-ramp of a highway: the on-ramp is engineered to provide a single choice.

## Technology and Fear

Fear of technologies is fear of what they will do to the life-world. To capture the sense of

public fear that emerges in certain situations and contingencies, we may look to the leg-ends of the past as incorporated in the darker fantasies of a time when Europe was a few cities and many villages and farming areas scattered within primal forest. What is the dark magic of legends, after all, but com-pletely opaque technologies in the hands of nonhuman beings for ends that are often not understood by humans? These are also char-acteristics of modern war technologies such as the blitzkrieg, which is designed to pres-ent such overwhelming technological force as to induce in a human population a state of awe and shock.

## Technology and Interest

All technologies embody interests. Although we think of technologies in terms of material devices, or of sets of instructions (or care-fully organized sequential operations) and sets of materials and specialized tools, tech-nologies inherently embody social interests. This is so because the choice to use a tech-nology (as opposed to leaving a task undone or using a different technology to address the task) shapes the future. Public articula-tion on this point is often not of high clarity, but public reaction and the crystallization of public opposition to specific technologies (or to specific technologies in specific situa-tions) inherently incorporates this knowl-edge, often expressed in ways that may seem irrational. Yet a bright and disciplined rationality does underlie public oppositions to fast food technologies, to use of antibac-terial agents to help "bulk up" cattle, to ge-netic alteration of the food supply, and, especially, to efforts of corporations and governments to keep information about such efforts secret.

## Natural Technologies

Strangely, although virtually all of our tech-nology has been developed or adapted for human use, we do see other species using technologies in prototypical fashion. Otters use rocks to open shells. Some monkeys can watch humans use doors and doorknobs, un-derstand causation, and then from that point forward copy the motions to open doors the same as humans would. Anyone who has watched a cat study a room for some hours and then suddenly take a series of jumps that place it on the top of the highest bookcase, looking down, sees a protoengineer who understands relationships but lacks speech and the ability to write things down. The impulse and the spatial sense are not differentiable from that of a young human with a bent toward engineering.

Everything involving intention that is done in the material world requires means. The means is always either a prototechnology or a technology, whether or not the intention is fully cognized.

## Evolutionary View of Technology

Finally, there is the matter of public understanding of technology and time on a historical scale. In the remote historical past, most members of the public would have had competency in the appropriate technologies that sustain human survival in a village and forest ecology. Such knowledge, however, has been lost except among primitive peoples who still live in isolated villages in primal forest. Technology evolves, and as it evolves, there is both gain and loss.

A relatively poor individual today who yet has some money may enjoy ice cream at any time of the year, receive successful treatment for a disease, or be healed of an injury, whereas none of these things would have been possible for the richest or most powerful individual of a previous century because the essential technology was not yet known.

Although the public is aware that technologies have side effects, the downstream implications of technologies are not understood and are not currently a matter of public focus and discussion. In the area of energy, for example, the public has virtually no awareness of the second law of thermodynamics. In the area of population ecology, the significance of problems of population burden on regional ecologies is not appreciated, and human populations and human settlement patterns are largely left to market mechanisms for resolution. In the area of recycling, public understanding of technology

remains superficial. The need for expression of collective social interest in product design has only begun to be discussed, and the problem of downcycling versus upcycling has only just been introduced by the sustainable architect William McDonough and the ecological chemist Michael Braungart.

As technology changes and evolves, so does public understanding of technology. Public understanding of technology is vastly operative in nature. When technology issues arise, public understanding is situational and contingent. Until limits are experienced, and as long as experts are virtually entirely in the pay of institutions known for short-term thinking such as for-profit corporations and national states (which define public interest first as service to corporate market interests), learning will be slow. The answers are likely to be in the opposite direction from the "brute force" thinking of globalization and liberalization of markets—rather, in the direction of the finalization movement in Germany, an attempt to bring well-thought-through social direction to applied science and technology, as well as the "technology procurement" effort developed in Sweden. It will be a long road with harsh Darwinian consequences.

*See also* **Public Understanding of Science**

**BIBLIOGRAPHY**

Böhme, Gernot, Wolfgang van de Daele, Rainer Hohlfeld, Wolfgang Krohn, and Wolf Schäfer. *Finalization in Science: The Social Orientation of Scientific Progress.* Dordrecht, The Netherlands: D. Reidel Publishing Company, 1983.

Ellis, Richard. *The Empty Ocean: Plundering the World's Marine Life.* Washington, DC: Island Press/Shearwater Books, 2003.

Engleryd, Anna. *Technology Procurement as a Policy Instrument.* Stockholm: Swedish National Board for Technical Development, Department of Energy Efficiency, 1995.

Greider, William. J. *One World, Ready or Not.* New York: Simon and Schuster, 1997.

Habermas, Jürgen. *The Theory of Communicative Action, Volume 1: Reason and the Rationalization of Society.* Boston, Beacon Press, 1984.

Marcuse, Herbert. *One Dimensional Man.* Boston: Beacon Press, 1966.

McDonough, William, and Michael Braungart. *Cradle to Cradle, Remaking the Way We Make Things*. New York: North Point Press, 2002.

Nilsson, Hans. *Market Transformation by Technology Procurement and Demonstration*. Stockholm: Swedish National Board for Industrial and Technical Development, Department of Energy Efficiency, 1992.

Nilsson, Hans. *Market Transformation, A Demand for Sustainability*. Stockholm: Swedish National Board for Industrial and Technical Development, Department of Energy Efficiency, 1994.

Odum, Howard T., and Elizabeth C. Odum. *A Prosperous Way Down, Principles and Policies*. Boulder: University Press of Colorado, 2001.

Olerup, Britta. *Good Energy Deeds: Renewable Sources, Efficient Use*. Lund, Sweden: Lund University, 1994.

Westling, Hans. *Technology Procurement in Swedish Construction*. Stockholm: Swedish Council for Building Research, 1991.

*Hugh Gilbert Peach*

# R

## RAIN FORESTS.

Rain forests are woodlands that are made lush by rainfall of at least 100 mm in each month of the year. Most of the world's rain forests are in the tropics of Africa, the Americas, Asia, and Australia. Temperate rain forests are found in coastal areas such as the U.S. Pacific Northwest and New Zealand. More than half the world's rain forest is in the Amazon River basin.

Rain forests play an important role in the global ecology despite occupying just 7 percent of the world's land area. They are home to at least 50 percent of all plant and animal species. Rain forests absorb more carbon than any other ecosystem, and they produce life-sustaining oxygen. Finally, the rain forests moderate the global climate system by distributing tropical heat to higher-latitude temperate zones in massive rain clouds. This increases biodiversity in these areas, enhancing agricultural productivity and increasing the range and comfort of human habitation.

Most indigenous inhabitants of tropical rain forests lived in overall balance with them prior to contact with European societies. Even the slash-and-burn agricultural techniques of native peoples had little impact, as their small populations and simple tools (compared to the vast array and enormous power of industrial technologies) kept resource use at sustainable levels.

European colonists brought more powerful technologies, but tropical rain forests remained relatively intact because of their remote locations, harsh conditions, and regenerative capacity. However, recent economic globalization and new production technologies (such as aerial logging and genetic engineering) have in-corporated the rain forests more directly into the world's industrial system. Consequently, some 20 percent of tropical rain forests were deforested in the last three decades of the twentieth century. In contrast to this recent deforestation, most of Europe's temperate rain forests had been eliminated by the end of the nineteenth century.

The primary source of this destruction is the demand that industrialization has generated for timber, oil, minerals, and other natural resources. In addition, the roads constructed for these extractive activities can be used to transport other types of goods to markets; this promotes further deforestation to clear land for new agricultural and ranching operations. These enterprises are often controlled by and for the wealthiest elements in society, who have the capital needed for costly projects and the political clout to push aside traditional inhabitants. The location of most tropical rain forests in formerly colonized nations with poorly performing socioeconomic systems has also made them an attraction to impoverished internal migrants.

Globally, deforestation reduces the absorption of carbon dioxide, thus exacerbating climate change. Locally, deforestation causes soil depletion and sedimentation of rivers, turning vast expanses of previously diverse ecosystems into simplified agricultural and cattle monocultures. Land and rivers rendered vulnerable in the transformation from rain forest to agriculture are frequently exhausted by growing numbers of poor migrants, who must then repeat the cycle in newly deforested areas just to survive.

Efforts to protect the rain forests have expanded significantly in the past two decades. Most actions fall into one or more of the following categories: laws and regulation, political mobilization, economic development, and science/technology. In the postcolonial nations where most tropical rain forests are located, governing institutions routinely lack the capacity to monitor and enforce laws and regulations effectively, and corruption is common. Political mobilization to confront government and economic elites is often met with repression, as in the famous case of Chico Mendes, the leader of a resistance movement in the Brazilian Amazon who was murdered in 1988 by ranchers who were appropriating land traditionally used by rubber gatherers.

Although regulation and political mobilization are more effective in the wealthy nations, violence there is not unknown. In one case, redwood forest activists Judi Bari and Derryl Cherney were seriously injured when a pipe bomb exploded in their vehicle in Oakland, California. Authorities alleged that Bari and Cherney were "ecoterrorists" who were knowingly transporting the bomb, but the charges were dropped for lack of evidence. Bari and Cherney later filed a civil suit against the Oakland Police and the U.S. Federal Bureau of Investigation and won a $4.4 million judgment. Many mobilizations have accomplished at least some of their goals, but most have failed to establish economic systems that meet the material needs of local inhabitants without depleting the forests.

Economic action focuses on acquiring rain forest land from private owners to place it in perpetual reserves, and on various forms of "green" business that seek to shift economic activities to sustainable enterprises such as ecotourism. Much land has been acquired and many businesses formed, but their economic significance remains small in comparison to such traditional activities as agriculture and logging. Some economic analyses indicate that the value of green business per hectare of rain forest significantly exceeds that of more exploitative activities, indicating that there is some promise in this approach.

Finally, much remains to be learned about the rain forests and new technologies for using them in productive activities. Biologists, agronomists, anthropologists, economists, and others at universities and research institutes around the world are studying ways to use the rain forests without destroying them.

No single approach among those previously mentioned is likely to suffice in protecting the rain forests, and many preservationists advocate combinations of them. Given the powerful forces behind the destruction of the rain forests, coalitions spanning wealthy and postcolonial countries will probably be required to advocate and implement effective reform, by connecting issues such as environmental conservation, economic justice, and human rights.

**BIBLIOGRAPHY**

Fimbel, Robert A., John G. Robinson, and Alejandro Grajal, eds. *The Cutting Edge: Conserving Wildlife in Logged Tropical Forest.* New York: Columbia University Press, 2001. This volume contains thirty chapters by international specialists on the impact of timber harvesting in the tropical rain forests and strategies for conservation.

Hill, Julia Butterfly. *The Legacy of Luna: The Story of a Tree, a Woman, and the Struggle to Save the Redwoods.* San Francisco: HarperCollins, 2000. The author advocated conservation in northern California's temperate rain forest by occupying a tall redwood for 738 days. Her experiences and observations about rain forest action are presented in a lively, accessible format.

Jukofsky, Diane. *Encyclopedia of Rainforests.* Westport, CT: Oryx Press, 2000. Written for the Rainforest Alliance, which advocates for sustainable rainforest uses, this book contains information on a wide range of rain forest topics and is highly accessible.

Revkin, Andrew. *The Burning Season: The Murder of Chico Mendes and the Fight for the Amazon Rain Forest.* Boston: Houghton Mifflin, 1990. A detailed account of the life of a humble rain forest inhabitant whose struggle to protect the rain forests and consequent murder focused international attention on the connections among deforestation, economic justice, and human rights.

*Richard Worthington and Elizabeth Merrill*

# RELIGION AND SCIENCE.

Populated with heroes and villains, often represented in terms of some cosmic conflict, the relationship between religion and science is increasingly a subject of popular as well as academic interest. When its master narratives are correctly recognized as products of a Western Christian culture in the last 150 years, however, "Science and Religion" assumes a different character than is reflected in much of the current literature.

## Reifying Terms

To reify a concept is to attribute concrete identity to an abstraction. From the outset, it should be evident that Science and Religion have both been reified, reflected in the common practice of denoting both terms with an initial capital letter. Yet an easy or neat definition is impossible; nor is there in historical perspective a definition that fairly represents the same concept in a variety of periods in the history of Western European culture. The phrase "Science and Religion" implies the existence of some trans-historical subject with terms clear enough in the abstract to be compared across cultures, but this is thus deceptively clear and simple. It is only by reifying the terms, and affirming the necessity of a dualistic approach, that we are able to create the mistaken perception of a comparative, trans-historical discourse.

Science and Religion is therefore not the result of some universal necessity, nor is it descriptive of some cosmic issue: it emerges from a specific historical, cultural, and social context. The operational definition of each term is determined by its relation to the other, like the overlapping area of a Venn diagram. The difficulty is not only defining Science and Religion, but identifying which aspects of Science relate to the aspects of Religion considered crucial in the discussion that takes place under the heading "Science and Religion."

Science is most easily identified as an activity. The social parameters of science considered as an institution enable its decomposition into those who "do science," where they are educated, what they need to know in terms of information and of skills, how the institution is funded, and where it fits within the goals or aspirations of a society. Using these institutional parameters, "science" understood as a social institution is a recent devel-

opment within Western European society in roughly the last 150 years. When terminology is considered—the names of disciplines, for example, the popular perception of science as observation, hypothesis, experiment, and theory (the "scientific method"), or even the term "scientist"—this chronology is confirmed. What preceded science in western European society was natural philosophy, and everything from the scope of its investigations (as likely to include alchemy as chemistry) to the information/skill set required, to the way it was funded and the language used to describe it, differs substantively from what we would recognize as modern science. Even if we restrict ourselves to an epistemic definition of "science," what constitutes "knowledge" changes from the natural philosophy of seventeenth-century Europe to the "Science" of late nineteenth-century America.

Thus, when we use the term science, and particularly when we use the reified term Science, we are talking about something that became a feature of Western society and culture after the Industrial Revolution. When we consider the practice of science in different places around the world, therefore, we need to recognize the Western character of its content and practice, regardless of the nationality of its practitioners. Cross-cultural comparisons of science, or even comparisons within the same culture at significantly different times, are therefore fraught with historical inaccuracies that do little to clarify the relationship between Science and Religion.

The definition of "Religion" is equally problematic. Understood as practice, its institutional expression is different in kind from its personal expression, often depicted in terms of spirituality or religious feeling. As a social institution, religion can be approached from a cultural perspective, as an integral part of whatever culture in which a particular religion is found. It serves political and ethical purposes, in binding together members of a culture; it serves as a means of social control or a vehicle of social cohesion. Whether expressed in terms of ritual, or philosophy, or morality, or ethical systems—and all of the above with differing interpretations depending on whether the interpreter is a believer—Religion (or, rather, the various religions) exhibits a multifaceted character. From a secular perspective, there would be competing anthropological definitions, sociological definitions, and psychological definitions. To say Religion deals with the metaphysical, defining it by exclusion from whatever is considered the "physical," is equally inadequate when actions and behavior are normally part of its expression. The indeterminacy of the term itself is overcome only if the definition is sought in the same context as that of "Science." In other words, it is not Religion *sui generis* that is at issue, but religion in nineteenth-century Europe and America—that is, Christianity. Attempts to extend the discussion of Science and Religion to other religious and cultural traditions tend, therefore, to be implausible, relating more to the encounter between Western science and indigenous religions.

Before leaving the problem of definition and reification, however, it should be noted that even reading "Religion" as Christianity is problematic. Only the elements of Western Christianity that intersect with certain elements of "Science" are part of the Science and Religion discourse. In brief, we need to consider the respective (and changing) perspectives on knowledge, meaning, authority, and experience that are found in the different discussions. Thus, to clarify the definitions of both Science and Religion, we need to turn to the master narratives of the last 150 years to understand how those narratives have caricatured both the institutions of Science and of Religion to suit the context in which they emerged.

## From Origins to the Origin

Debates in natural philosophy and in natural theology about the origins of life had been carried on for longer than the two hundred and fifty or so years between the work of Galileo Galilei and the appearance of Charles Darwin's *The Origin of Species* in 1859. Yet in these debates over what one could learn from the twin books of Scripture and of Nature, the focus was on how one should read the books each ultimately written by God.

The notion that one book was superior to the other, or that one book was necessarily in conflict with the other, remained an absurdity. There were arguments over the authority of the Church, both Catholic and Protestant, about knowledge and experience and what these might mean, just as there were heated arguments among the practitioners of natural philosophy as to the nature of experience and the meaning of observational knowledge. These arguments were selectively edited and interpreted in the latter part of the nineteenth century in Britain and America to illustrate an emerging polemic that focused primarily on the social authority of the Christian church.

The dominant master narrative of Science and Religion, setting out the scenario of the conflict between the two protagonists, arose from J. W. Draper's *History of the Conflict between Religion and Science* (1874) and A. D. White's *A History of the Warfare of Science with Theology in Christendom* (1896). Although both books were printed in America some time after the initial appearance of the *Origin of Species* in 1859, antagonism to Charles Darwin's ideas had become a contemporary example of the problem of the Church's authority. In Britain, it was T. H. Huxley who bannered the charge against the opponents of Darwinian theory who he felt were holding back the irresistible tide of the discoveries of modern science.

Draper, a professor at New York University, contended that the conflict resulted from the unbridgeable gulf between a divine revelation and the irresistible advance of human knowledge, making the whole history of Science into a narrative of two conflicting powers. He defined "religion" very narrowly, as Roman Catholicism, for his diatribe was sparked by actions of Pope Pius IX and the Vatican Council in establishing papal infallibility, something that he felt overstepped the bounds of Protestant common sense and reason. Andrew Dickson White began his own polemic in 1869, when as the first president of Cornell University he delivered a lecture at Cooper Union Hall after coming under attack for refusing to impose a religious test on students and faculty. He recounted the famous "battles" between religion and science in the "persecution" of Nicolaus Copernicus, Giordano Bruno, Galileo Galilei, Johannes Kepler, and Andreas Vesalius, and he cited his own position as the latest victim of religion's war on science. He defined "Religion" more narrowly as "ecclesiasticism" in *The Warfare of Science* (1876), and finally as "dogmatic theology" in *A History of the Warfare of Science and Theology in Christendom* (1896). White said in his introduction to the 1896 volumes that he continued to write on the subject after Draper's book appeared because he became convinced the conflict was between "two epochs in the evolution of human thought—the theological and the scientific." T. H. Huxley, as a public proponent of agnosticism, also challenged the social authority of the Church in the arena of intellectual activity, using his books and addresses to polarize the debate between the relation of the Church and State in Britain. Darwin's theories, and the response of some members of the Church, were enough to encourage Huxley to seize on this issue as another example of religious hegemony to be thwarted.

It is only in the last thirty years that a more critical eye has been cast both on the circumstances of their respective crusades and on the master narrative about the conflict between Science and Religion that has been used to buttress their claims. Scholars such as Colin Russell, who explored Huxley's antagonism to the Church of England as the reason for his championing of Darwin, and David Lindberg and Ronald Numbers, whose 1986 article on Draper and White exposed the reasons for the shape of their respective polemics, have contextualized the conflict scenario in a way that explains why Science and Religion came to be depicted in this specific fashion in the late nineteenth century in Britain and America.

The master narrative was initially conceived as a duality in perpetual conflict, a conflict that, for the good of humanity, Science had to win over the superstitions of the Church. The areas of conflict were identified as the nature and character of knowledge; the meaning of knowledge and thus of life itself; the source, justification, and boundaries of authority; and both the validity and epistemic content of experience. A resolute dualism is built into this initial master narrative, and heirs to the debate have arranged themselves on sides arguing for a limited set of possible relations. Where one side might maintain the inevitability of conflict, and the need for Religion to acquiesce in the supremacy of Science, the other would argue that there were two dimensions to human experience and that Science and Religion both needed to respect the boundaries of their authority. The conflict scenario would always find science ultimately victorious, as various superstitions and dogmas maintained by the Church were replaced by theories based on scientific evidence. Religion was represented, at least in its institutional form, as an impediment to the acquisition of knowledge about nature, and a barrier to scientific progress itself. In the language of materialism, religion dealt with the subjectivity of emotional experience, not the objectivity of the physical world. Whereas one could ascertain truth about the physical world by proper scientific method, no such truth could be forthcoming from a subjective emotional experience.

In the period after Darwin and before World War 1, European and American philosophy was dominated by the debate between materialism and idealism: how one could know anything about the external world, and how that knowledge might be obtained or verified. With the foundation of analytic philosophy in the work of such people as Bertrand Russell, a qualified materialism resulted, and attempts to derive knowledge from something other than experience was rudely dismissed as balderdash by the "logical positivists," as they came to be known.

Yet the philosophical implications of the new physics, the physics of Einstein's relativity theories and Niels Bohr's quantum mechanics, meant that certain statements about physical entities—the result of our knowledge of the external world—were impossible, and many Anglo-American thinkers explored whether there might be some other character of Religion, or some other relation between Religion and Science, that was more constructive than inevitable conflict.

For example, *Science, Religion and Reality* (1925), edited by Joseph Needham (later the famous scholar of Chinese science and culture) was intended to be an explication of the subject, not an apologetic for Christianity or an attack upon it. The authors of the various articles, all prominent in fields ranging from theology to anthropology and physics, realized the issue was far more complex than the conflict scenario permitted. Among a host of other publications, another anthology reflecting an even more congenial attitude toward Religion was published as *Science and Religion* out of a series of BBC lectures broadcast from September to December 1930, focusing on how the reality presented by science affected Christian theology. Bertrand Russell published his own rejoinder to this second anthology as *Religion and Science* in 1935, denying that Religion in any institutional form could lead to truth or knowledge of any sort, as he lampooned the philosophical musings of famous British physicists and astronomers Arthur Eddington and James Jeans.

After World War II, the development of the history of science as a discipline in its own right led to the critical exploration and assessment of the examples conventionally listed as evidence of the conflict between Science and Religion. Thus, work on the sixteenth and seventeenth century illuminated the circumstances of Galileo, and why his work might have got him into some trouble—and revealed the troubling fact that his most important work on heliocentrism was completed while under house arrest by the supposedly hostile Church. Similarly, work on the reception of the *Origin of Species* revealed many clergy supported Darwin, while secular figures did not; in addition, it was realized that it was the publication of *Essays and Reviews*—a collection of articles about the shape of a new critical and historical method for interpreting scriptures—that most concerned the religious establishment in 1859. The parallel work of Alfred Russell Wallace, and a realization of the intellectual context in which Darwin worked, also ate away at the

novelty—and shock value—of his discoveries about evolution, something that figured prominently in the initial master narrative.

One of the further philosophical implications of relativity theory was the reopening of the debate in the philosophy of science between realism and idealism. Although a strict materialism might not be possible, the positivist enterprise depended on the existence of some aspect of the physical world that was discernible and rational, and even if it had to be couched in the language of probability, it was preferable to a world in which there was no objective reality. Yet as the operations of science became more part of the domain of scholars asking questions, the nature of the scientific method, the way in which theories were developed, and the role of experiment in the discovery of knowledge yielded revelations unsettling for those who preferred the simple duality of the conflict scenario.

Ian Barbour's work in the period following Thomas Kuhn's *The Structure of Scientific Revolutions* (1962) set out some important characteristics of the changing master narrative. His *Issues in Science and Religion* (1965), as well as his *Myths, Models and Paradigms* (1974), critiqued the conventional depiction of Science and Religion. As the language and methods of modern science came under scrutiny, they were found to be increasingly less "scientific" and thus less superior to religious understandings of the nature and meaning of knowledge, or the validity and character of personal experience. Barbour's summary works—*Religion in an Age of Science* and *Ethics in an Age of Technology,* both from his Gifford Lectures in 1989–1991 (published before he received the 1999 John Templeton Prize in Religion)—established a more sophisticated dialectic than the original conflict scenario in which to consider the relations of Science and Religion. His fourfold model of relating Science and Religion included conflict, independence, dialogue, and integration.

Barbour's work, in particular, gave an alternative to the simplistic conflict scenario of Draper and White, but its inherent character—a dualistic representation, with indeterminate definitions of the terms—rendered it vulnerable not only to the history of science critique, but also to the charge that it was really about Anglo-American Christianity and Western science. The work of British historians of science John Hedley Brooke and Geoffrey Cantor, both separately and then jointly in their Gifford Lectures of 1995–1996, illuminated the personal and historical details that made up the mosaic of interactions between the institutions of Science and Religion, rendering Barbour's neat categories obsolete. Moreover, as sociologists of science further deconstructed the activities of scientists, and realized the elusive social character of the "knowledge" produced by the institutions of Western science, confident statements about the boundaries between these activities have come to reflect the personal convictions of different authors rather than any serious consensus or objective conclusion.

The work of the John Templeton Foundation in the 1990s to promote the teaching of Science and Religion bore fruit proportional to the enormous amount of money spent. Hundred of scholars around the world, primarily in the United States, received support to teach on the subject, and this (along with other incentives) encouraged the publication of a number of books on the relationship between Science and Religion unprecedented since the interwar period. Yet these initiatives are caught within the same net as earlier attempts to address the relationship between Science and Religion: they are culture-bound by the Western perceptions of knowledge, meaning, experience, and authority contained in the practice of Western Science, and by the parallel perceptions of the same topics in the (primarily Protestant) Western institutions of Christianity, writ large as Religion.

*See also* **Religion and Technology**

**BIBLIOGRAPHY**

Barbour, Ian. *Religion and Science: Historical and Contemporary Issues.* San Francisco: HarperCollins, 1997.

Bowler, Peter J. *Reconciling Science and Religion: The Debate in Early Twentieth-Century Britain.* Chicago and London: The University of Chicago Press, 2001.

Brooke, John Hedley. *Science and Religion: Some Historical Perspectives.* Cambridge, UK: Cambridge University Press, 1991.

Brooke, John Hedley, and Geoffrey Cantor. *Reconstructing Nature: The Engagement of Science and Religion.* Edinburgh: T&T Clark, 1998.

Denton, Peter H. *The ABC of Armageddon: Bertrand Russell on Science, Religion, and the Next War, 1919–1938.* Albany, NY: State University of New York Press, 2001.

Draper, John William. *History of the Conflict between Religion and Science.* 1874. Reprint, New York: D. Appleton and Company, 1897.

Lindberg, David, and Ronald Numbers. "Beyond War and Peace: A Reappraisal of the Encounter between Christianity and Science." *Church History* 55.3 (September 1986): 338–354.

Lindberg, David, and Ronald Numbers, eds. *God and Nature: Historical Essays on the Encounter between Christianity and Science.* Berkeley: University of California Press, 1986.

Moore, James R. *The Post-Darwinian Controversies: A Study of the Protestant Struggle to Come to Terms with Darwin in Great Britain and America, 1870–1900.* Cambridge, UK: Cambridge University Press, 1979.

Needham, Joseph., ed. *Science, Religion and Reality.* New York: Macmillan, 1925.

Numbers, Ronald. *Darwin Comes to America.* Cambridge, MA: Harvard University Press, 1998.

Russell, Bertrand. *Religion and Science.* Oxford: Oxford University Press, 1935.

Russell, Colin A. *Cross-Currents: Interactions between Science and Faith.* Grand Rapids, Michigan: Eerdmans, 1985.

*Science and Religion: A Symposium.* New York: Charles Scribner's Sons, 1931.

White, Andrew Dickson. *A History of the Warfare between Science and Theology in Christendom.* 2 vols. 1896. Reprint, New York and London: D. Appleton, 1926.

*Peter H. Denton*

# RELIGION AND TECHNOLOGY.

When considering the relationship between religion and technology, it is not always easy to separate technology from its close associations with science or with capitalist economic formations, both of which technology often uses or supports. Theories of modernization and secularization are likewise related. As a result, the present review will consider scholarly discussion of the religion–technology relationship even where the term *technology* itself may not play an explicit role.

The relationship between religion and technology first became an implicit theme for scholarly examination in the work of social scientist Max Weber (1864–1920), who focused on the contribution of Protestant Christianity to the development of capitalist industrialization. Before that, founders of the modern project, such as Francis Bacon (1561–1626) and René Descartes (1596–1650) had criticized Christian theology to justify their pursuit of science and technology, but did not think the religion–technology relationship itself worthy of extended argument. During the middle twentieth century, the historian of technology Lynn White Jr. (1907–1987) expanded discussion of the Christianity–technology relationship by pushing it back into the Middle Ages and relating it to other issues, such as the environmental crisis. This double expansion had the effect of globalizing the question, so that the issue of technology became a theme for consideration in relation to other religions as well. The present review of religion and technology scholarship is therefore divided into four sections: historico-philosophical background, the Weber thesis and its critics, the Lynn White thesis and its critics, and the globalization of the religion–technology discussion.

## Historico-Philosophical Background

Before the Renaissance, most political philosophies in both Europe and Asia were, either explicitly or implicitly, wary of what is now called *technology*. The arguments were fundamentally of two sorts: first, technological pursuits distract from higher things; second, technological change is socially destabilizing.

These ideas can be found expressed in Greek myths and in Judeo-Christian scriptures, as well as in Taoist and Buddhist teachings. The Greek stories of Hephaestus as the deformed worker of metals and of Icarus as the inventor of a flying machine that leads to his own destruction call attention to the spiritual dangers of technology. The accounts in Genesis of the conflict between the shepherd, Abel, and the builder of cities, Cain, and of humans' attempts to unite themselves through the technology of a tower construction that would detract from recognition of dependence on the divine likewise question any primacy of commitment to technology in human affairs.

A story from China documents similar attitudes in Asian traditions. One day on his travels, the Taoist sage Chuang Tzu (fourth century B.C.E.) saw an old man irrigating his garden by lifting water out of a ditch one bucket at a time. He explained how the laborer could construct a machine (a well sweep or shadoof) for making work easier. But the old man declined to adopt the machine, contending that "where there are ingenious machines, there are sure to be crafty actions [and] a scheming mind [that undermines] simplicity. . . . The unsettled spirit is not the proper residence of the Tao." An Indian story concerning the invention of "spirit-bearing engines" to guard relics of the Buddha has similar implications. The robots guarded the relics so well that devotees were unable to approach them. Only when the relics were freed from such technological control by King Asoka (third century B.C.E.) was the city of Pataliputra restored to cultural health.

With conscious opposition to such traditions, modernity arose during the long sixteenth century in association with arguments for the priority of technology in human affairs. Bacon, among others, called not just for the cultivation of nature but also its systematic control, in order to achieve magic-like power, adapting for his purposes an appeal to the Christian desacralization of nature. As paradigmatic inventions to be imitated he cited the printing press, gunpowder, and the compass as more beneficial than theological debates. Descartes likewise proposed a reconceptualization of the world that would make human beings "masters and possessors of nature." It was this inventive reconstruction of the world in thought that formed the basis two centuries later for the practical remaking of the world in the Industrial Revolution.

## The Weber Thesis and Its Critics

Such was the self-understanding of the Enlightenment, a self-understanding codified, for instance, in the idea of historical progress from theological, through metaphysical, to positive scientific culture as formulated by August Comte (1798–1857) and continued in numerous theories of secularization associated with sociological studies of religion and even the liberal theological demythologization of Christianity. The original challenge to the Enlightenment opposition between religion and scientific technology finds expression in the iconoclastic thought of Friedrich Nietzsche (1844–1900) whose *On the Genealogy of Morals* (1887) castigates the "slave morality" of Christianity as a primary source for the desire to achieve material comfort above all else. But it was Weber's *The Protestant Ethic and the Spirit of Capitalism* (1904–1905) that popularized the Nietzschean reinterpretation, although in a manner that gave technical development an economic camouflage.

According to Weber, cultures are characterized by the presence of numerous techniques. "There are," Weber writes, "techniques of every conceivable type of action, techniques of prayer, of asceticism, of thought and research, of memorizing, of education, of exercising political or hierocratic domination, of administration, of making love, of making war, of musical performance, of sculpture and painting, of arriving at legal decisions" (p. 65). In premodern cultures, these techniques were never analyzed or evaluated solely in their own terms, that is, in terms of effectiveness or what Weber terms their "rational" relation to the achievement of some clearly specified material end. The technique of butchering, for instance, was customarily oriented not only toward the preparation of meat, but also toward placating the gods or acting in harmony with various ritual prescriptions. Only coordinate with the rise of Protestantism did religion, economics, politics, the arts, and other aspects of culture get spun off into semiautonomous spheres, thus making possible an assessment of technique in strictly logical terms—and thereby giving birth to what we now appropriately call "technology."

The Weber thesis has been assessed by both historians and social scientists. Historians—suspicious of what they perceive as monocausal explanations and uneasy with giving such strong weight to "ideal types," ideas, and a measure of personal agency—have generally been more skeptical. Sociologists have been more sympathetic to precisely these same features of Weber's argument. The American social scientist Talcott Parsons (1902–1979), who first translated *The Protestant Ethic*, proceeded to develop a Weber-inspired general theory of social action. Subsequently, Weber's approach has contributed to the turn toward theory in culture studies, where the religion–technology issue has nevertheless taken second place to questions concerning the genealogy of desire and pleasure, as in Colin Campbell's *The Romantic Ethic and the Spirit of Modern Consumerism* (1987).

The most influential challenges to Weber are associated with two economic historians: Richard H. Tawney (1880–1962), an Indian-born Fabian socialist, and Amintore Fanfani (1908–1999), who served as a post–World War II premier of Italy. Tawney's *Religion and the Rise of Capitalism* (1926) argued the mutual influence of industrial capitalism on religion along with that of religion on industrial capitalism. Fanfani's *Catholicism, Protestantism, and Capitalism* (1935) made a case for Catholic influence on merchant capitalism in northern Italy. Both critics also questioned Weber's interpretation of Protestantism, which privileged popular and unorthodox interpretations of Calvinism over more orthodox theology. As the Christian theologian and sociologist Jacques Ellul (1912–1994) summarized this criticism in his study of *La Technique* (1954):

> *Christianity and Technique.* The East: passive, fatalist, contemptuous of life and action; the West: active, conquering, turning nature to profit. These contrasts,

so dear to popular sociology, are said to result from a different religion: Buddhism and Islam on the one hand; on the other, Christianity, which is credited with having forged the practical soul of the West. (p. 32)

In reality, according to Ellul, "The technical movement of the West developed in a world which had already withdrawn from the dominant influence of Christianity" (p. 35).

In order to defend Weber, his argument is thereby qualified so that it becomes not Protestantism but a debased form of Protestantism that provides the fundamental support for capitalist industrial development. An example would be the kind of Christianity exemplified in Bruce Barton's *The Man Nobody Knew* (1925), which argues that Jesus of Nazareth was the model for an industrial advertising executive. From this perspective, what Weber identified as the decisive contribution of "this-worldly asceticism" is not an authentic religious asceticism. Indeed, two hundred years earlier, John Wesley (1703–1791) had worried about how the Christian conversion experience, when incomplete, often led people to give up addictions such as alcohol only to replace them with others such as the pursuit of worldly riches. In the language of later sociopsychological analyses, one ersatz religion or religious substitute was replaced by another.

## The Lynn White Thesis and Its Critics

A new wave of discussion of the religion—technology relationship was initiated in the 1960s by the historian of medieval technology Lynn White. White pushed Weber's argument for a positive relationship between technical prowess and certain kinds of religious asceticism and theology back temporally into the Middle Ages and expanded it to include the influence not just of Protestantism but of Latin Christianity as a whole. He explicitly excluded only Eastern Orthodox Christianity because of its interpretation of Christian teaching as more a gnosis or spiritual enlightenment than a disciplining of the will, and its resistance to the contamination of sacred space with mechanical devices such as the organ or the clock. His

essay "The Historical Roots of Our Ecologic Crisis," originally delivered as a lecture at the annual meeting of the American Association for the Advancement of Science, blamed the Western Christian theological emphasis on active conformity to God's will and a sympathy for power technologies for environmental pollution. As such, it became a reference in such organs of popular culture as the *New York Times*, the *Boy Scout Handbook*, and the Sunday sermon; it was certainly White's most widely cited and criticized work.

But White's *Medieval Technology and Social Change* (1962) had earlier and with more nuance argued for significant technology-based changes in the military, in agriculture, and in power production. The utilization of such a simple device as the stirrup made mounted shock combat possible and thus contributed to Charles Martel's victory over the Moors at Tours/Poitiers in 732. The moldboard plow was part of a revolution in agricultural production. The medieval development of water and wind power were parts of a historically unprecedented flowering of mechanization. Once documented, such historical changes called for reflective interpretation on their causes. Given the existing association of the Middle Ages with Christianity, it was natural to place some of the weight there. As White already notes in passing, "It was in the works of the great ecclesiastic and mathematician Nicholas Oresmus, who died in 1382 as Bishop of Lisieux, that we first find the metaphor of the universe as a vast mechanical clock"—a metaphor destined to become a metaphysics (White, 1962, p. 125).

Finally, in a series of penetrating studies collected eventually in *Medieval Religion and Technology* (1978), White made his most persuasive case. In his essay "The Iconography of *Temperantia* and the Virtuousness of Technology," for instance, White argued that various manuscript illuminations and homiletic images from the ninth century onward revealed how the improvement and utilization of technological devices had become a major element in the Christian moral life. Temperance or discipline (which in earlier periods is alleged to have been one of the lesser, cardinal virtues) was elevated to a dominant position and symbolized by technology triumphant.

> Her essential character had become Measure, and, since we mortals dwell in time rather than eternity, her prime symbol came to be the clock. But then, toward the middle of the fifteenth century, Northern Europe clothed this supreme Virtue not simply with the clock but with the new technology of the later Middle Ages: on her head she wore the most complex mechanism of the era [the clock]; her feet now rested on the most spectacular power engine of that generation [the windmill]; in her hands she held eyeglasses, the greatest recent boon to the mature literature man. The new icon of Temperance tells us that in Europe, below the level of verbal expression, machinery, mechanical power, and salutary devices were taking on an aura of "virtuousness" such as they have never enjoyed in any culture save the Western. (White, 1978, pp. 201–202)

Before and parallel with White, other scholars in Europe were further exploring the medieval Christianity–technology relationship. For instance, Orthodox theologian Ernst Benz had previously developed an interpretation similar to White's. Other parallel work included M.-D. Chenu's *Nature, Man, and Society in the Twelfth Century* and Jacques LeGoff's *Time, Work, and Culture in the Middle Ages*. While concentrating on changes in theology, Chenu noted, like White, new favorable iconographic representations of artisans; a heightened naturalism in art; and the development of new sources of power, transportation, and military arms. LeGoff documented the rise of mechanized work in monastic communities, but argued that theology played more catch-up rationalization than initiating ideology.

The most immediate impact of White's argument, especially insofar as it was used to develop a pro-pagan environmentalism critical of Christianity (contrary to White's explicit intention, which was to reform the Christian tradition through an appeal to Franciscan respect for nature), was the stimulation of a two-pronged defense. One side challenged White's characterization of Christian theology, the other his assumptions about the environmental crisis. Against White's claim that Christianity promoted technological exploitation of nature and was thus to blame for the environmental crisis, a spectrum of theologians from the Catholic Paul Santmire to evangelical Calvin DeWitt replied, in arguments reminiscent of earlier responses to Weber, that White had failed to distinguish true Christianity from its cultural perversions. The authentic Christian teaching, such apologists maintain, is environmental stewardship. Against White's assumptions that technology was causing an environmental crisis, other defenders argued that technology had in fact done more to benefit humanity in ways consonant with the requirements of Christian charity than all environmentalists' criticisms—and that technology was actually overcoming the problems attributed to it. No book has better documented this pro-technology dimension of western Christian culture than David Noble's *The Religion of Technology*.

In North America, students influenced by White continued his explorations. Valuable additions to the religion–technology literature relation can be found, for instance, in George Ovitt Jr.'s *The Restoration of Perfection* and Elspeth Whitney's *Paradise Restored*. Ovitt provides a well-balanced examination of White and a large number of scholars working on similar themes in the history of ideas. Whitney offers a careful explication of relevant texts from late classical antiquity into the high Middle Ages. Both give the history of ideas about religion and technology a new depth.

## Beyond Christianity

Three factors contributed to a post-1970s expansion of religion and technology discussions. First was the sociological discovery of the persistence of religion and even its rebirth in high-tech cultures and in societies undergoing modernization. Peter Berger's *A Rumor of Angels* was an opening wedge here. The result was a questioning of secularization in ways quite different from Weber's.

Second was the transformation of the environmental crisis from a local to a global phenomenon. When the problem was not just the pollution of Lake Erie but the destruction of stratospheric ozone, then White's appeal for changes in Christianity's attitudes as a basis for dealing with the problem looked parochial. Third, the anti-Christian environmentalist appeal to other religions, such as Hinduism and Buddhism, not to mention paganism, invited investigation of the historical record with regard to these traditions.

The resultant globalization of the religion–technology question has yielded a growing body of literature in especially two areas: on modernist development and secularization, and on religion–technology relationships across time and cultures. In the face of this productive effulgence, some system for classifying and analyzing the results is needed. An early effort, operating still within the Christian–technology emphasis, was a typology developed by Carl Mitcham and Jim Grote based on the five basic theologies of culture identified by H. Richard Niebuhr's influential *Christ and Culture* (1951). For Mitcham and Grote, the spectrum of Christian responses to technology includes Christ as opposed to culture (for example, Tertullian, with contemporary exemplifications in Amish rejections of technology), as of culture (liberal theologians and mainline Protestant affirmations of technology), as transcending culture (Thomas Aquinas and Catholic judgments on technology), and culture in paradox (Luther and those who, such as Ellul, who argue a dialectical acceptance and rejection of technology), and as transformer of culture (Augustine and evangelicals inspired by faith to what they see as new forms of technology).

A more comprehensive framework has been developed by William Jones and Warren Matthews. Jones and Matthews begin by arguing that what is most important is not technology alone but technological innovation. Thus "the most fruitful and illuminating classificatory scheme(s) will reflect the central role of innovation [and] are likely to be cases of a religious tradition reacting to a technological innovation, . . . contributing to

a social climate fostering (or inhibiting) technological innovation, or some technological innovation having an impact upon a religious institution or value" (p. 5). The results of their taxonomical analysis, somewhat simplified, can be summarized in the following outline:

I. No technological innovation: Religion–technology relationships
 A. No interaction
 B. Religious support for an established technology
 C. Support from an established technology for religion
II. Technological innovation: Religion–technology relationships
 A. No interaction
 B. Religion influencing technological innovation
  1. Established religion facing new external technologies
  2. Established religion facing new internal technologies
  3. Indirect or unintended influences
 C. Technological innovation influencing religion
  1. Direct positive effects
  2. Direct negative effects
  3. Indirect effects

More than previous analysts, Jones and Matthews appreciate a wide spectrum of possible religion–technology relationships. The idea of differentiating religion–technology relationships in the contexts of established versus new technologies is especially significant insofar as it calls attention to the ways in which, for instance, inherited architectural techniques have been used to reinforced Catholic teachings whereas Baptists have opposed well-established technologies for alcohol production and the Amish have opposed electricity long since its introduction.

In the context of technological innovation, Jones and Matthews make another distinction between external and internal novelty. Cotton Mather's transferring vaccination from China and Turkey to eighteenth-century Boston is not the same as the legendary Ko Hung's promoting of herbal medical innovation within third-century China. Likewise,

Gandhi's criticism of industrial technologies imported into India from Europe is not the same as Catholic criticism of the European–North American development of nuclear weapons or human cloning. Finally, the ways in which technological innovation may impact religion ranges across forcing religious communities to examine their values, undermining social function, and developing alternative sources of authority. Developments in biomedical technology, for instance, challenge traditional religious values, reposition or marginalize prayers for the sick, and create new authorities to consult about issues of life and death.

## A Matrix of Possibilities

Following a century of scholarly and popular discussions, religion–technology interactions have become considerably more gradated than was originally the case. Combining distinctions introduced in relation to Weber and Lynn White, one may support the matrix at the bottom of the page to suggest four possibility spaces for further refinement.

Not only is it recognized that technology can both influence and be influenced by religion, but it would be useful for this two-way influence to be distinguished, however difficult it may be, in terms of authentic and inauthentic forms of religion. This is a position alien to anyone arguing for conceiving religion as technology or technology as religion. Insofar as religion is accepted as a player in the religion–technology dialogue, it is appropriate to allow debates within religious communities themselves to have a bearing on interpretation of the relationship. If one were to claim that religion influenced technology in the form of techniques as diverse as the spectrum originally postulated by Weber, a spectrum that would include techniques such as saying Mass, partisans of technology might well contest the reality of this as a technology. In the same way,

even though the distinction between authentic and inauthentic religion is more deeply contentious than that between real and illusory techniques (although this latter is not without disagreement), different interpretations of religion must themselves be allowed to enter into religion–technology discussions.

*See also* **Religion and Science**

### BIBLIOGRAPHY

Benz, Ernst. *Geist und Leben der Ostkirche.* Hamburg: Rowohlt, 1957.

Berger, Peter. *A Rumor of Angels: Modern Society and the Rediscovery of the Sacred.* Garden City, NY: Doubleday, 1969.

DeWitt, Calvin B. *Caring for Creation: Responsible Stewardship of God's Handiwork.* Grand Rapids, MI: Baker Books, 1998.

Ellul, Jacques. *La Technique ou l'enjeu du siecle.* Paris: A. Colin, 1954. English version: *The Technological Society*, translated by John Wilkinson. New York: Knopf, 1964.

Ferré, Frederick, ed. *Technology and Religion. Research in Philosophy and Technology*, vol. 10. Greenwich, CT: JAI Press, 1990. Includes fourteen papers followed by discussion sections and reviews.

Jones, William B., and A. Warren Matthews. "Toward a Taxonomy of Technology and Religion." *Research in Philosophy and Technology* 10 (1990): 3–23.

LeGoff, Jacques. *Time, Work, and Culture in the Middle Ages.* Translated by Arthur Goldhammer. Chicago: University of Chicago Press, 1980. A collection of eighteen closely related articles.

Mitcham, Carl, and Robert Mackey, eds. *Philosophy and Technology: Readings in the Philosophical Problems of Technology.* New York: Free Press, 1972. Paperback reprint, 1983. Includes a section of five "Religious Critiques" by Nicholas Berdyaev, Eric Gill, R. A. Buchanan, W. Norris Clarke, and Lynn White Jr.

Mitcham, Carl, and Robert Mackey. *Bibliography of the Philosophy of Technology.* Chicago: University of Chicago Press, 1973. An annotated

|  | Technology influencing | Technology influenced |
|---|---|---|
| Religion authentic | Technology influences authentic religion | Authentic religion influences technology |
| Religion inauthentic | Technology influences inauthentic religion | Inauthentic religion influences technology |

review that includes a section on religious critiques.

Mitcham, Carl, and Jim Grote, eds. *Theology and Technology: Essays in Christian Analysis and Exegesis.* Lanham, MD: University Press of America, 1984.

Mitcham, Carl, and W. Mark Richardson, eds. "Science, Technology, and the Spiritual Quest." *Technology in Society* 21.4 (November 1999): 343–486.

Muller, Max, ed. *The Sacred Books of the East,* Vols. 39 and 40: *The Writings of Chaung Tzu.* Translated by James Legge. Oxford, U.K.: Oxford University Press, 1891.

Newman, Jay. *Religion and Technology: A Study in the Philosophy of Culture.* Westport, CT: Praeger, 1997.

Noble, David F. *The Religion of Technology: The Divinity of Man the Spirit of Invention.* New York: Knopf, 1997.

Ovitt, George Jr. *The Restoration of Perfection: Labor and Technology in Medieval Culture.* New Brunswick, NJ: Rutgers University Press, 1987.

Panikkar, Raimundo. "The Destiny of Technological Civilization." *Alternatives* 10.2 (Fall 1984): 237–253.

Rossi, Paolo. *Francis Bacon: From Magic to Science.* Trans. Sacha Rabinovitch. London: Routledge and Kegan Paul, 1968. Pp. xviii, 280.

Santmire, H. Paul. *Brother Earth: Nature, God, and Ecology in Time of Crisis.* New York: Thomas Nelson, 1970.

Weber, Max. *Protestantische Ethik und der Geist des Kapitalismus.* 1905. English version: *The Protestant Ethic and the Spirit of Capitalism,* translated by Talcott Parsons (New York, 1930).

White, Lynn Jr. *Medieval Technology and Social Change.* New York: Oxford University Press, 1962. A history of the development of specific tools (such as the stirrup, heavy plow, horseshoe, horse collar, and mechanical crank) and the influences they had on social structures.

White, Lynn Jr. *Machina Ex Deo: Essays in the Dynamism of Western Culture.* Cambridge, MA: MIT Press, 1968. Reprinted under the title of one of its chapters as *Dynamo and Virgin Reconsidered: Essays in the Dynamism of Western Culture* (1971). A popular argument for Christianity as the key stimulus to medieval technological change.

White, Lynn Jr. *Medieval Technology and Religion: Collected Essays.* Berkeley, CA: University of California Press, 1978. Nineteen previously published scholarly studies with a synthetic introduction.

Whitney, Elspeth. *Paradise Restored: The Mechanical Arts from Antiquity through the Thirteenth Century.* Philadelphia: American Philosophical Society, 1990.

*Carl Mitcham*

# REPLICATION.

The possibility of replication of results seems to separate empirical science from the mundane world. If an empirical result is in doubt, others can test it independently. If the result is replicable by anyone, it is valid; otherwise it is not valid.

Replication is the experimental equivalent of induction; what is regularly reproducible is generalizable, and the test does not depend on where it is carried out or who conducts it. Of course, there is a philosophical "problem of induction," so there must be a "problem of replication" as well: Why should we consider that a sequence of replications will continue into the future? This problem, however, does not actually prevent us from generalizing.

A more immediate and sociologically interesting problem of replication comes into clear focus in cases where results are disputed. The problem is to know when one experiment counts as sufficiently like another, or possibly sufficiently *unlike* another, to count as a test. If experiments are too similar, they may embody similar mistakes, so apparent replication is merely the multiplication of error. On the other hand, if an experiment is too different from the one it is meant to replicate, it cannot count as an adequate test.

## Analytic Theory of Replication

Thinking this way we can develop an analytic theory of replication. First, the confirmatory power of an experimental replication grows from near zero as the second experiment becomes increasingly different from the first. Thus, the same experiment repeated immediately with the same apparatus by the same experimenter adds little in the way of reassurance, whereas if a similar but nonidentical apparatus is built by a different experimenter in a different place, then reproduction of the effect is more convincing. Should the second experimenter initially disbelieve the initial result (perhaps because he or she belongs to a rival group) but be forced to accept it nevertheless, then it is more convincing still. Thinking this way shows that the experimenter (including his or her internal states and social connections) is a part of the experiment. Nevertheless, this increase in confirmatory power ceases at the point when the second experimenter and his or her apparatus cease to have scientific credibility. Thus, should my measurement of the mass of the Higgs boson be confirmed by a newspaper astrologer using a crystal ball, that "experiment" will add no credibility at all. Indeed, in certain circumstances, replication by nonexpert groups can decrease the credibility of an initial claim. This has happened where the replicators were high-school students and the like.

To complete the analytic theory of replication, it should be noted that disconfirmation works the other way around: the more alike the experimental setup is to the original, the more powerful a disconfirmation is, whereas disconfirmations by very different apparatuses and types of person carry little or no weight.

## Replication in Practice

One can immediately see that the analytic theory of replication as set out in the preceding section invites a sociological analysis, because it turns on what counts as appropriate degrees of similarity and connectedness between apparatuses and persons. One scientist's expert is another scientist's fool; one person's "golden-handed experimenter" is another person's fraud. Cases of replication in scientific practice are next to be discussed.

Oddly, nearly all cases of replication in practice are not treated as tests of anything. Thus, in countless laboratories throughout the world, high-school students are regularly trying to repeat famous experiments and getting results that disagree with the original. These are dealt with unproblematically— they are simply treated as "wrong results."

Indeed, the test of whether a student is adequately skillful in respect of a certain experimental replication is whether they can reproduce the expected result. In these cases, then, instead of replication being treated as a test of a scientific result or principle, it is treated as a test of the experimenter. This reversal of conclusion applies to most replications in most laboratories most of the time, because most of the time the proper outcome of an experiment is undisputed, and scientists can hone their skills by testing their abilities against consensually agreed results. A different kind of sociological interest arises, however, where a result is disputed. Again, oddly, it is only when a result is disputed that the power of replication to separate science from the mundane is put to the test. Now, however, the analytic theory leads us to expect serious dispute about a result to turn into dispute about what counts as an adequate replicator and adequate replication. Where it is claimed that the outcome of one experiment disproves the result of another, the initial experimenter can dispute the adequacy of the second experiment or experimenter.

## The Experimenter's Regress

This transformation in the nature of the argument is known as the "experimenter's regress." The experimenter's regress starts with the observation that experiments, especially those on the frontiers of science, are difficult and that there is no criterion, other than the outcome, that indicates whether the difficulties have been overcome. In short, in the normal way, as just described, to know whether one has built and operated an experimental apparatus properly one checks to see that the findings lie in the expected range. To put this more graphically, suppose one tries to build an "$X$ detector." To know whether it is a working $X$ detector, one has to check and see whether it will detect $X$s. But suppose the existence of $X$s is in dispute. Then to find out whether there are $X$s, one has to build an $X$ detector to look for them. But to know whether one's $X$ detector is working, one has to use it to see whether it will detect $X$s. But to know whether it should detect $X$s when it is working, one has to know whether there are $X$s, and this means building an $X$ detector, and so on.

In practice, scientists have to turn away from replication itself in order to break out of the circle caused by the experimenter's regress. They have to find criteria for establishing that an experiment is working other than the outcome of the experiment itself. This is difficult because, in frontier experiments, no one knows exactly what the crucial variables are. As one scientist, discussing the problem of replicating gravitational wave detection experiments, said (quoted in Collins, p. 86):

> [I]t's very difficult to make a carbon copy. You can make a near one, but if it turns out that what's critical is the way he glued his transducers, and he forgets to tell you that the technician always puts a copy of *Physical Review* on top of them for weight, well, it could make all the difference.

This is just a colorful example of the problem of documenting the existence and nature of a skill, in this case an experimental skill.

In practice, to break out of the experimenter's regress, scientists turn in a variety of directions. Theory is one direction: It can be argued that an apparatus cannot be seeing such and such an effect because theory forbids it. But the very nature of science means that occasionally experiment will contradict theory, so this kind of argument changes probabilities but cannot settle an experimental dispute in a decisive way. Theories of how a detector must work can also be invoked, but, again, in frontier science such theories are not completely secure. In any case, complex theory can be subject to mistakes; there is a theoretician's regress that is parallel to the experimenter's regress and prevents theory from producing a definitive resolution to a dispute.

Calibration of the apparatus is another method that scientists try to use to circumvent the experimenter's regress. The argument is that a device designed to detect $X$s ought also to be able to detect well-understood $Y$s, so one can test the efficacy of rival $X$ detectors by using them as $Y$ detectors. Calibration works as long as no one questions the relevance of $Y$ detection for $X$ detection, but where a dispute is deep, the problem is likely to arise. For example, in the gravitational wave dispute just mentioned, the $Y$s were tiny impulses delivered to the detector by electrostatic drivers, but it was argued that these did not mimic gravitational waves either in their pulse shape or the way they interacted with the detector overall.

When these methods fail, scientists turn to a variety of criteria broadly similar to those used in making common-sense judgements in ordinary life. They will look to the experimenters' reputation, the reputation of their institution, their track record, previous personal interactions, surreptitiously obtained information about how the experiment was conducted, and so forth. What emerges from these considerations is that replication, the idea of which seemed to separate scientific findings from the mundane, eventually returns us to the mundane world we sought to escape.

## BIBLIOGRAPHY

Collins, H. M. *Changing Order: Replication and Induction in Scientific Practice*, 2nd ed. Chicago: University of Chicago Press, 1992.

Kennefick, Daniel. "Star Crushing: Theoretical Practice and the Theoretician's Regress." *Social Studies of Science* 30.1 (2000): 5–40.

Travis, G. D. L. "On the Importance of Being Earnest." The Social Process of Scientific Investigation, edited by Karin Knorr, Roger Krohn, and Richard Whitley. Sociology of the Sciences Yearbook, vol. 4. Dordrecht, The Netherlands: Reidel, 1980: 165–193.

*H. M. Collins*

# REPRESENTATION AND SCIENCE.

"Representation" is among the semantic nodes at which "science" and "culture" come together. The *Oxford English Dictionary* defines *representation* as "an image, likeness, or reproduction of some manner of a thing," "a material image or figure; a reproduction in some material or tangible form." *Representation* also refers to "the action of placing a fact, etc., before another by means of discourse; a statement or account, esp. one intended to convey a particular view or impression of a matter in order to influence opinion or action." Keywords latent in these definitions are central to the questions that science studies raises with regard to scientific representations, including notions of reference, correspondence, inscription, matters of fact, replication, and discursive practice. When we speak of "science and representation," are the representations at stake images on a screen or on a retina? Molecular models in a chemistry classroom? The graphs, paragraphs, and equations in a written report? Or the whole network of shared rules, standards, values, and exemplars constitutive of the "account" of the way the world is and works offered by a scientific paradigm? Do these representations ultimately "refer" to "nature," to other data, to configurations of bodies and machines, or to the cultural logic of the field in which scientific practice is carried out and the social context in which it participates? By what criteria does "science" attempt to establish the truth or falsity, consistency or inconsistency of its representations, and how might these criteria be revised to incorporate richer accounts of what "objectivity" might entail?

When we consider the denotative field of "representation" more broadly, the ambiguities multiply. Early modern usage deploys the term to denote one's "bearing" or "air"— the *manner* by which some one or some thing makes an "impression." Representation can also refer to dramatic performance: "the exhibition of character and action upon a stage." Perhaps most importantly, the *mimetic* sense of representation just defined—representation as a likeness *of* something—is troubled by a substitutional logic, in which representation denotes the fact of standing *for*, or on behalf of, some other thing or person. This apparently slight difference is consequential, for representation in the latter case does not imply a relationship of *imitation* but rather a formal agreement upon the symbolic value of a *simulation* that does not so much point back to some referent as replace it. For theories of scientific epistemology, this ambiguity

opens a passage between "representation" and "construction."

## Representing Construction

The ambiguous meanings of representation as imitation or simulation accounts for some of the semantic and taxonomic confusion in which debates about science and representation have occasionally been mired. The necessity to distinguish them is complicated by the ways in which they mutually coconstitute one another.

The issues at stake were involved in the self-described "hoax" by Alan Sokal, a physicist at New York University, who submitted an article entitled "Transgressing the Boundaries: Toward a Transformative Hermeneutics of Quantum Gravity" for publication in the "Science Wars" issue of *Social Text*, a leftist journal of cultural studies. Upon its publication in 1996, Sokal declared that the piece had been written as a parody, shot through with erroneous physics and intended to test the intellectual standards of certain precincts of the American academic humanities. Although Sokal later singled out the arrogance of what he called postmodernist *literary* theory, the fields that were most directly implicated were cultural studies and science studies. The ambiguity of the relation between representation and construction enabled Sokal to assemble a pastiche of theoretical claims that entangled science studies, cultural studies, and literary theory in his paper's keystone sentence (Sokal, pp. 217–218):

> It has thus become increasingly apparent that physical "reality," no less than social "reality," is at bottom a social and linguistic construct; that scientific "knowledge," far from being objective, reflects and encodes the dominant ideologies and power relations of the culture that produced it; that the truth claims of science are inherently theory-laden and self-referential; and consequently, that the discourse of the scientific community, for all its undeniable value, cannot assert a privileged epistemological status with respect to counter-hegemonic

narratives emanating from dissident or marginalized communities.

Beginning with a claim about the *construction* of "physical 'reality'," the sentence concludes by rejecting the *representational* privilege of "the discourse of the scientific community." The strategy of Sokal's pastiche is associatively to link together claims that have no necessary logical relation and that derive from radically different philosophical and theoretical programs. To say that science is socially constructed is to acknowledge that scientific practices take place within social contexts ranging from individual laboratories, to the national funding agencies that help to set research priorities, to international collectives that allocate research time on such scarce resources as particle accelerators. To say science is linguistically constructed evokes a different network of relations, including dominant metaphors that help to guide research programs, for example "invasion" and "warfare" metaphors for explaining disease that dominated early twentieth-century medical practice and that have recently been questioned as misleading simplifications of complex immunological processes. Further clauses in Sokal's thesis lump together critiques of scientific ideologies with reflections on paradoxes of self-reflexivity in scientific research, which again are research programs with different methodologies and aims.

For those interested in the topic of "science and representation," the relevance of Sokal's hoax lies primarily in its goad to clarify the distinctions that science studies have made and the relation between the projects that it has pursued in arguing for the socially embedded status of scientific knowledge claims. By erasing any meaningful distinction between representation and construction, Sokal's rhetoric made both seem absurd: The construction of scientific facts was hollowed of its materiality, while "linguistic" representation was endowed with world-making powers for which science studies had never really argued. The fires of the 1990s "science wars" were fueled by such misrepresentations, and the challenge for analyses of scientific fact and scientific cul-

ture is to articulate the way in which representation and construction are at once distinct and inexorably intertwined.

## Scientific Culture

The contemporary debate about science and representation that reached its nadir in the Sokal affair took its bearings from Thomas S. Kuhn's classic text, *The Structure of Scientific Revolutions* (1962), which set forth the central claim of science and technology studies: that science operates *as* a culture, *in* a culture. Kuhn's insight was that the capacity of science to operate at all—to generate the results that it offers as objective representations of natural laws—is predicated upon a prior set of representations that he called a "paradigm" or "disciplinary matrix." In this sense, one might say that science is "cultural" insofar as it participates in and is constituted by dialectic of representation and construction. Scientific facts are "constructed" insofar as they depend upon a set of shared representations that enable practice (equations, analogies, exemplary problem formulations and solutions), while the possibility of sharing those representations depends upon their reference to previously constructed matters of fact. Kuhn described as scientific revolutions those conceptual upheavals by which one paradigm is replaced by another; they occur, he argued, when the fit between representation and construction loosens and the cycle of their coimplication grows increasingly agitated by competing models of how they might best be realigned.

The process of negotiating such alignments of representation and construction is constitutive of scientific knowledge. Science offers representations of nature to a given culture; at the same time, its operations are also representative of the cultures in which they are embedded. For example, as noted by Steven Shapin and Simon Schaffer, the practice of "witnessing" and therefore verifying scientific experiments, initiated by the Royal Society in seventeenth-century England, depended on cultural assumptions about who was qualified to serve as a witness; aristocrats and men of means qualified, while women and working-class journeymen employed in the laboratory did not. Taking up these consequences of work by Kuhn and other "new" historians of science, the "sociology of scientific knowledge" (SSK) made it a methodological principle to bracket science's representations ("of nature") in order to investigate the manner in which science is representative ("of culture"). By bracketing reference, SSK did not attempt to *invalidate* matters of scientific fact or to espouse an epistemological antirealism. Rather, by investigating the "form[s] of life" and "patterns of activity" (in Shapin and Schaffer's phrases) upon which experiment is predicated, SSK offered thick descriptions of the manner in which the social organization of scientific communities delimited the range of matters that *could* be constructed as facts. On this model, science is cultural insofar as the normative practices that constitute sociality help to determine which constructions will be represented, and which representations will be accepted.

## The Representation of What?

If work in the sociology of scientific knowledge did not *directly* throw the philosophical gauntlet of antirealism down to defenders of scientific realism, it may nonetheless have begged the question of Ian Hacking's 1999 book, *The Social Construction of What?* Hacking's earlier masterpiece of conceptual ground clearing, *Representing and Intervening*, argued for two realisms: realism about theories and realism about entities. While realism about theories makes a claim about the truth of our representations (their correspondence to the way the world is), realism about entities makes a claim about the existence of the entities that are represented (though not necessarily about their precise correspondence to the representations). Arguing that "the final arbitrator in philosophy is not how we think but what we do" (p. 31), Hacking urged a realism about entities that would shift the focus of science studies from representation to intervention.

This emphasis on *practice* is exemplified by much of the stimulating work in science studies since the mid-1980s. Ethnographies of scientific communities, such as those by Sharon Traweek and Karin Knorr Cetina, have challenged the unity of "science" by offering rich accounts of institutions, laboratories, and research teams as particular scientific cultures characterized by specific methods of fact making. Correlatively, feminist programs of "strong objectivity" (in Sandra Harding's phrase) and "situated knowledges" (in Donna Haraway's) have urged understandings of scientific knowledge that do not give up on "objectivity" but rather situate it within particular cultures, rendering the standards of objectivity accountable to the contexts in which knowledge claims are produced.

Also contributing to the emphasis on practice is actor-network theory, which has been dedicated to examining the breakdown, in practice, of the theoretical distinctions upon which the scientific method is supposedly premised. Bruno Latour and Steve Woolgar's groundbreaking *Laboratory Life* (1978) focused on the centrality of "inscription" to the solidification of scientific matters of fact, excavating the technological and economic constraints that render particular *writings* of scientific knowledge valuable. Latour's subsequent work has traced the ways in which scientific practice undermines clean distinctions between humans and machines as agents and between nature and culture as separable spheres, even though the traditional rhetoric of scientific objectivity depend upon the maintenance of such binary models. A "strong" or "situated" understanding of objectivity incorporates these internal contractions within scientific cultures into the self-descriptions and methodological practices of those cultures.

In this vein, Andrew Pickering's recent work tracks the "dialectic of resistance and accommodation" as it unfolds in the course of scientific practice. For Pickering, "resistance" refers to the constraints upon knowledge construction occasioned not only by money, time, politics, technological configurations, and so forth but also by the gritty materiality of the phenomena under scientific investigation. In recognizing that the world pushes back, his work distinguishes itself from antirealist brands of constructivism. The dynamic's other component, "accommodation," refers to the compensatory maneuvers by which scientists and their support networks attempt to meet the resistance their projects encounter. Pickering's focus, then, is on the process by which the demands of objectivity, desire, and necessity are negotiated and made to fit. He argues that the dynamic between resistance and accommodation, which he calls "the mangle of practice," crucially involves temporality, for these are processes that happen in time. As a result, he argues for a shift in science studies away from what he calls a "representational idiom" to a "performative idiom."

Recall that the word "representation" can denote a dramatic *performance*: "the exhibition of character and action upon a stage." In this definition, representation *includes* a performative idiom. Pickering's work restores this denotation to our understanding of representation, enriching the emphasis on practice with an awareness of practice as a temporal process of performance. In so doing, he skewers on the arrow of time Hacking's formulation that "we represent in order to intervene, and we intervene in the light of representations."

The point about "science and representation" that should not be lost in "the mangle of practice" is that representations are *part of* the mangle in which we entangle ourselves when we intervene. Pickering's dialectic of resistance and accommodation can be further refined by noting that representation and intervention mutually entail each other: We intervene *by* representing, and representations are interventions. Representation, that is to say, operates on both sides of the dynamic between resistance and accommodation. If we acknowledge that co-operation, we will have come a long way toward a subtler understanding of the ways in which representation participates in the process of constructing scientific facts.

## Misplaced Concrete

In the Winter 2004 issue of *Critical Inquiry*, a prominent journal of literary theory, Latour published a lead article entitled "Why Has Critique Run Out of Steam? From Matters of Fact to Matters of Concern." Arguing for an approach to facts that would treat them not as fodder for critique but rather as gatherings of concern, Latour quotes Alfred North Whitehead as his exemplar of "a new respectful realist attitude" (Latour 2004, p. 244):

> For natural philosophy everything perceived is in nature. We may not pick and choose. For us the red glow of the sunset should be as much part of nature as are the molecules and electric waves by which men of science would explain the phenomenon.

Philosopher Brian Massumi, in his 2002 phenomenological study *Parables for the Virtual: Movement, Affect, Sensation,* cites the same passage, tacking two more sentences from Whitehead onto the end of his quotation: "The real question is, When red is found in nature, what else is found there also? Namely, we are asking for an analysis of the accompaniments in nature of the discovery of red in nature" (quoted in Massumi, p. 238). Massumi glosses Whitehead's question, "What else is found there also?", as a plea for an "expanded empiricism," his equivalent of Latour's "new respectful realism."

Along similar lines, feminist anthropologist Donna Haraway concludes her 1997 book, *Modest_Witness@Second_Millennium. FemaleMan©_Meets_OncoMouse™,* by defining "matters of fact" as "crucial points of contingent stability for possible sociotechnical orders, attested by collective, networked, situated practices of witnessing." Witnessing, she continues, "is seeing; attesting; standing publicly accountable for, and psychically vulnerable to, one's visions and representations" (Haraway 1997, p. 267).

The attitude towards "science and representation" evident in these recent formulations is not so much critical as it is additive, supplementary, amplificatory. Such an approach to scientific representation is also what feminist philosopher of science Sandra Harding had in mind when, in 1986, she called for a "strong objectivity" that could account for *more* of the ways in which science constituted its theoretical objects. Harding described "participatory values as preconditions, constituents, or a reconception of objectivity" (Harding, p. 249). For all these thinkers, proper conditions of objectivity (not the advocacy of relativism or antirealism) is the "matter of concern" that should be at the center of science studies.

The priorities of science studies have thus shifted in significant ways from the 1980s, when the sociology of scientific knowledge and actor-network theory concentrated on demonstrating that our experience of "Nature" was not antecedent but rather subsequent to its scientific construction; in this view science builds a simulacrum of Nature from its culture, one particle at a time. To insist, as does Massumi, that nature is partitive, giving infinitely of itself to us, is to realign the priorities of science studies, pointing the field in a direction very different from that in which it was headed twenty years ago. The priority now is not the disassembly of what science has constructed but rather the supplementary construction of what it has not yet represented.

If this continues to be the direction of further inquiries into science and representation, then we have finally arrived back at the dictionary's first, forgotten denotation for our title term: representation *also* means "presence, bearing, air." What best characterizes the current approaches to science and representation is not so much their "position" as their posture. Having learned the hard lesson that constructivist arguments can be put to work for reactionary agendas as well as for liberatory purposes, science studies has begun to "carry itself" differently, foregrounding collaborative projects and ethical obligations rather than deconstructive critique. Contemporary approaches to science and representation are less determined to stake out antagonistic, negative claims about the impossibility of objectivity than they are to elucidate the dynamic interactions in which imitation, simulation, *and* concern mutually coconstitute the nature of scientific representations.

*See also* **Representation and Technology**

BIBLIOGRAPHY

Haraway, Donna. "Situated Knowledges: The Science Question in Feminism as a Site of Discourse on the Privilege of Partial Perspective." *Feminist Studies* 14 (1988): 575–599.

Haraway, Donna. *Modest_Witness@Second_Millenium.FemaleMan©_Meets_OncoMouse™: Feminism and Technoscience.* New York: Routledge, 1997.

Harding, Sandra. *The Science Question in Feminism.* Ithaca, NY: Cornell University Press, 1986.

Knorr-Cetina, Karin. *Epistemic Cultures: How the Sciences Make Knowledge.* Cambridge, MA: Harvard University Press, 1999.

Latour, Bruno. *Science in Action: How to Follow Scientists and Engineers Through Society,* translated by Catherine Porter. Cambridge, MA: Harvard University Press, 1987.

Latour, Bruno. *We Have Never Been Modern,* translated by Catherine Porter. Cambridge, MA: Harvard University Press, 1993.

Latour, Bruno. "Why Has Critique Run Out of Steam? From Matters of Fact to Matters of Concern." *Critical Inquiry* 30.2 (Winter 2004): 225–249.

Latour, Bruno, and Steve Woolgar. *Laboratory Life: The Social Construction of Scientific Facts.* Beverly Hills, CA: Sage Publications, 1979.

Massumi, Brian. *Parables for the Virtual: Movement, Affect, Sensation.* Durham, NC: Duke University Press, 2002.

Pickering, Andrew. *The Mangle of Practice: Time Agency and Science.* Chicago: University of Chicago Press, 1995.

Shapin, Steven, and Simon Schaffer. *Leviathan and the Air-Pump: Hobbes, Boyle, and the Experimental Life.* Princeton, NJ: Princeton University Press, 1985.

Sokal, Alan. "Transgressing the Boundaries: Toward a Transformative Hermeneutics of Quantum Gravity," *Social Text* 46–47 (Spring-Summer 1996): 217–252. Available at http://www.physics.nyu.edu/faculty/sokal/.

Traweek, Sharon. *Beamtimes and Lifetimes: The World of High Energy Physics* Cambridge, MA: Harvard University Press, 1988.

*Nathan Brown and N. Katherine Hayles*

# REPRESENTATION AND TECHNOLOGY.

This entry includes one subentry

Imaging Technologies

Imaging techniques developed since the turn of the twentieth century have played a crucial role in Western conceptualizations of the body and the globe. Arguably, one of the greatest revolutions in medicine—and in the larger cultural imaginary—was the unveiling of internal physical structures. When Wilhelm Conrad Roentgen revealed his own x-rayed skeleton in 1895, he sparked a visual revolution that extended from medicine to literature to movies to art. Recent technological innovations have continued to change the way we see ourselves and our world. Ultrasound has allowed countless parents to visualize fetuses *in utero*; the Visible Human Project—an on-line, virtual representation of two human cadavers—has changed the way doctors understand, operate on, and manipulate the bodies of their living patients; and the Global Positioning System (GPS), originally developed for the military, has increased our ability to protect national boundaries and engage in site-specific warfare but also to explore and map terrain for peaceful purposes.

Each technology has certainly enriched the lives, security, and knowledge of Western citizens; yet, because each imaging technology is also embedded in cultural ideologies, economic circumstances, and national goals, representational images remain simultaneously promising and perilous. Photographs and sonograms of fetuses *in utero* have played a starring role in fetal-rights debates, the Virtual Human Project is steeped in ideologies of gender normalcy, and GPS is founded upon sometimes dubious national interests. Thus, while medical and scientific images are often valued for their realism—their ability to uncover the hidden mysteries of the natural world—they are rarely neutral or simply descriptive. Whether we examine microimages of cells, bones and muscle, an entire virtual cadaver, or the planet, imaging technologies determine not only what we see but how we see.

## X-rays: Pioneers and Popular Fascination

Before the discovery of x-rays, the invention of MRI, and the recent development of image-guided surgery (IGS), doctors and surgeons relied solely on superficial examinations, public anatomy theaters, and anatomical illustrations to visualize the human body and its various pathologies. Discovered in the fall of 1895 by German physicist Wilhelm Conrad Roentgen, x-rays changed medicine irrevocably. For the first time, physicians and surgeons could see into living bodies without prying back layer upon layer of skin and muscle. Such images offered new hope for diagnosing broken bones and even lead bullet fragments, but they also fundamentally altered the ways in which physicians, artists, and the lay public understood and visualized the human body.

Roentgen first detected x-rays by accident. He was experimenting with William Crookes's newly invented cathode ray tube, when he noticed that a screen several feet away began to fluoresce. At the time, cathode ray were known only to project a few centimeters; the faint glow seen by Roentgen was something new. Indeed, these new rays (named "X" for their unknown source) could travel through air, through opaque objects, and even through bodily tissue. More importantly for medicine, x-rays could produce spectacular images of the skeletal system, because bone absorbs more of the rays than soft tissue does, leaving a negative image (or shadow) on a photosensitive plate/film. On December 22, 1895, Roentgen produced the first and most famous of x-ray images: his wife's hand complete with an eerily "floating" wedding ring. These he distributed to some of the leading scientists of his time after submitting an article on the new rays to the *Proceedings* of the Physico-Medical Society of Würzburg.

Roentgen's documentary medium of photography, one of medicine's other nineteenth-century imaging techniques, helped to legitimate Roentgen's x-rays in the medical community and in the popular imagination. The photographs convinced several university physicians that x-rays could aid in the establishment of *scientific medicine,* a movement to standardize medical exams, diagnoses, and treatments. More importantly, Roentgen's photographs provided proof of x-rays that could be readily disseminated to the public. When the photo of Frau Roentgen's hand appeared on the front page of the *Neue Freie Press* (a Vienna newspaper) in January 1896, it sparked a popular fascination with x-rays that lasted for nearly a century. Public interest and involvement with the x-ray were due, in part, to the fact that Roentgen refused to patent his discovery. Thus, crude x-ray machines were produced throughout the Western world; component parts were ordered and apparatuses built in private homes. Sometimes, patients even took their own x-rays and shipped them to their physicians before ever being physically examined. Even when the Roentgen Society was founded in 1897, its meetings and membership were open to lay and scientific individuals interested in x-rays.

In the newspapers, in art, in educational films and in feature-length cinema, the x-ray was lauded, satirized, commended, and feared well into the 1960s. Women's x-rayed hands became fetish objects at the turn of the twentieth century; x-rays of tubercular-free chests were coveted in the 1940s; and Superman's newest power, x-ray vision, provided the foundation for several scandalous advertisements and editorial cartoons in the 1930s. The latter featured either bathing beauties whose skeletons were displayed under their swimming costumes or fully-clothed men and women whose underwear could be seen peeking through their formal outerwear. These images, and science fiction tales like *X: The Man with the X-ray Eyes*, often explore larger cultural fears about x-rays.

Public concerns about the new visual technology often revolved around conceptions of privacy and social difference. Given their ability to penetrate bodies and objects, x-rays promised to expose unseen body parts, unseat conceptions of personal bodily integrity, and dematerialize superficial markers of race and class. One particular editorial cartoon from 1897 depicts two people alongside their skeletons: a woman and a man who are very different in size and shape. Their skeletons, though, as the cartoon's caption notes, are relatively similar: "Whether stout or thin, the x-ray makes the whole world kin." The implications for this hidden sameness were many; indeed, if all people were the same "under their skin," then typical physical markers of race and class difference were vis-

ible only in superficial manifestations. For the public, and scientists specifically, x-rays shifted the value of human vision. Once associated with truth, realism, and direct knowledge of the world, sensory perceptions began to lose their purchase in science and culture. Unaided sight was deemed unreliable in the face of an unseen world of interiority exposed by x-rays.

The public frenzy over x-rays, though, was double-edged, not only because x-rays challenged established modes of visuality but also because the rays themselves proved to be physically dangerous, even deadly. As early experimenters quickly learned, x-rays caused harm as well as healing in exposed individuals. Some reported a superficial burn, others noticed swelling in joints and limbs, still others found tumors, and men like Clarence Dally (one of Thomas Edison's assistants) lost their bodies to amputation, limb by limb. Science fiction stories such as *X: The Man with the X-ray Eyes* explored similar personal tragedies for experimental scientists. In this particular book and film, a scientist who acquires x-ray vision eventually sees too much. While his new power appears harmless—if not beneficial—at first (he can see through clothing and can even see into sick patients to determine their illness), sleeplessness, skeletons, and visions eventually haunt his waking life. In the final scene of the movie, he plucks out his own eyes in search of relief.

## Medical Imaging: Ultrasound, CAT, and MRI

Ever since Roentgen first penetrated the body without scratching its surface, scientists have sought new techniques by which to image not only bone but soft tissue as well. Significantly, this search has not been limited to biology, chemistry, or physics laboratories. Indeed, twentieth-century imaging technologies are linked to the military. Ultrasound instruments used in obstetrics, for example, are based on sonar technology developed during World War I for use in submarine warfare. Medical use of these imaging technologies did not occur until they were demobilized and declassified. Without the advent of digital computers (first deployed in the military and public spheres)—including the related rise in public digital-image literacy—computed axial tomography (CAT) and magnetic resonance imaging (MRI) scans would not be possible. Regardless of their origin, though, the primary purpose of imaging technologies remains largely unchanged: to allow soldiers, scientists, and doctors to "see in the dark"—be it during a midnight military campaign or a physical examination of the body's internal structures.

Very basically, each imaging technique penetrates the body with sound or electromagnetic waves (x-rays in CAT and radio waves in MRI). When these waves encounter internal bodily structures, the energy they release or absorb produces signals (coded data) that are sent to and interpreted by computers. Unlike x-rays, which actually produce a negative image on a photosensitive plate, ultrasound, CAT, and MRI produce sets of data that are not immediately recognized as images until they are transformed by the computer. Moreover, because the information is in raw form and is often two-dimensional in nature, images constructed from the data can be manipulated in two and three dimensions; they can be enlarged, colored, coded, keyed, and annotated.

Because they necessitate the interaction of technician, machine, computer, and imaged body, ultrasound, CAT, and MRI scans demand another shift in visuality. As with the advent of the x-ray, unaided sensory acuity is devalued; perception is dispersed through an imaging machine that emits waves, collects signals, and processes data. Technicians are thought to be experts at decoding images, and images become valued as objective artifacts—objects more accurate and reliable than a patient's verbal explanation or a physician's visual or manual examination. A typical example occurs in hospital delivery rooms, where doctors have been known to look at fetal monitors rather than their patients to determine when a contraction is happening. In this scenario patients often become fragmented themselves: images are taken only of their uterus, their heart, their brain, or their knee and are evaluated as if each piece functioned independently of the entire body.

Even as they are fragmented and processed by machines, though, popular exposure to cinema, television, and computer monitors have provided lay audiences with the visual literacy to appreciate—if not altogether understand—the nature of the images produced by medical technologies. Most of us could recognize a skeleton and maybe even some of the muscles and tendons present in an MRI image. We have seen body scans countless times in movies like *Total Recall* and in television programs like *Star Trek*. Exposure to popular depictions of futuristic, mechanical vision has increased our reliance on and desire for imaging technologies. After sonograms became available to civilian physicians, for example, they also became one of the most requested and performed type of imaging procedure. In other words, many parents do not simply want to hear that their baby's data points indicate a healthy fetus; they would rather view the data on-screen while being told how to interpret the blurry shadow resembling a baby.

While fetal images, like those presented by sonograms, have provided many a baby's first portrait, their existence has also influenced debates about fetal rights and maternal responsibilities. Prior to the invention of ultrasound imaging, a pregnant woman might not have a sense that she was carrying a distinctly living being within her womb until she felt the fetus move (an event called "quickening"). With the advent of ultrasound, fetuses could be seen—and deemed to be living, human beings—much earlier. In many respects, this new vision helped to construct *in utero* fetuses as miniature people—children who existed autonomously from their mothers. In this scenario, pregnant women are positioned in a managerial role, if they do not disappear altogether. However, the way in which fetal images are understood is not merely the result or responsibility of medical imaging technologies. One of the most influential series of fetal images was not taken with an ultrasound or an x-ray machine, but with a camera. Lennart Nilsson's famous images of fetuses floating in a vast sea of blackness first appeared on the pages of *Life* magazine in 1965. In these photos, the mother is markedly absent; the fetus is presented as an autonomous being able, even, to suck his thumb. Ironically, as in many of the early x-ray films, the apparently living fetus—brought to light and life by the camera—is in fact dead. Eerily, Nilsson's images of life's beginnings return us to the strangeness experienced by early Victorian audiences viewing their skeletons for the first time.

## Virtual Representations: The Visible Human Project and Image-Guided Surgery

Over the past decade, scientists have sought various virtual representations of the human body that would be more accurate, individualized, and interactive than two-dimensional textbook images or static body scans. Some recent representations such as those used in image-guided surgery, provide real-time images of an individual patient's pathology; others provide wonderfully detailed, full-body anatomical models, such as the Visible Human Project (VHP), developed by the National Library of Medicine in 1994, which includes fully digitized sections of the bodies of two humans (Joseph Jernigan and an anonymous female). Each virtual technique demands the cooperation of several seemingly disparate disciplines and technologies: doctors must work with computer scientists, engineers, and—inadvertently—the entertainment industry to produce techniques and images once thought to be scientific fiction. In fact, science fiction, fantasy, and imagination have been crucial components in the development of virtual medicine.

Like ultrasound and MRI scans, virtual representations depend on computers to process data points; however, because virtual images are rendered in three dimensions—and typically include bony and nonbony bodily structures—they are often constructed by combining two or more different imaging techniques. Image-guided surgery, for example, is made possible by combining CAT scans, traditional x-rays, and MRI images. The resulting three-dimensional model is, essentially, a virtual patient: one that can be manipulated presurgically, allowing doctors to plan operations that will minimally harm healthy tissue and better isolate pathology.

Virtual patients are, of course, not a new concept in medicine—simulated patients and haptic models are often used to train medical students—however, the virtual patients of image-guided surgery are not only computerized but individualized. A generic model is helpful; a functional three-dimensional model of a specific patient is infinitely valuable. Some of the latest technology even allows surgeons to perform operations while patients are inside an MRI machine; thus, their patient's images are constantly updated and surgery can be performed with real-time, virtual feedback. Image-guided surgery may sound like something straight out of a science fiction novel; and it is, at least inspirationally. In fact, journalists and surgeons alike have compared virtual surgery to Superman's x-ray vision. Importantly, this comparison is not diminutive: It recognizes not only the imaginative foundation of medicine's newest technology, but also the power and prevalence of popular culture in the dissemination and explanation of medical innovations.

In fact, image-guided surgery and the Visible Human Project have foundations in the entertainment industry. Virtual images rendered in the medical industry are generally produced using the same software contained in video game cards available in retail electronics stores. In game machines and home computers, these cards produce realistic images of football players, car chases, and military combat; in medical imaging technologies, the same cards provide doctors with more accurate representations of human anatomy in virtual patients. Surgeons can even perform heart surgery on a patient in another room via video feedback and robotic arms controlled by a joystick-like apparatus. Medical imaging has been able to advance so rapidly—and with minimal financial and research investments—because the game industry has led the way.

People are quite willing to invest billions of dollars in the development of more realistic and entertaining video games; medicine has simply adapted these new, cultural technologies to the operating room.

## Global Positioning System

Compared to medical imaging technologies, which *create* bodies as much as they *describe* them, geographic images may appear less ideological, less metaphorical, and less invested in cultural and socioeconomic systems. However, mapping—be it medical or geographic—is often invested in ideals of progress, safety (personal or national), and territorial control. The Global Positioning System (GPS) is no exception. Developed in the 1970s by the U.S. Department of Defense, GPS is a group of thirty-seven orbiting satellites that transmit radio signals to ground-based receivers. While the European Union has plans to complete its own satellite constellation (Galileo) by 2008, and the Russian Federation has a group of functioning satellites (GLONASS) already in place, America's GPS remains the only functional satellite system accessible to the public. Available free of charge to anyone in the world who owns a GPS receiver, GPS has become an instrument for national security, globalization, and even art.

Nationally, GPS serves two primary purposes: to provide the U.S. military with a distinct advantage on a global scale, and to enhance what the National Security Council has termed the "Global Information Infostructure" (including terrain surveys, air traffic control, and global change research). When President Ronald Reagan first made GPS available for civilian use in 1984, the government degraded the civilian GPS signals, making them less accurate, while military systems had the capability to correct for the degradation. This degradation of GPS signals was termed "Selective Availability." After the termination of Selective Availability in 1996, public GPS receivers became ten times more accurate; users can now locate their bearings to within sixty feet. However, GPS is still funded and controlled by the U.S. government. Officials can, at any given time, discontinue signals over certain parts of the globe or reinstate Selective Availability—a decision that would reduce the accuracy of civilian handsets tenfold.

GPS is also a tool for globalization. As NASA administrator Sean O'Keefe points out in his Welcome Letter on the NASA Website dedicated to "GPS Applications Exchange," GPS signals "present an opportunity for the rapid transformation of a country's infrastructure in areas ranging from river and port navigation to aviation and telecommunications." In this scenario, GPS enables two globally progressive goals: first, GPS functions to soften national boundaries by encouraging the importation of foreign ideals, products, and projects—a maneuver that introduces and implements narratives of progress and globalization. Second, the availability of GPS technology solidifies national goals for other countries: If they have access to GPS and the drive to advance as a civilization, then the U.S. satellite constellation can help to "transform" their very "infrastructure" into one that is more modern.

Private use of GPS has sparked another kind of globalization: an international artistic movement. As several civilian users have discovered, handheld GPS receivers are a wonderful medium for creating unusual pictorial maps within various cities. Electronic drawings produced by this once-military technology are both challenging national boundaries and reconstituting the visuality of invisible radio waves. For Jeremy Wood and Hugh Pryor—two young men from the United Kingdom—the world has literally become their canvas. As they travel across roads, bridges, jogging trails, and streets (as well as in boats and airplanes) and record their successive positions, their GPS receivers act like paint brushes, tracing the shapes formed by their paths. Later, these data points can be better connected, visualized, and superimposed on city maps via computer software, somewhat as in an MRI scan. Wood, Pryor, and other GPS artists have discovered and created an ever-growing number of patterns such as an elephant in Brighton and a spider in Oxford. An entire Website is devoted to images produced by GPS artists.

Particularly significant, given the nationalistic and militaristic beginnings of GPS, is the fact that many of the artists are not American, and their drawings rarely reinforce national or even city boundaries. Instead, these drawings ask us to re-envision cities and nations as locations for artistic expression. In this respect—and perhaps despite the problematic nationalism embedded in GPS—the U.S. satellite system is providing for a global community. Furthermore, images drawn with GPS tend to challenge the very purpose of the satellite constellation. Instead of pinpointing an individual's location or mapping the fastest routes through busy city streets, GPS artists defy traditional mapping prerogatives. Whether pre-planned or free-form, their drawings re-envision spaces as shapes, ideas, words, or doodles. What matters is not the destination, but the journey. GPS, then—particularly GPS art—alters *how* we see the landscape, just as medical imaging technologies alter how we see and understand the human body.

*See also* **Representation and Science**

## BIBLIOGRAPHY

Blume, Stuart. *Insight and Industry: On the Dynamics of Technological Change in Medicine.* Cambridge, MA: MIT Press, 1992. A history of imaging technologies (including ultrasound, thermal imaging, CT scans and MRI) from an economic and health policy perspective.

Cartwright, Lisa. *Screening the Body: Tracing Medicine's Visual Culture.* Minneapolis: University of Minnesota Press, 1995. A cultural/science studies analysis concerning the convergence of cinema and medical imaging technologies. This book spans a period from 1895 to the 1940s. Note especially Chapters 5 and 6 on the cultural impact of x-rays.

Crary, Jonathan. *Suspensions of Perception: Attention, Spectacle, and Modern Culture.* Cambridge, MA: MIT Press, 1999. An analysis of modernist art, the science of attention, and visual literacy at the turn of the twentieth century. Particularly useful for anyone pursuing direct connections between art and science during a major visual revolution.

El-Rabbany, Ahmed. *Introduction to GPS: The Global Positioning System.* Boston: Artech House, 2002.

Franklin, Sarah, Celia Lury, and Jackie Stacey. *Global Nature, Global Culture.* London: Sage Publications, 2000. A feminist analysis of nature and culture in a world marked by globalization. Chapter 1, "Spheres of Life," and Chapter 6, "Life Itself: Global Nature and the Genetic Imaginary," may be relevant to anyone interested in icons of life (the fetus and the "blue planet") and the ways in which science changes our perceptions of bodies, respectively.

GPSDrawing.com. http://www.gpsdrawing .com. A Website devoted to GPS artwork, projects, and information about how to become involved and/or commission a project.

Kevles, Bettyann. *Naked to the Bone: Medical Imaging in the Twentieth Century.* New Brunswick, NJ: Rutgers University Press, 1997. A history of medical imaging technologies divided into two larger sections: Part One addresses the development of x-ray imaging, and Part Two takes on medical imaging after the dual revolutions of television and computers.

National Aeronautics and Space Administration. *GPS Applications Exchange.* http:// gpshome.ssc.nasa.gov/. Website describing many applications for GPS, whether actually implemented or merely suggested by users.

Treichler, Paula, Lisa Cartwright, and Constance Penley, eds. *The Visible Woman: Imaging Technologies, Gender and Science.* New York: New York University Press, 1998. A collection of essays divided into three sections that address what the editors term several paradoxes of visibility. Provides wide coverage of medical imaging technologies with a particular focus on women's health, reproductive and health policy issues.

Waldby, Cathy. *The Visible Human Project: Informatic Bodies and Posthuman Medicine.* London and New York: Routledge, 2000. A somewhat densely written cultural history and analysis of the National Library of Medicine's "Visible Human Project." Chapter 4, "Virtual Surgery: Morphing and Morphology," may be particularly relevant.

*Melissa M. Littlefield*

# Imaging Technologies

Each new kind of imaging technology is accompanied by fanfare that celebrates it for opening a new world to our senses. Indeed, one could almost say that technologies that produce pictures also have a tendency to create a certain kind of sensationalism. At one time, people marveled at photography's frozen stills of our seemingly familiar world; at Roentgen's X-rays, which showed us the bony insides of our bodies; and still later at swirled and glowing positron-emission tomographic scans of our brains. Now we are struck by the lifelike quality of 3D and 4D fetal imaging. Once a quack science, radiology has become a leading edge in medical science over the course of the twentieth century,. But it is only one invention that has changed the way in which we see our bodies, and the world we live in. Today it is ordinary to expect to be able to see not only into the deepest recesses of our bodies to diagnose illness, but also to use imaging technologies, such as satellites and night vision cameras, to see deep into the heart of enemy territories. Other technologies, such as fingerprinting and retinal scans, can help us not only to see our enemies, but also to control them. From the technologies glorified in spy movies to the trinkets that mesmerize viewers of TV crime shows, people seem to be enthralled and inundated by gadgets that make pictures.

Are these technologies that make pictures really so different from other kinds of technologies, such as bicycles and railroads? In some ways, probably not. For example, imaging technologies can also be said to be socially constructed. That is, they can be seen to embody a set of interests and social relations that are particular to a historical moment, just as was the bicycle or railroad. But there is at least one way that these kinds of technologies are not just like the bicycle or the railroad: they make pictures—that is, representations—of something else. Things like X-ray, fingerprint technologies, and satellite systems can all be considered forms of instrumentation in the sense that one means of defining how they "work" is in their capacity to produce knowledge about some other thing. Operators of railroads or reactors may need knowledge to operate these systems, but such machines themselves are not commonly thought of as mechanisms to produce knowledge. Because of the special relationship to knowledge production, imaging technologies raise a common set of questions about the relationship between technology and various social groups and

about the relationship between knowledge and power. While by no means exhaustive, some of those relationships concern popularization, professionalization, objectification, and transparency.

## Popularization

As mentioned in the foregoing, one thing that is different about imaging technologies is that they can be thought of as instruments; their role is to make images of something else and, in that sense, to help produce knowledge about something else. Many of these technologies are also intended to facilitate the rapid and mass production of images. Because of this, imaging technologies can give more people a wider range of possibilities to participate in certain forms of knowledge production. As such, they also frequently raise questions about the expertise needed to produce or interpret an image produced by one of these technologies. At the beginning of the nineteenth century, for example, only artists, painters, or sculptors could make a portrait. But once the camera was invented, the means necessary to produce a portrait were accessible to many more categories of people. Now the middle class could afford portraits also. But, as Walter Benjamin noted, it was precisely because everyone could utilize it that the aesthetic and eventually monetary value of the portraits produced by the camera diminished. Now that anyone can produce a portrait via photography, portraiture risks being considered a vulgar art. The phenomenon Benjamin pointed to can be applied to questions of aesthetics as well as to those concerning knowledge. From Abraham Zapruder's 26-second film of John F. Kennedy's assassination to the amateur video of the Rodney King beating and from lay observers of new stars to binocular-equipped backyard bird-watchers, imaging technologies seem to open up possibilities for lay participation in knowledge production. But what happens to the believability of a type of representation when everyone can produce it?

In many ways, the proliferation of a certain kind of imaging technology renders the representations it produces more familiar. But contrary to Benjamin's observations concerning aesthetics, there is no reason to suppose that familiarity with a technology makes the images it produces any less believable. In fact, familiarity could give those images a certain kind primacy and authority, and give them more seeming "objectivity" precisely because the images they produce are decipherable to everyone (or at least appear to be). The ease of accessing a camera, for example, can create an association among the general public between the appearance of a photograph and the expectation that what is familiar to our own eyes is more likely to be accurate.

Alternatively, we have all also seen photos of three-headed babies and aliens on the covers of tabloids at the checkout line of our grocery stores, too. Photographs removed from their context are rarely taken to be sufficient proof of anything, especially as we enter the twenty-first century. If the proliferation of imaging technologies makes the production of images accessible to almost anyone, so too does it make the alteration and manipulation of those images more easy to accomplish. In essence, the popularization of imaging technologies is a mixed bag. It may give laypersons the possibility of being more involved in knowledge production. But it also tends to make those lay contributions more suspect.

## Professionalization

One way to establish the potential for objectivity of a representational medium is for a group of experts to assert authority over that technology. That is, to take a technology circulating in the public sphere and to "professionalize" it. In the following, X-ray technology is offered as an example of what professionalization might mean here.

We often forget the history that goes into making a technology useful for science or medicine. Today it seems obvious that Wilhelm Roentgen's discovery in 1895 of rays that made the skeleton visible would have immediately been applied to medicine. But that was not the case. At the end of the nineteenth century, images produced with Roentgen's rays were generally considered to be just another type of "photography." Indeed

the technique we now simply call X-ray was then called Roentgen photography.

Photography, something that we now take to be a rather ordinary technology in that it only purports to image what our eyes can already see, was used as an aid to medical practice in the mid-nineteenth century. It was used to document rare medical cases, sometimes for diagnosis (for example, cases of smallpox), and even to solicit medical opinions from other doctors through correspondence. At that time, it was not in any sense obvious that the photographer was a medical professional. In fact, many hospitals had their own photography labs and photographers on hand well before they had radiology departments or "radiographers."

Just as the hospital had photography studios, at the turn of the century, most radiographs were taken not at the hospital but in privately owned portrait studios operated by photographers who also specialized in Roentgen photography. One common use of Roentgen's rays, especially popular among young women, was to make an image of one's hand to send as a memento to a betrothed or loved one. Initial steps to professionalize radiology involved distancing the field from photography. This was done at least in part by separating out the skills required to produce images from the interpretive expertise needed to read the images ("radiology"). A new subprofessional, the radiographic technologist, created a buffer between the X-ray specialist and the surgeon, and let the radiologist prove his status as a doctor in his own right. This move made a technique that had previously flowed between the spaces of the hospital and the portrait studio a strictly medical practice. As the profession of radiology took shape, there would no longer be anything entertainingly novel about X-rays; it was just that experts were required to read them.

Another common arena in which different social groups lay claim to an imaging technology by claiming expert interpretive skill over the images they produce is in the courtroom. Scholars in science studies have explored this aspect of technologies and representation in extensive detail. Some have studied the historical consolidation of certain professions, such as latent fingerprint scholars, as experts in courts of law. Others have shown how the lay perception of jurors or the television-watching public can be transformed in the courtroom through the assistance of expert testimony and the actions of the judge.

## Objectification

The examples referred to in the preceding section (photographs, X-rays, fingerprinting, video cameras, etc.) point to another characteristic of imaging technologies: their ubiquity and the seemingly enormous role they can play in our lives. Imaging technologies, it seems, can make the difference between our health and illness, or between our freedom and incarceration. The potential of imaging technologies to have such a monumental importance in our lives have led some to ask, "Do imaging technologies tend to produce a certain kind of relationship between the person creating an image and the person who is being represented?" Some would claim that they certainly do. For example, some feminists would argue that imaging technologies can objectify women's bodies.

What is meant by the term *objectification*? As social theorists often put it, the subject is the person or thing that has agency in the world, whereas the object is the person or thing that is acted upon. In the case of photography, for example, the subject would be the photographer and the object is the person depicted in the photograph. Some would argue that the ability to make a picture of someone else gives the photographer a certain power over the person who is being represented. A photographer can, for example, alter the lighting, or position the subject in a way that may evoke a certain mood. Perhaps the direction of the light can produce shadows that make the face seem imposing or somber. Others would argue that the very act of making pictures embodies a certain power differential. Because she takes the picture, the photographer is also in a position to produce knowledge about the person being photographed. She decides what to include in or exclude from that picture, whereas the person being photographed is not similarly able to produce knowledge about the photographer. The stakes of this unequal ability to produce knowledge about someone else are clear in the case of medical imaging.

Medical imaging technologies show how the ability to make images can reinforce unequal power relations in society. Some feminists would argue, for example, that the history of obstetrics is a case in which the technological means of imaging the fetus increasingly came to replace the knowledge that the mother and the midwife had previously had about the experience of pregnancy and about the well-being of the fetus. Technological developments such as ultrasound, along with such procedures as amniocentesis and fetal heart monitoring, gave the doctor more and more direct contact with the fetus and in turn reinforced the authoritative status of the doctor's knowledge over that of the mother because the former was "scientific" whereas the mother's was based only on experience. By the end of the twentieth century, the obstetrician had become the interpreter of the fetus to the mother, rather than the other way around. Now the mother experiences her pregnancy through the lenses provided by modern medicine.

Some feminists would argue that one stake of unequal power relations in imaging technologies is that they can transform what once were ordinary everyday experiences (like pregnancy) into medical experiences that now require technoscientific experts. Fetal imaging technologies are just one type of technology that gives such experts greater possibilities to intervene in our lives. In fact, some would argue that it is characteristic of all modern sociotechnical arrangements to enable experts to gain greater access to our lives by producing knowledge about them through imaging technologies. Emblematic of this view is Michel Foucault's account of the Panopticon.

The Panopticon was a prison plan designed by Jeremy Bentham at the end of the eighteenth century. The Panopticon is a circular tower in which the watchman of the central tower has total and perpetual view of the prisoners in the peripheral circles. Here, being observed is not just an unequal relationship between the person who can produce knowledge and the person being observed. Here being observed is a form of punishment. As Foucault describes it, in the sociotechnical arrangement of the Panopticon, even when the prisoner is not being watched, the prisoner begins to scrutinize his own behavior because he realizes that the possibility of being observed is ever present. It is through this process of self-observation, Foucault claims, that the interned learns to be self-aware and thereby becomes a conscious subject of modern civil society and an acceptable citizen of the modern nation-state. Foucault describes the Panopticon as an architectural arrangement that is ubiquitous in not only modern prisons, but also modern schools, factories, and hospitals, to name just a few. In this sense, Foucault's Panopticon is not just an isolated sociotechnical arrangement, but describes a whole system of governance that controls its citizens by producing knowledge about them.

Some types of technologies seem to reinforce Foucault's pessimistic view of the role of imaging technologies in modern societies. Hidden cameras and recording equipment, aerial and satellite surveillance technologies, and "total information awareness" defense projects today all make us vividly aware that "knowledge is power." But the question remains: do our feelings about these technologies come from their inherent characteristics? Or is it simply the political moment and social circumstances in which such technologies get deployed that enable them to "objectify" and control populations? Just as more imaging technologies give governments and those in power more opportunities to observe and produce knowledge about their citizens, devices such as digital cameras in our cell phones and camcorders can also offer ordinary citizens the opportunity to observe and publicize the activities of those in power.

## Transparency

If imaging technologies afford more opportunities for being surveilled, they also afford more opportunities for the public to hold authority figures accountable. We can all think of examples in which public officials have been caught in embarrassing or criminal acts, for example, via tape recordings of private phone conversations or video recordings of police beatings. In this sense, Foucault's sociotechnical imaging system can be understood not only as a mechanism to produce modern subjectivity, but also a system of political accountability. This argument is best articulated by Yaron Ezrahi.

Ezrahi argues that we should view the role of imaging technologies as not only controlling ordinary people but as affording the possibility of keeping track of public figures and those in power. Liberal democratic political culture, especially the American variant, Ezrahi says, is a "visual culture." What he means by this is that citizens' gazes have been instrumentalized and rationalized, at the same time that state actors rely on science and technology, including its metaphors, to make their actions appear visible and transparent to an "attestive" public. In particular, the mass-printed and electronic media act to sustain this political culture as "visual culture," rendering citizens as "witnesses" and politicians as ever cautious about the possibility of being observed. This arrangement essentially reverses the direction of the Panopticon, and suggests that imaging technologies, especially with the help of the mediating press, can also make it possible to produce knowledge about those in positions of power.

Because of the special relationship that exists between imaging technologies and knowledge production, they raise a common set of questions about the relationship between knowledge and power, and more generally about the relationship between technology and different social groups. As we go forward in the twenty-first century, we will no doubt again be amazed by technologies that make pictures, not just because they open new worlds to our senses, but because such technologies will continue to be battlegrounds over who has the authority to say what their representations show and over who has the right to say how such devices will be used.

## BIBLIOGRAPHY

Abbott, Andrew. *The System of the Professions.* Chicago: University of Chicago Press, 1988.

Benjamin, Walter. *Illuminations*. New York: Schocken, 1969.

Bijker, Wiebe, Thomas Hughes, and Trevor Pinch. *The Social Construction of Technological Systems*. Cambridge, MA: The MIT Press, 1989.

Burns, Stanley. "Early Medical Photography in America (1839–1883): III. The Daguerrean Era." *New York State Journal of Medicine* (July 1979): 1256–1268.

Burns, Stanley. "Early Medical Photography in America (1839–1883): VII. American Medical Publications with Photographs." *New York State Journal of Medicine* (July 1981): 1226–1264.

Cole, Simon. "Witnessing Identification: Latent Fingerprinting Evidence and Expert Knowledge." *Social Studies of Science* 28.5–6 (October/December 1998): 687–712.

Daston, Lorraine, and Peter Galison, "The Image of Objectivity." *Representations* 40 (1992): 81–128.

Foucault, Michel. *Discipline and Punish*. New York: Vintage Books, 1977.

Goodwin, Charles. "Professional Vision." *American Anthropologist* 96.3 (1994): 606–633.

Holtzmann Kevles, Bettyann. *Naked to the Bone: Medical Imaging in the Twentieth Century*. New Brunswick, NJ: Rutgers University Press, 1997.

Howell, Joel. *Technology in the Hospital*. Baltimore: Johns Hopkins University Press, 1995.

Jasanoff, Sheila. "Eye of Everyman." *Social Studies of Science* 28.5–6 (October/December 1998): 713–740.

Oakley, Ann. *The Captured Womb: A History of the Medical Care of Pregnant Women*. New York: Basil Blackwell, 1984.

Palmquist, Peter, ed. *Elizabeth Fleischmann: Pioneer X-Ray Photographer*. Berkeley, CA: Judah Magnes Museum, 1990.

Petchesky, Rosalind. "Foetal Images: The Power of Visual Culture in the Politics of Reproduction." In *Reproductive Technologies*, edited by Michelle Stanworth. Minneapolis, MN: University of Minnesota Press, 1987.

Stabile, Carol. "Shooting the Mother: Fetal Photography and the Politics of Disappearance." *Camera Obscura* 28 (Winter 1993): 179–206.

*Pauline Kusiak*

# RISK.

We live in a global age that was brought into being by a sequence of events over a fifty-year period—from the dropping of the atomic bomb in 1945, through the spanning of the globe by communication media and the first photograph of planet Earth from outer space, and to the acknowledgment by the United Nations in 1995 that climate change is due to human intervention. On account of events such as these, society has been reconceived in various ways. The atomic bomb compelled us for the first time to see the globe as a whole. The electronic media made us appreciate that world society is actually a "global village." The shrinkage of the earth to an image on a photograph, graphically expressing our sense of the proximity that nevertheless accompanies globality, called forth the name *the blue planet*. And years before the confirmation of global warming, the propensity of modern civilization to damage and, indeed, threaten to destroy itself by allowing human decisions and activities to give rise to dangers and hazards or risks on an unprecedented scale led German sociologist Ulrich Beck to coin the label *the risk society*.

Preceding the risk society, two distinct risk cultures were in evidence. Early modern capitalism, particularly fourteenth-century marine trade in northern Italy, brought forth the modern word *risk*, which depended on the theory and legal establishment of the insurance industry. From it stems the still-familiar Western culture of risk calculation, which, underpinned by an awareness of the precariousness of time, is aimed at the rational exploitation of available options and the simultaneous avoidance of loss. Risk taking in this sense found formal expression in the actuarial formula of risk as being equivalent to the probability of loss multiplied by the magnitude of loss divided by time. With the transition to industrial capitalism and accompanied by widespread poverty, risk and hence security underwent a change in the face of an unpredictable and uncertain future. Instead of the interests of the individual businessperson, dangers and hazards, such as unemployment and sickness, threatening the very existence of the working class required a different risk culture. In a combination of rational calculation with population and accident statistics, settlement formulas, and the principle of monetary compensation, risk became the central instrument in the creation of social security in the welfare state.

A radical transformation going far beyond these earlier risk cultures, however, occurred in the twentieth century. Rather than remaining confined to risk-taking individuals or a risk-averse social class, this reorientation signaled the global expansion of a contradictory culture of the production, perception, and communication of risk, which has become the signature of our own time. The welfare state was made possible by basing the economy on science and technology and the stabilization of the whole by state intervention. Reinforced by two world wars, this development by mid-century catapulted science and technology into the leading forces of production. On all accounts, these productive forces in their fused form of technoscience lie behind both the production of risks and the accompanying risk culture. Science and technology, in conjunction with capitalism, industry, and the state, are responsible for the alarming multiplication and even normalization of chemical and nuclear disasters (such as near Seveso, Bhopal, and Chernobyl) and of ecological catastrophes (such as the ozone hole and global warming). Simultaneously, their transformation of society into a laboratory, the so-called experimenting society, deliberately employing risk to advance knowledge, also feeds the spectacular upsurge in the pronounced consciousness of, the debates about, and the conflicts over real, suspected, and even hypothetical risks in a variety of fields.

Whereas risks were traditionally regarded first as individual and later as collective in a social sense, they have lately acquired a new quality. Today, risks are universal in the sense of threatening all living beings without exception, global in the sense of knowing no geographical boundaries, and irreversible in the sense of harboring adverse effects for all future generations. Under these conditions, the ontological security of the individual diminishes, with the result that both fear, dread, or anxiety and the desire for safety or security increase. Culturally, this experience is organized by a newly emerging set of concepts, rules of conduct, and meanings: the cognitive scheme of danger and safety. Three typical responses informed by this characteristic cultural model of our time reproduce the contradictory contemporary culture of risk. Some regard risk not as a threat to integrity but rather as the necessary starting point for advancement through the multiplication, specification, and exploitation of risks. Others who believe we are at the end of modern civilization project either a hedonistic or a religious apocalyptic culture. Beyond these extremes, still others insist on the inescapability of a new responsiveness to our situation, for instance, by developing new reflexive or discursive institutions, sustainable forms of life, a new sense of both collective and individual responsibility, and extending citizenship to include both ecology and technology.

## BIBLIOGRAPHY

Beck, Ulrich. *Risk Society: Towards a New Modernity.* London: Sage, 1992. The classical source of the concept and theory of the risk society, first published in German in 1986, and elaborated in a number of subsequent books.

Beck, Ulrich. *Ecological Politics in an Age of Risk.* Cambridge, U.K.: Polity Press, 1995. Sequel to the above in which modern technology is depicted as giving rise to ecological politics.

Bernstein, Peter L. *Against the Gods: The Remarkable Story of Risk.* New York: Wiley, 1996. An acclaimed work arguing that the moderns distinguished themselves from their predecessors who entrusted God with their fate by employing science and technology, particularly mathematical probability calculation, to transform dangers and hazards into manageable risks.

Douglas, Mary, and Aaron Wildavsky. *Risk and Culture: An Essay on the Selection of Technical and Environmental Dangers.* Berkeley, CA: University of California Press, 1982. A cultural theory leading the diversity of perceptions of risk in contemporary society back to the values entertained by different social groups.

Giddens, Anthony. *Runaway World: How Globalisation is Reshaping Our Lives.* London: Routledge, 1999. Includes a highly readable chapter on risk showing that this notion

illuminates some of the most basic characteristics of the world in which we now live.

*Piet Strydom*

# ROBOTS AND SOCIETY.

Robotics historically derives from two very
different origins. Designers of early mechanical automata sought to amaze viewers with
sophisticated machines that displayed lifelike
physical behaviors. The second origin is the
community of science fiction authors who
sparked the public imagination with an-
droids endowed with superhuman qualities
that, as often as not, possessed the capability
to do great harm or good to human society.
Commercial and research robotics did not initially realize these early explorations in social
robotics, but rather excelled on the factory
floor by providing highly accurate and reliable operations for repetitive tasks, such as
welding and painting. Robotics has more recently begun to return to its roots because the
science and engineering disciplines of robotics are enabling the creation of embodied automata that interact directly with people.
What was once science fiction is rapidly becoming a feasible, possible future.

Conventional robotics draws primarily
upon three fields of research: *mechanical engineering, electrical engineering, and computer
science.* Because robots have physical embodiment, mechanical engineers provide solutions
to the problems of chassis design and locomotion, from carlike rovers to anthropomorphic
walking machines. To enable robots to function
autonomously, computer scientists and, in particular, experts in artificial intelligence create
cognitive computational systems that implement the robot's control and decision-making
processes, which direct the robot's physical actions. Bridging computation and mechanism
are the electrical engineers, who deliver basic
competencies, such as intelligent power management and embedded processing.

By contrast, *social robotics* draws upon a
number of fields outside of engineering and
computer science to create social interactions
between robots and humans, which have
physical, emotional, and aesthetic qualities
and which together constitute a social structure. *Human–computer interaction* provides
evaluation methodologies and design heuris-
tics governing computer-based interfaces.
*Human factors* provide formal evaluation of
useful man–machine interfaces. *Design* provides methods and techniques to realize the
physical, emotional, and expressive qualities
in the morphology of a social robot. *Cognitive
psychology* and education lend predictive
models of cognition for human learning, thus
guiding both social robot cognition and
robot–human interaction analysis. *Sociology*
provides a system-level approach to understanding the social structure of human and
animal societies. Together the robot design
team must integrate perception, attention,
motivation, and emotion and must demonstrate these in intelligent and expressive behaviors on the part of the robot.

Social robotics is obviously a broad field,
encompassing both research projects that
aim to further the capabilities of robots functioning in human social spaces as well as research projects that aim to make use of social
robots to further our understanding of
human or animal cognition and social systems. Following are summaries of several
important areas of research endeavor.

## Multimodal Communication and Dialogue

Natural interaction between robots and humans requires all modalities of human–human
interaction to apply to robots as well. Basic perceptual competencies that are being developed
include gesture, facial expression, and speech
recognition; eye gaze interpretation; and visual
recognition of the location and pose of nearby
humans. At the cognitive level, robots must be
able to reason about human behavior and must
produce appropriate robotic responses. This research spans multimodal dialogue modeling
and control as well as the design of robotic ef-
fectory capabilities, including synthesized
voice modulation for affective dialogue, robotic facial expression generation, motion and
trajectory control, and manipulator control.

## Social Perception

In order for dialogue systems and other computational decision-making systems in a

social robot to function properly, the robot must recognize social context. Numerous research projects employ diverse robotic sensors, including thermal imagers, vision systems, laser scanners, and infrastructure-based environmental sensors to infer human social context. Laser scanners have been particularly effective in recognizing the trajectory of humans walking in an indoor environment. From this data, further reasoning enables extrapolation regarding the spatial goals of the humans. Ongoing projects aim to infer human goals through observation, responding with the most effective dialogue and motion control for a mobile robot to share close quarters in order to stand in line with humans, use an elevator, and share a narrow hallway. This work naturally draws upon theories of human behavior, because the robots must have sufficiently rich models of human activity to track the goals of each human and to draw rational conclusions regarding appropriate actions and communications to trigger.

## Social Learning

Although in some cases one can supply a social robot with a rich model of human social behavior for tracking and rational decision making, this approach has a significant limitation in that disagreement between the supplied model and reality can lead to dissonant robot behavior. Equally important, human responses to robots in their midst are not static but change over time; the responses vary from those of early adopters who embrace new technology to those of suspicious observers who would rather avoid the robotic encounter. An important open problem remains the design of computational metastructures for robots that are conducive to learning social behavior. The modest goal is termed *off-line learning*: A robot should, given massive data concerning social interactions, design the appropriate social regulatory control systems for future interactions. The ambitious goal is termed *on-line social learning*: A robot should, given continuous feedback regarding its own performance in social circumstances, make continuous adjustments to its own policies regulating human–robot

interaction in order to maximize the effectiveness of current and future interactions.

## Affective Modeling

There is significant debate over the role of emotion in robotic systems. At one extreme, some social roboticists argue that the path to effective interaction with humans is to model internal human mechanisms as accurately as possible, in effect producing a human replica and including affect. At the other extreme, researchers demonstrate relatively rich human–robot interactions using robot control systems that are deceptively simple, suggesting that rich social interaction may be an emergent phenomenon that does not require high-fidelity models of human sociocognitive mechanisms. The study of affective modeling offers a particularly fertile area for exploration of this spectrum. In terms of robot behavior, some researchers have designed affective internal models for social robots, so that their output is modulated not only by the robots' goals and by social context but also by internal drives, including frustration, happiness, and loneliness. System-level evaluations of deployed social robots have collected evidence suggesting that affective outputs may increase a robot's overall effectiveness by guiding humans to engage in the intended forms of interaction with the robot.

## Robot Morphology

Design plays a critical role in the creation of all robot systems, although that role is generally executed through the vision of mechanical engineers with an eye toward power efficiency, reliability, and mechanical dexterity. In social robots, the form, features, and aesthetics of the robot will have impact on the human's social perception of the robot. Members of the design, human factors, and human–computer interaction communities have increasingly adopted the social robotics design problem as a multivariable design challenge aiming to satisfy mechanical considerations and social considerations simultaneously. The largest investment in human-friendly form has emerged over the past thirty years in Japan. Recognizing the demographic trends in Japan, which portend a large elderly population in need of care, these companies have concentrated on domestic robot applications in which robots will provide assistance, companionship, and entertainment throughout the home environment. Both because homes have been designed to interface efficiently with the human form, and because of the hypothesis that humans may be more accepting of anthropomorphic robots, Japanese research has made significant strides in the creation of mass-producible anthropomorphic robots capable of bipedal locomotion.

The Japanese lead in the successful advancement in robotics, arguably a robotic revolution, is due to an important difference between Japanese social views of robots and those of the rest of the first world. Particularly with the advent of Astro Boy and other fictional characters, robots have been represented in Japan as empathetic, anthropomorphic creatures whose role is frequently to nurture and care for humans. Contrast this to the U.S. movie industry, for instance, which has frequently shown highly mechanized killing machines that are out to destroy humanity. Zoomorphic robot design has also achieved measurable success in Japan. For instance, one research project has designed a robotic seal pup, destined for nursing homes, with the intention of fostering increased communication between nursing home residents. The seal pup form is extremely successful because it is both friendly and unfamiliar. Its furry and soft exterior encourages petting and handling, while the public's unfamiliarity with seal pups ensures that users have no preexisting expectations regarding its behavior, thus mitigating the chance of disappointment.

Beyond Japan, two thrusts of research in robot morphology have also had significant traction. First, the study of facial expression and dialogue control has had impact both in demonstrating the power of the animated facial form and in presenting a foundation for research in sociocognitive development of caregiver–child relationships. Second, a number of projects aiming for productive short-term interactions between robots and untrained humans in public spaces have demonstrated that the right robot morphology can guide human first impressions appropriately. Notably, these deployed systems, seen both at the Swiss Expo and German Expo, are reminiscent in mechanical dexterity and expressiveness to the mechanical automata that founded the study of robotics several hundred years ago.

**BIBLIOGRAPHY**

Brooks, R., C. Breazeal, M. Marjanovic, B. Scassellati, and M. Williamson, "The Cog Project: Building a Humanoid Robot." In *Computation for Metaphors, Analogy, and Agents.* Lecture Notes in Artificial Intelligence 1562, edited by C. Nehaniv. New York: Springer, 1998: 52–87.

Fong, T., I. Nourbakhsh, and K. Dautenhahn. "A Survey of Socially Interactive Robots." *Robotics and Autonomous Systems* 42.3-4 (2003): 143–166.

Fujita, M., and H. Kitano. "Development of an autonomous quadruped robot for robot entertainment." *Autonomous Robots* 5.1 (March 1998): 7–18.

Paulson, E.   and J. Canny. "Social Tele-Embodiment: Understanding Presence." *Autonomous Robots* 11.1 (July 2001): 87–95.

Scassellati, B. "Investigating Models of Social Development Using a Humanoid Robot." In *Biorobotics,* edited by Barbara Webb and Thomas Consi. Cambridge, MA: MIT Press, 2000.

Siegwart, R., and I. Nourbakhsh. *Introduction to Autonomous Mobile Robots.* Cambridge, MA: MIT Press, 2004.

Steels, L. "AIBO's First Words: The Social Learning of Language and Meaning." In *Evolution of Communication,* 4(1), edited by H. Gouzoules. Amsterdam: John Benjamins, 2001.

*Illah R. Nourbakhsh*

# RUSSIA: PSYCHOLOGY OF SCIENCE.

In 1960, in Russia, an independent scientific direction—the psychology of science—was begun. This direction has been developing at the juncture of several sciences, including sociology, philosophy, psychology, the "science of science," and the history of science. The founder and initiator of this scientific discipline was Mikhail Grigoryevich Yaroshevsky (1915–2000), the author of textbooks on the history of psychology and social psychology of science. Professor Yaroshevsky created and, until his death, guided the department of psychology of scientific creative works at the Institute of the History of Science and Technology at the Russian Academy of Science.

Science can be considered in two ways: as a system of knowledge and as a system of activity that creates new knowledge. Understanding science as a system of activity, one can analyze it with the help of methodologies that are detailed and developed in social psychology.

The peculiarity of contemporary science presupposes the collective nature of scientific activity. Even when making a discovery or solving a problem individually, the scientist is dependent upon what was accumulated by other researchers. The scientist enters into discussions, arguing his or her own ideas with colleagues. Thus arises the phenomenon of scientific contacts: interpersonal relationships among scientists. This is why contemporary science is always a joint creative activity. For comprehensive analysis of this creative activity, it is very important to examine its social-psychological aspect.

Psychology of science is the study of the psychological peculiarity of people whose activity produces new knowledge; it is also the study of the psychological regularity of the development and functioning of scientific collectives.

The bases of a psychology of science in Russia arise from two methodological characteristics of the analysis of scientific activity. The first such characteristic is the study of any phenomenon of science that results from any of three groups of factors:

1. Objectively logical factors are those that deal with the internal logic of the scientific development—the development of its ideas as well as the analysis, elucidation, and discussion of a definite problem. They arise in connection with the requirements of practice.
2. Social-scientific factors involve the social environment in which science functions over a current cultural and historical pe-

riod. For example, from 1930 to 1940, the USSR prohibited research in some scientific disciplines, such as genetics, biological selection, and psychology. Many scientists' projects were repressed, and this ideological pressure was expressed through a significant deficiency in the development of some fields of Soviet science. In more recent times, we see the absence of a necessary level of government financing for fundamental sciences; this has led to the cessation of research in some scientific disciplines.
3. Personality-psychological factors include the personality and psychological peculiarities of a scientist, the traits of his or her character and temperament, the style and mode of thinking, and the system of values, standards, aims, and motivations of scientific activity.

The second sort of methodological approach that exists in Russia for the study of the psychology of science is the study of the scientific program and its scientific and social role. This notion of "program" designates the common scheme or project in which the scientist's activity of creating new knowledge plays a part. The program includes the general intention of the research, the scientist's social experience, the categories of the present domain of science, and psychological values and purposes.

Three types of scientific research programs, elaborated in research collectives, are distinguished:

1. Scientists are united in learning the same process but are interested in different objects. This division of scientific organization determines that the specialization of scientists is conducted according to the object in the place where they work and according to the methods of analysis.
2. Scientists study a common object but from different sides and levels. The specialization of scientists is determined according to the level at which the object is studied and according to the methods in use.
3. The members of a scientific group use a common methodological approach that can solve different problems and analyze various objects. For the realization of this

research program, the scientific collective differentiates among scientific functions; it selects scientific roles, such as the generator of ideas, the scholar, or the critic.

The personality of the individual scientist and the personality of the scientific collective as a whole are both the subjects of psychological analyses. Among the problems of the study of the individual scientist's psychology, the question of the effect of the intellect on creative ability stands out. This problem involves investigating the formation of the scientist's personality and the factors influencing the scientist's decision to enter the scientific profession—that is, the motivation behind scientific activity.

The study of the psychological peculiarities of scientific groups considers the problems of individual interactions and associations in science, the role of a scientific leader, the styles of leadership in a scientific group, the conflicts in research collectives, and the types of scientific groups. It is customary to distinguish formal and informal scientific groups. The formal groups, which include the department, section, laboratory, and institute, have a fixed social status, a legal address, a permanent staff, and an organized structure. The informal scientific group, which includes the scientific school, seminar, invisible college, and directions in science, does not have a formal status and is built on inter-individual scientific contacts, using the mechanism of cross-quoting. The subject, topic, and purpose of research unite the members of informal scientific groups. The leader of the informal group has a meaningful position. The scientist's perception of himself or herself as a member of this union and the identification with this scientific group are the basic psychological characteristics of informal groups in science.

Science in contemporary Russian society is in a crisis that was caused by society itself, outside of scientific factors. Psychology of science's focus has thus been transferred to a new domain: the interrelation between science and society. Members of the psychology of science profession in Russia discuss the following problems: the image of science and scientists in contemporary Russian society, connections among scientists, the psychological mechanisms that result in a brain drain, and the function of science in society.

We also analyze the representations of the contemporary Russian scientist that exist in both the public consciousness as a whole and the individual mind of the average Russian citizen. Four versions of the image are particularly relevant: the objective activity of the Russian scientist, the individual-personal characteristics of the scientist, the intellectual potential of the scientist, and the situation of the Russian scientist in society. The intellectual potential of the scientist and the contents of his or her professional activity are positively valued in contemporary Russian public consciousness, gaining admiration and respect for the scientist. However, the psychological peculiarities of the scientist and his or her social situation are negatively valued by the average Russian citizen, resulting in pity and scorn. The position of scientists in contemporary Russia can be characterized as "clever but poor."

*Elena Volodarskaya*

*See also* **Science in History: Eastern Europe and Russia; Technology in History: Eastern Europe and Russia**

# S

## SCIENCE AND TECHNOLOGY IN THE AGE OF UNCERTAINTY.

The idea that science must be assessed in the context of uncertainty is a very new one. It is not merely that quite suddenly the whole outlook for modern society has been discovered to be uncertain. The vision of the 1990s, an indefinite floating upward based on the dot-com revolution and unchallenged American power, is dead. Even before then, awareness had been growing that our totally science-based civilization was impinging on, or had already transgressed, its natural limits within the ecosphere. The easy assumption of a linear sequence in which research would discover, technology would apply, industry would produce, consumers would purchase, and happiness would increase is no longer plausible. In this age of uncertainty, we encounter a strange inversion of the classic distinction between hard, objective scientific facts and soft, subjective values because now we face decisions that are hard in every way, for which the scientific inputs are irremediably soft.

Readers of this article will find a broad perspective on the nature, functions, and future of science and technology in the emerging world order. Does the "nature" of science relate to the "world order"? This calls for a contextual analysis of science, of the sort that used to be called Marxist until it became obvious common sense. The perspective taken herein is to modify the specification slightly and, following the Marxist scholar Gerhard Harig, to refer to the "objects, methods, and functions" of science. The "methods" include both technical and social methods (thus, computer modeling on the one hand and peer review on the other), and the "func-tions" range from industry through to regulation, and include ideology.

Can the "nature" of science itself be dependent on the world order? Certainly, the objects of inquiry can be strongly influenced by fashions and commitments. We see this now in the debates over reductionist and holistic styles of work, dealing with simple elements or complex systems, respectively; the distinction is now most obvious in health care. Also, reality itself has boundaries that are culturally conditioned and negotiated. Consider acupuncture: Not long ago it was dismissed as an Oriental superstition, and now it is generally accepted as a complementary therapy. If we allow *nature* to stand for the synthesis of objects, methods, and functions, then we can say that the nature of science does indeed depend on the world order.

We must also consider the future of science. This can be done in two ways. One is to extrapolate what are seen as present trends and to assume that they will continue unchanged. The events of the end of the 1990s have shown the fallacy of that approach. The other approach is to attempt a longer perspective, to get a sense of the transformations that have occurred in the past, and to use these as clues to what might happen in the future. This approach does not yield predictions; in this age of uncertainty, predictions themselves might be out of fashion.

### Historical Phases and Their Styles

As early as the thirteenth century, Europe had already begun to expand against the Islamic societies, which were generally superior in technology, learning, and tolerance, when most of Spain suddenly fell to Christian forces. Even though the Ottoman threat was very real in eastern Europe through the seventeenth century, by the early fifteenth century Europeans were exploring the Atlantic and Africa. In this, the learned craft of shipbuilding and the state-supported mathematical sciences of cartography and navigation were crucial. Through the sixteenth century, "mathematical practitioners" emerged as independent professionals, serving rulers in warfare through the military arts (fortification, gunnery, organization, and supply of armies) and also in peacetime functions through architecture and civil engineering. The idea of pure science did not exist. If someone made a discovery in mathematics, he would secure credit through a challenge in which he announced the answer and invited others to obtain it.

Through the turbulent and troubled sixteenth century, all the sciences progressed steadily; the fifteen-year doubling time for the number of publications (which persisted for centuries after) was established in key fields such as mathematics textbooks. By the end of the century, fully matured disciplines had emerged, although the numbers of experts were still tiny by modern standards. Then, early in the seventeenth century, the study of nature was harnessed to philosophy. Profound thinkers such as Bacon and Descartes, and afterward Hobbes and Spinoza, found a source of certainty and progress in natural science, unlike in any other sort of study. Science and philosophy alike were involved in the "disenchantment" of common sense, in which magic and miracles steadily lost plausibility. Regular systems for intellectual property, the first patents for inventions, and the first journals for science were created. The atomic unit of knowledge reflected the atomized reality that it described.

In that period there also emerged distinctive national schools, with their characteristic styles and their characteristic cycles of growth and decline. The Italians were first, probably cut short by the atmosphere of fear after the condemnation of Galileo. The English came next; and Isaac Newton was both the greatest and the last of the luminaries that included Robert Boyle and Robert Hooke.

The eighteenth century was in some ways a pause, lacking great advances on any front. The main concerns during the Enlightenment were with "progress," creating the foundations of the Industrial Revolution, and "ideology," further extending the nonreligious approach into all areas of thought. In science, the French took the lead in the period before their revolution, promoting mathematics and the analytical approach, and enjoying unparalleled support from the state. French dominance came to an end with the downfall of Napoleon, and then the Germans took over. That was a victory of organization and zeal; through the nineteenth century, the Germans "professionalized" science across the whole range of fields. Well into the twentieth century, even after German science had been wrecked by war and dictatorship, students of chemistry needed to learn German in order to master the necessary literature.

The idea of pure science emerged in Germany during that time. It was always partly contradicted by practice, as scientists individually or collectively collaborated with the state and industry. But it was a powerful myth for university-based scientists, and it formed the basis of the *social contract* by which science was transforming the productive system while protecting its own autonomy and integrity. It fitted well with the conditions of twentieth-century America. Benefiting enormously from the flow of ethnic and political refugees from the Nazis, and enjoying a huge and continuing largesse from war work, American science and technology dominated the world for the latter half of the century.

The changes in size of the whole enterprise and of the scale of individual projects inevitably affected its character. Research became "industrialized" because it required huge capital investments for its projects and was ever more tightly bound to production and profit. And as the not-for-profit sector made increasing demands on the public purse, science became incorporated into the state apparatus, yet another claimant on public spending. Through the latter part of the twentieth century, elderly scientists mourned the passing of "little science," which was remembered variously as an idealistic endeavor or just good clean fun.

By the turn of the century, the social system of science was coming under increasing strain. The incursion of business practices and values into a previously not-for-profit enterprise caused confusion and sometimes corruption. The principles of openness and mutual trust, on which quality assurance and morale had depended, were eroded by the inevitable pressures for commercial secrecy on the one hand and the hyping of discoveries on the other. Leading academic scientists have voiced real outrage at the practices of those who move into public-sector research areas just when they are ripe for exploitation and thereby expropriate their achievements for private profit. The widespread patenting of results and techniques creates hazards for ordinary researchers, who need lawyers to trawl for restrictive patents before they can dare to start their research. The ethical problems of the new biotechnology, extending through molecular genetics and reproduction engineering, become ever more troubling.

Under these circumstances, it is not surprising that public disquiet about science, starting with the atomic bomb and increasing through environmental awareness, should have become aggravated. Evidence of the commercialization of knowledge, and the perils of runaway technology, steadily erodes public support. There are problems with recruitment at undergraduate and postgraduate levels in many countries. Worse, the use of leading scientific figures to provide patently false official reassurances has extended the public's cynicism from politicians and journalists to scientific spokespersons. There are now quite considerable attempts by governments to repair the damage through greater participation in decision making on science-related issues.

## The Future

The preceding survey has shown how great is the variety in the societal functions of science, and correspondingly in its "nature," in the various milieux where it has flourished. Underneath the steady growth in size and power, we have had the deep changes in the social enterprise of science, and also the successive rise and decline of various national centers. This enriched perspective on the past should enable us to be more bold and imaginative in confronting the future.

In some ways, the recent evolution of science represents a maturing. With the atomic bomb, said J. Robert Oppenheimer, science tasted evil. In retrospect, the heritage of the seventeenth century was of a disenchanted, sanitary magic, whereby we could achieve great good while at no risk of being bad. This illusion could not persist indefinitely; our task now is to grow beyond it.

Even the continued dominance of European science is not so certain. Through all the story of previous success, science outside Europe has been peripheral to the point of scarcely requiring mention. But this dominance may not persist. Just as European science is showing signs of senescence, China is early on a cycle of strong growth in scientific technology. Should they survive their current social and ecological perils, this ancient and proud civilization will be a formidable rival indeed.

Not long ago, the term *the new world order* would refer to a universal Pax Americana. But now we have experienced both the protest at the World Trade Organization's 1999 Seattle round and the terrorist attacks of 9/11, representing serious and permanent opposition, variously nonviolent and violent in principle. Military science can be used for creating terror in winning wars, but it cannot make peace. *Uncertainty* as a theme for the future is a reminder of the radical shift from the assumptions that have animated European science for nearly a half-millennium. Science now has a new societal function: coping with this uncertainty. What new methods and objects are appropriate for this? What does this uncertainty now mean?

It is not merely the circumstances of science that have become uncertain. A new popular concept describes "unintended consequences," which are inevitable whenever new technologies are introduced. They had always been there, of course, but they had been assumed to be either benign or negligible. Now they are serious, but, worse yet, they cannot even be predicted. No longer can the societal function of science be seen simply as speaking truth to power. We live in an age of policy-critical ignorance, where decisions about our future are thrust upon us, and our knowledge is swamped by our ignorance. We can express our new situation by comparing the iconic figures of the last century and the present. Albert Einstein was the man who created the twentieth century, for better and for worse. For ours, we can suggest Edward A. Murphy, Jr., the hero of Murphy's Law: that what can go wrong, will.

The primary societal functions of science in this next period can be characterized as cleanup and survival; reversing and then redressing the consequences of our well-intentioned but heedless rape of the planet. The appropriate methods would be those of post-normal science, where facts are uncertain, values in dispute, stakes high, and decisions urgent. The blinkered puzzle-solving of Thomas Kuhn's "normal science" can no longer be dominant, nor can we expect normality and predictability in the societal matrix of science. Scientists must become aware, primarily of the uncertainties in their work, but also of the values that direct and shape it. Otherwise they will be little better than hired hands, moving from job to job and awaiting redundancy as bioinformatics technology increasingly does their work for them. The categories of scientist and citizen cannot remain distinct and exclusive.

This optimistic vision is one of a science that finally comes to belong to a democratic polity. Its achievement will be necessary for a sustainable future. It is the only way for science to survive the present age of uncertainty and to evolve further in the service of humanity.

## BIBLIOGRAPHY

Bowler, Peter J., and Iwan R. Morus. *Making Modern Science: A Historical Survey.* Chicago: University of Chicago, 2004.

Funtowicz, Silvio, and Jerome Ravetz. "Science for the Post-Normal Age." *Futures* 25 (1993): 735–755.

Funtowicz, Silvio, and Jerome Ravetz. "Post-Normal Science: An Insight Now Maturing." *Futures* 31 (1999): 641–646.

Nowotny, Helga, P. Scott, and M. Gibbons. *Rethinking Science: Knowledge and the Public in an Age of Uncertainty.* Cambridge, UK: Polity Press, 2001.

Ravetz, Jerome R. *Scientific Knowledge and Its Social Problems.* New Brunswick, NJ: Transaction Publishers, 1996.

Restivo, Sal P., and Christopher K. Vanderpool, eds. *Comparative Studies in Science and Society.* Columbus, OH: Merrill, 1974.

*Jerry Ravetz*

# SCIENCE IN HISTORY

This entry includes nine subentries

> Global Perspective
> Africa
> China
> Eastern Europe and Russia
> India
> Japan
> Latin America
> Middle East
> United States and Western Europe

The account of science in history we are most familiar with places mathematics and measurable experimentation at the heart of scientific methodology: True science is equated with quantification. Some of the reasons for this equation are obvious; some of the less obvious reasons will reveal why we conceive of science in history as we do and how it might be conceived otherwise.

The most obvious reason we equate science with mathematics is that the word *science,* as we now understand it, was coined by a mathematician (William Whewell, 1794–1866). As the Nobel Prize–winning physicist Eugene Wigner (1902–1995) famously suggested several generations after Whewell, mathematics, and the institutions of modern science that are built upon it, are too "unreasonably effective" not to be in some fundamental way true. That is, if our capacity to quantify and to apply other forms of mathematical computation were not so well suited to reality, we would not expect such good results as we have obtained in comprehending, predicting, and manipulating reality. Since Whewell and Wigner there have been many similar expressions of this idea put forth by scientists, notably scientists whose work, like Wigner's, depends upon mathematics. The conviction is cogently expressed, for example, by contemporary cosmologist John Barrow. The titles of some of his books, *Pi in the Sky* and *Constants of Nature,* assert the essential idea: pi is indeed "in the sky"; there exist "constants of nature," and the prize of power and control goes to those who comprehend them.

It is important to notice the context in which this limited conception of science, with its emphasis on a potent, mathematical methodology and a consequent practical utility, was taking shape. If European and American scientists, from Descartes in the seventeenth century to Whewell in the nineteenth but especially from Whewell to Wigner, needed an external, albeit superficial confirmation of the worth and veracity of their science, they had to look no further than the dramatic subjugation of non-European peoples going on all about them in Africa, the New World, India, and Asia, most conspicuously during the nineteenth century—the period, not coincidentally, when science was achieving its distinction as *the* shibboleth of secular authority.

The actual causal relation between the expansion of modern science and our consequent ideas about science in history, on the one hand, and Europe's technological and imperial expansion, on the other, is, to say the least, a complex and inadequately understood subject. Yet, the very absence of a rigorous, empirically verifiable account of these historical relations has redounded to the benefit of those who would perpetuate an account of science in history that priviledges European achievements. In the absence of any clear understanding of the relation between science and the vast array of technological innovations, many of which were helping subdue and constrain the subjects of European imperialism—steam engines, medicinal cures for tropical diseases, new transportation, and military technologies—one might nevertheless assert that there had to exist some efficacious correlation. Though many of the important technological innova-

tions of the nineteenth century were made by craftsmen and not by what we would now call scientists, in the minds of Europeans and, significantly, non-Europeans alike the imagined association of science and technology was becoming sufficient proof of European science as the one true science. Technology and the sciences vaguely associated with it were becoming the new "measure of man," and the measure, too, of entire civilizations.

In the context of ongoing industrial revolutions and self-confident, unreflective global imperialism, science, by the proof of association, could be read as evidence of the superiority of the West or, in Latin terminology, the Occident (literally, the quarter where the sun sets). The implications for the "East," or Orient (literally, the quarter where the sun rises) were not difficult to project. The Orient came to be associated with all the Occidentalists imagined they were not; all the attributes they did not assoicate with their self-styled technoscientific culture, they gave to the Orient. Thus, "Orientalism," with its implication of the nonprogressive, traditionalist, nonrational, nonscientific cultures of the East, became a habit of thought among those contemplating science in history, a habit of thought suitably tailored to Occidentals' positive image of themselves. From this perspective, one can also see how conceptions of "science in history" and the "history of science" have reinforced each other. Temporally, the history of science has been conceived in the Occident (Europe, North America) as a tale of progress, a tale that achieved its critical turning point with the efforts of Europeans beginning about 400 years ago. Spatially, in a mirror image of the history of science, science in history has been conceived of as an enterprise that, upon its inception at the European "core," was gradually propagated to the cultural "peripheries."

Combining these inherited spatial and temporal notions, we obtain an account of science in history that may be simply characterized as "Western"—that is, it is an account from a western European or North American viewpoint. Accordingly, and with few exceptions, scholars have cast Middle Eastern science in the role of printers or scribes, conserving through translation the important discoveries of Greco-Roman antiquity. The allegedly unsystematic and practically unrealized ingenuity of the Chinese is allowed a more respectable cameo appearance. Indian civilization is given only a very slight part, limited usually to a few lines concerning their contribution to mathematics. Africans, as with other vast swaths of human culture, make no appearance, except by their conspicuous absence, in our "Western science."

To think more clearly about science in history, then, we must demystify the terms *science* and *history*.

"Science" suggests for many the idea of a body of incontestable fact arrived at by a consistent and reliable method; this idea is unfortunate and almost entirely wrong. A more accurate characterization of science is articulated by Saul-Paul Sirag: "The essential point in science is not a complicated mathematical formalism or a ritualized experimentation. Rather the heart of science is a kind of shrewd honesty that springs from really wanting to know what the hell is going on!" More prosaically still, science has been described as a "no holds barred" struggle for the truth. The point in both of these characterizations is that science cannot be reduced to a single, certain methodology, even less so to a static body of fact.

"History," not unlike popular but erroneous notions about science, connotes a body of fact that mirrors, more or less faithfully, the events of the past. However, history literally means "inquiry," and should in no way be read as a synonym for the past. The past refers to the totality of what was; history refers to our chronically incomplete, highly contestable interpretations of evidence about the past.

Our minds may balk at the suggestions that we have been misled by our education and that history, instead of referring to a definitive account of how the past progresses to what we are now, refers instead to an unremitting contest of interpretation. Accordingly, we embrace a caricature of history, and a caricature of science appropriate to it. Analogous to and apropos of a conception of history as a series of factual events leading toward us, we imagine science as something methodologically reliable and progressive. "Science" and "history" reinforce one another until we arrive at a conception of science in history as the application of a realiable methodology for practical and progressive ends. To conceive of "science in history" we must first begin to understand what "science"—literally, "knowing"—might mean for cultures outside the rigid framework of our "history." Clearly, many other cultures—present or past—have "science," that is, methods for obtaining knowledge. What are these methods? What is their knowledge used for? Is it "used" at all, in the sense we understand utility? In what temporal and spatial frameworks have non-Western cultures situated their ability to know?

## BIBLIOGRAPHY

Adas, Michael. *Machines as the Measure of Men.* Ithaca, NY: Cornell University Press, 1989,

Krebs, Robert E., and Carolyn A. Krebs. *Groundbreaking Scientific Experiments, Inventions, and Discoveries of the Ancient World.* Westport, CN: Greenwood Press, 2003.

Reingold, Nathan, and Marc Rothenberg. *Scientific Colonialism.* Washington, DC: Smithsonian Institution Press, 1986.

Teresi, Dick. *Lost Discoveries: The Ancient Roots of Modern Science, from the Babylonians to the Maya.* New York: Simon and Schuster, 2002.

*Peter John*

## Global Perspective

The *New Atlantis* (1627), Francis Bacon's answer to the utopia of his predecessor Thomas More, describes an institution located on an island in the Pacific Ocean for learning about the natural world. The inmates of Bacon's institution, the House of Solomon, tabulate and summarize knowledge from all parts of the globe, perform experiments to deepen understanding, and formulate general propositions and laws. Bacon's description became a vade mecum for the early Royal

Society of London, and his persistent imagery involving light helped bring apotheosis during the Enlightenment. For Bacon's successors, science was cumulative, interdiscursive, interventionist, and synthetic.

Bacon's successors imagined enlightened knowledge as a beacon for the world. Science was a new path toward inventions of practical use, to which savants and experimenters dedicated themselves as they delineated the qualities of the world. From their generalizations about nature they instigated improvements in the quality of life—for example, determining the time of day or protecting buildings from lightning strikes. Late in the eighteenth century, savants enjoyed mutually profitable relations with artisans; one site for this interchange was the First Industrial Revolution of steam, coal, and iron.

Possessing novelties produced by the activity of European scientists became a sign of edification and virtue throughout Christian Europe, whether maps of the world, telescopes and microscopes, thermometers and barometers, air pumps, or electrical machines; the novelties and institutions to interpret them spread with European expansion. Telescopes and heliocentrism found a place at the seventeenth-century court in Beijing through Jesuit ambassadors; European books on mathematics and surgery, along with European clocks and firearms, circulated in Japan from the port of Nagasaki. Houses of Solomon multiplied. Dozens of universities and colleges came to Spanish and English America, where "philosophical" apparatus was in evidence. One of the largest astronomical observatories in the Enlightened world was located at Batavia, on Dutch Java; one of the finest of the Enlightened world's libraries, the Palafoxiana, was at Puebla de los Angeles in Mexico. Scientific societies the equal of their European sisters graced Batavia, Philadelphia, Calcutta, Cap François on Haiti, and Cairo, Egypt. The Baconian imperative extended to studying knowledge that had been assembled before the arrival of Europeans. Chinese, Islamic, and Hindu astronomy received systematic attention from dedicated scholars, who also focused on local manufactures and medicines. Botanical gardens for plant experimentation and acculturation arose from St. Helena to Java to Philadelphia.

The Baconian discourse, inserted into industrial capitalism, became a justification for conquering the world. Imperialists spoke with one voice about the unique qualities of science, whose true method had been discovered in Europe. The effectiveness of their guide to truth seeking authorized them to raise the world's peoples up into smaller, darker copies of Europeans. A Jesuit in late nineteenth-century Ecuador spoke about the "divine gift" of science. A French physics professor in Indochina in the 1920s offered lessons from the history of science, notably "ideas of beauty and harmony that raise mankind well above immediate, practical interests and contribute to moral progress." A Dutch physicist on Java emphasized at this time that experimentation was "a fruit of spiritual culture," and for this reason a physics laboratory was a "cultural force of great import." He also noted that studying classics of the West, where science originated, was "an especially excellent way" for Indonesians "to be raised up to higher civilization" (Pyenson 1993, pp. 330–332). We may ask whether this sort of rhetoric is grounded in the history of science.

## The Truth-Seeking Enterprise

Truth seeking, the Socratic enterprise condemned by late-twentieth-century postmodernists, was certainly present in the Classical world, complete with theories, predictions, mathematical proofs, and experiments; nor were the collection, summary, and generalization of data foreign to Greek and Latin authors. Inventions were prized, for example, in civil and naval architecture, mechanics, or alchemy. These interests were maintained and elaborated over more than half a millennium of Islamic civilization, bringing algebra and astronomy to a form that could be used profitably by late-Renaissance scholars in Europe. The arts from Han to T'ang China include discoveries in printing and publishing and chemistry; by the Ming period, there are substantial innovations in mechanics. The signal discoveries Francis Bacon identified with Western Europe—printing, the mechanical clock, and firearms—came to Europe from Eastern civilizations. There is no reason to doubt that civilizations with nongraphic forms of record keeping, such as knotted strings or oral traditions, also made and transmitted innovations of many kinds.

Especially in the period 1850–1950, the justification for European superiority in the study of nature was often formulated in terms of rules for discovering new notions; the search for a logic of scientific discovery, indeed, became a central concern for academic philosophers, and it led a new branch of philosophy, epistemology, which dominated traditional inquiries in morality, aesthetics, and theology. During the second half of the twentieth century, epistemology came to conclude that creative activity depended as much upon whimsy as upon traditional logic—*fortuna* (in the form of working conditions) as much as *virtù* (in the sense of individual will) lay at the root of new scientific ideas. The observation invites us to consider the material conditions for prosecuting science in Europe and elsewhere.

In the two thousand years before Francis Bacon, there are clear records of scientific centers with a pedigree longer than any in, for example, North America. Professorships existed at many centers in the greater Mediterranean Basin under Greek and Roman rule—a tradition extending (with interruptions) into medieval Constantinople. There were smaller copies of the Lyceum and Academy of Athens, as well as copies of the Alexandrian Museum—possibly the greatest institution of higher learning. The museums, equipped with library and dining commons, were collections of individuals who could take on private pupils; the fellows or professors radiated the glory of a regime or city. In a different vein at roughly the same period of time, institutions in China combined instruction with codification of traditional learning, the notion being that the corpus of knowledge required for success in civil-service examinations and for the prestige of a ruler would be subject to continual revision. A vital part of the revision, undertaken by each

incoming regime, concerned calendar reform—hence the periodical interest among Chinese savants in hosting foreign astronomers. Academies such as the so-called Forest of Pencils could enjoy a run of a thousand years, but their purpose, even when associated with an observatory, was not to discover new knowledge. The institutions of medieval Islam are closer to Bacon's House of Solomon—the Hall of Wisdom in ninth-century Baghdad was just such a site of collection, translation, and innovation. Islam also formulated a mechanism for maintaining privately guaranteed institutions of higher learning—the charitable endowment, or testamentary trust. This device encouraged such new institutions as the hospital, the observatory, and the religious seminary to cultivate wealthy patrons. There does seem to have been an association between teaching and research at these institutions, although the relationship was not sanctioned by statute.

## Research Institutionalized

Europe improved on the Mediterranean tradition of public professors, schools of religious law, and medical lectures in hospitals by inventing the university—a complex corporation with guild privileges, notably the right to own property and to award guild certificates of competence. The sheepskin from one university was recognized in all Latin quarters—universities recognized fraternity just as bakers and drapers did, a circumstance not unrelated to a certain uniformity of standards in curricula. Students circulated along with professors, stimulating academic and municipal competition (the principal beneficiaries of increasing enrollment have always been town merchants and landlords); even if everyone read Aristotle and Galen, there was a premium on interesting lecturers and, one may imagine, free-wheeling discussions in the tavern.

Whether early-modern universities resisted new ideas in the manner of Humanist caricatures is a focus for research among historians of science. It seems clear, however, that there was no formal mandate for innovation. The mandate came following Bacon's prescription, where traveling merchants and artisans worked with university lecturers to generate new learning. Required were new and elite institutions, the scientific academies, which in succeeding centuries gave rise to more democratic and more specialized societies, a number of which had teaching as part of their charter. Between 1760 and 1830, largely in western Europe, universities and a host of educational institutions with the privilege of granting a diploma (*grandes écoles*, military academies, medical schools) came to absorb the research function of the academies and societies.

The union of research and teaching sustained the vigor of European science. Most students have always been interested in securing the means for a comfortable living, toward which end truth seeking is an inconvenience. If virtue were signaled by research ability, however, routine teaching could with a clear conscience be assigned to professors with no special talent in lecturing, and their material advancement would depend less on expository skill than on the published record. To state the matter in this way is to exaggerate the strength of the union between teaching and research. In most institutions then as now, pedagogy—that is, being nice to cash customers—was the watchword. But a stratification of institution emerged, and by the early decades of the twentieth century many of the research universities we recognize today were clearly visible. A sign of the prestige accorded to research is found in the transformation of the *Ecole normale supérieure*— a school in Paris for establishing standards of pedagogy—into a research institution under the direction of Louis Pasteur. Great educators in America—Andrew Dickson White, Charles William Eliot, David Starr Jordan—became infected with the research ethic.

Between 1800 and 2000, the European model of research-and-teaching institution spread around the world. Perhaps a better description would be that professors were everywhere recruited on the basis of research credentials, notably a research doctorate, even if in the course of their career they contributed hardly at all to the sum of human knowledge. Why should this criterion continue to hold, especially over the past fifty years, when most research is done beyond academic walls (in military enclosures, plant-breeding or animal-breeding stations, drug laboratories, mountaintop observatories, or particle accelerators)? The reasons for maintaining the improbable couple research-and-teaching in North America and Europe may be seen more clearly by examining how the model became established in areas into which Europe expanded.

## Science and the Civilizing Mission

Short of mass extermination, practiced with alarming regularity from Acadia to Argentina to the North American plains to Armenia (to mention only the beginning of the alphabet), imperial domination requires complicity between master and servant. Facilitators—pandits, *Gauleiter*, or a bewildering variety of *fonctionnaires*—can occupy the interface between the two groups, which may be called residents (the transplanted European invaders) and regents (the indigenous authorities). From the side of the European residents, it is useful to have a sense of rectitude. Because the day's work is easier for the imperialist resident with a righteous frame of mind, the autochthonous regent also tends to absorb the sensibilities behind the resident's sense of duty.

During the modern period, religious doctrine does not secure rectitude. The gods battle among themselves now and for all time, as sociologist Max Weber observed, and there is no general way to choose among the teachings of Lao-Tze, Ali Mohammed of Shiraz, and Ignatius de Loyola. Rectitude comes from nature. What emerged around 1800 was a transcendent set of beliefs tied to the workings of the natural world—that is, a belief system capable of embracing any number of religious doctrines. The principal advantage possessed by such a belief system is that it legitimizes the authority of the imperialist, who controls the means of transforming nature, without explicitly demanding an allegiance to the imperialist's god or the imperialist's sovereign. All that is required is the Baconian notion that civilization is characterized by domination over nature. Catholic and Presbyterian, Muslim and Jew, may in this way, and without sacrificing articles of faith, work for representatives of the King of England, or whichever agents are most adept at curing disease and building railways.

The proposition depends only on the proposition that science in the European sense is capable of doing what it claims to be able to do: prevent smallpox or puerperal fever, send messages by electrical cable, or manufacture artificial colors. The proposition provides a natural explanation for the civilizing mission. First, it explains the perplexing connection between religious missionaries *in partibus infidelium* and the prosecution of science. The heavens and the earth have been chronicled by Jesuit missionaries from Matteo Ricci in seventeenth-century Beijing to Pierre Lejay and Ernesto Gherzi in twentieth-century Shanghai; to say that astronomy and geophysics are impractical activities as awe-inspiring as a mass or a sermon is unhelpful, for one could say that same thing about a chamber-music concert or a football game (activities prominent at some American religious universities but not in missionary schools). Second, as President John Quincy Adams elaborated in his address to the U.S. Congress in 1825, observatories (and by extension laboratories) become symbols of civilization—in his words, "lighthouses of the sky." Third, facility in manipulating the laws of nature is a sign of moral superiority, an outward manifestation of being civilized. Practicality is less important than generality: the laws of nature are the justification, the steam engine the evidence. From this perspective there is a very large distinction between calculating longitude from the stars and memorizing a national myth. It has been contended that students in the French colonial ambit received lessons about "our ancestors the Gauls." These lessons were absorbed with as much conviction and taught with as much ardor as the lesson presented to today's schoolchildren about the founding fathers of the United States or of Russia.

The proposition also explains a *modus vivendi* among modern imperialist powers as they divided up China, Southeast Asia, Africa, and the Islamic world. Each one had to make a case for their common civilizing mission. The administration of a particular parcel of real estate—how much civilization was imparted—would warn off imperialist rivals. This system breaks apart with the Asian empire of Japan, facilitated by a mastery of European science and technology. But Japan is not an exception to the general proposition about science and the civilizing mission: Japan founded universities in Korea and on Taiwan, and it undertook archaeological and astronomical research on Java.

This way of looking at the civilizing mission applies equally well beyond and within Europe. The dramatic extension of universities, trade schools, observatories, and museums in nineteenth-century Europe reinforced the view that education meant knowing nature's laws. It is hard to imagine today that one hundred years ago quite the most exciting place for a young man or woman was a university physics or chemistry laboratory. The laws of the heavens, expressed in Western mathematics, were held to exemplify virtues useful in the home and family—whether on Java or in Boston.

## A Look Ahead

All the sites for the prosecution of truth past seem to constitute a class, or set of objects—just the sort of thing that appeals to law-seeking humanists. The sites are remarkably widespread, and the architecture wonderfully varied. This observation leads us to imagine that truth seeking is a different kind of enterprise from, say, the expressing of sensibility (in the form of art exhibits or symphony concerts), although the universe of laboratories is perhaps not so different from the universe of houses of worship. A postmodernist might say, "but then science is a system of belief." To which comes the answer, "Yes, of course, but it is a catholic or universalist system that in principles and application is stubbornly insensitive to local conditions." There is an interdiscursivity among scientists, notwithstanding their various languages, nationalities, and living circumstances. The kind of astronomy one does seems not to depend on whether one eats croissants or pickles for breakfast, whether one is married or celibate, or whether one professes the Nicean Creed or the Eightfold Way. The observation is a powerful refutation of the postmodernist contention that all knowledge is simply local knowledge.

Notwithstanding our enterprise in seeking truth, truth may be lying in wait for us, lurking around the corner to knock us on the head. Truth as a cudgel wielded by true believers lies at the heart of the mission to civilize barbarians—the general task of universities over the past several centuries. This is the great enterprise of Eleazar Wheelock at Lebanon, Connecticut; of Mark Hopkins facing a new student sitting on a log at Williamstown, Massachusetts; and of William Tecumseh Sherman bringing Protestant values to Alexandria, Louisiana: Dartmouth College, Williams College, and Louisiana State University. The task has become complex, George Steiner noted, because we no longer recognize a barbarian when we see one.

Henry Rosovsky observed in his memoirs of Harvard University that students are transient parasites and faculty are chronic tubercles in the life of the university. In reality, however, the institution is ephemeral—its enduring part is the enterprise of its people. Great schools of learning are no more: the Academy and Lyceum, the Alexandrian Museum, the monastery at Nalanda, the medical school at Salerno, the University of Harderwijk, Gresham's College, and Black Mountain College. But we recall Plato and Aristotle, Ptolemy, I-Hsing, Oldenburg, Linnaeus, and Buckminster Fuller—truth seekers who labored at the institutions.

When the American man of letters, Ralph Waldo Emerson, melancholy and disabused of religious doctrine, visited England to see a handful of the greatest British intellectuals, he traveled beyond the metropolises and universities to a Scots farm so remote that, in the words of the woman of the house, "one could hear the sheep graze." This was the home of historian Thomas Carlyle—the writer with the intellectual stature of Alfred Tennyson, Charles Dickens, and Elizabeth Barrett Browning placed one on top the other. It is useful to recall, in this context, that an institution, like a nation, is blameless; it has no conscience or soul; it is only as good as the people who inhabit it.

Thomas Carlyle, who would have become the greatest Scots astronomer had the University of Edinburgh the good sense to hire him, knew that the primary measure of civilization has become facility with manipulating nature's laws, notwithstanding an inflexible, discriminatory, and indifferent extension of sites for teaching about these laws. If the story ended here, scientists would rush to worship at the Positivist's Church of Humanity in London, where historian of science George Sarton lectured on New Year's Eve 1920. But truth seeking European-style served horrific ends. Economic exploitation and political oppression have been justified by the people who pay the tribe of truth seekers. This apprehension of reality is not grounds for throwing the baby out with the bath water, as some scholars are inclined to do, for the truth seekers have often struggled to keep themselves at arm's length from their paymasters: Max Planck under the Nazis is a sobering example.

Does the future hold hope or despair? Hope certainly, following the most astonishing fact of our lifetime: the absence of nuclear war, notwithstanding the widespread availability of nuclear weapons. Furthermore, at many of the sites for prosecuting science around the world have arisen vigorous communities of scientists who regularly make subtle and useful discoveries. From the point of view of abstract truth seekers in the nineteenth-century mold, however, there is a troubling corollary. Tremendous confusion has arisen between science and technology—so much so that mastery of nature's laws is once more equated with mastery over nature, in the sense of Francis Bacon. It is too early to say whether this recovery of an early sense of Western science is more than the necromantic inspiration behind revivalist movements, religious and secular.

*See also* **Technology in History: Global Perspective**

**BIBLIOGRAPHY**
Cosandey, David. *Le Secret de l'Occident: Du miracle passé au marasme présent*. Paris: Arléa, 1997.
Fernández-Armesto, Felipe. *Civilizations: Culture, Ambition, and the Transformation of Nature*. New York and London: The Free Press, 2001.

Floris Cohen, H. *The Scientific Revolution: A Historiographical Inquiry.* Chicago: University of Chicago Press, 1994.

Heilbron, John L. *The Dilemmas of an Upright Man: Max Planck as a Spokesman for German Science.* Berkeley: University of California Press, 1986.

Huff, Toby. *The Rise of Early Modern Science: Islam, China, and the West.* Cambridge, UK: Cambridge University Press, 1993.

*Itinerario.* Periodical of the Center for the History of European Expansion of the Rijksuniversiteit Leiden, Netherlands, the premier forum for studies about European colonialism and imperialism; published quarterly.

*Journal of the Netherlands Institute in Japan,* annual volume 3, 1991. This volume provides a good summary of opinion about science across cultures over the past millennium.

Lafuente, Antonio, Alberto Elena, and Marĺa Luisa Ortega, eds. *Mundialización de la ciencia y cultura nacional.* Madrid: Ediciones Doce Calles, 1993. Many of the contributions are in English.

Nakayama, Shigeru. *Academic and Scientific Traditions in China, Japan, and the West* (trans. Jerry Dusenbury). Tokyo: University of Tokyo Press, 1984.

Petitjean, Patrick, Catherine Jami, and Anne Marie Moulin, eds. *Science and Empires.* Dordrecht: Kluwer, 1992.

Pyenson, Lewis. "The Ideology of Western Rationality: History of Science and the European Civilizing Mission," *Science and Education* 2 (1993): 329–343.

Pyenson, Lewis, and Susan Sheets-Pyenson. *Servants of Nature: A History of Scientific Institutions, Enterprises, and Sensibilities.* New York: W. W. Norton, 1999.

Rosovsky, Henry. *The University: An Owner's Manual.* New York: W. W. Norton, 1990.

Steiner, George. *In Bluebeard's Castle: Some Notes Towards the Redefinition of Culture.* New Haven: Yale University Press, 1971.

*Lewis Pyenson*

# Africa

Africa is composed of 47 to 53 nations (depending on whether island nations are included in the count), with 14 percent of the world's population on 20 percent of the earth's surface. It is generally accepted that Africa is the motherland and home to the first humans. Our understanding of science and technology originates with humans living in communities, communicating, and altering their environment. Thus, science and technology have been present in African communities for thousands of years. Among the earliest examples of "science" is the manufacture of iron, since the first millennium B.C., at a high purity using forges made from clay. If we include knowledge of plants, water, wood, and other resources in our definition of science, it is clear that science has been alive and well in Africa as long as people have been living there. (The amount of information available in the West on science and technology in Africa is sadly very small compared to knowledge we have about science in Europe or science in China. This can be explained at least in part by what counts as science, which in turn depends on who is determining what is science, as well as perceptions of African history, culture, and knowledge on the part of European cultures.)

As early as the sixteenth century, travelers noted the types of tools, transportation, and agriculture used in Africa, However, these technologies were not respected as such by Europeans, who maintained that Africans lacked the knowledge, and intelligence, to utilize their natural resources and, therefore, that Europeans should take advantage of the African resources before the Africans did so themselves. The presence of the written word was taken as a measuring device to assess the existence of science and technology outside of Europe. The literate world pronounced the oral world not merely illiterate but also deficient in knowledge of science and technology, even though there was an indigenous science and technology that did not meet the Europeans' criteria.

If we define science as a way of understanding, rationalizing, and explaining the world and creating knowledge about the world, it is clear that science exists and existed in Africa. If we think of modern science as a social institution that originated with the Enlightenment and gained momentum and authority with the scientific revolution in Europe, then the questions of science's existence in Africa takes on a different meaning. For the purposes of this entry it is important to remember the distinction between science as a way of thinking, rationalizing, and explaining the world to produce knowledge and science as a social institution. It is on the latter that this essay will focus.

## Pre-Science, Colonialism, and Modern-Science

In the modern world, when science and Africa are placed in the same sentence, typical associations by industrialized peoples are that science in Africa is weak, nonexistent, prescientific, primitive, or lacking. In the early 1900s, Africa's science and technology was judged inferior to Western science and technology by its colonizers. This judgment reflected a political and economic agenda, because it is easier to justify colonizing a people if they are perceived to be inferior and in need of the colonizer's superior knowledge. Thus in the late 1960s when historian Robin Horton addresses African traditional thought and compared it to Western science, he was doing progressive theorizing. His goal was to demonstrate that African traditional thought precedes Western science in the evolution of theories. Horton (p. 68) recognized "the theoretical models of traditional African thought [as] the products of developmental processes comparable to those affecting the models of the sciences." He declares that, since African traditional thought shares so many theoretical characteristics with Western science, then it must be a *precursor* to Western science. He labeled it "prescientific," unable to interpret the similarities as proof that African traditional thought is a "science" or worldview in its own right. Horton's worldview or cultural lenses prevented him from seeing African thought as anything but prescientific. This example highlights how a term such as *science* carries a particular meaning for particular people that reflects certain historical and cultural values.

If African traditional thought is really prescientific, why has African modern thought not created its own "Western" science? David Bloor's answer would be that "ideas of knowledge are based on social images, that logical necessity is a species of moral obligation, and that objectivity is a social phenomenon" (Bloor, p. 157). Thus, Western science is a construct of Western culture and not a construct of African culture. There is no reason to expect that African culture will or should produce Western science, because its social institutions, power relations, authority, and knowledge structures are not Western. Not only is the idea of science culturally constructed, but so is the idea of Africa. Africa is often forgotten or overlooked by industrialized nations because it is perceived as separate, exotic, primitive, or unknown from industrialized ways of life. Recent scholarship on colonialism and postcolonialism highlights and captures the way Africa has been perceived, treated, and used by Europe/Western Nations (see, for example, the work of Susantha Goonatilake, V. Y. Mudimbe, Linda T. Smith, and Anne McClintock).

Colonialism is a good place to start our discussion of science and Africa, because when the European missionaries and explorers landed on the coast of Africa, they judged the local culture, people, technology and science to be "primitive" and inferior compared to their own. By the early 1900s, Europeans' judgments about science in Africa involved political and economic interests as well as ignorance and ethnocentricism. Western science has been one of the categories used to differentiate Europe and North America from the rest of the globe since the Enlightenment and to justify the former's economic global position. Being judged not to have science marked the latter as "savages" and was part of the institutionalization of colonialism.

Colonialism/imperialism introduced modern science to the continent of Africa through explorers, military conquest, commercial and political interests, and missionary activities. The first European sciences to be introduced through practice were biological sciences, navigational sciences, geographical sciences, and anthropology, where Africa and its inhabitants were the object of the scientific gaze.

As colonizers established their presence and authority in the new lands, they brought their social institutions, which included religion, education, and science. This begins phase two of the spread of European science, as described by George Basalla: "dependent science," in which the first colonial indigenous scientists begin to practice the "science" whose language, norms, laws, behaviors, methodologies, and manners they have learned in European scientific institutions. This is a dependent relationship, because the scientist works in the colony, away from the mainstream of scientific culture, and attempts to maintain the practice of science without the benefit of the informal scientific organizations that feed modern science. The hope is that colonial science will utilize the scientific tradition to create a new scientific tradition in the new location and ultimately create Basalla's third phase: an independent scientific tradition.

Paulin Hountondji argues that modern science in Africa has three stages: (1) the collection of data, (2) the interpretation of raw information, and (3) the application of theoretical findings to practical issues. His analysis shows that stage two did not occur or was not developed in African colonies and thus created a "theoretical vacuum." Africa has been seen as a source of data and a place to apply theory, but an undesirable place to encourage interpretation, by the powers that be. Or, in other words, "colonial science was knowledge *on* Africa" (Hountondji; italics in the orginal) or an extension of peripheral capitalism.

Today African modern science has grown with better laboratories, journals, and publishing houses, more universities, and more scientists. However, the content of African science is still geared toward the interest of the West. For example, export crops are more intensely studied than sustainable crops.

## What Is the Reality of Science in the Twenty-first Century?

As we look at science today in Africa, some may argue that it is still a dependent science, while others may argue that it has become independent science. One way to provide more information to this dialogue is to remember that science is a social institution with a particular culture with a means of reproducing itself, a common language, shared material culture, and solidarity activities. When we look at present-day Africa, we find that there are universities and centers that are doing scientific research and develop-

ment on such topics as malaria, rice, and AIDS, which clearly reflect the communities' concerns. There have been recent scientific symposiums and conference on biology, physics, women and science, and health. We find evidence of scientific professional societies in aeronautics, animal science, computer science, engineering, marine and fisheries, and public health and medicine, to name a few. Clearly there is a culture of modern science that is present in selected locations throughout Africa. Modern science has been transferred to Africa, and a scientific culture is present and taking shape. Something akin to modern science is being created, recreated, transferred, and transformed in Africa today.

In this process one common measure of African science has been to study the scientists who are working in Africa, or are African and trained abroad. Studies have been done comparing where scientists are trained and where they work and examining the flow of scientists from their countries of origin (often called the brain drain), collaborations between scientists in different countries, and NGOs and other developmental organizations funding and evaluations of "scientific activities" in less-developed nations. There has been no definitive answer to the question of whether science in Africa is dependent or independent. The Internet is seen by many as a potential boost to scientists working in isolated places and a major globalizing and equalizing force. However, the answers to how science, the social institution, is being practiced, defined, and augmented are still unclear and are an active area of scholarship for some.

The United Nations Educational, Scientific, and Cultural Organization (UNESCO) historically has been one of the biggest measurers of science and technology in Africa through papers, studies, and conferences on such topics as science and technology policy, brain drains, sustainable development, and economic development. Science has been identified as key to economic and social development throughout the twentieth century. These studies' general conclusion is that distributions of scientific capacity and its benefits are irregular throughout the countries of Africa because of a lack of formal education and an inability of local business, industry, and government to provide resources for research and development. The United Nations Conference on Environment and Development (UNCED) advocated that science's role in Africa is to address environmental stress and find ways to reduce damage to the environment and monitor changes in the environment. M. G. K. Menon, president of the International Council of Scientific Unions (now the International Council for Science), insists that Africa needs science to help solve its socioeconomic problems.

Thinking about science and Africa can challenge our notion of what science is. What knowledge is produced outside of a scientific culture? What is the role of culture on science? If we just train Africans to be scientists, then will they be scientists, practice science, and create a scientific culture? In addition, raising questions about science and Africa raises important ethical and economic issues. For example, indigenous knowledge of plants is being collected and used by Western industries for their own benefit. Who owns/controls this knowledge, and why? Or, typically, most published materials on science and Africa address issues of development, both technological and economic. Whom does this benefit? Who is advocating this development, and why?

New questions are being explored about science and Africa because of science studies. One example of this is Helen Verran's work on science and Africa that explores the meaning of logic in her work with Nigerian teachers and students that leads to connections between knowing, embodiment, and culture. Another example is Wes Shrum's (forthcoming) study of scientists' Internet use in Africa, which raises issues about globalization of science and African scientists' participation in global science.

See also **Technology in History: Africa and the Middle East**

**BIBLIOGRAPHY**
Adas, M. *Machines as the Measure of Men: Science, Technology and Ideologies of Western Dominance.* Ithaca, NY: Cornell University Press, 1989. An analysis of science, technology, and colonialism through the history of technology and science in Africa and Asia from the seventeenth century to the early twentieth century.
Basalla, G. "The Spread of Western Science." *Science* 156 (May 5, 1967): 611–621.
Bloor, D. *Knowledge and Social Imagery.* Chicago: University of Chicago Press, 1991.
Horton, R. "African Traditional Thought and Western Science." *Africa,* 37 (1967): 1–2.
Hountondji, P. "Knowledge Appropriation in a Post Colonial Context." In *Indigenous Knowledge and the Integration of Knowledge Systems: Towards a Philosophy of Articulation,* edited by Catherine A. Odora Hoppers, 23–38. Claremont, South Africa: New African Books Ltd, 2002.
Menon, M.G.K. (1993) "Introduction." In *Africa: In the Context of World Science.* 1–26. Kenya: Academic Science Publishers, 1993.
Pyenson, Lewis. *Civilizing Mission: Exact Science and French Overseas Expansion, 1830–1940.* Baltimore: The Johns Hopkins University Press, 1993. Pyenson's historical work traces the introduction of the "exact sciences" and scientists by French imperialism in Africa.
Shrum, W. "Reagency of the Internet, or How I Became a Guest for Science." *Social Studies of Science,* forthcoming.
Verran, H. *Science and African Logic.* Chicago: The University of Chicago Press, 2001.

*Wenda K. Bauchspies*

# China

The history of China is rich and as complex as that of any nation on Earth, yet for millennia it was virtually unknown to those in the West. Today it remains separated from most Westerners' acquaintance by language, customs, geography, and attitudes. This is no less true about the history of science in China than about any other aspect of its history.

The growth of the greatest river in China, the Yangzi (Yangtse), from its headwaters to its estuary, serves as a metaphor for the development of science in China, from the protoscience of small agricultural villages to the mighty stream of contemporary science and technology. Both the literal and the metaphoric river share their flows with no river system of the West. The dams under construction in the Three Gorges of the Yangzi also provide a metaphor for the barriers that blocked the flow of Chinese ideas after about 1300. Emperor Qian Long's letter to King George in 1793 regarding trade is the epitome of China's humiliating disdain toward the West. "There is nothing we lack. . . . We have never set much store on strange or ingenious objects."

Many Chinese ideas preceded those of Western science by dozens or more centuries. The West, however, experienced a scientific revolution, while China's advances sank slowly out of sight. The following sections will look at the flow of Chinese science from antiquity until it empties into the ocean of human knowledge, in a time line spanning twenty-six centuries.

## Origins of Chinese Science

Science organizes knowledge about the world to solve human problems and better enable humans to dominate nature. The early history of China records the combination of curiosity, observation, and a search for order that ultimately moved the Chinese from the hands of capricious deities toward a view more characteristic of today's science: an orderly and regular structuring of observations. Despite a strong retrograde movement during the days of Mao Zedong, the Chinese learned that useful science commands more of nature, and requires *unbiased* observations and systematic experimentation.

## Antiquity through the Third Century B.C.E.

The first steps in Chinese protoscience (circa 6000 B.C.E.) emerged in agriculture and medicine. Using the definition of science previously stated, the earliest known applications of observation and organization of nature came with growing crops in rows and plant cultivation, over two thousand years in advance of European agriculture. Concern with agriculture has been a Chinese obsession through the centuries.

Historians of science usually record that William Harvey in England discovered circulation of blood in the body. In fact, he was preceded by several natural philosophers going back to the writings of an Arab, Al-Nafis (d. 1288), who apparently got the idea from the Chinese. The Chinese thinking is recorded in great complexity in *The Yellow Emperor's Manual of Corporeal Medicine.*

Chinese contributions to astronomy also far surpassed those of Western science for thousands of years. The peculiar motions of the planets presented an irresistible intellectual puzzle to early Chinese, as they did to all peoples of developing civilizations. In the search for order and regularity, the Chinese seized upon the heavens as the paradigm of certain knowledge. By the fourth century B.C.E., astronomy emerged and remained the queen of the sciences for the next four thousand years.

As early as the fourth century B.C.E. the Chinese had recognized sunspots. Western people were as interested in the sun, regarding it as the source of life, but notions of the necessity of perfection of the heavens precluded recognizing sunspots as phenomena. The Chinese had no such predilections. They described and passed on just what they saw. Kan Te, a fourth-century B.C.E. astronomer, drew up the first star catalog. He also referred to sunspots when he incorrectly described a "solar eclipse" that started in the center of the sun and spread outward. The work of these early Chinese astronomers was in every way comparable to that of the Greek astronomer Hipparchus two hundred years later.

Chinese scholars early devised a calendar and methods of plotting the positions of stellar constellations. Since changes in the heavens augured important changes on earth (for the Chinese considered the universe to be a vast organism in which all elements were connected), astronomy and astrology were incorporated into the system of government from the very dawn of the Chinese state in the second millennium B.C.E.

Mathematics began to appear in China in the fourth century B.C.E. as well. The Chinese had a decimal system for counting as far back as the fourteenth century B.C.E., during the Shang Dynasty, and it was probably used long before that. A cogent reason for developing a decimal system is that Chinese writing used, first, drawings of objects and, later, characters derived from the drawings by simplifying and regularizing the pen strokes, to represent words and syllables. These characters do not fall into a fixed sequence as the letters of an alphabet do, so there is not the temptation to express numbers by matching the sequence of number word elements with the sequence of characters, as was done with the Greek alphabet, for example.

In physics the Chinese anticipated Isaac Newton by twenty centuries. In a collection of writings about natural philosophy, known as *Mo Ching,* a school of philosophers known as the Mohists (after their founder, Mo Ti) clearly stated what is now called Newton's first law: a body at rest, or in uniform motion along a straight line, remains in that condition unless an outside force is imposed upon it. The Mohists were the first and only of the ancient Chinese who considered dynamics. They did not last long as a school, thus did not have a major impact on Chinese history.

Joseph Needham, the greatest scholar of Chinese science, established that Europeans got the compass from the Chinese, not from the Arabs as was once thought. Lodestone is a naturally occurring magnetized material. The Chinese use of lodestone is found in *The Book of the Devil Valley Master* dating from the fourth century B.C.E. The book specifically states that jade collectors used the "south pointer" to distinguish directions.

## The Second Century B.C.E. through the Second Century C.E.

The second century B.C.E. brought advances in medical science in China as well. It is claimed that sex and pituitary hormones were isolated and extracted from human urine and used in treatment of disease as early as the second century B.C.E. This extraction was no simple feat; sublimation techniques were crucial to the extraction process. Another technique involved using a variety of chemicals that would cause the hormones to precipitate out of the urine.

Mensuration, the science of measuring, is fundamental to agriculture. Land is measured in *area*. Stocks of grain and reservoirs of water are measured in *volume*. Area and volume are represented by higher-order algebraic equations. By the first century B.C.E. the Chinese learned to solve algebraic equations of higher order and wrote them as such. Those equations would be unrecognizable to us today because they were written spelled out with characters, but they were valid equations. It was not until the mid-sixteenth century that leading mathematicians of the West were able to solve a cubic equation. The Chinese truly made a most fundamental contribution to mathematics in the theory of equations.

Given the importance of mensuration, and since the decimal system was in hand and had no competition, it is not surprising that decimal fractions were also introduced by the Chinese. An inscription dated 5 C.E. refers to a length "9.5" of a given unit. By the third century, decimal fractions were highly developed in China but, surprisingly, not often used. Chinese mathematics was so far advanced that scholars were quite skilled in manipulating ordinary fractions, and there was no obvious advantage in using decimals.

Advances in mensuration, the decimal system, and the need to move armies were driving forces for the Chinese development of cartography. By the second century the Chinese were using quantitative maps. Not only were scales preserved on the maps, but the Chinese also had a grid system and a well-codified set of rules on how to make a map. Special officials were responsible for developing and preserving maps, which were called *ditu* (face of the earth). The map was vital to agriculture, military planning, taxation, travel, and all the things we find it useful for today. All the old maps have been lost, and with the Han invasion in 111 B.C.E., cartography's advances sank back into the soft sand of history.

Use of maps and mensuration naturally led to a closer study of the precise "face" the earth presents. By the second century, a layered earth, composed in part of sedimentary rock and washed down by erosion from mountains, was an accepted fact, one thousand five hundred years before Clarence Edward Dutton, the Western "Father of Geology."

Science begins with a curiosity for the strange. In the second century a statesman and humanistic scholar, Chang Hua, compiled a book of "Strange Things" in which he recorded tales of all sorts of unexplained phenomena and attempted to study them in an organized manner. He recorded the destruction of the arsenal in the time of Emperor Wu (Jin dynasty, thirteenth century C.E.) that was caused by spontaneously ignited oil. It was not until 1757 that spontaneous combustion was recognized in the West.

### The Third to Fifth Centuries C.E.

For seventeen centuries the Chinese have protected orchard crops from predators by biological means. Close observation of Mandarin orange groves showed that when citrus ants were on the trees, other pests did not attack the oranges. This means of protection, the killer ants, is still used today, based on an observation made in the third century C.E.

Diseases caused by deficiencies of various vitamins and minerals (such as vitamin C) affected much of Western civilization, particularly during exploration. At the end of the nineteenth century, Western medical men recognized that many of these diseases (beriberi, scurvy, and rickets) were caused by diets that were monotonous and lacking in essential elements. The Chinese were aware of deficiency diseases as early as the third century. Chang Hi ("the Galen of China"), writing in *Systematic Treasury of Medicine,* gave vivid accounts of deficiency diseases and recommended dietary treatments.

Geometry and algebra remain two distinct areas of mathematics today. The field of analytical geometry combines the two areas and is fundamental to science and technology. These areas of mathematics were not combined in the West until the time of René Descartes in the seventeenth century. However, in China algebra was well known and applied to geometric figures by the third century C.E. It is odd that the Chinese were so far ahead in using the two together yet never developed analytical geometry. The study of conic sections (parabolas, ellipses, hyperbolae), for example, was never engaged. They failed to construct mathematical models of nature that relied on these curves.

In 264 Liu Hui presented a method for approximating the ratio of the diameter of a circle to its circumference, well known in the West as pi. He overtook the Greeks by computing pi to be 3.14159. By the tenth century pi was computed to ten decimal places. In the West, pi was not computed to as many as seven decimal places until the year 1600.

The Chinese used sound as an objective standard, particularly through sets of tuned bells. Concern with tonal and musical quality of bells led to concern for the timbre or quality of sound of all musical instruments. Timbre represents esthetically the instrument's vibration modes, studied exhaustively in China. By the third century, Chinese observations of vibrating stretched strings were just about as scientifically advanced as they were in the nineteenth century in Europe.

### The Sixth to Eleventh Centuries C.E.

Chinese astronomers were astute observers of the sky who were not confounded by preconceived notions of how things should look. In the sixth century, Chinese texts carefully recorded the positions of comets observed and the directions of their tails, and it was noted that the tails of comets always point away from the sun. The Chinese astronomers also realized that the light from comets was reflected light from the sun.

Chinese medicine continued to evolve. Much of it was based on trial and error, much on superstition. Still, a great deal of Chinese medical science far preceded Western medicine. Diabetes was not only described but also treated, to an extent, by the seventh century C.E., a thousand years ahead of European medicine. Chinese physicians noted that the urine of diabetics was characterized by the presence of sugar and that diabetes is often characterized by intense thirst, frequent and large amounts of urination, edema of extremities, impotence, and foot problems. By the seventh century C.E., goiter was another recognized condition, caused by a tumor of the thyroid gland. Advanced hormone extraction techniques allowed extraction of a thyroid hormone from rams that was used effectively in treating goiter.

Europeans incorrectly thought that *magnetic north* (the direction in which a compass points) was the same as *true north* (the direction to the earth's north pole) until the fifteenth century. The difference between true and magnetic north is the *magnetic declination.* If methods of navigation are to be successful, a correction must be made for declination. The Chinese discovered this by the ninth century C.E. As Joseph Needham states, "The Chinese were theorizing about the declination before Europe even knew of the polarity" (Needham et al., Vol. 4, Part 1, p. 333).

Producing maps that represent large areas of Earth's surface and that are useful for navigation presents a problem of topology. Earth's surface is shaped like a sphere. Maps are flat. How can a flat map be used for navigation? By the tenth century the Chinese had developed a method of projecting the sphere onto a flat surface. This was done in a manner such that at any intersection of longitude and latitude all angular relationships are maintained, and thus correct azimuths recorded. For a navigator, then, a straight line plotted from point A to point B on such a chart would represent a line that crosses all meridians at a constant azimuth. The navigator can follow such a course simply by maintaining a compass orientation. In the West this projection is known as the Mercator projection after the German cartographer who accomplished it only in 1512.

As already mentioned, the Chinese were well aware of lodestone and its magnetic properties. Some time before the eleventh century they discovered that magnetism could be removed from lodestone by heating it. Furthermore, magnetic properties could be induced into hot iron by rubbing it with a magnetic lodestone while the iron cooled. The Chinese also had the sophisticated insight that the induction process could magnetize iron only when it was held in alignment with Earth's magnetic field.

## The Eleventh to Fourteenth Centuries

Chinese astronomy advanced continuously owing to its importance throughout Chinese life. By the thirteenth century C.E. the Chinese had developed equatorial astronomical instruments (that is, instruments that were aimed with reference to the Earth's axis) and quite advanced their practical use. First, a calendar, far more precise than the Western Gregorian calendar, was developed. It showed not only the day and month, but also the position of planets at different times, and when one can observe an eclipse. Chinese astronomers discovered not only that the earth rotates around a tilted axis while revolving around the sun but also that the Earth's axis sweeps out a cone over a period of 26,000 years. Awareness of this gradual change, known as the *precession of the equinoxes* (and observed also by Hipparchus), helped greatly in fine-tuning the calendar. Chinese astronomical instruments of the thirteenth century were equatorial, not ecliptic (aligned with reference to the plane of the earth's orbit) as was the fashion in the West. The Chinese system was much more convenient and showed such great promise that Tycho Brahe adopted it in the seventeenth century for his astronomical observations.

After the thirteenth century, however, Chinese science began to languish. A scientific revolution came to Western Europe and with it the rise of modern science. This did not happen in China, despite the advanced state of Chinese science in antiquity and the Middle Ages. The reasons for this are not all clear. Chinese bureaucratic organization in its infancy clearly helped in Chinese scientific development. However, in its later stages, this bureaucratic organization became more of a roadblock to scientific advancement. What is clear is that the modern world owes a great debt to the Chinese for much of what we consider modern.

*See also* **Technology in History: China.**

**BIBLIOGRAPHY**
Dernberger, Robert F., Kenneth J. DeWoskin, Steven M. Goldstein, Rhoads Murphey, and Martin K. Whyte. *The Chinese: Adapting the Past, Building the Future.* Ann Arbor: University of Michigan Center for Chinese Studies, 1986.

Merson, John. *Culture and Science in China.* Sydney: ABC, 1981

Merson, John. *The Genius That Was China: East and West in the Making of the Modern.* New York: Overlook Press, 1990

Needham, Joseph, et al., eds. *Science and Civilization in China.* 21 vols. to date. Cambridge, UK: Cambridge University Press, 1954–1995. The massive bibliographies in each volume provide the fullest documentation of most topics.

Needham, Joseph, and Colin A. Ronan. *The Shorter Science and Civilization in China.* Cambridge, UK: Cambridge University Press. 1994.

Needham, Joseph. *Science in Traditional China: A Comparative Perspective.* Cambridge, MA: Harvard University Press, 1981.

Sivin, Nathan, ed. *Science and Technology in East Asia. History of Science: Selections from ISIS.* New York: Science History Publications, 1977.

Temple, Robert. *The Genius of China.* New York: Simon and Schuster, 1987.

Williams, Suzanne. *Made in China: Ideas and Inventions from Ancient China* (Dragon Books). Berkeley, CA: Pacific View Press. 1997.

*Fred L. Wilson*

# Eastern Europe and Russia

The rapid development of science in Italy, France, England, and Germany since the Renaissance cannot be compared to its development in the countries of Eastern Europe. Many lands in Eastern Europe were under foreign, often tyrannical, rule; their economies and cultures were in stagnation. The weak development of towns also prevented the nations' scientific progress.

There were Eastern European sovereigns who tried to change the economic and cultural situations of their countries. Western academies had been organized in Eastern Europe since the fourteenth century. The Russian Emperor Peter the Great (1682–1725) succeeded in drawing many talented scientists from the West to his academy's staff. The world society has recognized many prominent scientists with Russian, Hungarian, Polish, and Czech names, yet they largely represent the considerable scientific migration from Eastern European countries. For example, most of the Hungarian scientists who have been awarded Nobel Prizes have found employment abroad.

Lately, Eastern European countries have carried out reforms in the science and educational systems, while widening international cooperation. Although the work of each country is affected by certain problems, positive achievements are obvious. The countries of Eastern Europe have hope for the future.

## The First Universities and Scientific Centers in Eastern Europe

The appearance of medieval universities in Paris, Bologna, Cambridge, and other Western cities could not help attracting attention in Eastern Europe. Czech, Polish, and Hungarian young people left their home countries to study at the Sorbonne, in Padua, in Bologna, and later at the Protestant universities of Leipzig, Königsberg, and Basel. Charles I, king of Bohemia, named Prague his capital in 1346 and devoted a considerable amount of energy to the development of the city. In 1348 he founded Charles University, the first university in Central and Eastern Europe. The Bohemian King, who was also crowned the Holy Roman Emperor (as Charles IV), claimed that the University would attract scholars and students from throughout the Continent. This idea was put into practice, and in the second half of the fourteenth century the number of students in the University reached eleven thousand. Yet, with the appearance of similar schools in Kraków (1364), Vienna (1365), Heidelberg (1386), and Cologne (1388), this number decreased.

Beginning in the 1390s Czech magistrates at the University, under the leadership of Jan Hus (1369?–1415), who became rector of Charles University in 1402, opposed the dominating influence of the Germans and the Catholic Church hierarchy and expounded for equal rights of laymen and clergy, prefiguring the Protestant Reformation. Hus was condemned for his actions by the Council of Constance and burned in 1415. Four years later, a people's rebellion developed, leading to the Hussite war on Czech lands

(1419–1437). During that period, Charles University practically ceased to function and did not finally return to normal activity until the mid-1440s.

In 1556 the Czech King and Holy Roman Emperor, Ferdinand I, invited Jesuits to teach in the University. The main disciplines became theology and philosophy; the law and medicine faculties were absent.

At the end of the fifteenth century and the first half of the sixteenth a rise of activity took place in the University in Krakow, founded in 1364 by Casimir the Great, the King of Poland. The department of astronomy at the University of Krakow achieved great fame, thanks to Nicholas Copernicus (1473–1543) and his work on the heliocentric movement of planets. Copernicus was born at Torun in Poland and was educated at the University of Krakow from 1491. Then he studied astronomy at Bologna, medicine at Padua, and law at Ferrara. Although he spent most of his life as the canon of the cathedral at Frombork (Frauenburg) in western Poland, his main interest was always astronomy. Copernicus's main book, *De revolutionibus orbium coelestium* (On the Revolution of the Celestial Spheres), was first printed during the very year of his death, 1543, although he had been developing his theory for about a quarter of a century. In it, he explained the apparent motion of planets in the sky in terms of the motion of both the planets and the earth on orbits around the sun, rather than of the motion of the the planets and the sun around the earth as described by Ptolemy. Similar ideas had appeared as early as the times of the ancient Greeks, but it was Copernicus who brought about the modern outlook of the earth as a small part of the universe.

Academic science in Poland, however, did not at first adopt the revolutionary ideas of Copernicus. Because of academic, Protestant, and later Catholic opposition to the heliocentric movement of planets, astronomy as taught in Copernicus's homeland continued using Ptolemy's model, with the earth at the center of the universe, for two hundred years more.

The Copernican theory was confirmed at the scientific center that arose at the court of the Emperor Rudolf II in Prague. Rudolf II showed great interest in natural philosophy, alchemy, and astrology. He gathered more than one hundred alchemists, physicians, and astrologers in Prague. The prominent Danish astronomer Tycho Brahe was one of those who joined the institute established at the court of the Emperor. The great astronomer from Germany, Johannes Kepler, also came to the Prague Institute and assisted Brahe in his investigations. Kepler discovered that the planets orbited the sun in ellipses rather than in circles, and he also formulated two other laws that explained the speed of planets' motions in their orbits. Kepler's work eliminated the main astronomical objections to the Copernican heliocentric system.

Works by Czech scientists were recognized during that period. Physician, astronomer, and botanist Tadeáš Hájek was one of the exponents of Renaissance ideas. He wrote a book, *Dialex*, about a "new star" (now known to have been a supernova explosion) that appeared in 1572, and he contributed much to getting Tycho Brahe invited to Prague. Adam Zalužanský, who was born in Poland but became dean of the philosophy faculty and rector at Charles University and was well trained in classical philosophy and natural sciences, published the first Czech work on botany, *Methodi herbariae libri tres* (Three Books of Botanical Method). Slovak Jan Jesenský, or Jessenius, published in 1606 a book on anatomy called *Anatomia Pragensis* (The Prague Anatomy) and did much to raise the standards of teaching medicine at Charles University. Jan Marek Marci, or Johannes Marcus Marci, carried out investigations in physical optics. Finally, Jakub Dobřenský, a physician, marked the beginning of clinical medical teaching in the country.

In Hungary, János Apáczai Csere completed in 1653 the *Hungarian Encyclopaedia*, in which he described the advances in science, introduced the Hungarian equivalents still used today for mathematical terms, and popularized the Copernican ideas.

In Poland, Michał Sędziwój, or Sendi-vogius, was recognized as an expert in chem-istry. He worked in several European countries. In the middle of the seventeenth century, astronomer Jan Hewelke, or Hevelius, founded the first Polish observatory in Gdańsk.

In the eighteenth century, astronomic in-vestigations broadened as observatories were opened in St. Petersburg (1732), Prague (1751), Vilnius (1773), and Krakow (1791). The Russian scientist Mikhail Lomonosov, the Hungarian Miksa (Maximilian) Hell, and the Czech Josef Stepling succeeded in carry-ing out astronomical observations.

During the second half of the eighteenth century, Pál Mako in Hungary wrote a hand-book on differential and integral calculus, and István Hatvani contributed to statistical studies. The Czech inventor Prokop Diviš built the first lightning rod in Europe, and a professor at Charles University, Jan Boháč, used electricity for medical treatment. Jan Tesánek contributed to number theory in mathematics.

In 1825 the Hungarian Academy of Sci-ences was founded. János Bolyai worked at non-Euclidean geometry independently of the Russian Nikolai Lobachevsky. Physicist Loránd (Roland) Eőtvős invented the geo-physical pendulum, which was named after him, and performed careful experiments to confirm the equivalence of inertial and gravitational masses. Denes Kőnig became known through his works on algebra and his theory of sets, and Jósef Szabo investigated petrography and tectonics.

The Czech scientists, brothers Jan and Karel Presl, published in 1819 the first "Czech flora." Jan Presl, together with Josef Jungmann, edited the first Czech Encyclopedia in 1821. Juři Proháska, or Georg Prochaska, became one of the founders of the reflex theory widely used in neurophysiology. The well-known works by Jan Purkyně, or Purk-inje, influenced the development of cytology, anatomy, and embryology.

In Poland, Abraham Stern invented an original calculating machine at the beginning of the nineteenth century. The prominent astronomer Adam Praźmowski discovered in 1860 the effect of polarization in the solar corona. Works by Stanisław Jundziłł in botany and by Feliks Jarocki in zoology were highly appreciated within the scientific com-munity. Also, Polish scientists were recog-nized when working abroad. Filip Walter dealt with the chemistry of distilled products in France. The well-known chemist and min-eralogist Ignacy Domeyko worked in Chile; the mineral *domeykite* was named after him.

## At the Call of Peter the Great

At the close of the seventeenth century, inter-est in science spread far beyond the Euro-pean countries. It was a time when life itself demanded closer contacts between science and practice. The Emperor of Russia, Peter I, understood the importance of science and the people's enlightenment to Russia's de-velopment. The developing industries, such as transport and trade, needed scientific sup-port and a higher cultural level to gain both domestic and international ground. Peter the Great created an institutional project in Rus-sia that comprised an academy of sciences, a university, and some schools. The Imperial Academy of Sciences, later to be known as the Russian Academy of Sciences and, during the Soviet period, the USSR Academy of Sci-ences, was opened in St. Petersburg in 1724.

Peter I planned the Academy to fulfill con-temporary scientific requirements. Talented scientists from abroad, such as the mathe-maticians Leonhard Euler, Daniel Bernoulli, and Christian Goldbach, the astronomer and geographer Joseph-Nicolas Delisle, the physi-cist Georg Kraft, and others, were invited to live and work in Russia. Beginning in the 1740s, the Russian scientists who were trained at the university started to gain re-spect in the Academy. In 1748, Lomonosov suggested the principle of conservation of matter and motion. Bernoulli's kinetic theory of gases and Lomonosov's kinetic theory of heat also made contributions to the physics of matter.

In the nineteenth century the Academy be-came one of the leaders in European science. Russian scientists greatly influenced the de-velopment of mathematics: Lobachevsky, at the Kazan University, created non-Euclidean geometry; Pafnuti Chebyshev and his colleagues in the St. Petersburg Mathematical School made a weighty contribution to number theory, probability theory, and differential equations; and Sofia Kovalevskaya discovered a new case of the motion of solids, among many other mathematical contributions.

Russian physicists and chemists accomplished no less spectacular successes. Early in the nineteenth century, Vasili Petrov discovered the electric arc. Emili Lenz became known for his works in electromagnetism. Nikolai Zinin, Aleksandr Voskresensky, and others reached significant results in the field of organic chemistry. Aleksandr Butlerov's theory of chemical structure also increased this field of knowledge. Finally, Dmitry Mendeleyev's periodic law became a fundamental element of chemistry and proved to be an outstanding contribution to world science.

No less spectacular were successes in biology. Karl Behr laid the foundation in embryology. Ilya Mechnikov and Aleksandr Kovalevsky created comparative and evolutionary embryology. Ivan Sechenov and Ivan Pavlov contributed to the physiology of higher nervous activity. Pavlov (in 1904) and Mechnikov (in 1908) were awarded Nobel Prizes.

## Scientific Migration from Eastern Europe

Scientific travel to other countries became widespread during the nineteenth century. Visiting scholars ordinarily returned home after such voyages, but there were a few notable exceptions. The Hungarian scientist Fülöp Lénárd, or Philip Lenard, went to Germany and became a follower of the prominent scientist Heinrich Hertz, who investigated electromagnetic phenomena. Lénárd discovered important characteristics of cathode rays and was awarded the 1905 Nobel Prize for physics. Lénárd also became known as an ideologist of the "German physics" movement, harmonious with Nazism and antipathetic to modern theoretical physics, including the theory of relativity and quantum mechanics.

Marie Skłodowska-Curie, who left Poland in 1891, became the first great woman scientist. In France, she discovered polonium and radium, and carried out classical investigations on radioactivity. Skłodowska-Curie became a double Nobel laureate (1903 and 1910).

Another Eastern European scientist whose accomplishments were highly acknowledged abroad was the Russian Ilya Mechnikov. In the 1880s, the scientist moved to Paris and carried on his works related to comparative pathology and evolutionary embryology at the Institut Pasteur; as mentioned earlier, Mechnikov was awarded a Nobel Prize for biology. Pioneering works by Mechnikov contributed to the decision of the Institut Pasteur to admit a large group of Russian émigrés after the Bolshevik Revolution.

The escape of intellectuals from Soviet Russia between 1918 and 1925 was the first noticeable wave of scientific migration in twentieth century and was motivated by politics, not science. These Russian refugees then made a contribution to world knowledge, particularly in the humanities disciplines such as ancient and medieval history, paleography, and linguistics.

The Seminarium Kondakovium, founded in Prague by the outstanding art historian and Byzantine scholar Nikodim Kondakov, left a significant mark on the development of Byzantine and medieval Slavic studies in the West. Foundings by many prominent historians, including Georg Ostrogorsky, Henry Grégoire, and Georg Vernadsky, were much associated with the Seminarium.

In France a group of outstanding young scholars—Alexandre Koyré, Georges Gurvich, and Nadezhda Shchupak—found employment at the École des Hautes Études Prâtiques. The Université Libre de Bruxelles took on Professor Alexander Eck, the medieval historian, and the University of Vienna invited Nikolai Trubetskoy to the chair in Slavic philology.

In the United States, Mikhail Rostovtsev obtained a professorship at Yale University (1925), which gave him the opportunity to broaden his studies of ancient economic and social history. Feodosy Dobrzhansky, or Theodosius Dobzhansky, a prominent biologist, stayed in the United States after meeting with the outstanding geneticist Thomas Hunt Morgan at Columbia University in

1927. Morgan credited Dobrzhansky with introducing the creative air of the Russian school of population genetics into American genetics.

The next wave of migration from Eastern Europe, as well as from its central part, took place in the 1930s and was related mainly to the spread of Nazism. Many prominent scientists left Germany, Hungary, Poland, and adjoining countries at that time. Some of them crossed the ocean and settled in the United States. In 1939 Hungarian émigré scientists Leo Szilard, Eugene Wigner, and Edward Teller addressed the U.S. authorities and offered their services on the military application of nuclear physics. All of them, as well as the Hungarian János Neumann (well known as John Von Neumann after he became a U.S. citizen), the Russians George Kistiakowsky and George Gamow, and others, became active participants in the development of nuclear and thermonuclear weaponry.

Other scientists from Eastern European countries were highly acknowledged through their scientific achievements aimed at peace. The Russian Wassily Leontief, who lived in the United States from 1931, was awarded the Nobel Prize in economics in 1973 for the development of the input-output econometric method. Hungarians George de Hevesy and Dennis Gábor both became Nobel laureates, de Hevesy in chemistry in 1943 for the discovery of radioactive tracers and Gábor in physics in 1971 for the invention of holography. The Russian chemist Vladimir Ipatieff, who settled in the United States after a scientific trip abroad, performed investigations in organic chemistry that were highly appreciated by scientists and American industry, particularly the petroleum industry.

The next wave of emigration followed the end of World War II and also had a political character, as many scientists could not accept the social and political changes taking place with the Communist takeover of their countries. There were three great Hungarian scientists among the emigrants. Albert Szent-Györgyi, the 1937 Nobel laureate in physiology who discovered Vitamin C, left for the United States in 1947. He devoted many years of experiments to solving the secrets of cancer—the disease that claimed the life of his wife, his daughter, and his close friend von Neumann. Georg von Békésy, the experimental physicist, was awarded the Nobel Prize for the development of the physiological theory of hearing (1961). Békésy also contributed to physics, technology of communication, and physiology. Another prominent Hungarian émigré, John Harsányi, received the Nobel Prize for economics (1994). He revolutionized a branch of mathematics by applying the theory of games to social and economic processes. His work in mathematics and economics motivated him to publish his philosophical essays about the improvement of ethics.

## Science and Power in Eastern European Countries

The science that had been done in some Eastern European countries since the fifteenth century could not have a substantial development. The semifeudal political system, wars, loss of independence, and economic stagnation prevented the region from keeping pace with Western countries.

The Ottoman Turks and the Austrian Habsburgs ruled over Hungary, Bulgaria, and the Czech and Slovak lands from the sixteenth to the nineteenth centuries and prevented the development of these countries. Poland was weakened by endless wars, and its territory was divided among other countries at the end of the eighteenth century. Science in Eastern Europe continued to develop, but this process was not rapid enough because of the insufficient material base and other negative factors.

In the twentieth century the two world wars inflicted considerable loss onto the Eastern European countries. After World War II, Communists in Poland, Czechoslovakia, Hungary, Bulgaria, Romania, and Yugoslavia established pro-Soviet governments and joined the socialist bloc. Scientific development in these countries was organized mainly as it was in the Soviet Union.

Science in the Soviet Union was developed according to specific "endemic" laws. After 1917 the power of the Communist Party in the State was all-encompassing. According to the Communists, the role of the Academy of Sciences was to honor the Party's activity with scientific knowledge. In 1925 the Academy of Sciences of the USSR was confirmed as a State scientific institution, controlled by the Council of People's Commissars.

Scientific work became the planned labor of scientists controlled by the socialist state. Yet, as the well-known academician Vladimir Vernadsky and his supporters insisted, scientific activity had to be the free search for scientific truth by a human personality. The aim of the State should be to help the creative scientific work of a nation rather than to form a state organization of science. Though Vernadsky's statements received a hostile reception from the "new" Communist academicians, his point of view influenced the subsequent organization of science in the USSR.

The scale of the development of science in the USSR was unprecedented. In 1917 the Russian Academy of Sciences consisted of 40 full members and 100 personnel; by the end of the 1970s the Academy of Sciences of the USSR included 850 members and correspondent members and 52,100 doctors, doctoral candidates, and other researchers. Soviet science produced many well-known scientists: the Nobel Prize winners and physicists Lev Landau, Nikolai Basov, Alexander Prokhorov, and Pyotr Kapitsa; the chemist and physicist Nikolai Semenov; economist Leonid Kantorovich; and many others. And yet, the effectiveness of scientists on the whole, in the country, was low.

At the close of the 1980s the socialist bloc disintegrated and Communist rule ended. The Eastern European countries accomplished economic and technological developments under the conditions of a free-market economy. Reorganization affected the system of scientific research and development to a considerable extent. International cooperation broadened, and the system of financial support was changed. Scientists in Eastern Europe look forward to seeing the positive fruits of reform.

*See also* **Technology in History: Eastern Europe and Russia**

## BIBLIOGRAPHY

Graham, Loran. *Science in Russia and in the Soviet Union: A Short History.* Cambridge, UK: Cambridge University Press, 1993.

Hoensch, Jörg. *A History of Modern Hungary, 1867–1986.* London and New York: Longman, 1988.

Holy, Ladislav. *The Little Czech and the Great Czech Nation: National Identity and the Post-Communist Social Transformation.* Cambridge, UK: Cambridge University Press, 1996.

Kosáry, Domokos G. *A History of Hungary.* Cleveland and New York: The Benjamin Franklin Bibliographical Society, 1941.

McCauley, Martin. *The Soviet Union since 1917.* New York: Longman, 1981.

Palló, Gábor. "Scientific Migration from Hungary." *Journal of the International Committee for the History of Technology* 4 (1998): 210–219.

Russian Academy of Sciences. *Past and Present.* Moscow: Nauka. 1999.

Reddaway, William F., ed. *The Cambridge History of Poland.* Cambridge, U.K.: Cambridge University Press, 1941.

Wallace, William V. *Czechoslovakia.* Boulder, CO: Westview Press, 1976.

*Vasily P. Borisov*

# India

*Satyam eva jayate* (It is Truth that wins in the final analysis): This Upanishadic proposition is now inscribed in the national emblem of the Indian Republic.

## Science in Ancient India (3000 B.C. to 500 A.D.)

Two common, opposing attitudes exist toward ancient Indian knowledge. One view regards the Indians with almost superstitious reverence as having knowledge of cosmic mysteries and as superhuman geniuses whom contemporary scientists cannot hope to equal. The other attitude, inspired by the optimistic belief in scientific progress, considers reliance on ancient knowledge systems a mere obstruction to progress. Truth and falsity can be found in both attitudes throughout Indian history.

Some five thousand years ago, the Indian *rishis* (scientist sage) of the Vedic age propounded three fundamental principles of human knowledge: "Lead me from Darkness to Light" (*tamaso ma jyotir gamaya*); "Lead me from Falsehood to Truth" (*asato ma sad gamaya*); and "Lead me from Mortality to Immortality" (*mrityor ma amritam gamaya*). The three fundamentals, in turn, were aimed at three main areas of scientific inquiry. First, energy and light (*jyoti*), including fuel and fire, represented the basic support for life on this planet. Second, the search for truth-value (*sat/satya*) is developed through logical reasoning, philosophical studies, and analysis of grammar and linguistics. Third, the search for immortality (*amritam*) occurs through investigation of medicine, agriculture, and physical properties of life that exist on this planet.

Although many hypotheses propounded by the *rishis* cannot be demonstrated, it can be demonstrated that their process of hypothesizing was consistent with itself and with known facts. Modern philosophical thinking is, in fact, dominated by hypotheses and principles first appearing in the ancient records of the Vedic/Upanishadic period. Many scientific ideas are based on Vedic wisdom, and, regardless of how the ancient principles are depicted today, they have survived to help develop powerful ideas throughout history.

Many areas of mathematics, physics, and philosophy were first developed in India and then adopted by the ancient Greeks. The first atomic theories, for example, were asserted by Kanada and Pakudha Katyayana, contemporaries of Gautama Buddha (500 B.C.). These atomic theories are recorded earlier than those of the Greek philosopher Democritus and prior to knowledge exchange with the Greco-Roman world. While they had no physical experimental evidence, their atomic theories agree with those of modern physics in many respects. Many advances in algebra, logical reasoning, and astronomical calculations preceded those made in Western scientific development.

By the time Alexander the Great invaded India in 327 B.C., Indo-Greek knowledge exchange was well established. Great leaps had been made in developing astronomical calculations, as well as linguistic theories, theories of causation, and atomic theory. Panini, sometime between the sixth and fourth centuries B.C., founded the study of linguistics. His grammar, which contains a reference to Greek (*yavanani*, "Ionian") writing, shows an awareness of the logical and epistemological problems concerning *meaning* and the cognitive issues associated with making meaning of observations. Other scientific and technological advancements took place, particularly in animal husbandry and domestication techniques (such as elephants and lions), agriculture, invention of wheeled and other tools, and astronomical observation techniques.

The geometrical knowledge applied in building construction during this period is still a marvel today. The Ajanta and Alora caves were occupied around 200 B.C. to 200 A.D. by Buddhist monks and, at Alora, followers of the Hindu and Jain faiths as well. These caves were carved out from deep inside the monolithic mountain, and were big enough to house thousands of students; they include thirty-one three-story buildings. Building scale, design, and ornamentation all demonstrate the advanced technological levels of the time. The caves are adorned with painted and sculptured images. The exquisite ornaments and hairstyles of the female figures in these images are consistent with the high social status of women at the time. Motherhood was highly respected, and women were treated with deference and courtesy.

In the fourth century B.C., the sage Kautilya (also known as *Chanakya*, the Diplomat) wrote his politicoeconomic thesis *Arthashastra*. This work provides details about the socio-political life of a well-organized society and information about the level of advancement in scientific and technical knowledge. Kautilya, who was the ministerial advisor of the young Mauryan king Chandragupta, gives a good account of Indian civic society, comparable with the description of Greek civic life in Plato's *Republic. Arthashastra* also includes instructions on maintaining forests and public areas and suggests punishing anyone who pollutes public places. Environmental management of ponds, rivers, and other water resources for temples is yet another topic covered in *Arthashastra*. Political power during this period was held by a council of wise men and astronomers who predicted the seasons, announced the time for ceremonies, and encouraged intellectual pursuits.

Creative writings of poetry and fiction that flourished between 1000 B.C. and 500 A.D. include the great epics *Ramayana* and *Mahabharata*, along with eighteen collections of Puranas, all of which are revered by the Hindus as holy books, containing popular stories of heroic acts of gods and goddesses—superhuman beings who live in heavenly realms but still possess human characteristics and emotions (such as envy, anger, passion, compassion, hate, jealousy, greed, and violence). This literature contains an excellent mix of poetry, literary fancy, didactic thesis on civil society, norms for individual behaviors, social contract obligation, and ethical teachings and sex education as well. Interestingly, the Puranas resemble modern science fiction in that the characters routinely travel between Earth and heavenly worlds—frolicking, scheming, conspiring, attacking, stealing, robbing, kidnapping, and colonizing others as they do so. The heroes of these epics are described as having scientific and technological knowledge that could produce and use weapons of mass destruction comparable to those of today as a feat of divine engineering.

## The Decline of India's Knowledge Civilization (A.D. 500–1000)

Around the sixth century of our era, philosophical controversies about the nature of the universe and theories of creation abounded. The dominant philosophical debate between Buddhist and non-Buddhist thinkers was about Appearance and Reality. In the absence of experimental equipment such as particle accelerators, the philosophers advanced speculative theories and, at best, could apply epistemic principles to cosmic reality. Among the leading Buddhists were the logicians Dignaga in the sixth century A.D. and Dharmakirti in the seventh. The Buddhist reasoning was that "relational knowledge is never cognized through senses" (in Dignaga: *na sambandhah indriyena grhyate*) and that "the subject-predicate

relation is but dialectical" (in Dharmakirti: *samvrittisad eva dharmi-dharmalakshanam*).

At the end of the eighth century a young and visionary thinker, Shankaracharya or Shankara, challenged the Buddhists. He propounded the Vedanta philosophy, which was the beginning of the end for the Vedic investigative civilization. He rejected the material reality of existence and viewed the universe metaphysically, without scientific substance. He described physical phenomena as mere mirage (*maya*) or mere Appearance, devoid of real substance. Besides the Buddha, Shankara has been one of the most intellectually influential figures of Indian culture. In his teens the young Shankara renounced worldly life, took *diksha* (transfer of divine energy from a Guru to a disciple that destroys negative evil in the heart, body, and soul), studied philosophy, and, before the age of twenty-five, propounded the metaphysical thesis of the Vedanta. Hailing from Kerala in the South, over the next ten years Shankara traveled the length and breadth of the subcontinent. As a powerful orator he engaged the Buddhist opponents in debates, and established his centers of learning and teaching the doctrine of Vedanta; he founded four *Ashramas* (the Vedantic centers or seminaries) in the four corners of India. At the age of thirty-eight he died in the present state of Uttaranchal at Kedarnath, where the Himalayas reach a height of over 14,000 feet.

A great poet, orator, and yogi, Shankara is credited with many charismatic stories. In one story he left his body and entered the body of a recently deceased king. While gone, his body was preserved under the strict ayurvedic (medical) treatment by his devoted disciples. In this king's body, he learned the meaning of sex and passion with the palatial queens. Having enjoyed the company of princesses to satisfy the pleasures he was denied as a celibate monk, the yogi returned to his own body and spread his Vedantic philosophy.

The metaphysical and cultural significance of the Vedantic doctrine notwithstanding, its social implications proved catastrophic for the people of India. It suppressed the spirit of scientific investigative inquiry by declaring that all that appears is unreal (*maya*). Since the Vedantic doctrine taught that the ultimate reality (*Brahman*) is infinite, it could not be known by finite human knowledge, including scientific knowledge. Problems of life and death were attributed to the ignorance of I-Ego, which separates finite Self (*atman*) from the Cosmic Reality, Infinite Brahman (*aham brahma asmi, tad-tattvam asi*).

Questions of material well-being, social organization, and astronomical movements were of no use in Vedantic liberation of the soul. The aims of civic life no longer included building and improving upon material wealth or proper management of agriculture or governance. Even the rulers/kings and warriors were not to advance military science, and the tradition of valor was despised. Life at large became a mirage, because the only worthy pursuit was union of the self (*atman*) with Brahman. But paradoxically, union with Brahman was nonknowable, nonexpressible (*a-vyakta*), nonrecognizable, unnameable, and, of course, imperceptible. Being fixed on knowing the unknowable Brahman, the Indian people became fatalist under the Vedantic philosophy of *Maya* and *Mukti*.

It was at this point that India was invaded by Islamic hordes across its western passes. Travelers had carried tales of the riches of the Orient to Western countries including the Arabian lands and beyond. In A.D. 711, Islamic armies attacked India after passing through present-day Iraq, Iran, and Afghanistan. The Arabian cavalry forces quickly overcame the Indians, whom they found unsuspecting and militarily weak. They founded an Islamic state in the valley of the mighty river Indus, called *Sindhu* in Sanskrit and *Hind* in Persian, from which they called the people residing east of that river *Hindus*.

The iconoclastic invaders destroyed places of worship, shrines, and temples belonging to the Buddhists and the Hindu pantheon. The centers of learning and research at Taxila, Nalanda, Vaishali, and Mithila were destroyed. Handwritten philosophical texts, literature, and the holy books of the Hindus and the Buddhists were burnt in open fires.

## The Indian Response to Islam (900 to 1700)

During this time, seven hundred years before the arrival of European colonizers (Portuguese, Spanish, French, and English), India witnessed unprecedented social and political upheavals and civil wars along with internal and external insecurity. Introduction to the philosophy and faith of the newcomers from the near western lands was a new experience for the inhabitants of the South Asian subcontinent. For the first time Indians came into direct contact with worldviews they viewed as unprecedented in their hostility and intolerance.

For Indians, Nature consisted of that which was cared for and revered. Every atom of every element, every river, every mountain, and every ocean were divine and to be respected. Mountains were home to frolicking Apsaras (heavenly damsels), the oceans and lakes were full of golden lotuses, and the heavens above were full of bright stars where gods and goddesses continually performed mythological divine comedies.

The Indians (Hindus) believed in multivalued logic, in which diversity of viewpoints was an asset and allowed for exploration of Truth. In other words, Truth is described in many ways by pundits (*ekam sad viprah bahudha vadanti*). But for the Islamic newcomers, there was only *one* divine being, Allah, and one and only one holy place of deliverance (Mecca). Radical, revolutionary Islam offered no compromise for their only one truth and one exclusive path to Paradise. Nonbelievers in the Islamic Prophet were doomed to suffer here and hereafter.

Indians made many attempts to reconcile these diametrically opposite approaches to truth and reality. Indians engaged the invaders in their arts, dance, music, poetry, and agriculture. The Hindus even offered prayers at Muslim shrines and accepted the blessings of Muslim saints and Pirs. But Muslims remained mostly uncompromising conquerors. India witnessed no scientific and technological development during critical centuries after the Islamic invasions, for no rulers or rich people patronized scientific or technological pursuits.

A tolerant and liberal people were pitted against an intolerant and radical ideology. Even in the field of architecture, the near western people could have contributed greatly. Instead, just to take one example, the Muslim ruler of Hindustan, Shah Jahan (1592–1666), is said to have beheaded, blinded, or mutilated the artisans, craftsmen, and engineers who designed and built the Taj Mahal, lest they be able to build another Taj. His son, Aurangzeb (1618–1707), was a devout jihadi of his times. Like some modern military dictators, he murdered his brothers, and dethroned and imprisoned his father, accusing them all of infidelity to Islam. Aurangzeb imposed an Islamic tax (*jajia*) on the Hindus. While other Indian rulers were making trade concessions to the new East India Company (from the mid-seventeenth century), Aurangzeb attempted to curb the number of concessions he made, although the company set up trading bases wherever it could. Indian society underwent a drastic metamorphosis toward communal intolerance and conservatism. New ideas and scientific thinking became unwelcome intrusions, while violence and civil unrest overtook philosophical investigation.

The aggressive Islamic invaders claimed exclusive right to Allah and declared *fatwas* (rulings in Islamic law) against anyone questioning its *sharia* laws (derived from the Koran). Indian women were pushed into lower status and were forced to cover their faces according to Islamic rules. Upanishadic debates and philosophical discussions were now frowned upon by both sides. At this point, a new social and political denominator emerged in Indian history: *Hindu* and *Hindustan*. These terms were foreign in origin, as are the English names *India* and *Indian*.

The most authentic record of pre-Islamic Indian science and intellectual history is attributed to the writings of Ahmad Al-Biruni (973–1048), a native of Khwarizm (Khiva) in western central Asia, and a contemporary of the Muslim invader of India, Mahmud of Ghazni. A learned scholar, Al-Biruni mastered several languages, especially Arabic, Persian, and Sanskrit. His main interests were in astronomy, astrology, philosophy, religion, and the social customs of India. Though a sincere believer in Islam, Al-Biruni was tolerant toward the polytheistic ideology of India. As an associate of the Ghaznavid rulers, he wrote that the ruthless suppressions by Sultan Mahmud not only lost India its prosperity but also deprived foreign scholars of India's rich scientific heritage. According to Al-Biruni, Mahmud of Ghazni utterly ruined the country, destroyed their wonderful exploits, and turned the Hindus into "atoms of dust scattered in all directions," becoming an "old story in the mouth of the world." Al-Biruni's account of India describes pre-Islamic Indian knowledge in the northern region of the subcontinent, especially in the areas of astronomy, mathematics, and philosophy.

## Invasions and Civil Wars of the Middle Ages (1000–1700)

When Christopher Columbus (1451–1506) and Vasco da Gama (1469–1524) set their sails in search of the riches of the Orient, India was a center of commerce with good trading ports. Yet when it came to knowledge, the Indian people were anxious to protect their heritage from foreign aggression. While the Muslim rulers enforced religious conversion using threats or political and financial favors, the Hindus took to protecting their age-old belief systems. Passion for scientific investigation and acceptance of new ideas and discoveries had become alien to the defeated nation.

They did find applications for astronomical calculations in making astrological predictions and reading destiny in the stars. The *jyotishi* (astrologer) was honored in all high places, such as the courts of kings, the camps of military expeditions, and on all occasions of social importance such as childbirth, marriage, starting a new business, or laying the foundation for a new house. Matrimonial matchmaking was done in consultation with the *jyotishi*. Astrology thus became an application of the astronomical sciences for interpreting interrelationships and mutuality between natural forces on Earth and those of heavenly bodies. No experimental method was involved in the field of astrology, nor had telescopes reached the Indian pundits.

The ayurvedic medical profession was practiced by the Brahmins (those of highest social class), but, again, experimental methods had not been developed. Inflicting pain on others was regarded as social violence and not sanctioned by the Hindus. Therefore, bleeding and touching of blood and wounds were seen as inviting impurity. Menstruating women were considered impure and kept isolated so that they were not to touch others and not to enter the kitchen area. Personal hygiene was enforced ritualistically.

Several reform movements were led by writers who dared write in common dialects instead of in the Sanskrit language of the Brahmins. These leaders wrote poetry emphasizing human values, divine love, and common knowledge heritage. They spoke out against the exclusiveness of religion, condemning bigotry, the caste system, and the idol worship of the Hindus. These movements, popularly known collectively as the Bhakti (devotional) movement, had a good following among the masses, but the status of women and low-caste people remained unchanged.

The founders of the Sikh movement, Guru Nanak (1469–1538), and Tulsidas (1532–1623), were among these social reformers who wrote in popular dialects. For the first time, Sanskrit lost its commanding status as the medium of intellectual and scientific discourse. Influenced by the iconoclastic ideology of Islam, the Sikh gurus collected spiritual sayings of saints of various popular faiths and collected them into a guide book (called *Guru Grantha*, or Sayings of the Gurus). The Sikhs were required only to recite the sayings of the Grantha and serve the community without caste discrimination.

The term *Sikh* is derived from the Sanskrit *shishya*, meaning the discipline of the guru. A member of the Sikh faith was to dispossess his birth caste by joining a common caste of

the brave (whence the surname *Singh,* "lion"). Originally the Sikh was a radical people's movement ideologically opposed to the established Brahman theology. The Sikhs in the seventeenth century and the Arya Samaj in the nineteenth both challenged the cultural and social caste systems by claiming that they violated the principle of the spiritual oneness of Humanity. Divine grace was still the main aim of Sikh spiritualism, but service to the community was only to win the grace of God and His forgiveness.

The revolutionary thinkers of the Middle Ages seldom advocated scientific methods. To relieve pain and sickness, prayers and chanting of the holy book (Koran, Grantha, or Vedas) were prescribed, not the practice of ayurvedic (medical) science. Philosophical writings in the Middle Ages were rewritten old texts at best, with a few added interpretive explanations (*bhashyas*). They never questioned the theses of the old masters.

Education (*gurukulas*) was controlled by the Brahmins, whose goals were to teach and learn Sanskrit. Brahmin education mainly focused on preserving the old texts by memorization, along with learning grammatical axioms (*sutras*), medicinal plant names, or astronomical calculations that helped in astrological practices. Those who wrote in Sanskrit were mostly monks associated with various religious sects of Hindu philosophical schools. Intellectual activities during this period (900–1700) were predominantly preserved for high castes at theological centers. Women and low castes were not allowed to enter any place of knowledge. Nor could non-Hindus enter the Hindu centers of learning.

The social status of women under the influence of Islam was reduced to the lowest level. Women were to be kept out of public gaze and were to cover their faces. Even Hindu women were to cover their faces in public. Hindu women were forced to commit *sati,* a most heinous practice in which a surviving widow was burned alive along with the body of her husband, with full religious fanfare and chanting of holy mantras. Child marriage, which protected the honor of young girls kidnapped by invading forces, became a common practice. In case a young girl was widowed, even at ten to fifteen years of age, she was not allowed to remarry.

In early 1700, when Europeans reached India, no secular education was in vogue. In fact, Indians were not aware of the technological revolutions taking place in the West. Indians had turned inwards. Foreign travels were considered sinful and were forbidden for Hindus. In case a Hindu had to undertake a sea journey, he was, on his return, to undergo a special purification ceremony.

## Modern Science and English Rule (1800–1948)

After George Washington (1732–1799) defeated the English in the U.S. war of independence in 1783, the colonial British turned their attention fully to the newly acquired Indian empire. Warren Hastings (1732–1818) became the Governor General of India, and imperial colonization of India was fully established in 1857. The traditional Hindu caste system, which the English generally upheld and enforced, led to outright segregation of the lower castes and those born outside the castes (called untouchables, or now sometimes called Dalits, or downtrodden) from the upper castes. Segregation became a part of everyday life for all Indians; they were segregated at social events and facilities, at water wells, and on the river banks. Low-caste Hindus were not allowed to visit temples. Untouchables were to walk on the opposite side of the street in such a way to avoid their shadow from falling on higher-caste people. Christian missionaries established schools for untouchables, and the Mahatma Gandhi forced the entry of untouchables into Hindu temples (in the 1920s). As he did so, Gandhi was attacked by the Hindu priests, and it is, therefore, no surprise that he was assassinated in 1948 by a chauvinist Brahmin who believed that Gandhi was working against Hinduism. When the British arrived, the Indians were divided among small principalities ruled by Hindu and Muslim rulers. The natives of the Indian subcontinent were not unified, except in that they were either meditating on invisible Brahma or prostrating themselves five times a day, facing the distant Abode of Allah. The Mughals, Marathas, Sikhs, Holkers, Sindhias, Ranas, and Rajputs were just some of the large and small dynastic kingdoms that existed in India at this point. The lack of unification among them was a factor that allowed the highly motivated English to colonize and plunder the natural resources they found valuable.

The British came with modern weapons, skill in planning invasions, and experience with governing different cultures in conquered colonies. The East India Trading Company already had a firm hold with its trading bases. As the imperial language, English was also the language of new scientific institutions and diplomatic jargon against which the Indians found no means of common cultural defense. The Indians had ample valor and courage, but no modern weaponry, unification of purpose, or developed military science. British colonial rule in India (known as the British Raj) introduced secular education and "modern" scientific and technological methods, all of which proved critical for development and for constructive questioning of traditional belief systems.

The British divided Indians against each other. They bribed some, and supported the others. They offered weapons to the rich, along with an education in how to operate them. They dethroned some and enthroned others who supported the British.

Thus, British colonization brought good and bad policies. Voluminous studies have been written on the wrongdoings of the British Raj. Yet the colonization brought with it the technoscientific development that allowed Indians to participate in the intellectual movements igniting global civilization. The Industrial Revolution in England (1750–1850) intensified the Western search for raw materials and natural resources and created competition among the European countries. Colonization of South Asia gave them both rich natural resources to feed the mechanized industrial enterprises in Britain, and a huge, captive market for their manufactured goods under colonial control.

British rule in India is mostly attributed to their superior science and technology.

Sociocultural scientific movements flourished in Europe. Europe and England engaged in scientific study and research with full patronage and support by the state and foundations. Indians had yet to learn about modern scientific methods and European industrialization. To sustain the vast Indian Empire, the English required trained natives to govern the colonies.

## The Influence of the Western Orientalists (1800–1900)

English industrial professionals were anxious for information about the newly acquired Indian colony, its history, and its natural resources. Accordingly, many German and English scholars engaged in Sanskrit research on Indian heritage. To facilitate better communication, they founded the Asiatic Society of Bombay in 1804. They also began the Agricultural and Horticultural Society of India (1820) at Kolkata (Calcutta), and the Bombay Natural History Society (1883).

In many English, German, and other western universities, Oriental and Indian (Sanskrit) studies were introduced in regular curricula. This is when orientalism gained foothold in Western academic studies. Hundreds of ancient texts, dealing with astronomy, ayurvedic sciences, linguistics, epistemology, ancient and classical language and literature, religion and philosophy, along with antiques from religious monastic centers, were collected and taken away for critical examination by orientalists at the Western centers of higher learning.

Indian heritage was preserved at university libraries and museums in Europe and America. With the dedicated efforts of orientalists, India contributed to modern European development. European languages have incorporated thousands of words and expressions from languages of India, some of them not obvious. For example, the drink mixture *punch* gets its name from the Sanskrit word *pancha* (five); specifically, *panchamritam* is a tasty nectar or drink made of five ingredients. European linguists were enabled to understand the kinships among their own language families by the study of Sanskrit, to which most of the European languages are related,

forming what is now known as the Indo-European family. Indian influence on European arts, crafts, and personal hygiene has become so thoroughly incorporated into Western culture such that the influence is no longer recognized but instead taken for granted.

## Social Reform and Modernity

The discovery of the Indo-European heritage by the orientalists of the West offered Indians a radical perspective of the outside world. This view was not as closed, narrow, and exclusive as was the view introduced at the earlier arrival of Islam. It also gave Indians a renewed self-confidence as the inheritors of a great secular and scientific civilization.

Dayananda Sarasvati (1824–1883) founded a reform movement for Hinduism called Arya Samaj, or The Society of Aryan People. Along with the nationalist movement, Arya Samaj spread over most of northwestern India. A Vedic scholar, Dayananda was a prolific writer and an effective debater. Becoming a Swami (esteemed monk) at an early age, he studied the ancient Vedic literature from a blind guru, Swami Vrijananda Sarasvati. Having learned about the works of the orientalist Europeans, young Dayananda corresponded with a professor of Indic Studies at Oxford, Friedrich Max Mueller, who was to become renowned as a translator of and commentator on the Vedic and Sanskrit literature.

Originally from Germany, Mueller was the first orientalist to prepare and publish what was then the oldest document known in the history of the Indo-European language family: the *Rig Veda*. Mueller published the six volumes of the *Rig Veda* with commentary over the span of twenty-five years. He prepared and published a two-volume edition containing the *Samhita* (the form of the *Rig Veda* text with the words run together as they are sung) and *Pada* (the form in which the words are written as if pronounced separately). The volumes averaged 600 pages each. Mueller also published a history of *Ancient Sanskrit Literature* (1859) and *Sanskrit Grammar* (1866). He delivered lectures entitled "India: What It Can Teach Us" for English officers sent to India for colonial administration. In the summer 1898, at the age of seventy-five, he received requests from Indians to prepare and publish an Indian edition of the *Rig Veda*. Mueller died in 1900.

Swami Dayananda gave the call to Indians to "go back to the Vedas," and in response to the Judeo-Christian Ten Commandments, he formulated the ten principles of Arya Samaj, one of which was that "The source of all true knowledge is the Vedas." Another principle reads, "Always be ready to accept the truth and be ever ready to give up untruth." Arya Samaj was a modernist Hindu reform movement that was opposed to the caste system and recognized equal rights for women and untouchables. The organization set up schools for girls' education. The Samaj also condemned the image-worship of mythological Hindu gods and goddesses. Members of the Samaj were frequently arrested by the English administration charged with "disturbing the peace." Some Samaji leaders were assassinated by Islamic fanatics. Some fifty years before Bertrand Russell wrote *Why I Am Not a Christian*, Dayananda presented scientific arguments against all belief systems that perpetuate mythology as the final and revealed Truth.

Dayananda sincerely believed that modern scientific and technological inventions were known to the sages (*rishis*) in the Vedic Age, when the society was not yet burdened by blind faith and corrupt religious practices. He proposed to take the society back to the uncorrupted age of the Aryans. In his work titled "The Light of Truth" (*Satyarthaprakash*), Dayananda condemned traditional Hindu rituals and ridiculed the Brahmin priests who offer salvation by receiving gifts and offerings from the ignorant faithful. He also criticized the Christian and Islamic claims to having exclusive alliance with the heavenly Father, the Almighty who needs to send His Son and Messenger on Earth. With knowledge of astronomy, Dayananda questioned the description of planetary positions and astronomical claims of the Holy Books of the Hindus, Christians, and Muslims.

Swami Dayananda was the first reformer of Sanskritic tradition to condemn the evil practice of *sati* and the authority of the Vedas, and advocated marriage for widows.

As Christ threw the moneylenders out of the Jewish temple, Dayananda threw out stony icons of gods and goddesses from the Hindu temples. He authorized equal rights for women to read and write the forbidden Holy books of the Vedas. He declared that high caste status was not given by birth but was to be earned by qualification, profession, and noble deeds. On many occasions Dayananda was attacked by high-caste Hindus. Brahmin priests, finally, conspired against the Swami, and Dayananda was poisoned by his Brahmin cook.

The Arya Samaj saw English/Western education as an attempt to convert Indians to Christianity, which undermined the ancient Vedic foundation of Indian civilization. In response to the English educational system, the Arya Samaj set up *gurukulas*, education based on patterns established in age-old monastic schooling. Lessons were in Sanskrit, and the syllabus consisted of classical literature, Sanskrit grammar and linguistics, and a few old philosophical texts without much reference to modern secular knowledge. Popularly respected as the defender of Indian culture and nationalism, they were harbingers of sociopolitical movements that helped widen the Hindu-Muslim divide. Gurukulas encouraged the association of anti-Christian, anti-Western, and anti-science perspectives.

Muslim leaders organized *madrasas*, a system of schooling that reinforced Islamic education, devoid of any scientific inputs. The schooling in the madrasas mostly focused on Arabic and Koranic theological texts that glorified the Muslim history of expansion. Thus, in the nineteenth century, the Christian, the Hindu, and the Islamic modern reform movements were revivalist and devoid of scientific content.

This state of affairs turned around in the twentieth century. Encouraged by scientific and engineering developments in England, a dozen universities were founded in major cities. By the end of the first half of the twentieth century, India had roughly sixty-five science and engineering institutions and professional societies. The Indian Mathematical Society was founded in 1907, the Institution of Engineers in 1920, the Indian Botanical Society in 1921, the Geological Mining and Metallurgical Society of India in 1924, and the Indian Medical Association in 1928. The National Academy of Sciences was set up at Allahabad in 1930. The Indian Physical Society and the Indian Academy of Sciences were founded at Bangalore in 1934. The Indian National Science Academy was founded in 1935, and the Entomological Society of India in 1938. The colonization of India by the English introduced modernization, including scientific and technological systems, to the people of the Indian subcontinent.

## Independence and Indian Science Policy (1947–2000)

The first socialist (Labour) government in England was elected at the end of World War II. Ideologically, the Labour Party was committed to defending the democratic rights of colonized peoples. Although Sir Winston Churchill defended democratic freedom for the Western nations, he had opposed giving freedom to the Indian nation. Labour released Indian leaders from long years of imprisonment, and India gained independence from colonial rule. Scientific progress and acquisition of new technologies were deemed necessary pursuits to treat the socioeconomic ills of the country. At midnight on the August 15, 1947, Jawaharlal Nehru, speaking in the Indian Parliament, declared that "as the world sleeps, India is awakened" to its historical task of liberation of its multitude.

Educated at Eton and Harrow, Nehru experienced an English (Western) education through which he acquired new scientific perspectives. A committed leader of socialist democratic ideology under the leadership of Mahatma Gandhi, he led the Indian nation to freedom. Both Gandhi and Nehru had spent many formative years in England. Gandhi himself was educated at the Lincoln's Inns, London. Yet their steel was also tempered inside the British prisons, where they spent dozens of years, mediating on modern social and political scientific theories. They upheld the modern democratic rights of the subjugated peoples of India and the colonies. His-

torically, they defended modern, Western, scientific political value systems.

After gaining independence, India adopted the English parliamentary system. Nehru was committed to following scientific methods of development to make India a modern technoscientific nation. He encouraged the policy of propagating "scientific temper" as a way to guard against cultural and religious chauvinism. Where the traditional gods had failed, science was to succeed with human will and endurance to achieve social change. Technology, not gods, was to free the nation from poverty.

In the first half of the twentieth century, many Indian scientists carried out pioneering research in advanced scientific fields. Between 1930 and 1948, study and research in nuclear physics was conducted at several Indian universities. In 1930, Dr. Chandrasekhara Venkata Raman, for whom the Raman effect (the effect of a particle's quantum state on the light scattered from it) was named, received the Nobel Prize in physics. Dr. Satyendra N. Bose had established his claims to the Bose-Einstein statistic, describing the behavior of large numbers of particles with integer spin. Meghnath Saha was elected to the Royal Society in England, and was honored for his research in thermal ionization, stellar spectroscopy, propagation of radio waves, radio emission from the sun, and the aging of rocks. It was due to Saha's pioneering efforts that his brilliant student, Dr. Basanti Dulal Nag Chaudhuri, brought the first cyclotron to India in 1941. At age 24, Nag Chaudhuri received his doctorate at the University of California, Berkeley. Such luminaries emerged in Indian science as a result of the high priority placed on scientific research in Indian universities.

In the years leading up to India's independence, scientific activities depended mostly on the interest of individuals and required only small funds, so the gap between Indian and Western science was not excessively wide. In practice, science was open and universal. Publication of research was considered a scientist's most important function. This situation enabled Indian scientists to contribute directly to scientific advancement.

Science lost its innocence with World War II, nuclear fission, atomic bombs, and other weapons of mass destruction. The scientific enterprise that produced the bomb united science with technology. Scientific activities in advanced countries turned to "big war science" as a result of the synergistic effect of science, technology, and Cold War politics. The period from 1948 to 1958, the first decade of the Cold War, pushed secretive science and technological development in the West.

The people of India, on the other hand, freed after two hundred years of colonial domination, were involved in the task of restructuring postcolonial society. Because Indians were not directly engaged in wartime scientific activities, there was no infrastructure to support research in new scientific areas. Consequently, Indian science fell behind in the race of big scientific research projects, including war systems and manufacturing weapons of mass destruction.

A young, dynamic alumnus of Cambridge, Homi J. Bhabha, belonged to the large Tata industrial and engineering house. Tata engineering interests reaped rich dividends from the war. Bhabha set up the Tata Institute of Fundamental Research and was credited later with the first Atomic Research Centre (now named Bhabha Atomic Research Centre). The ambitious Bhabha had no experience working in an academic environment, and he went on to became the Science Policy Advisor to Nehru, the first Prime Minister of India, who appointed him the first Chairman of the Indian Atomic Energy Commission.

On August 1, 1948, the Indian government established the Atomic Energy Commission (AEC), aiming to protect the interest of the country in connection with nuclear energy. The Commission's directive was to increase the already active teaching and research facilities for nuclear physics in the Indian universities. C. V. Raman had become the first Indian director of the Institute of Sciences at Bangalore, and Meghnath Saha had established the first Indian Institute of Nuclear Physics (now named the Saha Institute of Nuclear Physics) in Calcutta in 1944.

In 1958, Bhabha refashioned the AEC in a manner that gave him a free hand in planning and executing India's atomic science policy. In a resolution dated March 1, 1958, the government of India gave him "full authority to plan and implement the various measures on sound technical and economic principles and free from all non-essential restrictions or needlessly inelastic rules."

The chairman of the commission was granted "full executive and financial powers" and was made the *ex officio* Secretary to the Department of Atomic Energy, responsible only to the Prime Minister. Thus, the AEC became a policy-forming body. The Department of Atomic Energy is the government agency responsible for executing missions decided upon by the commission. Both are formally under the Minister of Atomic Energy, and the Prime Minister always holds the office of the Minister of Atomic Energy. The chairman of the Atomic Energy Commission heads both the commission and the department.

The close relationship between nuclear research and the potential development of weapons made secrecy an integral component of nuclear research. All nuclear research activities, therefore, were covered in secrecy. Secrecy became the mantra with India's nuclear establishment, which was in itself a subgovernment.

Under the Atomic Energy Act of 1962, Bhabha secured an exteme level of personal autonomy within the formal constitutional framework of India. Bhabha was a close friend of Prime Minister Nehru, whom he could address as *bhai* (brother). Both were bachelors, belonged to aristocratic families, graduated from Cambridge, and enjoyed Western art and music.

In contrast, Meghnath Saha opposed secrecy in scientific research and criticized separation of fundamental research from mainstream science teaching in the universities. He was against the creation of an independent nuclear energy agency, insisting that nuclear physics and fundamental studies could continue as university research projects. Bhabha, however, using his closeness with Nehru, concentrated all big science research under his official domain of the Department of Atomic Energy. Indian universities were denied necessary funding for nuclear research and were unable to purchase equipment or attract young scientists from abroad. The question of social and economic issues in technological appropriateness was not a priority for Bhabha. He advised the Indian government to go for capital-intensive and energy-consuming nuclear technology for an Indian energy development strategy.

Under the Atomic Energy Act of 1962, the Department of Atomic Energy was the sole agency authorized to "initiate, explore, plan, and execute" nuclear research. A science policy resolution passed by the Indian Parliament in 1958 highlights the respective roles of technology, raw materials, and capital in India's quest for national prosperity and cultural heritage. The same resolution asserts that new scientific techniques can substitute for deficits in natural resources and capital.

The founder of India's science policy was not enthusiastic about peaceful use of the atom; he was not inclined to join or support the international scientists' peace movement. He did not sign the Russell-Einstein Manifesto (1955), calling for the banning of nuclear weapons. Nehru, being a follower of Gandhi, was opposed to nuclear weapons. He had agreed to host the first International Scientists' Conference Against Atomic Weapons in India. J. D. Bernal, Einstein, and Bertrand Russell had expressed concern about the use of science for evil purposes by the political powers. "India's leading official scientist," Bhabha, offered a "cold douche" to the proposal of Bertrand Russell. Recalling his efforts, the latter wrote: "Bhabha had profound doubts about any such manifesto, let alone any such conference . . . It became evident that I should receive no encouragement from Indian official scientific quarters" (quoted in Sharma 1983).

In India, Saha opposed Bhabha's policy for separation of nuclear-allied research from mainstream higher education in the country. Saha felt no necessity to allow the newly independent state to take over monopolistic control of open Indian scientific enterprise. He feared that the probability of misuse of science by the state would increase through governmental control of the atom, and that the free flow of information with mainstream knowledge would stop. Saha's fears proved true in time.

## State-Controlled Scientific Research and Development

Most research and development grants and funding in India (nearly 90 percent at the beginning of the twenty-first century) come from India's federal government. Priorities given to specific areas of science and technology are reflected in budgetary allocations by the government. The majority (almost 65 percent at the beginning of the twenty-first century) of the science budget is controlled by three departments: the Department of Atomic Energy, the Department of Space, and the Department of Defense Research and Development. The activities of these three departments have been conducted in total secrecy. (Interestingly, India adopted this policy of secrecy and nonaccountability by following the example of the United States of America, which, at the beginning of the 1990s, was spending nearly $350 billion a year on secret service activities.) Aside from the federal budget, there is no other independent source of funding available, nor another scientific community. The government agencies command the power of patronage and punishment over engineering graduates. Indian science has, thus, become captive of the State. Little scope has existed for independent and critical scientific work.

During the early Cold War decade (1950–1960), India's nuclear energy program received as much as 40 percent of research and development grants (data from the 1958–1959 budget). Although this percentage fell somewhat in the following decades, defense science departments received strong backing during the 1970s, when the U.S. military presence increased in the Indian Ocean.

In 1971, as India was engaged in the liberation of Bangladesh, the Nixon administration ordered a U.S. naval task force, headed by the nuclear-powered aircraft carrier *USS Enterprise*, into the Bay of Bengal. In response to what it viewed as the arrogant posture of the Republican administration, the Indian government stepped up the activities of its three big secret science organizations. On May 18, 1974, India tested its first atomic device, code-named "Smiling Buddha," at Pokhran in the Rajasthan desert.

Following this test, which came to be known as Pokhran I, United Nations sanctions were imposed on India. The country was denied access to advanced fields of science and engineering.

In response to the sanctions regime, India adopted policies of self-reliance and reverse engineering. From the mid-1970s to the mid-1980s, science and technology were aimed at self-reliance, but the country did not achieve that goal in socially useful sectors. According to the *Status Report on Science and Technology* for 1988, the percentages of India's budget allocated to the big three secret science departments (Defense, Space, and Atomic Energy) were 37 percent, 17 percent, and 12 percent, respectively, totaling 66 percent of science and technology funding. Funds left available for social sector research activities were insignificant. For example, the health and medicine funding for over a hundred million citizens represented only 2 percent of grants, just enough to meet the salary for a few hundred managerial staff. No funds were made available for necessary infrastructural, equipment, and research fellowships. Similarly, environmental research, agro-research, biotechnology, veterinary sciences, astronomical studies, high altitude research, remote sensing studies, waste management, water management, and other public utilities and social and public health services did not get high priority funding. Meanwhile, towns and cities lacked proper transport systems, and the citizens badly suffered from a lack of drinking water and no regular power supply. The gap between rural and urban had been widened to the extent that fewer than 1 percent of scientists working in India's science centers came from rural or underprivileged classes.

In the 1980s the government inaugurated a number of ministerial-level departments in various emerging fields of science and technology with the avowed aim to "close the gap" in fields of knowledge. Many areas of study and research that could have been part of university curricula instead became government departments without properly trained personnel: the Ministry of Environment; the Ministry of Non-Conventional Sources of Energy, the Ministry of Ocean Research, and the Ministry of Information Technology. A policy statement was made available that recommended that the country would adopt the strategy of reverse engineering, which involves disassembling an engineering device to see how the system works in order to duplicate it. India also obtained licensing agreements with multinationals for importing technology on a turnkey basis to manufacture goods not made locally.

During the twenty years between 1980 and 2000, India achieved self-reliance in many high-tech fields of weaponry, aerodynamics, and missile technology. Between May 11 and 13, 1998, five hydrogen bomb tests (named Pokhran II) were carried out, this time code-named "Shakti," after the goddess of power.

The secret service or war science research activities remained outside the purview of public scrutiny. Nor could any legislative authority look into these affairs. Neither the Controller nor the Auditor General of India, nor the Planning Commission nor the Regulatory and Environmental Agencies, could investigate the performance of the three "sacred cows" of secret science. The secrecy meant, among other things, that if these departments were to develop any new material with industrial application, Indian industry would generally not be able to obtain a patent. Moreover, Indian industry's requirements were not compatible with the high-tech systems and materials used and developed by the three departments.

After the Vietnam War, the United States increased its military presence in the Indian Ocean. An additional measure taken by the United States to counter the Soviet advances in the Central Asian region was to provide advanced military hardware to Pakistan and the Persian Gulf states. The United States upgraded its military presence by stationing B-52 bombers on the island of Diego Garcia off Indian shores, and increased the number of nuclear submarines patrolling in Indian backwaters.

The first Gulf War (1990–1991) and the Iraq invasion of 2003 demonstrated the vulnerability of India and other Third World countries when pitted against high military industrialized powers. The U.S. presence on Diego Garcia and its intervention in the Middle East aroused old memories of the imperialist invasions that were made possible because of advanced Western technological weapons. Budgetary increases for defense research invariably led to further loss of funds for social welfare, health, and educational programs (Sharma 1991).

Even if science is considered a vector of social change elsewhere, the Indian experience in the twentieth century was otherwise. Over fifty years, hardly any woman scientist appeared on the decision-making committees of the top scientific departments. No woman was a Commission Member of the Atomic, Space, or Defense Research Organizations. Most top positions at these three ministries (and at other higher education, engineering, and science centers) were occupied by high-caste Brahmins.

## Twenty-First Century Science and Technology Policy

On July 25, 2002, the scientist sage Dr. A. P. J. Abdul Kalam was elected the twelfth President of the Indian Republic. Under his guidance, the government of India announced its *Science and Technology Policy 2003,* which included almost doubling science and technology funding from 1.1 percent to 2 percent of the gross domestic product (GDP) by 2007. This science policy was formally made public by Prime Minister Atal Bihari Vajpayee. It further proposes to provide universities and research institutions with greater autonomy and to improve technoscientific infrastructure at centers for higher learning and research. Other objectives for twenty-first century science policy include making tax concessions for firms willing to invest in research and development, offering attractive perks to expatriate scientists to return to India, increasing funds for socially oriented study and research, researching natural disaster management tactics, promoting environmentally cleaner and more efficient technologies, and increasing government interface with industrial development interests. Admittedly, there is promise in such aims to make science and technology serve the civic society, and recognition of universities and industries in science and technology strategy. However, the document explicitly allocates funds on the old pattern of planning which had dominated throughout the years of Independence.

## Science and Religion

Over the centuries, scientific inquiry was aimed at overcoming fear of unknown natural forces. The power of scientific knowledge has now led us to a vision of self as paramount, not dependent on the sanctions of gods and goddesses who rule over nature. Yet, scientific knowledge did not liberate Indians from what are elsewhere considered to be theological falsities. Even those who are highly educated, familiar with modern (Western) developments, and influenced by contemporary debates (such as feminism, equal rights, and the secular democratic principle of treating people without discrimination based on gender, race, or religion), are culturally inclined to continue traditional social practices without questioning their truth value.

One example can be found in ongoing matrimonial practices. The most highly placed elite in India, for example, advertise matrimonial propositions in the daily newspapers and on the Internet to widen their choice. Surveying the matrimonial columns offers interesting reading on the attitudes of highly educated Indians. The very first word in each advertisement indicates the caste and the subcaste, specifically stating high-caste status. Even those who are Westernized, highly qualified, and well placed in the United States or working with multinational corporations, ask for horoscopes matching their own. The matrimonial negotiations are often conducted by the parents of the candidates. The national dailies index the matrimonial advertisements under the headings of castes, religion, gender, and region, and highlight financial status, social connections, and property.

Biosciences have given us the ability to detect a defective gene during gestation and take remedial steps to ensure a safe and healthy pregnancy. However, in a male-dominated society like India, begetting a son is a cultural necessity so as to continue the lineage for the salvation of forefathers. The birth of a female child is an unwelcome event. The cultural custodians of faith and nonviolence in India considered it justifiable homicide to commit an abortion or destroy a female child immediately upon its birth, claiming that the unfortunate child was born dead. In the olden days, the midwife became an accomplice in the crime. More recently, scientific methods of prenatal screening—amniocentesis and ultrasound—have come into play. These methods, originally designed to screen for congenital disorders, can also be used to detect the sex of the unborn child. The cultural pundits in India used these scientific methods to act out the age-old belief in the necessity of a son for the continuation of a family lineage. As prenatal scanning became increasingly popular in India, the rate of abortion rose dramatically. Alarmingly low female birth rates led to legislation banning sex detection during pregnancy.

Since scientific advancement is associated with the Western domination, there is a cognitive resistance factor evident in the sociocultural perspective of most Indians. Social thinkers, critics, writers, and professional scientists use double-speak about religious and scientific knowledge.

On June 20, 2003, in Jaipur (Rajasthan State), Arya Samaj performed a five-day *yagna* (fire worship) to pray for good rain in the drought-affected area. The President of Rajasthan, Arya Pratinidhi Sabha, Satyavrat Samvedi, is quoted in the periodical *The Hindu* (June 22, 2003) as saying, "Apart from its religious importance, *yagna* has a scientific basis as well. The fumes of ghee [purified butter] emanating from the *yagna* fire condense the vapours and cause rain." It is ironic that by using advanced technoscientific

systems, the cultural chauvinists want to propagate age-old unverifiable "spiritual" truth.

In a secular, democratic political system, every citizen in India is granted equal rights, irrespective of gender, race, religion, color, caste, creed, economic status, family, or dynastic peerage. The view that women, blacks, and the disadvantaged (poor) have an equal right in a civil society is an admittedly modern principle that is not recognized in any holy book, nor in any traditional part of the culture.

The history of science offers greater insight, indicating that the path to alleviate human suffering lies in evolving global scientific civilization. However, the most serious problem before us is how to liberate humanity from the cultural doctrines of divine caprice. But science must also be free from the political control of the State. Secret funding of science is antiscience and antipeople.

A paradox of Indian science is that Indian scientists invest professionally in scientific belief systems but personally invest in religious belief systems. At the beginning of the twenty-first century, India's most respected scientist-sage was Dr. Avul Pakir Jainulabdeen Abdul Kalam, who built the country's rocket systems, had been awarded the highest national title, *Bharat Ratna* (Jewel of India), and was also the top science advisor to the Indian government. As the Principal Scientific Advisor on Defense, Kalam was the architect of the Pokhran-II hydrogen weapons tests in 1998. Abdul Kalam was also, however, by faith a Muslim, teetotaler, and vegetarian as well as a thinker, philosopher, and poet. Though Kalam belonged to no political party, he was elected the twelfth President of India in 2002, with an almost unanimous vote by the electoral college.

The Missile Man of India symbolizes the nation's cultural synthesis approach to faith and reasoning, with a sense of professional duty, sacrifice, and achievement. In his personal life, Abdul Kalam is a Gandhian, so he has "no possession" and "no attachment" to worldly relations. His life story is that of a poor child, born in an underprivileged social setting, who sold newspapers and worked to achieve higher scientific posts with hard work and dedication. A secular believer, Kalam submits his entire work to Allah: "Be aware that God is with you and when He is with you who can be against you." Kalam has donated his entire pension for the education of disadvantaged children. His life story, he says in his writing, "will end with me, for I have no inheritance in the worldly sense. I have acquired nothing, built no (property), no family, sons, or daughters" (Kalam 1999). Kalam's paradox is the paradox of Indian scientists. They recognize no conflict between scientific ways of knowing and India's cultural belief systems.

In fact, most professional Indian scientists use scientific assertions to support age-old belief systems in their personal and social life. India's scientist President Kalam is a unique personality who successfully created a synergy between technoscientific, agro-industrial development and traditional ways of knowing and living. In the twenty-first-century science policy resolution "Vision 2020," Kalam provided a roadmap to make "India a developed country by 2020" by collective efforts of scientists at the centers of higher learning and research to work with industry, universities, agriculturalists, politicians, and the youth. After fifty-six years of Independence, the first scientist-President, Dr. Kalam, has given a new mantra to India's youth and to political parties: "The Nation Is above Politics and Religion."

## BIBLIOGRAPHY

Al-Biruni. *Alberuni's India* (1888). Translated and edited by Edward Sachau. New Delhi: Indialog Publications, 2003.

Basham, Arthur L. *The Wonder That Was India.* New York: Grove Press, 1959.

Council of Science and Industrial Research. *Status Report on Science and Technology.* New Delhi: CSIR Publications, 1988.

Darian, Steven G. *The Ganges in Myth and History.* Honolulu: University Press of Hawaii, 1978.

Foerstel, Herbert N. 1993. *Secret Science: Federal Control of American Science and Technology.* Westport, CN: Praeger Publishing.

Gosling, David L. *Religion and Ecology in India and South East Asia.* London and New York: Routledge, 2001.

Kalam, A. P. J. Abdul, and Arun Tiwari. *Wings of Fire: An Autobiography of A. P. J. Abdul Kalam.* London: Sangam Books, 1999.

Sharma, Dhirendra. *India's Nuclear Estate.* New Delhi: Lancers Publishers, 1983.

Sharma, Dhirendra. "India's Lopsided Science." *Bulletin of the Atomic Scientists* 47.4 (1991): 32–36.

Sharma, Dhirendra. "Science, Culture, and Conflict in India." *Cultural Dynamics,* 12.2 (2000)· 164–181.

*Dhirendra Sharma*

*See also* **Technology in History: India and Southeast Asia**

# Japan

Before the Middle Ages, China served as Japan's teacher in the ways of occidental knowledge. Since the second half of the nineteenth century, Japan has sought direct contact with the West for the purpose of introducing occidental science and technology. By the end of the twentieth century, Japan became a major world economic power. The Japanese people were able to assimilate Western science and technology in such a short while because critical foundations were already established in Japanese culture. In the following historical account, it is important to note the manner in which and methods by which the Japanese people coped with cultural incorporation of science and technology.

## Western Guns and Christianity: 1543–1600

Japan directly encountered occidental science and technology in 1543. A trade ship headed for China from southeast Asia was caught in a typhoon and drifted ashore on the coast of the Tanegashima, a small island located on the south of Kyushu. The Portuguese drifters gave two firearms to the islanders. This is the first delivery of such arms to Japan. The island of Tanegashima produced iron sand, and thus supported the island's excellent swordsmiths, who accomplished domestic production of firearms in the next year.

The techniques to manufacture firearms were brought without delay to Sakai, a city of commerce located in the south of Osaka. In the midst of a provincial warlord rivalry in Owari (presently Nagoya City), a feudal lord named Nobunaga Oda quickly heard about the distinguished performance of firearms. Nobunaga invited merchants from Sakai to visit and ordered them to deliver a large number of firearms. Availing himself of these new arms, Nobunaga started to change his battle tactics. He defeated the warlords of East Japan in a battle at Nagashino in 1575. East Japan's pivotal forces were conventional cavalrymen, while Nobunaga's forces held three thousand guns. By unifying this area, he made firm advancement toward unification of Japan, only 40 years after the first firearms arrived in the country.

Ieyasu Tokugawa, a feudal lord in Mikawa and once a vassal of Nobunaga, unified the whole land of Japan. In 1603, Ieyasu established shogunate rule in Yedo (today Tokyo) by appointing himself shogun. The Tokugawa Shogunate continued to govern the whole nation of Japan for over two and a half centuries. Fearing the spread of Christianity throughout the nation, which Portuguese and Spanish missionaries had been promoting since 1549, they prohibited Christianity in 1614. Publications with any description of Christianity were prohibited, regardless of whether they were written by Westerners or even by the Chinese. After that, a severe closure policy was adopted, which exclusively permitted the Dutch to make transactions with Japan, but only on a small artificial island called "Dejima" settled in Nagasaki, Kyushu. Thus the arrival of two firearms and encounters with Christianity effected great change in Japan half a century later.

## Mathematics and Medicine: 1601–1850

The process of unifying Japan made practical mathematics a necessity for facilitating infrastructural development (expansion of commerce, measurement technology, irrigation for agriculture, and so on). Mathematics (*wasan* in Japanese) was practiced and taught throughout Japan until after the Meiji Restoration in the mid-nineteenth century.

The development of *wasan* is attributed to Takakazu Seki (1642–1708), and his school's mathematics rivaled Western mathematics at that time. In fact, Seki's school determined how to solve monogenic equations of a higher degree earlier than Western mathematicians. Circles were an extremely important concept in geometry, and observations of circles were a focus of especially particular study. The circle ratio, known in the West as pi ($\pi$), was obtained with an accuracy of more than 40 digits. The concept of infinite series was taught, as well as definite integrals based on quadrature of parts. Because Arabic numerals had not been introduced in Japan at that time, numerical equations were still expressed in vertical columns using Chinese characters. *Wasan* emphasized practical results, with no importance attributed to theoretical processes of verification. Performing *wasan* was regarded as a kind of art, like tea ceremonies or martial arts. Multiple schools were formed, and the methods were taught exclusively to followers. *Wasan* was the first full-fledged scientific project realized by the Japanese in a state of isolation from other countries. However, mathematics as the basis for scientific inquiry was not developed, and no acceleration of experimental science occurred.

In the second half of the eighteenth century, two Japanese medical doctors, Genpaku Sugita and Ryotaku Maeno, independently came across a Dutch book on dissection of a human body, titled *Tafel Anatomia*. The doctors were surprised by the differences between schematic diagrams of the internal organs illustrated in the Dutch book and those they knew from Chinese medical books. The two were lucky enough to find out separately where executed criminals were being dissected, and met there. Fixing their eyes on the dissected internals in their pursuit of truth, the two Japanese doctors were deeply impressed with the precision of dissection diagrams in Dutch books.

The two doctors learned the Dutch language and laboriously translated Dutch medical literature into Japanese over the course of three years, completing the translation in 1774. This built up momentum for the spread of Dutch medical knowledge throughout Japan. By the end of the administration, the

Tokugawa Shogunate is said to have stocked no less than six thousand foreign books.

## Opening Borders: 1851–1900

When ships of foreigners eager to trade with Japan appeared close to the Japanese archipelago at the beginning of the nineteenth century, the Tokugawa Shogunate drove the ships away. This practice changed when Commodore Perry, an American admiral, came to Japan with his fleet in 1853 and demanded by force that Japan open herself to trade with the world. Keenly aware of its military weakness, the Tokugawa Shogunate decided to open the country by entering into the Japan–U.S. Treaty of Peace and Amity. The provincial feudal lords, who were becoming powerful enough to overwhelm the Shogunate, strongly opposed the opening of the country and the treaty. A state of national confusion ensued. The Shogunate established a naval training station in Nagasaki, where the Japanese could learn Western scientific techniques in areas such as navigation, shipbuilding, and gunnery. In this training station fundamental sciences such as physics, mathematics, and chemistry were taught, in addition to the Dutch language. This policy hearkens back to the Edo period, when foreign books were scrutinized to establish a base of Western learning and expertise in Japan. The training station later served as the foundation for the University of Tokyo, the most prestigious university in Japan today.

The nationwide confusion over opening Japan to trade raged for about ten years, until a coalition of feudal lords formed in the western part of Japan. In 1868 this development led the Tokugawa Shougunate to return its administrative rights to the Emperor, an event later called the Meiji Restoration. This should not be mistaken for a civil revolution but rather a revolution accomplished by the lower-class warriors.

The Meiji government aggressively introduced occidental culture. The government placed special emphasis on catching up with Western scientific and technological development. Education of the Japanese people in Western science became a top priority. The government employed several hundred foreign experts as teachers. They adopted a policy to send out hundreds of exchange students to Europe and the United States. Almost all of the professors at the University of Tokyo, the first university in Japan, were foreigners. Yet in twenty years these educational initiatives led to replacement of these foreign professors by Japanese scholars. Scientists belonged to a very high Japanese social class, and scientific research was considered an activity for the state.

## Science for Industry: 1901–1925

With its victories in both the Sino-Japanese War (1894–1895) and the Russo-Japanese War (1904–1905), Japan became an internationally recognized power. The Japanese government aimed to expand higher education gradually to enhance their national scientific and technological prowess. In this period, considerable research was done in areas of pure science, although many Japanese researchers conducted their work in foreign countries.

Jokichi Takamine is an example of one such scientist of the late nineteenth century and early twentieth century. After graduating from Tokyo University, he studied in Scotland and eventually founded a research lab in New York City. He is known for his enzyme and hormone isolation work, and he was the first to isolate and purify the hormone adrenalin at the turn of the century. He temporarily returned to Japan from the United States to insist firmly that Japan expand its industry to establish itself as a world power. He insisted that institutes for scientific research and development be founded, and he specifically proposed that a People's Science Institute be established. Finally, in 1917, the Institute of Physical and Chemical Research was established, and it has long played a pivotal role in Japanese scientific research.

World War I enabled Japan to establish industrial foundations and to stand up as an industrial country without any support from overseas. Thus Japan's status in the international community rose, and Japan joined the League of Nations established after the war.

## Science for War: 1926–1950

However, the prosperity after World War I did not last long, and within several years Japan entered an economic recession. The great 1923 earthquake in Tokyo and its surrounding areas claimed 150,000 lives and further deepened the recession. In an effort to escape the recession, Japan took a hint from Western countries and aimed to build up its military and, in the process, build up their scientific and technological areas in new universities and colleges.

Part of Japan's military goal was to advance on mainland China. The Japanese government's priorities and national budget centered on expanding scientific research to compete better in international science and technology markets. The Japanese Society for the Promotion of Science (JSPS) was established in 1932. Although scientific research funding provided through this agency was greater than science promotion funding through the Ministry of Education, the agency's research was designed to link industrial and military activities, emphasizing extensive, coalition-building research projects. Much of the national budget was allocated toward scientific research that would prepare for Japan to conduct colonial rule in mainland China, and a variety of such research agencies were established.

In December 1941, Japan levied war upon the United States by attacking Pearl Harbor, Hawaii, simultaneously declaring war on Great Britain through the U.S. link with the Allied Powers. Japan expanded its goals of conquest to include the resources in Southeast Asia. Around thirty institutes were thus founded to carry on what was referred to as "southern science," or "training of army doctors or naval surgeons," in efforts to further Japanese interests during World War II. By the time atomic bombs released on Hiroshima and Nagasaki in August 1945 accelerated Japan's surrender to the Allied Powers, Japan was developing enriched uranium, but development of atomic bombs was far from reality.

After the war, Japan was occupied and controlled by the Allied Forces guided by the United States. The occupation forces prohibited Japanese scientific research in areas connected to military and naval activities (such as atomic power, aircraft, and radar). The Cold War eventually restructured world politics, and as soon as war broke out in the Korean Peninsula, the United States changed its occupation policy in Japan. Expecting Japan to play a breakwater role, the United States permitted Japan to establish an army that was shortly converted into self-defense forces in Korea. In 1951, Japan regained her independence through the San Francisco Peace Treaty.

## Science for Economic Recovery: 1951 to the Present

Instead of being defeated by Western science and technology, however, Japan was defeated more by its hierarchical administration systems under an archaic regime, as seen in ministries and agencies, including universities and research institutes. For example, the Japanese army and navy research institutes were utterly ignorant of each other's projects to develop an atomic bomb. Japan's defeat provided an opportunity to remake such inefficient national administration systems. As the Cold War progressed, however, the United States changed its occupation policy, and Japan revived the conventional administrative bureaucratic system to expedite post-war recovery.

Japanese expenditure on scientific research after the war differs from that of the Western world in two ways. One is that, because Japan's military spending was minimal, governmental expenditure on science and technology accounted for a very small proportion of the whole budget. The country opted for security dependence on the United States through the U.S.–Japan Security Treaty, in order to focus resources on economic recovery. For this reason, private-sector research and research for practical use became two pillars upon which Japanese science and technology are based.

The second difference in Japanese science and technology expenditure was the classification of budget categories by three independently controlled government ministries and agencies. The system of specialization supported by such administrative arrangements bred poor interagency coordination and, as a consequence, led to rigid policies that were difficult to change. For example, Japan still supported a nuclear fuel cycle route already rejected by other countries for its economic inefficiency.

Japan's economic growth from the 1960s to 1980s made Japan the second greatest economic power in the world. However, excessive state control over the Japanese economy made it difficult to keep up with the changes at the end of the Cold War, inducing economic stagnation in the 1990s. To stimulate the economy, the government altered science policies and administrative structure. As of 2001, science and technology policies were made directly accessible by the Prime Minister through a new agency: the Council for Science and Technology Policy, Cabinet Office. The Ministry of Education and the Science and Technology Agency, which were formerly independent of each other, were unified into a single ministry. Meanwhile, Japan converted state-run universities into independent administrative corporations. By so doing, priority is given to developing applied science deemed useful to the industrial world. Now, Japanese vitalization begins with mobilization of scientists and technologists.

See also **Technology in History: Japan**

**BIBLIOGRAPHY**

Fortune, Robert. *Yedo and Peking*. London: John Murray, 1863. A book of travels written by an English horticulturalist who visited Japan as a plant hunter in the end of the Tokugawa Shogunate.

Mikami, Yosio. *The Development of Mathematics in China and Japan*. 2nd ed. New York: Chelsea Publishing Coporation, 1974.

Nakayama, Shigeru, Kunio Goto and Hitoshi Yoshioka (eds). *A Social History of Science and Technology in Contemporary Japan*. Melbourne: Trans Pacific Press, 2001. (English translation, covering the first three of the six volumes in the series as of late 2004). Outline works on the history of science and technology in Japan.

Sugita, Genpaku. *Dawn of Western Science in Japan* (*Rangaku Kotohajime, 1869*). Translated by Ryozo Matsumoto. Tokyo: Hokuseido Press, 1969.

*Harukazu Iguchi*

# Latin America

The great navigations from the sixteenth century onward exposed forms of scientific knowledge from different cultural environments to each other. The several ethnosciences involved in the encounters, among them European science, have been subjected to great changes as a result. This article will examine some of the consequences of this mutual exposure of cultures.

What is nowadays recognized as "science" comes from the European tradition developed in the Middle Ages and the Renaissance and imposed, after the sixteenth century, on practically the entire world, particularly in Latin America, in the processes of conquest and colonialism. Pre-Columbian cultures had different styles of doing their measurements and computations, of producing, of healing, and of performing the other tasks needed to cope with their environment. These practices are still prevalent in some native communities. The people from these cultures have no problems at all in assimilating the current European system.

The high prestige of science comes mainly from its recognition as the basic intellectual instrument of progress. It is recognized that modern technology depends on science and that the instruments of validation in social, economic, and political affairs are based on science and mathematics, mainly through the storing and handling of data.

The history of science in Latin America is better understood when we look into the specificities of the colonial administration and organization in viceroyalties: *New Spain* (roughly what is today the southwestern United States, Mexico, Guatemala, and Central America); *New Granada* (roughly what is today Panama, Colombia, Venezuela, and Ecuador); *Peru* (roughly what is today Peru and Bolivia); *La Plata* (roughly what is today Chile, Paraguay, Argentina, and Uruguay); and the lately created Viceroyalty of *Brazil*, under the Portuguese crown. Since the rise of the independence movements early in the nineteenth century, the political and cultural map of Latin America has remained roughly the same.

The influence of the navigators and chroniclers, particularly Portuguese, in building up the mode of thought that underlies modern European science is noticeable, essentially through the development of objective and serene curiosity, rigorous observations, and creative experimentation.

## Conquest and Early Colonial Times

Native scientists and learning were not recognized as such by the conquerors. One of the earliest European writers to register the culture of the conquered cultures was Fray Bernardino de Sahagún (1499–1590), who explains much of the flora and fauna, as well as the medicinal properties of herbs, of New Spain. But he writes (Sahagún, vol. 2, p. 478):

> The reader will rightfully be bored in reading this Book Seven [Which treats Astrology and Natural Philosophy which the naturals of this New Spain have reached], . . . trying only to know and to write what they understood in the matter of astrology and natural philosophy, what is very little and very low.

In the vision of the conquered, great impression was caused by the arms, tools, and equipment in general, as seen in pictures, for example in the *Codice Florentino*.

To acquire fluency in the language and numeracy of the conquered peoples was a priority. Grammars were a major contribution to this. As early as 1547, Fray Andrés de Olmos (ca. 1485–1571) published the *Arte de la lengua mexicana* on Náhuatl. This is followed in 1560 by the *Grammatica o arte de la lengua general de los indios de los reynos de Peru* on Quechua, by Fray Domingos de S. Thomas (ca. 1505–1570). A description of the number system was written by Juan Diaz Freyle: *Sumario compendioso de las quentas de plata y oro que en los reinos del Pirú son necesarias a los mercaderes y todo genero de tratantes. Con algunas reglas tocantes al arithmética*, the first arithmetic book printed in the New

World. Maps, essential in European actions, were not known as such in native cultures, whereas time measurement was done with remarkable precision. This has much to do with the differences in cosmological vision.

On the Atlantic coast, under Portuguese rule, efforts to master the languages were also important. In 1595 the Portuguese Jesuit José de Anchieta published the *Arte de grammatica da lingoa mais usada na costa do Brasil* on Tupi, and the Spanish Jesuit Antonio Ruiz de Montoya published the *Arte y bocabvlario de la lengva gvarani* on Guarani in 1640.

The *Historia natural y moral de las Indias*, published in Seville in 1590 by Joseph de Acosta, another Jesuit, is representative of the trend toward writing comprehensive studies of society, history, religion, and science.

Already in the first century after the conquest, practical books were published in Mexico, such as the *Arte menor de arithmetica* by Pedro de Paz in 1623 and *Arte menor de arithmética y modo de formar campos* by Atanasio Reaton in 1649. Also noteworthy is *Nuevas proposiciones geométricas*, written by Juan de Porres Osorio, in Mexico.

Upon the arrival of ships to and from the New World, it was necessary to provide for medical care of the crew, who in many cases were suffering from unknown diseases. The *Santas Casas* (Holy Houses), founded in Lisbon in 1498, proved a viable institution with these purposes. In 1543, Bras Cubas founded a *Santa Casa* in Santos, near São Vicente, one of the earliest urban settlements in the Portuguese colonies.

But soon there was a concern about creating formal medical classes in the colonies. In 1653 a Chair of Medicine was established in Nueva Granada, in the Colegio Mayor de Nuestra Señora del Rosario in Bogotá. This lasted until 1865 and was replaced by the Facultad de Medicina de la Universidad Nacional, in 1867.

Astronomy was a major area of interest in Latin America in the seventeenth century, and there were important discussions on the meaning of comets. In 1690, Don Carlos de Sigüenza y Góngora in México published *Libra astronómica y filosófica*, considered one of the most important works of Latin American science, in which he refuted prevailing astrological arguments about comets.

In Brazil the same tone of reflections can be found in the works of Valentin Stancel, a Jesuit mathematician from Prague who lived in Brazil from 1663 until his death in 1705. In the Viceroyalty of Peru, the first to be recognized as a mathematician is Francisco Ruiz Lozano (1607–1677), who wrote *Tratado de los cometas*, essentially a treatise of medieval mathematics explaining the comets.

## The Established Colonies

The religious orders were not immune to the new ideas coming from Europe, and several descriptive treatises of the world were written. A general geography of the continent, the *Geographia histórica de la América y de las islas adyacentes, y de las tierras arcticas y antárcticas, y islas de los mares del norte y del sur*, authored by the Jesuit Pedro Murillo Velarde, was published in Madrid in 1752.

King Carlos III (1716–1788) in Spain and the Marquis of Pombal (1699–1782) in Portugal were responsible for the intellectual movement known as the Enlightenment (*Ilustración* in Spanish), which had important consequences for the intellectual revival of the two countries, and consequently of their American colonies. This contributed to shaping the scientific development of Latin America.

The scientific expeditions of Amedée Frézier, published in 1716 as *Relación del viaje por el mar del Sur a las costas del Chile y del Perú realizada durante los años 1712, 1713 y 1714*, and those of Charles Marie de la Condamine (1701–1774), had important influence in determining the scientific development of the colonies. An important expedition was organized by Alejandro Malaspina (1754–1809), who, in five years, visited practically all the Spanish possessions and collected precious materials.

A number of intellectuals, well versed in a variety of areas of knowledge, emerged in the colonies. At the beginning of the nineteenth century, Juan Alsina and Pedro Cerviño lectured on infinitesimal calculus, mechanics, and trigonometry in Buenos Aires. In Peru, Cosme Bueno (1711–1798), Gabriel Moreno (1735–1809), and Joaquín Gregorio Paredes (1778–1839) are best known.

In Brazil, José Fernandes Pinto Alpoim wrote two books, *Exame de Artilheiros* (1744) and *Exame de Bombeiros* (1748), both focused on military mathematical problems and written in the form of questions and answers. (In contrast to the Spanish colonies, Brazil did not have a press, so these two books, although written in Brazil and focusing on Brazilian interests, were printed, in Portuguese, in Spain and Portugal.)

The Colombian José Celestino Mutis (1732–1808) was responsible for an unpublished translation of Isaac Newton's work and also for the introduction of modern mathematics in Colombia, mainly relying on the books by Christian Wolff. He founded the Observatorio de Bogotá in 1803. His most distinguished disciple, Francisco José Caldas (1771–1816), became the director of the Observatory. Caldas was deeply involved in the Independence War and was shot by the Spaniards.

In Chile the Universidad Real de San Felipe was inaugurated in 1747, in Santiago, and had a *catedra* of mathematics under the self-instructed Fray Ignacio León de Garavito.

Publications in Mexico with scientific value appeared in the second part of the eighteenth century. *Lecciones matemáticas*, by José Ignacio Bartolache, was published in 1769. In 1772 an anonymous inventor built a "calculating wheel," capable of performing the four basic operations for numbers up to eight digits. In the same year Benito Bails published the *Elementos de matemáticas*, which treated infinitesimal calculus and analytic geometry.

The complexity of problems related to water and to mining stimulated important technological development of the country. "Subterraneous geometry" became a major theme in Mexican science. The needs of mining led to the creation of mining schools, such as the Real Seminário de Minería, founded in Mexico City by the Spanish mineralogist Fausto de Elhuyar in 1792. Particularly important were the efforts toward urbanization that took place in all the colonies. The book *Comentarios a las ordenanzas de minas*, by Francisco Javier Gamboa, published in 1761, is most representative of these developments.

In Guatemala, which included Costa Rica, the most renowned scholar was José Antonio Liendo y Goicoechea (1735–1814). He taught at the Universidad de San Carlos de Guatemala and was responsible for modernizing its plan of studies, incorporating experimental physics to the project and introducing modern mathematics, again, like the efforts in Colombia, based on the texts of Christian Wolff.

## Independent Countries

The independence of the Viceroyalties of Nueva España, Nueva Granada, Peru, La Plata, and Brazil was achieved in the first quarter of the nineteenth century. In Costa Rica the colonial authorities had established the Casa de Enseñanza de Santo Tomás in 1814. Its most influential teacher was Rafael Francisco Osejo, born in 1780, who wrote, in 1830, *Lecciones de aritmética,* in the form of questions and answers. In 1843 the Casa de Enseñanza was transformed in the Universidad de Santo Tomás, and a curriculum in engineering was established.

Colombia soon attracted foreign mathematicians. The Frenchman Aimé Bergeron wrote the *Lecciones de matemáticas,* published in 1848, and introduced descriptive geometry in the country. The Italian Agustín Codazzi (1793–1859) was influential in creating the Colegio Militar. Lino Pombo (1797–1862), who was particularly influential in founding the Academia de Matemáticas de Venezuela, wrote a complete course of mathematics.

In Brazil, the transfer of the royal family of Portugal to escape Napoleon's invasion in 1808 was decisive in changing cultural life in the colony. They founded a major library, a museum, a botanical garden, and an Escola Militar, the first institution of higher learning in the colony. Joaquim Gomes de Souza (1829–1863), known as "Souzinha," was the first Brazilian mathematician with an European visibility. His results, contained in short notes to the Académie des Sciences de Paris and in the Royal Society, were posthumously collected as *Mélanges du Calcul Intégral,* printed by Brockhaus of Leipzig in 1889.

Argentina experienced a remarkable intellectual development after achieving independence in 1816. In 1822 the ephemeral Sociedad de Ciencias Físicas y Matemáticas was founded in Buenos Aires. The private library of Bernardino Speluzzi (1835–1898) is a landmark. Valentin Balbin proposed a new study plan at the Colegio Nacional de Buenos Aires in 1896.

In Peru the book *Ensayo de estadística completa de los ramos económico-políticos de la provincia de Azángaro . . .* by José Domingos Choquechuanca, published in 1833, raised an interest in statistics.

In Chile the Universidad de Chile was created in 1842, with a Faculty of Physical and Mathematical Sciences. A distinguished member of the faculty was Ramón Picarte, a lawyer, who had his paper *La división reducida a una adición* accepted and published by the Academy of Sciences of Paris in 1859. Fifteen German mathematicians, most with doctorates, emigrated to Chile in 1889.

The influence of Auguste Comte (1798–1857) towards the end of the nineteenth century was very important in all Latin America, particularly in Mexico and in Brazil.

## The Twentieth Century

As part of the enormous progress of technology at the beginning of the twentieth century, the Brazilian Alberto Santos Dumont, living in Paris, won the Deutsch prize in 1901 by piloting a balloon around the Eiffel Tower, and he flew a heavier-than-air device in 1904.

A major event in Argentina was the development, with German support, of the Observatório Astronómico de La Plata. Beginning in 1917, the Spanish mathematician Julio Rey Pastor (1888–1962) became a frequent visitor to Argentina and had a decisive influence in the development of mathematics in Argentina. José Babini (1897–1983), a disciple of Rey Pastor, founded the Unión Matemática Argentina and coauthored, with Julio Rey Pastor, a remarkable *Historia de la matemática,* in 1951. In the 1930s the distinguished Italian mathematician Beppo Levi emigrated to Argentina, established an important research center in Rosario, and founded an influential journal, *Mathematica Notae.* Another important immigrant was the Spanish mathematician Luis Alberto Santaló (1911–2001). The visit of Albert Einstein to Argentina and Brazil in 1925 was a great support to the emergent scientific establishment in both countries.

Very important in the health sciences was the discovery in 1881 of the vector of yellow fever by the Cuban Juan Carlos Finlay, who became the Republic of Cuba's first Director of Health in 1902 to implement programs of eradication of these vectors and of mass vaccination. In the same year in Brazil, Oswaldo Cruz became the director of the technical section of an institute for bacteriological research and disease control that has borne his name since 1908. Cruz was made Brazil's General Director of Public Health and successfully fought yellow fever, plague, and smallpox.

In Uruguay a tradition of mathematical research was established early in the twentieth century, mainly under the influence of Eduardo García de Zuñiga (1867–1951), who succeeded in creating an important mathematics library in Montevideo. At mid-century, Rafael Laguardia and José Luiz Massera were responsible for the creation of a most distinguished research group in the stability theory of differential equations at the Instituto de Matemática y Estadistica de la Facultad de Ingeneria de la Universidad de la Republica in Montevideo.

In Brazil, the proclamation of the Republic in 1889 reinforced the influence of Positivism (which, particularly in Brazil, was practiced as a religion, complete with liturgical calendar), and Einstein's 1925 visit was seen a sort of final coup in favor of it. At the beginning of the twentieth century a number of young mathematicians were absorbing the most recent progresses of Europe, among them Otto de Alencar, Manuel Amoroso Costa, Teodoro Augusto Ramos, and Lelio I. Gama. In 1916 the Academia Brasileira de Ciências was founded. With the inauguration of the Universidade de São Paulo in 1934, the first fully operational university in Brazil, a new direction in science and mathematics in the country can be observed. This moment can be said to be the beginning of academic research in Brazil. Young researchers were hired from Europe. Among them the most promising were Luigi Fantapiè, Giàcomo Albanese, and Gleb Wataghin from Italy, respectively in functional analysis, algebraic geometry, and physics; the chemists Otto Hauptman and Heinrich Rheinbolt from Germany; and the sociologists Pierre Monbeig and Claude Levi-Strauss from France. They were responsible for initiating an important research school in São Paulo.

## Developments after the End of World War II

A major recognition of Latin American science was the granting of the Nobel Prize in Physiology or Medicine in 1947 to Bernardo Houssay, an Argentina-educated scientist and Professor of the University of Buenos Aires, who was most influential in forming a number of Latin American scientists.

Further international recognition was earned by the Brazilian nuclear physicist Cesare Lattes (b. 1925), who played a major role in the discovery of the pi meson in 1947 at the University of California at Berkeley.

After World War II a number of European scientists and mathematicians emigrated to Latin America. In mathematics, the presence of Antonio Aniceto Monteiro from Portugal was particularly important; he was responsible for establishing important research centers in both Brazil and Argentina, in Rio de Janeiro and in Bahia Blanca, respectively.

The efforts of France and England to preserve their cultural presence in the former colonies led to the intensification of bilateral cooperation through organizations such as the British Council, the French ORSTOM (now the Institut de Récherche pour la Developpement), and the Coopération Française. This cooperation was extended to Latin America. Cultural and economic interest of the United States in Latin America were, at best, modest before World War II. This changed radically after the war, and the United States became the main destination of scientists and engineers pursuing doctorates abroad. UNESCO created the Oficina Regional de Ciencia y Tecnologia para America Latina y el Caribe in Montevideo, Uruguay, to support its activities, and the Organization of American States was activated toward the scientific and technological development in Latin America.

Practically every Latin American country has created a national research council such as Colombia's Consilio Nacional de Investigación Cientifica y Tecnológica (CONICYT) or Mexico's Consejo Nacional de Ciencia y Tecnología (CONACYT). An effort on the part of the American Association for the Advancement of Science (AAAS) to cooperate with homologous organizations in Latin America is also to be noted.

## The Challenge of Contemporary History of Latin American Science

To analyze contemporary history is a difficult task, because one must refer to processes still going on and risk stumbling into personal and political sensibilities. Several academicians were active in the period when military regimes took control of the governments of countries that were showing the strongest vitality in research in the sciences and mathematics. The military coups that occurred sequentially in the four countries that were active in research (Brazil in 1964, Argentina in 1966, Uruguay and Chile in 1973) originated an important migratory flux of scientists in all areas, initially among these countries, and soon directed to the few Latin American countries that were able to keep democratic regimes, particularly Mexico and Venezuela. After the redemocratization of Argentina (1983), Brazil (1984), Uruguay (1984), and Chile (1989), some scientists returned and reclaimed their positions. Others were able to maintain their positions during the military regimes and kept these positions after democratization. The dividing line between opponents and sympathizers and collaborators of the military regimes is very difficult to draw. Obviously, personal conflicts are still latent.

Recent developments in science and technology in Latin America acquire different characteristics because they are subordinated to more immediate international economic interests of giant industrial complexes.

This article has been an attempt to give an overall, although very incomplete, account of a vast subject. Although Mexico and other countries have advanced in the very attractive area of research of the history of science and technology, in Latin America as a whole

the field is still incipient. Practically all the names mentioned in this article, and several others, are open to investigation. There has been considerable research on the social studies of science and technology in Latin America, but the results are partial and dispersed.

*See also* **Technology in History: Latin America**

### BIBLIOGRAPHY

Azevedo, Fernando de, ed. *As Ciências no Brasil* (1955), Rio de Janeiro: Editora UFRJ, 1994.

Babini, José. *Páginas para una Autobiografia: Prólogo y notas de Nicolás Babini*. Buenos Aires: Asociación Biblioteca José Babini/Ediciones Letra Buena, 1992.

Catalá, José Sala. *Ciencia y técnica en la metropolización de américa*. Madrid: Theatrum Machinae, 1994.

Ferri, Mario Guimarães, and Shozo Motoyama, eds. *História das Ciências no Brasil*, 3 vols. São Paulo: E.P.U./Editora da Universidade de São Paulo, 1979–1981.

González Orellana, Carlos. *Historia de la Educación en Guatemala*. Guatemala City: Editorial Universitaria, 1985.

Mirador, Fernando Flores. *Tierra firme anticipada. El descubrimiento de América y las raíces arcaicas de Occidente* (to appear), Lund, 2003. An important study of the differences in cosmological vision between Europeans and native cultures.

Nicolau, Juan Carlos. "La Sociedad de Ciencias Fisicas y Matemáticas de Buenos Aires (1822–1824)." *Saber y Tiempo* 2 (1996): 149–160.

Nobre, Sergio, ed. *2° Encontro Luso-Brasileiro de História da Matemática e 2° Seminário Nacional de História da Matemática, Águas de São Pedro, São Paulo, Brazil, 23 a 26 de março de 1997. Anais-Actas*. Rio Claro, Brazil: Cinzeiro, 1997. The influence of Monteiro in both Brazil and Argentina was discussed at this conference.

Peset, José Luiz (ed.) *La Ciencia Moderna y el Nuevo Mundo*, CSIC/SLAHCT, Madrid, 1985.

Pyenson, Lewis. *Cultural Imperialism and Exact Sciences: German Expansion Overseas 1900–1930*. New York: Peter Lang, 1985. Contains a discussion of the Observatório Astronómico de La Plata.

Quevedo, Emilio, and Camilo Duque. *História de cátedra de medicina, 1653–1865*, Bogotá: Centro Editorial de la Universidad del Rosario, 2003. A detailed history of the Chair of Medicine in New Granada.

Rey Pastor, Julio, and José Babini. *Historia de la matemática.* Buenos Aires: Espasa-Calpe Argentina S.A., 1951.

Sahagún, Fray Bernardino de. *Historia General de las cosas de Nueva España,* 2 vols., Mexico: Alianza Editorial Mexicana, 1989.

Silva, Clóvis Pereira da. *A Matemática no Brasil. Uma História do seu Desenvolvimento.* Curitiba, Brazil: Editora da Universidade do Paraná, 1992.

Trabulse, Elías: *Ciencia y tecnologia en el nuevo mundo.* Mexico City: El Colegio de México/ Fondo de Cultura Economica, 1994.

Trejo, Jesús Galindo, Mario Alberto Torres García, María de la Luz Moreno García, José Ruiz de Esparza, Marco Arturo Moreno Corral, and Silvia Torres de Peimbert. *Lajas celestes: astronomia e história en Chapultepec.* Mexico City: CONACULTA-INAH, 2003.

Weinberg, Gregorio. "La Educación y los Conocimientos Científicos." In *Historia General de América Latina IV: Procesos americanos hacia la redefinición colonial,* edited by Enrique Tandeter and Jorge Hildalgo Lehuedé, 497–515. Paris: Ediciones UNESCO/Editorial Trotta, 2000.

Yepes, Ernesto, ed. *Algunos aportes para el estudio de la historia de la ciencia en el Peru.* Lima: CONCYTEC, undated.

Zuñiga, Angel Ruiz, ed. *Historia de las matemáticas en Costa Rica.* San José: Editorial de la Universidad de Costa Rica, 1995.

*Ubiratan D'Ambrosio*

# Middle East

The historian of the movement of Western science into and through the Middle East is faced with an immediate reason to rethink science, the West, and history itself. The history of this movement travels through the ages of exploration, colonialism, and imperialism through to the present. One must account, furthermore, for the indigenous contributions to science as a Western modern social institution. To proceed in this manner, however, would amount to speaking of a science without and outside history and to embody a wholesale Western posture. Perhaps the most pernicious assumption is that modern science is portable. Such a history would also naturally employ a fairly recent construct—the Middle East—as if it has always been there. Such a history would automatically erase the "before."

A concerted effort, commencing with the Abbasid caliphate in Baghdad and continuing over more than seven hundred years, with renewed surges of energy in various parts of the Arabic-speaking world, under various reigns, and involving Christian, Jewish, and Moslem Arab as well as non-Arab Moslem scholars equally as subjects in discovery, provided the momentum for the rise of a scientific West. Arab translators of the Middle Ages played a key role in the transmission of knowledge from other cultures to Europe. During this period, the Arab-Islamic tradition already epitomized the autonomous and universalistic ethos of modern science. Within this tradition, from Damascus to Andalusia and from Cairo to Samarkand, one can see over many centuries the development of theory, advanced mathematical tools, experimental methods, instrumentation, and a prototype observatory, hospital, and college. These inventions and discoveries were transmitted to Europe most intensively between the twelfth and sixteenth centuries, entering a West that was on the verge of becoming a global power. These transmissions paved the way for the breakthrough toward a relentless scientific revolution vital to a rising social order.

Through the use of methods rooted in my work (Makhoul 1978) to decipher the modern and drawing on my subsequent practice to establish during the 1980s a model of science that seeks commensurable theories of knowledge and development (Makhoul 2002), the Middle East suddenly becomes a uniquely illuminating site allowing for turning seemingly impossible propositions into obvious "of courses."

The dynamic of science in history neither begins with the movement of Western science into the region nor consists of a one-way flow. Rather, it is a somewhat cyclical course of complex processes that materialize in three distinct historical epochs that shape the intellectual landscape within this area's endlessly reconfigured political geography: first, an Arab-Islamic scientific revolution that delivered Europe's Renaissance; second, the appearance of Western scientific models a couple of centuries later in the provinces of a feudal Ottoman Empire on the verge of being supplanted by modern colonial rule; and third, the preeminence of Western science in postcolonial state policy, which takes us from the "Israeli exception" to the second American invasion of Iraq as scientific specter. This completes a full cycle in the movement of Western science into and through the Middle East.

## The Scientific Revolution of Medieval Arab Society

In the twelfth and thirteenth centuries, no Western astronomers or cosmologists could compare to such Arab-Islamic figures as Ibn Bajja (Avempace), Ibn Rushd (Averroes), Al-Bitruji, Al-Urdi, Al-Tusi, and Qutb al-Din al-Shirazi. The astronomical school of Maragha in western Iran made distinguished contributions to the development of modern science and astronomy, amounting to a scientific revolution before the Renaissance. The work undertaken at the Maragha observatory in the thirteenth and fourteenth centuries under the leadership of Nasir al-Din al-Tusi, and culminating in the work of Ibn al-Shatir (1375), the timekeeper of Damascus, gave us the first non-Ptolemaic (though still geocentric) models of the universe. Copernicus later adopted and duplicated those models, particularly in the case of the moon. In later centuries Europe adopted the Islamic observatory model and its instrumentation, as illustrated by Taqi al-Din's Istanbul observatory whose design is reflected in Tycho Brahe's Uraniborg observatory. Ibn Bajja (Avempace) enabled Galileo Galilei to generalize John Buridan's impetus theory and transform it into a general inertial dynamics.

The influence that Ibn Sina (Avicenna)'s *Canon* had on medical theory and practice from the fourteenth to the sixteenthth century illustrates the significant impact that Middle Eastern science had on the discussion of scientific method in the European Middle Ages. It took more than two centuries before Europeans (notably Realdus Columbo, 1510–1559) were able to improve on Ibn al-Quff's anatomical description of the heart and circulatory system and Ibn Nafiz's discovery of the lesser (pulmonary) circulation of the blood between the heart and the lungs.

The one field in the classical Arab-Islamic tradition considered to have been the most influential for the development of a discipline is optics. Continual advances, from Yaqub al-Kindi in the ninth century to Ibn al-Haytham (Alhazen) in the eleventh and Kamal al-Din al-Farisi in the fourteenth, laid the foundation for modern optics. This was vital for scientific theory in the West and for the development of the telescope and the microscope—two instruments that were to play major roles in the development of medicine and astronomy. In his *Optics*, Ibn al-Haytham showed how to solve "Alhazen's problem"—to find the location where a reflection of a given light source from a given mirror will appear to a given observer—and subsequently undertook the first systematic investigation of images produced by mirrors of various shapes. His theory of vision went beyond the scope of the Aristotelian and Ptolemaic programs to construct the concept of an optical image of the eye, which informed a psychological explanation of visual perception that has recently attracted the attention of present-day psychologists. His experimental theories of light and color and methods to establish mathematics as demonstrative science irreversibly superseded the hitherto prevailing, limited vision-oriented focus of the field. The extension of disciplinary boundaries left their imprint also on the subsequent development of a field, now called *Perspectiva*. Leading historians of science point out that Ibn Al-Haytham's optics directly influenced Kepler's theory of retinal image; that his experiment to explain the rainbow was replicated by Descartes in the seventeenth century; and that even Newton performed experiments identical to his regarding refracted light in vials of water.

This classical scientific revolution contributed to the core foundation of European university curricula: Ibn-al Haytham's *Optics*, al-Khwarizmi's *Algebra*, and Ibn Sina's *Canon* were among the decisive new thirteenth-century texts which, once translated into

Latin, laid the true foundations for the continuous development of science to the present in the West.

## Western Science in the Ottoman Empire and Beyond

How Western scientific models were received in the Ottoman provinces depended upon each province's relative power and geographic proximity to Europe. The Turkish center of the Ottoman world, however, would be the first environment to introduce Western science, reluctantly and selectively. Although translations of the modern European medical and astronomical works, including the Copernican system, had begun already by the middle of the seventeenth century, a printing press arrived in Istanbul only in 1727 and met with great resistance. It was not until the late eighteenth century that Ottoman Turkey began to import Western models of higher education. In nearby Iran, Shia orthodoxy dominated through the 1800s. Sufi mystics believed that study could reveal hidden meanings in the text of the Quran (Koran) and that religious authorities ought to keep "dangerous ideas from the West" out of Iranian society.

Emigrant Jewish scholars such as Ishak Efendi (d. 1836) were instrumental in the transfer of science to Ottoman Turkey. Although the introduction and spread of Copernicus's heliocentric concept into Ottoman Turkey was, in contrast with Europe, not at all contentious and even perceived as more consistent with religion than Ptolemy's geocentric system, such Western modes of thought as positivism and biological materialism were met with resistance.

At that historical juncture, the Ottomans needed immediate practical  transfers of technical knowledge to strengthen their military power. Theory, experiment, and research were disregarded, even in the program of the newly established European-type Imperial School of Engineering. In 1863 an American entrepreneur and businessman founded Robert College, now the University of Zoospores. Turkey's first university (Darulfunun), now Istanbul University, was not founded until 1900. In the second half of the eighteenth century, individuals started to establish learned societies and professional associations similar to the corresponding bodies in the West, which, with their legal statute and modus operandi, added to scientific culture in the region.

The Arab world, with the exception of Morocco and Yemen, was part of the Ottoman Empire, and it paradoxically entered the modern era from that precarious location. In this rather suppressed periphery, the appearance of Western models took on an entirely different course. It was only after the French occupation of Egypt and Palestine in 1798 that Western models of scientific and educational institutions suddenly appeared in the Arab world. Napoleon founded the first Arab learned society, L'Institut de l'Égypte. In retrospect, Napoleon's invasion can be viewed as the end of Ottoman attempts to impose physical and intellectual isolation on the Arab provinces, which only Lebanese and Syrian merchants could evade by maintaining contact with the West.

Immediately thereafter, the Ottoman ruler of Egypt, Mohammad Ali, vigorously adopted French models of engineering and medical schools, created a ministry of education, sent the first students to Europe, and invited mass translations of Western and Eastern scientific traditions into Arabic and Turkish. More pointedly, his son, Ibrahim Basha, in Syria, required officers and soldiers to learn geometry and mathematics and introduced a tolerant policy towards foreigners and indigenous Christians that materialized in the wake of missionary schools, as well as the development of scientific activity, including the field work of Saint-Simonians in Egypt and Algeria. Medical schools and hospitals were the first scientific institutions to be established in the Arab world. A fierce competition between two rival missionary groups—the American Protestants and the French Jesuits—in the middle of the nineteenth century sparked the creation of the first Western educational institutions in the Arab world, both in Beirut: the Syrian Protestant College (later the American University of Beirut) and the Jesuit-founded St. Joseph University.

Under the liberal rule of Khedive Ismail, freedom of the press inspired exponential growth in the number of periodicals. Along with the work of individuals such as Butrus al-Bustani, who published a seven-volume encyclopedia (*Dairat al-Maarif*) on modern knowledge in 1876, these new periodicals likewise provided the key mechanisms to diffusing Western scientific theories in the Arab world. The most committed of these periodicals was *Al-Muqtataf*.

Established in Beirut in 1876, and lasting (significantly) until 1952, when colonialism ended, *Al-Muqtataf* constituted the main channel for transmitting new ideas about the "technical aspects of scientific civilization" to the Arab world. In its first year of publication it introduced Darwinism, including the Huxley–Bishop Wilberforce exchange, to the Arab reader. It then served as a continuing forum for a vigorous debate over the materialist doctrine of natural selection, and it led the movement for the secularization of thought. New York University awarded its founding editors Sarruf and Faris Nimr—both graduates of the Syrian Protestant College—honorary doctoral degrees in 1890 after King Oscar of Sweden had awarded Nimr the Order of Golden Education for his contributions to educational development in the Arab world.

In contrast with *Al-Muqtataf*'s political purpose, another major journal, *Al-Mashriq*, which was launched by an Arab Catholic, Father Louis Cheikho, set out against a sweeping secularizing current to tackle scientific issues from distinctly religious and theological perspectives, thus admitting interpretations of Darwin's evolution in terms of Quranic authority. Scholars such as the Persian-born Jamal al Din al Afghani, who was expelled from Egypt by the British, as well as his Egyptian disciple Muhammad Abduh, who held that there was no contradiction between Islam and reason or between Quranic teaching and science theories, who were suppressed in the pages of *Al-Muqtataf*, were encouraged to speak their mind in *Al-Mashriq*.

At odds with the resistance of politically privileged Turkey, the Arab provinces enthusiastically received Darwin's biological materialism, even though Arab scholars categorically rejected the misapplication of Darwinism by Europeans in the pursuit of racist policies and for the justification of war. Nor were they particularly interested in the biological aspects of this theory. Arab scholars mainly yearned to strike at the foundations of repressive feudal Ottoman rule by means of the widespread transmission of the modern ideas of natural sciences.

Indeed, during the second half of the nineteenth century, intellectual life in the Empire's Arab province was invigorated by the encounter with Western science. For one thing, they readily recognized in the new Western civilization a hybrid of self and otherness—a combination of many cultures, their own civilization included. The challenge invited further vigilant revisiting of their dormant classical Arab-Islamic scientific tradition. This revisiting in turn facilitated the secularization of thought known in Arabic as *Nahda* (awakening), the by-product, however paradoxically, of all the aforementioned interventions.

The introduction and spread of Western science by an empire seeking to strengthen its military power against its subordinate populations also served to mobilize the politically underprivileged intellectuals to win, consciously or inadvertently, the ideological battle for modern colonial rule.

## Science in the Postcolonial Middle East

In the postcolonial epoch of independent modern nation-states we witnesses a dramatic shift from the predominance of the ideological that distinguished the former epoch toward the pre-eminence of the economic. Only in (post)colonial Israel does the ideological become the sublime object of the movement of science.

Foreign expertise, technology products, and modernization theories replaced missionaries and the spread of awakening ideas. The movement of Western science took on the form of commodities in search of expansive markets. A postcolonial Middle East with a new, geopolitical reconfiguration molded around a perpetual, violent Arab Jewish conflict and the imperatives of the modern oil economy presented a boundless potential market for Western scientific services (consultancies) and technology products.

It is not surprising, therefore, that in 1945, as the struggle for national independence from colonial rule peaked and the nation-state loomed on the horizon, colonialist Barton E. Worthington would endeavor to project the scientific needs of the Arab world, determining that these will be met by import of scientific manpower from foreign sources. The resulting colonial document, *Middle East Science*, became the blueprint for the science policy of the Arab "postcolonial" state.

Importation of scientific services and technology products was bound to emerge as the Arab state's only coherent science policy. Paradoxically, the demand for Western technology in African and Middle Eastern political economies reached its fullest intensity in the context of the policies of "economic indigenization," which sought to propagate the doubtful authenticity of the postcolonial state. During this epoch the Middle East became a grand market for the products of Western scientific activity. Within the primacy of its new, commodity form, the movement of Western science during this epoch served to sustain further a postcolonial political process reflecting the Cold War ethos and firmly remolding the area in theory and in practice around the demands of an arbitrary Arab-Jewish conflict. This conflict served as impetus for unrelenting demand for military technology transfer and provided experimenting grounds for new war technologies. A critical review of the Arabic, English, and Hebrew scholarly discourse on technology transfer into and through the Middle East and Africa has resulted in an annotated bibliography (Jerusalem Institute for the Study of Society, 1986) that vividly illustrates the intellectually stifling movement of Western frames of reference into and through the Middle East, serving to render this conflict eternal.

Complementing the constructivist landscape architecture of the modern national self, the importation of science involved also wholesale transfer of monumental science infrastructure, lucidly echoing the divisive identity politics of the global political cultural ensembles of the time.

In the early 1960s, as the superpowers' enmity rapidly spilled over to the field of science as did the missionaries' enmity in the preceding phase, the United States sponsored an "Atoms for Peace" program; the Western powers in the Baghdad Pact, later known as the Central Treaty Organization (CENTO), established the CENTO Research Institute of Nuclear and Applied Science in Tehran. Egypt and Iraq, on the other hand, acquired 2-megawatt nuclear reactors from the Soviet Union together with scientific and technical manpower. Perhaps the only source of encouragement of indigenous local research activity was United Nations agencies, but, sanctioned as indigenous, such research activity readily ceases to be science. The establishment of an Arab Regional Center for the Transfer and Development of Technology might inspire new directions, particularly in the context of the information technology revolution.

The Arab states' structural reliance on importation of scientific service has led to systematic discouragement effects, manifested in the near-total absence of research activity in universities and the plight of the Arab scientists who must shun their amply modernized Middle East and emigrate to a Western diaspora if they insist on doing science. This anomaly, which the nationalism and theoretical poverty of the existing interpretation invite positing as a lingering Ottoman legacy serving to obstruct a national enterprise, is actually rooted in the very purpose of the modern postcolonial state.

In sharp contrast with the "Arab" (postcolonial) state's policies of discouragement of a national enterprise of science, the "Jewish" (postcolonial) state goes out of its way to encourage its Jewish scientists and universities to engage vigorously in committed local research and international scientific activity, exclusively in service of national objectives. At the core of those national objectives, and the subsequent rewards and promotion criteria, is the creation of incentives for mobilizing new Jewish scientific immigration from the West. Israel is probably the only country with a special "Center for Absorption in the Sciences," itself serving as incentive for attracting continuous scientific immigration, however ethnically exclusivist. At odds with the global pattern of scientific immigration from South to North and from East to West, Jewish scientists are expected and keenly encouraged to shun their privileged locations in Western capitalism and immigrate to its periphery. The rationale for the "Jewish" state itself facilitates this unfamiliar process. This continuous free acquisition of ready-made Western producers of scientific knowledge also invites positing Israel as an instance of imperialism in reverse. The creation of Israel in a subsequently dismantled Palestine can be seen as the personification of the movement of Western science to flourish and strike roots and replicate itself outside its Western milieu, emerging as the nearest thing to a scientific settler colonialism. The Arab postcolonial states' technology import apparatus emerges as the functional equivalent of the Israeli postcolonial state center for absorption of scientific immigration: Both serve contrasting but complementary imperatives of the Western states' function in the accumulation of capital and in handling its distinctive problems of legitimation.

From a Palestinian point of perspective, this Western, scientific Israel is the sum of the total means of violence enlisted in substituting a reified, all-inclusive Jewish social entity for an inclusive Palestinian historical social entity. The singularity of Israel's political economy is often emphasized in order to render impractical its interpretation in other than Zionist value preference terms. However, when viewed through the lens of my original departure from the basic structure of Israeli sociology underpinning this analysis (Makhoul 1978), the amply revered enterprise of science in Israel appears as it really is: the key to securing the continuous relocation of the "Jew" as exists in the Western states' inherent problems of legitimation.

Such key Zionist figures as Theodore Herzl and David Ben-Gurion, whom apologetically critical approaches to the political economy of science in Israel unquestionably cite in support of an idealized notion of a harmonious marriage between Zionism and science, capture perhaps the essence of my basic thesis regarding the exceptional place of science in Zionism in expressively communicating Zionism's need for and commitment to science.

The above three-epoch model reveals the extent to which the actual history of science can be most fully understood when it is read sociologically, but through the lens of a critical theory of the society that undoes the superstructure-base dichotomy. Only during the classical Arab-Islamic tradition has science epitomized modern science's purported autonomy and universality, however supported by religious authorities. It formed the only unmistakably portable universal in a world of intensely incommensurable situated parochialisms.

Further, what in contemporary society is considered science cannot evidently develop in subordinate peripheral capitalist formations, Middle Eastern or other. Indeed, the Israeli exception, when decoded, most lucidly validates this thesis. Likewise, the notion of "indigenous contributions to science as a Western modern social institution," for which the historian of the movement of Western science must, as a matter of course, account, emerges as farfetched. The Arab scientists who must shun even the postcolonial Middle East to go into a western diaspora in order to do science must also cease to be indigenous. Conversely, what is sanctioned as indigenous automatically ceases to be science. Perhaps the only instance of indigenous contribution to science as a Western modern social institution can be found in the Israeli scientists' expected commitment to the Zionist ideal of creating incentives for continual Jewish scientific immigration from capitalism's Western center to its periphery.

By the same token, when looking at the contrasting movement of Western science into the Middle East through the lens of this early history, one cannot but wonder what had happened to the autonomy and universality of science in Europe.

With the Israeli exception deciphered, the actual history of the Middle East paradigmatically validates my theoretically informed thesis that modern science is uniquely *immovable*. Given the new location that it comes to inhabit in the distinctly modern historical context, science is impotent to move into capitalism's periphery to flourish. It can do only the tasks beyond science that capitalism has given to it. The war on Iraq is a collision with the perceived violation of this rule, which in turn brings us to the global erosion of boundaries and the promise that the computer revolution holds for the rise of a human science—a science that rises above parochialism and territoriality.

## BIBLIOGRAPHY

Crombie, Alastair C. "Avicenna's Influence on the Medieval Scientific Tradition." In *Avicenna: Scientist and Philosopher*, edited by George M. Wickens, 84–107. London: Luzac, 1952.

Duke, Dennis. *Ancient Planetary Model Animations*. Florida State University. http://www.scri.fsu.edu/~dduke/models. Interactive displays illustrate the relationships between the theories of Ptolemy, al-Tusi, Ibn al-Shatir, and Copernicus.

Faraj, Nadia. "*Al-Muqtataf* 1876–1900: A Study of the Influence of Victorian Thought on Modern Arabic Thought." *Ph.D. diss.*, Oxford University, 1969.

Hamarneh, Sami. "Arabic Medicine and its Impact on the Teaching and Practice of the Healing Arts in the West." *Oriente e Occidente* 13 (1971): 395–426.

Hogendijk, Jan P., and Abdelhamid I. Sabra, *The Enterprise of Science in Islam: New Perspectives*. Cambridge, UK: MIT Press, 2003.

Huff, Toby E. *The Rise of Early Modern Science: Islam, China, and the West*. Cambridge, U.K., and New York: Cambridge University Press, 2003.

Ibsanoglu, Ekmeleddin. "Ottoman Science." In *Encyclopedia of the History of Science, Technology and Medicine in Non-Western Cultures*, edited by Helaine Selin, 799–805. Boston: Kluwer, 1997.

Iskandar, Albert. "Ibn al-Nafiz." In *Dictionary of Scientific Biography*, edited by Charles Coulston Gillispie, vol. 9. New York: Charles Scribner's Sons, 1968.

Kheirandish, Elaheh. "Optics in the Islamic World." In *Encyclopedia of the History of Science, Technology and Medicine in Non-Western Cultures*, edited by Helaine Selin, 795–799. Boston: Kluwer, 1997.

Lindberg, David C. *Theories of Vision from Al-Kindi to Kepler*. Chicago: University of Chicago Press, 1976.

Makdisi, George. *The Rise of Colleges: Institutions of Learning in Islam and the West*. Edinburgh: Edinburgh University Press, 1981.

Makhoul, Najwa. "The Proletarianization of the Palestinians in Israel." *Ph.D. diss.*, Massachusetts Institute of Technology, 1978.

Makhoul, Najwa. "Agricultural Research and Human Nutrition: A Comparative Analysis of Brazil, Cuba, Israel and the United States." *International Journal of Health Services* 13.1 (1983): 15–31.

Makhoul, Najwa. "Technology Transfer and the Postcolonial Political Process: Middle Eastern and African Political Economies in World System Theories." Jerusalem Institute for the Study of Society, Working Paper No. 5, 1986.

Makhoul, Najwa. "Can Theory Be Leading History? The Portable Jerusalem Institute for the Study of Society as a Vehicle for Critical Inquiry." Second European Conference of the International Society for Literature and Science on "Experimenting Arts and Sciences," Aarhus, Denmark, May 8–12, 2002.

Moody, Ernest A. "Galileo and Avempace (Ibn Bajja): Dynamics of the Learning Tower Experiments." In *Roots of Scientific Thought*, edited by Philip R. Weiner and A. Noland, 176–206. New York: Basic Books, 1957.

Rashed, Roshdi. "Kamal al-Din al-Farisi." In *Dictionary of Scientific Biography*, edited by Charles Coulston Gillispie, vol. 7. New York: Charles Scribner's Sons, 1973.

Saliba, George. *A History of Arabic Astronomy: Planetary Theories during the Golden Age of Islam*. New York: New York University Press, 1994.

Sayili, Aydin. "The Emergence of the Proto-Type of Modern Hospital in Medieval Islam." *Studies in the History of Medicine* 4 (1980): 112–118.

Siraisi, Nancy G. *Avicenna in Renaissance Italy: The Canon of Medical Teaching in Italian Universities after 1500*. Princeton, NJ: Princeton University Press, 1987.

Smitha, Frank. *Worldhistory@fsmitha.com. 16th to 19th Centuries. Philosophy, Science and History.* http://www.fsmitha.com/h3/h45-psh.html.

Steinberg, Gerald, M. "The Political Economy of Science and Technology in Israel: Mutual Interests and Common Perspective." in *Scientists and the State: Domestic Structures and the International Context*, edited by Etel Solingen. Ann Arbor: University of Michigan Press, 1994.

Zahlan, Antoine B. *Science and Science Policy in the Arab World*. London: Groom Helm, 1980.

Ziadat, Adel A. *Western Science in the Arab World: The Impact of Darwinism, 1860–1930*. London: Macmillan, 1986.

*Najwa Makhoul*

## United States and Western Europe

Although it drew initially from important and sophisticated North African and Near Eastern traditions of mathematical, astronomical, and medical knowledge, Western science has traditionally been viewed as originating in classical Greece among thinkers whom Aristotle identified as *physicoi* rather than *theologoi*. What distinguished the attempts of these thinkers to account for natural phenomena from those of their near contemporaries between about 600 B.C.E. and 350 B.C.E. was the establishment of explicit criteria for judging the validity of knowledge claims regarding the natural world. No single early Greek speculator about nature incorporated all of these criteria, but by the time that Greek science reached its most sophisticated levels at the Alexandrian Museum during the first centuries of the common era, five criteria were widely accepted. First, science was to seek universal explanations rather than particular ones; that is, rather than seeking the reasons for a particular thunderstorm or flood, it sought general accounts of such classes of phenomena as storms and floods. Second, natural phenomena were to be accounted for in naturalistic terms, rather than by appeal to supernatural or divine agencies. Third, natural knowledge was expected to exhibit logical coherence, which, by 300 B.C.E., meant conformity to the syllogistic logic codified by Aristotle. Mathematical knowledge, in particular, demanded logical proofs. Fourth, the truth claims of a credible theory were expected to be consistent with sensory evidence and/or empirical results. Finally, though there were significant exceptions, there was a widespread expectation that natural phenomena could be described in the language of mathematics.

A few early Greek investigators of natural phenomena seem to have been interested in the application of that knowledge to practical purposes, including calendar making, medical therapies, and the creation of military technologies, but the two major natural philosophers of Greek antiquity, Plato and Aristotle, promoted a focus on knowledge for its own sake, overtly discouraging applications. This point of view predominated in the West during the medieval period. Its rejection in favor of a more utilitarian attitude toward natural knowledge, which emerged in the dynamic culture of Renaissance Italy and in connection with a Christian humanist ideology espoused by persons such as Tomasso Campanella in Italy, Johanne Andreae in Germany, and Francis Bacon in England, characterizes the transition to modern science in the West.

With the military collapse of the Roman empire and the concomitant deurbanization of Europe, the high level of scientific activity and natural knowledge characteristic of Alexandria between circa 300 B.C.E. and 200 C.E. declined rapidly. A small residue of natural knowledge, encouraged primarily for its role in biblical exegesis, was retained within the Western monastic tradition. The most sophisticated natural philosophy, astronomy, mathematics, and medicine from Alexandria were transmitted into the Near Eastern culture in Syria and subsequently into the Islamic tradition, largely through the efforts of members of the heretical sect of Nestorian Christians, who translated the advanced works of major Greek scientists first into Syriac and then into Arabic. This material found its way back into Europe during the twelfth and thirteenth centuries as a consequence of culture contact between Christian and Muslim scholars in the Iberian peninsula and in Sicily and as a consequence of the growth of a new European institution, the medieval university.

The university developed in Europe because rapid population growth, reurbanization, and the growth of a market-based economy created demands for civil and canon (church) lawyers, physicians, and educated clerics to serve a variety of functions in a society that increasingly demanded careful record keeping. These demands led first to the expansion of cathedral schools and eventually to the creation of institutions characterized by the existence of advanced faculties in law, medicine, and theology, all served by a common curriculum in the liberal arts (rhetoric, grammar, logic, geometry, arithmetic, astronomy, and music), which followed a pattern initiated by such Roman educators as Martianus Capella. As the sociologist Joseph Ben David and the historian Toby Huff have shown, the character of the medieval European university had a major impact on the rapid development of the sciences in Europe relative to that within Islamic cultures and in China.

## Why the Scientific Revolution Occurred in Europe

Within Islam and in China, training in the professions of law, medicine, and theology occurred principally in connection with apprenticeships to master teachers who specialized in one of the professions or in institutions devoted to a single profession. As a consequence, there developed no cultural specializations devoted to topics in the sciences. In Europe, on the contrary, where the arts curriculum preceded and served as a foundation for more specialized curricula, the teaching of arts subjects—including astronomy, arithmetic, and so on—within the university became an important profession in its own right. As a cadre of scholars whose job it was to study subjects that included the natural sciences developed, there was a drive to recover and understand the most sophisticated natural knowledge of antiquity. Aristotelian texts in natural philosophy soon became the core of the arts curriculum, but the more mathematically sophisticated works of Archimedes in mathematics and Ptolemy in astronomy, as well as some of the previously untranslated texts of Galen in anatomy, followed through the next three centuries.

Beginning in the late fourteenth century, the late medieval scholastic traditions of science were transformed into what we recognize as modern science. Starting in Italy, the traditions of Greek mathematics began to be appropriated by artisanal groups, including artists, architects, and engineers, for applica-

tions in the fine arts, architecture, and mechanical device design. This trend initiated a fusion of scholarly and artisanal interests that was unparalleled in other cultures and that was critical for the processes we identify with the term *scientific revolution.* If one looks at Galileo's foundational text in mechanics, *Discourses Concerning Two New Sciences,* of 1638, for example, one finds that it addresses problems faced by ship builders and machine builders in the Venetian Arsenal and by military engineers interested in the paths of projectiles fired from mortars and ballistae (machines for launching rocks, bombs, and the like).

The nearly simultaneous growth of a publishing industry grounded in movable-type technology, of humanism grounded in an admiration for action in the world and for the original writings of classical Roman and Greek authors, and of European commercial voyages throughout the globe—all of this growth served to further stimulate and redirect scientific activity. In fields such as astronomy, for example, printing stabilized the texts so that copying errors did not constantly creep in. As a consequence, reliable tables of planetary and lunar positions based on Greek astronomical theories could be produced and compared with observations with some assurance that lack of agreement was based on more than random arithmetical errors. When traditional Earth-centered theory diverged from observation, especially with regard to the dating of Easter, for instance, astronomers such as Nicolas Copernicus were driven to revive ancient sun-centered theories and to work out their details. Similarly, when Galen's anatomical works were recovered and certain claims were found to diverge from the results of human dissections, European anatomists such as Andreus Vesalius began to correct classical works.

The ability to print many copies of illustrations had a comparable impact on the sciences, as detailed illustrations of plants and animals, for example, allowed for the publication of herbals and other works of natural history that could actually be used in the field for identifying entities. This became particularly important as voyages of exploration located commercially valuable plants and animals—tobacco, corn, cane sugar, and potatoes from America, and teas and dyestuffs such as indigo from Asia, for example. The discovery of thousands of new kinds of plants and animals in turn called for the development of new systems of classification to keep the growing amount of information organized. As sea voyages increasingly departed from land, the publication of maps and charts as well as the need for trained navigators and methods for establishing longitude stimulated both the teaching of mathematics and astronomy and the development of new mathematical and astronomical knowledge. In consequence, the number of mathematically literate persons and the level of mathematical knowledge rose rapidly in the West.

## Institutional Support for Science in Early Modern Europe

The extent of natural knowledge grew, and its character became increasingly utilitarian. These trends promoted institutional support for scientific activity beyond the universities in secular courts. One example was that of the Holy Roman Emperor Rudolph at Prague, whose court mathematicians and mechanics included the astronomers Tycho Brahe and Johannes Kepler and the mathematician and mechanical genius Joost Bürgi, who built clocks that imitated the motions of the heavens and who invented logarithms to ease Kepler's calculations. Another was that of the Tuscan Prince Federico de Medici, who sponsored the Accademia del Cimento, which included Nicolas Steno, the author of the first serious theory of geological change; Alphonso Borelli, the student of Galileo who initiated an important tradition of iatromechanics; and Evangelista Torricelli, the inventor of the barometer. The sixteenth century also saw new forms of institutional support within higher education institutions. Beginning around 1565, for example, the Jesuits established positions as "scriptors," which lasted from two to six years, during which outstanding scholars were freed from their teaching duties in colleges or universities to focus on research.

Universities remained important centers of scientific activity, but in the seventeenth century two new patterns of institutional support for science emerged. In the commercial center of London, the Royal Society of London for the Promotion of Natural Knowledge was established in 1662 with a charter from the crown, although without state support. Members paid to belong to the organization, which initially offered demonstration experiments at each meeting and which soon established one of the first scientific journals, the *Transactions of the Royal Society of London*, to which any person could submit a scientific paper for publication on the advice of members of the society acting as referees. This society established a pattern followed by provincial societies that sprang up throughout Europe and the Americas over the following two centuries as well as professional societies that followed in the nineteenth century.

Four years later, the Paris *Academie des sciences* was established by the French monarchy. *Academiciens* were paid by the crown and were expected both to pursue their own scholarly work and to provide expert advice to the government on technical issues. Their *Journal des savans*, published papers that had been presented at *Academie* meetings. The French *Academie* provided the model for numerous national scientific academies, including those at Berlin and Saint Petersburg.

## Science in the Enlightenment and the Early Industrial Revolution

The publication of Isaac Newton's *Mathematical Principles of Natural Philosophy* (the *Principia*) in 1687 and many popularizations of Newtonian science by John Locke and others initiated a period of widespread interest in the natural sciences and of attempts to appropriate the methods of inquiry that had been successful in the natural sciences into moral philosophy and the investigation of human nature and society, giving rise to such disciplines as political economy, psychology, anthropology, and "philosophical history," which corresponds very roughly to modern sociology. In the salons of major continental cities, at popular science lectures given by itinerant lecturers almost everywhere in

Europe and the Americas, in the meetings of such international organizations as the Freemasons, and at local social clubs such as the Lunar Society of Birmingham, scientific ideas were discussed and applications envisioned among middle- and upper-class audiences.

The seventeenth and eighteenth centuries additionally saw a significant increase of women in science, which had largely been closed to them while universities (which allowed only male students) were the nearly exclusive centers of scientific activity. Wealthy aristocratic women, such as Margaret Cavendish (who was part of a family intellectual circle that included Thomas Hobbes), Emilie du Chatelet (who translated the technical portions of Newton's *Principia* into French), and the experimental natural philosopher, Laura Bassi (who actually became the first woman university professor of natural philosophy at Bologna in 1727), were educated at home privately by their families and entered into scientific activities. Not until the mid-nineteenth century were women allowed direct entrance into higher education, first into separate women's colleges and only later into coeducational universities. More frequently, middle-class women such as the astronomers Caroline Herschel and Maria Winkelman served as apprentices or helpers to scientifically active male family members before they developed independent reputations. At the craft level, increasing numbers of women served as apothecaries, scientific illustrators, collectors of specimens of plants and insects, preparers of anatomical models, and so on throughout the eighteenth century.

At the end of the eighteenth century, Europe was shaken by two momentous events: the Industrial Revolution, whose initial phases occurred in England, and a democratic explosion, whose original was the French Revolution. Many early commentators, including Edmund Burke, Joseph de Maistre, and William Blake, tended to see both revolutions as strongly dependent on scientific ideas and attitudes; and many postrevolutionary thinkers, including the economist Jean-Baptiste Say, and the early socialists Henri de Saint-Simon and Robert Owen, were convinced that at least the Industrial Revolution was heavily science-dependent. In France, the École Polytechnique, one of the premiere scientific institutions of the early nineteenth century, was established largely in the belief that its graduates would help France compete economically with British industries.

Most twentieth-century Western scholars tended to downplay the significance of science in fomenting both the French and industrial revolutions and argued that technological innovations, in particular, did not become closely linked to scientific knowledge until the mid-nineteenth century. But recent work by such historians as Margaret Jacob and Peter Mathias has emphasized once again the economic importance of science-based technologies—especially mechanical and chemical technologies—in the seventeenth and eighteenth centuries.

## The Nineteenth Century: The Century of Science

The nineteenth century saw the extension of basic literacy and subsequently of scientific literacy to all levels of society and to women as well as men throughout Western cultures. Organizations such as the Mechanics Institutes and Societies for the Diffusion of Useful Knowledge in nineteenth-century England, North America, Australia, and New Zealand, as well as the museum and zoological gardens movements, the World's Fairs (which began with the Crystal Palace exposition in London in 1851), and periodicals such as the *Edinburgh Review* and *Popular Science Monthly* promoted widespread interest in science in English-speaking countries, and there were comparable developments on the Continent.

Moreover, the nineteenth century was one in which scientific knowledge and technologies built upon that knowledge became the self-conscious foundations for economic and ideological developments alike. Modern industrial societies would be unimaginable without electrical and electronic technologies grounded in the sciences of electricity and magnetism developed by Michael Faraday,

André Marie Ampère, and James Clerk Maxwell. Similarly, without organic chemistry as developed by Justus Liebig and his many students, there would be no artificial fertilizers, no synthetic fabrics, no plastics, no vulcanized rubber, and no modern pharmaceuticals. Without the development of thermodynamics by Sadi Nicolas Léonard Carnot, Rudolph Clausius, Maxwell, and Ludwig Boltzmann, almost all modern, efficient, industrial processes would be unknown, not to mention automobiles and airplanes.

Although it was not immediately important for economic development, the theory of evolution by natural selection enunciated by Charles Darwin and Alfred Russell Wallace and built upon the work of such earlier theorists regarding the historical development of the earth and its inhabitants as Jean-Baptiste Lamarck, Charles Lyell, Herbert Spencer, and Robert Chambers, both refashioned our understanding of biological organisms and offered models of social and cultural development that became extremely important for socialist and capitalist ideologies alike.

## The Social Sciences and Social Policies

The Enlightenment attempt to appropriate scientific strategies for understanding human and social phenomena intensified through the nineteenth century, producing such social sciences as economics, sociology, psychiatry, anthropology, and political science, and grounding those in data analyzed through newly invented statistical methods. The most vigorous attempts at scientific understandings of social phenomena became the foundations for major ideologies and for political policies of great importance.

"Liberal" economics, as developed by Adam Smith and his followers, for example, promoted the intensification of European imperialism by insisting that economic progress could only be sustained by increasing and ensuring markets for the manufactured goods produced by advanced industrial nations. In the nineteenth century, that meant that European nations and the United States competed to ensure the control of markets in Africa, Asia, Oceania, the Caribbean, and Latin America. Coupled with an evolutionary ideology that accepted "the survival of the fittest" and justified the exploitation of the weak by the powerful, and linked to powerful military technologies built upon advanced scientific knowledge, a liberal economic doctrine justified Western nations in their effective exploitation of much of the rest of the world for their own benefit.

Even within many Western nations, including both England and the United States, free market economic doctrines allied with evolutionary ideologies justified both a pattern of increasing wealth accumulation by the few at the expense of the many and a rejection of humanitarian claims for charity to the poor. Social support for the poor was denied on the grounds that it would only encourage them to procreate and therefore to increase misery in the long run. Late in the nineteenth century and during the early decades of the twentieth century, notions of inheritance associated with evolutionary theories were also used to justify programs of eugenics and race hygiene that set out to improve the human species by discouraging the breeding of the unfit and encouraging the breeding of superior persons. These programs often included the enforced sterilization of persons deemed unfit, including those with diminished mental capacities or with moral defects, such as alcoholism or sexual license leading to illegitimate children.

At the other end of the political spectrum, ostensibly scientific economic arguments and social theories developed by such social theorists as Saint-Simon and Owen promoted cooperation rather than competition and led to arguments for providing important social services to all. As it became clear that capitalist economies were not going to follow the path proposed by what they deemed the "utopian" socialists of the early century, Karl Marx and Friedrich Engels sought to develop a more scientific socialism grounded in the detailed analysis of the growth of capitalism and of the character of productive processes. This analysis ultimately formed the core of communist ideologies that became the foundation for state governments in the Soviet Union, China, and satellite states in Europe and Southeast Asia.

## Science in Twentieth-Century Western History

Toward the end of the nineteenth century, there were a number of major changes in both the conceptual content and the social organization of the sciences in the West. The emergence of the "modern" physical sciences, with their emphases on both subatomic- and cosmological-scale phenomena were stimulated by the development of theories that abandoned many of the "common sense" notions associated with classical physics. Relativity theory involved the abandonment of traditional notions of time and space, and implied the intervertibility of matter and energy (leading to the development the atomic bomb and nuclear power). Similarly, quantum mechanics violated traditional notions of physical determinism. Later in the twentieth century, the discovery of the molecular basis of genetics offered the possibility of engineering organisms never produced within the natural order.

These developments in physics and biology have, on the one hand, immeasurably increased the scale of human impacts on the world—both for good and for ill. At the same time, they have increased the scale of support and cooperation necessary to do scientific work. Early in the century, such organizations as the Carnegie Foundation and the Rockefeller Foundation replaced private individuals as the principal supporters of what has come to be called *Big Science*. But the financial burden of such modern scientific operations as high-energy particle accelerators, extraterrestrial exploration, and the Human Genome project has placed primary responsibility for the support of big science in the hands of governments. In the case of some scientific fields that offer opportunities for rapid commercial exploitation, funding has come from major (usually international) corporations, who seek to limit access to the knowledge produced. As a consequence science, which had become almost entirely "public" knowledge in the early modern West, is being privatized to a significant degree.

*See also* **Technology in History: United States and Western Europe**

### BIBLIOGRAPHY

Ben-David, Joseph. *The Scientist's Role in Society: A Comparative Study with a New Introduction.* Chicago: University of Chicago Press, 1984. Excellent sociological introduction to the institutional development of science in the West from the time of the Alexandrian Museum to the twentieth century.

Boas, Marie. *The Scientific Renaissance, 1450–1630* (1966). New York: Dover, 1994. Subsequent works on this time period just do not match the combination of readability and attention to detail of this work, though it should be supplemented by Dear (see next citation) on issues of social context.

Dear, Peter. *Revolutionizing the Sciences: European Knowledge and Its Ambitions, 1500–1700.* Princeton, NJ: Princeton University Press, 2001. Best of a spate of short treatments of the "scientific revolution" that were produced between 1998 and 2001.

Greene, Mott. *Natural Knowledge in Preclassical Antiquity.* Baltimore: Johns Hopkins University Press, 1991. A most accessible discussion of "science" from prehistory to the time of Plato.

Huff, Toby, *The Rise of Early Modern Science: Islam, China, and the West.* Cambridge, U.K.: Cambridge University Press, 1995. Best comparative study of why the Scientific Revolution occurred in the West.

Jacob, Margaret C., *Scientific Culture and the Making of the Industrial West.* Oxford: Oxford University Press, 1997. Outstanding treatment of the cultural impact of the sciences in the eighteenth and early nineteenth centuries. Especially good on the spread of scientific knowledge and the application of scientific ideas to technological innovation.

Judson, Horace. *The Eighth Day of Creation: Makers of the Revolution in Biology.* New York: Simon and Schuster, 1979. Masterful story of the molecular revolution in biology during the mid-twentieth century.

Knight, David M. *The Age of Science: The Scientific World View in the Nineteenth Century.* Oxford: Blackwell, 1986. A very readable general introduction to science in nineteenth century culture.

Kohlstedt, Sally Gregory, ed. *History of Women in the Sciences: Readings from Isis.* Chicago:

University of Chicago Press, 1999. This volume introduces a group of outstanding writers on women in science through a series of delightful and path-finding articles. No other single volume offers nearly as many insightful perspectives covering the period from the Renaissance to the 1950s.

Lindberg, David. *The Beginnings of Western Science: The European Scientific Tradition in Philosophical, Religious, and Intellectual Context, 600 B.C. to A.D. 1450*. Chicago: University of Chicago Press, 1992. Best overall introduction to ancient and medieval Western science.

Mayr, Ernst. *The Growth of Biological Thought: Diversity, Evolution, Inheritance*. Cambridge, MA: Belknap, 1985. The most comprehensive history of modern biology up to the molecular revolution.

Merz, John Theodore, *History of European Thought in the Nineteenth Century*. 4 volumes. New York: Continuum, 2000. A reprint of the 1897–1902 original set. Although written over a century ago by a historian of philosophy, the first two volumes of this work remain indispensable to anyone interested in nineteenth-century science. The notes, which make up about 50 percent of the text, are especially valuable.

Olby, R. C., G. N. Cantor, J. R. R. Christie, and M. J. S. Hodge, eds. *Companion to the History of Modern Science*. London: Routledge, 1996. Incorporates excellent bibliographic guides to most topics and themes related to modern Western science.

Olson, Richard. *Science Deified and Science Defied: The Historical Significance of Science in Western Culture, Volume 2: From the Early Modern Age through the Early Romantic Era, ca. 1640 to ca. 1820*. Berkeley, CA: University of California Press, 1990. Unusually strong on the ideological uses of scientific knowledge and on the early development of the social sciences.

Rhodes, Richard, *The Making of the Atomic Bomb* (1986). New York: Simon and Schuster, 1995. Reprint edition. Beautifully written and comprehensive account of modern physics and the processes by which it became the most celebrated and feared science of the twentieth century.

Smith, Roger, *The Norton History of the Human Sciences*. New York: W. W. Norton, 1997. The only competent and comprehensive introduction the history of the topics that fall within the social sciences in American terms.

*Richard Olson*

# SCIENCE POLICY, DEVELOPMENT OF.

Science policy refers to government policy concerning the relationship between science, the state, and society. As a distinct policy area, it emerged with the atomic age during and after World War II. Before that, science policy existed only implicitly and in an ad hoc manner, as science often counted as part of the cultural life of a nation. In some countries this view was reinforced by a linguistic convention. In German, for example, the term *wissenschaft* even today includes scholarship in the humanities. The postwar period, however, narrowed the scope to mostly natural sciences, medicine, and engineering. In this article the background to the emergence of science policy is sketched and subsequent phases of development are reviewed.

## Marxist and Liberal Images

In the former Soviet Union science policy existed from the 1920s onward, and after World War II it was associated with a "science of science." Marxist circles in the West also argued that science can and should be planned. The case was forcefully put by the physicist John Desmond Bernal, who is usually considered the founder of science policy studies—that is, the academic field for analysis of resource allocation, organization, incentives, and priorities in science.

At first, research was considered a morally and politically neutral activity; good and evil lay only in the implementation of its results. Socialists held that science was frustrated and prostituted by the inherent inability, as they saw it, of capitalist society to utilize the fruits of research for the benefit of humankind. Only a new and socialist order of things would be able to unleash the full progressive potential of science. After the war, they emphasized the need to make science more efficient as a social institution and focused on planning, labor force, financial resources, instrumentation, and infrastructures. All these are factors belonging to science policy, which became identical with planning and state support.

The socialist view evoked strong opposition from more traditional scientists. In Britain, the chemist Michael Polanyi founded the Society for the Freedom of Science in 1941 on the principle that science by its very nature cannot be planned, and to try to do so will cripple creativity. What is needed is a self-regulated community of investigators who are entirely in control of setting research agendas and funding. Conflict between the idea of a socially responsible science serving explicit public goals and the ideal norm of a free, or autonomous, "Republic of Science" is a fundamental conflict that continually flares up in new guises. In practice, most scientists want something from both worlds.

### Autonomy and Military Entanglement

Compromise was also the line taken in the United States by Vannevar Bush, who drafted a report called *Science: The Endless Frontier* in 1945. It framed policy thinking in the West for much of the next forty years. An MIT engineer who played a crucial role in guiding the Los Alamos/Manhattan project that produced the first atomic bomb, Bush wished to free the scientific community from its wartime bondage to the military. He argued that basic research should be given institutional autonomy along the lines of the Republic of Science model, while only applied science and engineering development were linked to external goals set by industrialists, military men, and politicians.

The dualism carried over into two different kinds of public agencies: on the one hand a basic research council, and on the other, agencies to stimulate research for societal use. Bush's original plan sought to subordinate military interests to civilian organization of funding, but this plan was not realized. When the National Science Foundation in Washington was finally established in 1951, after several years of political turbulence and delay, its mandate was restricted to basic research, while the National Institutes of Health (NIH) and the three armed services

had control over money going to applied fields.

Vannevar Bush's insistence on autonomy for basic research reflected an attempt after the war to regain greater control on behalf of the scientific elite, a move that must be understood in the light of his and other scientists' wartime experience. The Los Alamos/Manhattan project represented a new way of doing science. It involved teamwork, large-scale funding, and masses of scientists and engineers, all of whom were part of an intimate entanglement with politicians, generals, and professional administrators. Today this is called "Big Science," to distinguish it from earlier, more individualist small-scale ventures characterized by a craftsmanlike mode of research in university laboratories with limited staff and one or two strong professors.

The bombs dropped on Hiroshima and Nagasaki momentarily shattered the image of a pure and morally neutral scientific community. Control of atomic energy, in both its belligerent and peaceful uses, became a top priority, together with the economic recovery of Europe and mass production of penicillin, vaccines, and other biologically based drugs. Lobbies of physicists in several countries, led by quantum physicist Niels Bohr from Copenhagen, who found himself snubbed by Winston Churchill, urged that atomic secrets should be shared among nations, atomic weapons banned, and nuclear power brought under the control of the newly founded United Nations. This attempt at international science and technology policy failed, thwarted by the Cold War and the iron curtain that divided East and West. Instead, the first twenty years of science policy—on both sides of the curtain—was marked by a continual buildup of military technologies, in think tanks involving applied mathematics, game theory, and systems analysis to tackle strategic problems relating to national security. In addition, there were schemes to use science to bolster the economic muscle of nations. Compared to the exponential rise of military budgets—accentuated by the surprise Soviet atomic bomb tests, the Korean War and in 1957 the launching of the first Sputnik-basic research councils had very little money to allocate.

In this context physicists continued to move forward into the most prominent positions as science advisers and opinion builders, and it was also from this community that critique of mainstream policies issued (in the *Bulletin of Atomic Scientists*, for example). The Pugwash movement, stemming from a manifesto by Albert Einstein and Bertrand Russell in the summer of 1955, renounced nuclear weapons and wanted to reduce armaments, calling them a threat to the continued existence of humankind. Physicists were joined by biologists, who could boast a morally purer reputation.

Pugwash played up the "neutrality of science" and forged a bridge between scientists in the East and West, thereby becoming something of a semi-official channel of communication between the United States and the Soviet Union. The idea of scientists as an elite got a further boost when C. P. Snow in his famous 1959 Rede Lecture at Cambridge coined the phrase "the two cultures," continuing a theme from one of his novels, The New Men. He claimed that physicists carried "the future in their bones," meaning that scientists ought to replace the older generation of civil servants, who had been trained in classics and literature, as an inevitable part of the modernization of society. Because of its technocratic policy twist, the "two cultures" thesis can still generate heat in debates between scientists and scholars in the humanities.

Actually, the scientists who had been responsible for the Bomb from conception to production were unable to prevent its use against Japan. It was the politicians and military men who decided, wanting it at least partially as a warning to the Soviet Union not to expand its sphere of influence into the Pacific. Fearful that the Bomb might nevertheless be seen as an inevitable result of physics, scientists found it convenient to adopt Vannevar Bush's separation of basic research from the messy affairs of state and industry. This theory of separation allowed them in good conscience to rationalize their continued pursuit of high science, even accepting research grants from the military. Solid state and nuclear physics were not the only games in town. In the 1950s and 1960s, such fields as oceanography and geophysics were also profoundly reconfigured thanks to direct military interest in the seabed (in case of submarine warfare) and the upper atmosphere and beyond (for missile warfare). Space exploration would not have been possible, either, without a combination of basic research, fundamental technological advances such as rocket propulsion, computers, advanced telecommunications, and Big Science support structures.

## Interacting Policy Cultures

From the foregoing it is clear that formulation of science policy is the outcome of interaction between different stakeholders and their interests. The main thrust of a given policy reflects the dominance of certain players. President Eisenhower, when he was about to retire in 1961, warned about the overwhelming power of what he called the "military-industrial complex." He was referring to the invisible network of relations between the Pentagon, scientific communities, and the weapons industry that drove each other into an unending spiral, beyond popular control: a military-scientific-industrial complex. The question is, who is on top? The ultimate objective of military commitments is destructive power and protection against it. For industry, the objective is economic gain, and in academic science it is the pursuit of knowledge. In accordance with these different objectives, each domain has its own ethos (values), incentives, and organizational characteristics—in a word, culture.

For the sake of analysis one can distinguish four "policy cultures" coexisting within each society, competing for resources and influence and seeking to steer science and technology in particular directions. These cultures have a bearing on how societal relevance and related criteria are defined. There is a bureaucratic policy culture, which has a top-down, rational decision-making ideal, based on following rules. This culture is representative of the mandate derived from state agencies and political decision-making bodies. A second policy culture, that of academe, is typically driven by the concern for academic freedom and self-regulation in the setting of priorities for research agendas. Third, there is a commercial policy culture with its sights set on "value-added" and monetary profit. It has its point of departure in the marketplace, firms, and multinational corporations. Finally, there is a policy culture that is often ignored, that of civil society. It can be seen to have emerged around social movements and related nongovernmental organizations (NGOs). Environmentalism, feminism, and animal rights movements are examples. They may be seen to work through, for example, ethical committees and other kinds of panels that among other things concern themselves with risk assessments and environmental impacts of new scientific projects as well as their societal relevance. Justice, equity, social benefits, and the desire to reduce the poverty gap between haves and have-nots, including the bulk of the population in the Third World, are often at issue here. These concerns call for a shift away from Eurocentric conceptions that are often found to be entrenched in science, both at the level of its ethos and in the very fabric of its notions of rationality. Bioethics calls for an even further shift of values, away from the dominance of anthropocentrist thought in science, toward a morality or ethos that takes the worth of the biosphere and all its members as key. The academic policy culture, with its tendency to put autonomy and growth of new science first, tends to contend in various ways with the other three policy cultures.

The bureaucratic policy culture, which in many countries is dominated by the military, is based on the state administration, with its agencies, advisory bodies, councils, and committees, which are concerned primarily with effective administration, coordination, planning, and organization. In this realm, science is of interest primarily for its social uses; the concern is with how science can be used for executing state policy or providing public policy making with a scientific base.

The academic culture has as its base scientific practitioners who often wish to act as an academic oligarchy free from outside interference. This is shown in their concern for the integrity of science, which they see as bound up with the preservation of traditional academic values of autonomy, objectivity, and their own control over funding and organization.

The economic culture is related to business and management, based in industrial firms, and focuses attention on the technological uses of science. At work in this culture is an entrepreneurial spirit or ethos to transform scientific results into successful innovations for the marketplace. Market demand replaces academe's principle of science-push as the motor of innovation.

The civic policy culture, finally, is very weak compared to the other three, but it does appear in pressures to introduce regulation and technology assessment called for by public interest organizations. While the dominant cultures tend to draw science and technology policy into a "technocratic" direction, the civic culture stands for a broad "democratic" inclination.

Struggles among these cultures may explain the way in which particular problems get framed differently in different countries and contexts. Historical and institutional interconnections also change, leading to bonding between, for example, academic and bureaucratic cultures when this mutually benefits both: administrators gain legitimacy in decision making by referring to the objectivity of experts, who in turn stand to gain more funding and academic authority for their scientific fields.

The history of science policy within the Organization for Economic Cooperation and Development (OECD) is a useful window for looking at other interactions and forces. Parallel to the military buildup under the North Atlantic Treaty Organization (NATO), OECD was created in 1961, joining rich nations to inherit the American Marshall Plan's economic part of the strategy of containment against Communism. For science policy it was much more important than the United Nations Educational, Scientific and Cultural Organization (UNESCO) founded in 1948, which was geared more toward helping developing nations. A trade union of scientists, the World Federation of Scientific Workers (WFSW), inaugurated in 1948 and interacting with UNESCO, tried to promote socialist approaches but ultimately with little effect in OECD countries.

## The 1960s: A Heroic Phase with a GNP-Growth Doctrine

The OECD report *Science and the Policies of Government* (the so-called Piagnol report) of 1963 formulated the distinction between "policy for science" and "science for policy." The 1962 OECD *Frascati Manual* (named after an Italian town where a workshop was held) codified categories for tracking funds to various activities: basic research, applied research, technological development, or, simply, development (altogether abbreviated as R&D). Higher education and research counted as long-term investments in economic growth, measured by a country's gross national product (GNP). Scientific performance was then taken to be the proportion of GNP reinvested in R&D (anything from 0.5 to 2.5 percent in those days; today 3 percent or higher is considered a good benchmark). The chain of events from new ideas and discovery of physical laws to marketable products or commodities defined a linear model of innovation. Recommendations to member governments included norms for establishing scientific advisory bodies to government.

Several debates followed. The Minerva debate (named after the journal that carried it) concerned criteria for choosing between different fields of science. The director of a big physics lab, Alvin Weinberg, came up with the idea of research as an insurance policy to guarantee a country a basic fund of resources for technological development as well as a set of general criteria for deciding upon socially relevant areas. The ambition was to combine scientific excellence with societal relevance.

A related debate in the mid-1960s concerned the linear model. Military and industrial leaders challenged the belief that coupled autonomy and investment in basic research to output of useful products. The

U.S. Defense Department commissioned a study called Hindsight. A large number of known innovations were identified in order to see how much purely curiosity-driven research lay behind them. The findings confirmed the military's suspicion that most new technology grew out of engineering development, and therefore too much money was being wasted on pure research. The National Science Foundation launched a counterstudy called TRACES. It found basic research was important at numerous points in the innovation chain, thereby confirming its role as a motor for innovation.

As the war in Vietnam grew unpopular, protest movements criticized military involvement at universities and insisted that military projects skewed research away from much-needed civilian technologies. Military spenders denied this and began to speak about the spin-off of military products to products useful in civilian life (such as teflon for frying pans). In the 1980s the bureaucratic culture shifted the frame of reference to a new catchphrase, "dual use," and made lists of technologies potentially useful in both military and civil spheres.

## The 1970s: Social Relevance and Accountability

New left critiques of modern capitalist society brought a growing "politicization," or "radicalization," of science. At the OECD in 1971, a panel chaired by Harvey Brooks also marked a turning point, calling for greater social accountability and attention to environmental protection and human welfare. Based on the findings of project Hindsight and TRACES, it also questioned the role of basic science in industrial innovation. In Britain the Rothschild Report demanded that science policy take on a broader range of objectives and advocated a strict contract between consumer and contractor for all commissioned research. Notions such as mission orientation, technology policy, and social relevance arose. Technology assessment and regulatory mechanisms also found more stable institutional forms, such as the Office of Technology Assessment (OTA), created by the U.S. Congress in 1972. Pressure from women's liberation groups raised issues of gender bias, and focus fell on areas of medical technology concerning reproduction and birth control. In the wake of the oil crisis, critical analysts pointed to alternative technologies and production systems.

## The 1980s: Orchestration Policy for Innovation

Under the influence of neoliberalism in the 1980s, however, many of the experiments of the 1970s were abandoned as emphasis shifted back to economic growth, promoted in part by the growing economic and technological challenge from Japan and the newly industrializing countries (NICs) of Korea, Taiwan, Singapore, and Hong Kong. Bureaucratic and civic cultures had failed to shape functional alternatives and devise appropriate techniques to manage and organize their science and technology priorities. They also lacked the power.

The new concern with "technological innovation" was also apparent at OECD, where a new formula was to integrate science policy with industrial policy. The point was to stimulate the growth of new technologies by means of active industrial policy as well as a more intimate relationship, and active partnership, between universities and industry. Technology transfer offices and science parks cropped up on campus, and the values of the economic culture made inroads into academe. Policy-oriented studies by the economist Christopher Freeman and his colleagues at the Science Policy Research Unit (SPRU) in Sussex, UK, became important. Freeman paid special attention to the innovation strategies of private corporations and went on to analyze the so-called "long waves" of scientific and technological development. The notion of "national innovation systems" came into vogue.

Several countries started research foresight studies to identify new and emerging technologies (ESTs) and predict their potential on the global market. "High tech" was characterized by its combination of broad social impact and a knowledge base in fundamental science. In policy documents a new term was coined: "strategic research," which is defined as basic research to be cultivated with an eye to social use in the long-term future. Another term was "generic technologies." Examples of these are the multitude of special subfields within microelectronics, biotechnology, and new industrial materials (now also referred to as nanotechnology). Carried by the biotech boom, life scientists started to replace physicists in advisory functions. In the new committees and programs, state agencies tried to follow the Japanese consensual model of decision making, bringing various actors together to look into the future on the basis of their specialized knowledge to advise governments in the selection of particular technological options. Foresight involved a strong element of social construction to shape consensus among representatives from the academic, economic, and bureaucratic cultures, while the civic culture was largely left outside this "orchestration policy."

## The 1990s: Globalization and a New Social Contract

The dominance of the economic culture increased even more in the 1990s, a trend again reflected in OECD. Discussion revolved around "a new social contract for science" and the older one retrospectively attributed to Vannevar Bush's *Endless Frontier* of 1945. Emphasis now fell on stronger integration of universities with the private sector, but with a simultaneous emphasis on basic research. The entrepreneurial university challenged traditional organization of knowledge production in disciplines, replacing it with interdisciplinary interaction governed by the context of application. Hybrid centers or consortia involving diverse fields and multiple actors from university, industry, and government became popular. In the science policy literature some authors, fascinated by new patterns of communication, stylized and exaggerated the interactive mode to beat the dead horse of the old linear model. Critical analysts preferred to emphasize power relations and how competition in the global marketplace was the major force behind the new social contract that defined relations among science, the state, and private corporations. This explains why "globalization" has become one of the new catchwords. Research agendas are increasingly set by international actors and influenced by international trade relations. The extent to which a region, a country, or a local site is drawn into globalization has become the new measure of its stage of "development."

At issue are two contending logics: competition and community. The former is evident in the policies of the World Trade Organization (WTO) with its Trade Related Intellectual Property rights (TRIPs), as well as the European Commission's idea of a European Research Area. Such initiatives at the supranational level are, from another point of view, criticized for disadvantaging locally based innovation and indigenous knowledge in developing nations. Scientific communities are divided on the idea of creating a European Research Council to balance the strongly applied thrust of the Framework Programs emanating from Brussels.

## Looking Ahead

In the years ahead there is a need for huge investment in research infrastructures and scientific mobility. High on the agenda will be intellectual property rights; international diplomacy for multicountry cost-sharing of megascience projects starting at a level of billions of dollars; common approaches to such global problems as climate change, biodiversity, AIDS, and poverty; and technologies for screening and controlling populations. Free market and deregulation talk suggests we are witnessing a withdrawal of the state, but in practice WTO and EU policies actually bring about stronger socioeconomic intertwining that goes beyond the state/market divide.

As the social impact of microelectronics, biotechnology, and new materials research and development—the core generic hightech areas of the 1980s—becomes more pervasive and transforms our lives in society, we are forced to rethink what it means to be human. At that point, existential, ethical, and legal issues multiply, demanding greater collaboration across faculty boundaries—between humanities and social sciences on the one hand, and fields in the natural sciences,

biomedicine, and engineering sciences on the other. If civil society's engagement in science policy is to prove meaningful, the broadening stakeholder participation from consumer groups, environmental organizations, and other users of knowledge in consensus conferences and public consultations will have to avoid the pitfalls of co-optation so prevalent in 1990s.

### BIBLIOGRAPHY

Bernal, John D. *The Social Function of Science.* London: Routledge, 1939.

Bird, Kai, and Lawrence Lifschultz, eds. *Hiroshima's Shadow.* Stony Creek, CT: Pamphleteer's Press, 1998.

Bush, Vannevar. *Science: The Endless Frontier.* Washington DC: U.S. GPO, 1945. Reprinted in *The Politics of Science: Readings in Science, Technology and Government,* edited by W. R. Nelson. New York: Oxford University Press, 1968. Nelson's reader contains many other classical articles.

De Solla Price, Derek. *Little Science, Big Science and Beyond.* New York: Columbia University Press, 1986.

Dickson, David. *The New Politics of Science,* 2d ed. Chicago: University of Chicago Press, 1986.

Elzinga, Aant. "Unesco and the Politics of Scientific Internationalism." In *Internationalism and Science,* edited by Aant Elzinga and Catharina Landström. Los Angeles: Taylor Graham, 1996: 89–131.

Elzinga, Aant, and Andrew Jamison. "Changing Policy Agendas in Science and Technology." In *Handbook of Science and Technology Studies,* edited by Sheila Jasanoff, Gerald E. Markle, James C. Petersen, and Trevor Pinch. Thousand Oaks, CA: Sage, 1995: 572–597.

Gibbons, Michael, et al. *The New Production of Knowledge: The Dynamics of Science and Research in Contemporary Societies.* London: Sage, 1994.

Horner, D. "The Cold War and Politics of Scientific Internationalism: The Post-War Formation and Development of the World Federation of Scientific Workers 1946–1956." In *Internationalism and Science,* edited by Aant Elzinga and Catharina Landström, Los Angeles: Taylor Graham, 1996: 162–198.

HMSO, *The Rothschild Report: A Framework for Government Research and Development.* London: Author, 1972.

Irvine, John, and Ben Martin. *Research Foresight: Priority-Setting in Science.* London: Pinter, 1984.

Jasanoff, Sheila. *The Fifth Branch.* Cambridge, MA: Harvard University Press, 1990.

Organisation for Economic Co-operation and Development. *The Measurement of Scientific and Technical Activities.* Paris: Author, 1962. (Frascati Manual, elaborated in many new editions since then).

Organisation for Economic Co-operation and Development. *Science and the Policies of Government.* Paris: Author, 1963. (Piagnol Report).

Organisation for Economic Co-operation and Development. *Science, Growth and Society: A New Perspective.* Paris: Author, 1971. (Brooks Report).

Organisation for Economic Co-operation and Development. *Science and Technology Policies for the 1980s.* Paris: Author, 1981.

Organisation for Economic Co-operation and Development. *Science and Technology Policy Outlook.* Paris: Author, 1992.

Ravetz, Jerome R. *Scientific Knowledge and Its Social Problems,* 2d ed. New Brunswick, NJ: Transaction, 1996.

Rose, Hilary. *Love, Power and Knowledge. Towards a Feminist Transformation of the Sciences.* Bloomington: Indiana University Press, 1994.

Rose, Hilary, and Steven Rose, *The Radicalisation of Science.* London: Macmillan, 1976.

Salomon, Jean-Jacques. *Science and Politics.* Cambridge, MA: MIT Press, 1973.Shils, E. ed. *Criteria for Scientific Development. Policy and National Goals.* Cambridge, MA: MIT Press, 1968. Records the Minerva debate.

Slaughter, Sheila, and Larry L. Leslie. *Academic Capitalism. Politics, Policies, and the Entrepreneurial University.* Baltimore: The Johns Hopkins University Press, 1997.

Snow, C. P. *The New Men.* London: Macmillan, 1954.

Snow, C. P. *Two Cultures and the Scientific Revolution.* London: Cambridge University Press, 1964.

Snow, C. P. *The Physicists. A Generation That Changed the World.* London: Macmillan, 1981.

Werskey, Gary. *The Visible College: The Collective Biography of British Scientific Socialists in the 1930s.* New York: Holt, Rinehart, and Winston, 1978.

Aant Elzinga

# SCIENTIFIC COMMUNITY.

This entry includes two subentries

Norms

Organization

Is *community* the right word to describe those people who do science for a living? Perhaps not. For sociologists, the prototype for community is a traditional village in some pastoral countryside, or maybe a stable urban neighborhood. Life in the village is characterized by intimate and diffuse associations with others who live contiguously and cooperatively, who share a common history and future, who emphasize the same values, and who routinely interact face-to-face. The "community" of scientists is sufficiently huge in number to preclude intimate or diffuse associations with more than a handful of peers, and it is spread all over the globe—with individuals more often connected by e-mail than face-to-face. Scientists surely share some common commitments to logic and empiricism, but they fight fiercely to defend their theories against those of a rival, and they compete vigorously among themselves for rewards and resources. *Scientific community* might be indeed be oxymoronic, but, to judge from its everyday usage, we seem to be stuck with it. There are other ways to describe those who make a career of science, and these terms may be more accurate and suggestive: *network, field,* and *profession.*

## Network

To see the scientific community as a network, imagine a set of dots (or "nodes") distributed unevenly in two-dimensional space, where some pairs of dots are connected by lines but many other pairs are not. The dots are scientists and the lines represent the various ways in which scientists are connected to each other: teacher and student; colleagues at the same university or research facility; collaborators in research or on a publication; or in informal communications, such as exchanging research papers before they appear in print. Sometimes, scientists are linked by indirect connections that do not depend on one scientist's actively seeking out another. For example, two scientists might be referenced (cited) together in a third scientist's research publication—an indicator, perhaps, that their work has something in common, even if they have never personally met.

If you were to unroll a gigantic map showing the nodes and links for the worldwide scientific community, several features would be immediately obvious. Most scientists will be found in relatively small and dense clusters, where almost every node is linked to every other node. These tight network clusters are scientific specialties, well-connected groups of scientists who are linked because they work on a common problem, or use similar observational/experimental instruments, or because they share a distinctive theoretical orientation to reality. An example of a scientific specialty might be the radio astronomical study of pulsars. These very dense clusters are themselves clustered, rather than distributed evenly. The resulting clusters of clusters are scientific disciplines, and much empty space is usually seen between disciplines (with few nodes or links). As a discipline, astronomy, for example, is no longer just a connected network of dense clusters; instead, scientific disciplines become institutionalized as departments within universities, as programs within funding agencies, and as distinctive journals within libraries.

These specialties and disciplines form the primary grounds for identity among scientists, and they create perhaps the most fundamental basis of differentiation in the scientific community. Now, if one could set the network map in motion to watch changes over historical time, one would see some specialties grow in the number of dots and links as they become less dense, even, in some cases, breaking apart into two new small clusters (with diminishing connections between them). Other clusters would disappear altogether as scientists move on to different problems or theories. In particular, you would see new clusters forming in the previously open spaces between disciplines: astrophysics" once emerged between astronomy and physics as a new interdisciplinary focus of research.

### Field

Although specialties and disciplines are sometimes referred to as *fields* of science, the word has taken on a different meaning in the work of French sociologist Pierre Bourdieu. As a field, in Bourdieu's sense, the scientific community is an arena of competitive struggles over scarce resources, structured by hierarchies of power. Scientists compete for research grants; faculty positions at prestigious universities; and honorific awards, such as the Nobel Prize, as well as for less-tangible resources such as esteem, visibility, recognition, influence, and credibility. The distribution of rewards and resources among members of the scientific community is remarkably skewed: a tiny elite enjoys a huge proportion of the benefits. Are they deserved? In principle, scientists should be rewarded for solving the research problems deemed most important to solve by members of a specialty, as measured crudely by the number of papers they publish and the number of times those papers are cited by other scientists. In practice, it is difficult to escape the conclusion that the scientific community is elitist, and that the distribution of rewards and resources is not always meritocratic or universalistic. Successful senior scientists actively promote the careers of their own students who give early evidence of promise, a pattern that occasionally results in the provision of fewer advantages and less encouragement to women (who remain underrepresented in the upper strata of the hierarchy). Moreover, the careers and reputations of "star" scientists rest in part on the labors of many who may never be in a position to compete: skilled technicians, failing graduate students, permanent post-docs.

### Profession

Even as scientists compete among themselves for prizes and prestige, they close ranks quickly and forcefully when the authority, autonomy, and collective investments in science at large are at stake. Members of the scientific community also function as professionals, as do physicians or lawyers; members work together to enhance the standing of science in society and to restrict the perquisites of membership to those deemed worthy. Organizations representing the professional interests of scientists publicize ideologies that legitimate public trust in the knowledge they produce and justify public expenditures for research, emphasizing, for example, the selfless or disinterested character of scientific assessments of candidate claims to knowledge. Public trust in science depends in part on the belief that the scientific community regulates the behavior of its members, so that the ideology of disinterestedness also becomes a norm prescribing appropriate behavior expected of the scientific role (and thus a basis for identifying deviance and meting out sanctions). As professionals, the scientific community spends considerable effort at patrolling its boundaries, marginalizing pseudoscientists who would usurp the authority of genuine science or poach its resources. Such exclusionary practices also serve as powerful tools of social control, enforcing conformity among scientists to some consensual moral code. No practicing scientist can risk banishment from the scientific community—and, indeed, few have been given the boot.

## BIBLIOGRAPHY

Bourdieu, Pierre. "The Specificity of the Scientific Field." In *The Science Studies Reader,* edited by Mario Biagioli. New York: Routledge, 1999: 31–50. About the competitiveness of scientists.

Crane, Diana. *Invisible Colleges: Diffusion of Knowledge in Scientific Communities.* Chicago: University of Chicago Press, 1972. The scientific community seen as a network.

Edge, David O., and Michael J. Mulkay. *Astronomy Transformed: The Emergence of Radio Astronomy in Britain.* New York: Wiley, 1976. Classic study of the emergence of a new scientific specialty.

Hagstrom, Warren O. *The Scientific Community.* New York: Basic Books, 1965. Describes the segmentation of science into specialties and disciplines.

Kevles, Daniel J. *The Physicists: The History of a Scientific Community in Modern America.* Cambridge, MA: Harvard University Press, 1971. Reprint edition. Describes the professional side of the scientific community.

Merton, Robert K. *The Sociology of Science.* Chicago: University of Chicago Press, 1973.

Considers the moral code of the scientific community.

*Thomas F. Gieryn*

# Norms

Science is the process of producing collectively verified knowledge. Science is a social activity because its practices, ideas, and outputs are shaped by social relationships. From the questions researchers choose, to the methods they use, and to the status that others accord to science, interactions among groups and individuals play an important role. Norms and values guide action and are tools that scientists and others use to demarcate science from, and liken it to, other social systems. Values are ideals that are important enough to place above other considerations. The practice of science, the knowledge produced through these practices, and the uses made of science have been associated with a variety of values throughout history. Norms, from the Latin *norma*, a carpenter's frame or rule, are a special kind of value. They are situation-specific guidelines for action. We may experience norms as imperatives that we follow because they compel us to do so, or as taken-for-granted ideas about what is appropriate in a given situation. Norms and values associated with science vary over time and place. Thus, an examination of their role in science is most fruitfully focused on identifying the contexts in which they operate and their influence on scientific ideas, actions, and on how they are strategically used to make claims about the relative autonomy of science from other social practices.

## Norms

In 1942, the American sociologist Robert K. Merton wrote the most influential formulation of the norms of science. He considered four central norms to be the "ethos of science," or "that affectively toned complex of values and norms which is held to be binding on a man of science" (p. 551–552). Merton argued that the ethos was a collectively binding framework that furthered the goal of producing reliable, certified knowledge, by promoting wide exchanges of ideas among large numbers of scientists, and allowing competing scientific claims to be settled by specialists. Merton derived the norms that made up the ethos by relying mainly on historical records of scientific practices from the seventeenth century onward. One of his motivations for identifying norms of science was that he wished to distinguish response to the uses of science by the Nazi government. In doing so, Merton wanted to establish that science was inherently a democratic system that had been perverted by politics. In formulating the ethos of science, he was both describing actions and ideas among scientists as well as encouraging them.

The four norms Merton identified are as follows.

1. *Universalism* is the imperative to judge claims whatever the social status of the speaker. In Merton's words, truth claims are subjected to "pre-established, impersonal criteria." Evidence and method, not the speaker, are the basis of judgments about the veracity of truth claims. Impersonal technologies of communication, such as the use of mathematics and review systems in which the identities of the reviewer and the claimant are unknown to each other, encourage universalism.

2. *Communism* (later termed *communality* or *communalism*) is the norm that scientific claims "are a product of social collaboration and are assigned to a community." Scientists are obligated to share their findings with each other in exchange for prestige and social recognition. Privatization of knowledge is antithetical to the production of shared truth claims. Awards and other honors for contributions to a field encourage communalism.

3. *Disinterestedness* insists that scientists avoid the temptation to make personal gains from their work, by directing their efforts to the collective benefit of other scientists. It does not mean that scientists cannot hope for a particular outcome of

an experiment, or that they cannot have multiple motivations for engaging in science, but that they engage in science for collective, rather than individual, benefit.

4. *Organized skepticism* is the process of subjecting all claims to the public, collective scrutiny of scientists, who then judge a claimant's methods, arguments, and evidence. The reliability of claims cannot be established until other scientists judge them publicly. Peer review, public demonstrations of experiments, and public presentations at professional meetings promote organized skepticism.

Merton argued that the four norms mutually reinforce each other to generate and sustain a distinct and functional social system. Scientists follow norms because socialization processes such as formal education and on-the-job training cause scientists to "internalize," or believe in them.

In the early 1970s, scientists and others began to question whether the four norms actually described contemporary and past scientific practice. Using interviews and laboratory ethnographies, researchers asked whether one set of norms could describe the actions and beliefs of scientists in all settings. In a study of Apollo mission scientists, Mitroff (1974) found that for every norm Merton identified, there existed a counternorm. Scientists, he found, sometimes valued individual over group advancement, practiced secrecy rather than communalism, were interested rather than disinterested, and paid closer attention to those scientists who had strong reputations than to those who did not. Other writers argued that science was not a closed social system but was closely tied to powerful groups such as the military. Over the next decade, social science researchers found that norms were specific to particular settings, groups, and activities.

Along with the recognition that norms in science are specific to particular settings, and that scientific practices were shaped by relationships with other powerful groups rather than internally generated by scientists, a third important development has been the rise of debates over whether methodological norms in science exist, and if so, whether they are distinct from norms that guide information exchange and relationships between scientists and other social groups. Some philosophers, anthropologists, and sociologists have argued that hypothesis testing, instruments, logic, mathematics, and empirical evidence are general methodological norms, although not every branch of science follows any norm equally. Theoretical physics, for example, relies heavily on hypothesis testing, logic, and mathematics, whereas primatology relies less on mathematics and quite heavily on empirical evidence. Like other kinds of norms in science, narrow methodological norms concerning measurement, depictions of nature, and methods of analysis are shaped not only by scientists' interests but also by the interests of research sponsors, regulators, and the public.

Beyond serving as guides for behavior in specific settings, norms play an important role in allocating social status and authority to science. The authority of science is based in part on the idea that it is a distinct social practice that is self-governing, autonomous, and has unique access to truth. Norms play an important ideological role in demarcating science from nonscience and scientists and nonscientists. Called *boundary work* by Thomas Gieryn, this is the process of drawing linguistic, legal, organizational, and cultural similarities and differences between science and other social activities. Scientists and their supporters must convince others that science is regulated by internal norms in order to preserve the autonomy of science. Thus, one of the functions of publicly punishing scientists for violating such norms as plagiarism or falsifying results is that it conveys to outsiders that scientists regulate the behavior of their peers. Regardless of whether norms exist, it is important for scientists to convey to nonscientists that they do exist. If nonscientists believed either that the norms of science were like those of other social practices or that they were highly variable, it would be difficult for scientists to maintain that they deserved social honor and autonomy from other social groups. At the same time, the authority of science is also based on the idea that science is a practice that is like other social practices, particularly democracy. To demonstrate the similarity between science and democracy, speakers can claim that both practices share similar norms, such as communalism, or the public sharing of ideas, whether or not such norms actually exist in either field.

## Values

Values motivate individual and collective actors by influencing their needs and aspirations, ordering their perceptions, and by providing criteria for judgments of social practices, groups, and individuals. Like norms, values can be explicit or implicit. Values are important in science because they suggest projects, questions, methods, and uses. They help to determine means to carry out projects, such as whether something is cheaper, faster, more helpful to certain groups, or easier. And like norms they can also be used to justify a certain course of action. Finally, the values that are collectively attributed to science are another form of the boundary work that helps to demarcate science from other social practices, thereby shaping the relative power that scientists have in controlling what is studied, who can claim to be a scientist, which methods are chosen, and how scientific products are used.

Thomas Kuhn's book *Structure of Scientific Revolutions* provided scholars with a vocabulary and perspective on how contextual factors influenced problem choice and what counts as knowledge. He showed that social contexts and mental frameworks shaped the knowledge claims that scientists made. Kuhn called the cluster of taken-for-granted ideas about which questions are worth asking, which methods are best, and what kinds of answers are reasonable "paradigms." Paradigms, he argued, were shaped by the social networks in which scientists were located. By examining cases of scientific discovery, Kuhn found that scientists do not choose from all theoretically possible ideas and practices but instead choose from those that were within the paradigm that existed in their network of fellow researchers, funders, and users.

Values influence the interconnected processes of choosing research questions, methods, interpretations, and uses of science. Funding systems are an important influence on problem choice. Most scientific research is paid for by governments, colleges and universities, and nonprofit organizations. Funders usually have specific problems they would like to investigate and make funds available for researchers to carry out research to address them. Funding agencies also make funds available for projects that have no immediate practical application. Problem choices in this kind of research are shaped by the field in which the researcher is working, which itself has been shaped by previous funding choices. The value inherent in funding this kind of research is that research without direct applications has often led to new breakthroughs that lead to new theories and applications in the future. For example, military agencies may fund research directed toward the development of weapons and research that have no currently known application. Similarly, a public health agency may fund research on a specific health problem, as well as biomedical research that is unrelated to known health problems. Thus, funding agencies shape questions by directing attention toward solutions to particular problems, and by allowing researchers to pursue questions with no short-term practical benefit.

The values of individual scientists can also shape their choice of research questions. A biologist who has family members who suffer from a particular illness may decide to find a cure for it, or a chemist who is also an environmentalist may choose to develop nonpolluting alternatives to chemical processes commonly used in industry. Values may drive scientists away from certain kinds of work, such as Quaker physicists who choose not to work on military research projects. The values of scientists that drive problem choice may not be derived from other roles they have in their lives. Scientists may value complex problems; others may value research questions that can be answered with relatively straightforward research programs.

Values can also shape the methods and interpretations that researchers use. One clear example of the relationship between values and methods is the use of animal vivisection in many countries. Although antivivisection campaigns existed as early as the nineteenth century, the practice did not disappear until the 1970s in many countries. The shift indicates that the avoidance of pain and suffering in animals became a more important value for a larger number of people, even though there are proponents of vivisection who value the possible benefits to humans over animal suffering. Values can also have less immediately obvious effects on interpretations and methods. The degree of accuracy in measurement, when experiments are considered to be over, and the categories into which results are placed can be shaped by the values of researchers, funders, and users. Implicit broader social values can shape interpretations of data as well. Values concerning gender and race have been shown to influence the interpretations of biological data.

Since many people may attempt to make claims about the nature of the world and values are bound to intrude, scientists have developed methods to reduce the influence of researchers' preferences. These include double-blind experiments, randomization in experiments, control groups, mechanical controls, and mathematics. The use of mathematics is considered to reduce the influence of values because it is less ambiguous than other languages, is widely shared, and has rule systems that are clearly articulated and thus can be followed by many kinds of people.

Values are important not only in the research process itself but also in the assignment of social status and autonomy to science. Although values associated with science vary widely, there is one value that has been consistently associated with science over the past four hundred years: that it is worthwhile to systematically understand the organization of nature through the use of rational methods because it will lead to material and moral progress. All cultures have created systems of understanding regularity in the world. Modern science is distinguished by systematic, collective manipulation and observation of nature by standardizing systems of evidence collecting and analysis. Such investigation has often valued because it is thought to lead to material progress through the improvement of human life through the creation of new ideas and products, and moral progress through the disciplining of the mind and body.

Science is sometimes popularly considered to be objective and therefore valuable. Objectivity refers to three distinct values: that there is a close or perfect fit between theory and world, that scientists avoid making all but the narrowest of judgments about knowledge claims, and that the methods to gather and evaluate evidence do not distort nature. These values are associated with a broader value, which is that science should be valued because it can tell us what is, rather than what ought to be. The attribution of objectivity to science did not develop until the late eighteenth century, and the idea that value-free science was possible and desirable did not appear until the late nineteenth century. As historian Robert J. Proctor has shown, scientists in the nineteenth century promoted the idea that science was value-free because it helped them to make the case that nonscientists ought not to interfere with their activities.

## Studying Norms and Values

The most obvious method for studying norms and values is to simply ask groups of scientists what kinds of norms they follow and what values they hold. This method however, is neither always feasible nor reliable. Even if particular actions are observed, the meanings associated with them can vary considerably. Thus, if we observe a group of scientists discussing the research of another group, they may not be engaged in organized skepticism but deliberating about when they should release their own research findings. Moreover, asking scientists which norms and values guide their work may produce misleading answers, for scientists may

not be able to articulate them clearly even if they follow them, or they may provide a list of norms and values that they do not actually follow.

Researchers sometimes examine the role of norms and values through direct observation of objects and interactions in laboratories, public settings, and other situations. Observation can give a researcher clues about what values underlie science as a whole or particular kinds of scientific practices and questions.

Another method used to study norms and values is to observe them in visual and written documents, such as rules, guidelines, laws, correspondence, newspaper and magazine articles, textbooks, and photographs. Codes of ethics or guidelines for reporting results to a funding agency, for example, are means by which norms and values can be made explicit. Just as there are methodological cautions against relying exclusively on behavioral norms and values, the use of laws and correspondence cannot always indicate which are actually in use. Laws may be out of date or reflect a concern that few actually follow, and correspondence between individuals and groups may not tell us much about actual behaviors.

Finally, norms and values can be observed when a community expresses moral outrage over their violation. Norms and values that had theretofore been implicit become explicit. For example, when groups of researchers indicated that they would attempt to clone humans, scientists and other groups expressed outrage. This expression indicates that those who would clone humans were violating a norm about how and when humans can shape biological processes. In general, it is best to use more than one evidence-gathering technique to assess the presence and significance of norms in a given setting.

## Contemporary Debates over Values in Science

As a social process, the values and norms in science are subject to debate and change. Three major contemporary debates are as follows: whether and how nonscientists should play a role in decision making about science and the production of scientific knowledge; the expansion of opportunities for scientists to profit individually from their research; and the extent to which humans should manipulate genetic material. In the first case, activists and policy makers have argued that they ought to be more frequently included in decisions about what research is funded, how it is carried out, and how it is used. Among the questions that participants in the debate have asked are as follows: How large a role, if any, should nonscientists have in making decisions about science? If nonscientists are to play a role, how should their values be included in debates about what problems to study and how knowledge should be used? Should nonscientists become involved in producing scientific knowledge itself? If so, how? Proponents of broader inclusion of nonscientists value the knowledge that nonscientists can contribute, the expansion of democracy, and the checks on particularism that a wide range of participants can provide. Some argue that it leads to greater objectivity, since it acts as a system of checks and balances that weeds out the prejudices of scientists. Opponents often value the benefits of specialization and expertise, efficiency, and autonomy that less inclusion of nonscientists' values can offer.

Science has long been associated with profit making, because scientists have long sought to understand problems associated with economic productivity. Recently, more individual scientists are able to profit from their own research. Governments, universities, and firms now offer scientists a share of profits from funded research. Fields of study are unequally affected by this change in values. Incentives for individual profit making are common in biotechnology, for example, and less common in astronomy. As scientists and their sponsors seek to protect knowledge that may be profitable, that knowledge is less available to other researchers, and problems whose solution has little economic value may be neglected. At the same time, profit making can be a useful incentive for scientists, thus leading to rapid and creative problem solving.

Finally, perhaps the greatest contemporary debate over the role of values in science is the debate over when and why humans should be involved in manipulating genetic material. The debates involve questions of who should participate in research and how, the role of profit making, the meaning of life, and who should decide on how knowledge is used. Because the methods and uses of genetic manipulation involve questions about race, gender, religion, and profit, they have drawn a wide range of participants who have strong interests in influencing the direction of genetic research. As a dynamic, changing social system, the values and norms that are associated with science will likely be influenced by the processes and outcomes of these debates.

## BIBLIOGRAPHY

Daston, Lorraine. "Objectivity and the Escape From Perspective." *Social Studies of Science* 22 (1992): 597–618.

Gieryn, Thomas F. *Cultural Boundaries of Science.* Chicago: University of Chicago Press, 1999.

Gould, Stephen J. *The Mismeasure of Man.* New York: W. W. Norton, 1996.

Harding, Sandra J., and Merrill B. Hintikka, eds. *Discovering Reality: Feminist Perspectives on Epistemology, Metaphysics, Methodology, and Philosophy.* Dordrecht, The Netherlands: D. Reidel, 1993.

Harding, Sandra J., ed. T*he Racial Economy of Science: Toward a Democratic Future.* Bloomington: Indiana University Press, 1993.

Kuhn, Thomas. *The Structure of Scientific Revolutions.* Chicago: University of Chicago Press, 1962.

Merton, Robert K. "Science and Democratic Social Structure." In *Social Theory and Social Structure,* edited by Robert K. Merton. Glencoe, IL: The Free Press, 1957.

Mitroff, Ian I. "Norms and Counternorms in a Select Group of the Apollo Moon Scientists: A Case Study of the Ambivalence of Scientists." *American Sociological Review* 39 (1974): 579–595.

Proctor, Robert N. *Value Free Science?: Purity and Power in Modern Knowledge.* Cambridge, MA: Harvard University Press, 1991.

*Kelly Moore*

## Organization

As with any large, complex social institution (for instance, religion, education, the family), science can be viewed through many lenses. Who is involved in scientific activity? Why do societies pay for it, and how do they understand their return on investment? How does science, as an institution, interact with other institutions—with religion, for example? Who trains scientists, and how? Does scientific knowledge accumulate, or is old knowledge eclipsed by new? Is science a matter of uncovering the truth about nature, or is it a question of politics, local beliefs, and mixtures of practice? Should science be seen as universal, international, and timeless—and if so, how can its spread be encouraged and activated (such as through development programs)? If not, how do we understand the meeting up of different traditions?

This article takes the point of view that science is not a universalistic, timeless sort of institution. Rather, it is heterogeneous and is growing both locally and on a wide scale; it can best be understood via a number of processes, including communication patterning, triangulation, and the articulation of disparate points of view. The specific focus of this article is that science is organized as a *system*, and as such science will be examined with respect to several of these processes simultaneously.

### What Is a System?

The idea of a social system exists in contrast to stand-alone, disconnected events, people, or discoveries. That is, to look at science as a social system means seeing how things interlock, mesh, triangulate, and consequentially affect each other. Along those lines, each process presented here should be seen as one part of the proverbial elephant being examined by blind people, each one grasping a different part: Each impression will be different, and *not* articulated with other aspects. Some of the emergent properties of science include continual updating of knowledge, conflicts between schools of thought, profound decentralization, and lack of a totalizing formal organization with an unchanging chain of command. These ensure that we can never really know the "whole elephant." This does not mean that science lacks systemic properties or that each view is some sort of isolated, anomic entity. What characterizes science is a continual interaction within, across, and between disciplines, institutions, and schools of thought. Scientists are constantly in communication with each other, trading techniques, data, databases, students, and materials.

### Larger and Smaller Lenses

From the beginning of contemporary STS (science, technology, and society) studies, scholars have attempted to do two things: (1) to answer, in their own right, the larger questions about the nature of science as a system and (2) to view the specifics of the organization of science through several types of analysis: such as case studies; themes; particular emphases, such as gender or race; comparative studies of national or regional traditions. Most of this article will focus on the second endeavor and will examine the various approaches that science studies researchers have taken toward the organization of science. The larger questions in the first endeavor, however, are not merely background for understanding the second. They thoroughly inform the undertakings of specific scholarship. Thus the article begins with an overview of how these larger questions inform the study of science as a system.

For many years, most scholarship about science subscribed to the view of science as universal, consensus-driven, and as revealing "the truth out there." This scholarship focused on Western science and on the occupational prestige of scientists as professionals. This structural-functionalist work was especially influential in establishing a view of science as an elite profession, with a strong public mandate to deliver the truth. Prominent Western figures in this sort of investigation were Robert Merton, Bernard Barber, and Joseph Ben-David, supplemented by the methods of Paul Lazarsfeld and his survey research. This was a systemic point of view of science, in that principles of consensus, ratification of the status quo, and the taken-for-granted prestige of scientists worked to constitute a system.

Important earlier exceptions to the structural-functionalist schools were the works of Joseph Needham and J. D. Bernal. Needham was one of the pioneers to investigate the comparative aspects of science (particularly as between China and the West), and both he and Bernal were centrally interested in questions of social justice, science, and "reality." As well, some of the pragmatist philosophers of science and Marxian historians and philosophers of science contributed to understanding science as a disputed territory, not a given or simply natural one. Pragmatism in particular played a powerful role in recent STS work taking the symbolic interactionist point of view. One of the central tenets of pragmatist philosophy has been that the consequences (not just the logic) of ideas are of primary importance.

Much recent STS work about scientific organization and its systemic qualities has moved away from the structural-functionalist perspective and its legacy. Instead, since roughly the late 1970s, many researchers have instead been asking questions focused on the process of making science; on heterogeneity throughout; on cooperation and criticism rather than consensus; and, especially, on the construction of scientific results embedded in a complex intersection of types of scientific organization. One of the most exciting aspects of this scholarship has been to understand the challenge of studying science as both open and systemic, as internally contradictory yet observable as a complex cultural pattern.

Among the types of scientific organizations and institutions that have been studied by STS researchers are the following: museums (both research-oriented and general public); the academy; schools and education; industrial research and R&D firms; the media, including the public understanding of science; hospitals; nongovernmental organizations (NGOs) such as World Wildlife Fund or Africa Action; and voluntary and professional scientific associations (such as the Canadian Sociological Association; the Nordic Scientific Council; the American Association for the Advancement of Science, and thousands more).

Rather than examine each of these forms of studying science, I will investigate the ways in which various lenses have recently been used to view the systematic nature of scientific organization. This means pointing to theoretical developments with a systemic cast—ones that attempt to answer the difficult, heterogeneous, and contingent nature of scientific work.

## "Give Me a Laboratory and I Will Raise the World"

This phrase comes from an article, later incorporated into a book concerning Louis Pasteur and how he was able to succeed in creating widespread belief in his germ theory. Pasteur, in the words of actor-network theory, *enrolled* a number of groups (statisticians, farmers, doctors) by playing to their specific interests in the germ theory. He then was able to act as a spokesperson across the groups, using them to bolster each other and to solidify the fact of germs. This system-building activity is often cited as key to actor-network theory—building a kind of pyramid of belief across different interests and thus passing across communities to make a transdisciplinary fact. Actor-network theory has itself passed from the world of STS into organizational theory, computer science, and other venues to become a key means of the analysis of complex, interlocking beliefs and their geneses.

## The Right Tools for the Job: The Role of Materials in the Scientific System

How does the material culture of science contribute to its systemic qualities? This question has become important in investigating another means of knowledge travel and the solidification of facts. Adele Clarke and Joan Fujimura collected a series of articles in an influential volume concerning the often-overlooked role of humble "stuff" in creating science. Contributors looked at fruit flies, Petri dishes, rats, taxidermy, and other aspects of the materials of life sciences to demonstrate that the production and distribution and, especially, the standardization of materials in science often directly affects the character of the results. The role of standardization in science has become increasingly important in STS work, as well. This has occurred in a number of ways: looking at how standards are constructed and how they spread historically and geographically; examining standards in information technology and what they reveal and conceal about work processes; and examining standards as a kind of technology in its own right.

## Social Worlds and Communities of Practice

A unit of analysis that crisscrosses scientific organizations and institutions is that of the social world, or community of practice. These are groups that form from interests that traverse formal boundaries. For example, many computer scientists are extremely active in the social worlds of science fiction, including fantasy games such as Dungeons and Dragons, and conventions ("cons") that run for days. From these sorts of groups, affiliations are formed that often lead to jobs, partnerships, and lifelong friendships. These communities help structure science as a system, although they are difficult to see without direct observation and knowledge of the local cultures. They are also an important means for learning membership as a trusted friend within scientific practice. They intersect, fission, and overlap in ways that have yet to be fully analyzed.

## Collaboratories and E-Science

Starting in the early 1990s, a number of government and private initiatives began to exploit the new possibilities offered by distributed networked computing and other electronic media. The idea of a "virtual laboratory," cutting across many geographically distant sites of work, meant new ways of building communication systems for scientists working on a common problem. These were sometimes called *collaboratories*. Among the problems to be solved were common indexing, learning about tacit knowledges and the impact of local practice, the limits of sharing, and how public scientists wished to be at different stages of their careers. Just as this effort was getting underway in several countries, the World Wide Web was also on the rise, roughly beginning in 1994 and building on earlier data-sharing tools, such as Telnet, Gopher, and group decision laboratories.

With the rise of the Web, new forms of sharing became possible, and many countries began to invest heavily in "e-science," or the conduct of science using the co-authoring and graphical capabilities of the Web and its associated applications.

## Boundary Objects and Trading Zones

In the new STS work that searches for ways to understand cooperation without consensus, two conceptual frameworks have been influential. Boundary objects are part of a concept developed to explain cooperation in museums and in clinical/basic research settings. Boundary objects are those that live in the conceptual center of several social worlds. In the center, they are malleable, epistemologically flexible objects. Locally, in different social worlds, they are tailored to meet local needs. Thus, they meet the needs of both cooperation globally and specificity locally. Trading zones are spaces within scientific institutions where, via proximity, scientists, technicians, and other workers share materials, ideas, architecture, and conversations.

## Infrastructure

Building on all of the foregoing issues (enrollment, boundary objects, trading zones, materials, and standardization), studies of infrastructure are becoming important to questions of systematic scientific organization. The work of Thomas Hughes on the building of networks of electrical power is foundational here, as is the work of JoAnne Yates on simple infrastructures in organizations (such as file folders) and their ultimate impact on both the moral and the technical aspects of communication.

## "Others"

The voices of those traditionally excluded from elite science are crucial to the new understandings of open scientific systems and organizations. Because of the desire ecologically to understand all of the knowledge practices that compose science, every voice "counts." This has resulted in a desire to be inclusive of voices traditionally deemed irrelevant to science. In turn, this has created important intersections of research in ethnic studies, women's studies, cultural studies, disability, and other venues where "the other" is represented. The intersections work in all directions; thus, for example, whereas feminism has often been desirous of including technoscience in its analyses, it is only recently that feminist STS scholars have been able routinely to include gender and race in their STS writing. And at the same time, feminist scholars (particularly in the life sciences and medicine), are now able and willing to draw on STS work to speak of social inequality, mobilization, and the construction of scientific facts.

## The Politics of Quantification and Formalization

One of the ways in which scientists and mathematicians have made claims about science being dispassionate and universal concerns mathematics and formal languages. On this view, it is often claimed that one can "see" the world as directly expressing mathematical or formal ways of knowing. Mathematics, it is claimed, transcends all cultures, all history, and all forms of local knowledge. Although our knowledge of mathematics is incomplete, it is becoming more complete all the time, and, thus, attaining a more perfect way of knowing the world formally is only a matter of time.

Some of the most difficult work in STS has been refuting these claims and instead learn-

ing to see mathematics as a specialized language, created for specific, often political, purposes. The rise of mathematics has been like the rise of an imperial language (such as Latin in earlier times or English in our own). Why use any other means of investigating the world when this more perfect means of representation is available?

Refutations of this argument about perfect representation have taken several shapes. One argues that the precision of mathematics was socially invited for particular military purposes (MacKenzie, 1990). Nancy Cartwright's work on physics (1989) also argues a social purpose for inventing some forms of accuracy. Some have argued that the mathematics found within computer science and mathematical proofs require a particular social setting and way of talking in order to work, making it not "above the fray," but rather part and parcel of the fray. One form of STS investigation that is growing rapidly is the investigation of formal systems of classification, examining them as treaties of sorts, a way of settling sometimes-bitter arguments about deep philosophical issues. For example, the issue of whether life begins at physical birth has been "settled" in terms of medical categories administered by the World Health Organization by reducing the questions to the number of breaths taken, or attempted to be taken, by a newborn. Dying after six or fewer breaths, the fetus is stillborn; should the baby die after seven, he or she becomes a part of statistics about infant mortality.

## Reflecting on and Investigating Our Own Ways of Knowing

The view of science as universal, formal, and independent of history depends on one further assumption: that the investigator is dispassionate and uninvolved in weighing science's findings. If science is above the fray, so too must be scientists. That is, scientists become neutral trusted witnesses, archetypically "gentlemen" whose words can be relied on to be free from bias or tawdry self-interest. However, if science is involved in everyday life, and if the languages of science are not neutral but instead are political, then the meaning of the ideal witness changes radically. We all have biases, partial knowledge, and political involvements.

## BIBLIOGRAPHY

Bowker, Geoffrey. and Susan Leigh Star. *Sorting Things Out: Classification and Its Consequences*. Cambridge, MA: MIT Press, 1999.

Cartwright, Nancy. *Nature's Capacities and Their Measurement*. Oxford: Clarendon Press, 1989.

Clarke, Adele, and Joan Fujimura. *The Right Tools for the Job: At Work in Twentieth-Century Life Sciences*. Princeton, NJ: Princeton University Press, 1992.

Clarke, Adele. and Susan Leigh Star."Science, Technology, and Medicine Studies." In *Handbook of Symbolic Interactionism*, edited by Larry Reynolds and Nancy Herman. New York: AltaMira Press, 2003: 539–574 .

Galison, Peter. *How Experiments End*. Chicago: University of Chicago Press, 1987.

Goguen, Joseph. "A Social, Ethical Theory of Information." In *Social Science, Technical Systems and Cooperative Work*, edited by Geoffrey C. Bowker, Les Gasser, Susan Leigh Star, and William Turner. Princeton, NJ: L. Erlbaum Associates, 1997: 27–56.

Dewey, John. *The Quest for Certainty*. New York: Minton, Balch, 1929.

Haraway, Donna. *"Modest_Witness@Second_Millennium.FemaleMan_Meets_OncoMouse"*: *Feminism and Technoscience*. New York: Routledge, 1997.

Hughes, Thomas P. *Networks of Power: Electrification in Western Society*, 1880-1930. Baltimore: Johns Hopkins University Press, 1983.

Latour, Bruno. *The Pasteurization of France*. Translated by Alan Sheridan and John Law. Cambridge, MA: Harvard University Press, 1988.

Lave, Jean, and Etienne Wenger. *Situated Learning: Legitimate Peripheral Participation*. Cambridge: Cambridge University Press, 1991.

MacKenzie, Donald A. *Inventing Accuracy: An Historical Sociology of Nuclear Missile Guidance*. Cambridge, MA: MIT Press, 1990.

Restivo, Sal. *Mathematics in Society and History: Sociological Inquiries*. Dordrecht, The Netherlands: Kluwer Academic Publishers, 1992.

Shapin, Steven, and Simon Schaffer. *Leviathan and the Air-Pump*. Princeton, NJ: Princeton University Press, 1985.

Star, Susan Leigh, and James Griesemer. "Institutional Ecology, 'Translations,' and Boundary Objects: Amateurs and Professionals in Berkeley's Museum of Vertebrate Zoology, 1907-1939." *Social Studies of Science*, 19 (1989): 387–420.

Strauss, Anselm. "A Social World Perspective." In *Studies in Symbolic Interaction*, Volume 1. Greenwich, CT: JAI Press, 1978: 119–128.

Verran, Helen. *Science and an African Logic*. Chicago: University of Chicago Press, 2001.

Yates, JoAnne. *Control through Communication*. Baltimore: Johns Hopkins University Press, 1989.

*Susan Leigh Star*

# SEX AND THE BODY IN SCIENCE.

There is perhaps no better trope than the intersexual child for destabilizing the myth of science as pure, that is, as removed from the values, commitments, and biases of the cultures of which it is a part. In the twentieth century the intersexual has become a site of scientific and technological concern that is infused with cultural values concerning nature, the two-sex economy, appropriate gender roles, and heterosexism.

Cultural understandings of individuals born with atypical sexual anatomy are neither uniform nor stable over time. By way of illustration, consider the range of attitudes within one cultural tradition. Plato's *Symposium* includes the myth of a time before humans sinned against the gods, in which there were three "natural" bodily forms, including one that included both male and female bodily characteristics. Later, in the time of Romulus, the Romans saw individuals with atypical sexual anatomy as "mutations" and bad omens and frequently killed them. Early European medical texts, viewing sex as arising from variations in amounts of heat, perceived sexual differentiation as biologically variable and as occurring on a continuum, a theory that provided a basis for various cultural attitudes toward individuals with atypical sexual anatomy—from acceptance in some instances to rigid regulation in others. It is thus important to understand our contemporary understandings and treatments of intersexuals as itself socially situated.

## Contemporary Scientific Accounts of Intersexuality

Intersexual is the term currently in use in activist movements in the United States to identify individuals who develop primary or secondary sex characteristics that do not fit neatly into scientific and societal definitions of female and male. Contemporary medical science has identified many conditions that can lead to intersex anatomy. In congenital adrenal hyperplasia, the adrenal glands produce abnormally high levels of virilizing hormones (androgens), leading to enlarged clitorises. Individuals with androgen insensitivity syndrome have XY chromosomes (typically regarded male genotype), but because they do not fully respond to testosterone, their bodily development runs on a continuum between male and female external anatomy depending on the degree of insensitivity. Individuals with complete androgen insensitivity syndrome will develop female external genitalia and, at puberty, breasts, but they will not develop a uterus or fallopian tubes. Individuals with 5-alpha-reductase deficiency have testes, which are typically undescended until puberty, as well as a vagina, labia, and a penis, which is classified as small. In such individuals male secondary sexual characteristics develop at puberty.

Chromosomes, like genitalia, cannot be used as the measure for differentiating between sexes if we are expecting to find two and only two sexes. In some cases individuals have neither XX (typically regarded female genotype) nor XY chromosomes. Individuals with Klinefelter syndrome have one or two additional X chromosomes in (XXY or XXXY), individuals with Turner's syndrome have a single X chromosome (XO), and individuals with sex chromosomal mosaicism have some cells with XX chromosomes and other cells with XY chromosomes.

The category of "intersexual" is a contested category. Intersexual activists, for example, criticize those who refer to intersexuality as an abnormality or a deformation, arguing that in many cases societal attitudes and practices are the cause of problems experienced by intersexuals. Even the question of who counts as intersexual is a topic of heated debate. Recognizing that the phenomenon of intersexuality is unknown to most laypeople, many scholars have worked not only to bring its existence to light but to raise concerns with the medical management of intersexuality. Anne Fausto-Sterling (1993) was one of the first to popularize this topic and to offer statistics on its prevalence. She modified her original estimate in 2000 based on an extensive review of medical literature, concluding that an estimated 1.7 percent of all births are intersexuals. This number was contested by Leonard Sax (2002) on the grounds that Fausto-Sterling's definition for *intersex* was too broad because it encompassed individuals who are phenotypically indistinguishable from "normal."

Sax, for example, disqualified all individuals with non-XX or XY chromosomes as intersexual on the grounds that many individuals with such chromosomal variations are phenotypically indistinguishable from "normal" males and females. Infertility is the most common presenting feature with Klinefelter and Turner syndrome, short stature in the case of Turner syndrome; mild mental retardation, and behavior problems are the most common impact of other chromosomal variants. Sax seems to insist that the category of intersexual be reserved for cases of "ambiguous genitals."

What is at stake in the disagreement between Fausto-Sterling and Sax is an understanding of sexual variation. Fausto-Sterling identifies the large range of variations in what counts as sex both to help to explode what she regards as the myth of two and only two "natural" categories and to reveal the social factors at work in viewing intersexuality as defective and in need of medical management. Sax's aim is quite different: to argue that "true" intersex conditions are pathological and that human sexuality is a dichotomy, not a continuum.

## Bodily Repair: Fixing Intersexuals

If intersexuality is seen as pathological and children with so-called ambiguous genitalia are viewed as having defects that will preclude their normal development, then medical management seems imperative. Genital cosmetic surgery for infants with atypical genitalia is a common medical response.

But it is important to ask why medical management seems so imperative. Physical health is usually not an issue with intersexual children. Some conditions that result in intersexuality have potentially fatal complications that clearly are in need of immediate medical treatment, but the majority of cases do not pose any physical threat to the health of the individual. Many scholars such as Fausto-Sterling, Alice Domurat Dreger, Bernice Hausman, and Suzanne Kessler, have argued that the primary reasons for genital surgery are cultural: namely, to ensure that the individual fits within the two-sex economy that limits choices to two—male or female—and the assumption that anyone who does not fit within this framework will experience negative psychosocial development.

Links between cultural values and genital surgery are illuminated in the process of "assigning" sex. "Atypical" is often demarcated on the basis of size: an enlarged clitoris or a small penis. But why should a clitoris that is larger at birth than the typical range of 0.2 to 0.85 centimeters or a penis that is smaller than 2.5 to 4.5 centimeters be considered a reason to perform genital surgery? Here heterosexism interacts with the two-sex economy to lead to an expectation not only of two (and only two) differentiated genitals but also of genitals that can (only) accomplish their "intended purpose," namely penetration for males, but not for females.

Sex assignment in the case of atypical genitalia involves a variety of factors including aesthetic, functionality, fertility, and future pubertal development. A common practice for those intersexual babies with atypical genitalia but with XX chromosomes is to preserve reproductive potential and perform "clitoral" reduction regardless of the severity of the virilization. But intersex babies with XY chromosomes are treated differently. Here the gender of assignment depends on the adequacy of the phallus size. For such infants technology intersects with cultural val-

ues, for it is far more difficult to surgically create or reconstruct a "functional" penis than a "functional" clitoris. The reasons here, however, are not only technological; they also have to do with what is held to matter in the arena of functionality. In the case of penises, both size and the location of the urethra are at issue, with the working assumption that "normal" males will require penises large enough for vaginal intercourse (the presumption of heterosexuality) with urethras that allow them to urinate standing up and to ejaculate semen at the tip. As a result, some intersex individuals who have XY chromosomes are assigned as female and genital surgery and hormone treatment determined accordingly. It is important to note that sensation and orgasmic capacity are *not* criteria of functionality.

The response of intersexual activists to such "management" focuses on the lack of medical necessity of such treatment, as well as the potential harm from the treatment. Genital surgery, in addition to having all the risks of surgery, can also seriously compromise orgasmic ability and damage general sexual response. Arguing that the prime motivation for imposing surgery on children is the unfounded assumption that psychological distress would result if a child's genitals were atypical or did not fit into a clear two-sex economy, intersexual activists have taken a stand against all but medically necessary surgery for children. Their view is that the decision as to what type of treatment is best—indeed any decisions concerning sexual identity—must be determined by the individual, not by physicians or parents. To do otherwise, they argue, is to risk irreversible harm and would constitute a violation of that child's human rights.

The rush to "correction" of children is the result of contemporary scientific accounts of gender identity. The work of theorists like John Money and Robert Stoller advanced the concept of gender identity defined as a person's inner experience of gender, a person's feelings of maleness or femaleness. Money argued that gender identity, although a psychosocial phenomena and thus not innate, is indelibly imprinted by early childhood rearing within the first four and a half years of age. Money's theory, combined with the assumption that genitalia "differences" would cause psychological distress, led to a medical emphasis on early surgical intervention.

A competing scientific conception of gender identity from that of Money and Stoller does not view individuals as psychosexually neutral at birth but rather contends that gender disposition is innate and based on biological factors such as chromosomes and hormones. This conception of gender identity does not correlate with a prohibition on genital surgery but rather on "correcting" to the genetically correct sex. The recent case of Joan/John, which was discussed in the press and was the topic of a BBC documentary in 1980, contributed to a strong swing in the direction of this position. Joan/John was a twin whose sex was reassigned following the loss of his penis during a surgical procedure when he was seven months old. Joan/John underwent "surgical repair" and was thereafter raised as a girl. At the age of 14, however, Joan/John convinced his physicians and psychiatrists to allow him to live as a boy. Milton Diamond, an endocrinologist who had disagreed with Money's views for years, argued that the Joan/John case demonstrated the hormonal basis of gender.

This swing to viewing gender as "natural" category carries with it other assumptions about nature of gender: namely, that there are two and only two genders, that one's gender is fixed and unchanging over the course of one's life, that all individuals can be classified as either masculine or feminine, and that genitals correlate with gender. The natural attitude toward gender also assumes heterosexuality. Again, we see how tightly scientific theories about sex/gender are interwoven with societal values. Interestingly, even Money's view that gender identity is not fixed at birth, while not seeing gender as biologically determined, nevertheless subscribes to variants of all the above views about the nature of gender.

However, as Judith Butler as argued, the case of John is far more complex. She notes that John takes hormones and has had surgical reconstruction to create a penis. Butler ar-

gues that John's "inner sex" required a transsexual transformation to achieve a sense of naturalness, to ensure that his outer body accords with his inner truth. The inner nature that Diamond and others insist upon often requires augmentation, reminding us that gender conformity is not simply an issue for intersexuals, but is a site for medical and technological intervention for those who wish to be in better compliance with the gender that they are or want to be; hence the turn to cosmetic surgery.

## Elective Sex Re/assignment Surgery

Interestingly, the history of elective sex reassignment surgery turned on the work of Money and Stoller. Arguing that one's core gender identity, which Money referred to as one's unquestioned certainty that she or he belongs to one of only two sexes, could be in tension with one's physical sex, transsexuals in the United States and Europe developed a rationale in the latter part of the twentieth century to justify sex reassignment surgery and hormonal sex change. Such individuals argued that their inner conviction of gender was in tension with their bodily manifestation, and thus they claimed that the same psychosocial harms medicine had used to justify medical intervention on intersexual children justified their own elective surgical and hormonal interventions. While transsexuals were able to successfully lobby for both surgical and legal sex change, they did so only by reinforcing the two-sex economy and all that it entailed. Transsexuals, for example, had to comply to a heterosexual norm to be candidates for surgery.

Toward the end of the twentieth century, thanks both to the feminist movement and its critique of essentialist theories of gender/sex as essential and to the development of Queer Theory and the realization of the performativity and plasticity of sex/gender, many individuals in the United States and Europe who were not comfortable with their assigned gender/sex system but who also questioned the two-sex economy developed the concept of "transgender" to identify a wide range of embodiments/performances of sex/gender. Part of the philosophy of the transgender movement is that there are far more variations of sex/gender then two: man/woman, feminine/masculine. Rejecting the either/or choice of gender/sex, many but not all individuals who are transgender turn to science and technologies of the body, including cosmetic surgery, hormone treatments, and, in some cases, sex reassignment surgery. The transgender movement not only departs from the two-sex economy, it also involves a critique of normative heterosexuality. Indeed, this critique is often reflected in the complexity of transgender embodiment. Is a male-to-female transsexual who is partnered to a transgender lesbian, an instance of heterosexuality or homosexuality? Here the use of science/technology and values merge in the lives of transgender individuals, at least for those who have the funds to afford the medical procedures.

But the phenomenon of transgender individuals reveals a long history of individuals using science and technology to "fit" better in terms of sex/gender. The use of elective "cosmetic" surgery to make the body accord more convincingly to one's inner sense of gender/sex began long before the movements of transsexuals in the twentieth century. Many individuals have felt that their bodily proportions did not adequately accord with societal norms. While not all want to "change sex" in the sense of transforming themselves from the body of one sex into that of the other, many individuals have wanted to be more "manly" or more "womanly" than they were. A sense of the plasticity of the body was in place long before the discipline of plastic surgery was perfected. As just one example, women in the 1920s were extolled to diet and exercise to fit the slender body that was then the feminine fashion.

Plastic surgery developed in the late nineteenth century, but World War I and the high numbers of soldiers in need of reconstructive surgery led to many refinements in surgical techniques. These techniques were quickly transferred from reconstruction to aesthetic enhancement. Sander Gilman documents how physicians in the early decades of the twentieth century justified elective cosmetic surgery by insisting on the psychological need for cosmetic surgery. Arguing that saggy breasts and hefty legs can result in psychic states ranging from unhappiness to madness and suicide, physicians argued that cosmetic surgery in such cases is a therapeutic necessity. The idea that plastic surgery could and should be used to treat depression stemming from a person's fixation on his or her bodily imperfections was widespread by the 1920s.

In the contemporary period, cosmetic surgery is most frequently used to maintain a youthful appearance, but it is also commonly used to "augment" one's gender. It is not only John/Joan who used cosmetic surgery and hormone therapy to fabricate a male body; men are turning to hormones and drugs, particularly steroids and testosterone, in addition to weightlifting, to bulk up their muscles by accelerating the rate at which the body can metabolize nitrogen into muscle. And although far from perfected, a vast array of technologies that promise to enlarge one's penis are promoted in men's magazines and over the Internet. A Web search on the words *penis enlargement* would result in over a million hits to sites all designed to sell drugs, herbs, vacuum pumps, and surgical techniques designed to increase both the width and length of a man's penis, and ads for these products are a notorious topic of unsolicited electronic mail. Erections, taken by many as a sign of manliness, are now artificially inflated and enhanced by a variety of technological devices and drugs. Indeed, Viagra and its more recent competitor, Levitra, have become the fastest-selling "recreational" drugs on the market, with the definition of "erectile dysfunction" becoming overblown to include the inability to maintain an erection for at least 30 minutes.

Women, of course, are not immune to the allure of these technologies. Although there has yet to be a successful version of Viagra developed for women, drug companies are still exploring drugs designed to cure "female sexual dysfunction" (FSD), and women's magazines and the Internet market numerous remedies for FSD, including "female arousal fluids," testosterone therapy, herb creams and pills, and mechanical clitoral stimulators (to name just a few), all designed to cure this elusive problem.

Many women turn to elective cosmetic surgery to refigure their bodies to reflect their image of feminine beauty more "accurately." Figures from the American Society for Aesthetic Plastic Surgery show that cosmetic procedures performed by plastic surgeons in the United States increased by 23 per cent per cent between 1997 and 2002. In 2002, nearly 6.9 million such procedures were performed. The top five were liposuction, breast augmentation, eyelid surgery, rhinoplasty, and breast reduction. The number of 18-year-olds who underwent breast-implant surgery nearly tripled from 2002 to 2003, from 3,872 in 2002 to 11,326 in 2003, according to the American Society for Aesthetic Plastic Surgery. The popularity of cosmetic surgery has received a boost from so-called reality TV shows like *Extreme Makeover* and *The Swan*, which literally track the surgeries of individuals who "change [their] looks completely in an effort to transform [their] life and destiny and make [their] dreams come true" (ABC, Inc.).

While far more attention has been placed on enlarging penises, efforts are being made to move female genitalia into the cosmetic surgery spotlight with the marketing of labiaplasty, the "neatening" of the external female sex organs. According to one firm, Lasertreatments, women request labiaplasty because "they want to alleviate discomfort caused by large and/or thick labia, which often interfere with biking, working out, wearing tight fitting clothes. Or many women want to enhance or improve appearance of 'butterfly' or asymmetrical labia" (Lasertreatments.com). Labiaplasty is often listed alongside vaginal rejuvenation designed to "tighten" the vaginal muscles.

Thanks to science and medical technologies, you can build a manly body with steroids and weightlifting or with sex reassignment surgery. Or you can choose a more womanly form with voluptuous breasts, fattened lips, and a flat, liposuctioned tummy. Here we find ourselves at the intersections of science/technology, cultural values, and embodiment, and here, again, we are confronted with the same types of issues that were raised by the intersexual individual. The human body and what we see as sex/gender is neither fixed nor static. Nor does it comfortably fit into two neatly labeled categories. But it is also not a completely plastic background that science/ technology and culture forms into particular structures. It is active, is productive, and is acted upon and produced. There is a materiality that must always be taken into account, but not separated from the values embedded in cultural practices, including the practices of science and technology. Perhaps the best conclusion is Michel Foucault's query, "Do we *truly* need a *true* sex?"

*See also* **Sex and the Body in Technology**

**BIBLIOGRAPHY**

ABC, Inc. *Extreme Makeover. Casting.* http://abc.go.com/primetime/extrememakeover/casting.html.

Blum, Virginia. *Flesh Wounds: The Culture of Cosmetic Surgery.* Berkeley: University of California Press, 2003.

Butler, Judith. "Doing Justice to Someone: Sex Reassignment and Allegories of Transsexuality." *GLQ: A Journal of Lesbian and Gay Studies* 7.4 (2001): 621–636.

Davis, Kathy. *Reshaping the Female Body: The Dilemma of Cosmetic Surgery.* New York: Routledge, 1995.

Dreger, Alice Domurat. *Hermaphrodites and the Medical Invention of Sex.* Cambridge, MA: Harvard University Press, 1998.

Dreger, Alice Domurat, ed. *Intersex in the Age of Ethics.* Hagerstown, MD: University Publishing Group, 1999.

Fausto-Sterling, Anne. "The Five Sexes: Why Male and Female Are Not Enough." *The Sciences* 33.2 (1993): 20–25.

Fausto-Sterling, Anne. *Sexing the Body: Gender Politics and the Construction of Sexuality.* New York: Basic Books, 2000.

Foucault, Michel. *Herculine Barbin: Being the Recently Discovered Memoirs of a Nineteenth-Century French Hermaphrodite*, introduced by Michel Foucault, translated by Richard McDougall. New York: Pantheon, 1980.

Gilman, Sander. *Creating Beauty to Cure the Soul: Race and Psychology in the Shaping of Aesthetic Surgery.* Durham, NC: Duke University Press, 1998.

Gilman, Sander. *Making the Body Beautiful: A Cultural History of Aesthetic Surgery.* Princeton, NJ: Princeton University Press, 1999.

Hausman, Bernice. *Changing Sex: Transsexualism, Technology, and the Idea of Gender*. Durham, NC: Duke University Press, 1995.

Kessler, Suzanne. *Lessons from the Intersexed*. New Brunswick, NJ: Rutgers University Press, 1998.

Lasertreatments.com. *Labiaplasty—Cosmetic Vaginal Surgery—Vaginal Rejuvenation (Tightening)*. http://www.lasertreatments.com/labiaplasty.html.

Money, John. *Sex Errors of the Body and Related Syndromes: A Guide to Counseling Children, Adolescents, and Their Families*, 2d ed. Baltimore: P. H. Brookes, 1994.

Sax, Leonard. "How Common Is Intersex? A Response to Anne Fausto-Sterling." *Journal of Sex Research* 39.3 (2002): 174–178.

*Nancy Tuana*

# SEX AND THE BODY IN TECHNOLOGY.

Technology and the body share a convergence around the narratives and interpretations offered about their respective limits in relation to culture. Historically, the body, like technology, is contradictorily understood as both a site of utopian possibilities for the transformation of society and as a material anchor to the reality that surrounds us. From silicone implants that bring the body into compliance with contemporary Western societies' notions of beauty and sexual empowerment, to DNA databases that delimit the body as the realization of predestined biological coding, the human body as a physical reality is in constant flux between cultural metaphors surrounding corporeality and the technologies that help manifest and inflect such metaphors. Such tension has often been characterized in terms of the competing concepts of materialism and idealism. That is, both technology and the body have been read on the one hand as material realities that reflect the unmediated natural world and on the other as malleable entities capable of realizing idealized fantasies about our control over nature.

The *speculum*, an ancient figure for subjective reflection that as early as first-century Rome was used as an instrument for medical observation, exemplifies the long-standing complex reflexivity between technology, bodies, and texts. A metaphor for mirroring or reflection (from the Latin speculum, "looking glass"), the speculum materialized as a technological object for scientific observation of the body—specifically, for the clinical inspection of female reproductive organs. The speculum is an instrument that contemporary feminist science and technology scholars in particular have viewed as early evidence of a historical bias within Western technology toward an aggressive control and invasion of the body, a body frequently feminized and characterized as passive terrain awaiting discovery. The tension between the speculum as both a conceptual abstraction and a technological material extends back to Greek and Roman societies and illuminates the two motivating questions in Western cultures regarding technology and the body: Do technologies and bodies simply manipulate and materialize more fundamental physical natures and needs, or do they help construct our very ideas about the connections to be made between culture, nature, and embodiment?

Theories of technology and culture have tended to favor the notion that technological interventions merely extend our natural bodily capacities and, further, that technological innovation enacts a concretization of preexisting physical laws. By the early twentieth century, critical studies had crystallized along these two competing interpretive trajectories: a pejorative view (exemplified by Martin Heidegger) of the mediation by technology of a more basic human existence and technology's distorting impact on human societies, and a more benign vision (exemplified by Marshall McLuhan) of technology as an extension of our sensory apparatus that augments the evolution of more efficient systems of communication and production. Both of these positions still largely embraced an abstract and deterministic understanding of technology as either an after-effect of culture or the hard ground on which culture is formed.

Yet we need only look to the cultural obsession with both eugenics and sexology throughout the nineteenth and twentieth centuries to demonstrate a more complex interaction among technological systems, cultural values, and bodies. To a great extent, biotechnology projects such as the Human Genome Project, an ambitious internationally coordinated effort to provide a full profile of human genetic material through databasing DNA sequences, have their cultural roots in the early efforts to scrutinize the microgenetics of human essence. While such research is often characterized as progressive, leading to a better understanding of congenital disease, the conceptual and historical motives behind such technologies are fraught and varied. The Human Genome Project shares a history with earlier eugenics research, technoscientific projects that sponsored programs for selective breeding through genetic analysis. In the nineteenth century, sexologists further employed the conceptual frameworks of hereditary predisposition in an attempt to define what was understood to be normal versus deviant sexual practices. The technologies of cataloguing physical difference were used to reinforce and enact cultural intolerance for sexual behaviors that existed outside expectations of heterosexual reproduction and the nuclear family.

In the twentieth century, eugenics became a principal weapon in institutionalized racial cleansing in both Nazi-controlled Europe and the 1930s American South. While the fascination with interpreting human embodiment through genes is no longer tied to political and social programs of racial and social cleansing, reducing the complexity of physical and embodied existence to a genetic magic bullet denotes a broader cultural investment in reading and managing the body through technology. Arguments in support of the Human Genome Project point to the implied capacity for preventing disease and even thwarting congenital defects, while critics emphasize the atomizing tendency in giving priority to technologies that work to efface more complex, systemic interactions between the human body and environment, including the social and political inequities that often influence perceptions of racialized and sexed bodies within culture.

The evolution of genetic technologies becomes even more complicated in realizing how the field of genetics has been appropriated by communities once targeted by such projects. A recent example is research done in the early 1990s on locating the "gay gene." Some social theorists and gay advocacy organizations saw DNA research not as a program of potential cultural exclusion but instead as an opportunity to generate evidence of the physical basis to homosexuality. In a society that values scientific and technological explanation as a form of natural objectivity that escapes cultural bias, such evidence can be thought to go a long way in securing acceptance of marginalized embodiments within society.

The economic issues surrounding specific forms of technology raise further questions relating to the body. With the onset of the industrial revolution, the relationship between the body and technology took on a new focus: the extent to which capitalist production could be enhanced and monitored under technological custodianship. The appearance of large-scale factories at the turn of the twentieth century, a period often referred to as the Fordist era in response to the assembly line techno-production introduced by Henry Ford's automobile plants, ensured that bodies and technologies would exist as prostheses for one another. To this end, bodies would employ new production technologies to enhance micro-labor, and in turn technologies would manage embodiment in the interest of greater global productivity. A new interest in technology and the sexed body emerged in this period as technological innovation implied both new possibilities for generating individual empowerment through technological systems, as well as distinctions or "divides" that further instantiated existing cultural assumptions about the body, sex, and technology.

The post-Fordist era of information technology is a case in point of how the divides that exist in society between male and female empowerment and material access can be both ameliorated and exacerbated by technology. Computers were initially marketed with metaphors promising a new democracy within a virtual environment free from cultural presuppositions. Role-playing and pseudo-personas in cyber-environments were underwritten with the possibilities for experimenting with alternate identities within online communities. Ultimately, however, computing terrains have tended to reproduce many of the very demarcations of embodied identity to be found in real-life communities. This was especially true throughout the 1990s, when the phenomenon of the "digital divide" materialized. The digital divide was a concept that developed in the wake of studies that pointed to dramatic disparities between males and females and between socioeconomic classes in accessing the material and skills required for computing and digital technologies. Though the digital divide has closed considerably in the late twentieth century between males and females in Western societies, even greater disparities have been identified in assessing a similar impact at the global level on socioeconomic and ethnic minorities. The cultural idealisms surrounding computing technologies encounter the same issues more traditional technologies have: negotiating the tension between the realities of embodied experience amidst economics and culture, and the need for soberly evaluating the idealization of technology as media for transcending such boundaries.

## Extensions and Interfaces

Much of the focus thus far has been on the ways in which technology is interpreted as an extension or prosthesis for the physical body or, in some cases, even for nature itself. Juxtaposed with the cultural divides that are exacerbated by technology are the more contemporary theories that embrace new paradigms of hybridity. Feminist science critics such as Donna Haraway render an entirely new framework for thinking about technology and the body via the cyborg. The concept of the cyborg, an organism that is part machine, part human, challenges us to think about the radical reconfiguration of the sexed body that is implied with representing technologies not as mere extensions of the "natural" body but as complex interfaces

that subvert conventional understandings about the very limits of embodiment, nature, and culture.

Returning to the figure of the speculum as both a technological instrument and a metaphor for technoscientific introspection, contemporary theorists in particular have begun to appropriate the symbol of the speculum as a new tool for recasting virtuality at the present moment. If technologies have been largely understood as tools that extend or intrude upon the body, viewing the body as a passive spectacle, is it possible to turn the figure of the speculum back upon the very narratives that are created about the origin and future of technology? The implications of newer technologies such as virtual reality (VR) and artificial intelligence (AI) would on the surface seem to indicate as much. VR in particular draws into question the premise that technology is based on creating difference around, and functioning as a prosthesis for, the body. In the case of VR, symbolic representation, via images and interactive spaces, interfaces directly with physiognomy and sensory response, blurring the conventional distinctions between real-time embodied experience and virtuality. Cyborg theorists such as Haraway point to the productive confusion created by newer digital and cyber-technologies in distinguishing the body/machine divide. Such hybridity implies an uncertain flux surrounding the idea of the "natural body" and imparts new means for empowering the concept of contextualized embodiment.

N. Katherine Hayles, a science and technology theorist who studies the relationship between textual representation and technology, highlights a competing complexity to virtual embodiment. Hayles has explored the tendency within the narratives about cyber and digital technologies to portray the flow of digitized information itself as form of corporeality, indeed even as its own form of eroticized embodiment. Along with any potential emancipation achieved by projecting humanness across the digital interface, Hayles finds a familiar reinscription of the history of male and female bodies.

A recent fascination with the construction of AI, a machine materialization of the human capacity to think in real time, exemplifies our continuing infatuation with the possibility of transcending the very flesh that binds us to the material world. Though premised on a new and scintillating vocabulary of the infinite variety to simulacra, these new objects bear a striking resemblance to a Platonic worldview. Although forged from silicon and binary digits, these objects still manage to reflect the airy promise of an ideal world not yet realized. Where we once saw "nature" and "culture" as the defining imprints of biological existence, we now project on the cave wall a palimpsest of "information" and "bit streams." And yet, the representations guiding such terms betray a more basic affiliation with embodied desire and reproduction. The replication of identity—the grounding conceptual frame for AI—by definition reflects back on the embodied reality of human sexual reproduction.

## Technological Power and Sexuality

The issues pertaining to human sexual reproduction inform some of the earliest institutionalized technologies. Rachel P. Maines (1999) demonstrates the close relationship between our most basic beliefs about inherited sexuality and the development of new technologies. The significance of the female orgasm, for example, has shifted along with historical interpretations of female sexuality. In the nineteenth century, the increased production of mechanical vibrators resulted in parallel with crystallizing definitions of gender and the pathology described as hysteria. Hysteria (literally, "womb disease") was a condition associated with women viewed as emotionally disturbed. For centuries the medical treatment of hysteria called for stimulation of the clitoris to orgasm. The appearance of commercially produced vibrators in the late nineteenth century coincided with a male medical community growing impatient with providing such stimulation as a sanctioned cure. The complicated interplay between technology, the female body, and the public and private spheres demonstrates the fraught intersection of technoscience and cultural systems. Worth noting here is the extent to which hysteria as a medical pathology

mirrored the cultural expectations of a Victorian era (1839–1901) that viewed unregulated female sexuality and mobility as threats to a conservative social order. Throughout the eighteenth and nineteenth centuries, the problem of female hysteria, constructed as a function of female sexual desire, was managed and codified within the scientific community. With a shift in the late 1800s and early 1900s in what constituted appropriate medical practice in relation to female sexuality, technology was pressed into service to facilitate a privatization of such issues. The link between hysteria and female masturbation was seen as an appropriate concern for the public scientific community up until the turn of the century, when such intervention in sexuality materialized as a threat to the role of the medical institution, a cultural system itself responding to the new pressures of an increasingly industrialized society. Though mechanical vibrators perhaps at first glimpse fail to meet our litmus test for advanced technology, it is clear that these machines were important conduits for moving a significant cultural aspect of female sexuality from the public to the private sphere.

The capacity for technologies to mediate public and private domains is even more striking today. The very enterprise of human reproduction has become a very public, and in many cases, disembodied phenomenon. The prevailing technologies of surveillance within the biomedical community translate what the cultural historian Michel Foucault defined as "biopower" into even more concentrated systems of control. Foucault's concept of biopower describes the awareness that developed, with the onset of modernity, of the capacity for medical, military, and scientific institutional practices to discipline and manage bodies at both the macro and micro levels. Today, ultrasound imaging techniques that provide highly detailed (even cinematic) representations of the *in vivo* fetus allow for a focus on the developing infant as an entirely autonomous living organism. Recent debates have centered on the extent to which such practices further endorse an androcentric view of medical intervention that works to efface and undermine the control women have over their bodies during reproduction. Indeed, some in the medical community have begin to argue for a prevailing concept of obstetric medicine that foregrounds the fetus as the primary patient, rather than the reproductive system of mother and fetus. Surveilling technologies that can simultaneously abstract and imagistically reproduce representations of the fetus *in vivo* contribute in a direct way to the policies organized around reproductive rights, embodiment, and autonomy.

The combined technologies of DNA analysis and "test tube" reproduction further complicate the blurring realities of individual embodiment and the public sphere. For the first time, reproduction can occur outside of the human body: Both male sperm and female eggs can be technologically harvested from donors and conception managed in the laboratory. A nascent program within the biomedical community entails sex selection technology: sperm-sorting and *in vitro* embryo DNA testing. In 1993 the British Human Fertilization and Embryology Authority (HFEA) issued a ban on sex selection, because it was feared that the process exceeded healthy family planning strategies and could be abused to the detriment of society. Yet, because of limitations on the ban, sex selection continues to be offered in Britain and is virtually unregulated in the United States. The very function implied by an organization entitled "Fertilization and Embryology Authority" points to broadening public and state interests in the ethics and practices of human reproduction, and such interests raise fundamental questions about both the regulation of an individual's body and societies' priorities in funding the future of certain technologies.

In other respects, technology's relationship with the body appears entirely to remap conventional beliefs about the origin of sexual identity. Research in the sciences, social sciences, and humanities around the end of the twentieth century has reconfigured the assumed fixity to an individual's biologically predetermined sex. With the aid of sexual reassignment surgery, those who find their sexed embodiment to exist in contradiction with their desires and sense of self can now physically redefine their sex to a certain extent. Such technological developments may appear as entirely new interventions that reflect societal revisions of the limits on biological sex in relationship to gender, but the significance of medical technology to sexed biology has a long and complex history. Anne Fausto-Sterling has examined how surgical techniques have been deployed at least since the early twentieth century to assist in making biological sex a more stable category. Our recent historical investment in the role of the X and Y chromosomes in determining sexual definition becomes uncertain in light of intersexed people; that is, individuals who carry physical traces of both sexes in addition to either a male or female identifier at the chromosomal level. For years, medical technologies have been employed to select the "right" sex for such individuals. Like most surgical interventions, such procedures are difficult and often physically mutilate or disable the person. These medical events signify a deep link between a prevailing cultural need for stability to the "nature" of biological sex and an endorsement of the technologies that can affect the appearance of stability to such categories. Fausto-Sterling emphasizes that such procedures— far more common throughout history, in one form or another, than is often acknowledged—reveal that biotechnologies are performances that strive to make certain where nature and the body end and culture takes over.

## Technology of Shifting Scales

Technology and the body have been reconfigured over the years as notions of nature and culture have come up against changing definitions of the technology-human interface and the various forms of hybridity that emerge in the wake of such changes. One area of study that still requires much work is the field of scalar studies. Scalar studies investigate how shifting perceptions and interpretations of space and time guide our transitions into new technologies.

This article began with an overview of theories of technology that have historically privileged the idea of technology as a material artifact that either responds to or places pressures on the more fluid categories of culture and society. Yet a survey of some of the most recent technological developments generates a definition of technology that is far more dependent on the reorganization of spatial and temporal scales. Digital imaging techniques gain their technological power through their capacity to refocus perceptual cognition in space and time; for example, a satellite-generated map (say, of the Western Hemisphere) that, when displayed by a computer program, allows for rapid sequencing of views, moving from the very small to the very large. Such events alter our cognitive expectations of space and time as embodied experiences. A definition of technology as the reorganization of spatial and temporal scales in relation to embodied cognition is as relevant to experimentations in the Renaissance with telescopes as it is to current work with the geometric image flows of VR environments. How we interpret shifts in scales of time and space has a direct impact on how we conceive of embodiment in the world. The scales attendant upon an increased atomizing of the world, a technoscience that abstracts holistic embodiment from the macroenvironment, speaks to a continuing Western fascination with technologies as systems that can dissect, discipline, and dominate the human body.

The Visible Human Project provides us with a final example of the importance of embodiment and scale to remaining attuned to the cultural implications of even our most progressive technological innovations. Sponsored and funded by the National Library of Medicine, the Visible Human Project is a digital anatomical database constructed from the dissected corpses of two human beings, one male and one female. To compile the database, researchers sliced each body into micro-slivered increments, and then scanned the pieces into digital data that could be analyzed and compared across various configurations. The declared objective of the project was to provide the most intensive array of data to date on the entire anatomy of the human body. Yet the cultural rhetoric surrounding these efforts is reminiscent of the Renaissance empirical philosopher Francis Bacon's characterization of the Western technoscience program: nature as a body to be scrutinized down to the very last atom. The first Website created to support the Visible Human Project described the two donated bodies as science's new "first parents," a remade Adam and Eve who would provide for new medical research. Though we have gone a long way down the path of technological development since the ancient speculum, the digital "mirrors" we use today reflect familiar metaphors of a persistent desire to "see into" the sexed body of nature.

*See also* **Sex and the Body in Science**

**BIBLIOGRAPHY**

Basalmo, Anne. *Technologies of the Gendered Body: Reading Cyborg Women*. Durham, NC: Duke University Press, 1996.

Davis-Floyd, Robbie, Joseph Dumit, Jennifer Croissant, and Sylvia Sensiper, eds. *Cyborg Babies: From Techno-Sex to Techno-Tots*. New York: Routledge, 1998. A balanced collection of readings that focus specifically on the impact of technology on human reproduction, embracing both progressive and critical perspectives on cyborg culture. The introduction offers a thorough and succinct overview of the background and development of metaphors and actualization of the figure of the cyborg.

Fausto-Sterling, Anne. *Sexing the Body: Gender Politics and the Construction of Sexuality*. New York: Basic Books, 2000.

Haraway, Donna. *Simians, Cyborgs, and Women: The Reinvention of Nature*. New York: Routledge, 1991.

Hayles, N. Katherine. *How We Became Posthuman: Virtual Bodies in Cybernetics, Literature, and Informatics*. Chicago: University of Chicago Press, 1999.

Heidegger, Martin. *The Question Concerning Technology and Other Essays* (1954), Translated by William Lovett. New York: Harper and Row, 1977.

Hopkins, Patrick, ed. *Sex/Machine: Readings in Gender, Culture, and Technology*. Bloomington: Indiana University Press, 1998. The most valuable volume of essays to date on the intersection of sex, gender, and technology. A collection of diverse insights that all converge on the power of contemporary technologies to undermine cultural assumptions about sexed and gendered bodies.

Ihde, Don. *Bodies in Technology*. Minneapolis: University of Minnesota Press, 2002. A more recent appropriation of the "technology as bodily-sensory extension" thesis that employs philosophical phenomenology to read the impact of digital and virtual technologies on the body.

Maines, Rachel P. *The Technology of Orgasm: Hysteria, the Vibrator, and Women's Sexual Satisfaction*. Johns Hopkins Studies in the History of Technology, New Series, vol. 24. Baltimore, MD: Johns Hopkins University Press, 1999.

McLuhan, Marshall. *Understanding Media; the Extensions of Man*. New York: Routledge, 1964.

Seltzer, Mark. *Bodies and Machines*. New York: Routledge, 1992. However limited in its fairly reductive definition of technology as a supplement to natural existence, this text offers a broader historical range for envisioning technology in relationship to the body and literary production within culture.

Terry, Jennifer, and Melodie Calvert, eds. *Processed Lives: Gender and Technology in Everyday Life*. London: Routledge, 1997.

Terry, Jennifer. *An American Obsession: Science, Medicine, and Homosexuality in Modern Society*. Chicago: University of Chicago Press, 1999.

Treichler, Paula A., Lisa Cartwright, and Constance Penley, eds. *The Visible Woman: Imaging Technologies, Gender, and Science*. New York: New York University Press, 1998.

Wacjman, Judy. *Feminism Confronts Technology*. University Park: Pennsylvania State University Press, 1991. Striking for its intervention in the sociology of technology of the late 1980s and early 1990s, this text offers an in-depth overview of the connection between technology and masculinity, with close attention to the imbrications of sexual division and the division of labor.

*Jen Boyle*

# SOCIAL SCIENCES.

Philosophical and speculative thought about society, politics, economy, and human nature has ancient origins, but most historians of social science trace economics, sociology, psychology, political science, and the other modern social science disciplines to nineteenth-century European origins. Throughout the history of the social sciences, periods of rapid political and economic change promoted efforts to think systematically about human nature and to seek to explain the sources of social order and disorder. The revolutionary upheavals in the American Colonies and France at the end of the eighteenth century, together with the conservative reactions to them, set the stage for the rise of the social sciences. These upheavals and the reactions against them were, in turn, influenced by eighteenth-century writings of the German idealist philosophers, the *philosophes* of pre-revolutionary France, and the Scottish empiricists and moral philosophers. Just how to characterize the Enlightenment is difficult and contentious, but *Sapere aud!* ("Dare to know!")—the motto quoted by Prussian philosopher Immanuel Kant (1724–1804) in his famous 1784 essay "What Is Enlightenment"—concisely expresses the challenge to the authority of clerics and hereditary elites in favor of secularism, popular enfranchisement, and individual autonomy. The political and economic ideals of the Enlightenment were embraced by the rising class of merchants and entrepreneurs. This challenge to received ideas and entrenched classes was an important precondition for the secular orientations and search for novel explanations that were characteristic of the social sciences.

## Social Engineering and the Search for Scientific Authority

It is not sufficient to trace the social sciences back to the intellectual and political ideas expressed by Enlightenment philosophy. Another major influence was the example provided by the natural sciences. In the early nineteenth century, social and political philosophers looked to the natural sciences and engineering for inspiration. French philosopher August Comte (1798–1857) coined the term "sociology," and he imagined that this new science would be a source of laws analogous to those in the physical sciences. Comte's new science appears highly speculative today, but he and some of his contemporaries imagined that the sciences of "man" not only would explain social order, but would also provide tools for social control and social engineering. Society and economy were viewed as transcendent fields, with laws of operation that could become intelligible only through the specialized methods of a science. Comte's followers went so far as to develop a secular cult that worshiped emergent "society" rather than a transcendent deity. Though often supported by quasireligious zeal, early efforts to develop "scientific" systems of social and historical analysis supported vastly different programs. Political economist Karl Marx (1818–1883) built a radical critique of capitalism and a design for socialist society on a positive "scientific" foundation of historical materialism. Other nineteenth-century thinkers developed "scientific" philosophies that supported conservative ideas and programs. For example, the Englishman Herbert Spencer (1820-1903) and the American William Graham Sumner (1840-1910) founded their social philosophies on evolutionary principles. The phrase "survival of the fittest"—often misattributed to Charles Darwin—actually was coined by the social Darwinist Spencer. Social Darwinist thinking was often used to support social discrimination along racial, class, or sexual lines. Cultural "backwardness" (from a Eurocentric point of view) was deemed to reflect a lower degree of biological evolution, and criminality and deviance were attributed to a lack of fitness or the persistence of lower animal traits. The eugenics movement in North America and Europe sought to implement evolutionary principles by promoting selective breeding of humans to enhance "fitness" and eliminate "unfit" individuals from the human population. Often discredited in association with Nazi ideology, eugenic programs were also supported well after World War II in more "progressive" nations such as Sweden and the United States.

Social Darwinism and eugenics represent a darker side of the scientism (the promotion of "scientific" ideas and "facts" that might better be viewed as ideologies and products of speculation) that has been so prominent in the history of the social sciences. For better and worse, scientism and social engineering

were deeply intertwined with progressive movements associated with the Industrial Revolution. In contrast to histories that stress intellectual lineages and foundational ideas, many historians of social science follow Michel Foucault's lead by stressing the utilitarian technologies of power for administering madhouses, clinics, and prisons; increasing worker productivity and military efficiency; promoting public hygiene; and controlling unruly social elements. The technocratic heroes of these regimes include English utilitarian Jeremy Bentham (1748–1832); Italian founder of rational criminology Cesare Beccaria (1738–1794); Belgian "moral statistician" Adolphe Quetelet (1796–1874); and French founder of "scientific" forensics Alphonse Bertillon (1853–1914). These innovators expressed an obsession with "objective" measures of individual bodies and collective identities, and a zeal for organizing society on a rational, scientific basis. Especially in the United States, the formation of the social sciences was encouraged not only by systematic reform efforts led by governments, but also by philanthropic societies, moral hygiene movements, and populist efforts to enfranchise women, limit child labor, and alleviate poverty. The political and intellectual orientations of such activist efforts ranged from radical Marxist to Christian fundamentalist, but regardless of such differences many participants hankered after objective evidence and scientific authority.

The sometimes confident, sometimes desperate effort to claim scientific status for the social sciences has not gone unopposed. Periodic disputes have occurred throughout the history of the social sciences between proponents of "scientific" methods and opponents who insist upon more "humanistic," "interpretative," or "qualitative" approaches. At times, the proponents of interpretative approaches have found allies among natural scientists and mathematicians who refuse to acknowledge the scientific status of the so-called hard social sciences or the validity of social statistics. Boundary disputes about scientific status intensified at times when social scientists attempted to secure a place in the modern university, and when they claimed—or seemed to be claiming—resources reserved for the "hard sciences." The most notable dispute—dubbed the *Methodenstreit* (strife about method)—occurred at the turn of the twentieth century in Germany. This dispute occurred at a time when aspiring social scientists were beginning to secure university chairs and to establish departments. One extreme position in the debate claimed that the social (and historical) sciences were, or could become, branches of natural science that put forward law-like generalizations and tested them with quantitative data. The other extreme insisted that the sciences of the "spirit" (*Geisteswissenschaften*) required methods that would be sensitive to the singular case, and adequate to the complex, often surprising, constellations of events that make up history. The debate was never resolved, though Max Weber (1864–1920) developed an "interpretative sociology" that borrowed elements of both positions and became highly influential on later developments in sociology. Later variants of the debate—such as the "two cultures" debate of the early 1950s, and the "culture wars" and "science wars" of the 1980s and 90s—also occurred at a time when the social sciences were securing increased support from research foundations or when they seemed to threaten public support for objective knowledge and the experimental method.

Despite a history of discord, boundary disputes, and periodic complaints about a lack of progress, the social sciences have succeeded in certain important respects. The postwar welfare states in Western Europe and North America provided employment for social scientists and demand for their research findings. A rapid expansion of universities resulted from a combination of increased public support for higher education and the demographics of the baby boom. Expanding sources of government funding for university-based scientific research also supported rapid growth of the social sciences. In the United States, the National Science Foundation extended support to the social sciences, but only after a fierce debate about their scientific status. Departments of economics, psychology, government, sociology, and anthropology became fixtures at universities, and they experienced a golden age of expansion in the 1950s and 1960s.

Support for the social sciences has been uneven, however. Economics, once dubbed "the dismal science," separated itself from the study of political economy, embraced the neoclassical paradigm, and became increasingly technical in the later part of the twentieth century. Professional economists enjoyed considerable influence in the public and private sectors, and their discipline attained a level of administrative and financial support that the other social sciences lacked. Psychology has had more checkered success, but today many psychologists claim to have moved beyond the subjective focus that once characterized the field (and continues to appeal to undergraduate university students) to become a branch of the neurosciences. The dominance of cognitive science in psychology departments, and of psychopharmacology in clinical psychiatry, is accompanied by the splitting off of psychoanalysis to become an individualized therapy in private practice and a prevalent source of "theory" in humanities and cultural studies departments. After rapid expansion through the 1960s, political science, sociology, and anthropology have enjoyed less support in recent decades. Conservative governments tend to curtail support for sociology (consistent with Margaret Thatcher's famous declaration that there is no such thing as society). Nevertheless, all of the major social sciences have remained entrenched in universities despite periodic budgetary crises and occasional hostility from governing elites.

In addition to experiencing uneven public support, political scientists, sociologists, and anthropologists have been unable to march to the tune of a single paradigm. These fields continue to support a pluralistic mix of theoretical orientations and methods. Particular orientations—such as functionalism in mid-twentieth-century anthropology and sociology, and rational choice theory in late- twentieth-century political science—sometimes dominate for a period of years, or even decades, but they never fully displace incommensurable orientations. For example, an approach in sociology that Herbert Blumer dubbed "variable analysis" has remained a fixture of North American sociology for decades, while the statistical techniques with which such analysis is associated have grown increasingly sophisticated. Nevertheless, despite its prevalence, variable analysis has faced powerful, and largely unanswered, criticisms by proponents of alternative interpretative approaches, such as Blumer's symbolic interactionism and Harold Garfinkel's ethnomethodology.

## The "New" Sociology of Scientific Knowledge

In the 1960s, sociology experienced fragmentation, as several avowedly radical approaches sought to displace the dominant alliance of structural-functionalist theory and statistical survey analysis. The alternative sociologies were characterized by explicit radical programs. These included intellectual efforts to raise reflexive questions about the very possibility of "scientific" sociology, as well as political efforts to effect fundamental change in contemporary institutions beyond the university. Although sociology was very popular with students in the 1960s, it also underwent a crisis. In part, this was an epistemological crisis that involved the very status of sociology's relation to its subject matter. One highly influential source of epistemological criticism—especially in the United Kingdom—was Peter Winch's *The Idea of a Social Science*. Winch insisted that sociology was unlike other sciences. Whereas physicists and engineers communicate with one another about "first order" properties of a material universe, social scientists investigate a "second order" world that is, itself, constituted through communicative actions. Winch insisted upon a sharp distinction between sciences that develop concepts and theories about an independent material world and "social studies," which borrow concepts that already have currency in social worlds studied. Thomas Kuhn, though for different reasons, also drew a sharp distinction between "normal science" characterized by a shared "paradigm"—a taken-for-granted matrix of techniques, principles, and tacit

knowledge—and "pre-paradigm" sciences, characterized by intractable standoffs between incommensurable theoretical and methodological programs. He made clear that sociology and other social sciences were, and would possibly remain, preparadigmatic.

Winch's and Kuhn's critical views of social studies had strong influence on a new sociology of scientific knowledge that emerged in the late 1960s. The "old" sociology of knowledge, represented by figures such as Robert K. Merton and Bernard Barber, adhered to the structural-functionalist research program that had dominated North American sociology through the 1960s. Science was construed as an institution serving distinctive functions, with a reward structure that set it apart from political, economic, and religious interests. Empirical investigations sought to measure, or graphically represent, the outlines of scientific communities without critically examining the status of the knowledge they produced. A group of upstarts who staffed a new Science Studies Unit at the University of Edinburgh—David Bloor, Barry Barnes, and David Edge—proposed a "strong programme" in the sociology of scientific knowledge (SSK) that would put aside the exceptionalist treatment often granted to the sciences. Scientific knowledge would be examined and explained much in the way that other beliefs would be investigated sociologically. Moreover, as subsequent developments in the field made clear, scientific laboratories would be treated as places of work, to be studied ethnographically much in the way that students of Everett Hughes, Howard Becker, Anselm Strauss, and Harold Garfinkel had studied work in machine shops, hospitals, clerical offices, and courtrooms.

"Social constructionism" (or "constructivism," as it is sometimes called) became the most commonly used name for the new programs in sociology of scientific knowledge. This name tends to be used in a vague and indiscriminate way by proponents and critics alike, but the general idea is that theories, facts, graphical data, and other elements of scientific knowledge arise from and are sustained through practical activity and cultural appropriation; they do not simply exist in an external world, the characteristics of which are hypothesized, observed, tested, and verified through the scientific method. Peter Berger and Thomas Luckmann's *Social Construction of Reality* was the first use of the term in a published title. Berger and Luckmann developed a general framework that drew upon an eclectic treatment of Karl Mannheim's sociology of knowledge, Alfred Schutz's phenomenological sociology, and Max Weber's historical approach to the rationalization of institutions. Berger and Luckmann expressed little interest in the natural sciences, and they were content to distinguish objective nature from the socially constructed realities constituted through social action and interaction. Weber, Mannheim, Kuhn, Winch, Schutz, and later theorists and philosophers such as John Searle and Ian Hacking all distinguish socially constructed realities (currency systems, customs, rituals, culture-bound syndromes, and so forth) from the "brute facts" of the natural world (the height of mountains, the speed of light, the existence of electrons, and so forth). Although these philosophers are often cited in support of social constructionist studies of the natural sciences and mathematics, the new research programs suspend judgment about any fundamental ontological distinction between social and natural realities. They also tend to turn away from the philosophical project (most famously associated with Sir Karl Popper) of developing criteria that would demarcate the methods of the natural sciences from those of the social sciences. Instead, the new sociologists of knowledge emphasize the way scientific research initially confronts a world that is not already demarcated in terms of nature-culture: it is a world in which objects cross back and forth between ontological and organizational boundaries. Instead of seeking normative criteria for demarcating science from non-science, the new research programs treat demarcations as instances of boundary work—historically specific polemical and organizational struggles to establish what counts as scientific or not—and focus on the intertwining or hybridization of natural and social realities.

Social constructionism became widespread throughout the social sciences and humanities in the 1980s and 1990s, and it was applied to a bewildering variety of subject matters. The "strong programme" and its offshoots in the emerging field of science and technology studies (S&TS) became especially prominent and controversial in the constructionist movement, partly because it addressed what appeared to many to be the most obdurate of realities: the facts and laws of natural science and the results of mathematics. If physical and mathematical realities could be shown to be constructed, then it would seem easy work to reveal the constructed nature of medical categories, technologies of all kinds, educational curricula, museum exhibits, race and gender, and whatever else one might take up for study. To understand what was at stake, consider a precursor to social constructionism: labeling (or "societal reaction") theory in the sociology of mental illness. Starting in the late 1950s and early 1960s, radical (antipsychiatrists such as Thomas Szasz and R. D. Laing, and sociologists such as Erving Goffman and Thomas Scheff) advanced an influential set of arguments and empirical studies that challenged the idea that mental illnesses (or, at least, some mental illness categories) were diseases.

Szasz denied that mental disorders were akin to infectious diseases, strokes, or cancers, in that they were characterized by the *absence* of any of the organic disease agents or lesions that medical science uses to explain clinical syndromes. Such "diseases" were, in Szasz's famous expression, "problems in living," which were compounded by social antagonisms. The now-defunct category of hysteria was Szasz's best case: the "disease" was characterized by symptoms that lacked medical explanation, but which mimicked genuine medical conditions (seizures, paralysis, digestive disorders, and the like). Moreover, the diagnosis and treatment of hysteria often expressed the misogyny that was entrenched in patriarchal societies. One could say, as Goffman did about a range of mental disorders, that the "patient" suffered not from a disease but from "contingencies": the circumstances of family, social location, and sheer accident that result in the diagnosis and commitment of a person as a mental patient.

In many respects, labeling theory was a precursor to the social constructionist approaches to science, technology, and medicine that later became prevalent. There was a key difference, however. Szasz, Goffman, and the others distinguished mental disorders from other conditions that they presumed to be appropriately classified as medical disorders caused by objective infections, lesions, or other bodily conditions. Historical accounts of the "construction" of mental disorder categories and ethnographic studies of how those categories are applied in individual cases were set off against a presumed "unconstructed" universe of medical categories and patient conditions. In principle, though not always in practice, social constructionist accounts of scientific facts do not make use of such a contrast. Although it is assumed that different entities are constructed differently—and that the differences are interesting and worthy of detailed attention—there is no assumption that some entities are constructed while others are not. This does not amount to a wholesale denial of "nature," or an effort to reduce all explanations to a set of social factors. Instead, it amounts to an open invitation to delve into the historical and situational origins of any established truth, without initially assuming that some truths reflect the nature of things while other *purported* truths do not. It also invites forms of explanation that do not trade upon a rigid distinction between natural or cultural factors, but which describe the entanglement, and ultimate inseparability, of nature and culture.

## From Topic to Resource and Back

The new sociology of scientific knowledge (SSK) arose from and reacted to developments in sociology and other social sciences in postwar Europe and North America, and it has become influential (often in connection with cultural studies and postcolonial themes and approaches) in the Far East, India, South America, and other regions of the world. In other words, it has become an international and interdisciplinary development from within the social sciences. How-

ever, contrary to what is sometimes assumed in the larger field of Science, Technology, and Society, SSK—as well as some other, more recent, developments such as Actor-Network Theory—is not an *application* of social science knowledge to the study of science and technology. It is as much an *investigation* of the social technologies (including those associated with the "sciences of man") through which natures and cultures are constituted. This is not a question of adding another S to SSK, so that it becomes the sociology of social scientific knowledge: an offshoot of SSK that produces case studies of the practices used for designing and administering social surveys, representing data with graphs and tables, or implementing social science knowledge in programs of social engineering. Instead, it is an unsparing investigation of the constitutive uses of social practices, representational devices, and instrumental complexes wherever they occur, and with whatever effects.

### BIBLIOGRAPHY

Barnes, Barry, David Bloor, and John Henry. *Scientific Knowledge: A Sociological Analysis.* Chicago: University of Chicago Press, 1996.

Berger, Peter, and Thomas Luckmann. *The Social Construction of Reality: A Treatise in the Sociology of Knowledge.* Garden City, NY: Anchor Books, 1967.

Blumer, Herbert. *Symbolic Interactionism: Perspective and Method.* Englewood Cliffs, NJ: Prentice Hall, 1969.

Foucault, Michel. *The Order of Things: An Archaeology of the Human Sciences.* New York: Vintage Books, 1970.

Gieryn, Thomas. *Cultural Boundaries of Science: Credibility on the Line.* Chicago: University of Chicago Press, 1999.

Hacking, Ian. *The Social Construction of What?* Cambridge, MA: Harvard University Press, 1999.

Kuhn, Thomas. *The Structure of Scientific Revolutions,* 2d ed. Chicago: University of Chicago Press, 1970.

Latour, Bruno. *Science in Action: How to Follow Scientists and Engineers through Society.* Cambridge, MA: Harvard University Press, 1987.

Smith, Roger. *The Norton History of the Human Sciences.* New York: Norton, 1997.

Snow, C. P. *The Two Cultures.* Cambridge, UK: Cambridge University Press, 1993.

Turner, Stephen, and Jonathan Turner. *The Impossible Science: An Institutional Analysis of American Sociology.* Newbury Park, CA: Sage, 1990.

Winch, Peter. *The Idea of a Social Science and Its Relation to Philosophy,* 2d ed. Atlantic Highlands, NJ: Humanities Press, 1990.

*Michael Lynch*

# SOCIAL STUDIES OF SCIENCE.

The social studies of science share some common concerns with literary criticism. No clear method in the social studies of science predominates, and no single standard bearer of "good" social studies of science exists. Social studies of science is a "field of research" in which any description of its object, given or implied, implicates a certain kind of politics. Less strongly, social studies of science, like literary criticism, either has no clear object of study or has an object of study in constant transition.

Unlike literary criticism, social studies of science (hereafter SSS) is growing, healthy, robust, and in demand. SSS is irretrievably linked with one of the dominant institutions of our time: technoscience. In turn, technoscience is tied to all the other major institutions, such as politics and economics. The work in SSS appears to be discussed well beyond its own field, in the halls of policy makers as well as by mainstream media.

The tradition of SSS is, in part, a critical one. SSS attempts to unlock the "black box" of the actual practices of scientific production. SSS studies how the social affects the cognitive in production of scientific knowledge; SSS documents, primarily though qualitative case studies, the negotiations that take place among scientists, the relationship between scientists and their artifacts, science boundary maintenance, and other organizational prerequisites. Some of these descriptions that show the "all too human" side of science meet, not surprisingly, with resistance and even outrage from some elements of the scientific community.

Science has not always been viewed as open to social explanations. Both Karl Marx and Karl Mannheim suggested that certain elements of human phenomena, such as the world view of mathematics and natural sciences, could not be accounted for by an analysis of class or generational interests. More recently, Robert Merton (1910–2003) noted that there are many social influences on the organization of science and its interactions with other institutions. However, he concluded that the dynamic between science as a certified, reliable body of knowledge and the social could not be explored through sociological analyses. The belief that natural science and mathematics were not amenable to traditional sociology of knowledge analyses meant that an important section of modern life was not available for examination.

One of the more exciting developments in SSS over the last thirty to fifty years was the opening up of domains of science left unexplored by traditional sociologists of science. The work of Thomas Kuhn had a tremendous impact, although not always for the better, on moving SSS forward. Kuhn, argued, in part, that science was not in a constant march toward the truth. Changes in our view of the physical world are more like "gestalt switches" than they are the product of accumulated evidence. Further, he suggested that these gestalt switches might be linked to such social processes as generational conflict; a younger generation of physicists may be more likely to adopt new theories than the older generation.

Some of the more interesting recent developments in our understanding of science as a social phenomenon have been the "strong program," social constructionism, feminist theories of science, evolutionary epistemology, social epistemology, and conflict and critical theories of science. The classifications of these developments seem to be strongly dependent on the traditions to which the classifier belongs. What these new developments in the social studies of science have in common is the continual challenge to a traditional understanding of science.

## Strong Program and Interests

The "strong program" sees the connection between social forms and knowledge production as dependent upon various "interest" models. Knowledge production, from this perspective, is a cultural phenomenon, rather than experiential. The strong program has four tenets that are methodological dicta on how to proceed in the analysis. The basic tenets of the strong program, as developed by Barry Barnes and David Bloor, are causality, symmetry, reflexivity, and impartiality.

Meeting the tenet of causality requires that knowledge production and beliefs be viewed as social phenomena; social accounts should be sought whenever possible. Causality asserts that knowledge has no special categorical status that allows it to be understood as spiritual or ethereal and not open to social accounting. The symmetry tenet requires that the same method of explanation be applied to both true and false beliefs. Because some beliefs turn out to be accepted as "true" in the future does not allow us to view the explanation of them as having some special characteristics different from the beliefs that failed. The reflexivity tenet means that the sociology of knowledge itself is open to the same kind of scrutiny, analysis, and explanation as any other kind of knowledge. Finally, the impartiality thesis refers to the notion that we should be value-neutral in our analyses.

One example of an interest model within that strong program is the *intellectual interest* model found in the work of Andrew Pickering, in which the intellectual combatants of the high-energy physics community debate whether charm or color is the best model to account for the behavior of certain elementary particles. Pickering argued that the charm model won because its proponents were able to mobilize more intellectual and organizational resources than were advocates of the color model. The charm model did not necessarily undergo the most rigorous trials. Rather, in Pickering's case study, elements of the charm model were more able than the color model to support the interests of an increasing number of subcultures within the high-energy physics community.

Another example of an interest model within the strong program is the *cultural interest* model articulated by Steve Shapin. Shapin argued that the gentlemanly practices in seventeenth-century England provided a context for the start of the social construction of truth. Within this gentlemanly society, individuals were trusted and could offer claims to truth and credibility.

## Social Constructionism and Ethnography

Social constructionism is another major development within SSS. Harry Collins and Trevor Pinch are identified closely with social construction. Their ethnographic work illustrates the negotiated reality that the so-called hard sciences have to overcome. According to them, science cannot be explained (that is, replicated) from written or even verbal accounts. In some cases, the scientist must go to the other research sites to learn how to set up and get the equipment to operate in the fashion needed.

Actor-network theory (ANT) is one variant of social constructionism. Bruno Latour, the most identifiable figure in ANT, argues that when viewing modern science from an anthropological perspective we see that modern science is not different from other everyday practices (cultural practices).

In his book *Science in Action*, Latour argues that science cannot be properly understood by examining the discourse of science in journals. Studying science requires going into the laboratory and following the scientists and engineers as they mobilize resources in the form of networks. Scientific claims must be examined in the context of the length of the network employed. Those scientists with the longer networks have advantages in the production of knowledge over those with shorter networks. For this purpose, networks are both material and social.

Latour argues that the study of networks among scientists is what is important. For example, scientists may supply each other with different things, and the primary ties of the laboratory may be extended to other ties in different networks. In systems with scarce resources, competition and collaboration develop. Some scientists may band together to acquire scarce resources, and others may compete for and fight over resources.

Latour attempts to dissolve the inside-outside (scientist-nonscientist) distinction. In *The Pasteurization of France*, he provides the example of Pasteur's lab in 1881. Large numbers of people who were not scientists were interested in Pasteur's lab, negating the view that the practice of science is removed from everyday life. He suggests that the methodology employed for examining the laboratory in society is the same as that for exploring the society in the laboratory. Science is part of normal motivational processes. Scientists are like the rest of us.

Latour suggests that we should not attribute any special method to the mind or method of the people involved in producing science. Looking at the rules of logic or the organization of society when charges of irrationality raise their ugly head will not help us understand their resolution. Rather, insight can be gained from knowing the angle and direction of the observer's displacement and the length of the network being built.

Finally, Latour also wants to dissolve the subjective and objective distinction in scientific accounts. He argues that claims are changed (translated) by others later. Thus, any claims or arguments have no peculiar cognitive status independent of what is attributed to them by others. Building on this argument, Latour suggests that the settlement of a controversy among scientists is the *cause* of how nature is represented—nature (consequence) cannot be used to explain how and why a controversy has been settled. Further, the settlement of a controversy is the cause of a society's stability; we cannot use society to explain how and why a controversy has been settled.

In a series of writings, Karin Knorr Cetina has argued for a *constructivist* interpretation of scientific reality. From her perspective, science is an artifact that is made to work. She claims that a constant decision-making operation occurs, which in turn becomes the basis for further decisions. She calls this process decision impregnatedness. Further, a strong contextuality, or contingent character, to modern laboratory research exists. This is sometimes described as idiosyncratic and unpredictable. Knorr Cetina argues strongly

against a problem-solving model of science because science creates a potentially increasing stock of problems that are sometimes ignored. One source of unpredictability lies in the fact that social action appears to be underdetermined by antecedent constraints.

Knorr Cetina rejects traditional macro approaches to the sociology of scientific knowledge. Her position is that this tradition is weak because the inferences regarding, for example, the link between class interests and scientific claims cannot be easily made. Her question asks how and at what junctures contextual factors such as social interests enter particular knowledge objects. She wants to examine the genesis and transformation of our objects of knowledge. She chooses to take the genetic and microlevel approach: examining scientific controversies at the actual site of scientific work.

Pickering's later work attempts to use a performative analogy to capture an understanding of the construction of science. Pickering claims that science has powers, capacities, and performances. In brief, science has material agency. Pickering argues that both humans and machines are influenced by each other in the performance of science. Machines tune humans, and humans tune machines. There is a complex dance of agency. The cultural context of scientific practice contains an element of uncertainty. The future of science exists only within the complex mangle of practice. Emergent forms of knowledge may become impure objects— the mixing of human and material agents.

## Feminist Theories of Science

The feminist critique of science has centered on a number of topics, such as the exclusion of women in the history of science, the lack of research on women or about women's concerns, the small number of women engaged in science, the interpretation of results that reinforce women's second-class status, and the problems that women experience in science careers. However, the most controversial challenge to science is the feminist argument that androcentrism is embedded in the very concepts and epistemic categories of science. For example, feminist standpoint theorists place the knowledge claims of women in a privileged position, arguing for epistemic relativism while guarding against judgmental relativism. Two recent feminist scholars who focus on science are Lynn Nelson and Helen Longino, both philosophers of science. They want to extend feminist concerns through a commitment to a modified empiricism. They believe that feminist values can be incorporated into science through the infrastructure of cognitive values that permeate science. Other major feminist scholars of science include Evelyn Fox Keller, Sandra Harding, and Donna Haraway. These scholars and others have rejuvenated the field of SSS with their feminist challenges and insights.

Lynn Nelson wants the incorporation of feminist criticism of science to lead to a different science—a science that will unabashedly include values subject to critical scrutiny. Unlike some feminist critics of science, however, Nelson argues that the justifications of feminists' claims have to be based on their empirical adequacy.

Nelson stresses the importance of the sex/gender system and the political context in the practice of science, noting that science is socially constructed within communities. However, she is *not* advocating that the only causes for beliefs are social causes (sociology of knowledge). Nelson maintains that the focus on communities does not preclude an emphasis on evidence as sensory. Rather, she suggests that sensory experiences themselves are dependent on membership in communities.

For Helen Longino, two types of values are fundamental to scientific inquiry. First, scientific knowledge rests on a bed of presuppositions, or constitutive values, about *what* questions are important and meaningful, about the general direction of causal relationships, and about what constitutes good explanations. These shared presuppositions, when viewed at all by the scientific community, will often be seen as self-evident truths. The second group of values is contextual values. These are the personal, social, and moral values about what ought to be.

According to Longino, since values have an important contextual role to play in sci-

ence, we must reject outright any commitment to the notion of a value-free science. Instead of a value-free science, Longino calls for a contextual empiricism viewed as a prescriptive theory. She believes that feminist scientists need to fashion research programs that are consistent with their contextual values and not split themselves into a "scientist here," and a "political actor" there.

Longino suggests that, with contextual empiricism, the objects of inquiry are not given in nature, but rather they are instantiated in part by social needs and interests embedded in the assumptions of research processes. Because scientific methods generated by constitutive values cannot guarantee independence from contextual values, Longino says we must work harder to discover the limits of present interpretive frameworks and build more inclusive and useful frameworks.

Evelyn Fox Keller writes that combining feminist thought with SSS promises both radical new insights and political dangers. One danger she sees is that, by viewing science purely as a social product, science becomes ideology and any sense of objectivity is lost. The arbiter of truth becomes politics. In this way, science would lose any emancipating function.

Keller posits that science should be characterized by a plurality of views of what scientific inquiry should be. Nevertheless, one image that has come to predominate in modern times is objectivity, where objectivity means both emotional distance and control over the object of knowledge. For Keller, this results in reductionist and mechanistic models (explanations) of natural phenomena.

Keller identifies historical and psychological reasons for why we have a reductionist and mechanistic model. She argues that in the seventeenth century there was a fierce struggle between two methods of scientific inquiry. The one that ultimately won is a heterosexual fantasy of control and submission that makes science a masculine endeavor and makes women and nature appropriate objects of domination.

The androcentric nature of scientific knowledge is not due, says Keller, to most scientists' having been and being male. Rather, psychological processes are reflected in the notion of static objectivity. Keller's feminist object relation theory asserts that there are two sorts of autonomy, and correlatively there are two sorts of objectivity. Static autonomy results when the self is created in opposition to another, primarily the mother. Static autonomy is characterized by constant anxiety over the self's boundaries, and anxiety can be relieved only by attempting to control all those who attempt to break the boundaries. Dynamic autonomy is created through differentiation but also through relatedness to others. It is characterized by tolerance for ambiguous boundaries and a sense of self in a context of other selves. Static objectivity is predominately male; it radically separates the subject from the object of knowledge.

Keller favors complex and interaction models of nature. Dynamic objectivity, by her account, can provide more accurate and reliable representations of nature than are possible through static objectivity.

In her book *Whose Science? Whose Knowledge?*, Sandra Harding attempts to synthesize standpoint and postmodern feminist epistemologies. She argues that feminism, as well as science, contain both regressive and progressive tendencies and that ways are needed to advance the progressive sides and inhibit the regressive natures of science and feminism. Harding suggests that, by combining feminist standpoint theory with postmodernism, we can overcome some of the limitations of the former. She believes there are many links between postmodernism and standpoint theory.

Harding, in a more recent defense of standpoint theory, argues that marginalized lives should be the subject of scientific inquiry, not the solution to an epistemic problem. The subjects of knowledge in standpoint theory are embodied and visible in their social locations as communities of knowers that are multiple, heterogeneous, contradictory, and even incoherent.

Donna Haraway's work is important because she combines a concern for theories of the body, the body politic, and science. In particular, her focus on the biobehavioral sciences illustrates how the "social" can be shown to have an influence on observations as well as interpretations of scientific phenomena.

Haraway focuses her attention on the study of animals where animals are looked at as raw materials, subject to exact laboratory discipline. Animal sociology has been central in making the argument that patriarchy is natural. In her work, she attempts to expose weaknesses of the reduction of society to sexual physiology.

Haraway is well known for her critique of Yerkes' and Carpenter's studies of apes. Like some other animal researchers, Yerkes believed that primates are like humans; if we study them, we will discover something about ourselves. Based on his observations of apes, Yerkes concluded, in a highly influential study, that primate intelligence allowed sexual states to stimulate the beginnings of human concepts of social rights and privileges. Following Yerkes, Carpenter concluded from his study of the effects of removing the alpha male ape from the social group that domination was a natural property with a physical-chemical basis. Haraway provided a powerful critique of the sexual reductionism and the androcentric nature of these conclusions.

As a whole, feminism has had and continues to have a thriving and fertile relationship with the study of science, technology, and medicine.

## Evolutionary Epistemology, Social Epistemology, and Critical Theories of Science

*Evolutionary epistemology* is mostly associated with Donald Campbell, past president of the American Psychological Association and well-known methodologist who actually coined the term. His goal was to extend Popper's ideas of falsification to knowledge processes everywhere: biological, psychological, and social (see the extended bibliography in Cizko and Campbell, 1990).

As an orientation toward knowledge, evolutionary epistemology argues that knowledge is a product of variation and selection that characterizes other living, natural processes. Knowledge is an aid to survival and reproduction, and organisms with more useful knowledge than other organisms will be in a better position to survive and reproduce. Thus, knowledge helps phylogenetic evolution. In addition, an evolutionary epistemology perspective notes that knowledge pieces are subject to variation and selection. Again, as ideas emerge from variations, some are selected as more useful than others.

Campbell's orientation toward evolutionary epistemology rests on three ideas: (1) The social and biological processes that generate novel insights cannot predict what they will find (because if they could, it would be known already), and only some of the ideas discovered will be useful; (2) once some idea or piece of knowledge is retained in memory, it can be vicariously selected into the internal memory of the organism before it undergoes further environmental testing; and (3) the selection process itself can undergo alteration, pieces of the selective apparatus may be changed, the whole selector can be altered and new criteria introduced, thus altering higher-level selector systems. This last idea may involve a metatransition of the selection process. This development allows the use of a more adaptive and perhaps more intelligent system.

*Social epistemology* is not new. However, it has only recently been creatively applied in the social studies of science by Steve Fuller. In Fuller's work, social epistemology is not confined narrowly to description and explanation. Rather, he sees the activity as thoroughly normative. What kind(s) of science do we want? How should science be organized? Fuller argues that we need an ethic of accountability. Fundamentally, what SSS should teach us, from Fuller's perspective, is that one does not need to be an expert to understand expertise and that it is not clear that experts live up to their own standards of expertise.

The list of social epistemologists can be extended to Barnes and Bloor, Helen Longino, Michel Foucault, and many others. They all more or less share the same aversion to classical epistemological questions and focus more on what passes for belief. They ask, how are beliefs maintained individually and institutionally and, following Campbell, what selectors make some beliefs more likely to succeed?

Arguments from *critical and conflict perspectives* are embedded in much of the work discussed above. Other key figures from this tradition are Langdon Winner, Sal Restivo, Randall Collins, Jurgen Habermas, and Paul K. Feyerabend.

Sal Restivo has refused to examine science within a narrow context. He even theorizes science as a social problem.

Randall Collins has attempted to articulate a social theory of intellectual life. In his more recent writing, he tries to account for the social factors that influence the production of philosophy and for how some philosophies get passed down along generational chains. His work meets with resistance among intellectuals because he suggests that their work has a very tiny chance of being read hundreds of years from now.

The work of Jurgen Habermas represents an ongoing tradition of critical theory. Habermas attempts a systematic critique of modern science as it is implicated in the evolution of modern social systems. He argues that we a need a "critical" theory to deal with this complexity.

The basic Habermasian position is that two systems are operating on us simultaneously. The first is the life world that is understandable from the perspective of everyday actors, and the second system (society) is understandable from the perspective of the other (observer). Neither system has an ontological priority over the other, but they exist in tension, an evolutionary tension, with each other. Further, Habermas argues that their continued existence does not result in some higher unity of the spirit, revolution, or transcendental unity of apperception (in the subject). He claims that on an evolutionary scale, the two distinct spheres were collapsed together earlier in time. This meant that the performance of a task was also a system maintenance function for some institution. Moreover, the actor's understanding was also a comprehensive understanding at the level of system. Thus, ritualistic behavior could also function with economic consequences and theological content.

The uncoupling of system and lifeworld is experienced in modern society as a particular kind of objectification: The social system escapes the horizon of the lifeworld and is separated from the intuitive knowledge of everyday communicative practice. Thereafter, knowledge of the system is known only through the counterintuitive knowledge of the social sciences developed from the eighteenth century. The more complex social systems become, the more provincial lifeworlds become. The lifeworld appears to shrink under a bureaucratic management regime.

In particular, the subsystems of political administration and economic production and distribution fueled by power and money have broken off from the lifeworld. If social reproduction (that which must take place in the lifeworld—cultural reproduction, social integration, socialization) is constantly subjected to money and power, social pathologies occur. According to Habermas, only a critical social theory can help us prevent the outbreak of social pathologies.

Habermas convincingly argued that Marx's distinction between the infrastructure (material forces) and superstructure (ideology) was inadequate for analyzing the plight of modern societies. Rather than using Marxian analytic categories, the critical and conflict theorists within SSS focused their analyses on three subsystems of action: work, language, and power. When the cognitive interests of science are analyzed from these three subsystems of action (work, language, and power), three approaches of sciences with distinct interests are found. The approach of the *empirical-analytic sciences* incorporates a *technical* cognitive interest; that of the *historical-hermeneutical sciences* incorporates a *practical* interest; and the approach of the *critically oriented sciences* incorporates the *emancipatory* cognitive interest. This classification of the sciences also illuminates whether some sciences are repressive or liberative in their social-historical context. In addition, the discovery of these interests allows us to see a new form of domination in the modern world: scientism.

Critical social science determines the meaning or validity of propositions through self-reflection. A critical social science differs from the empirical-analytic or hermeneutic sciences in that it presupposes that all self-conscious agents can become aware of the self-formative processes of society and with this knowledge achieve a historically conditioned autonomy. Thus, the character of a critical science is unique insofar as it is concerned with the assessment of the socially unnecessary modes of authority, exploitation, alienation, and repression. The interest of a critical social science is the emancipation of all self-conscious agents from the seemingly "natural" forces of nature and history.

Finally, the extensive critique of science provided by the philosopher of science, Paul K. Feyerabend completes this coverage of conflict and critical perspectives. Feyerabend challenged the defenders of science to subject their work to democratic review and critique.

## BIBLIOGRAPHY

Barnes, Barry. "On the Extension of the Concepts and Growth of Knowledge." *Sociological Review* 30 (1982): 23–44.

Barnes, Barry, David Bloor, and John Henry. *Scientific Knowledge: A Sociological Analysis.* Chicago: Chicago University Press, 1996.

Bloor, David. *Knowledge and Social Imagery.* Chicago: Chicago University Press, 1976.

Bloor, David. "The Strengths of the Strong Programme." *Philosophy of the Social Sciences* 11 (1981): 199–213.

Callon, Michel. "Some Elements of a Sociology of Translation: Domestication of the Scallops and the Fishermen." In *Power, Action and Belief: A New Sociology of Knowledge?* edited by John Law, 196–229, London: Routledge, 1986.

Cizko, Gary, and Donald T. Campbell. "Comprehensive Evolutionary Epistemology Bibliography." *The Journal of Social and Biological Sciences* 13 (1990): 41–81.

Collins, Harry. *Changing Order: Replication and Induction in Scientific Practice.* London: Sage, 1985.

Collins, Harry. "What Is TRASP? The Radical Programme as a Methodological Imperative." *Philosophy of the Social Sciences* 11 (1981): 215–224.

Collins, Harry, and Trevor J. Pinch. *The Golem: What Everyone Should Know about Science.*

Cambridge, UK: Cambridge University Press, 1993.

Collins, Randall. *Conflict Sociology: Toward an Explanatory Science*. New York: Academic Press, 1975.

Collins, Randall. *The Sociology of Philosophies: A Global Theory of Intellectual Change*. Cambridge, MA: Harvard University Press, 1998.

Creager, Angela, Elizabeth Lunbeck, and Londa Sciebinger, eds. *Feminism in Twentieth-Century Science, Technology, and Medicine*. Chicago: University of Chicago Press, 2001.

Feyerabend, Paul K. *Farewell to Reason*. London: Verso, 1987.

Feyerabend, Paul K. *Science in Free Society*. London: New Left Books, 1978.

Fuhrman, Ellsworth R. "STS and Utopian Thinking." *Social Epistemology* 13 (1999): 85–93.

Fuller, Steve. *Philosophy, Rhetoric, and the End of Knowledge: The Coming of Science and Technology Studies*. Madison: University of Wisconsin Press, 1993.

Fuller, Steve. *Social Epistemology*. Bloomington: Indiana University Press, 1988.

Fuller, Steve. *Thomas Kuhn: A Philosophical History for Our Times*. Chicago: University of Chicago Press, 2000.

Gieryn, Thomas F. *Cultural Boundaries of Science*. Chicago: University of Chicago Press, 1999.

Habermas, Jurgen. *Knowledge and Human Interests*. Boston: Beacon Press, 1971.

Habermas, Jurgen. *Theory of Communicative Theory of Action*. Boston: Beacon Press, 1987.

Heylighen, F. "Evolutionary Epistemology." http://pespmc1.vub.ac.be/EvolEPIST.html, 1995.

Keller, Evelyn Fox. "Feminism and Science." In *Feminist Theory: A Critique of Ideology*, edited by Nannerl Keohane, Michelle Rosaldo, and Barbara Gelpipp, 113–126. Chicago: University of Chicago Press, 1981.

Keller, Evelyn Fox. "Gender and Science." (Ed.) *Feminist Research Methods*, edited by Joyce McCarl Nielsen, 41–57. Boulder, CO: Westview Press,1990.

Keller, Evelyn Fox. *Reflections on Gender and Science*. New Haven: Yale University Press, 1984.

Haraway, Donna. *Simians, Cyborgs, and Women: The Reinvention of Nature*. New York: Routledge, 1991.

Haraway, Donna. "Situated Knowledges: The Science Question in Feminism and the Privilege of Partial Perspective." *Feminist Studies* 14 (1988): 575–599.

Harding, Sandra. "Rethinking Standpoint Epistemology: 'What Is Strong Objectivity?' In *Feminist Epistemologies*, edited by Linda Alcoff and Elizabeth Potter, 49–82, New York: Routledge, 1993.

Harding, Sandra. *Whose Science? Whose Knowledge?: Thinking from Women's Lives*. Ithaca, NY: Cornell University Press, 1991.

Knorr-Cetina, Karin. "The Ethnographic Study of Scientific Work: Towards a Constructivist Interpretation of Science." In Science Observed, edited by Karin Knorr-Cetina and Michael Mulkay, 115–140, Beverly Hills, CA: Sage, 1983.

Knorr-Cetina, Karin. *The Manufacture of Knowledge: An Essay on the Constructivist and Contextual Nature of Science*. Oxford: Pergamon Press, 1981.

Latour, Bruno, and Steve Woolgar. *Laboratory Life: The Social Construction of Scientific Facts*. London: Sage, 1979.

Lator, Bruno. *The Pasteurization of France*. Cambridge, MA: Harvard University Press, 1988.

Latour, Bruno. *Science in Action: How to Follow Scientists and Engineers through Society*. Cambridge, MA: Harvard University Press, 1987.

Longino, Helen. "Can There Be a Feminist Science?" In *Feminism and Science*, edited by Nancy Tuana, 45–68. Bloomington, IN: University of Indiana Press, 1989.

Longino, Helen. *Science as Social Knowledge: Values and Objectivity in Scientific Inquiry*. Princeton: Princeton University Press, 1990.

Lynch, William, and Ellsworth Fuhrman. "Recovering and Expanding the Normative: The New Sociology of Scientific Knowledge." *Science, Technology and Human Values* 16 (1991): 233–248.

MacKenzie, Donald. *Knowing Machines: Essays on Technical Change*. Cambridge, MA: MIT Press, 1996.

Merton, Robert K. *The Sociology of Science: Theoretical and Empirical Investigations*. Chicago: University of Chicago Press, 1973.

Nelson, Lynn. *Who Knows: From Quine to a Feminist Empiricism*. Philadelphia: Temple University Press, 1990.

Pickering, Andrew. *Constructing Quarks: A Sociological History of Particle Physics*. Chicago: University of Chicago Press, 1984.

Pickering, Andrew. *The Mangle of Practice: Time, Agency, and Science*. Chicago: University of Chicago Press, 1995.

Pinch, Trevor. *Confronting Nature*. Dordrecht: Reidel, 1986.

Restivo, Sal. "Modern Science as a Social Problem." *Social Problems* 35 (1988): 206–225.

Restivo, Sal. *The Social Relations of Physics, Mysticism, and Mathematics.* Dordrecht: Reidel, 1983.

Schiebinger, Londa. *The Mind Has No Sex: Women in the Origins of Modern Science.* Cambridge, MA: Harvard University Press, 1989.

Shapin, Steve. *A Social History of Truth: Civility and Science in Seventeenth-Century England.* Chicago: University of Chicago Press, 1995.

Shapin, Steve, and Simon Schaffer. *Leviathan and the Air Pump: Hobbes, Boyle, and the Experimental Life.* Princeton: Princeton University Press, 1985.

Winner, Langdon. *The Whale and the Reactor.* Chicago: University of Chicago Press, 1986.

Zuckerman, Harriet. "The Sociology of Science." In *Handbook of Sociology*, edited by N. J. Smelser, 511–574, Beverly Hills, CA: Sage, 1988.

*Ellsworth R. Fuhrman*

# SOCIOLOGY OF PHILOSOPHY.

Science and Technology Studies (STS) has many origins due to its interdisciplinary nature, but one of its most central origins is the so-called "strong program." While the efficacy of the program and its subsequent modifications are still a source for much debate, the original sociology of scientific knowledge (SSK) tenets (impartiality, causality, symmetry, and reflexivity) serve as an important springboard for critical thought regarding mainstream epistemology and sociology. Within sociology itself, however, similar critical stances are emerging. Randall Collins's *The Sociology of Philosophies* stands out as one exemplar. It has a great deal in common with the original SSK position and suggests an even stronger sociological position on the nature of knowledge.

Both Collins and SSK are concerned with the formation of truth products, and both assert that they are the outcomes of networks of intellectuals, rather than separate objective realities or existing objects as natural kinds. Intellectuals themselves identify their occupation as creating "timeless" products that are immune to human miscommunication or are otherwise "freestanding." It is this very self-perception, Collins asserts, that intellectuals have in common, and this forms the basis of solidarity in their moral order. Intellectuals rely on their community to indoctrinate members (and to appreciate legacies), participate in the current state of ideas (with a certain set of "rules" established by the moral order), and eventually to create ideas as intellectual products (intellectual creativity). Collins has comprehensively reviewed Greek, Chinese, medieval, Renaissance, and modern philosophers. By charting networks of both affiliation and rivalry and the breakthroughs that result therein, Collins paints a compelling portrait of human creativity and intellectual affairs as outcomes of social networks.

It would be a mistake, however, to read Collins's work as a specific history of philosophy or as a study of the impact of a particular philosopher on his or her network, though it is possible to do so. Indeed, Collins's emphasis is not on individual "great" philosophers but on the minor philosophers that surround and support them. At the heart of this position is Collins's general theory of *interaction ritual chains* (IRCs), which arises out of all the details of philosophical legacies. Discarding the conventional sense of "ritual," Collins redefines it as a sort of machine, with three components: (1) A group is assembled; (2) there is a common referent, either a symbol, individual, or the group itself; and (3) a common mood or emotion shared by the group. Philosophers make a special point of the sharing of intellectual products as their common referent and the reverent production of those products as a common mood. It is this ritual machine that unfurls macrolegacies out of the micro-social moments that produce them.

Several key ideas result from this position. One is the "law of small numbers." As intellectuals compete for attention, knots of argument surround the ones that are successful at sparking the ritual machine. Since the crowd willing to participate limits attention space, there is an eventual threshold of competing arguments. A historical outcome of this process is that the number of competing networks in the field at any given time is usually limited to about three to six, and the success or failure of an individual career must take this into account. This is not meant to be a precise, hard number, but a general empirical observation, perhaps still operating among contemporary intellectuals.

Let us not concern ourselves only with groups, however. In a Western society that prizes the individual, free will, and its geniuses such as Kant, Kierkegaard, and Hegel, Collins weighs in with perhaps his most controversial thesis: *Thinking is internalized conversation.* This must be understood in terms of emotional energy (EE) and cultural capital (CC). From the moment they are born, individuals constantly seek higher levels of EE. It is earned as a product of solidarity and participation in social rituals. The more central an individual is to a ritual, either by leading or participating in a critical way, the more EE is earned. CC is a byproduct of earning EE and is fuel for further participation in rituals, and thus as a means toward further EE.

EE levels diminish over time, and without constant contact with others, individuals gradually lose their ability to function, both in a social sense and, at extremes, in a biological sense as well. In order for the individual to sustain itself, and also as a means of "practicing," self-talk is engaged in as a survival strategy. Since the stuff of self-talk is previous interaction in the rituals of others, unique combinations of cultural capital may occur in individuals who are privy to disparate networks, and such combinations may result in that most sought-after intellectual product: a unique, creative idea. Such results, however, owe as much, if not more, to the alliance of minor philosophers (and perhaps their major mentors) who support creative individuals as to the outstanding creative individuals themselves.

Collins has given us a new understanding of what it means to do philosophy, indeed, what it means to think at all. The sociology of intellectuals has radical implications for epistemology, for sociology, for cognitive science, and even for radical new directions in artificial intelligence. It moves current tentative efforts toward a social view of mind and brain in a strongly sociological direction based on strong theories and empirical studies.

**BIBLIOGRAPHY**
Collins, Randall. *The Sociology of Philosophies.* Cambridge, MA: Harvard University Press, 1998.

*Colin Beech*

# SOCIOLOGY OF UPPER LIMITS.

Most scientific experiments and observations set out to find some new phenomenon or make a measurement of a known effect. Some, however, either through accident or design, do neither; instead, they set an "upper limit" on the incidence or flux of a phenomenon.

An example is the famous Michelson-Morley experiment of 1887. Albert Michelson and Edward Morley set out to measure the speed of the earth through the ether by looking at the effect of the "ether wind" on the speed of light. But they could not measure any significant change in light velocity according to whether the earth was traveling with or across the supposed wind. They set an upper limit on the variation of the measured speed of light governed by the sensitivity of their apparatus, and this was well below the earth's speed (or they would have done so if they had completed the experiment in the way that many people think they did). This upper limit was sufficiently in conflict with the theory of the ether to cause serious trouble (the result would later be explained by the theory of relativity). Incidentally, the Michelson-Morley result is often described as though it showed that the speed of light was a constant, but it did not; it set only an interestingly low upper limit on velocity variation under particular circumstances.

Not all upper limits are as interesting as that set by the Michelson-Morley experiment, however. An upper limit can be as uninteresting as you like and can be set in principle by almost anyone. There are three interlinked features of the "logic" of upper limits that distinguish them from positive claims.

The *first* feature is that *to set an upper limit is to see nothing,* and one can see nothing no matter how insensitive the apparatus. The very encyclopedia you are holding is, in virtue of its stability, setting upper limits on all kinds of phenomena (the temperature of the air, the amount of hydrofluoric acid in the atmosphere, the flux of all manner of radiations, our proximity to a black hole, and so forth).

The *second* feature is that, although the sensitivity of a detector needed to make a positive observation is (social constructivism aside) set by nature—it depends on how strong the phenomenon to be observed is— the observer can choose the level of an upper limit. Observers can make upper limit claims that are as high and undemanding (and uninteresting) as they like.

The *third* feature that is interlinked with these two is that, to set an upper limit, signal and noise in the detecting apparatus do not have to be separated. One can set an upper limit by claiming only that the size of the signal is no greater than the *sum* of the signal and the noise. If there is a lot of noise, the upper limit thus set will constrain the world less than it otherwise would, but since the signal cannot possibly be larger than the signal plus the noise, the claim cannot be wrong. Of course, the more noise that can be eliminated, the stronger will be the upper limit. Thus, if the Michelson-Morley experiment had been so noisy that it could show only that the measured variation in the velocity of light was less than some number higher than the speed of the earth in its orbit, it would have set a valid upper limit but one that was scientifically vacuous. It might be thought that the only kind of upper limit that would be scientifically interesting was one that engaged with established theories, but this supposition, it turns out, is not true.

The *sociologically* intriguing cases of upper-limit setting are those on the cusp of *scientific* interest, such as those produced by gravita-tional wave detectors at the turn of the millennium. In 1999 an upper limit was set for gravitational radiation emitted by inspiraling binary neutron star systems in our galaxy. The observations showed that (with 90 percent confidence) there was less than one such event every two hours, or fewer than 4,400 per year. Astrophysical observations and cosmological modeling, however, suggested that there should be no more than one such event every million years. Thus, the upper limit exceeded the prediction by nine to ten orders of magnitude, yet it was still published in a high-prestige journal.

The resolution of this enigma is that upper limits can be interesting when technology is new, because even a scientifically meaningless result can demonstrate the functioning of new experimental technologies or data analysis techniques that may one day reach scientifically interesting levels of refinement. Such results are important to assure funders that a program of research is worth pursuing. Otherwise, meaninglessly high upper limits can also be given scientific significance if they confront otherwise unbelievably high positive results or tenuous theories. In sciences such as sociology, where technologies are simple, and where there are few firm results to confront, upper limits are almost never published.

## BIBLIOGRAPHY

Allen, Bruce et al. "Observational Limit on Gravitational Waves from Binary Neutron Stars in the Galaxy." *Physical Review Letters* 83 (1999):1498–1501.

Collins, Harry. *Gravity's Shadow: The Search for Gravitational Waves.* Chicago: University of Chicago Press, 2004.

Collins, Harry, and Trevor Pinch. *The Golem: What You Should Know About Science.* Cambridge, UK: Cambridge University Press, 1993.

Swenson, Lloyd S. *The Ethereal Aether: A History of the Michelson-Morley-Miller Aether-Drift Experiments, 1880–1930.* Austin: University of Texas Press, 1972.

*H. M. Collins*

# T

## TECHNOLOGICAL SYSTEMS.

Many contemporary technologies are systemic in character. Railroads, electricity utilities, the telephone network, the Internet, the air transport networks and technologies, food manufacture and distribution, surveillance technologies of all kinds from city-center closed-circuit television to satellite systems, global positioning, electronic battlefield technologies, patient record systems, and methods for tracking the movement of people across national frontiers are just a few of the technologies that take the form of networks of components distributed across time and space. Indeed, it is often argued, as has Manuel Castells, that we live in a "network society."

It is, however, a mistake to imagine that technological systems are purely technical. All the technologies mentioned above are heterogeneous, being simultaneously technical, social, organizational, economic, and political. Technological systems are produced in "heterogeneous engineering"—that is, the artful combination of the human, the social, and the material. The failure of any part of the system, technical or otherwise, may lead to its failure. Technologies are thus folded into the human, and vice versa. Indeed, the extent to which it makes sense to distinguish the "technical" from the "social" if we want to understand such systems is limited.

The exemplary studies of technological systems explore their character in several different ways. Some writers, such as Wiebe Bijker, have considered how social and economic factors shape technologies. Others, such as Thomas P. Hughes, have stressed the importance of "holistic" innovators who are able to think outside conventional categories and link together heterogeneous elements in science, technology, economics, and the law. Yet others, such as Michel Callon, have stressed the way in which networks of relations shape or, indeed, create not simply the technical but also the social elements that they bring together. Just as technologies themselves are created and shaped within technological systems, so too new social roles and realities are brought into being.

Different technological systems reveal different configurations. Some are highly centralized, such as electric utilities, which need to make overall decisions about routing and distribution. Others, however, are not. The Internet was created without a single center in order to increase its resilience in the event of failure or attack. Other technologies that are equally systematic, for instance indigenous agricultures, may be similarly decentered while again revealing their adaptability and resilience.

While there is no agreement about how to think about this, it is obvious that technological systems have political implications. Arguments about technology and politics come in a number of forms. First, some suggest that there are inherent connections between particular technologies and specific forms of politics. For instance, it has been argued that nuclear power is an intrinsically authoritarian technology because its reactors and their waste products are so dangerous that they require the political control of a strong, even quite authoritarian, state for a period of tens of thousands of years. Correspondingly, it can be argued that some versions of alternative power generation are intrinsically democratic because they are decentralized and resist centralized control.

Second, if it is the case (as argued above) that technological systems are simultaneously technical and social, then it follows that whatever the intentions of those involved, such systems necessarily have political effects. This is because they make and remake social roles and relations. But, contrary to the argument about inherently authoritarian (or democratic) technologies, the character and implications of such implicit politics may be both unclear and complex. For instance, new technological systems may deskill workers or render them redundant, as with the creation of electronic typesetting. They may also lead to work intensification (as with word processing). Conversely, they may actually empower previously less privileged groups. Finally, as is obvious, they may also be used for overt political and social purposes. (There are many examples of this, where, for instance, automation is used to secure the docility of labor, state control, or patriarchal privilege.) The general message is that there is no general message. Or, to put it differently, it is that users are "configured" in technological systems (in Steve Woolgar's terminology), but the political implications of this have to be taken case by case.

Third, notwithstanding the widespread argument to the contrary, it does not appear to be the case that new technologies and technological systems necessarily "dehumanize" people. They may indeed do so, but once again the argument has to be taken case by case. For instance, recent work on women receiving assisted conception shows that the intensive medical procedures involved are found to be invasive and dehumanizing by those who fail to conceive but not by those who are successfully implanted. More generally, it seems that all people, certainly in rich countries, are caught up in and remade by technological systems in endless and subtle ways. Arguably our humanity is bound up and made within such systems.

Finally, many authors are cautious about the hype surrounding technical systems. Some media and academic commentaries offer dramatic utopian or dystopian narratives about technological change. Much talk about the "network society" or genomics is of this kind. Such narratives need to be treated cautiously. Although technologies and technological systems are profoundly important, technological systems have existed since the dawn of agriculture about 10,500 years before the present, and most of the new technologies that have appeared over the last 200 years (for instance, the telegraph or the railroad) have been greeted with similarly exaggerated utopian or dystopian accounts. Arguably the importance of technological systems will become clear only once scholars find ways of describing them in ways that avoid exaggeration and reflect their social, political, and technical subtleties.

*See also* **Energy Systems; Internet, The; Networks and Large-Scale Technological Systems**

## BIBLIOGRAPHY

Abbate, Janet. *Inventing the Internet.* Cambridge, MA, and London: MIT Press, 1999.

Bijker, Wiebe, and John Law, eds. *Shaping Technology, Building Society: Studies in Sociotechnical Change.* Cambridge, MA: MIT Press, 1992.

Callon, Michel. "The Sociology of an Actor-Network: The Case of the Electric Vehicle." In *Mapping the Dynamics of Science and Technology: Sociology of Science in the Real World,* edited by Michel Callon, John Law, and Arie Rip, 19–34. London: Macmillan, 1986.

Castells, Manuel. *The Rise of the Network Society.* Oxford: Blackwell, 1996.

Cockburn, Cynthia. "The Material of Male Power." In *The Social Shaping of Technology,* edited by Donald MacKenzie and Judy Wajcman, 177–198. Buckingham, U.K., and Philadelphia: Open University Press, 1999.

Cussins, Charis. "Ontological Choreography: Agency for Women Patients in an Infertility Clinic." In *Differences in Medicine: Unravelling Practices, Techniques and Bodies,* edited by Marc Berg and Annemarie Mol, 166–201. Durham, NC, and London: Duke University Press, 1998.

de Laet, Marianne, and Annemarie Mol. "The Zimbabwe Bush Pump: Mechanics of a Fluid Technology." *Social Studies of Science* 30.2 (2000): 225–263.

Haraway, Donna. "A Manifesto for Cyborgs: Science, Technology and Socialist Feminism in the 1980s." In *Feminism/Postmodernism,* edited by Linda J. Nicholson, 190–233. New York: Routledge, 1990.

Hughes, Thomas P. *Networks of Power: Electrification in Western Society, 1880–1930*. Baltimore: Johns Hopkins University Press, 1983.

Latour, Bruno. *Science in Action: How to Follow Scientists and Engineers Through Society*. Milton Keynes, U.K.: Open University Press, 1987.

Law, John. "Technology and Heterogeneous Engineering: the Case of the Portuguese Expansion." In *The Social Construction of Technical Systems: New Directions in the Sociology and History of Technology*, edited by Wiebe E. Bijker, Thomas P. Hughes, and Trevor Pinch, 111–134. Cambridge, MA: MIT Press, 1987.

McNeill, John R., and William H. McNeill. *The Human Web: A Bird's-Eye View of World History*. New York and London: Norton, 2003.

Scott, James C. *Seeing Like a State: How Certain Schemes to Improve the Human Condition Have Failed*. New Haven, CT, and London: Yale University Press, 1998.

Winner, Langdon. "Do Artifacts Have Politics?" *Daedelus* 109 (1980), 121–136.

Woolgar, Steve. "Configuring the User: The Case of Usability Trials." In *A Sociology of Monsters? Essays on Power, Technology and Domination*, edited by John Law. Sociological Review Monograph, 38. London: Routledge, 1991: 58–99.

*John Law*

# TECHNOLOGY IN CULTURE.

Rather than existing in a separate realm from culture and society, technology is a constitutive part of social life, shaping and being shaped by culture. Moreover, as a product of culture, understandings of technology are themselves situated and contingent. In other words, the same technology will be viewed differently by different people, depending on one's social location, cultural background, political inclinations, and historical context. Some of the main themes in understanding the relationship between technology and culture can be traced by considering the following frameworks: narratives of technology as an agent of salvation (or of doom), and questions of agency and accountability in technological development.

Underwritten by the philosophies of the Enlightenment and the Scientific Revolution, technological innovation has long been associated with progress and change for the good. From the early beginnings of the Industrial Revolution in the eighteenth century to the present, there have always been those promising better living through technology (such as electric lights, washing machines, cars, or computers) and chemistry (such as medicines, vaccines, and, more recently, genetically engineered food and bioproducts). This tendency has been especially evident since the rise of cultural modernism, beginning in the late nineteenth century and picking up intensity in the early to mid-twentieth century. Signaling the sense of a break with the past and evident across the fields of industrial relations, agriculture, architecture, and the arts, cultural modernism is associated with favoring innovation over tradition, machine production over hand production, homogenization over uniqueness, and new over old. Within this value system, technology has come to occupy a privileged position.

The modernist response to social problems has typically been to look to technology for a quick fix. This approach is motivated by the belief that technology can "clean up" messy social problems. One example is responding to problems of urban poverty by destroying housing stock in underprivileged neighborhoods and building in its place high-rise housing projects (typically with higher densities than the housing they replaced); another is responding to the problem of widespread obesity by developing ever greater numbers of pharmaceuticals to manage the un-health produced by too much eating and not enough activity. This approach, termed the "technological fix," typically relies on a one-size-fits-all approach that does not acknowledge local history or traditions and ignores the multiple nature of a social problem's causes. As evidenced by rhetoric around the potential of new information and communication technologies such as the Internet and personal computers to transform society, hopes and dreams regarding the emancipatory potential of technology persist into the twenty-first century.

At the same time, some have always questioned the view that technological innovation is a certain path to social improvement. Like slavery and in some ways colonialization, the Industrial Revolution has long served as an engine of cultural and economic imperialism both between states and between classes in the same state. As the Luddites illustrated in early nineteenth-century England by laying to waste the machines they saw as threatening their way of life, technological innovation has long been interpreted differently by workers than by those who make a profit from workers' labor. The mechanization of death in the form of automatic weapons and the development of the atomic and hydrogen bombs led many, by the mid-twentieth century, to a further loss of innocence regarding technology. Their critiques remind us that the way technological innovation is interpreted is not singular or monolithic.

As there is disagreement over the material consequences of technological innovation, so is there a diversity of opinion over the question of agency *vis-à-vis* technology and historical change. Traditionally, technology has been posited as *driving* history. In this formulation, technology is imagined to be politically neutral and beyond human control or guidance. To Langdon Winner, the failure to recognize our agency in shaping how technology develops—accepting instead whatever technological innovation happens to occur—can be likened to sleepwalking ("technological somnambulism").

In the past two decades alternative conceptualizations of the relationship between technology and society have brought questions of agency to the fore. There are two main theories in this vein. The first, referred to as the social construction of technology (SCOT), stresses how social forces shape technology. The second, referred to as actor network theory (ANT), suggests that in addition to human actors, nonhuman entities in social-technical systems—such as machines—can also be seen as having agency. Although ANT does not deny human agency, it does seek to challenge the androcentricity of narratives of technological change.

Contra technological determinism, social constructionism views technology as a product of social action, and as such imbued with cultural, economic, and political motivations. This approach encourages us to ask, what political and social factors shape technology, and who gains and loses when a given technology develops in a certain way? Perhaps most importantly, the recognition that technological development *can* be shaped encourages us to consider how this approach can provide answers. Efforts to shape technological development have included holding to account businesses in industries that use high-risk technologies, as well as advancing design models that encourage higher levels of public participation. Large industrial disasters in the late twentieth century such as the oil spill from the *Exxon Valdez* and especially the mass death at *Bhopal* brought global attention to the dangers of high-risk technology, and in their wake citizens and activists have begun to demand higher levels of accountability from corporations engaged in high-risk practices. In turn, based on the model of Dutch science shops and Danish citizen councils, technology scholars and activists of different orientations have encouraged design processes that solicit the perspectives and concerns of end users and others who will ultimately be affected by a technology. Such "participatory design" is undertaken in the hope of avoiding such disasters. Recognizing our ability to affect how technological development will unfold, interventions such as these reflect the desire to claim agency in shaping how technology and culture will interact in the future.

**BIBLIOGRAPHY**

Bijker, Wiebe, Thomas Hughes, and Trevor Pinch. *The Social Construction of Technological Systems: New Directions in the Sociology and History of Technology.* Cambridge, MA: MIT Press, 1987.

Fortun, Kim. *Advocacy after Bhopal: Environmentalism, Disaster, New Global Orders.* Chicago: University of Chicago Press, 2001.

Hess, David. *Science and Technology in a Multicultural World: The Cultural Politics of Facts and Artifacts.* New York: New York University Press, 1995.

Sclove, Richard E. *Democracy and Technology.* New York: Guilford Press, 1995.

Smith, Merritt Roe, and Leo Marx, eds. *Does Technology Drive History?: The Dilemma of Technological Determinism.* Cambridge, MA: MIT Press, 1994.

Staudenmaier, John M. *Technology's Storytellers: Reweaving the Human Fabric.* Cambridge, MA: MIT Press, 1985.

Winner, Langdon. *The Whale and the Reactor: A Search for Limits in an Age of High Technology.* Chicago: University of Chicago Press, 1986.

Kate Boyer

# TECHNOLOGY IN HISTORY

This entry includes eight subentries

Global Perspective
Africa and the Middle East
China
Eastern Europe and Russia
India
Japan
Latin America
United States and Western Europe

## Global Perspective

Technology is the means by which humans accomplish results in the material world, so it is inherently global. Every project in which we engage is engaged using technologies. In a global perspective, technology incorporates practices and technical knowledge; technology is also embedded in our tools from a stick or a stone to a computer program or a spacecraft. On a world scale, technologies incorporate both a social-organizational basis (social technology) and hardware.

### Human Society Is Made Possible through Technology

Technology is a part of virtually all human activity. The range of human activity incorporating technologies includes shelter (housing and the built environment); protection from the ravages of nature (flood control, weather forecasting, air conditioning); protection from humanity (the technologies of insurgency and revolution as well as the technologies of war and suppression); food technologies (from development of GMO-free and pesticide-free foods to chemical and biological engineering);

and the media of communication (ships flags to cell phones, television, and radio). These are but a few of the vast and highly diverse human projects that incorporate technologies to engage the material world. Similarly, the arts are accomplished through technologies; for example, music and painting use instruments to inscribe the intention of artists in the material world. Even thought is technically assisted, using forms of argument, experimental designs, statistical tests and the technologies represented by mathematics and analytic methods.

The material world has primacy over humans, and we must engage it successfully to prosper or even to continue to exist. Where there is a person, there is technology in use. In global human society there is a massive interaction of multitudinous technologies that interlock and sometimes compete, showing vast and intricate and synergies and interrelations on a global scale.

### Technology Is Globally Integrated

Historically, those in the West thought of the world in terms of separate peoples or separate nations. It did make practical sense, for example, to speak of the superior products of the technologies and skills of the weavers of India or of the specific technologies of wine-making regions. The eventual dominance of the globe by the West followed as much from the development of weapons technology as from ideas of natural superiority and an ethic the permitted easy use of violence against other peoples. The perception of advanced technologies as a property of the West and of distinct technologies of Asia and other regions had, until recently, a material basis. Although it is still possible to identify human groups typified by different technological levels, speaking of the technological levels of peoples has acquired a flavor of quaintness usually associated with ethnocentric and racist perceptions. Today, human interconnection is a much stronger feature of global existence than separateness, and the integration of peoples and societies is increasing, facilitated by easy-to-use, inexpensive, and universalized technologies. Technologies that were once localized are now more correctly assessed as globally available and globally integrated.

## Global Technology Is Shaped by Political Economic Organization

Technologies can be placed at virtually any place on the globe. Because of the ascendancy of business and commerce, technologies are often placed to maximize economic advantage in a business form of social organization of the planetary society (that is, a capitalist system). The elements of a productive system are intentionally dispersed. Consider shoes, for example. A manufacturer of sport shoes will typically be legally organized with headquarters in a country with an advanced economy. The manufacturer will focus sales into advanced economies and into other economies typified by relative wealth, as well as to wealthy nodes of global elites in countries across the face of the globe. The entire manufacturing and assembly operations, however, may very well be located in diverse countries with less-developed economies. This form of globally integrated technology is possible with technical capabilities of electronic communication and shipping developed in the last half century. It is also dependent on assembly-line methods that have been developed over the past century and a half. Computer-assisted production and assembly with associated forms of social control in the factories and in the societies of less-developed host countries make it possible to open successful computer-assisted production plants in the traditional economies. With step-by-step work tasks, members of a responsible and attentive new work force recruited from a traditional economy are often young women in extended family systems who are bright and are also trained in traditional deference to authority. They may not even need to know the language of the dominant element of the corporation that has sited the plant in their city or village in order to produce highly technical goods of uniformly superior quality using methods of computer-assisted production.

## Centers of Technology Are Shifting to the East and South

The difference in wage rates and the ability of the global corporation to dominate local governments in economically weak countries (in such matters as benefits and worker protection and banning of unions) provide a large cost advantage over a similar plant located in an advanced economy. If over the past fifty years manufacturing work has moved from Europe and North America to Asia and Latin America, that shift was made possible by technology no less than by political economics. If white-collar work such as accounting, back-office functions, insurance, and computer programming are now moving toward India and China, it is the result of both technology and political economics.

These changes are supporting the growth of highly technical centers in India and China. These centers excel at developing consumer products of all kinds as well as production goods, including machinery and computer programs. Just as in the 1930s an automobile manufacturer in Detroit may have had several plants around the city or in nearby cites and a separate engineering and administrative center, today a company may operate plants and offices around the globe. This change brings new living standards and participation in a high-level productive and consumer society in places where new plants are sited.

The traditional cultures of China, India, and much of the rest of Asia have for thousands of years placed a premium on knowledge and respect for a learned class. This class, under current conditions of dispersal of technology, is creating the new centers of advanced technical production and innovation. The new centers in China and India extend with links to Latin America—for example, the Chinese-Brazilian space alliance, in which there is also some African participation. There is a blending of culture and opportunity that produce material control of both current and new technologies.

## Technology in the West Is Declining

The dispersal of technology is not simply dispersal from the West and new beginnings in the East and South. It is a loss to the West. It lowers living standards, increases stress on individuals and on families, and interferes with the intergenerational transmission of culture in communities and regions from which plants and jobs are dispersed. Following World War II, a young person in the West could expect that hard work, college education, and deference to authority could lead to a lucrative career in engineering or applied science. The links in this chain have been broken, and there is no longer a guarantee of technical employment, or even a guarantee that, if technical employment is possible, it will offer career opportunities. Perhaps for this reason many young persons today in the West are disaffected and avoid mathematics and engineering. As manufacturing and white-collar work is dispersed, older regional cities of the West are losing the character of corporate centers and entering a cycle of decreasing employment and a shrinking tax base. For example, major tele-marketing firms and call centers are shutting down operations in the advanced economies as new units in India with employees who have superior training and attentiveness are able to handle the core calling plus the over-flow. As core production is dispersed, an accompanying factor is the concentration of capital through mergers and acquisitions. In the West, the major regional cities have fewer corporate headquarters and CEOs. The vibrancy and initiative in city life and public services decreases as middle and lower managers become the top business leaders of regional cities.

Corporate centers are retreating from many locations throughout the West through mergers and acquisitions. This devolution of regional corporate leadership through concentration of capital is devolving the territory of the West: in terms of the "center" and the "periphery" economies, the resource-rich center is getting small and farther away. The periphery, from which value is pumped, is getting much larger. This "change in structure"—caused as much by technology as by political economics—is casting a long shadow over the twenty-first century.

## The Global Technology and the Globalization Phenomenon

Technology as a global perspective cannot be separated from the globalization phenomenon. The "efficiency principle" is the core doctrine of both the North American Free Trade Agreement (NAFTA) and the General Agreement on Trade in Services (GATS). This principle states that local law must give way to "least trade-restrictive" practices as determined by specialized panels representing primarily corporate interests and in which most ordinary public concerns are not admissible. The emergence of institutions to enforce GATS is a development of social technology on a world scale. In a global perspective, it is not only the tools, machines, and the narrowly defined technologies of production , but also the technologies of social organization and enforcement on a world scale that change the quality of human society. For example, the inherent tension between the "efficiency principle" and its mode of decision making and enforcement and the principle of "public interest" as locally defined by communities or national states implies a potential to remove elements of relative autonomy that have been essential to city life since the emergence of cities in the early Middle Ages in Europe.

The potentials of technologies are multitudinous, but the forms in which technologies are employed tend to follow the forms of the dominant institutions of the age. Technologies that could facilitate economic democracy and local control are typically installed to facilitate hierarchical control from remote centers instead. The set of technologies that make the Internet possible provide an illustration of this tension. For example, the creative use of the Internet by nongovernmental organizations (NGOs) may have protected an indigenous social movement and political rebellion. This success may create the fear in some government and elite military circles that by using the Internet, a great number of people might decide independently to do something suddenly (so-called swarming, which a great number of people are independently experimenting with as a cultural expression).

## The Character of Global Technology

What would the shape of global technology look like from the outside, that is, from an off-planet perspective? If we could assess global technology for the planet, we would see the dramatic emergence, proliferation, extension, die-off, and transmutation of technologies at an accelerating pace over the past three centuries. We would see technological development governed in part by existing levels of technical knowledge (the technical base, without which extension of new knowledge is not possible) but also by particular, often entrepreneurial interest rather than by generalized social interest. We would also see an explosion of technical development with each wave of expansion potentially setting the stage for further expansion. And we would see a lack of thinking things through. Because both expansion of technology and expansion of population are largely left to market forces, resources are consumed rapidly and without much concern for any limiting conditions that might exist until they are actually approached.

From the outside, global technology and humanity are largely inseparable. Our global technology takes on the character of the dominant "species being" of the planet. Its content and shape constitute a direct expression of the nature of the human species under conditions of capitalist production linked to the thoughtless excesses of an individually centered consumer culture. For example, if we consider radio services today, outside of the public networks such as the BBC, radio is a mere caricature of its potential. Television is generally a set of commercial messages with weakly developed programming set to the lowest common denominator, produced at the least cost to draw viewers to the brainwashing functions of incessantly repeated commercial mantras. Food production is generally sited such that it could not sustain certain populations if transportation systems should fail. We are experiencing global warming, but the technologies for attempting to turn the situation around are often not implemented. In spite of occasionally successful local efforts, air and water (including the oceans) are increasingly polluted and we are seeing both intensive over-harvesting of the oceans and a radical die-off of important fish stocks. Energy remains under market control while world oil production has peaked and identified harvestable natural gas reserves will diminish over a finite number of years.

From a distance, global technology would be seen as a distinguishing feature of human use of the planet. Currently, the uses of global technology generally follow the needs of a capitalist economic system in which the "efficiency principle" dominates the principle of public interest. As our thought experiment in viewing global technology from the outside shows, human use of human technology appears immature and unsustainable.

## The Future of Global Technology

The rising centers of global technology are in Asia. Within the next fifty years the primary centers of technology will likely be in Asia and Latin America. Exactly how the "limits" dilemma will play out is unclear, but the pattern of concentration of capital on a world scale is likely to subordinate the peoples of the West, except in a very few very rich nodal centers. The pattern of technological development as governed by the capitalist political economy is hollowing out the West. The future centers of new technology lie in new nodal centers in the East, primarily China and India, as well as in certain nodal centers of the West. Normal, everyday technologies will continue to be integral to all human societies.

*See also* **Science in History: Global Perspective**

**BIBLIOGRAPHY**

Croissant, Jennifer, and Sal Restivo, eds. *Degrees of Compromise: Industrial Interests and Academic Values.* Albany, NY: State University of New York Press, 2001.

Deffeyes, Kenneth S. *Hubbert's Peak: The Impending World Oil Shortage.* Princeton, NJ: Princeton University Press, 2001.

Fixico, Donald L. *The Invasion of Indian Country in the Twentieth Century: American Capitalism and Tribal Natural Resources.* Niwot, CO: University Press of Colorado, 1998.

Georgescu-Roegen, Nicholas. *The Entropy Law and the Economic Process.* Cambridge, MA: Harvard University Press, 1999.

Greider, William. J. *One World, Ready or Not.* New York: Simon & Schuster, 1997.

Hymer, Stephen Herbert. *The International Operations of National Firms: A Study of Foreign Direct Investment.* Cambridge, MA: MIT Press, 1976.

Kiernan, V. G. *The Lords of Humankind: Black Man, Yellow Man, and White Man in an Age of Empire.* Boston: Little Brown, 1969.

Odum, Howard T., and Elisabeth C. Odum. *A Prosperous Way Down: Principles and Policies.* Boulder, CO: University Press of Colorado, 2001.

Ronfeldt, David, John Arquilla, Graham E. Fuller, and Melissa Fuller. *The Zapatista Social Netwar in Mexico.* Santa Monica, CA: RAND Arroyo Center, 1998 (prepared for the United States Army).

Schmidt, Jeff. Disciplined Minds: *A Critical Look at Salaried Professionals and the Soul-Battering System That Shapes Their Lives.* London: Rowman & Littlefield, 2000.

*Hugh Gilbert Peach*

# Africa and the Middle East

Our awareness of technological development in Africa has been enhanced by several new findings in various parts of the continent. These include the Paleolithic tools, found mostly in Ethiopia, dated to 2.5 million years ago, and more sophisticated double-pointed blades with carved barbs in the Congo region, dating as early as 173,000 years ago. Equally significant for historians of technology was the discovery in that same region of what may well be the world's earliest mathematical artifact, the Ishango bone, presently in the Museum of Natural History, Brussels; a bone about 29,000 years old, similarly marked with 29 defined notches, which may have been calendrical, has been discovered in Namibia. When a villager of northern Nigeria stumbled on what turned out to be the remnants of an 8000-year-old boat, a new chapter in the history of navigation was opened, not only for Nigeria but for Africa as a whole. The town of Dufuna, after which the boat was named, has never been the same since then, given the attention that it has attracted from scholars around the world. But Africa's technological history has not been confined to ancient tools, boats, or mathematical objects. We have also become aware of multiple suspension bridges in Gabon and the Ivory Coast and a wide range of irrigation terraces in areas of northeast Nigeria and Cameroon. A comprehensive history of engineering and building technology in Africa is yet to be written, but when it does emerge, it will have to take into account

not only the numerous pyramids, stelae, and obelisks from Egypt and Sudan and the numerous sculptured churches from Egypt, Sudan, and Ethiopia but also a great variety of walled enclosures, fortifications, underground aqueducts and irrigation channels. These structures have been found in a wide range of city-states, kingdoms, and empires in the Nubian-Egyptian and Ethiopian zone; in the Puntite, Swahili, and Zimbabwean regions; and in West Africa in the Sahelian-Sudanic complex. Successive West African empires such as Ghana, Mali, and the Songhai Empire, and Hausa city-states such as Kano and Zaria, emerged in the context of adobe structures including elaborate mosques and walled cities.

This article first discusses two main areas of science and technology development in Africa and then proceeds to address, briefly, Mesopotamian ingenuity and intellectual prowess in selected fields.

## Science and Technology in Africa

This discussion of Africa focuses on historical trends in medicine, agriculture, and agro-processing in various parts of the continent. Our references are largely to developments in the continent before the twentieth century, although it must be recognized that very many of the concepts, values, and systems of ideas continue into the contemporary era in various forms. Brief reference is made, first of all, to the British Imperial Institute, so as to highlight the significance of African indigenous science and technology to the Empire.

To justify European colonization the traditional systems of knowledge were often marginalized by colonial legislation. The production of local soap or local textile, locally based steel, or even indigenous glass threatened European export markets. Moreover, traditional practitioners were viewed as hostile to colonial rule, and indeed some spearheaded anticolonial revolts during the era of British, French, German, Belgian, Portuguese, Spanish, and Italian occupation as they evolved in the mid-nineteenth century and afterward. An important link between indigenous African knowledge and the British scientific establishment was made through the creation of the British Imperial Institute in the early twentieth century. The institute specified the need for industrial espionage and the acquisition of information with respect to "both natural and manufactured products." There was a call for experimental research and technical trials and commercial valuation of medicinal products "held in high repute" locally. Active research into samples of African silk and varieties of guinea corn, iron ore, and dyestuff was encouraged. There were documented references to coal from Southern Africa; mica from Somalia; onion samples, palm nuts, grain, silk, iron ore, and guinea corn from Nigeria; and palm nuts from Sudan. In June 1911 a sympathetic colonial official pointed out in writing that "the native was unprotected by such patent laws as obtained in European countries" and that the practitioner in general "had to risk the secret of his trade." The Institute soon became a series of well equipped research laboratories, with galleries for display to merchants and industrialists on the lookout for new natural and processed products. Africa was an important source of information for the British Imperial Institute and certainly a major, though publicly unacknowledged, contributor to the Institute's success.

Researchers on Obafemi Awolowo University and the University of Ilorin, Nigeria, as well as institutions in Zimbabwe and post-apartheid South Africa, have begun to spearhead the rejuvenation of African traditional medicine (ATM) and other local indigenous systems of science, technology, and knowledge in general. The AIDS pandemic, as well as IMF/World Bank-derived structural adjustment programs (which demand the removal of government subsidies in health and education), have led to inward-looking, survivalist strategies.

*African Traditional Medicine.* Before the twentieth century, African medicine consisted of an accumulated body of diagnostic techniques, causative models, and personalistic and naturalistic explanations.concerning the prevention and cure of disease.

In West Africa, significant medical texts include those written by the nineteenth-century Nigerian scholar Muhammad Bello (1781-1837), caliph of Sokoto, whose writings include *Al-Tibb al-Hayyin, Al-Qawl al-Manthur,* and *Al-Qawl as-Sanna* on the eye, hemorrhoids, and laxatives, respectively. He wrote a total of sixty-five texts in the sciences in general, ten of which specialized in medicine. One of the most important medical texts of the nineteenth century was the *Masalih,* which dealt with diagnostic procedure, including urine examination and screening. Another notable scholar of medicine in West Africa was Muhammad Tukur, whose major texts focused on faith healing and aphrodisiacs, in addition to general healing procedures. Included amongst his writings are the *Qira' al-Ahibba,* the *Mu'awanat al-ikhwan,* and the *Talkhma.* It is not inconceivable that the thousands of manuscripts discovered in Mali in recent times will prove to include medical writings, which would shed light on medical practice before the fifteenth century.

In the case of northeastern Africa, numerous medical writings are known to exist in the context of Aksumite and post-Aksumite Ethiopia, although many works were lost during military invasions and internal civil strife. Legendary are the ancient Egyptian medical texts such as the Edwin Smith, Ebers, Hearst, Chester Beatty, and Kahun papyri, some of which go back to 5000 years ago. Information in such papyri includes the naming of major vessels, neuroanatomy and neurological symptoms, and the treatment of compound fractures and ailments such as epilepsy and scurvy. In the Edwin Smith papyrus, so called because it was acquired by Edwin Smith of Connecticut in 1862 and later donated to the New York Historical Society, there are twenty-seven cases of head injuries and six cases of cervical injuries. It has been pointed out that the first ever written reference to the brain is made in that document.

Indigenous African pathology and illness in Africa were explained in terms of naturalistic, empirical, and behavioral accounts that were very often embedded in supernaturally centered discourses. Such accounts proposed that spiritual forces pervade all forms of life; that there are negative and positive energy forms; that the universe is alive with unseen intelligence, consciousness, and energy; and that spirit travel, spirit possession, and out-of body experiences are real. The naturalistic aspect of African traditional medicine intersected with a supernatural model on several occasions. A theoretical distinction may thus be made between ATM and the biomedicine evolved in the West. Absent from the African traditional discourse was the Cartesian-Newtonian perception that the body is a machine and that disease is the consequence of the breakdown of that machine. For the African traditional practitioners, no part of the human body could be disembodied from the whole, nor could human anatomy be conceptualized in terms of body parts and fragments. ATM was holistic and cumulative.

Health care practitioners included itinerant pharmacists, sedentary pharmacists, specialists in cataract removal, bone setters, orthopedic specialists, and midwives. Diagnostic procedures included observation, the questioning of family members, clinical examination through pulse and body temperature, as well as divination, water gazing, and sand divination. Intervention included the use of infusions, elixirs, and enemas, and there are various historical accounts of surgery, administered in some regions with the use of herbally based anesthetics. Among such surgical interventions were scarification, circumcision for both males and females, the removal of inflamed tonsils, cataract removal, and the extraction of incisors. In some parts of the continent Caesarean childbirth was common as well. Interventions also included cranial trepanning, the resetting of pulmonary hernias, laparotomies in the treatment of abdominal wounds, removal of ovaries, and forms of plastic surgery. There is generous reference to the use of hydrotherapy in the Northeast, and inoculation seemed to have been a common practice in West Africa. Some scholars believe that the development of inoculation procedures in conventional Western medicine is indeed traceable to enslaved Africans in the Americas. Spinal manipulation and massage were sometimes combined with heat therapy and hydrotherapy in the treatment of fractures and spinal

disorders. The World Health Organization has recognized the efficacy of some traditional treatments for mental disorders, which also included the use of group therapy, family-based therapeutic sessions, and even, in some cases, mind-altering drugs, some of which were also used in religious-based ceremonies. One of the most well-known of these drugs is iboga (*Tabernanthe iboga*), now the basis of treatment for heroin addiction in some rehabilitation centers in the United States and Western Europe.

Traditional medical practitioners seemed very ineffective, however, in the treatment of waterborne diseases such as schistosomiasis and several insect-borne diseases. Practitioners seemed also to guard their knowledge with secrecy and often failed to standardize their medicinal preparations. Their client base has continued to grow in the contemporary era for a variety of economic and sociocultural reasons. Their more recent clients have been Western-based pharmaceutical companies in search of cures for ailments such as hypertension, cancer, and dietary disorders. Elusive has been a cure for AIDS, but the quest continues.

*African Agricultural Science.* Indigenous experimental science in agriculture included a variety of activities. The identification and naming of various botanical species took place in various parts of the continent in various time periods. Sacred groves were at the center of ecological reserves in various parts of the continent, intersecting with religious belief systems such as Orisa. The Niger basin and Lake Chad would be the center of production of African rice (*Oryza glaberrima*) in West Africa, whereas grasses such as sorghum emerged in the Nile region and the Sahara and date as early as 15,000 B.C.E., according to recent findings. The early domestication of plants within Africa was wide- ranging and included fruits, vegetables, legumes, roots, and tubers. Among the cultivated fruit were a wide range of melons, date palms (*Balanites aegyptica*), bush-butter fruit or safou (*Dacryodes edulis*), and baobab (*Adansonia digitata*). Domesticated vegetables have included the African eggplant (*Solanon macrocarpon*), amaranths (*Amarantus* spp.), bitterleaf (*Vernonia amygdalina*), fluted pumpkins, okra, gherkin, and various species of edible flowers. Legumes such as Bambara groundnuts, cowpeas, locust beans, pigeon peas, and sword beans were well known throughout the continent. Roots and tubers included a wide variety of yams, including yellow guinea yam (*Dioscorea cayensis*) and the potato yam (*Dioscorea esculenta*). We know that there were pragmatic responses to drought throughout the continent in zones such as the Kalahari and the Sahara, where the domestication of drought-resistant plants such as *Avena abyssinica* and *Hoodia* took place alongside the use of underground channels and aqueducts in some areas. Maps of the world related to the origin of food crops seldom insert information that is common knowledge to Africanists, namely, that Africa is the original home of a wide range of internationally known food items that include, in addition to the aforementioned items, coffee, kola nuts, and specific varies of cotton.

An important aspect of African technology is the transformation of legumes and tubers into edible pulverized products and beverages. Historically, the general trend in this field was empiricist and experimental, and the basic objective was to detoxify food, preserve it, enhance digestibility, and enhance its flavor. To achieve these purposes there were chemical, biological, and mechanical interventions. The discovery of beer-making facilities in the Northeast, dating back to 3500 B.C.E., has given us more information on the antiquity of brewing on the continent. Sorghum beers continue to dominate the continent at the present time, and some modern industries have benefited from ancient, indigenous brewing technologies and expertise.

Richard Okagbue has pointed out that women were in the forefront of food-processing technology in Africa and that fermented drinks such as *ogogoro* in Nigeria and *ekpeteshi* in Ghana are of great antiquity and derive from West African palm wines. Fermented milk products were also developed in various parts of the continent in Ethiopia, Kenya, Nigeria, Zimbabwe, and Sudan. African culinary tastes, no less than the expertise that evolved over time, grew out of specific ecological and environmental situations. Dried fish and dehydrated pulverized grain preparations, along with vegetable- based soups, predominated at a meal in various parts of the continent. River systems such as the Nile and the Niger ensured a generous supply of varieties of fish.

In the sphere of animal husbandry there were some striking developments, not only in the knowledge and ability to stimulate the behavior of cattle and other livestock, but also in the area of agro-meteorology. The Masai and other East African pastoralists accumulated information on the interconnections between weather, ecology, and climate. They also applied veterinary treatments for a wide range of animal-based diseases such as rinderpest. Masai outrage at the mass slaughter of infected cattle by the British in the late 1990s stemmed from their conviction that the disease could have been cured effectively by traditional veterinary methods.

## Science and Technology in Mesopotamia

The massive looting of ancient Iraqi artifacts in April 2003 caused a major outcry around the world. Stephen Fidler spoke of "launderers and looters, mobsters and museums" in an illuminating analysis of the trade in stolen antiquities. Some have spoken of the links between organized crime, looted antiquities, and the funding of terrorist activities given the tightened restrictions on money laundering. Research into science and technology in ancient Mesopotamia has undoubtedly been challenged by recent events, but fortunately we do have at our disposal the works of previous researchers in the field. There is also the hope that the remaining looted antiquities will surface and be returned to their rightful owners over time. This discussion will focus primarily on aspects of scientific development in the area of mathematics, but the region must be situated briefly in the context of local and regional politics.

*The Near Eastern Historical Context.* Mesopotamia was home to a succession of city states, kingdoms and empires. Sumerians, Akkadians (including the later Assyrians and Babylonians), Amorites, Hurrians, Hittites, Arameans, Medes, and Persians dominated the region prior to Greek conquest in 331 B.C.E. The region has been recognized as a major center of wheel technology, sailboats of various types, and early writing systems (though not necessarily the earliest script as previously thought). There are claims that the first underwater tunnel was constructed in this area, and also the world's first battery, called the Baghdad battery. The Euphrates and the Tigris had as its core about 700 miles of territory, between the two rivers. Recent discoveries at Hamoukar, Syria, have led to the modification of previous assertions about the unidirectional flow of intellectual and technological knowledge from southern Mesopotamia (Sumer) to the rest of the region, Hamoukar shows features of urban civilization at least as old as the oldest Sumerian cities, suggesting a multidirectional movement of ideas and technologies within the region.

The peoples of Mesopotamia interacted commercially with Northeast Africa: Egyptians, Nubians, and Aksumites were important players in the regional and inter- regional balance of power over the millennia. Mesopotamians and Egyptians had significant trade relations with Persians and South Asians. Amorites, Hurrians, Hittites, and Israelites flowed into Canaan (modern Israel and Palestine), and confrontations ensued between these peoples and the Egyptians for the control of the region. One of the major contentions for successive empires was access to the Red Sea and the control of the caravan routes plying the region. Egyptians coveted the cedar trees of Lebanon for the construction of buildings and ships. Nubian gold also attracted envious eyes in Egypt and the countries with which Egypt had contacts.

Rival theological systems arose in the region, between 500 B.C.E. (or earlier) and 700 C.E., namely, Judaism, Christianity, and Islam. Common to the three systems of ideas were propositions about divine intervention, prayer, fasting and other mortifications, and pilgrimage as well as moral traditions.

There were a variety of reasons for the development of mathematics in ancient Mesopotamia. Of great relevance was the calculation of interest on loans and prop-

erty taxes and the compilation of accounting ledgers for commercial purposes. The strict observance of religious festivities merited precision, and so, too, did planting and harvesting. While Northeast Africans were erecting their numerous pyramids, the Mesopotamians were also building pyramidal structures (ziggurats) of various dimensions. In the course of such activity, precise systems of measurement evolved, and so too did general formulae based on empirical data and mathematical tables.

The region has become well known for the development of the sexagesimal system, with numbers classified in groups of 60 for the calculation of time, space, weights and measures, and geometrical shapes. The system of positional numeration, or place value, means that the value of a number is determined by its position in a sequence of numbers. To shift a number one place to the left or right is to change its value.

Mesopotamian astronomy developed in the context of data collection over long periods of observation. Propositions emerged about the occurrence of eclipses, meteors, and comets. Planets were identified, and time was perceived in terms of a lunar year periodically adjusted to the solar year by insertion of an extra month. The Babylonians used mathematics in the calculation of motion and distance and in the assortment of their vast data.

The discovery of thousands of cuneiform clay tablets at Ashur and the more recent unearthing of the Sumerian necropolis at Umm al-Ajarib mark a major boost to our understanding of various aspects of daily life in the region. We hope that historians and archeologists of the region will be able to resume the analysis of those artifacts that have survived the Second Gulf War.

*See also* **Science in History: Africa; Science in History: Middle East**

**BIBLIOGRAPHY**

Abdalla, Ismail. *Islam, Medicine, and Practitioners in Northern Nigeria.* Lewiston, NY: Edwin Mellen Press, 1997.

Andah, Bassey. *Nigeria's Indigenous Technology.* Ibadan, Nigeria: Ibadan University Press, 1992.

Ben-Sasson, Haim *A History of the Jewish People,* Cambridge, MA: Harvard University Press, 1976.

Burstein, Stanley, ed. *Ancient African Civilizations: Kush and Axum.* Princeton, NJ: Markus Wiener, 1998.

Dunbar, Robin. *The Trouble with Science.* London: Faber and Faber, 1995.

Eglash, Ron. *African Fractals: Modern Computing and Indigenous Design.* New Brunswick, NJ: Rutgers University Press, 1999.

Emeagwali, Gloria. "African Indigenous Knowledge Systems: Implications for the Curriculum." In *Ghana in Africa and the World: Essays in Honor of Adu Boahen,* edited by Toyin Falola. Trenton, NJ: Africa World Press, 2003.

Fidler, Stephen. "Launderers, and Looters, Mobsters and Museums." *Financial Times,* May 2003.

Finch, Charles. *African Background to Medical Science: Essays on African History, Science and Civilizations.* London: Karnak House, 1990.

Msigani, K., ed. *Traditional Medicinal Plants.* Dar es Salaam, 1991.

National Research Council, Board on Science and Technology for International Development. *Lost Crops of Africa.* Vol. 1. *Grains.* Washington, DC: National Academy Press, 1996.

Nunn, John. *Ancient Egyptian Medicine.* Norman: University of Oklahoma Press, 1996.

Okagbue, Richard. "Food Technology in Africa." In *Encyclopedia of the History of Science, Technology and Medicine in Non-Western Cultures,* edited by Helaine Selin. Dordrecht, The Netherlands: Kluwer Academic, 1997.

Postgate, Nicholas. *Early Mesopotamia: Economy and Society at the Dawn of History.* London and New York: Routledge, 1992.

Some, Malidoma Patrice. *The Healing Wisdom of Africa: Finding Life Purpose through Nature, Ritual, and Community.* New York: Putnam Publishing Group, 1999.

Weiss, Harvey, ed. *The Origins of Cities in Dry-Farming Syria and Mesopotamia in the Third Millennium BC.* Guildford, CT: Four Quarters, 1986.

Wesler, Kit, ed. *Historical Archaeology in Nigeria.* Lawrenceville, NJ: Africa World Press, 1998.

*Gloria T. Emeagwali*

# China

The English Renaissance scholar Francis Bacon was a champion of modern science. In his work *Novum Organum,* he notes

Again, it is well to observe the force and virtue and consequences of discoveries, and these are to be seen nowhere more conspicuously than in those three which were unknown to the ancients, and of which the origin, though recent, is obscure and inglorious; namely, printing, gunpowder, and the magnet. For these three have changed the whole face and state of things throughout the world. (Aphorism I.129)

Bacon did not know that each of these inventions was Chinese and that many more extraordinary inventions were part of Chinese life long before their appearance in the West. It is folly indeed to think that the West is the source of all modern progress and to be ignorant of the incredible accomplishments of the Chinese.

In the following discussion, the evolution of China's rise to technological virtuosity will be traced through the centuries. China reached its acme of development in approximately the thirteenth century C.E. and began to decline thereafter.

## Antiquity through the Fifth Century B.C.E.

Early in prehistory, Chinese peoples learned to organize resources to produce food and provide shelter and protection. However, these early technological innovations were not recorded. Yet, archeological artifacts and later records show tools, devices, and techniques requiring a considerable level of sophistication. Technical innovation in ancient times was the result of an apparent need to be satisfied. A long history in agriculture led to important tools and a need for measurement led to devices for mensuration and the development of standards. Arts—graphic, decorative, and musical—required advancing technology to allow them to serve social needs. Thus, we see an evolution of increasingly sophisticated technology, anticipating the West by centuries.

The production of food requires cultivated land. The use of a hoe probably came first and was followed by animal-drawn hoes, or the crude plow. Western plows were wooden, evolving from all wood to one with a moldboard in England with a 1720 patent. Plows made with cast-iron moldboards and shares originated in Scotland in 1785, used by James Small. Much earlier, the Chinese freed themselves from plowing with crude and inefficient devices by developing iron plows. When Europeans learned of the Chinese plow, they copied the design, thus revolutionizing common European agriculture practices.

Just after 1900, archaeologists found curious sets of bronze bells in tombs throughout China. The bells ranged from 2,000 to 3,600 years old, and they looked like truncated slightly squashed cones.

These bells were developed in the sixth century B.C.E. Each bell gives two tones, which are separated in pitch by an interval of either a major third or a minor third. Every bell is, in fact, two bells in one. A closer look reveals that many bells even have marks on the side where one should strike to get one note or the other.

The importance of this discovery grows when we realize that it took the West a thousand years to develop the cathedral bell, which surfaced during the Middle Ages. Yet China had these remarkable bells during the Golden Age of Athens. The bells produce a rich tone, take far less bronze than a cathedral bell, and deliver two sounds instead of one.

Acousticians just now understand how the bells work. For consistency of tone, the bells were tuned against a standard string. The consistency of their shape made them a standard of volumetric measure. The amount of bronze in each was measured according to a carefully controlled weight standard. Each set of bells amounted to a miniature bureau of standards in ancient China—probably the only civilization to base its measurements on the notes sounded when bells are struck.

## The Fourth to the Second Century B.C.E.

An incredible development in the fourth century B.C.E. was the use of the horse as a power source. When a Malthusian competition for food exists, a farmer could cultivate more land and produce more food if only the horsepower were efficiently harnessed. Ancient systems of harnessing were unsuitable.

The earliest forms of harness were based on the yoking system—a system designed for oxen, with heads set low, at the fronts of their chests, rather than on the high, fairly upright necks of horses. A rigid yoking/harnessing system depending primarily on a harness around the neck would choke the horse. An efficient means of harnessing horses would not appear in China until about the fourth or third centuries B.C.E.: a breast-strap system of harness, still in use today.

Attaching a horse to a plow is not quite as difficult as matching the plow to the horse's strength. If the plow smashes into a rock, the plow is ruined. Iron plows were in use, but meanwhile the Chinese experimented with heat to treat metals. Methods of producing high-temperature ovens were available to the Chinese for casting bronze and firing clays. The achievement of higher temperatures required new technology. The double-acting piston bellows was developed, providing for a continuous flow of oxygen for the fire. This bellows was known and was in widespread use by the fourth century B.C.E.; the Chinese philosopher Lao Zi (Lao Tse) even described it in his writings. Through trial and error, the Chinese developed blast furnaces to make cast iron in the fourth century B.C.E., whereas the blast furnace did not reach Europe until 1380.

A further advance in the third century B.C.E. led to the discovery that annealing could allow the production of malleable cast iron. Plow points made from such iron survived after striking large stones. Weapons of war were improved, and perhaps because it could be done, several enormous architectural structures were built using cast iron.

The crossbow, not appearing in Europe until the time of mounted knights, had been the standard weapon of Chinese armies since the fourth century B.C.E. A crossbow is an incredibly difficult weapon to build. Some method for storing energy (a stretched leather strap, for example) must be used. The bow must be ready to use, necessitating a trigger mechanism bearing the entire stored energy load, yet simply released allowing the bolt (projectile) to leave the crossbow immediately. Even artillery pieces were built on the design of the individual soldier's crossbow.

Two major innovations in civil engineering highlight the technology of third-century B.C.E. China. The first of these was the making and using of relief maps, especially for military purposes. Relief maps are also essential to agricultural planning, irrigation, and hydraulics. The Chinese usually constructed relief maps from carved wood.

Knowledge that water seeks its own level is ageless. Undoubtedly Chinese engineers knew this in earliest times. Once contour maps of a sort were constructed, the idea of canal waterways would be an obvious solution to the many problems of navigating on natural waterways. Steep declines make navigation without portage impossible on natural rivers. Rapids and sand bars are seasonal, limiting traffic. Canals could be built on contours linking points of commerce and avoiding seasonal vagaries of natural rivers.

The Grand Canal is the greatest of these artificial waterways. The first segments of the Grand Canal were begun in 486 B.C.E. during the Wu Dynasty. It would eventually become the world's longest canal. Remnants of it are still in use, as can be seen near the city of SuZhou.

The Chinese invented paper in the second century B.C.E. Modern paper is mostly made from a slurry of wood fibers, whereas the earliest known Chinese paper was made from pounded hemp fiber. Apparently, it was not used for writing until much later. The earliest example of a bit of paper with writing on it dates from 110 C.E. With printing technology, and indeed, even with the need to transmit business and governmental messages, paper was finally recognized as the most natural thing to use.

Steel manufacture starts with cast iron. Cast iron is malleable to a degree but remains brittle because of its carbon content. Reducing the content of carbon in iron yields steel. The Chinese learned how to blow air into molten iron (burning the carbon). Centuries later (1856) William Kelly, from Kuttawa, Kentucky, brought four Chinese steel experts to Kentucky to demon-strate the methods of steel production. Ironically, in 1987 the president of the American Metallurgical Society was brought to China to advise on steel making for the large BaoShan steel plant near Shanghai.

## The First Century B.C.E. to the Second Century C.E.

The first century C.E. set the stage for a number of devices that reduced human labor and supported transportation. For the first time, water power was used to replace human energy, driving the bellows for blast furnaces producing iron. A driving belt, with buckets attached, had one end placed in a water source as a means for bringing water to a reservoir. The device could also be used to empty water from sites or to get water from a river to irrigate a field.

Steering a ship properly requires a rudder. As the Chinese transported heavier loads over longer voyages, they realized that oars were too unreliable. One person could easily control a lever system using a flat surface at the rear of the boat. Such rudders were used in the first century C.E. but did not appear in the West until the twelfth century C.E. The idea was adopted in the West from the Chinese. In sailing, two advances were highly important: the development of a strong mast to support a sail for sea voyages, and watertight compartments to guard against swamping and sinking from vessel damage.

An ingenious device developed during this time was the seismograph, or seismoscope. China borders the Pacific, with much of its coast in the "ring of fire," plagued by earthquakes. Chang Heng, an extraordinary scientist in the Han Dynasty, devised a seismoscope in 132 C.E. Sensitive to Earth vibrations, a bronze ball would fall from the mouth of a dragon into the open mouth of a bronze toad. The ball's loud clunk into the toad's mouth signaled an earthquake event. A set of dragons was arranged around the circular seismoscope, and the dragon from which a ball fell signaled the direction to the epicenter of the earthquake. Because they incurred huge losses of life over centuries, the Chinese continue to set earthquake prediction as a high priority for science and engineering today.

## The Third to the Fifth Centuries C.E.

Early Chinese compasses had no needles. In the third century c.e., the Chinese introduced an inscribed platform over which a suspended needle (or dial) could rotate. The instrumental value to be read was inscribed on the table. One interesting device was a "south-pointing" machine. A representation in jade of a sage was mounted on a device that had a differential gear mechanism beneath it. No matter how the device was moved on a surface, the sage always pointed to the south.

The invention of the stirrup in the third century C.E. was undoubtedly the most important invention of the time. The effectiveness of a war horse is vastly reduced without stirrups. Without them, the rider must hold onto the horse with one hand, limiting his ability to strike with a weapon. With stirrups, the rider and horse become one, allowing for speed, maneuverability, and vastly increased shock capability.

The basic design of a reciprocating water engine marked the fifth century. It did not have a shaft to rotate, and the Chinese never used the crankshaft. Waterpower was used to drive pistons for a variety of uses, including forcing air into smelters for steel making. This allowed for a new form of steel to be made, composed in part of steel being mixed with wrought iron. The resulting metal was incredibly well suited for weapons of war, namely sabers and swords.

During the Qi Dynasty (479–502 c.e.) some additions were made to the canal system that would eventually be linked and extended to form the Grand Canal. The widespread use of canals in China stemmed from a desire to connect rivers, which generally flow from west to east in China, so as to enable boat traffic between north and south as well.

## The Sixth to the Eleventh Centuries C.E.

The Grand Canal reached its full length under Emperor Yangdi of the Sui Dynasty, who united existing canals and built new ones in a six-year crash program from 605 to 610 c.e., until Beijing north of the Yellow River was linked with Hangzhou on the coast south of the Yangzi by a single waterway 1,795 km (1,114 miles) long.

Complementary to canals, which enable travel in boats across land, are bridges, which enable travel by pedestrians, animals, and land vehicles across water. Hence, it was natural to consider a variety of bridge types; the Grand Canal eventually was crossed by some sixty bridges. For centuries a kind of arch bridge was used. These bridges often leaped to soaring heights because the arches formed complete semicircles. Li Chun, a genius and engineer of the seventh century c.e., recognized that a load-bearing arch could be built on a considerably shorter arc than a full semicircle. Such a segmented arch required less material to build and still allowed for boats and flood waters to pass. Li Chun's remarkable Great Stone Bridge over the Jiao River, built in 610 c.e., survives today.

The inventions of the eighth century C.E. were no mere novelties. The first major work was a mechanical clock. Clocks were developed by the Chinese 500 years before they were introduced to Europe, but the reasons for clock development in China and in the West were poles apart. In the West, John Harrison worked all his life to develop a clock that would keep accurate time aboard a ship for navigation. In China, an accurate measurement of time was desired for astrology.

The other astonishing development of the eighth century C.E. (to be further perfected in the eleventh century) was woodblock printing on paper. The prototypes for printing were the ubiquitous seals used by Chinese as official signatures. A set of characters, usually indicating an individual's name, was carved in a block of wood or stone. The "white space" of the character is the part carved out of the stone in intaglio. Whenever it was necessary to put one's name on a legal document, the stone or wood seal was inked and stamped. These seals still bind legal signatures in China today.

In the ninth century, many innovations found their way into public life. Playing cards, paper money, and lamps with wicks were all developed in this period. Another invention, however, profoundly affected the world's people: gunpowder. By the late twelfth century, gunpowder was realized in the West. By this time the Chinese had gone through several stages of gunpowder development and had a developed lore concerning gun tubes of all sizes. The essential ingredient to gunpowder is saltpeter (potassium nitrate), not widely available in the West. In fact, a major contributing cause to the French Revolution was the habit of sending French soldiers to raid farmers' barns to scrape saltpeter off the stone walls, where it had crystallized from cattle urine.

The chain drive for transferring power was invented in 976 C.E. for use in a clock. It provided an example for many other uses of the continuous transmission of power. A far more important invention was a canal lock system. China's canal system was extensive and supported commerce as well as the taxation system: Taxes were paid in grain that was delivered to collection points via canals. The canal lock, introduced in about 984 C.E., conserved water and reduced the dangers of manually hauling boats to higher levels. Twenty-four locks were built along the Grand Canal.

One of China's enduring problems has been trying to unify all its parts into a nation. The tenth through the thirteenth centuries in China were characterized by provincial states at war with each other. Weapons of war were certainly developed during this period and included the first true flamethrower (based on the continuous piston device described earlier) that could produce a constant stream of fire. A number of explosive devices, including flares, bombs, mines (both land and naval), were also developed. Gunpowder, originally used for fireworks, soon turned to devices of deadly effect.

## The Eleventh to Fifteenth Centuries C.E.

During the eleventh century C.E. more practical devices were introduced, including utilization of buoyancy for salvaging artifacts sunk underwater. The spinning wheel for making thread also was developed. Joseph Needham believes that the spinning wheel was first brought to Europe by Marco Polo, who saw its use in China.

Two more devices of war were also introduced: rockets and repeating guns. Rockets were developed as parts of Chinese fireworks centuries earlier. Their use in war was twofold. A main problem with a rocket is guidance because its path is unpredictable in practice, owing not to gravity but to aerodynamics. The Chinese fixed rockets to what were formerly incendiary arrows. These large arrows had much more predictable paths. The military significance of rockets is the shock effect. Single rockets cause little concern, so the Chinese bundled several rockets together in a cluster launcher, thus directing an enormous shock effect when as many as 300 rockets were launched almost simultaneously. Another form of rocket was a "ground rat," a rocket launched along the ground. It sped along a serpentine path, perhaps igniting flammable materials along its way.

Firearms were introduced in China in the thirteenth century C.E. Firearms presented enormous problems for military uses. For larger guns, such as mortars and howitzers, the barrels could be cast, but they often burst, killing the gunnery crews. True shoulder-fired weapons, capable of directing firepower on an enemy, required many more years to develop.

China's inventive virtuosity dulled as a result of the wars ongoing between the warlords and the Mongols' invasion soon after. China began to become modern China in the fifteenth century. The inventive genius of the preceding centuries was never recaptured.

"Look to the West for things useful; look to China for things spiritual," has been a well-known slogan in China since Deng Xiaopeng's time. Perhaps if the two were reversed, China might still be leading the world in technological innovation.

*See also* **Science in History: China**

**BIBLIOGRAPHY**

Dernberger, Robert F., Kenneth J. DeWoskin, Steven M. Goldstein, Rhoads Murphey, and Martin K. Whyte. *The Chinese: Adapting the Past, Building the Future.* Ann Arbor: University of Michigan Center for Chinese Studies, 1986.

Merson, John. *Culture and Science in China.* Sydney: ABC, 1981

Merson, John. *The Genius That Was China: East and West in the Making of the Modern.* New York: Overlook Press, 1990

Needham, Joseph, et al, eds. *Science and Civilization in China.* 21 vols. to date. Cambridge, U.K.: Cambridge University Press, 1954–1995. (The massive bibliographies in each volume provide the fullest documentation of most topics.)

Needham, Joseph, and Colin A. Ronan. *The Shorter Science and Civilization in China*: Cambridge, UK: Cambridge University Press. 1994.

Needham, Joseph. *Science in Traditional China: A Comparative Perspective.* Cambridge, MA: Harvard University Press, 1981.

Sivin, Nathan, ed. *Science and Technology in East Asia. History of Science: Selections from ISIS.* New York: Science History Publications, 1977.

Temple, Robert. *The Genius of China.* New York: Simon and Schuster, 1987.

Williams, Suzanne. *Made in China: Ideas and Inventions from Ancient China* (Dragon Books). Berkeley, CA: Pacific View Press. 1997.

*Fred L. Wilson*

# Eastern Europe and Russia

The economic, technological, and cultural developments of Eastern European countries have been greatly dependent on the historical background of the region. On the one hand, all of these countries have experienced, to a certain extent, the progressive influence of the principal European states. On the other hand, the ambitious and aggressive intentions of neighbors very often prevented the Eastern European countries from achieving normal development.

In the middle of the thirteenth century a devastating invasion by the Mongols brought disaster to the populations of Russia, Hungary, Poland, and other countries. Consequently, Russia suffered under the Mongol yoke for more than two centuries. Incursions by the Turks from the fourteenth to the eighteenth centuries led to stagnation in Bulgaria, Hungary, Romania, and Serbia. The Austrian Habsburg dynasty, who replaced the Ottomans in the main part of Hungary and also ruled over the Czech and Slovak territory, had an uneven effect on development in these countries; development advanced in the Czech country but was retarded in Hungary and Slovakia. Moreover, Poland was weakened by the wars with Turkey during the seventeenth century. Later, the territory of modern Poland was divided between Prussia, Austria, and Russia. As a consequence of such obstacles, the technological revolution that has affected the industrial development of the principal European states since the second half of the eighteenth century touched the Eastern European region only to a small extent.

Two world wars inflicted major losses on the Eastern European countries. After the end of World War II, Communists in Poland, Czechoslovakia, Hungary, Bulgaria, Romania, and Yugoslavia established pro-Soviet governments and, together with the Soviet Union and East Germany, formed a socialist bloc. At the close of the 1980s, the socialist bloc disintegrated and the Communist rule ended. The Eastern European countries then proceeded with economic and technological developments under the conditions of a free-market economy.

## Russia

The development of towns in Russia during the Middle Ages necessitated the defense of such settlements from enemy attack. Potential invasions could easily come from Lithuania, Poland, or the Mongol and Tatar khanates.

In the beginning, the defensive structures for towns consisted of moats and walls. The moats were 4 to 10 meters deep and about 8 meters or more wide. A stone wall built in the fifteenth century as a defense for the Kremlin, the central part of Moscow, was more than 2 kilometers long, had an average height (with foundation) of about 15 meters, and was about 4 meters thick.

At the end of the fourteenth century, Russian weapon makers began manufacturing artillery. By the sixteenth century Moscow had no less than two thousand guns of various caliber. J. Fletcher, the English ambassador to Russia in 1588–1589, wrote that not one Christian sovereign had a stock of firearms equal to that of the Russian Tsar. Fletcher noted that the cannons he had seen in Moscow Armory strongly suggested skillful copper casting.

The "Tsar Cannon," manufactured by Russian craftsman Andrei Chokhov in 1586, was an unsurpassed achievement in gun production by size. The inner barrel had an inner diameter of 890 millimeters and an outer diameter of 1200 millimeters. The total length of the cannon was 5.34 meters.

Although the country was well equipped with fighting installations, Russia fell behind advanced nations in machine building and instrument making as a whole. The aims of Russia's radical transformations initiated by Peter the Great at the end of the seventeenth century were the development of industry and transport and the mastering of Western techniques by Russian industry. Peter the Great planned Russian works to meet the modern technological requirements, thus deeming it necessary to invite scientists and skilled craftsmen from Germany, Holland, France, and other countries.

During the second half of the nineteenth century Russia began competing with Western European standards through the creation of the Technological Institute, the Institute for Railway Engineering in St. Petersburg, and the Imperial Technical School in Moscow, among other institutes and schools. Alexander Popov, a lecturer at Navy Torpedo School near St. Petersburg, was the first to construct a radio receiver (1895), originally intended for use as a lightning detector. Professor Nikolai Zhukovskii was the founder of the science of aerodynamics, that was essential for the development of aircraft. Aircraft designer Igor I. Sikorsky developed multi-engine airplanes that surpassed other models in technical characteristics (1913–1914). Engineer Vladimir Shukhov designed original metal constructions for petroleum pipelines, the Radio Moscow transmission tower, and other industrial facilities. By the beginning of the twentieth century, the metalworking, textile, chemical, electromechanical, and other Russian industries were expanding.

World War I and the subsequent October Revolution (1917) and civil war (1918–1920) disrupted the period of economic growth and damaged industry, science, technology and culture. Many scientists and engineers left Russia during that period, seeing no opportunities for work and fearing Bolshevik persecution. Some of them have been well recognized abroad thanks to their achievements in technology: Vladimir K. Zworykin's fundamental inventions made electronic television possible; Igor Sikorsky established himself as a prominent designer of fixed-wing aircraft but is renowned as the developer of the helicopter; Stephen P. Timoshenko became a leading theorist of American applied mechanics; and Vladimir I. Yourkevitch was the designer of the French liner SS Normandie, the fastest transatlantic passenger ship of the 1930s.

Soviet Russia needed more than ten years to reach the same level of output as tsarist Russia had enjoyed in 1913. In 1925 the Bolshevik Party Congress decided to turn Russia into an industrial country, manufacturing all of the necessary machines and equipment, in order to overcome the country's dependence on the capitalist surroundings.

Plans for industrialization and a technological revolution demanded huge labor and material expenditures. During the process of fulfilling the first Five-Year Plan (1928–1932), the Union of Soviet Socialist Republics (USSR) created a tractor and automotive industry and a machine tool and aircraft manufacturing industry, and developed a metallurgy and chemical industry. A system of higher and secondary technical schools was formed to prepare qualified specialists for the industries. Within the same period, the government formed the Institute for Non-ferrous Metals and Gold, the Petroleum Institute, the Machine Tool Institute, the Polygraph Institute, and many others.

New generations of people with higher and secondary education were forced to support the Bolshevik Party's realization of the industrial and social plans. "Old engineers," Soviet newspapers wrote, "are 90 per cent counterrevolutionary." Thus, in the 1930s the work of many scientists and engineers was suppressed by political intentions.

In spite of the losses from political repression and planning errors, industrialization brought significant results. At the beginning of World War II (1939), based on the nation's large productive capacity, the USSR developed a system of research institutes and design offices and extensive scientific and engineering establishments. This system became an important factor through which the USSR could withstand Nazi aggression during World War II. It also played a decisive role during the postwar period, when the Soviets began solving the complicated problem of making nuclear weapons.

English scientist John Bernal wrote in the 1940s that that the main achievements of the twentieth century (up to that point), including the automobile and the aircraft, were based more on the science of the nineteenth century than on that of the twentieth. A lag between scientific discovery and its use in technology characterized most of the technological innovations in the past. However, work on the atomic bomb broke this tradition. The discovery of nuclear fission with possible release of massive energy was published for the first time in 1939. Within only ten years, the United States and the Soviet Union built nuclear reactors and developed technical devices using the chain nuclear reaction. This result strongly suggested the possibility that the Soviets would gain scientific achievements and advantages in technology.

The concentration of resources on certain scientific and technological initiatives allowed the USSR to achieve another important goal. The launch of the first artificial Earth satellite by the Soviet Union (1957) was a shock to many people. American newspapers named the event "a bloodless Pearl Harbor" to sound the alarm in the United States. In 1959 one Soviet probe became the first artificial planet around the sun, and another sent back the first images of the far side of the moon. In 1961 the USSR flew the first orbital mission with cosmonaut Yuri Gagarin on board the Vostok I spacecraft. In 1965 Alexei Leonov performed the first human extravehicular activity in space. In 1975 spacecraft from two countries, a Soviet Soyuz and an American Apollo, docked with each other in space for the first time. During the 1970s and 1980s the Soviet Union launched 100 to 120 space vehicles annually.

Near the close of the 1980s, it became clear that centralized government had to become less rigid, and excessive military development had to slow its growth, to permit social progress. The Soviet Union, and after 1991 the Russian Federation, began political and economic reforms aimed at shifting to an open society and participating in world markets.

The import of foreign goods in the 1990s made clear the inefficient standards of Russian manufacturing and production for competition in the world marketplace. Many enterprises had to dismiss most employees and master new production and labor systems. During its history, Russia has had many achievements in science, technology, and culture. Time will uncover the results of Russia's movement in a new direction.

## The Czech Republic

During the medieval period, the Czech Kingdom (known as Bohemia) was one of the most influential states, playing a notable role in the economic, political, and cultural life of Europe. For instance, there was much handcraft industry in the country's towns, such as cloth manufacturing and metalworking. During the thirteenth and fourteenth centuries, silver mining and the production of silver articles were developed.

The mass settlement of German colonists greatly influenced life in Czech towns and villages during the fourteenth and fifteenth centuries. German settlers brought with them many new machines and technological methods for mining and the manufacturing of various articles. For several years the prominent German mining expert Georgius Agricola lived in the city of Joachimsthal, now known as Jáchymov, in the Erzgebirge (Krušné Hory, "Ore Mountains"). A large silver coin was minted there, the Joachimsthaler, from whose name comes the word *dollar*.

In the course of almost four centuries, beginning in the year 1526, the Czech lands were under the Austrian Habsburg power. The food industry, light industries including textile, leather, and porcelain manufacturing, and the heavy industries of arms and transport were developed in the country. During that time, the Czechs became the most industrially developed people of Eastern Europe. In 1707, the Prague Engineering School was opened.

Noticeable progress in technology and technological sciences took place in the nineteenth century. The Institute of Engineering Education, founded in Prague in 1717, was soon reorganized into the Polytechnic Institute in 1806. Professor František Gerstner's work on mining lifts became well known, and Josef Ressel made an original screw propeller for ships in 1826. Moreover, in 1845 the first railroad line from Prague to Olomouc, designed by the Austrian engineer Alois Negrelli, was put into operation. A portion of the railroad between Prague and Pardubice was, to a great extent, the work of the Czech engineer Jan Perner (1815–1845). He died when passing through a tunnel at Choceň.

Since 1860 the Polytechnic Institute has taught in both the German and the Czech languages. Czech scientists and engineers contributed to the investigation of the ore deposits (František Pošepný), metallurgy (František Wald), and electrical engineering (František Křižik). French scientist Joachim Barrande, who lived for many years in the Czech lands, wrote a fundamental work on geology and paleontology.

In 1918 the Czechs were released from the power of Austria-Hungary and formed an independent state. The machine and chemical industries gained rapid development. Machine manufacturing then became the main branch of the Czech industry.

In 1939 the Czech lands, part of Czechoslovakia, was occupied by Nazi Germany. At the end of World War II in 1945, the Soviet Army defeated the German troops, and Czechoslovakia then became part of the bloc of socialist countries.

In the second half of the twentieth century Czechoslovakia developed an engineering industry, specializing in power-generating plants (large boilers and engines for electrical power stations), precision instruments, chemical engineering, and transportation system design and construction (railway locomotives, cars, road vehicles, and ships).

Since the end of the 1980s, industry in Czechoslovakia and, since 1993, the Czech Republic has been in a controlled shift to "market-driven" development.

## Hungary

Hungary was a principal state in east central Europe during the twelfth century. A devastating invasion by the Mongols destroyed the nobility of the country in 1241. From the beginning of the fourteenth century until 1918, Hungary was ruled mainly by foreign dynasties. The central part of the territory fell under the Ottoman Turks for three hundred years. In the late seventeenth century, the Austrian Habsburgs replaced the Ottomans as Hungary's rulers.

The tyrannical foreign rule led to a semifeudal system and stagnation in most of Hungary's industries. Handicraft manufacturing and mining production were developed. In the 1740s the prominent land surveyor and hydraulic engineer Sámuel Mikoviny worked at the Štavnica (Selmec) mines. He was the first to make a consistent use of the triangular method and led a number of hydraulic construction projects.

In 1849 Hungary declared its independence from Austria. The country continued to contribute to the increased development of national industry, even after the establishment of the Austro-Hungarian dual monarchy (1867). Consequently, the Hungarian physicist A. Jedlic invented a construction of the electrical generator (1861). The engineer Donát Bánki designed a float carburetor at the close of the nineteenth century (independently of the German Wilhelm Maybach). Finally, in 1912, Bánki greatly contributed to the design of compressors for combustion engines.

Austria-Hungary's defeat in World War I resulted in the dismemberment of Hungary. Romania, Czechoslovakia, Yugoslavia, Austria, Poland, and Italy received parts of Hungary's territory. In an attempt to regain independence, Hungary sided with Germany during World War II. After the war, the pro-Soviet Hungarian People's Republic was formed (1949). The national industry achieved considerable development during that period.

Hungary is one of the major world producers of bauxite, though most of its mineral and raw material resources are limited. The country's principal products include coal, cement, iron, steel, and aluminum. Hungary's main industrial exports now include electronic wares, electrical machinery, buses, locomotives, and railroad cars, and medical instruments. Since 1989, Hungary has been a multiparty state that has favored a gradual shift to free-market development.

## Poland

Over the past thousand years Poland's territorial base has changed many times. At one point, in the sixteenth century, the Union of Lublin, comprising Poland and Lithuania, was the largest state in Europe; its territory stretched from the Baltic to the Black Sea. The unified elective monarchy of Poland exported grain to France, England, Spain, and other countries, and towns developed shop craftsmanship. In the sixteenth and seventeenth centuries Poland was called the "granary of Europe." At other times in history, however, Poland as a state did not exist at all.

The Russo-Polish and Polish-Turkish-wars of the seventeenth century weakened the country. At the end of the eighteenth century, the territory of modern Poland was divided between Prussia, Austria, and Russia.

In the nineteenth century textile production increased place in the Lódź region. Silesia exported coal, iron, and zinc, and Warsaw developed craft manufacturing. The prominent geologist and geographer Stanislaw Staszic compiled the first geological map of Poland (1806).

A Polish educated man, Baron Karl Glave-Kolbielski, became known as an expert in a well-organized system of industrial espionage practiced by many European countries. Baron Kolbielski spent the beginning of his life in Poland and then lived mainly in Vienna. He succeeded in the clandestine purchasing of English textile machine drawings and hired the necessary experts to reproduce similar machinery (1800–1801).

Poland re-emerged as an independent state at the end of World War I in 1918. The country had mainly agrarian production; the only developed heavy industry was mining. Many Polish engineers left their homeland in the nineteenth and twentieth centuries and were recognized for their work while abroad. Wladyslaw Kluger and Ernest Malinowski built channels, roads, and railways in Peru. Stanislaw Kierbedz and Andrzej Pszenicki distinguished themselves by designing bridges in Russia. Stefan Drzewiecki worked at submarine projects and, later, aircraft in Russia and then in France, developing a general theory of propellers. Gabriel Narutowicz became well known from his works of hydraulic engineering in Switzerland; he was the first president of the newly independent Poland after World War I but was assassinated after two days in office.

A distinguished expert in metal welding was Stefan Bryla, who designed several bridges in Poland and the Ukraine in the 1920s and 1930s. During World War II he was accused of an anti-Nazi conspiracy and was shot.

After World War II, Poland's eastern and western borders shifted westward. In 1952 Poland became a People's Republic, and its economic, scientific, and technological developments were closely connected with the Soviet Union. Many industries, including coal and lignite mining, electrical power engineering, ferrous and nonferrous metallurgy, machine building, and chemical and textile production, were developed considerably during that period.

After leaving the socialist bloc in 1989, the Republic of Poland continued its scientific and technological development. Poland is now among leading producers of many industrial wares. It is in the world's top ten coal-, lignite-, copper-, and sulfur-mining countries and one of the top twenty countries that produce cement, television sets, cars, trucks, and electrical power.

## Bulgaria

Bulgaria, a country with an ancient history, was invaded by the Ottoman Turks at the end of the fourteenth century and lost its independence for almost five hundred years (1396–1878). While the Bulgarian peasantry was under Turkish masters, science and technology were at a standstill. Craftsmanship, such as harness making, blacksmithing, and jewelry manufacture, had been previously developed in many Bulgarian towns, such as Gabrovo, Sliven, Sopot, Samokov, and Kotel.

After the Russo-Turkish War (1877–1878), Bulgaria, covering more than half the Balkan Peninsula, became independent. The country was then on the way to capitalist development and industrial revolution. In the period from 1900 to 1909 one hundred new enterprises appeared in the textile, mining, food, and metalworking industries, among others.

Bulgaria's defeat in World War I caused the country to lose part of its territory. The country also had to pay reparations and occupational debts. In 1923 a fascist-inspired coalition seized power, which served the important foreign financial interests. From the 1930s to the end of World War II, Bulgaria's economy was dominated by German monopolistic capital.

After World War II, the Bulgarian Communist Party determined the social, scientific, and technological policy in the country. Economic development was organized for almost half a century, using the same methodology as was used the Soviet Union. The country focused on five-year plans aimed at rapid industrialization. As a result, Bulgaria achieved a predominantly urban and industrial society.

Bulgaria's present gross national product per capita is similar to that of other developed countries. Industry (including mining and public utilities) employs more than one-third of the labor force. Bulgaria's main exports are machinery and equipment, fuel, mineral ores, and metals.

See also **Science in History: Eastern Europe and Russia**

BIBLIOGRAPHY

Endrei, Walter. "The Kolbielski Project." *Journal of the International Committee for the History of Technology* 3 (1997): 9–23.

Hoensch, J_rg. *A History of Modern Hungary, 1867–1986*. London and New York: Longman, 1988.

Holy, Ladislav. *The Little Czech and the Great Czech Nation: National Identity and the Post-Communist Social Transformation*. Cambridge, U.K.: Cambridge University Press, 1996.

Kosáry, Domokos G. *A history of Hungary*. Cleveland and New York: The Benjamin Franklin Bibliographical Society, 1941.

Lyashenko, Peter I. *History of the National Economy of the USSR*. Vols. 1–3. [In Russian.] Moscow: Gospolitizdat, 1956.

Mc Cauley, Martin. *The Soviet Union since 1917*. New York: Longman, 1981.

Reddaway, William F., ed. *The Cambridge History of Poland*. Cambridge, U.K.: Cambridge University Press, 1941.

Wallace, William V. *Czechoslovakia*. Boulder, CO: Westview Press, 1976.

*Vasily P. Borisov*

# India

The study of the history of science and technology in a given area of the world, such as the contemporary Indian subcontinent, is usually undertaken through archaeological findings and literary records. In the case of India, understandable written records are not available until the Vedic age, which dates back to 1700 B.C.

## The Pre-Vedic Period: 8000–1700 B.C.

Archaeological evidence of ancient India suggests that at the close of the pre-ceramic Neolithic period (ca. 8000–5000 B.C.) mud brick architecture; cultivation of wheat and barley; domestication of cattle, sheep, and goats; and the first evidence of the cultivation of cotton had already developed at Mehrgarh, the earliest settlement in the Indian subcontinent. Around the first half of the fourth millennium B.C., evidence of a greater use of pottery and the first introduction of copper tools is also found. Development of the Indus Valley civilization during the second half of the fourth and early part of the third millennium B.C. contributed to the social and cultural foundation for the later classical and modern Indian civilization. Bridget and Raymond note that "the gradual build-up of population, and its spread through the Indus plains . . . ; the growth of technology and agricultural know-how; and the establishment of a socio-economic interaction sphere over an enormous area" (Allchin and Allchin, p. 165) had been the major forces for the transition from incipient urbanism to the mature urbanism. In fact, the mature Indus Valley civilization, comprising a number of well-known settlements such as Harappa, Mohenjo-daro, Kalibangan, and Lothal. lasted for about five centuries. The Allchins argue that the technical level achieved over so great an area, as is demonstrated in the Harappan technology, is probably unique in the ancient world.

Excavated urban centers of the mature Indus Valley civilization have unearthed a wide range of potter's kilns, dyer's vats, metal tools, deposits of beads, and other artifacts. There is ample evidence of local transportation networks in terms of streets, roads, and extensive cart tracks. In addition to terra cotta models of bullock carts, models of vehicles in metal—copper and bronze models of carts with seated drivers and a number of elaborate solid cast copper models of various transportation devices from the late Indus period (ca. 1800–1500 B.C.)—have also been discovered. These artifacts not only show a high degree of metallurgical skill in casting and designing but also provide some indication of the level of technological sophistication attained during that period.

### The Vedic Period and Post-Vedic Buddhist Period: 1700 B.C. to A.D. 500

The Indus Valley civilization slowly disintegrated around 1700–1500 B.C. No civilized settlements appeared till the advent of the Aryans, who first settled in the regions of the former Indus valley and later migrated to the fertile land on the banks of the Ganges River. With the breakdown of the Indus Valley civilization, city life seems to have disappeared for several centuries before emerging afresh in the Ganges Valley. The emergence of city life in the Ganges Valley marks the beginning of the classical period of Indian civilization. The literary evidence from the Vedic period (1700 to 500 B.C.), provided by the religious texts of the Aryans, the *Vedas* (such as *Rgveda, Yajurveda, Samaveda,* and *Atharvaveda*), suggests regular use of cosmetics, glass, gold, silver, copper, tin (*trapu*), lead (*sisa*), bronze, and a metal called *ayasa*, which is recognized as iron by most scholars. Tanning of leather and dyeing of cotton were also found to be practiced during the period. "The Aryans made a particular kind of polished grey pottery known as P.G. ware, which was dated to 1000–800 B.C. by the technique of radio-carbon dating. Other va-

rieties of pottery, for example, red or northern black-polished N.B.P. ware, were also made by them" (Deshpande, p. 133). Use of intoxicant beverages—*Somarasa*, the fermented juice of the plant *Soma* (held by many to be *Sarcostemma viminalis* or *S. acidum* (Roxb) Voigt); *Sura*, barley beer; and *Madhu*, honey wine—is mentioned in the Vedas. It is to be noted that, although these materials originated in the Vedic period, they developed significantly in the post-Vedic or Buddhist period (500 B.C. to A.D. 500).

### The Medieval Period: Fifth to Fifteenth Centuries

There are indications of improvements in techniques and implements and development in knowledge relating to agriculture in the subsequent medieval period, but these technologies did not differ radically from what they had been in the Buddhist period until the advent of modern technology and mechanical tools (Gopal, 1999, p. 314). The Turco-Afghan sultans and Mughal emperors of the medieval period did not make any effort for significant change in the techniques of agriculture. They pursued a policy of subsistence agriculture even up to the end of the seventeenth century. During this period, however, India excelled in technologies for the products of everyday use, which were often made from agricultural/forest produce. Some of these achievements, as noted by A. Rahman, include the development of grafting technology and production of different fruits; specialization of making herbal medicines such as distillation products, herbal extracts, and oxides of metals; cosmetics, in particular a variety of perfume, used by the kings and nobility; various types of press-mill for extraction of oil from oilseeds and juice from sugarcane; and utensils such as copper vessels with *kalai* (tinning) extensively used for cooking purposes and ceramic utensils with intricate engravings or designs of written text and floral patterns. The system of production was mostly decentralized, with each household producing for itself, using mostly solar energy; there were also karkhanas, or small-scale production units, built to meet the demands from kings, nobility, rich landholders, and traders.

*Textiles.*  All aspects of textile technology, from production of cotton fiber to manufacture of textiles and their dyeing and printing, were highly developed. Another related area of clothing designs was also developed. Depending on seasons and ceremonial occasions, new patterns and designs were developed, and styles also underwent many changes during this period. Using simple textile tools such as looms, carpets were made of cotton and silk as well as wool. Although this technology in all probability came from outside, it took root in India, particularly in Kashmir. Apart from carpets, mention may also be made of mats made of grass and reeds and decorated with beautiful designs and colored patterns.

*Metalwork.*  Another important area around which a number of technologies were developed in medieval India was metal technology. Following Rahman (pp. 256–258), this area can be subdivided into armament technology, coinage technology, and jewelry technology.

War was a preoccupation in the medieval period, and the armies of the various kingdoms had to be equipped with weapons—arrows, swords, shields, and armor and, later, guns, cannons, and rockets. The industries and *karkhanas* for the manufacture of these items had been established with proper mechanism for mining and smelting operations and for supply of raw materials, which led to the development of armament technology in medieval India. Smelting was carried out in furnaces that, although they appeared crude, were fabricated with precision using the cubit as the standard unit of measurement, with a length:width ratio of 6:5 being maintained throughout.

A considerable literature is available on coinage technology. B. N. Mukherji (chapter in Roy and Bagchi 1986) gives a comprehensive description of not only the processes of refining metals (copper, silver, and gold) for making coins but also the types of coins made in India since antiquity. He notes four types of coins: cast, repoussé, die struck, and archaic die struck.

As a major item of consumption of precious metals, jewelry was a small-scale industry, with goldsmiths spread all over India producing tailor-made ornaments for their clients with different social status. Jewelry technology covered not only the fine decorative designs and inlaying of precious stones but also the cutting of precious stones in different shapes and designs to suit specific ornaments. The manufacture of jewelry thus integrated technology with art and culture.

*Construction Technology.*  The historical developments of demonstrative and construction technology in India were, as Rahman (p. 249) argues, essentially aimed at meeting the needs and fancies of the kings and nobles, who patronized the master craftsmen. This can be evident in the case of construction technology. The statues of Buddha and Mahavira show the high degree of skill acquired by the stone cutters. The iron pillar at the Qutab Minar complex and the stone pillar at Firoze Shah Kotla, both in Delhi, demonstrate metallurgical and casting capabilities as well as the high degree of skills of masons and stone cutters. In addition to its high level of stone-cutting capabilities, a high degree of inlay work can be seen in the Taj Mahal. The skill of making delicate patterns of different types of grilles, as can be found in Fatehpur Sikri, is also noteworthy. Some of these acquired skills clearly exhibit the integration of knowledge in optics and geometry.

In the case of other types of construction technology, with relatively more utility, mention can be made of three types of structures: places of worship, palaces, and mausoleums. Although the final construct in most cases had been highly sophisticated and beautiful, the idea of replicating such products on a large scale did not happen. As a result, the productive resources and capacities were directed mainly toward meeting a whole range of specific and distinctive needs, not toward further development of resources and systems of production.

*Paper.*  The very important area of paper technology is regarded as the main vehicle for a knowledge revolution. Although historians have found evidence that the paper-making industry had already been established in India during the early Sultanate period, it was not until the reign of Sultan Zainul Abedin of Kashmir (A.D. 1417–1467), when the establishment of paper industry got a big boost.

> The Kashmiris learnt paper-making and improved upon the technique to such a degree that, within a few years, Kashmiri paper earned considerable reputation for its excellence. . . . Kashmiri paper was in much demand in the rest of the country for writing manuscripts. It was used by all who wished to impart dignity to their correspondence. The pulp from which the paper was made was a mixture of rags and hemp fibre, obtained by pounding these materials. Lime and some kind of soda were used to whiten the pulp. (Rahman, p. 265)

With the rapid diffusion of technical know-how and increase in demand for paper, a number of paper-making centers were soon established in different parts of India.

## Impediment during Colonial Rule: 1757–1947

During British colonial rule in India, there was a well-organized and directed effort to destroy the indigenous industries. "Most of the state sponsored scientific and technological activity was geared toward the agricultural sector, and toward the engineering colleges and technical institutes established for the training of engineers to undertake the construction of irrigation systems" (Baber, p. 220). The main aim of such initiatives was to make the Indian subcontinent a source of raw materials for British factories and a market for their finished products. In order to secure raw materials and provide effective control, surveys were made to map the country with its geological features and natural resources. These were followed by extensive mining operations and the establishment of a network of roads, railways, and communication systems. In 1854 the Public Works De-

partment (PWD) was constituted in order to centralize the operations of these systems. The major impact of the above developments/underdevelopment was the creation of two rival segments of the Indian society: one based on the *traditional* handicraft and handloom system, and the other based on *modern* technology imported from Europe.

Despite the clearly defined interests and motives for specific projects as just mentioned, there was no explicitly formulated science and technology policy during the initial phase of British colonialism under the East India Company. After the Industrial Revolution and the consequent rapid growth of science and technology in Europe, the colonial Indian state began several experiments in the application of science and technology around the mid-nineteenth century. When India came under direct Crown rule in 1858, science and technology became an integral component of state policies. The engineers associated with the PWD got the opportunity to work in various fields of applied science and technology, such as forestry and mining research, locomotive design, and the manufacture of iron rails. By the late nineteenth century, many Indians were discouraged "by the heavy emphasis on applied technical education and the visible neglect of theoretical scientific research and teaching" (Baber, p. 228). In fact, the creation of the Indian Association for the Cultivation of Science (IACS) in 1876 was a response of Indian scientists to the existing situation. Recurring famines during the last quarter of the nineteenth century were instrumental for the creation of a Board of Scientific Advice (BSA) in 1902. The BSA, in liaison with the Royal Society, had been engaged in offering the government any scientific and technical advice whenever needed.

Meanwhile, the process of deindustrialization was practically complete, with the banning of Indian textile exports to Great Britain and of Indian ships for the East India Company's trade. Although the colonial administration felt the necessity of producing munitions for the allies during World War I, and many committees and royal commissions had been urging the colonial government to stimulate industries, no definite attention had been paid to India's industrial development. A group of Indian leading scientists, along with the Indian National Congress, started playing an active role in formulating the science and technology policy of preindependent as well as independent India. In 1938 a National Planning Committee was formed, with Jawaharlal Nehru as its chairman. In the final debates prior to independence, one group, spearheaded by Mohandas K. (Mahatma) Gandhi and P. C. Ray, advocated the revival of indigenous industries based on handicrafts, seeing both modern science and technology as alien to the Indian tradition and as instruments of European domination and exploitation. The opponent group, led by Nehru and Meghnad Saha, was in favor of modern science and technology and advocated heavy import-substitution plans to aid India's industrialization.

## Technology Policy for Development during the Postindependence Period

Impressed by the success of both planned industrialization in the Soviet Union and the Tennessee Valley Authority project in the United States, Nehru, the first Prime Minister of independent India, along with India's leading scientists, endorsed economic development programs for the country through the adoption of modern science and technology in the 1950s and 1960s. The basic development strategy of the Second Five-Year Plan (1956–1961) was "Growth in Heavy Industry." A Scientific Policy Resolution that was passed by the Indian Parliament in March 1958 stressed:

> The key to national prosperity, apart from the spirit of the people, lies, in the modern age, in the effective combination of three factors, technology, raw materials and capital, of which the first is the most important, since the creation and adoption of new scientific techniques can, in fact, make up for a deficiency in natural resources, and reduce the demands on capital. (Government of India)

Because the modern science and technology based institutions neglected cottage, small-scale, and artisan-based industrial structures, there was very little integration of the traditional segment into the modern segment during the 1950s and 1960s. Critics started questioning the validity of the development strategy of "Growth in Heavy Industry" and the related science and technology policies. For instance, the Green Revolution, which is portrayed as a success story by many technocratic advocates, is found to be responsible for the dysfunctional consequences such as the incidence of marginal farmers turning into landless laborers, long-term harmful impact on soils, and the cultural dislocation of peasant communities.

By the early 1970s India had apparently achieved near total self-sufficiency in its capacity to produce most of the standard modern capital goods required by Indian industry. However, the achievement of the manufacturing sector was not accompanied by any significant acquisition of design capacity. In the early 1970s a separate ministry for science and technology was created. Science and technology cooperation agreements with many international organizations and countries were also signed. In the mid-1970s, the first Science and Technology Plan (1974–1979) was launched in order to assist technology absorption, assimilation, and development of indigenous capabilities. As a result of very slow progress in these directions, the socioeconomic critique of science and technology increased its intensity. As Sukhamoy Chakravarty notes (p. 66), it was also increasingly felt during the early 1980s that the estimated technology gap in India's capital goods sector was of the order of around ten to fifteen years.

The realization that India was lagging behind in technology growth in key sectors of the economy led the Government to issue the Technology Policy Statement of 1983. The Technology Policy Implementation Committee (TPIC) in the Planning Commission then recommended setting up Science and Technology Advisory Committees (STACs) in all socioeconomic Ministries. The Department of Science and Technology (DST) took charge of coordinating all the committees and began to act as focal point for evolving multipartner Joint Technology Projects. The main objectives were to make sure not only that emphasis is placed on the development of internationally competitive indigenous technology, with export potential, and stress on efficient absorption and adaptation of imported technology, but also that a greater role is assigned to in-house industrial research and development (R&D) units for a closer interface between industry and other technology users, national laboratories, and the educational sector. Under the Research and Development Cess Act of 1986, a cess of 5 per cent of all technology import payments by the industrial concerns was imposed. Using this cess fund, the Industrial Development Bank of India (IDBI) established the first Indian venture capital fund (VCF) in 1987 to promote new technologies and entrepreneurs. The Technology Policy of 1983 also paved the way for several initiatives including the establishment of a number of national and regional organizations and laboratories.

With the beginning of the era of liberalization in the mid-1980s, the central government regulatory agencies and state-owned enterprises have made conscious attempts in the area of high-technology industry to increase state actions aimed at complementing and promoting private entrepreneurship. Realizing a huge export potential of the information technology (IT) industry, particularly its software and services sector, the government had established Software Technology Parks (STPs) in 1990 in order to provide shared support facilities. The number of such parks subsequently increased, and they were registered in June 1991 as a new society called the Software Technology Park of India (STPI). Towards the end of the 1990s the government of India has designed a series of major policy initiatives to enhance competitiveness in the IT sector. A new Ministry of Information Technology was set up in October 1999 as the nodal agency for facilitating all initiatives taken by all levels of government, academia, the private sector, and successful Indian IT professionals abroad. As of December 1999, 170 Indian software companies had acquired international quality certification. According to a survey in 2000 made by India's quasi-governmental software industry promotion organization, the National Association of Software and Services Companies (NASSCOM), more than 185 of the *Fortune* 500 companies (that is, almost two out of every five global giants) outsourced their software requirements to India during 1999–2000.

With its world-class IT industry, large pool of scientific talent, and vibrant pharmaceutical sector, India is currently well positioned to emerge as a significant player in the global biotech area as well. India's strength in selected areas of biotechnology includes its capacity in bioprocess engineering, skills in gene manipulation of microbes and animal cells, capacity in downstream processing and isolation methods, and competence in recombinant DNA technology of plants and animals.

*See also* **Science in History: India**

## BIBLIOGRAPHY

Allchin, Bridget, and Raymond Allchin. *The Rise of Civilization in India and Pakistan.* London and New York: Cambridge University Press, 1982.

Baber, Zaheer. *The Science of Empire: Scientific Knowledge, Civilization, and Colonial Rule in India.* New York: State University of New York Press, 1996.

Chakraborty, Chandana, and Dilip Dutta. "Indian Software Industry: Growth Patterns, Constraints and Government Initiatives." In *Indian Economic Reforms,* edited by Raghbendra Jha, 317–333. Hampshire, U.K., and New York: Palgrave-Macmillan, 2003.

Chakravarty, Sukhamoy. *Development Planning: The Indian Experience.* Oxford and New York: Clarendon Press, 1987.

Deshpande, Vijay Jayant. "History of Chemistry and Alchemy in India from Pre-historic to Pre-modern Times." In *History of Indian Science, Technology and Culture: AD 1000–1800,* edited by Abdur Rahman, 129–170. Oxford University Press, 1999.

*Economic Survey: 2000–2001,* New Delhi: Ministry of Finance, 2001.

Gopal, Lallanji. "Indian Agriculture—A Historical Perspective." In *History of Indian Science, Technology and Culture: AD 1000–1800,* edited by Abdur Rahman, 314–333. Oxford: Oxford University Press, 1999.

Government of India. *A Scientific Policy Resolution.* Indian Parliament, March 1958.

Krishna, Venni V. "Science, Technology and Counter Hegemony: Some Reflections on the Contemporary Science Movements in India." In *Science and Technology in a Developing World*, edited by Terry Shinn, Jack Spaapen, and Venni Krishna, 375–411. Dordrecht, Boston and London: Kluwer, 1997.

Mahmud, (Sayed) Jafar. *Metal Technology in Medieval Indi.* Delhi: Daya Publishing House, 1988.

Rahman, Abdur, ed. *History of Indian Science, Technology and Culture: AD 1000–1800,* Oxford: Oxford University Press, 1999.

Roy, Aniruddha, and S. K. Bagchi, eds. *Technology in Ancient and Medieval India.* Delhi: Sundeep Prakashan, 1986.

*Dilip Dutta*

# Japan

The anchoring of an American gunboat in Tokyo Bay in 1853 marked the beginning of modern Japanese technological development. Priority was placed on military technologies, leading up to the American occupation of Japan in 1945. The following years involved a shift from defense-oriented technological development to a market-oriented paradigm. International competitiveness following this market-oriented approach peaked in the 1980s. This article will describe both the defense-oriented technological development strategy and the market-oriented development strategy throughout modern Japanese history.

Crucial background for this history includes Japanese importation of technology from Korea and China and other international influences. Imported bronze swords have been found in burial areas and provide contrast to Japanese steel swords, made using iron sand and renowned as cutting weapons. Matchlock guns were introduced from a wrecked Portuguese boat in 1543, adding a weapon to the wars of the sixteenth century. Establishment of the Tokugawa Shogunate regime in the seventeenth century began more than two hundred years of peace. In such a long peaceful period, gunsmiths found few work opportunities and generally turned into fireworks pyrotechnicians.

## Defense-Oriented Technological Development

At the time of Commodore Oliver Hazard Perry's arrival in Japan in 1853, Japan was ruled by the samurai class, traditional warriors of the highest social class, composing the top 5 percent of Japanese society. Despite the long peacetime and lack of real war experience, the samurai responded professionally to the Western threat, each wearing two traditional swords. Their first move was to position cannons along the coastline. They quickly realized that their seventeenth-century matchlock weaponry was no match for post-Napoleonic Western military technology. All available experts in Western science and technology gathered at Saga clan and attempted to cast a steel cannon using Western methods. Differing translations of Western military text led to different weapon-making approaches to fight the West. These attempts proved fruitless, when some of the technologically "advanced" clans were defeated in 1863 by a Western gunboat armed with a cannon firing with superior range. These events opened Japan to Western imperialists and eventually to the regime change in 1868, giving way to the Meiji oligarchy government.

It thus comes as no surprise that the most urgent concern of the Meiji oligarchy leaders was how to defend Japan from foreign military aggression. But at the same time, they recognized the industrial background behind Western military superiority. In order to survive, their new priorities were summarized in their slogan: "Rich state and strong military." They established the College of Technology within the Ministry of Technology in 1870. Emphasis on science and technology in Japanese education was evident in the high percentage of graduates in scientific disciplines from Tokyo University (85 percent in the 1880s).

The Meiji oligarchy established modern bureaucracy, modern institutions, and infrastructure such as military works, telephone and telegraph lines, and railroads. The defense priority was designed into the structure of the Imperial University, founded in 1886. Besides familiar departments such as mechanical engineering and civil engineering, Imperial University had departments of military works and ammunition, unlike Western universities of the 1880s. The department of shipbuilding technology was established at the demand of the Japanese Navy.

Modern Japanese scientific and technological professions were created by the new, Western-oriented government. Former samurai warriors were the main practitioners of these new professions, and during the Edo period they became primarily administrative bureaucrats. In the past they had inherited stipends from their family in exchange for their loyalty to the Shogunate or local feudatories. They were the class long accustomed to thinking in terms of public affairs and holding public office.

In the 1870s the Meiji government, in an effort to abolish class difference between samurai and commoners, curtailed the inherited family stipends of the samurai class. Whereas other classes (farmers, artisans, and merchants) could continue their inherited vocations, the samurai lost their traditional source of revenue. They had to find a new way to make their livelihood. Science and engineering jobs were new and promising because experts were needed for the building of a modern nation. Infrastructure development required telegraph networks and defense production. Medical schools were occupied by sons of traditional doctors, agricultural schools by sons of hereditary landowners. Thus, modern Japanese scientific and technological professions were, at the beginning of their development, very "samurai-spirited."

Samurai graduates found their posts in the government, as their fathers found positions in Shogunate or clan bureaucracy. As they did then, they filled their roles as technocrats. They designed the technological infrastructure of a modern nation by translating Western books into Japanese practice. Combining the high status of samurai with scientific and technological development yielded a different class effect in Japan than in Europe, where the rise of lower and middle classes occurred. Samurai were quite suited to civil and

military engineering pursuits and made their technology "public technology" in contrast to traditional artisans' "private technology."

Despite their samurai spirit and scholarly expertise in public technology, many of their attempts ended in failure because there was no precedent to fit economic considerations to both a local and a capitalistic system. Private market-oriented technology, including a modern textile industry, was encouraged by the government. Experimental government factories turned out to be economically disastrous and were later taken over by the private sector. These early attempts at scientific and technological infrastructure development ended up with the dissolution of the Ministry of Technology in 1885.

Modern Japanese engineering and technology may be said to have two origins. One is "public-centered" engineering, best exemplified by the military engineering taught and practiced at the French École Polytechnique. The other is profit-making, capitalistic engineering practiced in the private sector, such as electrical engineering and pharmaceutical technology.

The construction of late nineteenth-century Japanese technological infrastructure was one of the earliest examples of a process now common in the history of countries reaching a modern level of development. The process typically moves from development in the public sector by an underdeveloped dictatorship oligarchy, until operations are suitable for profit making by the private sector, which then takes over development. In the process of transferring a technology to the private sector, questions about continuing development of imported technologies in the public sector arise. First, does the public sector have a sufficiently developed infrastructure to support a private development sector? Next, who can support the initiation and transplantation of private scientific and technological development? The technology practiced in the private sector (or "private technology") was maintained by the traditional artisan classes, such as carpenters and fishermen. Their motivation was profit, as contrasted with samurai focus on public technology.

In the early Meiji period the government encouraged invention by sponsoring domestic industry-fostering fairs and conferring prizes for the best inventions. Industrial designers and inventors were not legally protected in Japan, however, and the ideas of inventors were easily stolen. It was also the time when international industrial property regulations were formulated. Westerners claimed that the Japanese, who so far had freely appropriated Western inventions, should subscribe to the international patent regulations. But those who welcomed the enforcement of international regulations were, of course, the Westerners, not the Japanese.

Around 1885, a major political and diplomatic issue was the adjustment of unequal treatises between Japan and major Western powers. The Japanese government made a preliminary draft of a domestic patent system and intended to give patents to foreigners, in exchange for Western denunciation of unequal treaty items. The treaty amendment was postponed to 1899, when Japan subscribed to the international industrial property regulations. Until then, Japanese inventors could obtain domestic patents by imitating Western inventions. Such was the case of Sakichi Toyoda (1867–1930), of artisan origin, who founded what began as a textile firm using his loom designs and eventually became Toyota Industries and the Toyota Motor Corporation. During that time, patent assessors were generally graduates of engineering departments, whereas inventors were mostly traditional artisans.

This contrast may be explained by supposing that, while elite graduate engineers were busy introducing and translating the technology imported from the West, innovation in the private sector was demonstrated in local technological practices. College graduates were engaged in public works, where it would be less proper (as in the samurai spirit) to apply for profit-making patent rights, whereas private-sector inventors were eager to make a profit. Only in the twentieth century, after Japan had subscribed to the international industrial property regulations, were high-level engineers concerned with patent rights.

When Japan adopted the international industrial property regulations at the end of the nineteenth century, electricity generation and corresponding equipment were sorely needed. The Department of Electrical Communication was created within university already in the 1870s, but it was the establishment of a national telegraph system in the public sector that proved to be quite useful in the Japanese Civil War of 1877. At the end of the century, however, the private-sector electricity needed for lighting demanded quite a different power generation strategy. Some of the early graduates of the Imperial University traveled to the United States as workers to learn about these new private technologies, not yet taught and established at Japanese schools. This facilitated the transfer of electrical engineering from the public to the private sector, while civil and military engineers remained in the public sector. Thereafter, samurai engineers moved gradually into the private sector.

Thus, to characterize prewar Japanese industrial development in one phrase, it was a constant process of transfer from the public to the private sector, often interrupted by the effects of the government arms industry. It was a rather slow process of privatization, completed only after World War II.

## Market-Oriented Technological Development

Japan's surrender in 1945 marked the beginning of another transformation in Japanese science and technology, from military-oriented to market-oriented development. In the beginning of the Occupation that followed, the Occupation army's task was to demilitarize and demobilize Japanese wartime science and technology. Accordingly, the two Japanese cyclotrons were thrown into the sea. Some strategic areas, such as nuclear and aircraft research, were forbidden by the Occupation forces.

The civilian officers of the science and technology division of the Economic and Scientific Section (ESS/ST) of the Occupation forces soon realized that the Japanese scientific community had no intention of recovering military-oriented technology. Science and technology officers promoted science and technology for the sake of the economic recovery of impoverished Japan. They promoted the export trade through development of quality control measures in Japanese industries.

After the Occupation was over in 1952, the Japanese were free to engage in nuclear and space research, two major areas of Cold-War public science. However, Japan's late start made closing the gap between their research and development and that in the West difficult. The Japanese Ministry of International Trade and Industry (MITI) pursued an industrial policy to protect and encourage industry, first for the domestic market and later for the export market. It did not have a comprehensive science policy at all, as often believed, but acted as advisor to industry. This ministry designed science and technology promotion in the 1950s and early 1960s, exercising the power of scarce currency allocation for purchasing machineries and devices needed for scientific and technological development.

The Japanese Science and Technology Agency (STA), created in 1956, sought to develop big national projects of public science, such as nuclear technology and oceanography, but it remained a small office. STA complained in its "White Paper" on science and technology (an annual publication) that the Japanese government did not allocate as much to science and technology as other advanced countries did. This constituted an indirect criticism of the Ministry of Finance, in charge of allocating funds for the national budget. The Ministry of Finance persistently responded that the major reason for Japan's relatively small science and technology budget is that Japan does not conduct military research and emphasizes only economic goals for research and development. After a subsequent *White Paper* included an international comparison of the ratios of public/private expenditure excluding military research, Japan's government expenditure on research and development was similar to the research and development budgets of other nations.

The causal correlation between economic growth and research and development invest-

ment remains a popular, but false, belief. Differences between Western and Japanese culture are apparent when basic concepts driving scientific and technological development are compared. When Japanese funding levels are examined relative to growth, the Japanese case disproves the Western belief that a positive relationship exists between investment and economic growth. Japanese high growth was attributable to other factors entirely.

Within the worldwide science policy community, Japan represents an unexplainable and embarrassing example during the high-economic-growth period of the late 1950s through the 1960s. The discovery of such negative correlations in various settings frustrated the whole science policy-making community and led to policy change and drastic budget cuts in the United States, England, and France in the latter part of the 1960s. Japanese funding for research and development was much lower than that in major Western countries at the beginning of the postwar period. The total percentage of government expenditure on research and development remained between 2 and 3 percent throughout the 1960s, compared to 11 percent of the total government budget spent in the United States and France around the same time.

Unlike England or France, Japan has had no grand research and development plan, yet it has been able to realize rapid economic growth. Thus, the Organization for Economic Cooperation and Development (OECD), the stronghold of international science policy makers, in their Conditions for Success in Technological Innovation (Paris, 1971), concluded that "there is no observed correlation between the proportion of national resources devoted to research and development and the rate of growth in productivity."

Since then, technologically advanced nations have taken lessons from the Japanese experience, as evidenced in their gradual reduction of total investment in public-sector research and development. The United States' nondefense public obligation was reduced from 47.2 percent in 1965 to 28.9 percent in 1987, although defense research and development was not significantly reduced. The French nondefense commitment was reduced from 62.3 percent in 1965 to 32 percent in 1986, although its defense research and development spending increased.

This poses a problem in that it disputes the belief on the part of Western technocrats that increased research and development investment yields increased economic growth. This was an idea entertained by postwar science policy makers as a "linear program" that asserted that investment in basic science eventually leads to technological innovation. The belief gained prominence after Vannevar Bush's *Science, the Endless Frontier* was published immediately after World War II by the United States Office of Scientific Research and Development.

Third World science policy makers observe Japan's secret to success in terms of how assumptions about science and technology differ from traditional Western notions. First, Japan avoided costly basic science development, which does not guarantee economic return to society. Instead, Japan imported basic ideas and research results from other countries and then used them in economically viable lines of development without expending large amounts of money. Japan has, as a result, been able to reconstruct its economy and attain high economic growth.

## Secret of Japanese Technological Success

During the 1970s, Japanese industries first faced a nationwide antipollution movement, then the oil crisis of 1973, which stopped economic growth (but only for a single year). Growth increased again, and priorities shifted from a heavy chemical industry to one focused on microelectronics. This shift is captured in the popular Japanese slogan from "heavy-thick-long-large to light-thin-short-small," referencing not only product orientation but also the value of minimizing energy use and pollution.

The 1980s brought the stamp of international quality on Japanese market-oriented products. Quality control in Japan received considerable attention as the world searched for reasons behind Japan's economic success. They found that what made Japanese private industry so successful was not the outcomes of large research and development budgets but instead the improvement of production process quality. This includes process innovation rather than product innovation, such as the introduction of robotics and the quality circle movement in the 1970s. The latter is perhaps more a clever labor management measure to reduce alienation at the workplace.

In the late 1980s Japan was involved in an intellectual property ownership conflict with the United States. American science policy bureaucrats were internally blamed for the United States' spending more public money on research and development yet still being less competitive in producing quality technology than Japan. In turn, the American bureaucrats blamed the Japanese for getting a "free ride" on American basic science. Japan responded to these charges by more aggressively competing in the IT and biotechnology markets while finally increasing its research and development spending after 1995 with the "Basic Plan." Ironically, the Japanese science policy community now makes the same complaint as late 1980s American science policy makers: that Japan spends more on research and development but gains less competitive economic growth.

*See also* **Science in History: Japan**

BIBLIOGRAPHY
Low, Morris, Shigeru Nakayama, and Hitoshi Yoshioka. *Science, Technology, and Society in Contemporary Japan.* Cambridge, U.K.: Cambridge University Press, 1999.
Nakayama, Shigeru. *Science, Technology, and Society in Postwar Japan.* London and New York: Kegan Paul International, 1991.

*Shigeru Nakayama*

# Latin America

It is of interest to consider two distinct periods of the history of technology in Latin America: the period before the arrival of the Europeans to that part of the world and the period after the arrival. The ways people lived in what would become Latin America and in Europe at the time of the arrival of the Europeans, were quite different. For one thing, in America there was no bronze or iron metallurgy before the arrival of the Europeans. Although the Europeans did have a technological edge over the native people, however, that should not be understood simplistically; in comparison with the demands of local life, local technology was quite efficient and developed. If we understand technology as sets of ideas and techniques, as well as artifacts, that different cultures have developed to improve their own lives and, in some cases, to dominate the lives of others, we will see that the native people of Latin America were doing well by the time the Europeans arrived.

## Technology of the American Peoples at the Time of the Conquest

In 1492 Christopher Columbus, an Italian sailing for the Spanish kingdom of Castilla, arrived at the Caribbean. In 1500 Pedro Álvares Cabral, a Portuguese, arrived in Brazil. In 1519 Hernan Cortés, a Spaniard, reached what is today known as Mexico. Who were the people living in these regions then, and how had they lived before that time? None used the wheel or had metal tools, but there were some significant differences among the peoples.

In the region known today as Brazil, large centralized societies never developed, although around 1500, an estimated 3 million people inhabited the land. The main group, the Tupi-Guaraní, had spread to different parts of the region, including the shore, coming from the Amazon region, and lived a life based on hunting, fishing, and fruit and root extraction (particularly manioc), which led them to move constantly. Native Brazilians have possessed, since that time, a substantial knowledge of the medical uses of herbs and plants. Significant social differentiation, coercive power structures, tax systems, and institutionalized forms of religion were not present.

Three major civilizations developed in Central and South America before the colonization: the Maya, in the Yucatán Peninsula and part of south Mexico and Central America; the Aztec, in a region around what is today Mexico City; and the Inca, who spread from the south of Ecuador and Colombia to all of Peru and Bolivia and to parts of Chile and Argentina, in the western portion of South America.

*The Maya.* There is evidence that the Maya, the oldest of the three civilisations, was already building small burial mounds as early as some 3000 years ago. This simple architecture eventually evolved, along the centuries, to the construction of spetacular pyramids, palaces, temples, and observatories. In most cases, those were made of limestone, a quite adequate type of stone for builders who did not have metal tools. In earlier times a mixture of crushed limestone ($CaCO_3$) and burnt limestone (containing quicklime, $CaO$) was used as mortar to hold the pieces together; modern concrete contains these ingredients as well. With time and practice, the cutting and fitting of the stones became much more precise and did not always require the use of mortar; the use of this mixture as a "plaster" finishing, however, remained, as well as in the construction of more delicate structures. In simpler constructions such as houses, wood and a mixture of mud and straw (adobe) was used. Only in the Maya city of Comalcalco have remains of fired clay bricks been found; that region lacks any stone deposits.

Because the civilization was based on centralized government, the issues of territorial extension and population size, involvement in long distance trade, and frequent military campaigns naturally led to a need for people with specialized knowledge to conduct the governement affairs: priests and scribes, belonging to the higher layers of the society and even trained, in some cases, in specialized schools such as the one in Mayapán. Mayan mathematics, calendars, astronomy, and astrology developed and were deeply interrelated. That knowledge was used in agriculture, everyday business—including legal judgments of individuals—and the planning of military campaigns, and resulted in the development of calendars involving different kinds of cycles; some authors conjecture that the Mayan calendar system was possibly the most complex ever to be found anywhere in the planet. Observatories were constructed to allow the study of the movements of celestrial bodies, especially the moon and Venus; one of the best known is Chichén Itzá, in the Yucatán Peninsula. In association with the calendar system, the Mayans developed a mixed-base number system that was sophisticated to the point of having a positional zero, and a writing system that combined phonetic and pictographic elements. A huge amount of texts were produced, but only some 5000 were carved in stone. Most of the rest, which were written on a kind of bark fabric, were burned on the orders of priests, shortly after the Spanish conquest. Over 85 percent of the surviving Mayan texts had been deciphered by 2004.

It seems that a combination of factors contributed to the decline of the Classic Maya culture, a decline that began around the ninth century A.D. but lasted for many centuries. Authors list two factors as the most important. One was continuing war involving different city-states (which formed the basis of the political structure). The other was difficulty in feeding the population, a problem due not only to a not-so-efficient food production system but also to some two hundred years of drought in the period from 800 to 1000. Recently researchers have suggested that deforestation due to the huge amounts of wood used to burn limestone for making mortar and "plaster" finishing, might well have had an impact, leading, for example, to substantial soil erosion and its consequences for agriculture. The slow deaggregation process that folowed eventually led to the end of the Maya Empire. The last Maya city-state was taken by the Spanish only in 1697.

*The Mexica (or Aztec).*   The name *Aztec* is somewhat inadequate to name the people who built a civilization that, after a nomadic start, established itself in the region around present day Mexico City. The explorer Alexander von Humboldt (1769–1859) made the suggestion to apply to those people the term "Aztec," meaning "people from Aztlán," their legendary land of origin, somewhere in what is now northwestern Mexico or the southwestern United States. They referred to themselves, however, as *Mexica* (pronounced "mesh-ee-kah"), from which the name of the country was derived. The city of Tenochtitlán was built in 1325, on

land skilfully gained from Lake Texcoco. Floating platforms, called *chinampas* by the Spanish, were built using canes and mud, and on those *chinampas* the Mexica would grow their crops—a technique first developed by the Mayans before A.D. 1000. As roots from the platform reached the lake bottom, some of them would become actual islands. At one time the total cultivated area reached 30,000 acres. The water of the lake was salty, so engineers were called in to create a separate area within the lake to hold fresh water coming from springs, using a dike built across it. This was complemented by mechanisms to control the water level and by aqueducts. The whole enterprise, undertaken as a series of state-managed construction works, set up the infrastructure for quite successful agriculture (corn, beans, pumpkins, chilies, and cocoa grains; the latter were used as currency) and fishing, more than sufficient to feed the city's large population (estimated at between 200,000 and 300,000 at the time of Cortés's arrival in 1521). It is worth noticing that the success in food production allowed more time for the Mexica to concentrate on crafts, arts, and military matters. Corn was for the Mexica, as much as for other Mesoamerican cultures, part of their staple food. They had developed a technique of treating corn with limewater (*cal* in Spanish, $Ca(OH)_2$, made by dissolving burnt limestone), which made the corn much easier to digest and so a much better source of nutrients. When corn was taken to Europe, people started eating it as a substitute for wheat, without this treatment; because of the resulting vitamin deficiency illnesses, it was decided in Europe that corn would be used exclusively to feed animals. Today, corn is treated with cal in Mexico and other places to produce the flour from which tortillas are made.

But the Mexica were not only accomplished hydraulic engineers, farmers and fishermen. According to James E. McLellan and Harold Dorn, they "posessed sophisticated botanical and medical knowledge," with priests acting as medical specialists who also conducted empirical research, leading to a body of knowledge and techniques that were "at least the equal of their Spanish conquerors." It is estimated that at the time of the Conquest, life expectancy in the Mexica Empire was at least 10 years higher than that in Europe. In the late twentieth century, public hospitals in Mexico began to reintroduce the *herbolaria* (traditional herbal medicines) for the treatment of some illnesses because of the much reduced costs.

But this sound medical knowledge was not sufficient to prevent the defeat of the Mexica by an (unintentional) episode of biological war. Even taking into consideration the power of the military equipment that Hernán Cortés had brought to Tenochtitlán and the support that he had from natives fighting the domination of the people in Tenochtitlán, it is well known that the contamination of the natives with smallpox played a major role in allowing Cortés to finally defeat Emperor Cuauhtémoc and his people. A similar effect also had an important impact on the conquest of the Incas by Pizarro.

*The Inca.* The Inca Empire existed from about 1200 to 1533, when an expedition led by the Spaniard Francisco Pizarro defeated the forces of Emperor Atahualpa; its capital was Cuzco ("navel" of the world, in the Quechua language), in what is today Peru. Stretching almost 2000 miles from its northern to its southern limits, the Inca territory was home for many different cultures and languages, and the Inca state served to articulate and control all those cultures. The needs of such a large and complex empire asked for some kind of technology through which information (on demographics, taxes and commerce, for instance) could be kept. In contrast to the Maya, who developed quite sophisticated writing and written number systems, the Inca developed the *quipu,* a sophisticated record-keeping device, which was also used to record historical information.

The *quipu* consists of strings of cotton, tied to a stick, and in those strings complex patterns of knots are made to represent the required information. Although much work has been done on decoding the *quipu*, it is believed that a considerable amount of information is still hidden in them. The absence of

traces of a written Incan language (although Quechua is still spoken throughout the former empire) does not help in the attemps to decipher the *quipus*. They were transported around the empire by messengers on foot (*chasqui*), who chewed coca leaves to increase their strength during the journeys; this organized system of messengers, who also carried packages, functioned as a postal service.

Inca astronomers observed the sky with the help of special constructions, such as the Coricancha temple in Cuzco, or natural markers, like the mountains of the Andes. Because of the lack of written records, it was common practice to adopt different calendar systems in different regions of the empire.

The expansion of agriculture to the highland areas of the Andes was done through the use of terraces in which such crops as potatoes, tomatoes, cotton, peanuts, and coca were grown. With the development of a complex and efficient system of irrigation canals; the whole system provided enough food for the population. The llama, a mammal related to the camels, was used for the transportation of goods but also for meat (as were guinea pigs as well) and wool; in no other culture of Central and South America did a domesticated animal play a role as important as that of the llama for the Inca. The engineering ability shown in the building of canals is also witnessed by the construction of temples and fortresses, rope bridges, and a road system; even today, specialists marvel at how the Inca cut, moved, and fitted blocks of stone weighing up to 100 tons. After the end of the Inca Empire in 1533, the terraces and canals were either destroyed by the Spaniards or gradually abandoned, but in 1977 British archaeologist Ann Kendall founded the Cusichaca Trust, a nongovernmental organization dedicated to restoring the ancient Inca agricultural system, to help raise food production in the Andean villages.

The medical and botanical knowledge of the Inca was substantial and, in some cases, quite sophisticated for the time. For instance, they performed amputation and trepanation (cutting of a hole in the skull of a patient, supposedly to prevent further damage in cases of brain swelling). A major contribuition of the

Inca to mankind is the use of quinine to treat malaria. Mummification was skilfully performed; as recently as March 2004, dozens of exquisitely preserved mummies were found and recovered in the outskirts of Lima, the capital of Peru.

In 1532, Francisco Pizarro and his 150 men met Emperor Atahualpa, who refused to convert to Christianity, eventually leading to his execution by the Spanish, after a series of military confrontations. The Inca army had some 40,000 people, but was defeated by a combination of three factors: the massive number of deaths as the result of smallpox infection (as had happened with the Mexica), Pizarro's association with groups that wanted freedom from the rulers, and the superior warfare technology of the Spanish. By 1535 the conquest was established, and Pizarro moved the capital from Cuzco to Lima.

## Technology in Latin America after the Conquest

After the conquest many of the traditions of the native peoples disappeared or were transformed and adapted to life in the villages, but the impact of technology developed by them was felt in many places, through their agricultural knowledge (the potato, for instance, helped to prevent famine in China and in Europe, and maize is a staple food in many places) and their knowledge of medical properties of herbs and plants (pharmaceutical laboratories continue, centuries later, to study the "folk" knowledge of Central and South American native people in search of herbs and plants that can lead to new drugs). Unfortunately, it is quite likely that much has also been lost in the colonization process.

What happened after the arrival of the Europeans is probably better understood not in terms of cultures and civilizations but in terms of technologies and economic activities.

The first wave of colonizers came to Latin America in search of gold and silver, but soon an interest in agriculture developed. Until the arrival of the Europeans in Latin America, sugar was produced in small scale in the West and bought at high prices from the East. In his second voyage to the Caribbean, Columbus brought sugar cane from the Canary Islands and planted it in what is today the Dominican Rebublic. Around 1530, driven by the need to occupy and to defend its new territories, the Portuguese created large-scale sugar cane plantations in the northeast of Brazil. According to Brazilian historian Gilberto Freyre, for a long time the history of Brazil was the history of sugar. For almost three centuries after the conquest, sugar cane was, for Europe, the most important agricultural product, being central in the economy of the Caribbean as well. In the nineteenth century, steam engines were used instead of water wheels where sufficient water was not available, and until the Brazilian Abolition Act of 1888, intensive use was made of slave labor of African origin. Labor division and the use of specialized machinery made this economic activity somewhat similar to manufacturing.

The tradition of the sugar plants and the huge availability of sugar cane led to plans to use ethanol, made by fermenting the sugar, as fuel. In 1923 Brazil produced its first systematic study of such use, and in 1975, two years after the oil crisis, the National Ethanol Program was created, leading to the use of pure hydrated ethanol in cars in 1980 (previously it had been added to gasoline), and to the development of a specific technology of car engines that are better resistant to the corrosion caused by ethanol. In addition to being a renewable source of energy, the burning of ethanol is also less damaging to the environment than that of fossil fuels. The use of ethanol to produce electricity directly in fuel cells is also being studied in Brazil.

Oil production is a major economic activity in some Latin American countries, particularly Mexico, Venezuela, Brazil, Argentina, Colombia, and Ecuador. In 1878 extraction began in Ecuador, and it began in Colombia in 1909. Venezuela started commercial extraction in 1917 and Brazil in 1939. In Mexico, oil was found in 1901. Brazil, with some 5000 miles of shoreline, has most of its oil extraction done in off-shore platforms; this has led Petrobras, the state-owned oil company, to develop what is today considered the most advanced technology for drilling and extraction in deep and ultradeep water. Natural gas is gradually becoming more important in energy production in South America, especially because of new reserves recently found in Brazil and the construction of a gas pipeline connecting Bolívia and Brazil.

A widespread problem with education and a low level of investment in research and development are two of the factors that have led Latin America to be a consumer rather than a producer of new technology, but political and geopolitical factors are also significant. Mexico and Brazil are also affected by problems in education and investment, but the sizes of those economies make it less difficult for the two countries to develop original technology. Mexico, for instance, develops technology based on soil dynamics, to be used in the tracking of seismic events (earthquakes) and in the design of earthquake-proof skyscrapers. Brazil is a world-leader in the field of bank services automation (hardware and software); it supported the development of a fully electronic voting system, unique in the world; and it has developed a biodiesel that will be added to normal diesel, to reduce oil dependence and to reduce pollution. On the design and production of small and medium-sized airplanes, the Brazilian company Embraer is an international competitor.

*See also* **Science in History: Latin America**

**BIBLIOGRAPHY**

Brasil H$_2$ Fuel Cell Energy. *Portal Célula a Combustível.* 2004. http://www.celulacombustivel.com.br/sobreabh2/sobre-english.htm. A Brazilian firm "devoted to the study and spreading of the revolutionary technology of energy generation, the Fuel Cell."

Couto, Jorge. "Las gentes de la tierra" [in Spanish]. Camões: *Revista de Letras e Culturas Lusófonas* 8 (1999). Instituto Camões. http://www.instituto-camoes.pt/revista/revista8es.htm. Reflections on the European conquest of Brazil on the occasion of the 500th anniversary of the arrival of Pedro Cabral.

Galeano, Eduardo H., and Cedric Belfrage. *Open Veins of Latin America: Five Centuries of the Pillage of a Continent.* New York: Monthly Review Press, 1998.

Gama, Ruy. *Engenho e Tecnologia.* São Paulo, Brazil: Livraria Duas Cidades, 1983.

McClellan III, James E., and Harold Dorn. *Science and Technology in World History: An Introduction.* Baltimore: The John Hopkins University Press, 1999.

Pacey, Arnold. *Technology in World Civilization.* Cambridge, MA: MIT Press, 1991.

Sheridan, Richard. *Sugar and Slavery: An Economic History of the British West Indies, 1623–1775.* Baltimore: The Johns Hopkins University Press, 1974.

Schwartz, Stuart B., and Alan Knight, eds. *Sugar Plantations in the Formation of Brazilian Society: Bahia, 1550–1835.* New York: Cambridge University Press, 1986.

Taylor, Kit Sims. *Sugar and the Underdevelopment of Northeastern Brazil, 1500–1970.* Gainesville: University Presses of Florida, 1978.

Topik, Steven C., and Allen Wells. *The Second Conquest of Latin America: Coffee, Henequen and Oil during the Export Boom, 1850–1930.* Austin: University of Texas Press, 1998.

Vargas, Milton. *História da técnica e da tecnologia no Brasil.* São Paulo, Brazil: Editora da UNESP, 1994.

von Barloewen, Constantin. *History and Modernity in Latin America: Technology and Culture in the Andes Region.* Oxford and New York: Berghahn Books, 1995.

*Romulo Lins and Marcos V. Teixeira*

## United States and Western Europe

An unprecedented degree of technological change accompanied by an extraordinary growth of wealth frame the history of Western technology. Some of the threads woven into this history include the story of the continuous harnessing of more and more energy; the transformation of largely rural societies to urban ones; providing billions of people with access to complex technologies and technological systems; the replacement of human with nonhuman labor power; the development of major industries, such as textiles, chemicals, railroads, robotics, and energy; and the development of medical technologies that have expanded the human life span by decades. Other threads include the technologically induced devastation of indigenous and socially marginalized peoples; the replacement of jobs by machines; the increasingly centralized and monopolistic control of technology by private corporations; the dominant role of the nation-state and military ambition in subsidizing and promoting technologies of control; and continuing environmental devastation.

As seemingly diverse as these threads are, we can perhaps connect them by viewing the history of Western technology as the continuing application of knowledge and practice to create technological systems designed to accelerate the rate of capital accumulation (see Table).

The foundations of the remarkable technological and societal changes that were to take place in the nineteenth and twentieth centuries in the West were in place by 1800. These include the rudiments of the scientific method, the use of money as a measure of value, and a culture and society dependent on perpetual economic growth and capital accumulation. But it would not be until the Industrial Revolution that these foundational devices would begin to transform the West and, ultimately, the rest of the world.

## The Technology of Everyday Life in the West

We can appreciate the history of the role of technology in capital accumulation by examining the technologies of food, work, and travel. Each illustrates some of the ways in which technology promotes the accumulation of money: first, technology accelerates natural processes, devising ways to make plants and trees grow faster and animals grow bigger faster; second, it accelerates work and reduces labor costs by substituting nonhuman for human labor, increasing the pace of labor, and downgrading skilled to unskilled labor; third, it compresses space, accelerating both travel and communication in the interests of capital accumulation; and, finally, it redefines social relations and institutions in the interests of economic growth.

Agriculture and animal husbandry were, perhaps, the first areas to which technology was applied to accelerate production. Virtually all of the food crops and domesticated animals that make up the Western diet existed 2000 years ago. Today, Western farmers and ranchers just grow and raise them faster and with less human labor. In the nineteenth

century, the advent of mechanized agriculture—the steel plow; the mechanical reaper; and the combine, which combined the machines for reaping and thrashing wheat—cut the labor time for each bushel of wheat harvested from sixty-one to three hours. In cereal production, in general, output per worker more than tripled between 1840 and 1911, with 60 percent of the increase attributable to mechanization. In the latter half of the twentieth century, the so-called green revolution further reduced labor time by substituting nonhuman energy (largely subsidized by nation-states) in the form of chemical herbicides, pesticides, and fertilizers for human energy. Technology has also contributed to the accelerated growth of domesticated animals. A grass-fed steer in 1750 would not attain maximum weight until it was five or six years old. By 1950, grain feeding would bring the steer to market in two or three years. But with today's use of herbicides to increase pasturage, protein supplements, growth hormones, forced feeding of grain, and antibiotics to ward off infections, a typical steer can reach its slaughter weight of some 1,200 pounds in 14 to 16 months. Now, with the advent of genetic engineering, researchers are further accelerating the growth and size of plant as well as animal species, such as salmon, trout, catfish, and oysters.

In addition to accelerating the pace of nature, technology has also altered the pace and rhythm of work, continually devising ways for more work to be done in less time. Up until the late 1700s, trained artisans worked at home or in small workshops, turning out items by hand with relatively simple tools. But increased demand for more rapid production led to two revolutionary changes in the organization of work in the second half of the eighteenth century: the introduction of the factory system and development of systems of mass production. The factory system first emerged in the British textile industry as increased demand for products led to the development of new technologies—the spinning jenny in 1765, the water frame in 1769, the spinning mule in 1779. The size of these machines and the power sources required to run them led to

**Table 1.  Selected Technological Developments and Per Capita GNP for Selected Countries 1820–1998 and the Contribution of Technologies to U.S. GNP in 1996**

| Technology Category | | | | | Energy | Manufacturing (organization of work) | Communication | Travel | Military | Medicine | Agriculture, forestry, fishing |
|---|---|---|---|---|---|---|---|---|---|---|---|
| Approximate contribution and percent of total to GNP of United States in 1996 (in billions)† | | | | | 283.2 (4.2%) | 1316.0 (19.4%) | 695.3 (10.2%) | 455.5 (6.7%) | 270.0 (apx) (4.0%) | 459.1 (6.8%) | 130.4 (1.9%) |
| Per Capita GNP by Country‡ | | | | | | | | | | | |
| Year | USA | UK | France | Germany | | | | | | | |
| 1998 | $27331 | 18714 | 19558 | 17779 | | Computerization of the workplace Offshore assembly plants | Personal computer | | "Smart" bombs | Genetic manipulation Advanced reproductive technologies | Genetically engineered crops |
| 1973 | 16689 | 12022 | 13123 | 11966 | Nuclear | | | Space travel Jet travel | ICBMs | Transplant surgery | "Green" revolution |
| 1950 | 9561 | 6907 | 5270 | 3881 | | | Television | | Nuclear weapons Rockets Airplanes Tanks | Penicillin | Insecticides Herbicides |
| 1913 | 5301 | 4921 | 3485 | 3648 | Electricity | Fordism Taylorism | Radio Wireless Phonograph Telephone | Airplane Automobile | Maxim rifle Submarine Ironclad ships | Purification of water supplies X-rays | Tractor |
| 1870 | 2445 | 3191 | 1876 | 1821 | Oil | Factory production | Telegraph | Railroad | | | Norfolk-four-course rotation Steam-driven threshing McCormick reaper |
| 1820 | 1257 | 1707 | 1230 | 1117 | Steam | Mass production | | | Mass production of weapons | | |

*This does not imply a causal relationship between technology and economic growth, but, rather, that the two are interrelated.

†GNP figures (the dollar total of all goods and services produced) by area are from the Bureau of Economic Analysis of the U.S. Department of Commerce (http://www.bea.gov/bea/dn2/gpoc.htm). The figures are approximate and based on BEA categories and do not include government expenses.

‡Data from Maddison 2001.

the centralization of production into factories. This permitted greater control over workers and, in effect, turned skilled artisans into "unskilled," poorly paid laborers. Manufacturers also accelerated production by introducing production methods—initially to the production of armaments—thus allowing the standardization and interchangeability of parts. A French gunsmith, Le Blanc, first applied mass production to armaments in 1785, but the first large-scale application occurred in 1798 in the United States when a contract to make guns was given to Eli Whitney, the inventor of the cotton gin. Whitney claimed that he could substitute machinery for the work of highly skilled artisans; produce a uniform product; and, in the process, increase the speed with which guns could be produced. Although Whitney's claims for speed and exactness were not realized, *armory practice,* or the *American System,* as it was called in Great Britain, spread quickly to civilian industries, such as clock making and sewing machine manufacture. Traditional clock makers, who worked by hand in both wood and brass, could make four or five clocks a year, each selling for about $50. By 1850, an average clock factory was turning out 130,000 to 150,000 clocks annually, and the price of a clock had fallen to $15.

Frederick Taylor's contribution to management marked the next milestone in the acceleration of work. Taylor's system, first publicized in 1895 and elaborated in his book *Theory of Scientific Management,* analyzed every stage of the worker's task carefully to determine how quickly it could be done, and whether it and/or the worker could be eliminated. By applying his methods in one steel plant, he increased the capacity of the pig-iron handler from 12.5 to 47.5 tons per day. In 1912, Henry Ford introduced assembly line production for the Model T to further accelerate the speed at which laborers performed their tasks. Within a year, assembly time of the flywheel magneto of the Model T dropped from 20 person-minutes to 5 minutes; that of the engine from 594 to 226 minutes; and that of the chassis from 12.5 hours to 93 minutes. Ford, however, had to double the daily wage rate to keep workers at their

mind-numbing tasks. With the continued application of labor-accelerating and labor-saving technologies, the production of the average automobile today requires less than 20 hours of human labor.

As technology has accelerated the world of nature and the world of work, so it has also greatly accelerated communication and travel, along with the money that each generates. In the early nineteenth century, if one were to travel by coach, it took 3 days to go from London to Birmingham, 4.5 days to Manchester, and nearly a week to Edinburgh. A trip from Boston to Washington over the best of the roads took 10 days traveling 15 to 16 hours a day. Today a trip from Washington, DC, to Boston would take 7.5 hours by automobile and 1.5 hours by air.

The acceleration of transportation and communication over the past 200 years required large amounts of capital, far more than what was available to solely private interests. Thus, one of the earmarks of the history of Western technology has been the involvement of the nation-state as a provider of investment capital. For example, to facilitate trade, the United States in the early 1800s embarked on government-funded projects of road and canal building to speed the movement of goods and services to market. Moving a ton of grain, salt, or ore overland from Buffalo to New York City cost roughly $100 in 1820. Five years later, the same load could have been shipped via canal and river barges to the same destination for $9.

Rail transport was next to revolutionize the movement of goods and people. Originating in German mines in the sixteenth century, human, and later horse- or mule-drawn railways were constructed in Great Britain and the United States. The first steam locomotives were built in Wales in the early 1800s. Robert Stephenson designed the ancestor of the conventional steam locomotive in 1830 for the Liverpool & Manchester Railway. By 1835, trains were running in Germany, Belgium, and the United States. By 1900, the United States had some 250,000 miles of track, Great Britain some 21,000 miles, France some 31,500 miles, Germany some 40,000 miles, and Russia some 44,000 miles, which had promoted the accumulation of vast fortunes in industries such as meat packing and petroleum. Today in France, Germany, and Great Britain, high-speed trains transport people and goods in excess of 175 miles per hour.

The automobile was the next major innovation in transport, leading directly and indirectly to the development of an industry that remains one of the largest contributors to capital accumulation. First developed in Europe, the automobile used the internal combustion engine designed by Otto and modified by Wilhelm Maybach, Gottlieb Daimler, and Karl Benz. The first cars were sold in Europe in the early 1890s and shortly thereafter in the United States, where, by 1910, there were 458,500 registered automobiles in an industry that had jumped from 150th to 21st in product sales in less than a decade. By the end of the twentieth century, there were 226 million vehicles in the United States, with the rate of ownership growing at twice the rate of the population.

The development and application of household technology illustrates how western technology gradually redefined household labor and altered patterns of social relations. A housewife prior to 1800 would have cooked and baked galore, but her husband would have done much of the preparation—chopping wood, shelling corn, and pounding grain into meal; and the children would have helped as well. But in the nineteenth century, two technologies would begin to change that: milled flour and the cast-iron stove. Before flour was commercially milled, men ground the wheat or corn or hauled the grain to the local mill. Since milled wheat, rye, and corn do not keep long, grinding or hauling grain to the mill had to be done at least weekly. The commercial milling of flour (which kept longer than other ground grains) freed men of these chores, but, in the meantime, increased the labor of women because milled grains require more work to turn into breads, porridges, cakes, and the like, than do coarse grains.Before industrialization, only city dwellers or the very wealthy used fine milled flour. But as fine white flour became cheaper, yeast breads replaced the coarse breads. Cakes also were difficult to make and required long, laborious steps, further increasing a housewife's labor.

The introduction of the cast-iron cooking stove, which Ruth Schwartz Cowan calls "the single most important domestic symbol of the nineteenth century," also reduced labor, but, again, largely of men. It halved the amount of fuel that a household needed and halved the amount of work men had to do in cutting, hauling, and splitting wood. But it changed the labor of women little, if at all, and may have increased their work. Cooking before an open hearth, the most common meal was the one-pot dish; the stove allowed many dishes to be cooked at once. Furthermore, stoves had to be cleaned at the end of each day, unlike fireplaces, a task generally performed by housewives.

The introduction of other household innovations followed much the same pattern. Outhouses were built and cleaned by men (or children), but cleaning the modern bathroom fell to housewives. Even the automobile increased women's labor by assigning to her, without-pay, jobs that had been done by paid laborers. Thus, household services that had been provided by businesses (such as grocery delivery, laundry pickups, and milk delivery) now fell to the housewife. In sum, in spite of the revolution in "labor-saving" household technology, the hours that women, of all social classes, are expected to work in the homes has changed little in the past century, while, at the same time, generating billions of dollars of sales of household technologies.

There have been, of course, many other ways that technology has served as a means of capital accumulation, from the development of armaments (currently $850 billion) to video games (some $10 billion annually). But the story of the remarkable growth and success of Western technology would be incomplete without examining the unprecedented protest and resistance that accompanied it.

## Resistance to Technological Change

Whereas the application of technology and technological systems has dramatically increased the speed of capital accumulation, this technological proliferation, often characterized as "progress," comes at a price. For in the process of accelerating capital accumulation, technology also serves as a tool of *capital conversion*, turning nonmonetary capital into money. This process is facilitated by a cultural value system in which money is the measure of all things and in which environmental resources, political processes, and social relations are considered valueless unless they can be commodified. Thus, it would make no sense to convert a forest into lumber unless a greater value is placed on the finished lumber than on the trees; it makes no sense to kill thousands of people unless the result (such as the opening of markets or the acquisition of a resource critical to capital accumulation) is valued more than the lives lost; it makes no sense to convert family functions into capital-generating services unless the monetary capital is valued more than the family interactions they replace. There is no mystery to the process. Money (in itself a technology), replaces natural capital as people use technology to draw on more and more environmental resources. When water is used to grow crops more rapidly, a natural resource is converted into food and profit. When excess carbon is released into the atmosphere, a portion of the environment is exchanged for the burning of carbon-based fuel in the interests of capital growth. Money replaces political capital when technology requires greater capital investment, and, consequently, greater concentration of power thereby making people more dependent on technological systems and the people who control them. And money replaces social capital as the application of technological systems disrupts social networks, patterns of reciprocity, and collective bonds.

Because technology often serves to convert one set of valued things and activities to monetary resources, protest has accompanied virtually all technological development throughout Western history. In November 1811, in the English shire of Nottingham, a group of men, faces blackened, gathered to march to the homes of weavers with mechanical looms; they entered, and destroyed the machinery. Taking the name of their mythical founder, Ned Ludd, the Luddites were opposed to more than just the expansion of machines; the protestors were fearful for what the machines meant to their lives and particularly their livelihoods, which were rapidly being supplanted, and to their family lives, which were being undermined.

The initial uprising lasted some 15 months, and hundreds of mechanical looms were destroyed. But the name *Luddite* has been resurrected to signify those opposed to the continued technologizing of people's lives. Such protests fall along a spectrum that might place the Amish of the United States, who carefully evaluate the effects of technology on their society, on one side and the Unabomber, whom some view as an extreme example of the protest movement, on the other.

We can illustrate the evolution of protest with three items that make up the technological inventory of everyday life—the automobile, computer, and television. Together, on the average, they occupy one-quarter of the lives of members of Western society; the average person in the West spends about 180 minutes a day watching television, about 90 minutes a day at a computer, and 70 minutes a day driving an automobile. Each makes demands on the environment; each has political consequences in terms of the concentration of power, and each reduces autonomy and increases access of others to the private lives of users.

Television, for example, has been a major engine of economic growth and consumption; but, in addition, critics claim that it is the technology most responsible for the decline of social capital, particularly in the United States. Critics claim that the increase in television watching reduces political participation and face-to-face social and civic activities.

Automobiles, of course, are responsible for air pollution, urban sprawl, loss of natural landscapes, and growing dependence on corporate monopolies. Yet 25 to 30 percent of the energy needs of the United States, an amount greater than that of China and India combined, are required to move personal vehicles beyond the speed of a bicycle.

The personal computer also destroys social capital, opens up personal lives to whoever can gain access to computer files, and creates vast garbage heaps of toxic materials, while, at the same time, accelerating consumption and contributing billions of dollars to national economies. The end result of each of these technologies, along with all the others, has been an acceleration, not only of capital accumulation, but of the pace of everyday lives in the West. Thus, through technology, and in the interest of capital, people in the West must work, play, exercise, eat, and interact faster than ever before.

*See also* **Science in History: United States and Western Europe**

**BIBLIOGRAPHY**

Cowan, Ruth Schwartz. *More Work for Mother: The Ironies of Household Technology from the Open Hearth to the Microwave.* New York: Basic Books, 1983.

Cowan, Ruth Schwartz. *A Social History of American Technology.* New York: Oxford University Press, 1997.

Derry, T. K., and Trevor I. Williams. *A Short History of Technology from the Earliest Times to A.D. 1900.* New York: Clarendon, 1960.

Ellul, Jacques. *The Technological Society.* New York: Random House, 1967.

Heidegger, Martin. *The Question Concerning Technology and Other Essays.* New York: Harper and Row, 1977.

Hounshell, David. *From the American System to Mass Production, 1800 to 1932.* Baltimore: Johns Hopkins University Press, 1984.

Illich, Ivan. *Tools for Conviviality.* New York: Harper and Row, 1973.

Levine, Robert. *A Geography of Time.* New York: Basic Books, 1997.

Maddison, Angus. *The World Economy: A Millennial Perspective.* Paris: OECD, 2001.

McNeil, Ian, ed. *An Encyclopedia of the History of Technology.* London: Routledge, 1990.

Mumford, Lewis. *Technics and Civilization.* New York: Harcourt Brace, 1934.

Noble, David. *America by Design: Science, Technology, and the Rise of Corporate Capitalism.* New York: Oxford University Press, 1977.

Pacey, Arnold. *Technology in World Civilization: A Thousand-Year History.* Cambridge, U.K.: Basil Blackwell, 1991.

Polanyi, Karl. *The Great Transformation.* Boston: Beacon Hill, 1957 [1944].

Postman, Neil. *Technopoly: The Surrender of Culture to Technology.* New York: Vintage, 1992.

Pursell, Carroll. *The Machine in America: A Social History of Technology.* Baltimore: Johns Hopkins Press, 1995.

Putnam, Robert D. *Bowling Alone: The Collapse and Revival of American Community.* New York: Simon & Schuster, 2000.

Rosenberg, Nathan. *Inside the Black Box: Technology and Economics.* Cambridge, U.K.: Cambridge University Press, 1982.

Sale, Kirkpatrick. *Rebels Against the Future: The Luddites and Their War on the Industrial Revolution.* New York: HarperTrade, 1996.

Sclove Richard E. *Democracy and Technology.* London: Guilford, 1995.

Singer, Charles, E. J. Holmyard, and A. R. Hall, eds. *A History of Technology. Volume I: From Early Times to Fall of Ancient Empires.* Oxford, U.K.: Oxford University Press, 1954.

Singer, Charles, E. J. Holmyard, A. R. Hall, and Trevor I. Williams, eds. *A History of Technology. Volume II: The Mediterranean Civilizations and the Middle Ages.* Oxford, U.K.: Oxford University Press, 1956.

Singer, Charles, E. J. Holmyard, A. R. Hall, and Trevor I. Williams, eds. *A History of Technology. Volume III: From the Renaissance to the Industrial Revolution, c. 1500–1750.* Oxford. 1957

Singer, Charles, E. J. Holmyard, A. R. Hall, and Trevor I. Williams, eds. *A History of Technology. Volume IV: The Industrial Revolution, c. 1750–1850.* Oxford, U.K.: Oxford University Press, 1958.

Singer, Charles, E. J. Holmyard, A. R. Hall, and Trevor I. Williams, eds. *A History of Technology. Volume V: The Late Nineteenth Century, c. 1850–1900.* Oxford, U.K.: Oxford University Press, 1958.

Turkle, Sherry. *Life on the Screen: Identity in the Age of the Internet.* New York: Simon & Schuster, 1995.

Williams, Trevor I. *A Short History of Twentieth-Century Technology, c. 1900–1950.* Oxford, U.K.: Oxford University Press, 1982.

*Richard H. Robbins*

# TELEPHONE, THE.

In 1877, the *Revue et Gazette Musicale de Paris* defined a novel mode of communication called *telephony:* "Telephony, as every one knows, is that new process which permits the transmission of the human voice by means of electricity" (quoted in Martin 1991). Though it was presumptuous to assume that "everyone knew" what telephony was, the sobriety of the definition reflects the simplicity of the technology, which did not change much until the coming of the computer. Early telephones involved three components: a system of transmission for carrying each conversation; a signaling system for warning the subscribers that someone was "on the line"; and, later, a discriminatory system enabling individual calling and including a secrecy device to keep eavesdroppers at bay and achieve privacy. The amazement with which this technology was received was due not to its complexity but to its potential for extending the range of the human voice over many miles.

The transmission of voice tones and characteristics was the important asset in the new artifact, because long-distance transmission of information already existed with the telegraph system, which, during the half century of its existence, had been worked up into a complex and sophisticated network, especially in large cities. In fact, the technique of telegraphy had indirectly led to that of telephony, because Alexander Graham Bell developed the telephone while working on the harmonic telegraph. An application for a patent that made Bell the sole inventor and only developer of the telephone was filed on February 14, 1876. As a result, for years Bell Telephone Company (BTC) in Canada and American Telephone and Telegraph (AT&T) in the United States had a quasi-monopoly of telephone development and manufacturing. It was only when the American patent expired in 1893 that independent companies could buy their equipment from other manufacturers. Still, for years BTC refused to have telephone lines put up by these companies connected to its network.

Researchers from various countries contested the invention of the telephone by Bell. In the United States, for instance, Thomas Edison and Elisha Gray were in competition with Bell for securing a patent, while in Germany the physicist Philipp Reis was the first to apply electric transmission to music through an instrument called the "tone telephone." Other countries (such as Russia and China) also claimed the invention of telephony. However, the real interest of the telephone does not lie in the identification of its inventor but rather in its socioeconomic and technological development and in its cultural impact on some social groups.

## Uneven Telephone Distribution

The telephone involved "limitless possibilities of development," according to F. B. Jewett, vice-president of AT&T in the 1930s. Yet the technical development of telephone systems was oriented toward a specific kind of expansion that considerably narrowed down these possibilities. Indeed, although it was seen as a "universal" means of communication due to the low skill level necessary for its use, the people who controlled its expansion chose a pattern of distribution that limited its accessibility nationally and internationally. For decades, Canada and the United States were the two leading countries in world telephone expansion, not only in terms of the ratio telephones/population but also of the number of calls put through the systems. In 1920 they were still at the top of the list, with an average of twelve telephones per 100 inhabitants, while Germany, the country with the most extensive telephone system in Europe, had only three, and Great Britain, two.

Among the Western countries, France was one of the slowest in developing an efficient telephone system. In fact, when American soldiers came to France in 1917, they had to bring their own material to build the lines required for their military maneuvers. The French telephone system was still underdeveloped in 1966, when it took three years for a household to get a line, while only three days were needed in North America. The problem was resolved only in the early 1980s, when France Telecom developed the Minitel system. Nevertheless, the telephone system was not necessarily more democratic in United States, since it took thirty years to have the lines extended to all the towns and villages of the country. The telephone also developed slowly in Asian countries; Japan, for instance, had only 2.6 telephones per 100 inhabitants in 1953. Even today, with the coming of satellite systems since the 1960s, the telephone is far from being universally accessible, because 15 percent of the world population owns 85 percent of the telephones. Countries such as Russia, China, and India have a very low ratio, while large portions of the African population have still no access to telephone lines.

So, despite what most companies' managers claimed, the telephone did not expand naturally, either in terms of access or in terms of uses. Though various possible uses were gradually introduced through technological development, only those profitable to telephone companies were retained. For these companies, the most important technical issue to be coped with was that of privacy. The initial system was based on party lines, with several people connected to the same circuit that was provided with one general signal for incoming calls. This meant that all subscribers on the circuit had to answer every call and hang up when it was not for them. The lack of privacy that such a system permitted had to be overcome by telephone companies if they were to attract businesses as subscribers. To create the desired level of privacy on party lines, various devices were invented. The first was a discriminatory signaling system that provided an individual signal for each subscriber. Nevertheless, individual signaling was not enough to prevent eavesdropping, so other devices were developed. The most sophisticated, invented by an American during World War I, had three functions: to warn the users that someone was eavesdropping, to identify the eavesdropper, and to measure the length of the telephone call, because the device could not be used for more than four minutes. All these devices were user managed and unsatisfactory to telephone company managers, who wanted to keep control over their lines.

Therefore, despite the high cost of this type of expansion, telephone companies decided to install private lines. This brought important changes in the way in which the telephone system expanded and in the ways in which telephone workers operated it and subscribers used it. Because private lines were much more expensive to install than party lines, companies increased their phone rates so that only some classes of subscribers could afford the lines. The most promising customers for telephone managers were

small entrepreneurs. Though the telegraph already existed for long-distance communication, the telephone was ideal for instantaneous business transactions. So the telephone business targeted these consumers and the wealthy classes in its policies of expansion. To justify expansion based on income distribution, companies such as Bell Telephone spread the myth that low-income people were not interested in acquiring telephones. However, the case of the Beaches exchange in Toronto, Canada, clearly proves BTC wrong. In 1902 the Beaches were mainly inhabited by farmers, low-income families, and a small group of bourgeois families during the summer. Bell's local manager provided the company with a long list of families who wanted telephones, but BTC set its subscription rate so high that only a few families could afford it. So the company claimed that it had received no encouragement from the residents, and it installed a summer telephone service to accommodate the wealthy families.

## Telephone Operators

An essential element in this type of expansion was the workforce, particularly the telephone operators. In the first stage of telephone development, boys who were telegraph messengers were hired to be telephone operators. However, telephone companies soon considered them unsuitable for the work, and in the early 1880s, they began to hire female operators. Apparently, a woman's upbringing in Victorian society gave her all the necessary qualities to be a "perfect" operator, "gifted" with courtesy, patience, and skilful hands as well as a good voice and quick ear. She was alert, active, even-tempered, adaptable, amenable, and easy to discipline in the companies' rules and regulations. As such, female telephone operators were instrumental in the transformation of telephone systems from manually operated to highly mechanized and finally to automatic connection. The demands made on them caused a change in their work that led to the construction of a specific telephone structure and to the making of a telephone culture.

Operators were an important asset for the telephone business, because they were mediators between the users and the telephone companies. As such, their work was closely related to both groups: They had to be productive enough to satisfy companies' managers, and they had to be reasonably amenable to the subscribers. Operators responded to every technical transformation in a manner specific to women and attractive to a large number of subscribers. Yet, they were placed in a paradoxical situation where they represented both a necessary element and an obstacle to the production of instantaneous private telephone calls. Indeed, before the adoption of the automatic switchboard, they were essential to making connections between subscribers. However, they could also delay telephone calls or intrude on the privacy of callers, despite the rules and regulations imposed on them by telephone companies toward the end of the 1880s. They were particularly accused of connecting wrong numbers, deliberately delaying connections asked by disliked subscribers, or eavesdropping on users' conversations. In response to subscribers' complaints, telephone companies tried to form operators into particular habits, skills, and attitudes in schools especially opened for that purpose. The combination of their formation with the technical transformation of the telephone system contributed to the elimination of the operators' indiscretions (which were replaced by another type of eavesdropping: wiretapping). As operators gradually adapted to technological change, their relationship with the subscribers and users became increasingly impersonal. Their role went from that of a community worker on a party line to that of a connecting voice on a mechanized switchboard.

## Early Telephone Services

The accommodation of telephone companies to private communication to satisfy the business and bourgeois classes affected the types of interaction between users and differentiated those interactions from those occurring on party lines. In their nineteenth-century telephone practices, these classes wanted to preserve privacy at two levels: secrecy in telephone communication and protection of the household from intrusive phone calls. Indeed, many subscribers thought that the phone was a terrible invader of domestic intimacy, a "nerve-wracking" technology that had the capacity to intrude on one's privacy at any time of the day. At the beginning, the system was organized in domestic circuits, with several households and businesses connected to a collective line that formed a group of people who knew each other and who created various telephone activities. In early 1880s, however, exchanges, which consisted of central offices equipped with telephone switchboards to which several subscribers' lines were connected, were opened. Thus, when a call was made through an exchange, all subscribers connected to it could hear it and listen to it. This did nothing to promote the use of the telephone, so that even those who could afford it did not necessarily adopt it.

Doctors and lawyers were not immediately drawn to the telephone, partly because of this lack of privacy and partly because they were afraid to lose money with over-the-phone consultations. Even railroad stations were reluctant to shift distant communication from the telegraph to the telephone, because they found the latter unreliable. The fact is that the service was available only at certain periods of time. Daytime full telephone service appeared in cities and towns toward the end of the 1890s, with the extended use of copper wires, and night and Sunday service was provided to all exchanges with more than 100 subscribers. Although the night service work raised relatively few objections, though it continued to be assumed by boys for more than a decade, the Sunday service caused much controversy. The churches argued that there was no justification to make the operators work on God's day. This limited the utility of residential telephony considerably.

Moreover, though developers and newspaper reports and advertisements claimed that the use of the telephone required no specific skills or training, the same media repeatedly emphasized that people needed to be "educated" in the "art of telephoning." This education was in the form of instructions and rules published in telephone books explaining in detail how to use the technology. For example, in 1899, instructions addressed the voice and elocution of users, telling them how to pronounce a telephone number and suggesting some elements of "telephone etiquette." Discipline was imposed in several ways. Discourteous users were cut off or were given a fine, while very serious offenders had their telephones removed or could even be tried in a court of law and go to jail for up to thirty days.

## Early Telephone Uses

The different kinds of telephone lines provided to subscribers influenced the form of long-distance interactive communication in a community, creating various telephonic practices, some of them unexpected. Early uses of the telephone differed from contemporary uses, partly as a result of the technology itself and partly as a result of other factors. For example, the fact that transportation was often difficult, especially during winter and spring in some parts of a country, accounted for certain specific uses such as "meeting on the lines." Furthermore, because the radio in most Western countries was commercialized only around 1920, the telephone was used for various types of "broadcasting:" sometimes to "bring God to the home of the sick and invalid," but more often for entertainment such as concerts, recitations, poetry readings, and even newspaper transmission. Each user used a telephone receiver to listen to the desired program. A newspaper called *Telefon-Hirmondo*, transmitted in Budapest, began in 1893 and lasted for over 18 years. It concentrated on news, and the handling of the information was said to be quite sophisticated and rapid. Another, started in 1911 in Newark, New Jersey, lasted only a year, though the subscribers connected to a special circuit said that they received the news of the day more promptly than with the printed newspaper. This telephonic newspaper was called *The Telephone Herald*, "broadcast" from 8:00 a.m. till 10:30 p.m., and cost five cents per day. For households that could not afford such a means of entertainment, the operators supplied, on demand, useful information from the name of a butcher to football scores,

as well as updates of returns on election days. The diversity in the early uses of the residential telephone, often imagined by women operators and users, was astonishing. Most have since disappeared.

Meanwhile, developers considered the telephone as a means of facilitating business activity or as a link between homes and businesses. Women's activities were not of prime importance for telephone entrepreneurs. Yet, in that uneven distribution of telephone services, female users played a significant role, forcing the industry into accepting new and unexpected uses of the system. The apprenticeship of late Victorian women in the use of the telephone was class related, because working-class telephone operators were in direct contact with bourgeois and petit bourgeois women, who were the main users. The telephone development had not been planned to accommodate working-class women but to make money, and though public phones appeared rather early in cities, they were expensive. At first, women's access to the telephone for personal use was very much subjected to their husbands' trust in the technology and to the uses suggested by the companies. Women were invited to shop over the phone, more so in that the department-stores' ads encouraged its use to order merchandise and the telephone directory provided various shops' phone numbers. Some people called it the "buying guide." Another proposed use was for "nightly protection" against illness, fires, and thieves.

By the early 1900s, women who could afford it, used the phone regularly for such purposes but also for activities of their own. Among the latter, phoning friends and relatives was certainly one of the most popular. Conversing over the telephone was seen as "taking the place of visiting." Although it is impossible to determine the percentage of residential calls made just to chat, complaints published in newspapers and magazines about women's habit of talking on the phone for "futile motives" suggest that the telephone was regularly used for that purpose. Motives for making calls included chatting, courting, discussing, gossiping. In fact, the coming of the telephone had mixed impacts on women, operators, and users. While it granted the former some emancipation by opening a new occupation, this occupation was so nerve wracking that, generally after two years, an operator had to either be promoted to another type of job or quit altogether. For the users, the phone helped in some ways to decrease domestic chores and to free women from some social restrictions in mixed relationships, but it also had the effect of reducing the number of visits they made to friends and of outings to concerts, as some of them were transmitted over the phone.

In rural areas the independent telephone companies, which developed party line systems, applied much looser rules to the use of the technology and charged much lower rates, so in many areas almost everyone could afford a phone. Moreover, rural communities were more closely knit socially than urban ones, although they were more sparsely distributed geographically. All of these elements helped generate different types of telephone activities. Actually, for years in the code of rural party line practices, listening to others' conversations was not seen as eavesdropping by subscribers, but rather as participation in community life. In fact, one of the most important characteristics of rural party lines was that they were regularly used for "meeting on the lines," such as by women to organize church activities or by farmers for business purposes.

All these unexpected uses of the telephone practiced by women influenced the companies' notion of its value. This technology, which had been conceived exclusively for business, seemed to have alternative uses that were worth considering for profit. Indeed, the development of a technology in society is a complex issue involving many interrelated elements. It is clear, for instance, that women's telephonic activities and practices influenced the distribution of telephone systems as well as their use as a form of communication. The pressures exerted by the companies to orient women's use toward planned and expected practices met with unexpected responses. Women drew the attention of telephone industries

to sociability as a use for the telephone, and thus they indirectly helped not only to enlarge its expansion but also to develop new cultural practices that resulted in a "telephone culture."

## BIBLIOGRAPHY

Bernard, Elaine. *Long Distance Feeling.* Vancouver, Canada: New Books, 1982. A work on the occupation of telephone operator in western Canada.

British Telecom. *Britain's Public Payphones (A Social History).* London: British Telecommunications, 1984.

Bruce, Robert V. *Bell: Alexander Graham Bell and the Conquest of Solitude.* Boston: Little, Brown, 1973.

Carmagnat, Fanny. *Le téléphone public. Cent ans d'usages et de techniques.* Paris: Hermès, 2003. A book on the expansion of public telephone booths in France and the impact of the cell phone on their survival.

Carré, Patrick A. *Le téléphone. Le monde à portée de voix.* Paris: Gallimard, 1993.

Chapuis, Robert J. *100 Years of Telephone Switching (1878–1978).* Amsterdam, New York: North-Holland, 1982.

Fischer, Claude. *America Calling. A Social History of the Telephone to 1940.* Berkeley: University of California Press, 1992.

Green, V. "Goodbye Central: Automation and the Decline of 'Personal Service' in the Bell System, 1878–1921." *Technology and Culture* 36 (1995): 912–949.

Lefebvre, Antoine. *Les conversations secrètes sous l'occupation.* Paris: Plon, 1993. An analysis of the different means of communication, including the telephone, used to communicate during the German occupation of France in World War II.

Lipartito, K. "When Women Were Switches: Technology, Work, and Gender in the Telephone Industry, 1890–1920." *American Historical Review* 99 (1994): 1074–1111.

Martin, Michèle. *Hello Central? Gender, Technology and Culture in the Formation of Telephone Systems.* Montreal: McGill-Queen's University Press, 1991. A book on women's contribution, as telephone operators and users, to the development of the telephone in Canada.

Martin, Michèle. "Capitalizing on the 'Feminine' Voice." *Canadian Journal of Communication* 14 (1989): 42–62. A comparative study on the political economy of the use of women's voices as operators and as radio speakers.

Moyal, Ann M. *Clear Across Australia.* Melbourne: Nelson, 1984. A work on the expansion of the telephone in Australia.

Pool, Ithiel de Sola, ed. *The Social Impact of the Telephone.* Cambridge, MA: MIT Press, 1977. A collection of articles that discuss the impact of the development of the telephone from different perspectives in various countries.

Rakow, Lena F. *Gender on the Line: Women, the Telephone, and Community Life.* Champaign, IL.: University of Illinois Press, 1991. A contemporary analysis on the use of the telephone by women in the United States.

*Réseaux, Dossier: Les usages du téléphone.* 55 (Sept.-Oct.). Paris: CNET/France Télécom, 1992. Special issue on the history of different uses of the telephone in various countries.

*Michèle Martin*

# TRANSPORTATION TECHNOLOGIES.

A person walking at a normal speed can cover about 5 km (3 miles) in an hour. Although this pace enables most people to cover a distance of only 25 to 50 km (15.5 to 31 miles) a day, depending on terrain and the load carried, it allowed the human species to populate every habitable region of the earth. Today, in many parts of the world, walking still constitutes the main means of getting from one place to another, and several hours may be expended each day traveling to work, school, and other destinations. People on foot also carry sizable loads of water, firewood, and other items. Throughout history, and in many parts of the world today, one's two feet have provided adequate transportation, but the inherent limitations of pedestrianism have resulted in severely restricted mobility, confining vast populations to the immediate area in which they were born.

## Draft Animals and the Wheel

The domestication of large animals, which first occurred about 9,000 years ago, considerably improved speed and range, and in various parts of the world people have used donkeys, mules, llamas, oxen, camels, yaks, elephants, reindeer, and even dogs to haul themselves and their possessions. The result has been much greater mobility. A rider on horseback can easily traverse 80 km (50 miles) a day; in some instances, when high speed was essential, as was the case during the short-lived existence of the Pony Express in the American West, a rider could cover 300 km (186 miles) a day by changing horses every 10 to 30 km (6.2 to 18.6 miles).

Speed comes at the expense of carrying capacity, and most uses of animal power have emphasized the latter over the former. Range is similarly reduced by carrying a load; in late medieval times, a packhorse could carry 120 kg (265 pounds ), but only for a distance of about 25 km (15.5 miles). The key transportation parameters—speed, range, and carrying capacity—are all enhanced when the animal providing the power is hitched to a wheeled vehicle. The wheel was probably invented in Mesopotamia around 5,000 years ago, and the idea subsequently diffused into Europe, Africa, and Asia. However, for reasons that remain unclear, wheeled vehicles were unknown in pre-Columbian America, even though the indigenous people were familiar with the concept. Wheeled vehicles represented a major step forward in making transportation more efficient; a single horse-drawn cart can accommodate a payload of 500 kg (1110 pounds) while covering a distance of 30 km (18.6 miles) in an eight-hour day. Oxen also have been widely used as draft animals in many parts of the world. Although they move at half the speed of a draft horse when pulling the same load, they require less water and can maintain their pace for days on end. Oxen also have the advantage of being easy to hitch to a wagon, requiring only a simple harness of leather straps. This arrangement is far less satisfactory when used on a horse because the straps press down on the creature's windpipe, sharply limiting its ability and willingness to pull against a heavy load. A padded-yoke harness that obviated this problem was invented in China no later than the first century B.C.E., but it was not until the ninth century C.E. that it finally diffused into Europe, where it resulted in a threefold increase in a horse's pulling power.

More than pedestrians and pack animals, wheeled vehicles require decent roads for effective travel. Road building in some parts of the ancient world was quite impressive, with the Roman Empire being particularly noted for its well-constructed system of roads, which at its peak extended a total of 80,000 km (49,700 miles). Rome's collapse was paralleled by the deterioration of Europe's roads, a situation that began to be remedied in the eighteenth century when the French government's *Corps des Ponts et Chaussées* developed techniques for making solid, long-lasting roads. These techniques were adopted, refined, and popularized in England. During the latter half of the nineteenth century, concrete and asphalt began to be used for binding the stones that formed the roadbed, and eventually for the road surface itself, greatly improving durability and the prospects for all-weather travel.

Road improvements along with the breeding of stronger draft animals produced significant improvements in land transportation during the immediate preindustrial era. Nevertheless, moving goods and people by boat was far more efficient because much greater loads could be carried and no roadways were required. According to one nineteenth-century estimate, it cost as much to ship 1,000 kg (2,204 pounds) of freight across the Atlantic Ocean as it did to haul it over 50 km (31 miles) of road. Ocean-going sailing ships also had the great advantage of not requiring an onboard source of energy, and they could remain at sea for months on end. Inland water transportation rarely had the opportunity to make use of wind power, so horses and other draft animals provided the motive power. When a team was hitched to a barge, each horse could pull a load of up to 50,000 kg (110,231 pounds), although at only half the speed achieved while pulling a cart.

## Canals

Horse-drawn boats made use of navigable waterways, and where these were not available, a canal might be built. Canals for transportation were built in ancient Babylonia, Egypt, and India, but until the eighteenth century the most advanced center of canal building was China. One of the greatest feats of preindustrial civil engineering was the Grand Canal, which, after it was completed in the fourteenth century, stretched from Hangzhou to Tianjin, a distance of 1,770 km (1,100 miles). Canal building in Europe accelerated during the late eighteenth and early nineteenth centuries, with Britain leading the way. Speculators poured capital into numerous projects in the hope of gaining substantial profits, whereas local businessmen sponsored them in order to stimulate local economies. In the United States, the Erie Canal, which linked the Hudson River with Lake Erie, a distance of 584 km (363 miles), was built between 1817 and 1825. The technical and economic success of the Erie Canal touched off what could best be described as a canal-building mania; by 1840, 5,353 km (3,326 miles) of canals traversed various parts of the United States. All of this construction entailed heavy expenditures, $125,000,000 by 1840, and an additional $75,000,000 in the years that immediately followed. A significant portion of this capital was provided by the federal and state governments through bond sales, stock purchases, and grants of land.

Traffic on many canals peaked in the years immediately leading up to the Civil War, and some continued to serve as significant transportation arteries for a decade or two thereafter. By this time, however, the railroad had become the dominant technology for land transportation.

## Railroads

The use of rails to support wheeled vehicles goes back to their use in late medieval times, and in 1804 Richard Trevithick built what is usually described as the first steam locomotive. Credit for being the first common-carrier railroad goes to England's Stockton and Darlington Railway, which began to carry passengers and freight in 1825. Echoing the canal boom, the nineteenth century was the scene of massive building and overbuilding of railroad systems in Europe and North America, as governments and private investors eagerly supported networks of railroad lines.

In addition to greatly facilitating the movement of people and things, the railroad set the pattern for the modern corporation. Its appetite for great amounts of capital gave a powerful stimulus for the issuance of stocks and bonds, and for financial markets in general. Most investors played no role in railroad operation, nor could they, given the organizational demands of running a far-flung railroad system. Meeting these demands necessitated the development of a cadre of professional administrators, further separating ownership from management.

Outside England and North America, railroads often were state creations, as governments saw them as forces for national integration and economic development. Railroads in the United States were nominally the products of free enterprise, but they too absorbed substantial government grants of cash and land. At first, state and local governments were the prime sources of public financial assistance, but with the building of the first transcontinental railroad (the Union Pacific westward from Nebraska and the Central Pacific eastward from Sacramento), the federal government took on a considerable share of the burden. Much of the financial assistance to the two railroads came in the form of land grants, 5,180 ha (12,800 acres) per mile of track laid. This government largesse was eventually more than offset by low rates for carrying the U.S. Mail, but the fact remains that without it the two railroads might never have raised the capital necessary for the construction of the transcontinental line.

Along with helping to generate capital, governments had a significant influence on railroad administration, even in the United States, where private ownership was universal. In many parts of the country, individual railroads were monopoly enterprises, and they often took full advantage of their position. Consequently, hostility against the railroads was intense, especially on the part of farmers and other shippers who had no other means of getting their wares to market. Pressure from these quarters resulted in the formation of the Interstate Commerce Commission (ICC) in 1887. The ICC was not the first agency of the federal government to reg-ulate an industry; that had occurred in another sector of the transportation industry, when explosions of steamboat boilers led to federally mandated safety standards and inspections. The ICC was given the responsibility for ensuring that freight rates were "just and reasonable." At first, any action required a court order, but this requirement was eliminated by the 1906 Hepburn Act. As can be imagined, the interpretation of this mandate was hardly self-evident, and it can be argued that in the years that followed the passage of the act, the ICC's involvement in the setting of rates and other matters contributed to the railroads' slow decline.

## The Bicycle

To be sure, government actions were not the only source of the collective woes besetting the railroads. Although they were efficient movers of people and goods, railroads were inherently inflexible, both spatially and temporally. At a time when the railroads were the dominant mode of transportation, the bicycle was introducing millions of people to the delights of personal mobility. The bicycle's origins can be traced to the hobbyhorses of the early nineteenth century. Little more than wooden frames with two wooden wheels, these contrivances went down the road as the rider pushed with his feet. Bicycles that used pedals to turn the front wheel were created in France in the 1860s. Efforts to increase speeds resulted in the classic high wheeler, which thrilled sporting riders and produced spectacular accidents. Considerably less hazardous was the "safety bicycle," which used chain-and-sprocket gearing to generate high speeds while keeping the wheels of reasonable diameter. For the first half of the twentieth century, the bicycle served as a key means of personal transportation in many parts of the industrialized world, and it remains so in today's less-developed countries. China, for example, produced more than 40 million bicycles annually during the mid-1990s.

The dominance of the railroads resulted in the neglect of road construction and maintenance, a situation that began to change as cyclists clamored for better roads. Equally important, during the bicycle boom that began in the 1880s, bicycle producers made major contributions to industrial manufacture through the development of metal stamping techniques, brazed tube frames, tension-spoke wheels, electric welding, ball bearings, and chain drive. These technologies were readily adapted to the production of automobiles, and in fact many automobile producers started out as bicycle manufacturers.

## The Automobile and the Truck

At first little more than a rich man's toy, during the first two decades of the twentieth century the automobile rapidly evolved from being a cantankerous but exhilarating source of adventure to being a key means of getting people from one place to another. At this time, the dominant figure was Henry Ford, who with his associates developed the industrial techniques that relentlessly drove down the price of automobiles, putting them within the financial reach of a citizenry Ford called "the great multitude."

Although it was an individualistic rather than collective means of transportation, in many ways the automobile accelerated the economic, social, and cultural effects that had been initiated by the railroad. Above all, the orbit within which people lived their lives expanded dramatically.

At first, the automobile's influence was most strongly felt in the countryside, especially in the United States. In 1911, the U.S. Department of Agriculture counted about 85,000 cars on American farms; by 1920, that figure had increased twenty-five times to nearly 1,150,000. In that year, 30.7 percent of farms included an automobile. People who had largely lived within the range of a horse and buggy could now travel to large towns and even metropolitan centers for a day's shopping, their children could attend better-equipped consolidated schools, and religious services were no longer limited to worship at a local church served by an itinerant minister. At the same time, the ties that bound local communities together began to unravel as the general store, the country church, and the one-room school house began to fade from the scene. And for rural dweller and urbanite alike, automobile excursions offered novel recreational opportunities, not all of which were in accordance with traditional moral values.

Ironically, just as the automobile was transforming traditional rural life, it began to play a key role in the development of a suburban lifestyle that attempted to substitute a semirural way of life for an urban one. For this to happen, people had to have a means of traveling from their exurban residences to their jobs in urban offices and factories. At first, this need was met by extensive networks of trolley lines, but these confined suburban development to areas close to their tracks. Commuting by automobile removed this restriction and opened up new tracts of land for suburban development. The result was, and continues to be, a built environment that often has been criticized for its inefficiency, excessive resource use, and anomic culture.

Continuing a cultural shift that began with the railroads, automobiles also offered the visceral excitement of traveling at high speeds, a sensation that was augmented by being in control of a speeding vehicle, not a passenger just along for the ride. Although automobiles have the practical task of moving their occupants from one place to another, this is by no means the only source of their appeal. What one journalist observed in 1906 is still valid today: "the sensation which arouses enthusiasm for the automobile comes almost solely from the introduction of the superlative degree of speed and from the absence of effort or fatigue. The automobile is a vehicle that touches a sympathetic chord in all of us."

The automobile's cousin, the motor truck, did not have the visceral appeal of travel at high speed, but its effects on the economy and society were almost as great as the automobile's. Early trucks showed their mettle during World War I, when they moved great quantities of men, equipment, and supplies under very difficult conditions. The diesel engine, which became a practical vehicular power source in the 1930s, was well suited to trucks, and its use significantly contributed to economical operation. As the twentieth century drew to a close, railroads still moved more freight when measured in ton-miles; however, when the value of shipments is taken into account, intercity trucks accounted for dollar outlays that were more than seven times greater.

As with all wheeled vehicles, cars and trucks are of little use when they lack adequate roads. Italy can claim the first limited-access divided highway, a privately built toll road that linked Milan with Varese and Lake Como in 1925. Germany's famous autobahns began to be built a few years later when the Nazi government effectively used their construction as a means of alleviating unemployment. In the United States, the Pennsylvania Turnpike, completed in 1940 offered a similar driving experience, and the first of California's legendary freeways, the Arroyo Seco Parkway, opened to traffic in the same year. The transformation of America's highway system began in earnest with President Eisenhower's signing of the Interstate Highway Act in 1956. Characterized as the greatest civil engineering project in human history, the resulting system eventually grew to more than 67,600 km (42,000 miles) of high-speed, limited-access highways.

## The Airplane

While the automobiles, trucks, and improved roads were dramatically changing earthbound transportation, humanity's long dream of taking to the skies had become a reality. In 1903 (the year that the Ford Motor Company was established), Orville Wright made what is usually credited as the first powered flight. Eleven years later, airplanes were thrown into the cauldron of World War I, and, although they proved to be of little strategic or tactical importance, wartime pressures accelerated technological development. Modest commercial service began in the years that immediately followed. As with the early development of canals and railroads, government aid, this time in the form of lucrative airmail contracts, supplied an essential stimulus to the development of an emerging industry. It was not until the 1930s that a new generation of aircraft, most notably the Douglas DC-3, made the transport

of passengers a commercially viable proposition. Again, the role of the federal government was significant; some key advances in aviation technology stemmed from research sponsored by a government agency, the National Advisory Council on Aeronautics (NACA). Commercial aviation also benefited from advances made in military aircraft. Supercharging, jet engines, radar, and aircraft pressurization made postwar commercial air travel faster, safer, and more comfortable.

Together, the automobile and the airplane dealt a blow from which passenger trains and ocean liners never fully recovered. Travelers to overseas destinations no longer took ocean voyages, and passenger trains required substantial governmental subsidies for their survival. On a more positive note, railroad haulage of freight in the United States received a substantial boost from the abolition of government rate regulation in 1980. A similar pattern of deregulation transformed the airline industry in the United States and abroad, ushering in an era of lower fares, along with diminished amenities and services.

While jet aircraft made the airline industry the glamour sector of the transportation industry, land transportation was being revolutionized by the homely freight container. These standardized metal boxes are the key elements of intermodal transportation, a system that allows goods to be shipped by ship, rail, and truck without the need for slow and expensive loading, unloading, and reloading. Intermodal transportation has lowered shipping expenses to such an extent that items manufactured in low-wage countries across the globe can be sold more cheaply than goods made and sold in the same city. More generally, containerization has greatly expanded the size and scope of international trade, making it a key component in the process of globalization.

As this brief survey has indicated, advances in transportation have been intimately connected with ongoing economic and social changes, and at the same time they have been the product of purposive actions taken by individuals, business firms, and governments. Not all of these changes have been beneficial; job losses, accidental deaths, pollution, sprawl, the depletion of resources, the evisceration of local economies, and the loss of cultural anchorages can all be attributed to advances in transportation. A tally of gains and losses is beyond the scope of this article, and it would be a highly subjective enterprise were it to be attempted. Advances in transportation have been of crucial importance in the creation of the modern world, and how one feels about the latter will heavily influence one's evaluation of the former.

*See also* **Automobile, The**

**BIBLIOGRAPHY**

Berger, Michael. *The Devil's Wagon in God's Country: The Automobile and Social Change in Rural America, 1893–1929.* Hamden, CT: Shoe String Press, 1979.

Bilstein, Roger E. *Flight in America: From the Wrights to the Astronauts.* Baltimore: Johns Hopkins University Press, 2001.

Flink, James J. *The Automobile Age.* Cambridge, MA: MIT Press, 1988.

Foster, Mark, S. *From Streetcar to Superhighway: American City Planners and Urban Transportation, 1900–1940.* Philadelphia: Temple University Press, 1981.

Johnston, Paul F., *Steam and the Sea.* Salem, MA: Peabody Museum, 1983.

Lay, M. L. *Ways of the World: A History of the World's Roads and the Vehicles That Used Them.* New Brunswick, NJ: Rutgers University Press, 1992.

McShane, Clay. *Down the Asphalt Path: The Automobile and the American City.* New York: Columbia University Press, 1994.

Miller, Ronald, and David Sawers. *The Technical Development of Modern Aviation.* New York: Praeger, 1970.

Owen, Wilfred. *Transportation and World Development.* Baltimore: Johns Hopkins University Press, 1987.

Post, Robert. *Technology, Transport, and Travel in American History.* Washington, DC: American Historical Association, 2003.

Rae, John. *The Road and Car in American Life.* Cambridge, MA: MIT Press, 1971.

Schivelbusch, Wolfgang. *The Railway Journey: The Industrialization of Time and Space in the 19th Century.* Berkeley, CA: University of California Press, 1986.

Taylor, George Rogers. *The Transportation Revolution, 1815–1860.* New York: Harper Torchbooks, 1968.

Vance, James E., Jr. *The North American Railroad: Its Origin, Evolution, and Geography.* Baltimore: Johns Hopkins University Press, 1994.

*Rudi Volti*

# TUSKEGEE PROJECT, THE.

The infamous "Tuskegee Study of Untreated Syphilis in the Negro Male," sponsored by the United States Public Health Service (PHS) from 1932 to 1972, has been described as "the longest non-therapeutic experiment on human beings in medical history" (Jones, p. 91). When the Tuskegee experiment began, penicillin had not yet been discovered. Approximately 15 years into the Tuskegee study, penicillin was found to be effective in treating syphilis; nevertheless, physicians and nurses involved in the Tuskegee study took extraordinary measures to prevent the "subjects" from being treated, despite knowing that untreated syphilis caused excruciating pain and debilitation. For example, PHS officials prevented subjects who were drafted from taking their physical examinations so that they would not know that they had syphilis and be treated.

The very absence of rules governing ethics in medical research was used to "excuse" the study and enabled individual practitioners to determine the conduct of human experimentation on a case-by-case basis. Although PHS physicians reached a "consensus" that the Tuskegee study was worth doing, there is no evidence that they discussed the advantages and disadvantages of the study before reaching that consensus. In fact, it is not so much a consensus as an example of "groupthink": The PHS physicians were largely white, middle-class males sharing the same perspectives. Therefore, what appeared to be a group discussion was actually an affirmation of an existing, implied consensus among the group.

Nevertheless, the experiment breached a fundamental principle in the physician's code of conduct: "Above all, do no harm." Clearly, the Tuskegee study did a great deal of harm to subjects and their families. Because there was never a formal protocol for the experiment, both the rationale and procedures evolved as the study continued. Ironically, the very fact that the study had persisted for so long became the rationale for its continuation.

There is no consensus on the number of African-American men involved in the Tuskegee Study, because record keeping was unsystematic. According to Fred Gray, there were 623 men; according to James Jones, there were 600, of whom 399 had a latent form of syphilis; the remaining 201 men served as controls. The period of latency could last from "a few weeks to thirty years" (Jones, p. 3). When twelve men in the control group contracted syphilis, they were moved to the "treatment" (that is, observed but not treated) group. All of the people studied were African-American men who were poor, were uneducated, had little access to medical care, and lived in rural Alabama. PHS officials said that the study's subjects were limited to blacks because a similar study had been conducted already in Sweden using only white subjects. The PHS doctors used as makeshift clinics schools and churches—institutions respected by the black community. This, in conjunction with offers of free medical care and burial expenses, made it most unlikely that the men would refuse to participate in the study. A trusted black nurse, Eunice Rivers, helped to recruit and monitor the subjects. Like the male subjects, Nurse Rivers had little choice insofar as she was a black nurse in a profession dominated by white male physicians. Neither the male subjects nor Nurse Rivers had any real choice about participating in the study; they made what Susan Reverby calls a "choiceless choice."

One of the most egregious ethical violations of the study concerns what is now called "informed consent." Not only were subjects never told the real purpose of the study; they were deliberately deceived. The doctors told them that they were being tested—and treated—for "bad blood," a term at that time used by blacks to refer to a variety of conditions. The doctors told the men that they were getting "free treatment" including "spinal shots," which were spinal taps without anesthesia—although anesthesia was available.

Although PHS officials began discussing issues related to research on human subjects in 1945, they did not issue guidelines on clinical research until more than 20 years later. Revisions of the 1966 Surgeon General's guidelines on clinical research and training grants established a system of peer review at each investigator's home institution. The most important revision was that peer review panels were to have members with nonscientific backgrounds; in 1971 this was revised again to mandate that peer review panels include people able to judge projects by community standards. The most important recommendation from the government's blue ribbon panel convened in 1972 was for Congress to create a permanent entity to regulate all federally sponsored research on human subjects.

Results from Congressional hearings on human experimentation, public outrage, and pressure from both the civil rights and consumers' rights movements compelled the government to review existing regulations of research involving human subjects. As a result, Department of Health, Education, and Welfare guidelines were revised to specify criteria for research involving human subjects and require a larger role for ethicists and community members on institutional review panels. "The ultimate lesson that many Americans learned from the Tuskegee Study was the need to protect society from scientific pursuits that ignored human values" (Jones, p. 14). In sum, the Tuskegee study was a catalytic event in medical research history, the ramifications of which have not yet been thoroughly explored. The Tuskegee study created a deep-rooted mistrust of the medical establishment that still persists among African Americans.

*See also* **Human Subjects in Medical Experiments; Medical Values and Ethics**

BIBLIOGRAPHY
Brandt, Allan M., and Lara Freidenfelds. "Commentary Research Ethics after World War II: The Insular Culture of Biomedicine." *Kennedy Institute of Ethics Journal* 6.3 (1996): 239–243.
Gray, Fred D. *The Tuskegee Syphilis Study.* Montgomery, AL: New South Books, 1998.

Jones, James H. *Bad Blood: The Tuskegee Syphilis Experiment*. New York: The Free Press, 1993.

Reverby, Susan. *Tuskegee Truths: Rethinking the Tuskegee Syphilis Study*. Chapel Hill: University of North Carolina Press, 2000.

Rutecki, Gregory W., Michael Youtsey, and Bernard Adelson. "The Institutional Review Board: A Critical Revisit to the Protection of Human Subjects." *Ethics and Medicine* 18.3 (2002): 134–144.

Tuskegee Syphilis Study Legacy Committee. *Final Report*. May 20, 1996. University of Virginia Medical School Library. http://www.med.virginia.edu/hs-library/historical/apology/report.html.

*Cheryl B. Leggon*

# W

## WOMEN AND MINORITIES IN THE SCIENTIFIC COMMUNITY.

Gender, racial, and ethnic minorities have become an integral part of U.S. scientific communities since the late nineteenth and early twentieth centuries, when avenues to careers in the sciences were closed to them. Representational parity for women and minority men, however, remains an elusive goal. Less than a quarter of all U.S. scientists and engineers are women; and African Americans and Latinos together comprise a mere 8 percent of all positions in science and engineering, a third of their combined representation in the general population. Although the limited presence of minorities in scientific communities in the academy and industry mirrors the stratification and inequalities present in American society at large, the factors shaping their current composition are also specific to the field and include:

- Disparate educational attainment and access
- Difficulty for underrepresented groups in establishing social networks that help to secure the employment experience and research collaborations necessary for success in the sciences
- Micro- and macro-level discrimination in the academy and industry

At present, fewer scientists are being trained in the United States than are needed to sustain research and industry. A 2000 Congressional Commission on the Advancement of Women and Minorities in Science, Engineering, and Technology Development (CAWMSET) study diagnosed a critical scarcity in scientists, predicted that this shortage would become more acute, and suggested that the shortfall in

human capital could be ameliorated if the percentage of women and minorities in scientific communities was increased. In the interest of social equity (as well as national security and economic development), the public and private sectors have partnered to develop initiatives aimed at increasing the presence of women and minority men in the sciences. The demand for scientific expertise in the United States has also been filled by student and researcher recruits from abroad, especially from countries in Asia. Although Asian/Pacific Islander Americans (API) and other scientists of Asian descent are now employed in the sciences at levels higher than their African-American and Latino counterparts, and higher than their percentage of the total population, they also face hurdles to full membership in scientific communities.

### Education

Education continues to be both the largest obstacle to, and the central device for, the increased participation of women and minorities in the sciences. A few historical details about the Massachusetts Institute of Technology (MIT), one of the oldest and most renowned institutions of science education in the United States, are instructive of the barriers to education that, with the rise of academic science, has been the most important threshold to membership in scientific communities. Although discriminatory practices prevented minorities from matriculating at MIT when it was established in 1861, a woman and an African American enrolled at the institution within a few decades of its founding: Ellen Swallow Richards, in 1873, who was permitted to take classes but not to earn a doctoral degree, and Robert Taylor, in 1888, who became the college's first black American graduate. These achievements are significant, but misleading in isolation. Despite these auspicious beginnings, gender and racial minorities did not constitute a significant proportion of the MIT student community until the late twentieth century. For example, in 1972, more than a century after the founding of MIT, underrepresented minorities (defined as African Americans, Hispanics, and Native Americans) constituted just 3 percent of its student body. In 2002, these minorities amounted to 17 percent of MIT's incoming undergraduate class—a fivefold increase in thirty years, but a figure still well below their combined percentage of more than a quarter of the national population.

MIT is a microcosm of the general status of underrepresented minorities in scientific communities, which reflects these groups' historically restricted access to the fields of science, engineering, and technology. Indeed, the educational pipeline that places students on the path to membership in scientific communities has been described as a "leaky" or "funnel-shaped" for women and minorities, the numbers of whom precipitously decline in negative correlation with the level of educational attainment. Presently, women receive more than half the bachelor's degrees conferred in the United States, as they have done since the late 1980s. For ex-

ample, in 2001, 536,023 men graduated with bachelor's degrees, compared to 721,625 women. However, as the level of education rises, so too does the disparity between male and female recipients of the doctoral degrees often necessary to a career in science; men earned 16,162 doctoral degrees in scientific disciplines, compared to 9,303 women, in 2001. The representation of underrepresented minorities in science education shows a similar funneling effect from undergraduate to doctoral education (see Tables 1 and 2). (This trend is somewhat reversed among African Americans; black women surpass their male counterparts in postgraduate degree attainment.)

This funneling has been attributed to differential access to educational resources; socialization at school and home that may place horizons on educational aspirations; a paucity of women and minority role models among students and teachers in postbaccalaureate science education; and a normative peer culture that reinforces existing gender inequalities and values extreme competition among scientists and laboratories, an ethos with which men are more likely to be comfortable than women. Educational and economic disparities that correlate with race and ethnicity also are a rough indicator of the likelihood of entering a scientific field; substandard education and rel-

**Table 1.  Bachelor's Degrees, by Race/Ethnicity, Gender, and Field, 2001**

| All Fields | | 1,257,648 | |
|---|---|---|---|
| | | | Percentage of S&E |
| All science and engineering (S&E) | | 400,206 | |
| | Male | 197,623 | 49.4 |
| | Female | 202,583 | 50.6 |
| | White | 267,848 | 66.9 |
| | Asian/Pacific Islander | 36,398 | 9.1 |
| | Black | 33,290 | 8.3 |
| | Hispanic | 28,321 | 7.1 |
| | Native American/ Alaskan Native | 2,796 | 0.7 |

*Source:* National Science Foundation, Division of Science Resources Statistics, 2004, Table C-4, "Bachelor's Degrees, by Sex and Field: 1994–2001," and Table C-6, "Bachelor's Degrees, by Field, Citizenship, and Race/Ethnicity: 1994–2001."

**Table 2. Doctoral Degrees, by Race/Ethnicity, Gender, and Field, 2001**

| All fields | | 40,744 | |
|---|---|---|---|
| | | | Percentage of S&E |
| All science and engineering (S&E) | | 25,509 | |
| | Male | 16,162 | 63.5 |
| | Female | 9,303 | 36.5 |
| | White | 14,905 | 58.4 |
| | Asian/Pacific Islander | 6,540 | 25.6 |
| | Black | 864 | 3.4 |
| | Hispanic | 1,181 | 4.6 |
| | Native American/ Alaskan Native | 84 | 0.3 |

*Source:* National Science Foundation, Division of Science Resources Statistics, 2004, Figure F-1, "Doctoral Degrees Awarded in S&E and Non-S&E Fields, by Sex: 1966–2001," and Table F-5, "Field Distribution of Doctorate Recipients, by Race/Ethnicity and Citizenship: 2001."

ative poverty result in lower aspirations in science and technology for minority individuals. Last, women and minorities may be simply unable to imagine themselves as members of scientific communities owing to the fact that successful scientists from these groups may be uncelebrated or unknown. For example, Lewis Latimer, an African American, is better known as the drafter of Alexander Graham Bell's telephone patent application and as Thomas Edison's first drafter than as a successful inventor in his own right. Similarly underacknowl-

edged, until fairly recently, was British X-ray crystallographer Rosalind Franklin, whose images were essential for revealing DNA's double-helix structure to James Watson and Francis Crick, who received Nobel Prizes for the discovery.

## The Workplace

As in education, women and minorities are underrepresented in the workplace (Table 3).

Women and underrepresented minorities are virtually absent from upper levels of

**Table 3. Employed Bachelor's or Higher Degree Recipients, 2000**

| | Category | Percentage all S&E occupations |
|---|---|---|
| By gender: | | |
| | Male | 74.6 |
| | Female | 25.4 |
| By race/ethnicity: | | |
| | White | 76.4 |
| | Asian | 14.0 |
| | Black | 4.4 |
| | Hispanic | 3.4 |
| | American Indian/ Alaska Native | 0.3 |

*Source:* National Science Foundation, Division of Science Resources Statistics, 2004, Table H-1, "Employed Bachelor's or Higher Degree Recipients, by Occupation, Sex, Race/Ethnicity, Country of Birth, and Disability Status: 2000."

scientific research in industry and the academy. Moreover, they are more likely than their nonminority counterparts to fill the ranks of the lower-status jobs in scientific communities. As a case in point, the fields of architecture and engineering include the occupational subcategories of architects, aerospace engineers, chemical engineers, civil engineers, computer hardware engineers, electrical and electronics engineers, industrial engineers, mechanical engineers, drafters, and engineering technicians, and surveyors, among others. In 2003 women were more likely to be engineering technicians than full-fledged engineers; similarly, blacks were most often engineering technicians and least often aerospace engineers. Hispanics or Latinos were most likely to be drafters and least likely to be computer hardware engineers (Table 4). Thus, there are patterns of stratification within scientific communities, with women and underrepresented minorities more likely to be in occupational subcategories that are lower in status than those filled by their nonminority counterparts.

Workplace characteristics also shape a scientist's status in academic communities. Women and minorities tend to be relatively over-represented in public universities, colleges, and community colleges that are not as highly valued as the prestigious, private institutions, where their majority peers are more likely to be employed. In addition, black women faculty are more likely to teach at junior colleges than black men. African-American male faculty are more likely to teach the "hard sciences" such as physics and engineering, whereas black women faculty are most common in the "soft" social, health, and biological sciences. In several studies, women and underrepresented minorities have reported spending more time on teaching than research, a fact that is a function both of the requirements of the institution at which they are employed (such as the required teaching load and whether the institution is a teaching college or a doctoral degree-granting university) and of the available opportunities and financial and social resources required to advance a research agenda.

A well-known study on the status of women faculty in the sciences at MIT sheds some light on the disparities present in the academic workplace and on the role of discrimination in shaping the composition of scientific communities. The report found that women scientists entering the junior faculty ranks were initially optimistic about their jobs, expressed high job satisfaction, and had little experience with discrimination. As they progressed in their careers, these same women began to feel undervalued in the profession as demonstrated by disparities in compensation, promotion, and prestige. They also received little support for their efforts to balance career and family. Senior faculty women, despite their

**Table 4. Employed Persons in Science Fields by Occupation, Gender, Race/Ethnicity, 2003**

| Profession | Total Employed (in Thousands) | Women (Percent of Total) | Black or African--American (Percent of Total) | Asian (Percent of Total) | Hispanic or Latino (Percent of Total) |
|---|---|---|---|---|---|
| Computer and mathematical occupations | 3,122 | 28.8 | 8.1 | 12.9 | 5.5 |
| Architecture and engineering occupations | 2,727 | 14.1 | 4.4 | 8.7 | 5.2 |
| Life, physical, and social science | 1,375 | 43.0 | 6.3 | 10.3 | 5.9 |

*Source:* U.S. Department of Labor, Bureau of Labor Statistics, 2003, Table 10: "Employed Persons by Detailed Occupation, Sex, Race, and Hispanic or Latino Ethnicity," http://www.bls.gov/cps/cpsaat11.pdf.

tenured status, reported feeling invisible and disempowered to play leadership roles in their respective departments. This marginalization was manifest in pay differentials and access to fewer institutional resources (in the form of office space and research support, for example) than their male counterparts, despite demonstrating comparable talent and success in the profession. In sum, even women who, by virtue of having secured tenure at a top-ranked institution, would be considered a success by almost any standard expressed significant job dissatisfaction and attributed some of it to structural forms of discrimination in the workplace.

## Asian/Pacific Islander Americans and Visiting Scientists

Because there are not enough American-born scientists to fill the need for scientific and technical expertise, the United States has increasingly relied on non-native-born scientists, particularly Asian immigrants. Although API Americans and Asian immigrants constitute a racial/ethnic minority group, they are not underrepresented in the scientific community; in fact, proportionally, they have more representation than any other minority group (Tables 1, 2, and 4). APIs and Asians of both genders participate in the sciences at rates equal to, or in excess of, their percentage in the general population.

In addition to the shortage of native-born scientists, other factors driving the influx of non-native-born scientists are levels of educational attainment and convenience. Taking the field of engineering as an example, as a general rule, immigrant engineers have achieved higher levels of education than their native-born counterparts. A 1999 National Science Foundation (NSF) study found that among engineers working in the United States, three times as many immigrants as native-born engineers had doctoral degrees. Many non-native born engineers obtain their graduate education in the United States. Following their training, they are frequently recruited to join the American workforce, based on their expertise, advanced degree, and existing knowledge of the country and of the field.

Although APIs and Asians constitute a significant proportion of American scientists, some have expressed dissatisfaction with their treatment in the industry. Language barriers and citizenship status have been cited as the source of feelings of marginalization from scientific communities. In response, Asian-American scientific researchers have begun to organize academic and technical labor unions to address their concerns.

The pros and cons of American labs being increasingly staffed with noncitizens is an issue of debate. Many argue that it is in the best interest of neither the U.S. economy nor national security to draw such significant numbers of scientists from abroad. Others claim that more restrictive immigration standards for foreign students and researchers implemented following the terrorist attacks of September 11, 2001, may imperil the nation's role as an innovator in science and technology. It is undisputed, however, that immigrants have made exceptional contributions to the fields of science, technology, and engineering in the United States for generations.

## Initiatives to Increase Representation of Women and Minorities in Scientific Communities

Several stakeholders are advocating for increased recruitment and retention of women and minorities as a way to increase the ranks of scientists in the United States, including discipline-specific organizations and group-specific professional associations. Supported by data demonstrating that proactive outreach can foster diversity in scientific communities, and validated by legislation, these groups have developed specific programs and initiatives to address the underrepresentation of racial and gender minorities.

Together, the U.S. Supreme Court's *Brown v. Board of Education* decision in 1954 and the enactment of the Civil Rights Act in 1964 opened avenues for minorities to enter scientific communities—most notably, by outlawing segregation in public schools and discrimination based on race, color, religion, or national origin. Several years later, Title IX of the Education Amendments Act of 1972 prohibited sex discrimination by public and private colleges and universities. The federal government has also backed more recent legislation to increase numbers of minority scientists explicitly. Over a decade ago, the Excellence in Mathematics, Science and Engineering Education Act of 1990 was passed to promote interest in these disciplines among students. The Act included a number of special provisions intended to encourage greater participation by women and minorities in scientific fields.

On October 16, 1997, President Clinton signed H.R. 3007, the Advancement of Women and Minorities in Science, Engineering, and Technology Development Act, into law. Introduced by Representative Connie Morella of Maryland, the bill established a commission to study the barriers that women, minorities, and persons with disabilities face in the sciences. The resulting body was the Commission on the Advancement of Women and Minorities in Science, Engineering, and Technology Development (CAWMSET). Given a mandate to advise the state and federal government, academia, and private industry, CAWMSET recommended the adoption and implementation of high-quality state education standards in math and science, teacher training, and facilities. It also suggested aggressive interventions to prepare students for postsecondary education and expanded federal support for college scholarships and fellowships. It advocated holding employers accountable for the career development of underrepresented groups. And, in recognition of the that fact that popular imagery shape young people's ideas about which members of society are best suited to work in the sciences, the commission also called for a coordinated effort to improve the public image of scientists and engineers. Many of CAWMSET's suggestions have been taken up in the public and private sectors.

There are several academic organizations for minorities and women that help facilitate their socialization in scientific fields. For example, the National Organization of Black Chemists and Chemical Engineers (NOBCChE), the National Society of Black Engineers (NSBE), the Committee on Women in Science and Engineering (CWSE), and American Women in Science (AWIS) have made recruitment and retention of minorities and women a central component of their respective organizational missions. As part of such efforts, these organizations, often in partnership with educational institutions and nongovernmental organizations, provide financial incentives to students who remain in the sciences in the form of fellowships and stipends; such programs also supply minority students with internship and apprenticeship opportunities. In addition, social and academic support is facilitated through structured programs that socialize students in a scientific community. For example, senior scholars and researchers in the sciences are encouraged to serve as mentors to fellowship recipients; and professional conferences are held in dynamic social settings believed to increase networking among attendees.

Some educational institutions have been especially active in creating opportunities for women and minorities to thrive in the academic community. Since 1992, Lehigh University has participated in the CHOICES (Charting Horizons and Opportunities in Careers in Engineering and Science) program for middle-school girls. Each spring, 60 girls come to Lehigh to spend one day working in teams in the lab under the supervision of Lehigh engineering faculty and students. In 2002, the first CHOICES alumna enrolled as an engineering student at Lehigh. The student says her decision to become an engineer was influenced by her experience with CHOICES as a seventh-grader. As well, educational institutions with small minority enrollments in the sciences desiring to increase student diversity in their ranks frequently enter into bridge and partnership programs with colleges and universities with significant minority student constituencies. These programs aid majority universities and institutions in accessing a larger pool of minority students than exists on their own campus; such collaborations are common between elite, private institutions and historically black colleges and universities.

The combination of sustained legislation and recruiting and retention has shown promise as a tool for diversifying scientific communities. However, government oversight is also critically important to ensure that the playing field in science and technology workplaces remains equitable and accessible. The NSF, the National Institutes of Health, and other government organizations have been proactive in promoting increased representation and retention of women and minorities in science. The NSF, in particular, has funded thousands of graduate students through initiatives that assist racial, ethnic, and gender minorities in science education and training. However, according to a report issued in 2004 by the Government Accountability Office, the NSF, along with the National Aeronautics and Space Administration (NASA) and the Department of Energy, have been failing to monitor grantee institutions to check whether they were complying with Title IX.

The recent decision in *Grutter* v. *Bollinger* has placed some limits on the consideration of minority status in higher education placement and may stall the progress that has been made in the last several decades toward the diversification of scientific communities. As a result, programs such as CHOICES are no longer exclusive to minorities and women. The impact of these changes remains to be seen; but the ability to use such programs to encourage the participation of underrepresented groups in higher science education may be diminished.

*See also* **Gender, Race, and Class in Science; Gender, Race, and Class in Technology**

## BIBLIOGRAPHY

Bhattacharjee, Yudhijit. "GAO Faults Science Agencies on Title IX Compliance." *Science* 305 (2004): 589.

Bradburn, Ellen M., and Anna C. Sikora. *Gender and Racial/Ethnic Differences in Salary and Other Characteristics of Postsecondary Faculty: Fall 1998,* NCES 2002-170. Washington, DC: U.S. Department of Education, National Center for Education Statistics, 2002. http://nces.ed.gov/das/epubs/2002170/index.asp.

Burton, Lawrence, and Wang, Jack. National Science Foundation, Division of Science Resources Statistics. *How Much Does the U.S. Rely on Immigrant Engineers?* Data Brief NSF 99-327. Arlington, VA: National Science Foundation, 1999. http://www.nsf.gov/sbe/srs/issuebrf/sib99327.htm.

Congressional Commission on the Advancement of Women and Minorities in Science, Engineering, and Technology Development. *Land of Plenty: Diversity as America's Competitive Edge in Science, Engineering and Technology,* 2000. http://www.nsf.gov/od/cawmset/report/cawmset_report.pdf.

Fouché, Rayvon. *Black Inventors in the Age of Segregation: Granville T. Woods, Lewis H. Latimer, and Shelby J. Davidson.* Baltimore: The Johns Hopkins University Press, 2003.

Hanson, Sandra. *Lost Talent: Women in the Sciences.* Philadelphia: Temple University Press, 1996.

Hill, Susan T., and Jean M. Johnson, project officers. *Science and Engineering Degrees, by Race/Ethnicity of Recipients: 1992-2001,* NSF 04-318. Arlington, VA: National Science Foundation, 2004. http://www.nsf.gov/sbe/srs/nsf04318/sectb.htm.

Lawler, Andrea. "Silent No Longer: 'Model Minority' Mobilizes." Science 290 (2000): 1072–1077. Levin, Sharon, and Paul Stephan. "Sociology of Science: Are the Foreign Born a Source of Strength for U.S. Science?" *Science* 285 (1999): 1213–1214.

Maddox, Brenda. *Rosalind Franklin: The Dark Lady of DNA.* New York: HarperCollins, 2002.

Massachusetts Institute of Technology. "A Study on the Status of Women Faculty in Science at MIT." *The MIT Faculty Newsletter* 11.4 (1999). http://web.mit.edu/fnl/women/women.html.

Merton, Robert K. *The Sociology of Science: Theoretical and Empirical Investigations.* Chicago: University of Chicago Press, 2000.

National Science Foundation, Division of Science Resources Statistics. *Women, Minorities, and Persons with Disabilities in Science and Engineering: 2004,* NSF 04-317. Arlington, VA: National Science Foundation, 2004; updated May 2004. http://www.nsf.gov/sbe/srs/wmpd/.

Nettles, Michael T., Laura W. Perna, and Ellen M. Bradburn. *Salary, Promotion, and Tenure Status of Minority and Women Faculty in U.S. Colleges and Universities,* NCES 2000-173. Washington, DC: U.S. Department of Education. National Center for Education Statistics, 2000. http://nces.ed.gov/pubs2000/2000173.pdf.

Traweek, Sharon. *Beamtimes and Lifetimes: The World of High Energy Particle Physicists.* Cambridge, MA: Harvard University Press, 1988.

U.S. Department of Labor. Bureau of Labor Statistics. *Labor Force Statistics from the Current Population Survey.* Washington, DC: U.S. Bureau of Labor Statistics, 2003. http://www.bls.gov/cps.

Williams, Clarence, G. *Technology and the Dream: Reflections on the Black Experience at MIT, 1941–1999.* Cambridge, MA: MIT Press, 2001.

*Alondra Nelson and Lindsey J. Greene*

# Topical Outline of Articles

The entries and subentries in *Science, Technology, and Society: An Encyclopedia* are conceived according to the general conceptual categories listed in this topical outline. The list of articles that follows reflects the sequence of overview essays and alphabetical entries; for particular terms not noted here, refer to the index.

# List of Contributors

Atsushi Akera
*Rensselaer Polytechnic Institute*

Barbara L. Allen
*Virginia Polytechnic Institute and State University*

Stanley Aronowitz
*City University of New York*

Wenda K. Bauchspies
*Pennsylvania State University*

Charles Bazerman
*University of California, Santa Barbara*

Colin Beech
*Rensselaer Polytechnic Institute*

Deborah Blizzard
*Rochester Institute of Technology*

David Bloor
*University of Edinburgh*

Vasily P. Borisov
*Russian Academy of Sciences*

Lise Bouchard
*Runajambi (Institute for the Study of Quichua Culture and Health), Otavalo, Ecuador*

Geoffrey C. Bowker
*Santa Clara University*

Kate Boyer
*Rensselaer Polytechnic Institute*

Jen Boyle
*Hollins University*

Leslie Brothers
*University of California, Los Angeles*

Nathan Brown
*University of California, Los Angeles*

David T. Browne and Sheila E. Browne
*Mount Holyoke College*

Grace Budrys
*DePaul University*

Lawrence Busch
*Michigan State University*

Nancy D. Campbell
*Rensselaer Polytechnic Institute*

W. Bernard Carlson
*University of Virginia*

Daryl E. Chubin
*American Association for the Advancement of Science*

Harry M. Collins
*Cardiff University*

David Conz
*Arizona State University*

Susan E. Cozzens
*Georgia Institute of Technology*

T. Hugh Crawford
*Georgia Institute of Technology*

Jennifer L. Croissant
*University of Arizona*

Ubiratan D'Ambrosio
*State University of Campinas (UNICAMP), São Paulo, Brazil*

Marianne de Laet
*Harvey Mudd College*

Peter H. Denton
*University of Winnipeg*

Gary Lee Downey
*Virginia Polytechnic Institute and State University*

Rachel Dowty
*Rensselaer Polytechnic Institute*

Joseph Dumit
*Massachusetts Institute of Technology*

Dilip Dutta
*The University of Sydney*

Aant Elzinga
*University of Gothenburg, Sweden*

Gloria T. Emeagwali
*Central Connecticut State University*

Paul Ernest
*University of Exeter, U.K.*

Henry Etzkowitz
*State University of New York, Purchase*

Jill A. Fisher
*Rensselaer Polytechnic Institute*

Mark S. Frankel
*American Association for the Advancement of Science*

Ellsworth R. Fuhrman
*Virginia Polytechnic Institute and State University*

Steve Fuller
*University of Warwick, U.K.*

Brent Garland
*American Association for the Advancement of Science*

Thomas F. Gieryn
*Indiana University*

Priska Gisler
*Federal Institute of Technology, Zurich, Switzerland*

Chris Hables Gray
*Union Institute and University, Cincinnati*

Lindsey J. Greene
*Yale University*

Nathan Greenslit
*Massachusetts Institute of Technology*

Edward J. Hackett
*Arizona State University*

Sandra Harding
*University of California, Los Angeles*

N. Katherine Hayles
*University of California, Los Angeles*

David J. Hess
*Rensselaer Polytechnic Institute*

Linda F. Hogle
*University of Wisconsin*

Rachelle D. Hollander
*National Science Foundation*

Richard T. Houang
*Michigan State University*

Harukazu Iguchi
*National Institute for Fusion Science, Japan*

Mario Incayawar (Maldonado)
*Runajambi (Institute for the Study of Quichua Culture and Health), Otavalo, Ecuador*

Sarah S. Jain
*Stanford University*

Peter John
*University of California, San Diego*

Elaine Kamarck
*Harvard University*

Margaret Kerr
*University of Pittsburgh*

Nicholas B. King
*University of Michigan*

Karin Knorr Cetina
*University of Chicago; University of Konstanz*

Emily S. Kolker
*Brandeis University*

Pauline Kusiak
*Northwestern University*

Marcel C. LaFollette
*American Association for the Advancement of Science*

Frank N. Laird
*University of Denver*

John Law
*Lancaster University, U.K.*

Joan Leach
*University of Pittsburgh*

Cheryl B. Leggon
*Georgia Institute of Technology*

Les Levidow
*Open University, U.K.*

Romulo Lins
*State University of Rio Claro, Brazil*

Melissa M. Littlefield
*Pennsylvania State University*

Andrew S. London
*Syracuse University*

Juan C. Lucena
*Colorado School of Mines*

Michael Lynch
*Cornell University*

Rachel Maines
*Cornell University*

Najwa Makhoul
*Harvard University*

Brian Martin
*University of Wollongong, Australia*

Michèle Martin
*Carleton University, Ottawa,
Canada*

Elizabeth Mazzolini
*Pennsylvania State University*

Bari Meltzer
*University of Pennsylvania*

Elizabeth Merrill
*Pomona College*

Martina Merz
*University of Lausanne*

Carl Mitcham
*Colorado School of Mines*

John Monberg
*University of Kansas*

Kelly Moore
*Brooklyn College, City University of
New York*

Lynn M. Morgan
*Mount Holyoke College*

Thomas H. Murray
*Hastings Center, New York*

Linda Muzzin
*University of Toronto*

Nancy A. Myers
*Northeastern Ohio Universities
College of Medicine*

Shigeru Nakayama
*Institute for Advanced Studies on
Science, Technology and Society*

Alondra Nelson
*Yale University*

Illah R. Nourbakhsh
*Carnegie Mellon University*

Richard Olson
*Harvey Mudd College*

Nelly Oudshoorn
*University of Twente, The Netherlands*

John Parker
*Arizona State University*

Heather Paxson
*Massachusetts Institute of Technology*

Hugh Gilbert Peach
*H. Gil Peach & Associates*

Trevor Pinch
*Cornell University*

Susanna Hornig Priest
*Texas A&M University*

Lewis Pyenson
*University of Louisiana at Lafayette*

Jerry Ravetz
*Research Methods Consultancy Ltd., U.K.*

Richard H. Robbins
*State University of New York*

Barbara Katz Rothman
*City University of New York*

William H. Schmidt
*Michigan State University*

Ernst Schraube
*University of Roskilde*

Darleen V. Schuster
*University of Southern California*

Dhirendra Sharma
*Centre for Science and Industrial
Policy Research, India*

Elizabeth P. Shea
*Northeastern University*

Susan Leigh Star
*Santa Clara University*

Linda M. Strauss
*University of California at San Diego*

Piet Strydom
*University College Cork, Ireland*

Karen-Sue Taussig
*University of Minnesota*

Albert H. Teich
*American Association for the
Advancement of Science*

Marcos V. Teixeira
*Universidade Estadual Paulista, Brazil*

Stefan Timmermans
*Brandeis University*

Jeanette Trauth
*University of Pittsburgh*

Paula A. Treichler
*University of Illinois, Urbana-Champaign*

Nancy Tuana
*Pennsylvania State University*

Thomas W. Valente
*University of Southern California*

Mary Virnoche
*Humboldt State University*

Elena Volodarskaia
*Institute of History of Science and Technology, Moscow*

Rudi Volti
*Pitzer College*

Wendy E. Wagner
*University of Texas*

Judy Wajcman
*Australian National University*

Angela Wasunna
*Hastings Center, New York*

Ron Westrum
*Eastern Michigan University*

Fred L. Wilson
*Rochester Institute of Technology*

Edward J. Woodhouse
*Rensselaer Polytechnic Institute*

Steve Woolgar
*University of Oxford*

Richard Worthington
*Pomona College*

# Index

# Z

JOHNSTON COMMUNITY COLLEGE

3 3319 00043 5920

## DATE DUE

| | | | |
|---|---|---|---|
| | | | |
| | | | |
| | | | |
| | | | |
| | | | |
| | | | |
| | | | |
| | | | |
| | | | |
| | | | |
| | | | |
| | | | |
| | | | |

GAYLORD No. 2333      PRINTED IN U.S.A.

LRC/LIBRARY